Rock Mechanics and Engineering

Rock Mechanics and Engineering

Volume 1: Principles

Editor

Xia-Ting Feng

*Institute of Rock and Soil Mechanics, Chinese Academy of Sciences,
State Key Laboratory of Geomechanics and Geotechnical Engineering,
Wuhan, China*

CRC Press is an imprint of the
Taylor & Francis Group, an **informa** business

A BALKEMA BOOK

CRC Press/Balkema is an imprint of the Taylor & Francis Group, an informa business

© 2017 Taylor & Francis Group, London, UK

Typeset by Integra Software Services Private Ltd

Printed and bound in Great Britain by Antony Rowe (A CPI-group Company), Chippenham, Wiltshire

All rights reserved. No part of this publication or the information contained herein may be reproduced, stored in a retrieval system, or transmitted in any form or by any means, electronic, mechanical, by photocopying, recording or otherwise, without written prior permission from the publisher.

Although all care is taken to ensure integrity and the quality of this publication and the information herein, no responsibility is assumed by the publishers nor the author for any damage to the property or persons as a result of operation or use of this publication and/or the information contained herein.

Library of Congress Cataloging in Publication Data

Applied for

Published by: CRC Press/Balkema
P.O. Box 11320, 2301 EH Leiden, The Netherlands
e-mail: Pub.NL@taylorandfrancis.com
www.crcpress.com – www.taylorandfrancis.com

ISBN: 978-1-138-02759-6 (Hardback)
ISBN: 978-1-315-36426-1 (eBook)

Contents

Foreword		*ix*
Introduction		*xi*

Discontinuities

1
Characterization and modeling of the shear strength, stiffness and hydraulic behavior of rock joints for engineering purposes — 3
N.R. BARTON & S.C. BANDIS

2
Statistical fracture toughness study for rocks — 41
M.R.M. ALIHA & M.R. AYATOLLAHI

Anisotropy — 63

3
Rock damage mechanics — 65
F.L. PELLET & A.P.S. SELVADURAI

4
Experimental and numerical anisotropic rock mechanics — 109
K-B. MIN, B. PARK, H. KIM, J-W. CHO & L. JING

5
Characterization of rock masses based on geostatistical joint mapping and rock boring operations — 139
M. STAVROPOULOU & G. EXADAKTYLOS

Rock Stress — 181

6
Hydraulic fracturing stress measurements in deep holes — 183
D.R. SCHMITT & B. HAIMSON

7
Hydrofracturing — 227
S.O. CHOI

vi Contents

8 Methodology for determination of the complete stress tensor and its
   variation versus depth based on overcoring rock stress data   245
   D. ASK

9 Measurement of induced stress and estimation of rock mass strength
   in the near-field around an opening   267
   Y. OBARA & K. SAKAGUCHI

Geophysics   297

10 Compressive strength–seismic velocity relationship for sedimentary rocks   299
    T. TAKAHASHI & S. TANAKA

11 Elastic waves in fractured isotropic and anisotropic media   323
    L.J. PYRAK-NOLTE, S. SHAO & B.C. ABELL

Strength Criteria   363

12 On yielding, failure, and softening response of rock   365
    J.F. LABUZ, R. MAKHNENKO, S-T. DAI & L. BIOLZI

13 True triaxial testing of rocks and the effect of the intermediate principal
    stress on failure characteristics   379
    B. HAIMSON, C. CHANG & X. MA

14 The $MSDP_u$ multiaxial criterion for the strength of rocks and rock masses   397
    L. LI, M. AUBERTIN & R. SIMON

15 Unified Strength Theory (UST)   425
    M.-H. YU

16 Failure criteria for transversely isotropic rock   451
    Y.M. TIEN, M.C. KUO & Y.C. LU

17 Use of critical state concept in determination of triaxial and
    polyaxial strength of intact, jointed and anisotropic rocks   479
    M. SINGH

18 Practical estimate of rock mass strength and deformation parameters
    for engineering design   503
    M. CAI

Contents vii

Modeling Rock Deformation and Failure   531

19 Constitutive modeling of geologic materials, interfaces and joints using
    the disturbed state concept   533
    C.S. DESAI

20 Modeling brittle failure of rock   593
    V. HAJIABDOLMAJID

21 Pre-peak brittle fracture damage   623
    E. EBERHARDT, M.S. DIEDERICHS & M. RAHJOO

22 Numerical rock fracture mechanics   659
    M. FATEHI MARJI & A. ABDOLLAHIPOUR

23 Linear elasticity with microstructure and size effects   699
    G. EXADAKTYLOS

24 Rock creep mechanics   745
    F.L. PELLET

    Series page   771

Foreword

Although engineering activities involving rock have been underway for millennia, we can mark the beginning of the modern era from the year 1962 when the International Society for Rock Mechanics (ISRM) was formally established in Salzburg, Austria. Since that time, both rock engineering itself and the associated rock mechanics research have increased in activity by leaps and bounds, so much so that it is difficult for an engineer or researcher to be aware of all the emerging developments, especially since the information is widely spread in reports, magazines, journals, books and the internet. It is appropriate, if not essential, therefore that periodically an easily accessible structured survey should be made of the currently available knowledge. Thus, we are most grateful to Professor Xia-Ting Feng and his team, and to the Taylor & Francis Group, for preparing this extensive 2017 "Rock Mechanics and Engineering" compendium outlining the state of the art—and which is a publication fitting well within the Taylor & Francis portfolio of ground engineering related titles.

There has previously only been one similar such survey, "Comprehensive Rock Engineering", which was also published as a five-volume set but by Pergamon Press in 1993. Given the exponential increase in rock engineering related activities and research since that year, we must also congratulate Professor Feng and the publisher on the production of this current five-volume survey. Volumes 1 and 2 are concerned with principles plus laboratory and field testing, i.e., understanding the subject and obtaining the key rock property information. Volume 3 covers analysis, modelling and design, i.e., the procedures by which one can predict the rock behaviour in engineering practice. Then, Volume 4 describes engineering procedures and Volume 5 presents a variety of case examples, both these volumes illustrating 'how things are done'. Hence, the volumes with their constituent chapters run through essentially the complete spectrum of rock mechanics and rock engineering knowledge and associated activities.

In looking through the contents of this compendium, I am particularly pleased that Professor Feng has placed emphasis on the strength of rock, modelling rock failure, field testing and Underground Research Laboratories (URLs), numerical modelling methods—which have revolutionised the approach to rock engineering design—and the progression of excavation, support and monitoring, together with supporting case histories. These subjects, enhanced by the other contributions, are the essence of our subject of rock mechanics and rock engineering. To read through the chapters is not only to understand the subject but also to comprehend the state of current knowledge.

I have worked with Professor Feng on a variety of rock mechanics and rock engineering projects and am delighted to say that his efforts in initiating, developing and seeing through the preparation of this encyclopaedic contribution once again demonstrate his flair for providing significant assistance to the rock mechanics and engineering subject and community. Each of the authors of the contributory chapters is also thanked: they are the virtuosos who have taken time out to write up their expertise within the structured framework of the "Rock Mechanics and Engineering" volumes. There is no doubt that this compendium not only will be of great assistance to all those working in the subject area, whether in research or practice, but it also marks just how far the subject has developed in the 50+ years since 1962 and especially in the 20+ years since the last such survey.

John A. Hudson, Emeritus Professor, Imperial College London, UK
President of the International Society for Rock Mechanics (ISRM) 2007–2011

Introduction

The five-volume book "Comprehensive Rock Engineering" (Editor-in-Chief, Professor John A. Hudson) which was published in 1993 had an important influence on the development of rock mechanics and rock engineering. Indeed the significant and extensive achievements in rock mechanics and engineering during the last 20 years now justify a second compilation. Thus, we are happy to publish 'ROCK MECHANICS AND ENGINEERING', a highly prestigious, multi-volume work, with the editorial advice of Professor John A. Hudson. This new compilation offers an extremely wide-ranging and comprehensive overview of the state-of-the-art in rock mechanics and rock engineering. Intended for an audience of geological, civil, mining and structural engineers, it is composed of reviewed, dedicated contributions by key authors worldwide. The aim has been to make this a leading publication in the field, one which will deserve a place in the library of every engineer involved with rock mechanics and engineering.

We have sought the best contributions from experts in the field to make these five volumes a success, and I really appreciate their hard work and contributions to this project. Also I am extremely grateful to staff at CRC Press / Balkema, Taylor and Francis Group, in particular Mr. Alistair Bright, for his excellent work and kind help. I would like to thank Prof. John A. Hudson for his great help in initiating this publication. I would also thank Dr. Yan Guo for her tireless work on this project.

Editor
Xia-Ting Feng
President of the International Society for Rock Mechanics (ISRM) 2011–2015
July 4, 2016

Discontinuities

Chapter 1

Characterization and modeling of the shear strength, stiffness and hydraulic behavior of rock joints for engineering purposes

Nick R. Barton[1] & Stavros C. Bandis[2,3†]

[1]*Director, Nick Barton & Associates, Oslo, Norway*
[2]*Professor, Department of Civil Engineering, Aristotle University of Thessaloniki, Thessaloniki, Greece*
[3]*Principal, Geo-Design Consulting Engineers Ltd, UK*

Keywords: joint characterization, roughness, wall-strength, peak strength, shear stiffness, normal stiffness, physical and hydraulic apertures (Quantification of parameters: JRC, JCS, ϕ_r, K_s, K_n, E and e).

1 INTRODUCTION

The term 'characterization' will be used to describe methods of collection and interpretation of the physical attributes of the joints and other discontinuities, in other words those which control their mechanical and hydraulic properties, and the behavior of jointed rock as an engineering medium. Rock discontinuities vary widely in terms of their origin (joints, bedding, foliation, faults/shears, etc.) and associated physical characteristics. They can be very undulating, rough or extremely planar and smooth, tightly interlocked or open, filled with soft, soil-type inclusions or healed with hard materials. Therefore, when loaded in compression or shear, they exhibit large differences in the normal and shear deformability and strength, resulting in surface separation and therefore permeability. Such variability calls for innovative, objective and practical methods of joint characterization for engineering purposes. The output must be quantitative and meaningful and the cost kept at reasonable levels. The practical methods to be described will be biased in the direction of quantifying the non-linear shear, deformation and permeability behavior of joints, based on the Barton-Bandis (BB) rock engineering modeling concepts. The term 'modeling' will be used to introduce the basic stress-displacement-dilation behavior of joints in shear, and the basic stress-closure behavior when joints are compressed by increased normal stress. These are the basic elements of the (non-linear) behavior, which are used when modeling the two- or three-dimensional behavior of a jointed rock mass. They are the basic BB (Barton-Bandis) components of any UDEC-BB distinct element numerical model (used commercially and for research since 1985). The BB approach can also be used to determine improved MC (Mohr-Coulomb) strength components for a 3DEC-MC three-dimensional distinct element numerical model. In other words for acquiring input at the appropriate levels of effective stress, prior to BB introduction into 3DEC, believed to be a project underway. Due to space limitations, constant stiffness BB behavior of rock joints is given elsewhere.

2 BASIC GEOMETRIC INPUT FOR ROCK MASS REPRESENTATION IN MODELS

ISRM has recommended the following key attributes for the characterization of rock discontinuities:

(a) *Physical attributes affecting the engineering properties of discontinuities:*
   1) Roughness
   2) Strength of rock at the discontinuity surfaces
   3) Angles of basic and residual friction
   4) Aperture of discontinuities
   5) Infilling material

(b) *Geometrical attributes defining the spatial configuration of discontinuities:*
   1) Joint orientations (dip & dip direction)
   2) Spacing
   3) Number of sets
   4) Block shape and size
   5) Joint continuity

In this chapter we will be addressing the characterization and quantification of the first 'smaller-scale' set *(a)* of *Physical attributes* in detail, and the effect each of them can have on the physical behavior of the joints. We can use photographs to introduce *(b) Geometrical Attributes* without going into further detail about these larger-scale structural-geology attributes of rock masses, which determine modeled geometries with UDEC-MC, UDEC-BB and 3DEC-MC. (MC Mohr-Coulomb, BB Barton-Bandis).

Figure 1 Characteristics of joint sets as observed in a Finnish open pit and at the portal of an old unsupported road tunnel in Norway (100 years prior to Q-system tunnel support guidance). We see variable orientation (dip and dip direction), variable spacing within each set, variable numbers of joint sets (two to three), variable block shape and size, and variable joint continuity (e.g. 1–10m and discontinuous).

In order to arrive at a credible final output, namely the mechanical properties of discontinuities, the 'characterization' of the physical and geometrical attributes must adopt integrated approaches by combining *observations, measurements* and *judgment*.

Observations will cater for the intrinsic heterogeneity and variability and thus contribute to reducing 'sampling bias'. *Measurement* of the physical and geometrical attributes requires credible techniques that can be applied in the field and/ or in the laboratory in a standardized manner. Several techniques are available including index tests, laboratory tests and in situ tests. *Index tests* are simple, empirical methods, amenable to standardization and easily executable for measuring fundamental 'indices', such as friction, rock strength, roughness, etc. *Laboratory tests* (*e.g.* direct shear, uniaxial compression) are useful for confirmation of engineering properties predicted by index testing, notably when special types of discontinuities are involved (*e.g.* infilled or intensely pre-sheared). *In situ tests* may also be used for deriving parameters at representative geometrical scales and to study behavioral trends of particular critical discontinuity types, such as major weak features (*e.g.* fault zone materials).

Geometrical and other factors such as continuity, block size, history of displacements, etc. need to be taken into account when interpreting the characterization data in order to derive engineering properties. It is at that stage of characterization that expert *engineering judgment* acquires a special role.

3 CHARACTERIZATION AND QUANTIFICATION OF JOINT PROPERTIES

A convenient assembly of the recommended index tests needed for applying the Barton-Bandis BB model is shown in Figure 2. These tests, including the direct shear tests, were used by Barton & Choubey (1977) in their comprehensive research and developments using 130 joint samples collected from road cuttings near Oslo, Norway. The sketches were developed in the form of colored 'over-heads' for lecture courses, and bought together in one figure in Barton (1999).

Fortunately for the more rapid development of the BB model, Bandis (1980) used the same methods for characterization and description of his numerous joint replicas (used in his scale-effect studies) and for his natural joint samples (used for his normal stiffness studies). The suggested parameters from Barton (1973): JRC, JCS and ϕ_b were expanded to include the potentially lower ϕ_r for weathered joints because of the sometimes slightly weathered joints tested by Barton & Choubey (1977). Following Bandis' 1980 Ph.D. studies, the combined techniques for modeling both shear and normal loading were published in Bandis *et al.* (1981) (mostly concerning shear behavior and scale effects) and in Bandis *et al.* (1983) (most concerning normal stiffness behavior). In Figure 2 histograms can be seen for (suggested) presentation of variability within each index test. For example JRC is given with subscripts JRC_0 and JRC_n. These represent nominal 100mm long or larger-scale values, which might be obtained by the a/L method of Barton (1981). This is also shown in Figure 2, and expanded upon later in this chapter.

Since direct shear tests may be performed as part of the site characterization studies, some short notes are provided, which may or may not perfectly conform with suggested methods. However they are the result of collectively performing many hundreds of direct shear tests on rock joints, rock joint replicas, or rougher tension fractures.

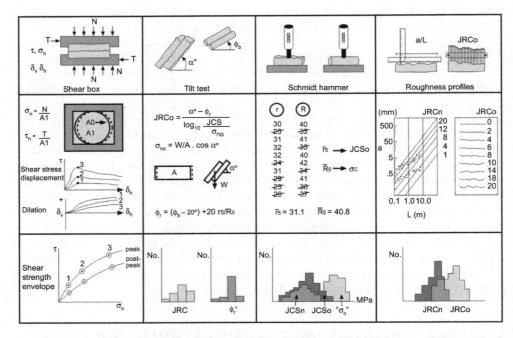

Figure 2 Four columns of diagrams showing 1. direct shear tests principles (Note: apply shear force T 'in-line' to avoid creating a moment), 2. tilt test principles, 3. Schmidt hammer test principles, and 4. roughness recording principles. Each of these simple methods are described in the following paragraphs.

1. *Direct shear tests:* The joint samples may consist of (cored) nearly circular or elliptical, or (sawn) square or rectangular samples, *i.e.* prepared from core, or from sawn blocks recovered from adits or from freshly excavated rock slopes. A strong recommendation is to recover sufficient numbers of representative samples of each joint set of interest, so that multiple testing of the same sample is avoided. The latter tends to 'rotate' the shear strength envelope, when tests at low stress are succeeded by tests at higher stress. An (even more) artificial 'cohesion' intercept is thereby obtained. (See discussion in Barton, 2014). Shear stress-displacement curves and dilation-displacement curves are plotted, and may look similar to the sketches in panel 1.2. The third Panel 1.3 shows 'peak' and ultimate' strength envelopes which will tend to be curved if joints have significant roughness and/or if a significant range of normal stress is applied, such as 0.5 to 5MPa, or 1 to 10MPa. Note that residual strength envelopes are highly unlikely to be reached with just a few millimeters of joint shearing ($\approx 1\% \times L$ may be needed to reach peak, or 1mm in the case of a 100mm long sample. This 1% reduces when testing longer samples). A method of estimating an approximate residual strength based on Schmidt-hammer tests is shown in Figure 2 (combine Panel 3.2 with Panel 2.2). It will be found that $\phi_r < \phi_b$, usually by several degrees if joint weathering (r < R) is significant.

2. *Tilt tests:* It is believed that Barton & Choubey (1977) were the first to apply tilt tests in a 'scientific' way to determine specific 'designer-friendly' joint strength properties, since they showed how both ϕ_b and JRC could be obtained from tilt tests. Because a

sound empirical non-linear shear strength criterion is used (Panel 2.2), the tilt test result from gravity shear-and-normal loading at a failure stress as low as 0,001 MPa can be extrapolated by three to four orders of magnitude higher normal stress. We will of course reproduce the 'standard set' of 100mm JRC profiles very soon, but in the meantime emphasize that many who concentrate (in the last decades) on the exclusive use of 3D-laser profiling of roughness, may be missing some important details of shear behavior by never performing (3D) tilt tests and 'always' criticizing 2D roughness profiles. (The latter were always intended just as a rough guide, and some 400 could have been selected to represent the typical (direct shear tested) JRC values of Barton & Choubey (1977), since 3 × 130 tilt tests were performed and 3 × 130 2D profiles were recorded. (The representative JRC values were however selected from among the single DST tests on the same 130 joint samples). Panel 2.1 represents the tilt test principle for testing the natural joints for back-calculating JRC, shown in Panel 2.2. Panel 2.1 also shows tilt tests on core sticks (these could be sawn blocks). The way in which the basic friction angle ϕ_b is utilized is shown in the last equation in Panel 2.2. In the case of using artificially 'prepared' surfaces for ϕ_b it is important to avoid using 'polished' samples due to slow drilling or slow diamond sawing. Brief sand-blasting should be performed to expose the mineralogy, without adding roughness. If ridges are present across either type of sample then grinding away of the ridges followed by sand-blasting should be sufficient. Values of ϕ_b tend mostly to be in the range 25° to 35°, and most frequently 28° to 32°. However if a single rock type like chalk or limestone is of interest, values may be consistently close to the upper values. Please be aware that 'so-called ϕ_b values' obtained by subtracting dilation angles from peak shear strength may be (dangerously) over-estimated (by as much as 10°), due to neglect of the asperity failure component a_s (which is of similar magnitude to the dilation angle). This will be illustrated later.

3. *Schmidt-hammer tests (for JCS).* Panel 3.1 illustrates, in diagrammatic format, the use of Schmidt-hammer rebounds (respectively r or R) when measuring on natural joint surfaces, and when measuring on artificially 'prepared' surfaces (core-sticks or sawn blocks). In each case, a flat concrete laboratory floor and clamping to a steel 'V-block' base is advised, so that the impact and rebound are not affected by unwanted 'rocking' or other movements. However, to be on the safe side and in order not to have even the effect of crushing a loose mineral grain, the mean of the top 50% of measurements is found to be superior to the normally recommended mean values. This simple technique is shown in Panel 3.2. Artificially low vales are thereby removed as unwanted 'noise', and the remaining 50% tend to be more uniform and therefore more representative. So finally, the mean values of r_5 and R_5 are used to represent, respectively, the *JCS (joint wall compressive strength)* and an approximate measure of UCS (unconfined compression strength). Of course more direct measurement of the latter is usually a part of the site investigation.

4. *Roughness measurement (for JRC).* Panel 4.1 of Figure 2 illustrates the two principal methods for recording joint roughness, and estimating JRC. Panel 4.2 shows in symbolic format, the a/L method and the JRC-profile matching method. A nearly full-scale set of roughness profiles of characteristic 100mm length, with associated JRC_0 estimates, from nearly smooth-planar JRC = 0 to 2, up to extremely rough, undulating JRC = 18 to 20, is reproduced on the next page for ready reference. However tilt testing where possible, or amplitude/length (= a/L) measurements are

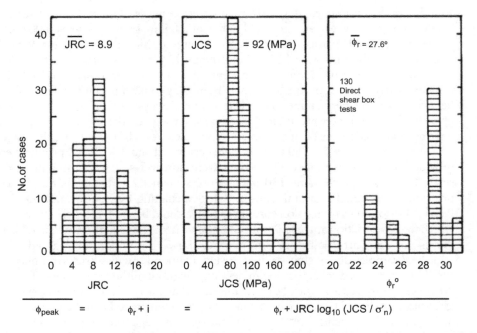

Figure 3 The results of characterization of JRC and JCS and testing (tilt test and DST) one hundred and thirty rock joint samples. Note that the 'i-value' of Patton (1966) is replaced by a stress-dependent logarithmic function incorporating variable (and scale-dependent) roughness, and the ratio of normal stress and joint wall strength, the latter also scale-dependent. The resulting strength envelope is non-linear.

recommended, in addition to profile 'matching', because the latter is inevitably subjective. This was pointed out not only by the first authors, but probably by each of the researchers responsible for a reported 49 equations for JRC (seen tabulated in a 2016 paper review). None were interested in performing tilt tests it seems. Figure 3 reproduces the original results of JRC, JCS and ϕ_r from Barton & Choubey (1977), and Figure 4 provides typical roughness profiles. Several index tests, and test samples are illustrated in Figure 5.

A practical and economic design for a tilt-test apparatus is shown in Figure 6. This was developed while the first author worked in TerraTek, with various joint characterization and testing projects. Today the company is owned by Schlumberger, and one may guess that this petroleum service company is less oriented for fracture characterization of reservoir rocks due to scarcer sources of samples, and the unfortunate tendency (from a rock mechanics point of view) of sectioning core, thereby losing the possibility of testing circular or elliptical samples. Furthermore, there is the remarkable tendency of those practicing reservoir geomechanics of only using linear friction coefficients (from Byerlee, 1978) and linear Mohr-Coulomb strength envelopes for the matrix rock. Both methods have severe limitations in terms of interpreting reservoir behavior because effective stresses rise by tens of MPa with increased production.

Figure 7 illustrates the way that the residual friction angle (ϕ_r) correlates approximately with the Schmidt hammer rebound (r). It is estimated from the empirical

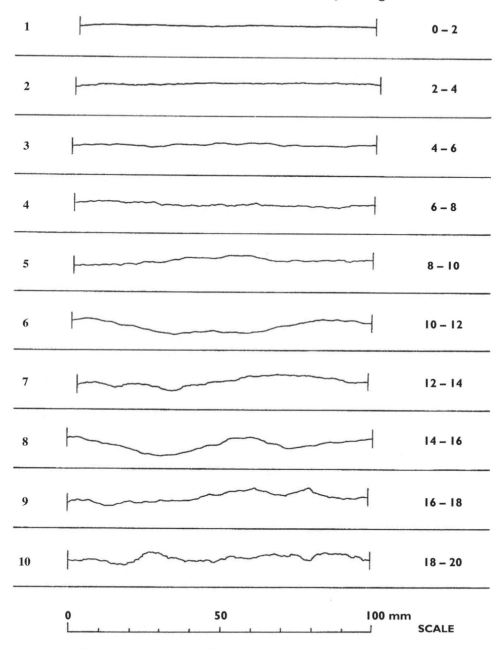

Figure 4 The 100mm long roughness profiles were associated in each case with back-calculated JRC values in the given range, based on direct shear tests of the individual joint samples. In each case, three roughness profiles were recorded on each sample, and one was chosen as representative. In each case, three tilt tests were performed on each sample, so as to predict the DST. Barton & Choubey (1977).

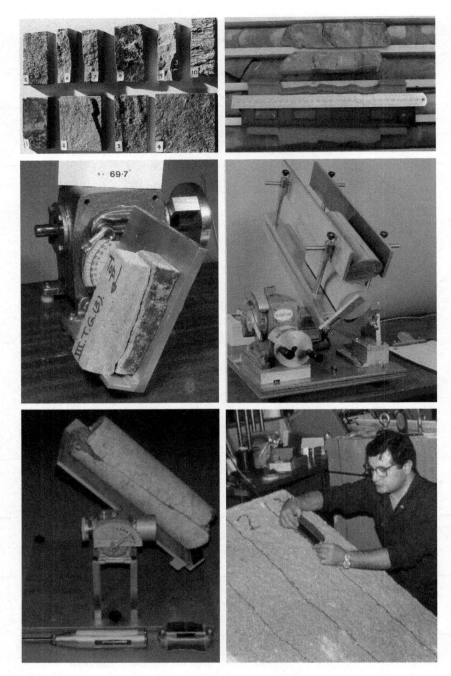

Figure 5 Top-left: a) The ten joint samples profiled in Figure 3. Top-right b) During core-logging (for Q-parameters) JRC is estimated using the profilometer and the a/L method (note magnets holding steel rule). c) One of the tilt-tests performed by Barton and Choubey, 1977. d) An electric-motor driven tilt test of φb using core-sticks (no ridges, no polish). e) Tilt test on large core showing Schmidt hammer and roughness-profiling comb. f) Roughness recording at 150mm and 1300mm scales, on fractured 1m3 blocks, prior to 1.3 ton tilt-tests, followed by biaxial flat-jack shear test (Bakhtar and Barton, 1984).

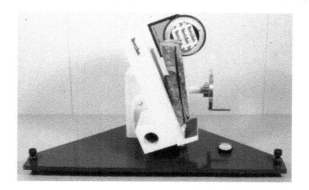

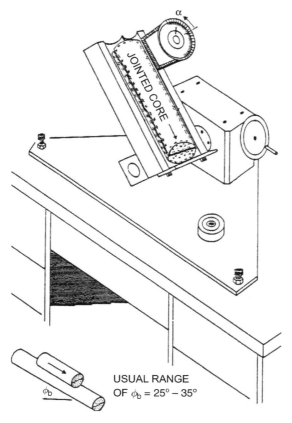

Figure 6 An economic tilt test apparatus, which consists of a triangular steel base-plate with three leveling screws, a circular spirit level, a tilt-angle recorder, a heavy 1: 200 reduction gear (with rotating handle), and a core-shaped or V-notch shaped 'tilt-table' which must be screwed to the gear axle at one end, so that the turning moment remains anti-clockwise (when viewed from front) throughout the tilt test. (One must avoid vibration occurring due to 'gear-slack' just before sliding occurs, as then the correct tilt angle will be missed).

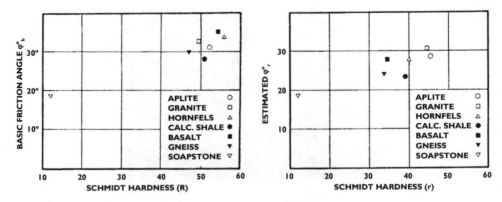

Figure 7 The Schmidt L-hammer is a useful way to register the degree of weathering in a rock joint. This method was used long ago by Richards, 1975 for registering the weathering grades (and low residual friction angles) of joints in sandstones, where values as low as $\varphi_r = 12°$ and r as low as 15 were recorded.

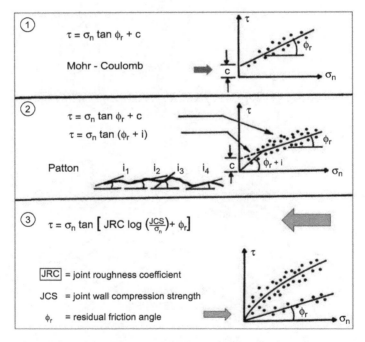

Figure 8 Diagrams and equations representing the most common shear strength criteria for rock joints: Mohr-Coulomb, Patton (1966) and Barton & Choubey (1977), equations 1, 2 and 3.

equation shown in Figure 2 (bottom of Panel 2.2) The reduction from (R) to (r) due to weathering effects was illustrated in Figure 2 (Panel 3.2). As a result of the three components JRC, JCS and ϕ_r we see the non-linearity sketched in Figure 8, in contrast to Mohr-Coulomb linearity, or Patton (1966) bi-linearity.

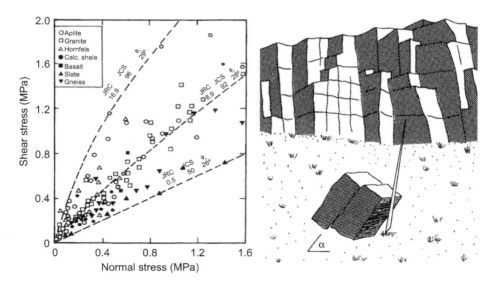

Figure 9 The DST measured peak shear strengths of the 130 joint samples, whose JRC, JCS and φ_r statistics were summarized in Figure 3. The symbolic tilt-test block is designed to emphasize that the shear strength of rock joints can be estimated from index tests carried out on recovered core or blocks of rock, if the latter are freshly exposed in e.g. rock cuttings or open-pit benches. 3D laser and long equations are not needed.

When block-size is taken into account, and we move beyond nominal laboratory $L_0 = 100$mm samples (where JRC_0 and JCS_0 apply), then input data applies to block-size L_n (the mean cross-joint spacing of each set) and we will refer to JRC_n and JCS_n. It is then correct to refer to the Barton-Bandis criterion, as scale effects are accounted for following the block-size scale effect adjustments suggested by Bandis *et al.* (1981). Note the red arrow in Figure 8, next to the actually non-existing cohesion (c) which is an artifact of the M-C linearity assumed. Only joints with steep steps, such as cross-joints, have real cohesion. When testing at very low normal stress (see small blue arrow) the friction angle may become very large, and the 'limit' is the tilt test, commonly performed at 1000 to 10,000 times lower stress than in rock engineering designs. Since envelope curvature is correct, a good estimate of engineering performance is achieved, as verified in Barton & Choubey (1977) who studied and proved the validity of the normal stress 'jump' from 0.001 MPa to 1 MPa (approx.) This is also discussed in Barton (1999), where the important topic of stress transformation errors is introduced: as applying to 45° loaded-direction shear test apparatuses. (This will be introduced later.)

The direct shear test results for joints recovered from the seven different rock types are shown in Figure 9. Deliberate choice of highest and lowest JRC, JCS and ϕ_r values allow the curved upper-most envelope and the lowest, almost linear envelope to be drawn. In the latter JRC is only 0.5 ('smooth, planar') compared to JRC = 16.9 ('rough, undulating') for the upper envelope. The mean results of the three parameters (JRC = 8.9, JCS = 92 MPa and $\phi_r = 28°$, are shown by the central envelope.

Regrettably in *geomechanics for petroleum*, there is an almost universal tendency (oil companies on both sides of the Atlantic and in the Middle East) to use the so-

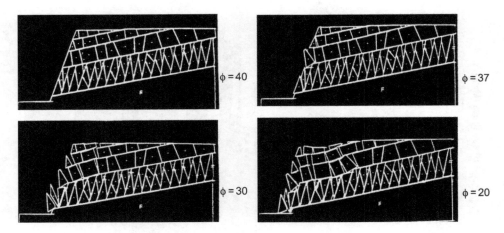

Figure 10 The importance of variable frictional angles on rock slope stability (and indeed on slope deformation characteristics) are nicely illustrated by these early distinct element (μDEC) models which was a method developed by Cundall, and culminated in UDEC (UDEC-MC and UDEC-BB) and 3DEC-MC. These four slope models are from Cundall et al. (1977). The given friction angles applied to all joints in these cases. Today we can model deformable blocks (in UDEC) and differentiate the (possible) non-linear response of the different joint sets (the latter with UDEC-BB, commercially available since 1985).

called 'Byerlee law', in which a (linear) friction coefficient of 0.85 ($\phi = 40.4°$) is assumed to represent 'critically stressed' fracture sets or joint sets. Byerlee was clearly not happy with three joint parameters (JRC, JCS etc.), and generations of (Stanford) researchers and professors have followed his simple (and sometimes very inaccurate) linear and limited friction coefficient approach. The need for more accuracy and acknowledgement of the actual important role of rock type and roughness are nicely emphasized by the following classic μDEC (pre-UDEC) result from Peter Cundall and a former Ph.D. student Mike Voegele. Only the most stable case shown in Figure 10 corresponds to the Byerlee 'law'. The linear 'belief' has been further spread by Zoback (2007). Cross-Atlantic research (on non-linear description for rock joints) does not seem to be popular in geomechanics, despite 65% of remaining petroleum in naturally fractured reservoirs (NFR) with guaranteed non-linear behavior. See further discussion in Barton (2015, 2016).

4 QUANTIFICATION OF JOINT PROPERTIES AT LARGER SCALE

There have been various stages in the profession's acknowledgement of the need for scale-effect adjustment concerning the shear strength of rock joints. In particular, the studies at different scales by Pratt *et al.* (1977) (using in situ tests), by Barton & Choubey (1977) (see Figure 11), by Barton & Hansteen (1979) (using studies with different block sizes in 250, 1,000 and 4,000 tension-fracture block-assemblies) and especially by Bandis (1980) and Bandis *et al.* (1981) (from work with different size replicas of rock joints), leaves one in no doubt about the importance of scale effects. We will also see tilt test results and roughness profiling in relation to typical large-scale JRC_n values, from 130 cm long fractures tested

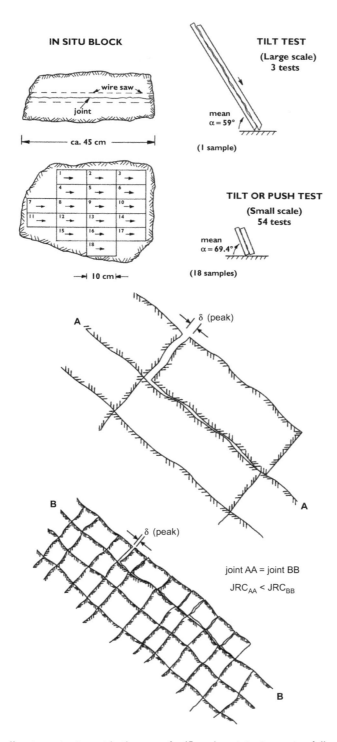

Figure 11 Scale-effect investigation with tilt-tests of a 45 cm long joint in granite, followed by individual tilt tests on eighteen component samples from the same joint, sheared in the same direction. The JRC value increased from 5.2 to 8.8, and eight of the eighteen samples now had to be push-tested as 'so rough'. Shear strength is lower for the largest sample, and the displacement to reach peak (δ_{peak}) is also larger.

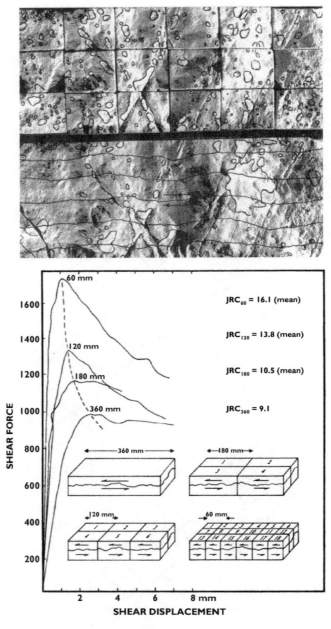

Figure 12 Some of the scale-effect studies of Bandis (1980) which were journal-reported in Bandis et al. (1981). Replicas of natural rock joints could be reproduced many times, and therefore could be tested at different scales. This particular set shows some specific changes in contact areas, reduced JRC and increased δ_{peak}.

by Bakhtar & Barton (1984). JRC_n is lower than JRC_0, and as we shall see, this has a significant effect on shear strength-displacement behavior.

The joint replica tests performed by Bandis (1980), and parallel sets of normal-closure tests on natural rock joints (see later) were each described by the JRC, JCS and ϕ_r parameters developed in Norway in the preceding years. An important summary of the scale-effects observed by Bandis is given in Figure 13. Here we see use of the asperity failure components S_A of Barton (1971), which may be of the same (angular) magnitude as the peak dilation angle. It is not correct to assume that this component is zero, and by subtracting the dilation from peak strength to assume one has reached the 'basic' (flat surface) friction angle. The error may be at least 10°.

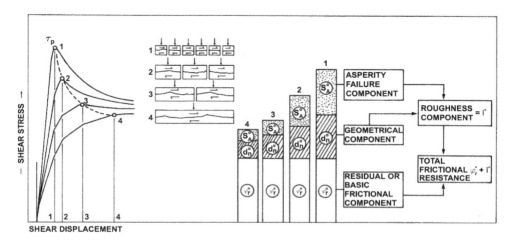

Figure 13 The principal scale-dependent components of the shear strength of rock joints as summarized by Bandis (1980) and Bandis et al. (1981). Note that S_A has almost the same magnitude as the peak dilation angle, at various scales, so subtracting dilation from peak strength leaves two components remaining, not φ_b as assumed by Hencher on various occasions.

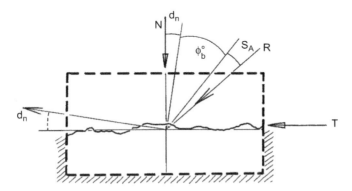

Figure 14 The angular components of shear strength for a non-planar rock joint. Barton, 1971. Note that the presence and influence of dilation d_n requires adjustment of the classic stress-transformation equations for all shearing-and-dilating geotechnical materials. (See Bakhtar & Barton, 1984; Barton, 2006). This important topic will be discussed later in this chapter.

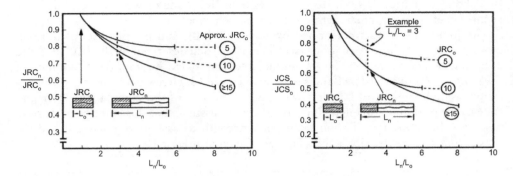

Figure 15 When representing the shear strength of rock masses with the parameters JRC and JCS, it is necessary to consider the mean spacing of joints crossing the joint set of primary interest, in order to apply the scaling Equations 4 and 5 of relevance to this block size (say 2m). The spacing of joints crossing the joint set of secondary interest (say 3m) will define the mean block size for the secondary set.

The scale effects illustrated in Figure 13 are the 'product' of individual scale effects on JRC and JCS, and as a result of combining the results of Barton & Choubey (1977), Barton & Hansteen (1979) and Bandis (1980), the following suggestions for reductions of JRC and JCS with increasing block size were given by Bandis *et al.* (1981). The following equations were recommended:

$$JRC_n \approx JRC_o \, [L_n/L_o]^{-0.02\,JRC_o} \qquad (4)$$

$$JCS_n \approx JCS_o \, [L_n/L_o]^{-0.03\,JRC_o} \qquad (5)$$

The tests shown in Figures 16 and 17 were performed by Barton on tension-fractured brittle model materials, using a double-bladed guillotine for generating intersecting sets of equally-spaced fractures. The tests (which were pre-UDEC, and therefore pre-UDEC-BB) were performed prior to large-span cavern modeling (for underground *nuclear power plant* purposes), using various fracture configurations and stress levels. Although the deformation moduli of the smallest-block models were lowest, the greater freedom for block rotation in these cases gave them higher shear strength, and induced kink-band formation.

A further study of scale-effects was conducted by Bakhtar & Barton (1984) using samples of the type illustrated in Figure 5f and Figure 18. These studies are incorporated in the next topic as they also gave insight into the need for a re-think about stress-transformation (from principle stresses σ_1 and σ_2 onto an inclined plane, in the form of the geotechnically important shear stress (τ) and normal stress (σ_n).

5 STRESS TRANSFORMATION, JRC (MOBILIZED) AND SHEAR STIFFNESS

An important subject that goes beyond the more common distinction that we make between constant normal stress and 'constant' normal stiffness shear testing of rock joints (the latter actually not constant in reality), is the correct transformation of stress.

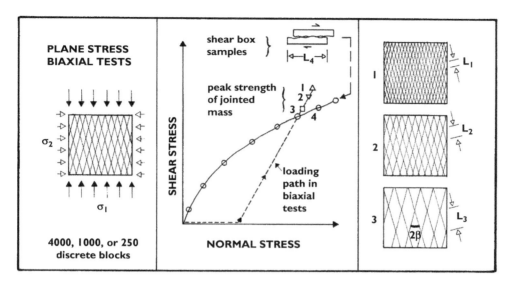

Figure 16 Biaxial shear strength testing of (guillotine-tension) fractured block assemblies consisting of 250, 1,000 and 4,000 distinct blocks. The first set of parallel fractures were continuous, while the second set crossing the first had potential cohesion, due to the steep steps created (Barton & Hansteen, 1979).

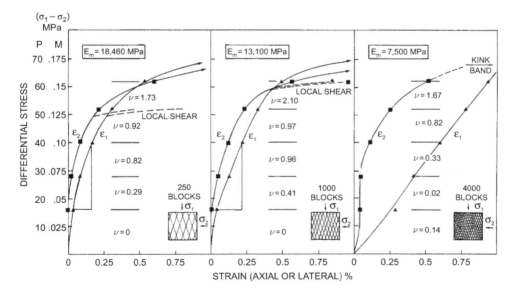

Figure 17 The reductions in block size shown symbolically here, had a consistent tendency to cause higher shear strength. However, in the case of the smallest blocks, whether with 60°/60° or 45°/45° fracture orientation, there was a remarkable 'linearization' of the stress-strain diagrams. Note the strong increases in the 'lateral expansion coefficient' (jointed Poisson's ratio) due to joint shearing (and violation of continuum behavior) (Barton & Hansteen, 1979).

The subject of concern is the transformation of stress from a principal (2D) stress state of σ_1 and σ_2 to an inclined joint, fault or failure plane, to derive the commonly required *shear and normal stress components* τ and σ_n. If the surface onto which stress is to be transformed does not dilate, which might be the case with a fault at residual strength, or for a thickly clay-filled discontinuity, *then* the assumption of co-axial or co-planar stress and strain is no doubt *more* valid.

If on the other hand dilation is involved (as in Figure 14), then stress and strain are no longer co-axial. In fact the plane onto which stress is to be transferred should be an *imaginary plane* since continuity is assumed. Non-planar rock joints, and failure planes through dense sand, or through over-consolidated clay, or through compacted rockfill, are neither imaginary *nor are they non-dilatant* in nature. This problem nearly caused a rock mechanics related injury, when Bakhtar & Barton (1984) were attempting to biaxially shear a series of ten $1m^3$ samples, applying shear and normal stress to the 130 cm long diagonal fractures (which were without weathering effects).

The experimental setup and a tilt test are shown in Figure 18. The sample preparation was unusual because of principal stress (σ_1) controlled-speed-tension-fracturing. This allowed fractures to be formed in a controlled manner, with less roughness than typical for (laboratory-formed) tension fractures. Figures 19 and 20 show the stress application and related stress transformation assumptions, presented in three stages.

The rock mechanics near-injury occurred when a (σ_1-applying) flat-jack burst at 28 MPa, damaging pictures on the laboratory walls and nearly injuring the writer who was approaching to see what the problem was. The sample illustrated in Figure 18 (with the photographer's shoes, pre-test stage) was transformed into ejected slabs, and ejected high-pressure oil, as a result of the explosive flat-jack burst.

These 1.3m long tension fractures gave tilt angles varying from 52° to 70°, and large-scale ($L_n = 1.3$ m) joint roughness coefficients (JRC_n) varying from 4.2 to 10.7. A clear scale effect was exhibited in relation to the 100mm long JRC_0 profiles shown in Figure 4.

The conventional stress transformation Equations 6 and 7, and the dilation-corrected Equations 8 and 9 are given below. It will be noted that a mobilized dilation angle is needed. A dimensionless model for mobilization of roughness (JRC_{mob}) is used, and is seen to have wider application in the BB modeling.

$$\left. \begin{aligned} \sigma_n &= \frac{1}{2}(\sigma_1 + \sigma_2) - \frac{1}{2}(\sigma_1 - \sigma_2)\cos(2\beta) \\[2mm] \tau &= \frac{1}{2}(\sigma_1 - \sigma_2)\sin(2\beta) \end{aligned} \right\} \qquad \text{(6) and (7)}$$

$$\left. \begin{aligned} \sigma_n &= \frac{1}{2}(\sigma_1 + \sigma_2) - \frac{1}{2}(\sigma_1 - \sigma_2)\cos[2(\beta + d_{n\,mob})] \\[2mm] \tau &= \frac{1}{2}(\sigma_1 - \sigma_2)\sin[2(\beta + d_{n\,mob})] \end{aligned} \right\} \qquad \text{(8) and (9)}$$

Angle β is the acute angle between the principal stress σ_1 and the joint or failure plane. The peak dilation angle and mobilized dilation angle can be written as:

$$d_n^0(\text{peak}) = \frac{1}{2}JRC(\text{peak})\log(JCS/\sigma_n') \qquad (10)$$

An estimate of the mobilized dilation angle $d_{n\,(mob)}$ for adding to the joint angle β, is as follows:

Characterization and modeling of the shear strength 21

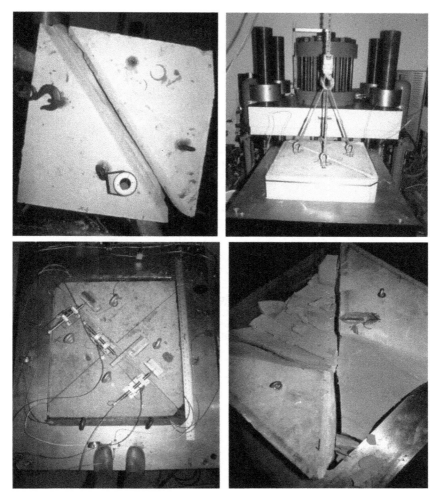

Figure 18 Sample loading test setup and tilt testing of 1m³ samples of rock, hydrostone and concrete by TerraTek colleague Khosrow Bakhtar in the early 1980s (in a pre-Schlumberger era). Note the tilt testing (at 1m³ scale), lowering a lightly clamped sample into a test frame, LVDT instrumentation, and a (rare) sheared sample of an undulating fracture in sandstone (Bakhtar & Barton, 1984).

$$d_n^0(\text{mob}) = \frac{1}{2}JRC(\text{mob})\log(JCS/\sigma_n') \qquad (11)$$

$JRC_{(mob)}$ is an important component of the Barton-Bandis joint behavior criterion. It is shown in Figure 21. It was developed by Barton, 1982 while analyzing the results of TerraTek's ONWI-funded 8m³ *in situ* heated HTM (hydro-thermo-mechanical) block test, which was performed by colleagues Hardin *et al.* 1982, at the Colorado School of Mines experimental mine.

The $JRC_{(mob)}$ concept illustrated in Figure 21 has the effect of 'compressing' a series of shear-displacement curves obtained from DST at widely different normal stresses (*e.g.* see Panel #1.2 in Figure 2) into a narrow band of behavior. Conversely, from the single

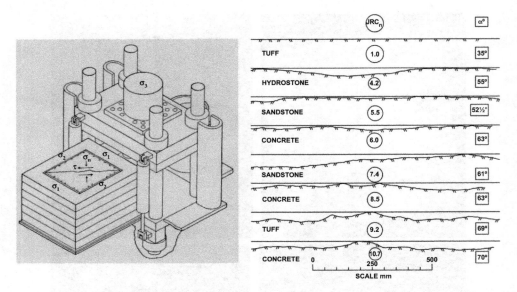

Figure 19 The TerraTek 1m³ 'cube-machine' biaxial (or poly-axial) loading facility. (This facility is now owned by Schlumberger). The roughness profiles were obtained by the simple techniques shown in Figure 5b. The JRC_n values were obtained by back-analysing the large-scale (1300mm long) tilt tests which were performed on the 1 m³ diagonally-fractured samples (Bakhtar & Barton, 1984).

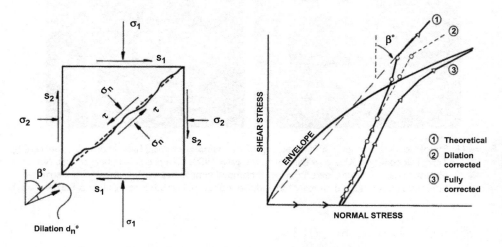

Figure 20 The assumed stress transformation components. In order to minimize boundary friction, molybdenum-greased double-teflon sheets (virtually fluid boundaries) and pairs of stainless-steel, 0 to 30MPa flatjacks were used on all four sides of the 1m³ blocks. It was assumed at first that failure was 'long overdue' when apparently reaching the location #1 in the right-hand diagram. In fact we were only just approaching the strength envelope, at location #3, so shear failure could not yet have occurred (Bakhtar & Barton, 1984).

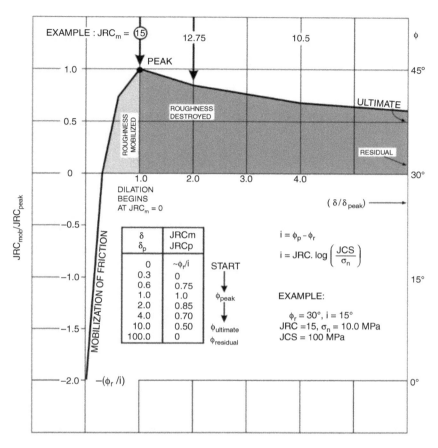

Figure 21 The dimensionless JRC$_{(mob)}$ concept was designed to match the details of joint or fracture behavior during direct shear testing. An example of this is shown in Figure 22. Note the different level of information compared to one 'peak' friction coefficient μ (top point of figure only) as used at present in petroleum geomechanics (Barton, 1982).

JRC$_{mob}$/JRC$_{peak}$ versus displacement δ/δ$_{peak}$ curve shown in Figure 21 we can generate (by hand if necessary) shear stress-displacement (and dilation-displacement) curves for widely different input data (JRC, JCS, ϕ_r) and widely different boundary (stress) conditions.

Figures 22 and 23 demonstrate how the JRC$_{(mob)}$ concept is used to generate stress-displacement (and also dilation displacement) diagrams, for joints of any roughness, or any normal stress level. These were readily generated by hand (Barton, 1982) *i.e.* demonstrating the simplicity of the concept, devoid of 'black-box' software needs, as common in today's commercial software.

Since we now have a simple method of generating shear stress-displacement (and dilation-displacement) curves, we can take the method one stage further and generate shear stress-displacement (and dilation-displacement) curves for rock joints (or jointed blocks of rock) at various scales, using the JRC and JCS scaling Equations 4 and 5 listed earlier, below Figure 14. In fact in the next section, considering joint apertures and joint

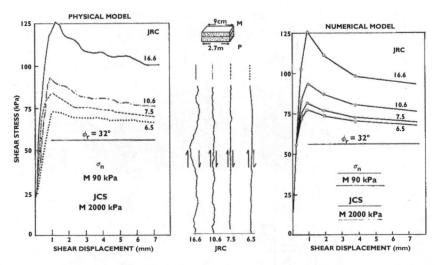

Figure 22 Generation of shear stress-displacement curves for four types of rock joint, as DST tested by Bandis (1980). The 'numerical model' was generated by hand using the JRC$_{(mob)}$ concept, Figure 21 (Barton, 1982).

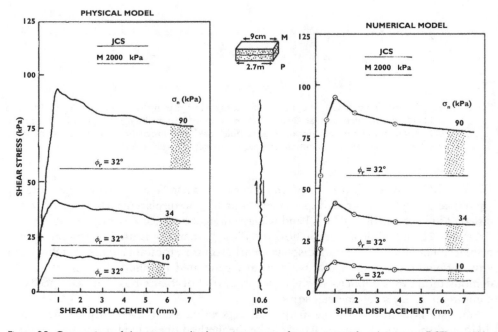

Figure 23 Generation of shear stress-displacement curves for one type of rock joint, as DST tested by Bandis (1980) at three different levels of normal stress. The 'numerical model' was generated by hand.

Characterization and modeling of the shear strength 25

conductivity, we will also be able to see how shearing and dilation affect the conductivity. We are then close to seeing the coupled nature of the Barton-Bandis model, which can be used for modeling rock mass deformation (*i.e.* caused by tunneling) and the joint-related flows (simplified and idealized in 2D) toward the same tunnel.

We may experience that the set of joints suffering (slight) shearing and dilation may not be the same set that conducts most flow to the tunnel. It all depends on the magnitudes of JRC, JCS, ϕ_r and on the magnitudes of the initial hydraulic apertures prior to deformation. In the BB model we convert the dilation-induced aperture $(E + \Delta E)$ into the less altered $e + \Delta e$ hydraulic aperture, using JRC_0 or $JRC_{(mob)}$ depending on whether opening/closing or shear/gouge production is occurring. This is described later in this chapter.

When estimating the values of JRC_n of most relevance to the given joint sets forming the rock mass (example block-sizes of 1m and 2m are shown in Figure 24) it is helpful to utilize the a/L method, which was sketched at small scale in Panels 4.1 and 4.2 of Figure 2. In Figure 25 this simple guide to scale effects is shown at a scale which can be used in practice. The black circles show imaginary a/L data that might have been collected by measuring convenient exposures of the joint set in question. In the case illustrated there is most data at L = 0.2 m and at L = 0.5 m scales. There is only one data point for the imaginary mean block size of 2m. Nevertheless we must make the conservative extrapolation and use an estimate of $JRC_n = 3$ in this case, and check from Equation 4 if this is consistent with the small scale (L = 15 to 20cm) JRC_0 estimates of approx. 5 to 10 seen down to the left-hand side of Figure 25. The final value will be a question of engineering judgment, and may also include a look at the 130cm long profiles in Figure 19.

As noted in the figure caption of Figure 24, and in the inset to this same figure, an estimate of δ_{peak} is required in order to derive appropriate shear stress-displacement (and dilation-displacement) curves. A collection of some 600 DST results for block sizes from 10 cm to more than 3m assembled in Barton (1982) indicated a rather wide spread of data for δ_{peak}. The statistics suggested the following formula as a workable approximation:

$$\delta_{peak} \approx L/500 \, (JRC/L)^{0.33} \tag{12}$$

where δ_{peak} is in meters and L is the block-size in meters.

Examples:

Lab. sample: $L_0 = 0.1m$, $JRC_0 = 15$. Equation 12 gives $\delta_{peak} = 0.0011m$ or 1.1mm.
In situ block: $L_n = 1.0m$, $JRC_n = 7.5$. Equation 12 gives $\delta_{peak} = 0.0039m$ or 3.9mm

As summarized in the figure caption to Figure 24, the double strength *and* δ_{peak} scale effect have a quite dramatic effect on the shear stiffness Ks. Many hundreds of DST data were assembled in Barton (1982) and gave the trends shown in Figure 26 a. Data for clay-filled discontinuities, natural rock joints and model-material joint replicas are shown.

In Figure 26, right-hand figure, a shear-stiffness prediction exercise is performed, using two widely different joint samples as a starting point. The top right-hand corner shows the parameters assumed for a very rough joint in hard unweathered rock (15, 150 MPa, 30°). The bottom left-hand shows the parameters assumed for a more planar weathered joint (5, 50 MPa, 25°). As we can see it is likely that shear stiffness Ks will often lie within the range of 0.1 to 1 MPa/mm for typical in situ rock block sizes and moderate (civil engineering) stress levels. In a later section of this chapter such low values will be

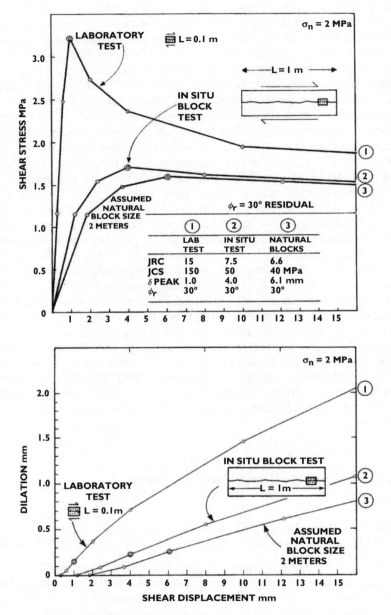

Figure 24 Generating shear-displacement-dilation behavior, for three different block sizes (Barton, 1982). Note the inset showing the scaling assumptions from Bandis et al. (1981) Equations 4 and 5. It will immediately be noted that, as experienced in practice, there is an increase in δ_{peak} as block size increases. Since there is also a reduction in peak shear strength, the peak shear stiffness Ks suffers a double scale-effect, and is lower than the values most numerical modelers are familiar with. This is shown later.

Characterization and modeling of the shear strength 27

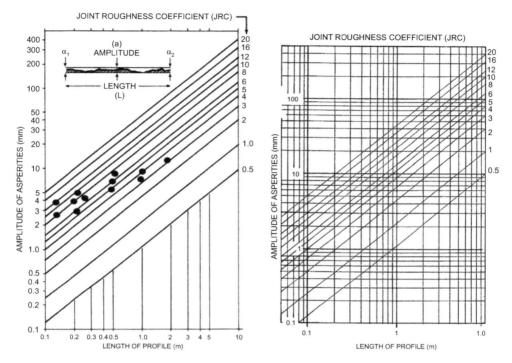

Figure 25 An illustration of the a/L method for estimating the large-scale JRC$_n$ value. Note that at L = 10cm scale JRC$_0$ ≈ 400 a/L, while at L = 1m scale, JRC$_n$ ≈ 450 a/L. A field-logging sheet is given on the right, with interpolation down to 5cm scale for use when core-logging and needing to estimate JRC$_0$ (Barton, 1981).

contrasted with the much higher (x 50?) values of normal stiffness Kn, which of course emphasizes the fundamental anisotropy of real rock masses, a property lost and forgotten in most continuum analyses, and even lost by some UDEC modelers.

6 THE CONDUCTIVITY OF ROCK JOINTS AND THE EFFECTS OF DEFORMATION

The theoretical Hagen-Poiseuille equation for the hydraulic conductivity (K) of a smooth parallel-plate, during laminar flow is:

$$K = (\rho g/\mu) e^2/12 = ge^2/12v \tag{13}$$

where K is in units of velocity (LT^{-1}), g = gravity acceleration 981 cm/s/s (LT^{-2}), e = equivalent parallel plate (smooth wall) aperture (L), v = coefficient of kinematic viscosity of the fluid (L^2T^{-1}) (where $v = \mu/\gamma$ = viscosity / density has units of $ML^{-1}T^{-1}/ML^{-3} = L^2T^{-1}$).

There are (at least) two types of joint aperture that need be considered when modeling the effect of joint deformation, namely the physical aperture (E) and the (theoretical) hydraulic aperture (e). A key issue is how to correlate the hydraulic apertures (e) to the generally larger physical apertures (E). The following empirical conversion formula was

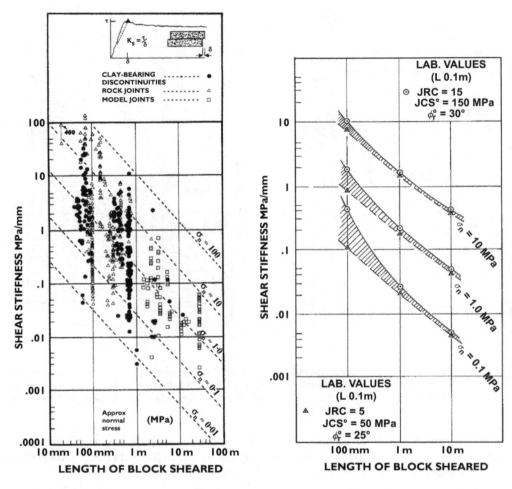

Figure 26 Left: Several hundred DST data for a variety of joints, filled-discontinuities and model-material replicas of joints. Note the strong scale effect, and also the approximate influence of the effective normal stress. Right: Predicted block-size scale and stress effects for a rough joint in hard rock, and for a smoother joint in weaker weathered rock. Care is needed when selecting Ks for modeling, as instruction manuals usually have the ratio Kn/Ks too small, even 1.0. (Kn = normal stiffness). This is incorrect (Barton, 1982).

developed by Barton, 1982 and more widely published by Barton *et al.* 1985. It is for correlating E and e in relation to roughness (JRC). This formula applies to normal closure effects. Olsson and Barton, 2002 extended the modeling of E and e to the case of (potentially) gouge-producing shearing effects, in this case involving joint aperture conversion using $JRC_{(mob)}$. This is shown later, in Figure 28.

$$E = (e \cdot JRC^{2.5})^{1/2} \tag{14}$$

where e and E are in units of μm.

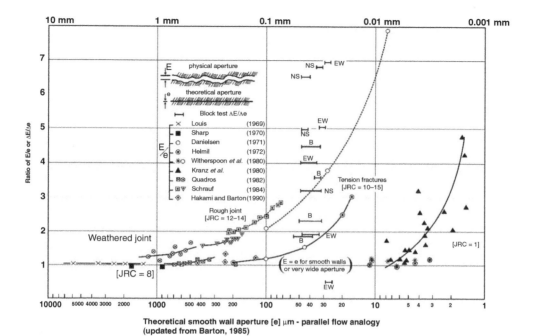

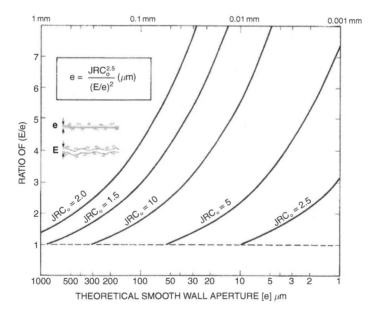

Figure 27 Top: Test data from which apertures e and E (or Δe and ΔE) were measured (Barton et al., 1985, updated by Quadros in Barton & Quadros, 1997. Bottom: The empirical model for converting between these two apertures was developed in Barton (1982) and became a part of UDEC-BB in 1985.

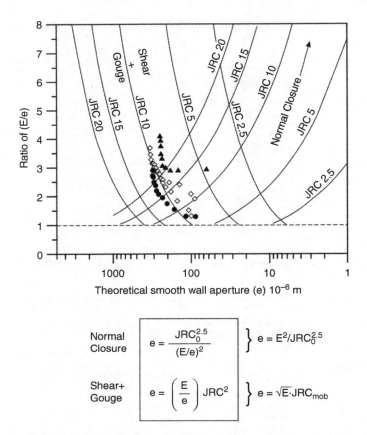

Figure 28 Coupled shear-flow tests performed by Olsson, which were summarised in Olsson and Barton (2001). In this figure the modification required to account for possible gouge-production during shear is shown in the 'left-hand' curves together with some data points for CSFT with samples having JRC$_0$ values of 7.2, 8.8, 9.7 and 12.2. The 'right-hand' curves are for closure/opening modeling, as in Figure 27. This figure shows Equation 14 ('normal closure') while 'shear + gouge' involving JRC$_{(mob)}$ can be listed as Equation 17.

The kinematic viscosity ν of water is 0.01 at 20° C and γ = 1 gm/cm^3. Checking units again we see that ν = μ/γ = viscosity / density, with units $ML^{-1} T^{-1}/ML^{-3} = L^2/T$). It follows that:

$$K = ge^2/12\nu = 8175\, e^2 \,(cm/s) \tag{15}$$

Substituting e ≈ $E^2/JRC_0^{2.5}$ yields the following formula for hydraulic conductivity:

$$K \approx 8175\, [E^2/JRC_0^{2.5}]^2 \times 10^{-10}\, m/s \tag{16}$$

Observations published by Makurat *et al.* (1990) indicated that CSFT (coupled shear flow tests) could cause gouge-production during shearing if stress levels were high in relation to JCS wall strengths. This could compromise both the physical (E) and

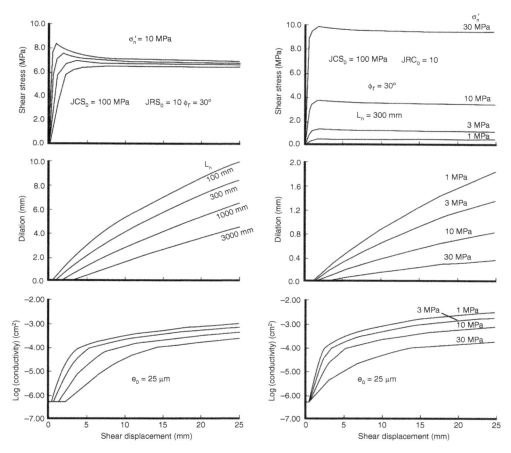

Figure 29 Examples of 'coupled' shear-dilation-conductivity modeling with the Barton-Bandis modeling assumptions. When block-size variations are involved (left) the delayed dilation and therefore delayed conductivity change can be noted. These curves were produced in 1983 by Bakhtar using a programmable HP calculator and the BB equations by now assembled in Barton (1982). ONWI and AECL funded work were responsible for the 'finalization' of the BB model prior to its programming (by Mark Christiansson of Itasca) into the distinct element code UDEC-BB (Barton & Bakhtar, 1983, 1987).

hydraulic aperture (e) as interpreted from the 'cubic law'. Logically speaking there would be a whole range of stress/strength ratios (high strength, low stress) in which gouge or damage would be minimal. In those cases, the conversion involving the empirically derived roughness factor $JRC_0^{2.5}$ could be used in place of the Olsson and Barton equation with $JRC_{(mob)}$.

Examples showing the (BB-predicted) effect of shearing and dilation on the conductivity of joints of variable size, or on single samples tested at varied normal stress, are shown in Figure 29. Note that for simplicity a 'starting' aperture of e = 25µm has been assumed in each case. As we shall see in the next section concerning normal stiffness, this assumption of an unchanged 25µm is likely to be erroneous when changes of

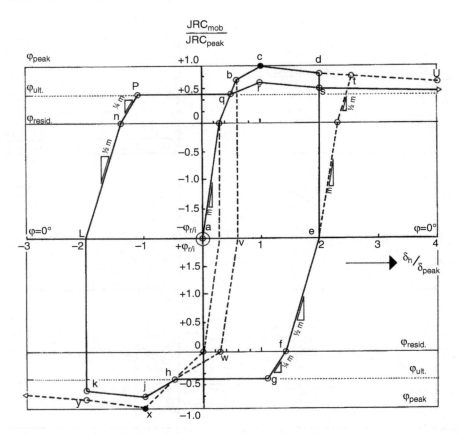

Figure 30 Reversed shear logic based on the dimensionless $JRC_{(mob)}$ concept. This model was developed in Barton, 1982 and shows gradual degradation of shear strength. A related model for the associated dilation has the shape of a 'saucer' which is thickening at its base by decreasing amounts with each cycle of reversal.

normal stress are being modeled. Of course in the UDEC-BB program, the 'starting' apertures are calculated more correctly from the following normal stiffness behavior.

Although unlikely to be noticed by petroleum geomechanics modelers who are satisfied with linear friction coefficients and linear Mohr-Coulomb, these 'delay's may have influence on what should be the desired effect of massive hydraulic fracturing concerning the 'accompanying' microseismic evidence of shearing of natural fractures in the gas shales. This occurs at larger distances from the central elliptic regions of sand-propped fractures (Barton, 2015).

As a result of seismic loading involving potential *reversals of shearing direction*, or as a result of some particular rock engineering excavation sequences which might result in a reversed shearing direction, one needs to consider what is likely to happen to shear strength. In Figure 30, having due consideration of some reversed-shear DST appearing in the literature some decades ago, Barton, 1982 formulated the 'degrading of roughness' JRC related model shown in Figure 30. Progressing through some reversed cycles of shear would of course have the result of compromising some of the dilation-related permeability increases, and the 'gouge-production' adjustment shown in Figure 28 would obviously apply with successively renewed strength, due to accumulating damage and inevitable gouge-debris accumulations.

7 NORMAL CLOSURE OF JOINTS AND HOW TO MODEL IT

The diagrams and equations assembled in Figure 31 show how Bandis (1980) formulated the normal closure behavior of rock joints, using the JRC and JCS parameters previously detailed in Barton & Choubey (1977) and now easily acquired by performing the index tests shown in Figure 2. An important detail to note about normal closure behavior is that every sample tested has been unloaded and disturbed during the recovery period. This applies more to core than block samples, if the latter are 'banded' with steel belts prior to transport. This of course also applies to samples which will be tested in shear. However in this second case one is concerned about behavior of millimeter scale, while the closure of tight rock joints might be measured in a few tens of microns. So near-removal of the effects of sample recovery by performing load-unload cycles becomes an important part of the testing procedure. The arbitrary but practical assumption is made that after about three to four load-unload cycles there is so little change that behavior can be considered as representative of undisturbed behavior.

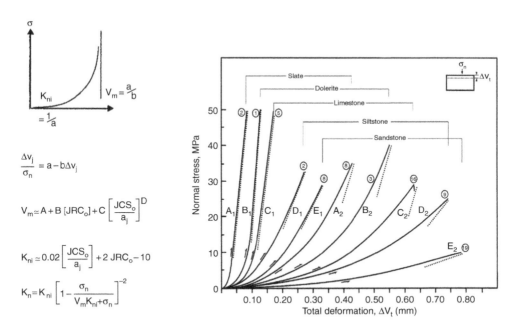

Figure 31 Left: The hyperbolic function (Equation 18a) and related equations (Equations 18b,c,d) chosen by Bandis (1980) to describe the highly non-linear behavior of closing interlocked rock joints. (This method and numerous experimental results were published in Bandis et al. (1983)). The normal stiffness expression is a derivative of the hyperbolic function. Right: Experimental evidence acquired by Bandis show strong rock-type/rock joint-type dependence, but of course some of this can be explained by rock-type-characteristic values of JCS and JRC.

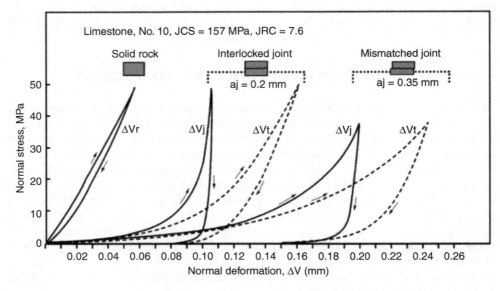

Figure 32 It is appropriate to 'define' normal loading behavior of rock joints by indicating the subtraction that has to be made of the elastic portion of the deformation associated with the intact matrix on either side of the joint. The net joint deformation is highly non-linear, but is made less so by shear deformation, or by mismatching (absence of interlock) (Bandis et al. 1983).

Figure 32 shows how normal stiffness behavior is interpreted as a net deformation of the joint, by subtracting the intact-rock deformation from the monitored behavior of load-unload cycles (sets of three are shown in Figure 33) from the 'rock-plus-joint' overall deformation. Wide differences in behavior from rock type to rock type and from joint type to joint type are indicated in Figure 33. A 'reservoir drawdown-and-injection' scenario is also demonstrated in Figure 33 (right-hand diagram).

8 CASCADING PROGRESSIVE FAILURE OF JOINTED ROCK MASSES

There is an almost 'universal' belief (by those not using the non-linear JRC/JCS model) that the shear strength of rock joints consists of cohesion and friction, and that one can add $c + \sigma_n \tan \phi$. The assumed cohesive strength is actually a purely 'arithmetic' construction due to linearized strength envelopes, and represents something hypothetical in relation to the reality of increasing curvature (friction angle plus dilation angle increase) experienced close to zero normal stress. Barton (1971) measured this lack of actual cohesion, even for the case of extremely rough tension fractures. The 'total friction' angle may reach 80° or more when normal stress is extremely low. The 'cohesion intercept' is an arithmetic convenience, but potentially exaggerates the shear strength actually available.

When on the other hand we consider the possible shear strength of *rock masses*, the same Mohr-Coulomb equation automatically comes to most people's minds. It is

Characterization and modeling of the shear strength 35

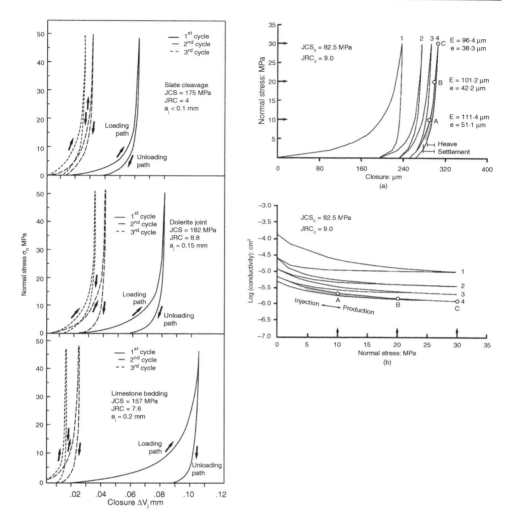

Figure 33 Left: Examples of normal stress-closure cycles for a cleavage joint in slate, for a rougher joint in dolerite, and bedding in hard limestone. Right: Using the BB-plotting routine programmed by Bakhtar on an HP calculator, to demonstrate the consolidating cycles seen on the left, and the assumed 'undisturbed behavior' 4th cycle, onto which the modeled apertures e and E have been written at 10, 20 and 30 MPa. A 'reservoir drawdown and injection' effect have been shaded on the conductivity curves (Barton et al., 1985).

assumed that 'c' can again be added to '$\sigma_n \tan \phi$'. In this case there may be a genuine cohesive strength component due to the 'necessary' failure of 'intact bridges', where joint sets do not 'line-up' as potential failure surfaces. However this time there is another type of problem. While the cohesion is no longer hypothetical but real, the problem is that it fails at much smaller strain than the mobilization of frictional strength along the newly formed fresh and rough fracture surfaces. These in turn may be stiffer than the lower strength (and possibly weathered) natural joints (*i.e.* those

36 Barton & Bandis

Table 1 A crude (and probably non-conservative) way to account for three of the components illustrated in Figure 33. Cohesion is ignored (this is conservative) but the three remaining fracture, joint, and filled-discontinuity components are imagined, for ease of hand calculation, to mobilize at the same strains.

Feature	Joint #1	Minor fault	Joint #3	Intact bridge (failed at small displacement (*)
Small-scale strengths		$Jr = 1.5$		
JRC_o	14	$Ja = 4$	12	18
JCS_o	75		55	150
Phi r	29°	$\tau/\sigma_n = 1.5/4$	28°	32°
Large-scale		$Jr = 1.5$		
strengths(est.)		$Ja = 4$		
JRC_n	9		6	12
JCS_n	45	$\tau/\sigma_n = 1.5/4$	30	90
Phi r	29		28	32°
Block-size Ln (m)	2 to 4 m		3 to 5 m	1m
Large-scale undulation (?)	(+ i = 4°)	(+ i = 0°)	(+ i = 2°)	(+ i = 4°)
Partial safety factors	1.2	1.3	1.2	1.1
Final strength components	45°/1.2 = 37.5°	21°/1.3 = 16°	37°/1.2 = 31°	56°/1.1 = 37.5°
Length (m)	50	25	30	20

Weighted mean strength 34°

$$\frac{37.5° \times 50 + 16° \times 25 + 31° \times 30 + 51° \times 20}{50 + 25 + 30 + 20}$$

(*) Clearly it is extremely conservative to ignore an intact rock cohesion of e.g. 25 MPa

capable of shearing because of adverse orientations. There may also be clay-filled shear zones or faults, with their smaller-and-later response.

A crude and non-conservative way to make allowance for the different components, and assume (when estimating by hand) that contributions can be added as if occurring at the same strain, but with the smallest-strain cohesion ignored, is illustrated in Table 1. With the aid of a computer one could mobilize the various components at their respective strains. How often do we see this done? Since the answer is clear we must conclude that it is time for change.

9 A BRIEF COMPARISON OF MOHR-COULOMB AND BARTON-BANDIS MODELING

Rock masses may range from almost intact, through well jointed, to heavily crushed, due to increased proximity to fault zones. The result is variable geometrical patterns resulting from several types of joint sets with their variable roughness and continuity. Notwithstanding an implied need for engineering rationalization, the assessment of strength for such complex media as rock masses cannot be approached on the basis of a single generic strength criterion.

The type, frequency and orientation of the jointing and faulting define the likely modes of deformation, and some indication of the likely ultimate failure mechanism. In significant volumes of rock there may be two or three classes of discontinuities (natural and stress-induced fracturing) which can become involved in the pre-peak and post-peak deformation and failure. Rock mechanics practitioners to date have generally adopted one of the following two approaches for the characterization and engineering study of *jointed* rock.

- A *discontinuum* approach, in which the geologic structure is explicitly represented and in turn controls the modes of deformation and mechanisms of ultimate failure (prior to modeling of appropriate rock mass reinforcement).
- A *continuum* approach, which involves a semi-empirical simulation of the rock mass, transforming the *in situ* (actual) discontinuous state into a hypothetical continuous medium, in which the weakening and softening influence of jointing is allowed for implicitly

Due to the complexity we must resort to numerical UDEC-BB or 3DEC simulations. A useful starting point, and a demonstration of the fundamental differences between M-C and B-B can be gained by performing simulations of large scale biaxial and triaxial tests. These give a useful insight into mechanisms at failure, and comparisons of shear strength estimates based on the above non-linear strength criteria, with linear M-C criteria are quite revealing. Of course predictions from "global" continuum strength criteria (an example would be GSI-based H-B Hoek-Brown) are quite different to both. Contrasting B-B and M-C behavior for an equally jointed 'rock mass' sample are shown in Figure 34.

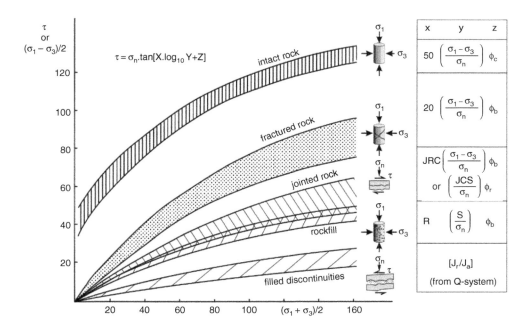

Figure 34 The components of the shear strength of rock masses, as assembled in Barton (1999). When considering the shear strength of rock masses with possible intact portions of rock along the potential failure surface(s) it must be realized that these four components (excluding rockfill) are unlikely to resist shear failure at the same shear strain. A 'cascading' progressive mobilization is involved from the upper to the lower curve at successively increasing shear strain (assuming filled discontinuities are also present).

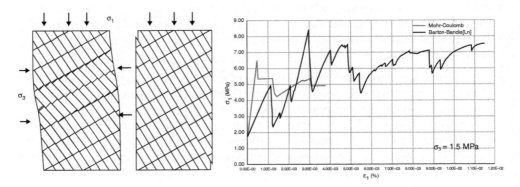

Figure 35 Contrasting stiffness and strength behavior, assuming non-linear (BB – left block plot) and linear (MC – right block plot). Due to the non-linearity of BB simulations compared to the linear (or bi-linear 'ramping') in the case of MC, significant block rotations are seen with BB which tend to be absent with MC. A comprehensive set of comparisons with numerous geometric and boundary conditions are given by Barton & Bandis (2017, in preparation).

Such diverging approaches, coupled with the inherent complexity of jointed rock behavior, have created unfortunate barriers that, to date, have prevented comprehensive and generally accepted approaches to rock mechanics, as needed for solving rock engineering problems. Discontinuum and continuum mechanics approaches are applicable to specific rock mass conditions, and *cannot be used interchangeably*, as summarized below:

– The *continuum* (implicit) approach is appropriate where the frequency and orientation of jointing are such that no preferential paths of stress-strain responses are present. Such conditions are present in an estimated <10% of rock masses.
– The *discontinuum* (explicit) approach is appropriate when the geological structure controls anisotropy, deformation modes, and strength. Such conditions are applicable to the vast majority (>90%) of rock masses.

10 CONCLUSIONS

1. The authors are aware of some 50 equations for evaluating JRC. While this should be considered gratifying, the great majority are rather complex descriptions of a topological nature, or linked to 3D laser profilometric analysis. The simple performance of tilt tests seems mostly to have escaped those analyzing roughness and 'improving' JRC.
2. This chapter addresses the recommended methods of describing and performing index tests of rock joints by means of the two basic parameters JRC and JCS suggested by Barton, 1973.
3. Due to the considerable further work of the two authors (including that with Choubey and Bakhtar) and subsequent correction for block-size scale effects, mostly from Bandis, the methods of predicting non-linear joint behavior (stiffness,

strength, dilation, conductivity, each with deformation) have become known as the Barton-Bandis or BB model. We must thank Bandis for his unmatched contribution during too short a lifetime (1951–2016 †)

4. At the end of this chapter we have shown the fundamental differences between linear (Mohr-Coulomb, MC) and non-linear BB. The differences are worthy of attention, and will be illustrated in detail in our book.

REFERENCES

Bakhtar, K. & Barton, N. 1984. Large scale static and dynamic friction experiments. Proc. of 25th US Rock Mech. Symp., Northwestern Univ., Illinois.

Bandis, S. 1980. Experimental studies of scale effects on shear strength, and deformation of rock joints. Ph.D. Thesis, Dept. of Earth Sciences, Univ. of Leeds.

Bandis, S., Lumsden, A. & Barton, N. 1981. Experimental studies of scale effects on the shear behavior of rock joints. *Int. J. Rock Mech. Min. Sci. Geomech. Abstr.* 18: 1– 21.

Bandis, S., Lumsden, A.C. & Barton, N. 1983. Fundamentals of rock joint deformation. *Int. J. Rock Mech. Min. Sci. Geomech. Abstr.* 20: 6: 249–268.

Barton, N. 1971. A relationship between joint roughness and joint shear strength. Proc. of Symp. of Int. Soc. Rock Mech. Rock Fracture, Nancy, paper I.8, 20p.

Barton, N. 1973. *Review of a new shear strength criterion for rock joints.* Engineering Geology, Elsevier, Amsterdam, Vol. 7, pp. 287–332. Also NGI Publ. 105, 1974.

Barton, N. & Choubey, V. 1977. The shear strength of rock joints in theory and practice. *Rock Mech.* 1: 2: 1–54. Springer, Vienna. Also NGI Publ. 119, 1978.

Barton, N. & Hansteen, H. 1979. Very large span openings at shallow depth: Deformation magnitudes from jointed models and F.E. analysis, 4th Rapid Excavation and Tunnelling Conference, RETC; Atlanta Georgia, Vol. 2: 1131–1353. Eds A.C. Maevis and W.A. Hustrulid. American Institute of Mining, Metallurgical, and Petroleum Engineers, Inc. New York, New York, 1979.

Barton, N. 1981. Shear strength investigations for surface mining. 3rd Int. Conf. on Stability in Surface Mining, Vancouver, Ch. 7: 171–192, AIME.

Barton, N. 1982. Modelling rock joint behaviour from in situ block tests: Implications for nuclear waste repository design. Office of Nuclear Waste Isolation, Columbus, OH, 96 p., ONWI-308, September.

Barton, N. & Bandis, S. 1982. Effects of block size on the shear behaviour of jointed rock. Keynote Lecture, 23rd US Symp. on Rock Mech., Berkeley, California.

Barton, N., Bandis, S. & Bakhtar, K. 1985. Strength, deformation and conductivity coupling of rock joints. Int. J. Rock Mech. Min. Sci. Geomech. Abstr. 22: 3: 121–140.

Barton, N. & Bakhtar, K. 1987. Description and modelling of rock joints for the hydro-thermal mechanical design of nuclear waste vaults. Atomic Energy of Canada Limited. TR-418. Vols I and II. (Published following TerraTek, Salt Lake City contract work in 1983).

Barton, N. & Bandis, S.C. 1990. Review of predictive capabilities of JRC-JCS model in engineering practice. Int. Symp. on Rock Joints. Loen 1990. Proceedings, pp. 603–610.

Barton, N. 1995. The influence of joint properties in modelling jointed rock masses. Keynote Lecture, 8th ISRM Congress, Tokyo, 3: 1023–1032, Balkema, Rotterdam.

Barton, N. & Quadros, E.F. 1997. Joint aperture and roughness in the prediction of flow and groutability of rock masses. Proc. of NY Rocks '97. Linking Science to Rock Engineering. Ed. K. Kim, pp. 907–916. Int. J. Rock Mech. and Min. Sci. 34: 3–4.

Barton, N. 1999. General report concerning some 20th Century lessons and 21st Century challenges in applied rock mechanics, safety and control of the environment. Proc. of 9th ISRM Congress, Paris, 3: 1659–1679, Balkema, Netherlands.

Barton, N. 2014. Shear strength of rock, rock joints, and rock masses: problems and some solutions. Keynote lecture, Eurock, Vigo, 14p.

Barton, N. 2015. Where is the non-linear rock mechanics in the linear geomechanics of gas shales? Keynote lecture. China Shale Gas Conf., Wuhan.

Barton, N.R. 2016. Non-linear shear strength descriptions are still needed in petroleum geomechanics, despite 50 years of linearity. 50th US Rock Mech. Houston. ARMA 16, paper 252, 12 p.

Barton, N. and Bandis, S. (2017 inpress). Engineering in jointed and faulted rock. Taylor and Francis, est. 750p.

Cundall, P.A., Fairhurst, C. &Voegele, M. 1977. Computerized design of rock slopes using interactive graphics for the input and output of geometrical data. Proc. of 16th US Symp. on Rock Mech. (1975), ASCE, New York.

Makurat, A., Barton, N., Rad, N.S. & Bandis, S. 1990. Joint conductivity variation due to normal and shear deformation. Int. Symp. on Rock Joints. Loen 1990. Proceedings, pp. 535–540.

Olsson, R. & Barton, N. 2001. An improved model for hydro-mechanical coupling during shearing of rock joints. *Int. J. Rock Mech. Min. Sci. and Geomech. Abstr.* 38: 317–329. Pergamon.

Patton, D.F. 1966. Multiple modes of shear failure in rock and related materials. Ph.D. Thesis, University of Illinois, 282p.

Pratt, H.R., Swolfs, H.S., Brace, W.F., Black, A.D. & Handin, J.W. 1977. Elastic and transport properties of *in situ* jointed granite. Int. J. Rock Mech. Min. Sci. 14: 35–45. UK: Pergamon.

Zoback, M.D. 2007. Reservoir geomechanics. Cambridge Univ. Press. 461p.

Chapter 2

Statistical fracture toughness study for rocks

M.R.M. Aliha[1] & M.R. Ayatollahi[2]

[1]Welding and Joining Research Center, School of Industrial Engineering, Iran University of Science and Technology (IUST), Narmak, Tehran, Iran
[2]Fatigue and Fracture Lab., School of Mechanical Engineering, Iran University of Science and Technology (IUST), Narmak, Tehran, Iran

Abstract: The presence of heterogeneity, bedding planes, porosity, inherent defects etc. in rocks inevitably generates large scatters in the strength and fracture toughness values which are obtained from fracture tests on rocks. Hence, statistical analyses may provide better estimations for investigating the mechanical behavior of rocks. In this research, using several fracture toughness tests, mode I and mode II fracture toughness was studied statistically for two different rocks (*i.e.* limestone and marble) tested with different geometry and loading configurations. Disk type specimens including straight through center crack Brazilian disk subjected to diametral compression and straight edge cracked semi–circular bend and chevron notched Brazilian disk specimens which are among common rock fracture toughness testing configurations were employed for experiments. It was observed that the shape and geometry of a specimen and also its loading mode can affect significantly the scatter in fracture toughness test data. It is shown that the observed differences between the experimental results can be estimated well if the effects of higher order stress terms (T and A_3) are considered via a modified maximum tangential stress criterion. The obtained statistical results are then predicted using two- and three-parametric Weibull distribution models. The failure probability curves of tested specimens are also evaluated successfully in terms of the failure probability curve of a reference mode I sample.

I INTRODUCTION

In rock mechanics, strength and integrity evaluation of rock masses is a very important task for investigating performance and durability of rock structures. In comparison with other engineering materials like metals or polymers, rock masses usually contain numerous cracks, inherent discontinuities, flaws, bedding planes and natural fractures which make them very susceptible to sudden failures at loads much lower than those expected for intact rocks. Hence, despite the availability of the conventional strength criteria like Mohr–Coulomb (Coulomb, 1773; Labuz & Zang, 2015) which are used for evaluating the load bearing capacity of intact rock masses, more precise analyses of naturally fractured or cracked rock masses require that the influence of cracks and discontinuities inside rock structures should be taken into account. Fracture mechanics has been known as a suitable framework for integrity assessment of cracked rock masses.

Fracture toughness is an important parameter for classification of rocks, design of rock structures and analysis of rock related problems. This parameter which defines the resistance of rock material against crack propagation is determined experimentally using suitable test configurations. A review of rock fracture mechanics literature indicates that most of the specimens used for rock fracture toughness testing have disk or cylindrical shapes. This is mainly due to the convenience of specimen preparation from rock cores. Meanwhile, because of difficulties related to introducing a fatigue pre-crack in rocks, chevron notch cracking is a suitable and preferred technique instead of straight cracking for these materials. Accordingly, till now four cylindrical shape fracture test specimens, namely the cracked chevron notch short rod tension specimen (Ouchterlony, 1998), the chevron notched cylindrical specimen subjected to three-point bending (Ouchterlony, 1998), the cracked chevron notched Brazilian disk (CCNBD) subjected to diametral compression (Fowell, 1995) and the semi-circular bend (SCB) specimen subjected to symmetric three-point bend loading (Kuruppu *et al.*, 2015) have been suggested by the International Society for Rock Mechanics (ISRM) for determining a versatile and reliable value of mode I fracture toughness (K_{Ic}) for rock materials. It is widely accepted that the tensile type mode I fracturing is the most important mode of fracture in rock masses. Many researchers have therefore investigated the mode I fracture behavior of rock materials either using ISRM suggested rock samples (Khan & Al-Shayea, 2000; Aliha *et al.*, 2006; Ingraffea *et al.*, 1984; Nasseri *et al.*, 2010; Zhou *et al.*, 2010; Siren, 2012; Whittaker, 1992) or other mode I test configurations (Funatsu *et al.*, 2004; Xeidakis *et al.*, 1997; Krishnan *et al.*, 1998; Tutluoglu & Keles, 2011). However, in practice the pre-existing cracks in rock masses and rock structures are usually subjected to complex loading conditions and due to arbitrary orientation of flaws relative to the overall applied loads, cracks often experience shear mode deformation (K_{II} component) as well in addition to the opening mode (mode I or tensile type) deformation. For example, hydraulic fractures initiated from the wall of inclined and horizontal wells propagate under mixed mode tensile-shear loads (Haddad & Sepehrnoori, 2015). For another practical application, most of the observed cracks in gravity dams, wall of tunnels or rock slopes are prone to both modes I and II fracture (Kishen & Singh, 2001; Arslan & Korkmaz, 2007; Aliha *et al.*, 2012a). The shear type loads and waves during earthquakes also create predominantly mode II fracture patterns in rock structures. Therefore, for practical situations it is necessary to know the value of mode II fracture toughness (K_{IIc}) of rocks as well. Some test specimens such as punch through shear specimen (Backers *et al.*, 2002), anti-symmetric four-point bend loading (Aliha *et al.*, 2009), inclined center cracked Brazilian disk (Aliha *et al.*, 2012b) and inclined edge cracked semi-circular bend (Aliha *et al.*, 2010) specimens are among frequently used configurations for mode II fracture toughness testing of rocks.

For brittle and quasi-brittle materials, the failure loads obtained from fracture tests on a specific specimen may not be the same even under identical testing conditions. This is related to the distribution of physical flaws and micro-cracks inside the body of brittle materials and initiation of brittle failure process from these weak points. For example, when a cracked rock material is subjected to mechanical loading, a damage zone ahead of the crack front called fracture process zone (FPZ) is formed which experiences non-linear behavior due to random distribution of flaws in different sizes and along arbitrary orientations and locations. Indeed, the FPZ is developed gradually by initiation, coalescence and propagation of micro cracks in front of the crack tip by increasing the level of applied load. When the density of micro cracks in this region reaches a critical value and FPZ is saturated by micro cracks, a larger macro crack is formed in this region from the tip of initial crack which eventually results the overall brittle fracture of cracked rock body. The shape, size, orientation and type of micro cracks and consequently the saturation stage of this region which is not necessarily the same for identical testing conditions may be considered as possible source of scatter in maximum load bearing capacity of rock materials and thus it is reasonable to see some variations in their fracture toughness values. The variations in fracture toughness can be expressed in terms of a failure probability model. Weibull (1951) proposed a stochastic approach for describing the strength distribution of brittle materials based on the weakest link concept. This concept assumes that the ultimate strength or crack growth resistance of a body involves the products of the survival probabilities for the individual volume elements. A fair comparison can be made by observing the strength of a chain, in which the strength is determined by the weakest link. Once this element is broken the next weakest element will determine the strength of the remaining parts.

On the other hand because of the inherent heterogeneity, porosity, humidity and water content, bedding planes, texture, crystal boundaries, secondary phases, fissures, pores, inclusions, defects caused by environmental factors, composition and anisotropy of rock materials, large scatters in experimental fracture toughness data of rocks is inevitable. Hence, to obtain versatile and reliable data for rocks (especially for brittle and porous rocks), it is necessary to investigate the fracture toughness of rocks statistically using a larger number of test samples. Indeed, average fracture toughness value obtained from three or four tests cannot necessarily provide reliable fracture toughness values for rocks. A few researchers have investigated crack growth response of rocks using statistical approaches. For example, Donovan (2003) studied the fracture toughness values for a number of rocks including metabasalt, siltstone and granite obtained from Virginia, USA, using edge notch disk (END) specimen and with at least 10 samples for each type of rock and reported upper and lower bound values and also mean fracture toughness value for the investigated rocks. In another work, Chang *et al.* (2002) tested some Korean rocks such as marble and granite using BD and SCB specimens under both modes I and II. They showed statistically that the fracture toughness of rocks is nearly independent of specimen's thickness. Using edge notch bend beam specimen, Iqbal and Mohanty (2007) studied the effect of rock material orientation and bedding planes on mode I fracture toughness using a large number of samples with similar geometries. Amaral *et al.* (2008) has also employed Weibull statistical analysis to study the bending strength of a granite rock. However, theoretical predictions for statistical data obtained experimentally for fracture toughness of rocks have been rarely presented in the past. Hence, in this research, brittle fracture behavior of some rock materials with different test configurations is investigated statistically using Weibull probability analyses. It is shown that the relatively large scatter in the fracture toughness data for the investigated rocks can be predicted well by the Weibull analysis. For each rock material the corresponding Weibull parameters are determined and using a fracture theory, the fracture probability curve of different geometries and loading modes are estimated from the mode I fracture data of a reference specimen.

2 TWO- AND THREE-PARAMETER WEIBULL MODEL

If we conduct fracture toughness tests N times for a certain brittle or quasi brittle rock using a given test sample, probably N different fracture loads (*i.e.* fracture toughness values) would be obtained due to natural scatter in the results. By sorting the obtained data and using rank and cumulative probability method, the statistical fracture toughness behavior can be evaluated (Wallin, 2011). The failure probability P_f for total number of N test samples can be written as:

$$P_f = \frac{i - 0.5}{N} \tag{1}$$

where i is the number of tests, sorted in order of increasing fracture load or toughness. Wallin (1984) proposed the Weibull statistical distribution (Weibull, 1951) as a probability function for brittle failure. This function (which has a power law type in general) can be written as following two- and three-parameter models:

$$P_f(K_C) = 1 - EXP\left(-\left[\frac{K_C}{K_0}\right]^m\right) \text{- Two-Parameter model } (K_0, m) \tag{2}$$

$$P_f(K_C) = 1 - EXP\left(-\left[\frac{K_C - K_{min}}{K_0 - K_{min}}\right]^m\right) \text{-Three-Parameter model } (K_0, m, K_{min}) \tag{3}$$

where K_c is the applied stress intensity factor at the onset of fracture, K_{min} is threshold fracture toughness below which the probability of fracture is zero, K_0 is a normalization factor which is equal to the value of K and stands for the failure probability of 0.623 and m is a fitting parameter that describes the magnitude of scatter. If the variations of test results from sample to sample are small, the calculated m value will be high and conversely for those cases where the scatter of results is large the calculated m would be small. As seen from Equation 3 by setting K_{min} as zero, three-parameter Weibull distribution will be identical with the two-parameter one. The application of the Weibull distribution was introduced by Basu *et al.* (2009) and Todinov (2009) as one of the appropriate models for predicting the probability of fracture in the brittle materials. Other researchers (Smith *et al.*, 2006; Curtis & Juszczyk, 1998; Diaz & Kittl, 2005; Aliha & Ayatollahi, 2014; Danzer *et al.*, 2007) have also investigated the applicability of this method for different types of brittle materials such as rocks and ceramics based on the Weibull statistical model. According to the previously published papers the distribution of rock fracture toughness is thought to be well approximated by the three-parameter (K_0, m, K_{min}) Weibull distribution. Since the crack growth behavior (*i.e.* the mode I and mode II fracture toughness of rocks) is critically dependent on microstructure and defects, it is required and preferred to perform a stochastic analysis for modeling and characterizing brittle fracture in rocks. Therefore, in the forthcoming sections of this paper the Weibull statistical analyses are used for evaluating the experimental fracture toughness in some types of rocks tested under different conditions.

3 STATISTICAL FRACTURE TOUGHNESS DATA FOR ROCKS

In this section, the experimental program employed for investigating the statistical behavior of rocks is outlined. As mentioned earlier the disk shape specimens such as BD and SCB are among the suitable and commonly used configurations for conducting rock fracture toughness experiments. Hence, using the mentioned specimens the statistical study of this research was performed in these two subjects:

(1) Statistical mode I fracture toughness study of straight cracked BD and SCB specimens made of a soft rock (Guiting limestone) to investigate the influence of specimen's geometry on mode I fracture data.

(2) Statistical study for mode I and mode II fracture toughness in chevron notched BD specimen made of a marble rock (Harsin marble) to investigate the influence of loading mode on the statistical parameters of K_{Ic} and K_{IIc} results.

For investigating the influence of specimen type and its geometry in mode I fracture, a sedimentary soft limestone (Guiting limestone) was used for the experiments. This rock is a homogenous material composed of calcite. It is a porous limestone that is beige in color which is widely found in the UK. Also, for conducting mode I and mode II tests using the chevron notched specimens, a white, coarse grain and relatively homogenous marble (Harsin marble) excavated from the west of Iran was selected. For mode I fracture toughness testing on Guiting limestone, a BD specimen and two SCB specimens (labeled by SCB1 and SCB2) were used. For conducting both pure mode I and pure mode II fracture tests on marble a chevron-notched Brazilian disk was employed. Figure 1 shows the geometry and loading conditions of the employed test specimens, where R and t are the disk radius and thickness, respectively. F is the applied load and $2S$ shows the span length for three-point bend loading.

For the sake of comparison, the overall dimensions of disk and semi-disk specimens (*i.e.* BD and SCB specimens) were considered to be the same for both rocks and were as follows: $2R = 100$ mm and $t = 30$ mm. However, the type and length of crack was different for two sets of limestone and marble samples. While for limestone samples a straight crack of length $a = 15$ mm (for BD and SCB1) and 30 mm (for SCB2) was created using a very narrow fret saw of thickness 0.4 mm, two chevron notches were cut in the center of marble disks from each side using a thin rotary diamond saw having diameter and thickness of 80 mm and 1 mm, respectively. It is seen from Figure 1 that three geometrical parameters a, a_0 and a_1 describe the geometry of chevron notch in the CCNBD specimen, in which the following dimensionless parameters are often used for characterizing the geometry of chevron notch in the CCNBD specimen:

$$
\begin{aligned}
\alpha_0 &= a_0/R \\
\alpha_1 &= a_1/R \\
\alpha_B &= t/R \\
\alpha_S &= D_S/R
\end{aligned}
\tag{4}
$$

where D_s is the diameter of cutting rotary saw blade. As stated earlier the chevron notch Brazilian disk was used in this research for conducting both pure mode I and pure mode II fracture toughness tests. The state of deformation mode in this specimen can be controlled easily by rotating the crack direction (*i.e.* α) relative to the applied diametral

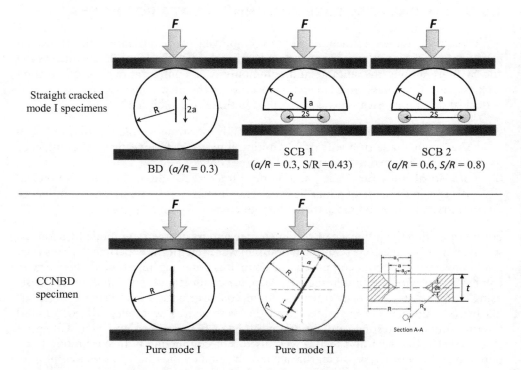

Figure 1 BD and SCB specimens employed for mode I and mode II fracture toughness experiments.

load as shown in Figure 1. When α is zero, the CCNBD specimen is subjected to pure mode I condition. Pure mode II inclination angle in this specimen depends on the crack length ratio (a/R). Some researchers have determined this angle for the Brazilian disk specimen using numerical and theoretical methods. For example, Ayatollahi and Aliha (2007) analyzed the Brazilian disk specimen by the finite element method and obtained the corresponding pure mode II crack inclination angle (α_{II}) for different a/R ratios.

For preparing the chevron notches, the cutting depth of disk was 17 mm from each side of CCNBD specimen. Using the measured values of a_0 and a_1 the average crack length a is found to be about 22.5 mm. Thus, the crack length ratio a/R is approximately about 0.43 for pure mode II tests in this research and the corresponding value of α_{II} is found from Ayatollahi and Aliha (2007) to be about 24°.

For the straight cracked BD and SCB specimens used for mode I fracture studies, the critical mode I fracture resistance (K_{If}) at the onset of fracture is determined from the following equations:

$$K_{If} = Y_{BD} \frac{F_c}{Rt} \sqrt{\frac{a}{\pi}} \qquad \text{for BD specimen} \qquad (5)$$

$$K_{If} = Y_{SCB} \frac{F_c}{2Rt} \sqrt{\pi a} \qquad \text{for SCB specimen} \qquad (6)$$

where F_c is the critical fracture load, Y_{BD} and Y_{SCB} are the geometry factors of the BD and SCB specimens, respectively. For the BD and SCB specimens these geometry factors

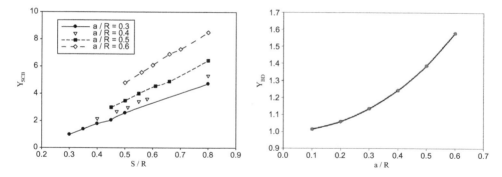

Figure 2 Variations of geometry factors Y_{BD} and Y_{SCB} for different geometry and loading conditions under pure mode I (Ayatollahi & Aliha, 2007).

are functions of the crack length ratio (a/R) and the span to diameter ratio (S/R). Some analytical and numerical solutions are available for Y_{BD} and Y_{SCB} (Ayatollahi & Aliha, 2007; Lim et al., 1994). For example, Ayatollahi and Aliha (2007) computed numerically these geometry factors using finite element method for a wide range of geometry and loading conditions. Figure 2 shows the mode I geometry factors for BD and SCB specimens with different a/R and S/R ratios.

The crack length ratio a/R was 0.3 for the straight crack BD specimen. For the SCB specimens, the crack length ratio and loading span to diameter ratio were as follows: ($a/R = 0.3$ and $S/R = 0.43$) for SCB1 and ($a/R = 0.6$ and $S/R = 0.8$) for SCB2 samples. From the graphs of Figure 2, corresponding geometry factors for the BD and SCB specimens of this research was found as: $Y_{BD} = 1.135$, $Y_{SCB1} = 2.013$, $Y_{SCB2} = 8.8$.

For determining pure mode I fracture toughness using the CCNBD specimen, ISRM (Fowell, 1995) has suggested the following equation:

$$K_{Ic} = \frac{F_C}{B\sqrt{D}} Y^*_{min} \tag{7}$$

where F_c is the fracture load of CCNBD specimen. The critical non-dimensional stress intensity factor Y^*_{min} is obtained in terms of $\alpha_0, \alpha_1, \alpha_2$ from the following equation (Fowell, 1995):

$$Y^*_{min} = u e^{v\alpha_1} \tag{8}$$

where u and v are the constant parameters given by Fowell (1995) in terms of α_0 and α_B. The average value of u and v were 0.277 and 1.784, respectively for the tested mode I CCNBD specimens in this research. Pure mode II fracture toughness K_{IIc} value was also determined from the following equation:

$$K_{IIC} = \frac{F_C}{\sqrt{\pi R} B} \sqrt{\frac{a}{R}} \sqrt{\frac{a_1 - a_0}{a - a_0}} Y_{II} \tag{9}$$

where Y_{II} is pure mode II geometry factor which depends on the crack length ratio (a/R) in the Brazilian disk specimen. Mode II geometry factor of Brazilian disk specimen has been calculated earlier by Ayatollahi and Aliha (2007) using finite element method for

Table 1 Experimental fracture loads and fracture toughness results obtained for CCNBD, BD, SCB1 and SCB2 specimens.

Specimen number	F_c (kN)	K_{IIc} (MPa.m$^{0.5}$)	F_c (kN)	K_{Ic} (MPa.m$^{0.5}$)	F_c (kN)	K_{Ic} (MPa.m$^{0.5}$)
	CCNBD (marble-mode II)		CCNBD (marble-mode I)		BD (limestone-mode I)	
1	15.017	2.430	10.658	1.003	3.412	0.1784
2	10.740	1.737	14.548	1.305	3.729	0.1950
3	9.451	1.594	10.559	0.939	4.52	02119
4	16.244	2.500	13.946	1.244	4.192	02192
5	9.939	1.807	12.571	1.120	4.327	0.2263
6	12.688	2.075	11.259	1.012	4.573	0.2372
7	16.864	2.658	12.361	1.120	4.704	0.2460
8	12.703	1.935	12.108	1.150	4.717	0.2467
9	15.611	2.361	14.554	1.278	4.757	0.2488
10	12.940	2.012	14.369	1.275	4.838	0.2530
11	15.294	2.360	14.080	1.216	4.896	0.2560
12	16.962	2.576	12.686	1.069	4.972	0.2600
13	14.209	2.117	14.841	1.367	5.074	0.2653
14	19.703	2.892	12.292	1.091	5.164	0.2700
15	15.988	2.543	14.125	1.221	5.470	0.2860
16	14.297	1.983	10.172	0.893		
17	17.546	2.634	11.200	1.036		
18	14.841	2.266	13.749	1.197		
19	12.368	1.927	10.867	0.949		
20	15.195	2.383	11.166	1.030		
21	16.782	2.529	11.867	1.075		
22	14.179	2.107	14.125	1.210		
	ave.= **14.526**	ave.= **2.2466**	ave.= **12.641**	ave.= **1.1273**	ave.= **4.589**	ave.= **0.2400**

Specimen number	F_c (kN)	K_{Ic} (MPa.m$^{0.5}$)	F_c (kN)	K_{Ic} (MPa.m$^{0.5}$)
	SCB1 (limestone-mode I)		SCB2 (limestone-mode I)	
1	1.968	0.2920	1.476	0.3407
2	2.056	0.3051	1.542	0.3808
3	2.132	0.3163	1.599	0.3550
4	2.150	0.3190	1.616	0.3745
5	2.163	0.3210	1.622	0.3690
6	2.167	0.3215	1.625	0.3509
7	2.204	0.3270	1.653	0.3952
8	2.244	0.3330	1.683	0.3601
9	2.279	0.3382	1.709	0.3847
10	2.309	0.3425	1.732	0.3874
11	2.325	0.3450		
12	2.341	0.3473		
13	2.377	0.3526		
14	2.463	0.3654		
15	2.493	0.3699		
	ave.= **2.245**	ave.= **0.3331**	ave.= **1.625**	ave.= **0.3698**

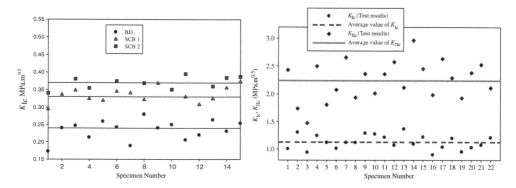

Figure 3 Experimental results obtained for pure mode I and pure mode II fracture toughness of tested limestone and marble rocks.

different a/R ratios. The corresponding value of Y_{II} for the tested CCNBD specimen with $a/R = 0.43$ was found to be 2.04 from Ayatollahi and Aliha (2007).

The prepared circular and semicircular specimens made of limestone and marble were then placed inside compression and three-point bend fixtures and loaded monotonically to obtain the critical peak load (F_c) of each specimen. The straight cracked BD and SCB specimens were tested by constant loading rate of 1 mm/min. For the chevron notched Brazilian disk samples, ISRM has suggested that the total time for each fracture test should be less than 20s. Therefore, the loading rate was chosen equal to 200 N/s to satisfy the ISRM requirements. The load–displacement curves of the whole samples were linear showing the validity of LEFM for the tested rock samples. By replacing the fracture loads (*i.e.* peak load obtained from each test) into related Equations 5, 6, 7 and 9, the corresponding values of K_{Ic} or K_{IIc} were determined for the tested materials. Table 1 and Figure 3 present the obtained fracture loads, corresponding fracture toughness values and the mean values for the whole investigated specimens.

As shown in Figure 3, there is a natural scatter in fracture toughness of tested rocks. These results reveal that the scatter of fracture toughness data may be dependent on the type of rock, configuration of test specimen, loading condition and also the crack type. For example, the scatter in the test results obtained for mode II fracture toughness is greater than the mode I fracture toughness of CCNBD specimen, or the scatter of SCB fracture toughness data is less than the BD specimen under mode I loading. The average values of K_{Ic} and K_{IIc} for the tested Harsin marble were about 1.12 MPa.m$^{0.5}$ and 2.25 MPa.m$^{0.5}$, respectively which gives the average K_{IIc}/K_{Ic} ratio of about 2 for this marble using the CCNBD tests. Meanwhile, the average K_{Ic} values obtained from BD and SCB1 and SCB2 samples made of Guiting limestone were 0.24, 0.33 and 0.37 MPa.m$^{0.5}$, respectively which shows the influence of geometry and loading conditions on pure mode I fracture toughness data. However, these findings are not in agreement with the predictions of conventional fracture criteria. Indeed, the available fracture criteria like the maximum tangential stress (Erdogan & Sih, 1963), the minimum strain energy density (Sih, 1974), the maximum energy release rate (Hussain *et al.*, 1974) and the cohesive zone model (Gómez *et al.*, 2009) suggest that the ratio of mode II over mode I fracture toughness (K_{IIc}/K_{Ic}) is a figure less than one and also the value of K_{Ic} for a given material would not be dependent on the shape and configuration of test specimen. These discrepancies between the experimental data and theoretical predictions are mainly because in the conventional fracture criteria, the influence of singular stress terms is only considered. But, it will be shown in the next section that the obtained statistical fracture toughness data can be predicted much better if a more precise description for stress field is used via a more generalized fracture criterion.

4 THEORETICAL FRACTURE CRITERION FOR MODE I AND II

A generalized stress-based fracture criterion is employed here to investigate the reason for the differences observed for the statistical results of tested limestone and marble rocks. The crack tip elastic stress field can be expressed as an infinite series expansion outlined by Williams (1957):

$$\sigma_{ij} = \sum_{n=1}^{\infty} A_n r^{\frac{n-2}{2}} f_{ij}^{(n)}(\theta) + \sum_{n=1}^{\infty} B_n r^{\frac{n-2}{2}} g_{ij}^{(n)}(\theta) \quad i,j = r, \theta, z \tag{10}$$

where σ_{ij} are the stress components and r, θ are the crack tip polar coordinates. Also A_n and B_n are the constant coefficients of the nth terms in the series expansion and $f^{(n)}_{ij}(\theta)$ and $g^{(n)}_{ij}(\theta)$ are the symmetric and anti-symmetric angular functions. The first term in this series expansion is singular which is related to the stress intensity factors (K_I and K_{II}). The other terms are non-singular which the first non-singular term is known as T-stress. Based on the maximum tangential stress (MTS) criterion (Erdogan & Sih, 1963) under any of modes I and II loading, brittle fracture occurs from the crack tip along (θ_0) *i.e.* the direction of maximum tangential stress (see Figure 4). For pure mode I loading, the tangential stress component in the vicinity of crack tip can be written as:

$$\sigma_{\theta\theta} = \sum_{i=1}^{3} \frac{i}{2} A_i r^{(\frac{i}{2}-1)} \left[(2 - \frac{i}{2} + (-1)^i)\cos(\frac{i}{2} - 1)\theta + (\frac{i}{2} - 1)\cos(\frac{i}{2} - 3)\theta \right] + O(r) \tag{11}$$

where A_i are the constant coefficients of the terms in the series expansion. Accordingly, A_1 is the coefficient of the singular term related to the mode I stress intensity factor by $K_I = A_1(2\pi)^{0.5}$, A_2 is the coefficient of the first non-singular term (corresponding to the T-stress which is independent of the distance from the crack tip r) and A_3 is the coefficient for the second non-singular term. For mode I cracks where the fracture path is self-similar and along the line of initial crack, the direction of fracture initiation (θ_0) is zero and crack growth occurs when the tangential stress $\sigma_{\theta\theta}$ at a critical distance r_c from the crack tip reaches a critical value of σ_c. Accordingly by ignoring the effects of higher order terms $O(r)$, based on the GMTS criterion the onset of mode I fracture is obtained from Equation 11 as:

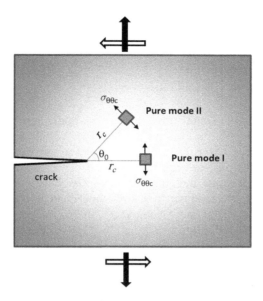

Figure 4 Schematic representation of brittle fracture based on the maximum tangential stress criterion for both modes I and II loading.

$$\sigma_c \sqrt{r_c} = \frac{K_{If}}{\sqrt{2\pi}} + 3r_c A_{3c} \qquad (12)$$

where K_{If} is the critical mode I fracture resistance of any given test sample which can be dependent on its geometry and loading conditions. By replacing $\sigma_c\sqrt{2\pi r_c}$ with K_{Ic} (i.e. pure mode I plane strain fracture toughness) in Equation 12 the following equation will be obtained:

$$\frac{K_{If}}{K_{Ic}} = \frac{1}{1 + 3\frac{A_{3n}}{A_{1n}}\left(\frac{r_c}{a}\right)} \qquad (13)$$

where A_{1n} and A_{3n} are the normalized forms of A_1 and A_3, respectively. Based on this equation, the mode I fracture toughness ratio K_{If}/K_{Ic} depends on the specimen type and its geometry through the geometry factors A_{3n} and A_{1n} and also the type of material which is related to r_c. The ratio of A_{3n}/A_{1n} has already been determined numerically for different values of a/R and S/R in the BD and SCB specimens (Aliha et al., 2012c).

Meanwhile, the tangential stress term for pure mode II case can be written in terms of singular and non-singular (K_{II} and T) terms as:

$$\sigma_{\theta\theta} = \frac{-3K_{II}}{2\sqrt{2\pi r}} \cos\frac{\theta}{2} \sin\theta + T\sin^2\theta + O\left(r^{1/2}\right) \qquad (14)$$

Unlike pure mode I, fracture propagation of pure mode II loading is not along the direction of initial crack. The direction of mode II crack growth (θ_{0II}) and the onset of pure mode II fracture can be obtained from:

$$\left.\frac{\partial \sigma_{\theta\theta}}{\partial \theta}\right|_{@pure\,mode\,II} = 0 \quad \Rightarrow \quad 3\cos\theta_o - 1 - \frac{16}{3}\hat{T}_{II}\sqrt{\frac{2r_c}{a}}\sin\frac{\theta_o}{2}\cos\theta_o = 0 \tag{15}$$

$$\frac{K_{IIc}}{K_{Ic}} = \frac{1}{\hat{T}_{II}\sqrt{\frac{2r_c}{a}}\sin^2\theta_o - \frac{3}{2}\sin\theta_o \cos\frac{\theta_o}{2}} \tag{16}$$

where $\hat{T}_{II}$ is the non-dimensional form of T-stress under pure mode II loading (T_{II}) defined as $\hat{T}_{II} = T_{II}\sqrt{\pi a}/K_{II}$. In Equations 13 and 16, r_c is a critical distance from the crack tip which is often taken as the size of fracture proses zone (FPZ) for rock materials. Based on previous research studies the size of r_c for different rocks typically varies in the order of a few millimeters (Schmidt, 1976; Labuz et al., 1987; Aliha et al., 2008). The value of r_c for tested limestone and marble has been reported about 2.6 mm (Aliha et al., 2012c) and 14.7 mm (Aliha et al., 2006), respectively. By ignoring the effects of A_3 and T terms in Equations 13 and 16, the predictions of conventional MTS criterion are obtained for pure modes I and II fracture. But it will be shown that considering the influence of higher order terms (T and A_3) via a generalized form of MTS criterion (i.e. GMTS) can provide very good estimations for the statistical fracture toughness data of limestone and marble rocks with different geometry and loading conditions.

5 RESULTS AND DISCUSSION

The probability of fracture (P_f) and the fitted curve for the obtained fracture toughness data have been shown in Figure 5 for each set of statistical results. The Weibull parameters can be obtained by fitting a curve to the obtained fracture toughness

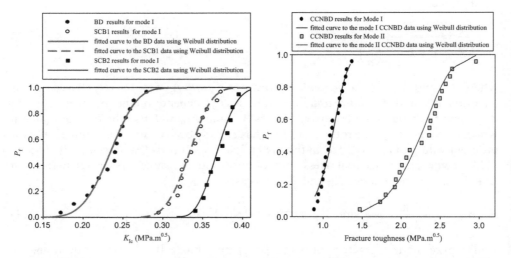

Figure 5 Probability curves plotted for mode I and mode II fracture toughness of limestone and marble rocks tested with the BD and SCB specimens.

data. For example, m and K_0 can be determined for two-parameter Weibull model, by rewriting Equation 2 in the form of following logarithmic relation:

$$\ln(\ln\left(\frac{1}{1 - P_f}\right) = m \ln K_0 - m \ln K_c \tag{17}$$

and then fitting a linear equation (straight line) to the statistical fracture toughness data. Figure 6 shows rock fracture toughness data of this research in a $\ln(\ln(1-P_f)^{-1})$ versus $\ln(K_c)$ and the corresponding m and K_0 values obtained for each set of test data.

In general, the Weibull parameters for two- and three-parameter models can also be determined by fitting a curve to the obtained statistical fracture toughness results (*i.e.* P_f versus K data). For example, the following equation can be minimized using the least square method to obtain three-Weibull distribution parameters (m, K_0, K_{min}):

$$f(m, K_0, K_{min}) = \sum_{i=1}^{N} \left\{ P_F(i) - \left(1 - EXP\left(-\left[\frac{K_c(i) - K_{min}}{K_0 - K_{min}}\right]^m\right)\right)\right\}^2 \tag{18}$$

Table 2 summarizes the Weibull parameters determined from curve fitting of rock fracture data using the mentioned method in MATLAB code for two- and three-parameter statistical models.

From the practical consideration, it is important to predict the statistical behavior of a given case in terms of other available statistical results. For example, although a large number of test data are available for K_{Ic} of rocks using CCNBD specimen, limited mode II fracture toughness K_{IIc} data have been reported for K_{IIc} of rocks using the same specimen. Therefore, it is useful to examine whether the fracture toughness data obtained from the ISRM-suggested specimen (*i.e.* mode I CCNBD) can be used for predicting the experimental results of CCNBD specimen tested under pure mode II loading condition. Meanwhile, it is important to examine whether the mode I fracture toughness test data obtained from one specimen (such as BD specimen) can be used for predicting the experimental K_{Ic} results of another test specimen (*e.g.* SCB specimen). This subject is of practical importance since by knowing the effect of specimen geometry and loading conditions, one can use the test data obtained from laboratory specimens for estimating the mode I fracture resistance of real cracked bodies made of rocks with other geometry and loading conditions.

Using the GMTS criterion, the mode I fracture resistance of SCB1 and SCB2 can be predicted in terms of mode I fracture toughness data of BD specimen. Furthermore, the mode II fracture toughness of tested Harsin marble can be predicted in terms of its K_{Ic} value obtained from mode I CCNBD specimen. For using this criterion, the value of A_3 and T (or their normalized parameters A_{3n}/A_{1n} and $\hat{T}_{II}$) should be known for mode I and mode II test samples, respectively. These normalized parameters have been determined using extensive finite element analyses by Ayatollahi and his core searchers (Ayatollahi & Aliha, 2007; Aliha *et al.*, 2012c) for the BD and SCB specimens. The ratios of A_{3n}/A_{1n} for the tested BD (with $a/R = 0.3$), SCB1 (with $a/R = 0.3$ and $S/R = 0.43$) and SCB2 (with $a/R = 0.6$ and $S/R = 0.8$) were found from (Aliha *et al.*, 2012c) to be about 0.24, –0.28 and –1, respectively. In addition for the Brazilian disk specimen with $a/R = 0.43$ subjected to pure mode II loading, the value of $\hat{T}_{II}$ was –0.95 (Ayatollahi & Aliha, 2007). Accordingly, by replacing the corresponding values of A_{3n}/A_{1n} for the tested mode I

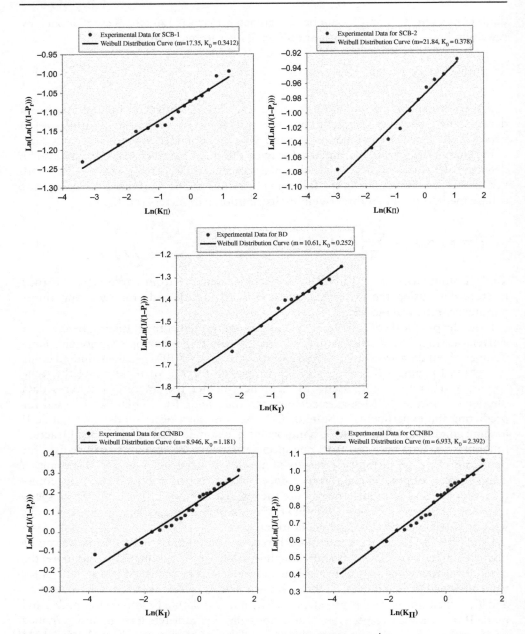

Figure 6 Extracting m and K_0 values by fitting a linear curve to $\ln(\ln((1-P_f)^{-1}))$ versus $\ln(K_c)$ graphs for the tested rock materials.

limestone samples the fracture toughness ratio K_{If}/K_{Ic} can be estimated from Equation 16. As a material constant the value of K_{Ic} for this limestone rock was determined experimentally as 0.26 MPa.m$^{0.5}$ using an ISRM suggested chevron notched bend beam method. The average ratio of K_{If}/K_{Ic} obtained experimentally for the BD, SCB1

Statistical fracture toughness study for rocks   55

Table 2 Weibull distribution parameters for the tested limestone and marble rocks.

Material	specimen	m		K_0 MPa.m$^{0.5}$		K_{min} MPa.m$^{0.5}$
		2 parameter	3 parameter	2 parameter	3 parameter	3 parameter
limestone	BD (mode I)	10.61	4	0.252	0.242	0.141
limestone	SCB1 (mode I)	17.35	4	0.341	0.345	0.256
limestone	SCB2 (mode I)	21.84	4	0.374	0.378	0.282
marble	CCNBD (mode I)	8.946	1.81	1.181	1.16	0.843
marble	CCNBD (mode II)	6.933	1.81	2.392	2.336	1.696

Table 3 Comparison of experimental fracture toughness ratios (K_{If}/K_{Ic} & K_{IIc}/K_{Ic}) with theoretical predictions of the GMTS criterion for the tested rocks.

K_{If}/K_{Ic}						K_{IIc}/K_{Ic}	
BD		SCB1		SCB2		CCNBD	
Fracture Test	GMTS prediction	Fracture Test	GMTS prediction	Fracture Test	GMTS prediction	Fracture Test	GMTS prediction
0.89	0.86	1.31	1.22	1.43	1.39	2.01	1.906

and SCB2 specimens and the theoretical predictions of K_{If}/K_{Ic} from the GMTS criterion (Equation 16) has been compared in Table 3 that indicates good consistency between the experimental results and theoretical predictions. Similarly, for the tested marble rock the mean fracture toughness of mode II CCNBD specimen can be predicted via the GMTS criterion, in terms of the results obtained from mode I tests on this specimen. As presented in Table 4, the predicted K_{IIc}/K_{Ic} ratio of 1.9 obtained from the GMTS criterion is in good agreement with $K_{IIc}/K_{Ic} = 2.01$ obtained experimentally from the tested CCNBD specimens under pure mode I and pure mode II loading conditions.

In addition to the mean fracture toughness values, it is very useful to predict the statistical results of each material in terms of its mode I Brazilian disk data by using the GMTS criterion. In other words, the probabilistic fracture curves for the SCB1 and SCB2 test data are predicted in terms of the Weibull parameters of the BD specimen made of limestone. Also, mode II statistical fracture toughness results of CCNBD specimens can be predicted in terms of the mode I Weibull parameters of the ISRM suggested CCNBD specimen. Based on the GMTS criterion, under pure mode I loading the effect of A_3 is considered to be responsible for shifting the mode I (K_{If}) results. Based on the assumption that at $r = r_c$, $\sigma_{\theta\theta c}(BD) = \sigma_{\theta\theta c}(SCB\,1) = \sigma_{\theta\theta c}(SCB\,2)$, Equation 13 can be used for making the following relation between the BD and SCB specimens:

$$K_{If}(SCB\,i) - K_{If}(BD) = 3\sqrt{2\pi r_c}([A_{3c}(SCBi) - A_{3c}(BD)]) \quad i = 1, 2 \tag{19}$$

The fitted parameters of K_{min} and K_0 for the BD specimens are modified to determine the corresponding values for the SCB specimens. Thus,

$$\frac{K_{If}(SCBi)}{K_{If}(BD)} = \frac{K_{I\,min}(SCBi)}{K_{I\,min}(BD)} = \frac{K_{I0}(SCBi)}{K_{I0}(BD)} \quad i = 1, 2 \tag{20}$$

$$\frac{K_{\text{I min}}(SCB\,i) - K_{\text{I min}}(BD)}{K_{\text{I min}}(BD)} = \frac{K_{\text{I}0}(SCB\,i) - K_{\text{I}0}(BD)}{K_{\text{I}0}(BD)}$$

$$= 3\sqrt{2\pi r_c} \frac{([A_{3c}(SCBi) - A_{3c}(BD)])}{K_{\text{If}}(BD)} \qquad i = 1, 2 \quad (21)$$

Equation 21 shows the relationship between A_3 and the Weibull parameters for the BD, SCB1 and SCB2 specimens. Therefore, by knowing A_3 and the Weibull parameters for one specimen type and employing the value of r_c for the investigated rock the values of $K_{\min}$ and K_0 can be predicted for other specimens from their corresponding A_{3c} values. The predicted curves for the SCB results in terms of mode I BD data are shown in Figure 7. It is seen from this Figure that the proposed statistical model based on the mode I BD data provides reasonable predictions for the statistical mode I fracture behavior of SCB1 and SCB2 test data.

The mode II Weibull parameters can also be predicted in terms of mode I statistical data, since the micro-mechanism of brittle fracture in rocks for mode I loading is expected to be similar to that for mode II loading and for both modes, fracture initiates along the direction where the tangential stress is maximum. Therefore, a model similar to pure mode I can be used to explore a statistical description of the mode II fracture

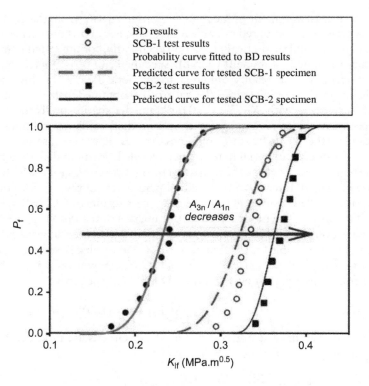

Figure 7 Prediction of mode I probability curves for statistical results of SCB1 and SCB2 specimens in terms of BD test data.

toughness data. The failure probability for mode II loading condition was determined using an extension of the Wallin model developed by Fowler et al. (1997) and Hadidimoud et al. (2002). The failure probability for mode II is given by

$$P_F(K_{\text{IIc}}) = 1 - \text{EXP}\left(-\left[\frac{K_{\text{Ic}}}{K_{0,\text{II}}}\right]^m\right) \quad \text{two-parameter } (m, K_{0,\text{II}}) \text{ model} \quad (22)$$

$$P_F(K_{\text{IIc}}) = 1 - \text{EXP}\left(-\left[\frac{K_{\text{Ic}} - K_{\min,\text{II}}}{K_{0,\text{II}} - K_{\min,\text{II}}}\right]^m\right)$$
$$\text{three-parameter } (m, K_{0,\text{II}}, K_{\min,\text{II}}) \text{ model} \quad (23)$$

Using the $K_{\text{IIc}}/K_{\text{Ic}}$ ratio, the mode I fitted parameters $K_{\min,\text{I}}$ and $K_{0,\text{I}}$ were modified to directly estimate the mode II parameters $K_{\min,\text{II}}$ and $K_{0,\text{II}}$ from:

$$\frac{K_{\text{IIc}}}{K_{\text{Ic}}} = \frac{K_{\min,\text{II}}}{K_{\min,\text{I}}} = \frac{K_{0,\text{II}}}{K_{0,\text{I}}} \quad (24)$$

accordingly mode II Weibull parametrs of the tested CCNBD specimen can be found from the mode I ones as:

$$K_{0,\text{II}} = \left(\frac{K_{\text{IIc}}}{K_{\text{Ic}}}\right) \times K_{0,\text{I}}$$
$$K_{\min,\text{II}} = \left(\frac{K_{\text{IIc}}}{K_{\text{Ic}}}\right) \times K_{\min,\text{I}} \quad (25)$$

where $K_{\text{IIc}}/K_{\text{Ic}}$ in Equation 25 (already determined from Equation 16) depends on the T-stress. The values of $K_{\min,\text{II}}$ and $K_{0,\text{II}}$ calculated using Equations 25 and 21 were replaced in Equations 22 and 23. The corresponding values of $K_{0,\text{II}} = 2.336$ MPa.m$^{0.5}$ and

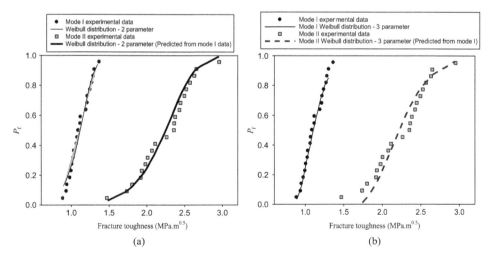

Figure 8 Prediction of probability curves for mode II fracture toughness of CCNBD specimen using the mode I data; (a) two-parameter Weibull distribution curves and (b) three-parameter Weibull distribution curves.

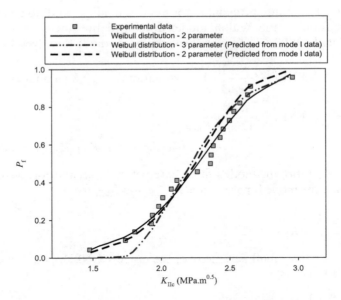

Figure 9 Comparison of two- and three-parameter Weibull probability curves for mode II fracture toughness of CCNBD specimen.

$K_{min,II} = 1.696$ MPa.m$^{0.5}$ predicted from Equation 25 are in very good agreement with the obtained Weibull parameters through curve fitting of mode II results (see Table 3). The curves predicted for the mode II results of CCNBD specimen using mode I statistical parameters are shown in Figure 8. It is seen from this Figure that the proposed statistical model based on the mode I data provides good predictions for the mode II fracture toughness data. Figure 9 compares the statistical curves of two- and three-parameter Weibull models for mode II data that shows the reasonable accuracy of both models for predicting the K_{IIc} data in terms of K_{Ic} results in the tested CCNBD specimens. Therefore, according to the findings of this research the mode II fracture behavior for the CCNBD specimen (*i.e.* the mean fracture toughness and the statistical parameters) can be estimated well by knowing only the pure mode I fracture toughness results of CCNBD specimens and employing the GMTS criterion. Consequently, there is no need to perform rather complicated pure mode II tests to obtain K_{IIc} from the CCNBD experiments.

6 CONCLUSION

- Mode I and mode II fracture behavior of two rock materials (limestone and marble) was studied experimentally using a large number of test specimens. The effects of specimen geometry, loading condition and type of loading were noticeable on the observed scatters in the test results.
- For each set of obtained fracture toughness results, the two- and three-parameter Weibull models were determined by fitting suitable curves to failure probability of each specimen.

- The fracture toughness ratios (*i.e.* K_{If}/K_{Ic} for straight cracked BD and SCB samples and K_{IIc}/K_{Ic} for chevron notched Brazilian disk specimens) were predicted very well using the GMTS criterion by taking into account the effects of higher order stress terms.
- The mode II fracture toughness probability curve of CCNBD specimen and also the mode I fracture toughness probability curves of SCB1 and SCB2 specimens were predicted successfully in terms of the mode I statistical Weibull probability distribution curves of Brazilian disk specimen.

REFERENCES

Aliha, M. R. M., Ashtari, R., & Ayatollahi, M. R. (2006) Mode I and mode II fracture toughness testing for a coarse grain marble. *Applied Mechanics and Materials*, 5, 181–188.

Aliha, M. R. M., Ayatollahi, M. R., & Pakzad, R. (2008) Brittle fracture analysis using a ring-shape specimen containing two angled cracks. *International Journal of Fracture*, 153(1), 63–68.

Aliha, M. R. M., Ayatollahi, M. R., & Kharazi, B. (2009) Mode II brittle fracture assessment using ASFPB specimen. *International Journal of Fracture*, 159(2), 241–246.

Aliha, M. R. M., Ayatollahi, M. R., Smith, D. J., & Pavier, M. J. (2010) Geometry and size effects on fracture trajectory in a limestone rock under mixed mode loading. *Engineering Fracture Mechanics*, 77(11), 2200–2212.

Aliha, M. R. M., Mousavi, M., & Ayatollahi, M. R. (2012a) Mixed mode I/II fracture path simulation in a typical jointed rock slope. In Crack Path (CP2012).

Aliha, M. R. M., Ayatollahi, M. R., & Akbardoost, J. (2012b) Typical upper bound–lower bound mixed mode fracture resistance envelopes for rock material. *Rock Mechanics and Rock Engineering*, 45(1), 65–74.

Aliha, M. R. M., Sistaninia, M., Smith, D. J., Pavier, M. J., & Ayatollahi, M. R. (2012c) Geometry effects and statistical analysis of mode I fracture in guiting limestone. *International Journal of Rock Mechanics and Mining Sciences*, 51, 128–135.

Aliha, M. R. M., & Ayatollahi, M. R. (2014) Rock fracture toughness study using cracked chevron notched Brazilian disk specimen under pure modes I and II loading–A statistical approach. *Theoretical and Applied Fracture Mechanics*, 69, 17–25.

Amaral, P. M., Fernandes, J. C., & Rosa, L. G. (2008) Weibull statistical analysis of granite bending strength. *Rock Mechanics and Rock Engineering*, 41(6), 917–928.

Arslan, M. H., & Korkmaz, H. H. (2007) What is to be learned from damage and failure of reinforced concrete structures during recent earthquakes in Turkey? *Engineering Failure Analysis*, 14(1), 1–22.

Ayatollahi, M. R., & Aliha, M. R. M. (2007) Wide range data for crack tip parameters in two disc-type specimens under mixed mode loading. *Computational Materials Science*, 38(4), 660–670.

Backers, T., Stephansson, O., & Rybacki, E. (2002) Rock fracture toughness testing in Mode II—punch-through shear test. *International Journal of Rock Mechanics and Mining Sciences*, 39(6), 755–769.

Basu, B., Tiwari, D., Kundu, D., & Prasad, R. (2009) Is Weibull distribution the most appropriate statistical strength distribution for brittle materials? *Ceramics International*, 35(1), 237–246.

Chang, S. H., Lee, C. I., & Jeon, S. (2002) Measurement of rock fracture toughness under modes I and II and mixed-mode conditions by using disc-type specimens. *Engineering Geology*, 66(1), 79–97.

Coulomb, C. A. (1773) Sur une application des règles de maximis et minimis à quelques problèmes de statique relatifs à l'architecture. *Mémoires de mathématiques et de Physique, Académie Royale des Sciences*, 7, 343–382.

Curtis, R. V., & Juszczyk, A. S. (1998) Analysis of strength data using two-and three-parameter Weibull models. *Journal of Materials Science*, 33(5), 1151–1157.

Danzer, R., Supancic, P., Pascual, J., & Lube, T. (2007) Fracture statistics of ceramics–Weibull statistics and deviations from Weibull statistics. *Engineering Fracture Mechanics*, 74(18), 2919–2932.

Diaz, G., & Kittl, P. (2005) Probabilistic analysis of the fracture toughness, KIc, of brittle and ductile materials determined by simplified methods using the Wiebull distribution function. In *Proceedings' of International Conference on Fracture (ICF 11)*, Torino, Italy.

Donovan, J. G. (2003) *Fracture toughness based models for the prediction of power consumption, product size, and capacity of jaw crushers* (Doctoral dissertation, Virginia Polytechnic Institute and State University).

Erdogan, F., & Sih, G. C. (1963) On the crack extension in plates under plane loading and transverse shear. *Journal of Basic Engineering Transactions of ASME*, 85, 519–525.

Fowell, R. J. (1995, January) Suggested method for determining mode I fracture toughness using cracked chevron notched Brazilian disc (CCNBD) specimens. *International Journal of Rock Mechanics and Mining Sciences & Geomechanics Abstracts*, 32, 57–64.

Fowler, H., Smith, D. J., & Bell, K. (1997) Scatter in cleavage fracture toughness following proof loading. Advances in fracture research. *Proc. 9th International Conference on Fracture (ICF 9)*, 5, 2519–2526.

Funatsu, T., Seto, M., Shimada, H., Matsui, K., & Kuruppu, M. (2004) Combined effects of increasing temperature and confining pressure on the fracture toughness of clay bearing rocks. *International Journal of Rock Mechanics and Mining Sciences*, 41(6), 927–938.

Gómez, F. J., Elices, M., Berto, F., & Lazzarin, P. (2009) Fracture of U-notched specimens under mixed mode: Experimental results and numerical predictions. *Engineering Fracture Mechanics*, 76(2), 236–249.

Haddad, M., & Sepehrnoori, K. (2015) Simulation of hydraulic fracturing in quasi-brittle shale formations using characterized cohesive layer: Stimulation controlling factors. *Journal of Unconventional Oil and Gas Resources*, 9, 65–83.

Hadidimoud, S., Sisan, A. M., Truman, C. E., & Smith, D. J. (2002) Predicting how crack tip residual stresses influence brittle fracture. *Pressure Vessels and Piping*, 434, 111–116.

Hussain M. A, Pu, S. L., & Underwood J. (1974) Strain energy release rate for a crack under combined mode I and Mode II. Fracture Analysis, ASTM STP 560. American Society for Testing and Materials, Philadelphia, 2–28.

Iqbal, M. J., & Mohanty, B. (2007) Experimental calibration of ISRM suggested fracture toughness measurement techniques in selected brittle rocks. *Rock Mechanics and Rock Engineering*, 40(5), 453–475.

Ingraffea, A. R., Gunsallus, K. L., Beech, J. F., & Nelson, P. P. (1984) A short-rod based system for fracture toughness testing of rock. *ASTM STP*, 855, 152–166.

Khan, K., & Al-Shayea, N. A. (2000) Effect of specimen geometry and testing method on mixed mode I–II fracture toughness of a limestone rock from Saudi Arabia. *Rock Mechanics and Rock Engineering*, 33(3), 179–206.

Kishen, J. C., & Singh, K. D. (2001) Stress intensity factors based fracture criteria for kinking and branching of interface crack: application to dams. *Engineering Fracture Mechanics*, 68(2), 201–219.

Krishnan, G. R., Zhao, X. L., Zaman, M., & Roegiers, J. C. (1998) Fracture toughness of a soft sandstone. *International Journal of Rock Mechanics and Mining Sciences*, 35(6), 695–710.

Kuruppu, M. D., Obara, Y., Ayatollahi, M. R., Chong, K. P., & Funatsu, T. (2015) ISRM-suggested method for determining the mode I static fracture toughness using semi-circular bend specimen. *In The ISRM Suggested Methods for Rock Characterization, Testing and Monitoring: 2007–2014*, 107–114. Springer International Publishing.

Labuz, J. F., Shah, S. P., & Dowding, C. H. (1987, August) The fracture process zone in granite: Evidence and effect. *International Journal of Rock Mechanics and Mining Sciences & Geomechanics Abstracts*, 24(4), 235–246.

Labuz, J. F., & Zang, A. (2015) *Mohr–Coulomb failure criterion*. In: *The ISRM Suggested Methods for Rock Characterization, Testing and Monitoring: 2007–2014*, 227–231. Springer International Publishing.

Lim, I. L., Johnston, I. W., Choi, S. K., & Boland, J. N. (1994, June) Fracture testing of a soft rock with semi-circular specimens under three-point bending. Part 2—mixed-mode. *International Journal of Rock Mechanics and Mining Sciences & Geomechanics Abstracts*, 31(3), 199–212.

Nasseri, M. H. B., Grasselli, G., & Mohanty, B. (2010) Fracture toughness and fracture roughness in anisotropic granitic rocks. *Rock Mechanics and Rock Engineering*, 43(4), 403–415.

Ouchterlony, F. (1988) ISRM Suggested methods for determining fracture toughness of rocks. *International Journal of Rock Mechanics and Mining Sciences & Geomechanics Abstracts*, 25, 71–96.

Schmidt, R. A. (1976) Fracture-toughness testing of limestone. *Experimental Mechanics*, 16(5), 161–167.

Sih, G. C. (1974) Strain-energy-density factor applied to mixed mode crack problems. *International Journal of Fracture*, 10(3), 305–321.

Siren, T. (2012) *Fracture toughness properties of rocks in Olkiluoto: Laboratory measurements 2008–2009*. Posiva Oy, Helsinki (Finland).

Smith, D. J., Ayatollahi, M. R., & Pavier, M. J. (2006) On the consequences of T-stress in elastic brittle fracture. *Proceedings of the Royal Society A: Mathematical, Physical and Engineering Science*, 462(2072), 2415–2437.

Tutluoglu, L., & Keles, C. (2011) Mode I fracture toughness determination with straight notched disk bending method. *International Journal of Rock Mechanics and Mining Sciences*, 48(8), 1248–1261.

Todinov, M. T. (2009) Is Weibull distribution the correct model for predicting probability of failure initiated by non-interacting flaws? *International Journal of Solids and Structures*, 46 (3), 887–901.

Wallin, K. (1984) The scatter in KIC-results. *Engineering Fracture Mechanics*, 19(6), 1085–1093.

Wallin, K. R. (2011) Simple distribution-free statistical assessment of structural integrity material property data. *Engineering Fracture Mechanics*, 78(9), 2070–2081.

Weibull, W. (1951) A statistical distribution function of wide applicability. *Journal of Applied Mechanics*, 18(3), 293–297.

Whittaker, B. N., Singh, R. N., & Sun, G. (1992) *Rock fracture mechanics*. Elsevier.

Williams M. L. (1957) On the stress distribution at the base of a stationary crack. *Journal of Applied Mechanics*, 24, 109–114.

Xeidakis, G. S., Samaras, I. S., Zacharopoulos, D. A., & Papakaliatakis, G. E. (1997) Trajectories of unstably growing cracks in mixed mode I–II loading of marble beams. *Rock Mechanics and Rock Engineering*, 30(1), 19–33.

Zhou, Y. D., Tang, X., Hou, T., & Tham, L. (2010, 23–27 October) Fracture measurement of marble specimens from Maon Shan, Hong Kong by Chevron-notched bend test: some discussions based on nonlinear fracture mechanics approach. In: *International Symposium 2010 and 6th Asian Rock Mechanics Symposium – Advances in Rock Engineering*, New Delhi, India.

Anisotropy

Chapter 3

Rock damage mechanics

F.L. Pellet[1] & A.P.S. Selvadurai[2]

[1]MINES ParisTech, Geosciences and Geoengineering Department, Fontainebleau, France
[2]McGill University, Department of Civil Engineering and Applied Mechanics, Montreal, Canada

Abstract: Damage mechanics of rocks is a concept that requires a comprehensive presentation as it involves multiple physical and sometimes chemical phenomena. Mechanical damage is inextricably linked to cracking and the resulting loss of mechanical properties. A collateral effect of damage is the alteration of the fluid transport characteristics of rocks that can have geoenvironmental consequences. This chapter outlines damage characterization at the scale of the microstructure of a rock specimen and then presents different analytical approaches available to mathematically describe the evolution of the mechanical properties. These approaches essentially rely on Fracture Mechanics and Continuum Damage Mechanics. Finally some examples of numerical modeling are presented to illustrate their application in the field of rock engineering. The chapter also presents a brief review of the techniques that can be adopted to integrate results of damage mechanics to fluid transport phenomena.

I INTRODUCTION

Damage is a consequence of physical degradation that can impair or progressively weaken a structure. In geological formations, damage can affect the rock matrix, the rock discontinuities or the assembly of both: the rock mass. In rock engineering, any damage can be of concern to the serviceability and stability of structures such as rock slopes, underground openings, deep wellbores, etc.

Environmental issues related to transport and diffusion of harmful substances (radionuclides, CO_2, etc) may arise due to the development of damage zones in rock formations. Damage to rock or engineered structures unavoidably generates micro to macro-cracks, which increases the rock permeability. This has been extensively investigated over the last few decades for deep geological disposal of radioactive waste (Hudson *et al.*, 2009). Rock formation damage is also an important feature for the extraction of unconventional resources such as shale gas (Zimmerman, 2010), or geothermal energy (Li *et al.*, 2012).

Rock damage can occur due to either monotonic or repeated mechanical action (alteration of the in situ stress field), changes in temperature, transformation resulting from chemical processes or a variation in the pore pressure.

Today, several different methods have been adopted to address damage from a mechanical point of view. This chapter aims to clarify the advantages and disadvantages of the approaches currently used to study mechanically-induced damage.

2 EXPERIMENTAL EVIDENCE

2.1 Investigations at the laboratory scale

The problem of damage will first be addressed from an engineering point of view. Let us imagine an intact rock specimen, void of any defects, subjected to a uniaxial state of stress in compression. Initially, the specimen will store the energy provided by the loading system. When this energy exceeds the specific potential energy associated with the inter-atomic bonds of the material, these will be progressively broken and micro-cracks will appear. The final damage stage occurs when the rock specimen can no longer sustain additional stress and the cracks coalesce to form a large macro-crack. Between the incipient loading and the final collapse, the specimen undergoes successive transitional states of damage, from slight to moderate and finally severe damage.

Figure 1a illustrates a typical stress–strain curve for a rock specimen axially loaded in compression. Measurement of the axial strain ε_1 and lateral strain ε_3 allows the computation of the volume variation of the specimen, referred to as the volumetric strain, ε_v. After the crack closure (σ_{cc}), the onset of damage is evidenced by the loss of linearity in the variation of the lateral deformation. This corresponds to the nucleation of cracks, which results in a slower decrease in volume change (σ_{ci}, for crack initiation). The specimen is then slightly damaged. With a further increase in the loading, the crack will grow steadily until the energy provided by the loading system leads to unstable crack propagation. The transition point between these two propagation modes (σ_{ci}, for crack damage) is indicated by the change of the volume variation passing from compaction to dilation. At this time, the material is severely damaged. The loading process then continues until total failure of the specimen is observed (σ_{cf}, for crack failure). Figure 1b shows a thin section extracted from a shale rock specimen after failure. Macro-crack networks as well as micro-cracks are clearly visible.

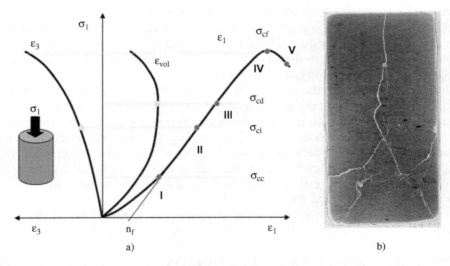

Figure 1 (a) Typical stress–strain curve for a rock specimen axially loaded with characteristic thresholds defining different damage intensity; (b) Macro-crack and micro-crack networks in a shale specimen (80 mm in height) loaded in mono-axial compression (Pellet, 2015).

The damage process described above has been reported and interpreted by several authors (*e.g.* Martin, 1997; Patterson & Wong, 2005) since the pioneering works of Brace *et al.* (1966, 1972) and Bieniawski (1967). At the present time, standard laboratory tests can be supplemented by additional measurements such as the Acoustic Emission activity (Stanchits *et al.*, 2006) or changes in ultra-sonic wave velocity induced by the degradation of the specimen (Fortin *et al.*, 2007; Pellet & Fabre, 2007). Indeed, as shown by Keshavarz *et al.* (2009), such data can also help to delineate the different damage thresholds.

2.2 Physical interpretation of cracking of rock at the micro-scale

Rocks are heterogeneous materials with many flaws and micro-defects such as micro-pores or micro-cracks. The latter play a significant role in the changes that occur in the mechanical properties of a rock during the loading process. Micro-defect characteristics have been widely studied; for example, Kranz (1983) distinguished four types of cracks:

- Cracks associated with grain boundaries that separate two alike or different crystals or crystals and cement.
- Inter-crystalline cracks (or inter-granular) initiated at the contact between the grains and pass along one or more grains.
- Intra-crystalline cracks (or intra-granular) which propagate in the grain; the length of these cracks is often much shorter than the grain size and the opening can be very small ($< 10\ \mu$).
- Cleavage cracks are particular intra-crystalline cracks separating the cleavage planes of a crystal; they occur as parallel networks to the symmetry planes of the mineral.

Further information on the rock fabric and petro-physical characteristics is given by Davis and Reynolds (1996).

During the loading process, micro-cracks will nucleate or propagate, eventually coalescing to form macro-cracks (Figure 1b). In order to be considered a macro-crack, several grains or crystals have to be involved. The images presented in Figure 2, taken with an optical microscope, show different types of cracks. Figure 2a illustrates cracks in the clay matrix of a Callovo-Oxfordian marl showing the debonding of calcite crystals (Fabre & Pellet, 2006). Figure 2b shows trans-granular cracks in a gabbro specimen subjected to very high stresses, up to 1.7 GPa (Pellet *et al.*, 2011).

The theoretical strength of a crystal, R_{th}, is related to the inter-atomic bonding forces (Figure 3). It can be computed using the following expression (Dorlot *et al.*, 1986):

$$R_{th} = 2\sqrt{\frac{E\gamma}{a_0}} \tag{1}$$

with:

E the Young's modulus of the crystal,
γ, the energy required to debond surfaces per unit area,
a_0, the equilibrium distance between atoms.

The theoretical strength is between one-third and one-tenth of the Young's modulus E. However, due to the presence of flaws and micro-defects, the usual values for the tensile

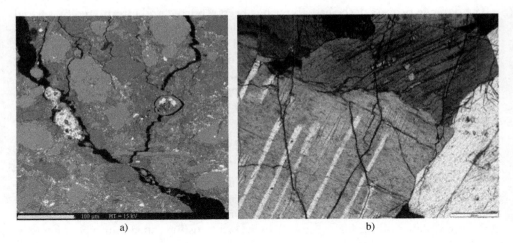

Figure 2 Crack damage in rocks specimens: (a) Inter-granular cracks in Callovo-Oxfordian marl (Fabre & Pellet, 2006); b) Trans-granular micro-cracks in plagioclase crystals of a gabbro (Pellet et al. 2011). Note that the development of micro-cracks crosses the original crystal twinning.

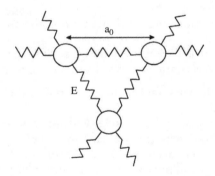

Figure 3 Schematic of crystal atomic bonding.

strength of rocks are much lower, in the order of 1/100 to 1/1000 of the Young's modulus.

In summary, since most rocks are formed by an assemblage of various types of crystals, fracturing can occur at different scales. At the scale of the micro-structure, damage could result from inter-granular bonds breaking or sometimes rupturing within the crystal. This type of rupture is referred to as dislocation mechanics (McClintock & Argon, 1966).

3 CONSTITUTIVE MODELING

3.1 Introduction

Constitutive modeling of damage evolution has motivated numerous studies in recent decades. Basically, there are two approaches: The first is based on the theories of fracture mechanics while the second is derived from continuum damage mechanics. In the following sections the main aspects of both approaches will be reviewed and the domain of validity of each method will be identified.

3.2 Linear fracture mechanics

3.2.1 Energy balance consideration

It is important to review the basic concepts involved in fracture mechanics. Griffith (1924) formulated a rupture criterion for a material based on energy balance. He acknowledged that defects (pore spaces, cracks or structural defects such as dislocations in the crystal lattice or grain boundaries) can exist in any material and stress amplification at such defects will govern the failure stress.

Considering an elementary volume subjected to a uniaxial tensile stress (Figure 4), and based on the principles of thermodynamics, equilibrium is achieved when the total potential energy of the system is at a minimum. The energy balance of the system leads to the expression of the total energy, W, as the sum of the energy of the external forces, the elastic strain energy, and the energy dissipated during the cracking process:

$$W = W_{ext} + W_{el} + W_s \qquad (2)$$

with:

W: total energy
W_{ext}: energy due to external forces
W_{el}: elastic strain energy
W_s: dissipated energy

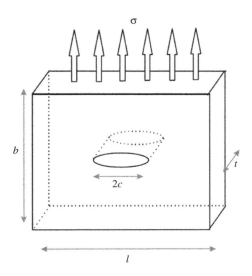

Figure 4 Elementary volume of rock with a plane crack loaded in tension.

During extension, the crack length increases from $2c$ to $2(c + dc)$ and the conservation of energy gives:

$$\frac{\partial W}{\partial c} = \frac{\partial W_{ext}}{\partial c} + \frac{\partial W_{el}}{\partial c} + \frac{\partial W_s}{\partial c} = 0 \tag{3}$$

The energy release rate G is therefore defined as:

$$G = -\frac{1}{2}\frac{\partial}{\partial c}(W_{ext} + W_{el}) \tag{4}$$

The energy dissipated by the crack propagation is assumed to vary linearly with the increase in crack length, *i.e.*:

$$dW_s = 4\gamma\, dc \tag{5}$$

where γ is the specific energy of the material.

By substituting the expressions 4 and 5 in Equation 3, the crack static equilibrium is obtained when:

$$G = 2\gamma \tag{6}$$

Consider the particular case of an elastic plate of unit thickness with a through crack of length $2c$ and subjected to a uniaxial tensile stress (Figure 4). The change in elastic energy caused by the introduction of the crack in the elastic plate (Jaeger *et al.*, 2008) is given by:

$$W_{el} = \frac{\pi\sigma^2 c^2}{E} \tag{7}$$

where E is the Young's modulus of the plate material and c is the crack half-length.

According to the virtual work principle and for a constant load, it can be shown that:

$$W_{ext} = -2\, W_{el} \tag{8}$$

Thus, according to Equations 4 and 7, the rate of energy release G (Equation 4) can be expressed as a function of the stress and the crack length as follows:

$$G = -\frac{1}{2}\frac{\partial W_{el}}{\partial c} = \frac{\pi c \sigma^2}{E} \tag{9}$$

The equilibrium condition (Equation 6) is then used to calculate the critical failure condition (Equation 10) and the critical stress, σ_c is given by:

$$\pi c \sigma^2 = 2\gamma E \tag{10}$$

$$\sigma_c = \sqrt{\frac{2\gamma E}{\pi c}} \tag{11}$$

Note that the critical failure stress depends not only on the intrinsic material properties (Young's modulus E, and the specific surface energy γ) but also on the crack length, $2c$. In particular, the critical value of the stress is lower when the length of the crack is larger. The expression of the critical stress in the case of an arbitrarily oriented crack in a biaxial tensile stress field is presented by Eftis (1987) and extensive discussions are given in many texts on crack and fracture mechanics and in the volume by Sih (1991).

Griffith's analysis was extended to the case of an inclined crack in a biaxial stress field taking into account the friction on the crack faces (Jaeger et al., 2008). Under these conditions, the critical failure condition becomes:

$$\pi c \left(|\tau| - \mu\sigma_n \right)^2 = 2\gamma E \tag{12}$$

where σ_n and $|\tau|$ are, respectively, the normal stress and shear stress acting in the crack and μ is the friction coefficient between the two faces of the crack.

In this analysis, Griffith states that the energy dissipated by the propagation of the crack is solely due to the extension of the crack surface. In fact, this analysis reflects crack initiation rather than crack propagation. Other forms of energy dissipation must be taken into account during the crack propagation; for example, plastic energy dissipation, which is associated with high stress concentrations at the crack tip, and kinetic energy, related to the acceleration of the crack propagation. In the case of a purely brittle process, the plastic deformation energy can be neglected. On the other hand, it seems incorrect to neglect the effects of kinetic energy dissipation even in the case of quasi-static loading. This will be discussed in the next paragraph.

3.2.2 Steady and unsteady crack propagation

Extending Griffith's analysis, considering energy dissipation due to inertia, leads to the calculation of the total energy (Equation 2), by including a term for the kinetic energy, W_{cin}:

$$W = W_{ext} + W_{el} + W_s + W_{cin} \tag{13}$$

Crack propagation will be steady if the kinetic energy remains constant during the cracking process:

$$\frac{\partial W_{cin}}{\partial c} = 0 \tag{14}$$

From the equilibrium equation (Equation 13) and the definitions provided by Equations 4 and 5, we obtain:

$$G = 2\gamma \tag{15}$$

We thus recover the Griffith equilibrium condition for a static crack (Equation 6). However, the significance of Equation 15 is that it represents the condition of crack propagation at a constant rate; but crack propagation will be unsteady if the kinetic energy increases:

$$\frac{\partial W_{cin}}{\partial c} > 0 \tag{16}$$

Therefore, from the balance Equation 13 and the definitions (4) and (5):

$$G > 2\gamma, \tag{17}$$

Resuming the system of uniaxial tensile elementary volume (Figure 4), the axial deformation is then expressed by:

$$\varepsilon = \frac{\sigma}{E_{eff}} \tag{18}$$

with E_{eff}, the effective Young's Modulus:

$$E_{eff} = \frac{E}{1 + 2\pi c^2/bt} \tag{19}$$

$$\varepsilon = \frac{\sigma}{E}\left(1 + 2\pi c^2/bt\right) \tag{20}$$

By introducing the stability condition (15) into Equation 20, we obtain the stability criterion in the stress–strain plane:

$$\varepsilon = \frac{\sigma}{E} + 8\frac{\gamma^2 E}{\pi b t \sigma^3} \tag{21}$$

Representation of Equation 21 in the stress–strain plane is shown in Figure 5 (curve AB). For the undamaged material (a plate without a crack), Hooke's law is satisfied with a modulus of elasticity E (line OA). For a plate with a crack half-length $c = c_1$, the material also follows Hooke's law but with a lower modulus E_{eff} (Equation 19) than the virgin material (line OP', Fig. 5). When the stability curve AB is reached, the crack steadily grows to a point P" where unloading is achieved. According to Equation 19, unloading shows a weaker effective modulus of elasticity (line P"O). The surface area

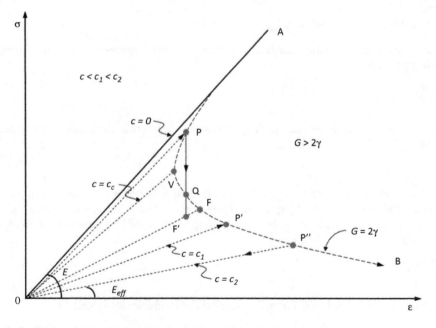

Figure 5 Griffith's stability criterion for an elementary volume in the stress–strain plane.

between segments OP', P"O and P'P, corresponds to the energy expended for crack propagation.

It should be noted that if the initial half-length of the crack is less than $c = bt/6\pi$ (corresponding to the point where the tangent to the curve is vertical, point V), then the propagation of the crack will be initially unstable (line PQ). The excess energy represented by the area PQV, is then transformed into kinetic energy. Part of this energy enables the crack to propagate beyond the stability limit (point Q) to point F'. The area QF'F represents the surface energy needed to propagate the crack, although some of the energy is dissipated as heat. The remaining kinetic energy is emitted in the material as elastic waves. In addition, for a stress-imposed loading, the crack propagation can be unstable since stability is only possible if there is a decrease in the stress. For a strain-imposed loading, two types of post peak behavior are observed depending on the initial length of the crack. The first (Type I) occurs when the crack length is such that $c < c_c$, and requires an increase in the energy transmitted to the specimen to obtain a steady propagation. In contrast, when the initial crack length is such that $c > c_c$, energy has to be extracted from the specimen to stabilize the crack propagation (Type II). Point V, where $c = c_c$, represents the case where the energy of the specimen is in equilibrium with the energy necessary for crack propagation. Wawersik and Fairhurst (1970) experimentally observed the presence of these two post-peak behavior patterns by controlling failure using a servo-controlled testing machine. In the case of a behavior of type II, known as "snap–back", only the control of transverse deformation allows the control of rupture.

The case of overall compressive stress was studied by Cook (1965). The stability curve has the same characteristics as that shown in Figure 5. Following Cook's approach, Martin and Chandler (1994) found, using Lac du Bonnet granite, that the change of stress marks the initiation of unstable crack propagation during progressive damage of the Griffith criterion type.

The Griffith global energy approach shows that the essential phenomena governing the behavior of a cracked body lie near the crack tip. Fracture mechanics investigates issues of the initiation and development of cracks by analyzing the stress in the vicinity of the crack. In the case of a brittle elastic solid, the presence of cracks leads to stress singularities. The study of these singularities allows the definition of stress intensity factors that correspond to the particular kinematics of the crack propagation. These stress intensity factors control the behavior of the crack.

When the stored elastic energy is close to the specific surface energy ($G = 2\gamma$), cracks propagate sub-critically (Atkinson, 1984); in other words, the propagation rate is lower than the sonic velocity. In contrast, when the stored elastic energy is much higher, cracks propagate super-critically.

In all cases, the cracks develop in the direction of the minor compressive stress or perpendicular to the major principal stress.

3.2.3 Griffith crack initiation criterion

In the case of complex loadings, stress analysis at the vicinity of the crack tip is required to determine the crack propagation criterion, the length increase and the orientation change. For this purpose, the discipline of Fracture Mechanics has been developed and widely used since Griffith (1924).

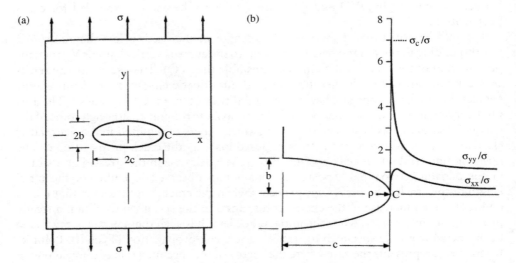

Figure 6 (a) Geometry of the plate loaded in uniaxial stress; (b) Elastic stresses distribution in the crack tip area.

In order to study the stress field around an elliptical cavity in a plate subjected to a uniaxial tensile stress, as shown in Figure 6a, the solution for the stress distribution at the periphery of the crack has to be calculated using the theory of elasticity (Inglis-Kirsch type solution) (Timoshenko & Goodier, 1970; Little, 1973; Barber, 2010; Selvadurai, 2000a).

Referring to Figure 6, the stress σ_{yy} is largest at the wall of the cavity, where the ellipse curvature radius is the smallest (point C). Thus:

$$\sigma_{yy(c)} = 1 + 2\sqrt{c/\rho} \qquad (22)$$

Subsequently, the stress σ_{yy} gradually decreases to the lowest value corresponding to the applied field, σ.

The stress σ_{xx} increases rapidly to reach a maximum near the end of the ellipse and then gradually decreases to approach zero. The area of influence of the cavity is of the order of the length c and the stress gradients are very high in the length zone of the curvature radius of the cavity ρ. According to Equation 22 for a semi-axis cavity of half-length c, the greater the radius of curvature the smaller the stress.

3.2.4 Stress field and fracture modes

Consider plane problems (2D) for which all the components of the tensors of stresses and deformations depend only on two Cartesian or polar coordinates. The crack is assumed to be in a homogeneous medium, which is isotropic linear elastic.

Recall that two basic modes of fracture exist (in fact, there are three fracture modes but the third is not relevant to massive bodies) (Figure 7):

The extension mode (Mode I) corresponds to a discontinuity of the normal displacement field to the plane of the crack. The plane shear mode (or Mode II) corresponds to a

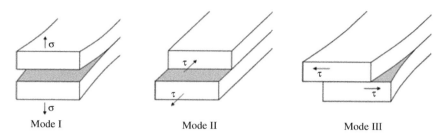

Figure 7 The three main fracture modes; in rock mechanics only Mode I and Mode II are relevant.

shift of the edges of the crack parallel to the plane of the crack along a sliding direction normal to the crack front. For any load, several basic fracture modes may overlap. This is referred to as a mixed fracture mode.

Solving the equilibrium equations with the compatibility conditions leads to a biharmonic equation for a stress function. The exact solution of this problem was established by Muskhelishvili (1953) based on the solution by Westergaard (1939) and expresses the stress fields and displacement in the crack for each of the three modes in coordinate systems.

The solution introduces the notion of stress intensity factors K_{Ic} and K_{IIc} for Mode I and Mode II fracture, respectively. Knowledge of these stress intensity factors will allow the determination of the stresses and displacements in the fissured structure. Conversely, if we know the stresses and displacements, it is possible to determine the stress intensity factors. For example, in the simple case of a plane infinite medium containing a crack of length $2c$ loaded in Mode I by a stress σ, the stress intensity factor can be expressed in the form:

$$K_I = \sigma\sqrt{\pi c} \qquad (23)$$

The stress intensity factor is measured in units $\text{N.m}^{-3/2}$ or $\text{MPa.m}^{1/2}$.

A similar approach can lead to the establishment of the stress intensity factor K_{II} for Mode II fractures (Irwin, 1957).

3.2.4.1 Failure criterion in the open mode (Mode I)

In the tensile mode, stress at the tip of the crack is infinite. Therefore, a crack initiation criterion can be introduced based on the concept of a critical threshold for the factor K_I. This criterion postulates that:

$K_I \leq K_{Ic}$ No crack propagation

$K_I = K_{Ic}$ Crack propagation onset

The critical value, K_{Ic}, is the toughness, a physical characteristic of the material (Atkinson and Meredith, 1987). Toughness values obtained from the literature (Table 1) show a significant scatter, which results from the high parameter sensitivity to the investigative method used and the environmental conditions (pressure, temperature, fluid). Note that toughness increases rapidly with the pressure applied.

Table 1 Toughness of different types of rocks, after Atkinson and Meredith (1987).

Rock Type	K_{Ic} [MPa.m ½]
Berea Sandstone	0.28
Carrara Marble	0.64 – 1.26
Westerly Granite	0.60 – 2.50

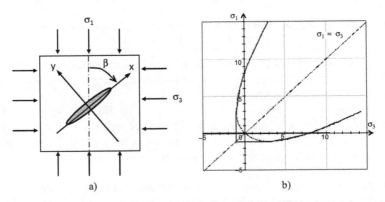

Figure 8 Griffith onset crack propagation criterion in the principal stress plane (adapted from Jaeger et al., 2008).

In terms of energy, crack growth requires that the released energy is higher than the specific energy, which leads to the establishment of the following equation:

$$W = W_{el} - W_s \tag{24}$$

Therefore, when $\frac{\partial W}{\partial c} \geq 0$, stress at the crack tip is:

$$\sigma \geq \sqrt{\frac{2\gamma E}{\pi c}} \geq \sigma_t \tag{25}$$

Remember that the energy release rate G controls the crack behavior:

$G > 2\gamma$ instability
$G = 2\gamma$ controlled steady cracking (the kinetic energy does not increase)

3.2.4.2 Failure criterion in the mixed mode

Now we consider the same plate as previously, but loaded in compression on all four sides (Figure 8a). The development of the crack is no longer along its plane and the contact of the crack surfaces can therefore withstand stress. The direction of the extension of the crack must be defined and the work dissipated by friction must be taken into account.

In the mixed mode, the definition of a failure criterion is therefore more difficult. However, linear fracture mechanics makes it possible to find a relationship between the stress intensity factors and the rate of energy release, G, when the crack grows in extension. This relationship is known as the Irwin formula:

$$G = \frac{(3 - 4\nu) + 1}{8\,G}\left(K_I^2 + K_{II}^2\right) \qquad \text{(24) In plane strain}$$

where G is the shear modulus.

The Griffith onset crack propagation criterion (Equations 25 and 26) thus generalizes the original criterion:

$$(\sigma_1 - \sigma_3)^2 - 8 \cdot \sigma_t \cdot (\sigma_1 + \sigma_3) = 0 \quad \text{if } \sigma_1 > 3 \cdot \sigma_3 \qquad (25)$$

$$\sigma_3 = -\sigma_t \qquad\qquad\qquad\qquad \text{if } \sigma_1 < 3 \cdot \sigma_3 \qquad (26)$$

Note, however, that the growth assumption of fissure in extension is a serious limitation to the application of the Irwin formula. In fact, this assumption is only valid in the case of a failure mode in pure opening (Mode I); in the case of a mixed mode, the crack deviates from its plane. Other onset failure criteria in the mixed mode are available in the literature, namely the criterion of maximum normal stress, the energy criterion of minimal elastic deformation, the criterion of maximum energy of restitution, etc.

3.3 Continuum damage mechanics

Continuum damage mechanics was first developed for structural engineering (Kachanov, 1958), with a special emphasis on structural elements loaded in tension. This approach has attracted particular interest in the design of concrete structures (Chaboche, 1988; Bažant, 1991; Lemaitre, 1996; Selvadurai, 2004).

The motivation for this approach was based on the evidence that, under certain loading conditions, the development of micro-cracks that spread throughout the material does not necessarily cause a macroscopic fracture. However, gradual deterioration of physical properties such as strength and stiffness are observed, sometimes well before rupture occurs. A comprehensive understanding and modeling of these phenomena that precede rupture is therefore of great practical interest.

Continuum Damage Mechanics was developed within the framework of Continuum Mechanics. This method enables the analysis of how the development of micro-defects influences the overall behavior of the material using a phenomenological approach, which describes the evolution of the material structure with internal state variables.

Let us suppose that the effects of micro-debonding in a Representative Elementary Volume (REV) of a solid body can be described using mechanical variables called damage variables. Since such damage variables are macroscopic variables that reflect the state of the loss of mechanical integrity, they can be considered as internal state variables from a thermodynamic point of view. With these assumptions, damage problems can be analyzed based on the principles of thermodynamics of irreversible processes. Therefore, the phenomenological damage model can be defined by:

- A law of transformation,
- A principle of equivalence,
- The nature of the damage variable,
- The evolution law of the damage variable.

3.3.1 Law of transformation

Phenomenological models appeal to the effective macroscopic quantities whose role is to define the damage state of the material. Thus, a heterogeneously damaged material obeys the same mechanical constitutive laws as its homogeneous undamaged equivalent; however, the states of stress and strain undergo changes from the nominal configuration. This change is made via a law of transformation, defined by the tensors $\overset{=4=}{\mathbf{M}_\sigma}$ and $\overset{=4=}{\mathbf{M}_\varepsilon}$, linking the effective variables $(\tilde{\sigma}, \tilde{\varepsilon})$ to nominal values (σ, ε) as follows:

$$\tilde{\sigma}(\sigma, D) = \overset{=4=}{\mathbf{M}_\sigma}(D) : \sigma \tag{27}$$

$$\tilde{\varepsilon}(\varepsilon, D) = \overset{=4=}{\mathbf{M}_\varepsilon}(D) : \varepsilon \tag{28}$$

The tensors $\overset{=4=}{\mathbf{M}_\sigma}$ and $\overset{=4=}{\mathbf{M}_\varepsilon}$ are linear operators that apply a symmetric tensor of the second order on itself and which are functions of the damage variable, formally denoted by D. When there is zero damage, the actual magnitudes coincide with the nominal values, and the transport tensor coincides with the unity.

Zheng and Betten (1996) demonstrated, by analyzing the general form of the transformation law, that the transport tensor must be an isotropic function of its arguments. Assuming a damage variable as a tensor of order two, this mathematical condition is:

$$\overset{=4=}{\mathbf{M}}\left(\mathbf{Q}\,\mathbf{D}\,\mathbf{Q}^T\right) = \mathbf{Q}\overset{=4=}{\mathbf{M}}(D)\,\mathbf{Q}^T \quad \forall \mathbf{Q} \in \Theta \tag{29}$$

where Θ is the full group of orthogonal transformations $\mathbf{Q}$.

3.3.2 Principle of equivalence

The principle of equivalence can be expressed in terms of deformation, stress or energy. The principle of equivalence in deformation implies that the effective deformation is equal to the nominal deformation, thus:

$$\tilde{\varepsilon} = \varepsilon \tag{30}$$

$$\tilde{\sigma} = \overset{=4=}{\mathbf{M}_\sigma}(D) : \sigma \tag{31}$$

More explicitly, it means that any representation of the deformational state of a damaged material can be described by the constitutive laws of an undamaged medium by simply replacing the nominal stress by the effective stress. This leads to the concept of the effective stress, which is not to be confused with the Terzaghi effective stress, which can be defined through reciprocal analysis, as the stress to be applied to the undamaged volume element to achieve the same strain as that caused by the nominal stress applied to the damaged volume element.

The principle of equivalence in stress is defined by assuming that the effective stress, $\widetilde{\sigma}$, is equal to the nominal stress, σ, *i.e.*, the effective deformation causes the same stress as the application of the nominal strain to the damaged volume:

$$\widetilde{\sigma} = \sigma \tag{32}$$

$$\widetilde{\varepsilon} = \overset{=4=}{\mathbf{M}_\varepsilon}(D) : \varepsilon \tag{33}$$

Finally, the equivalence energy principle is established assuming that the effective elastic energy density, $\widetilde{W}$, is equal to the nominal density of elastic energy, W. This assumption is expressed by:

$$\widetilde{\sigma} = \overset{=4=}{\mathbf{M}_\sigma}(D) : \sigma \tag{34}$$

$$\widetilde{\varepsilon} = \overset{=4=}{\mathbf{M}_\varepsilon}(D) : \varepsilon \tag{35}$$

$$\widetilde{W} = \frac{1}{2}\widetilde{\sigma} : \widetilde{\varepsilon} \tag{36}$$

$$W = \frac{1}{2}\sigma : \varepsilon \tag{37}$$

$$\widetilde{W} = W \tag{38}$$

3.3.3 Damage variable

Since the work of Kachanov (1958), a new variable is available to describe the internal state of a damaged material in the context of continuum mechanics. As indicated before, the notion of damage is closely linked to micro-structural aspects. However, the damage variable introduced by Kachanov described the presence of micro-defects in a comprehensive manner. The advantage of this homogenized approach is that it allows the indirect determination of the influence of the state of damage on the overall strength of the material through simple measurements.

Although the authors have not detailed the physical significance of this variable, it can be easily understood by using the concept of a Representative Elementary Volume (REV). Let us consider a REV in a damaged solid (Figure 9). Assuming that S_{total} is the area of a section of a volume element indicated by the normal $\vec{n}$, and S_{flaws} is the area of all micro-cracks and micro-pores, the classical damage variable can then be expressed by:

$$D_{\vec{n}} = \frac{S_{flaws}}{S_{total}} \tag{39}$$

From a physical point of view the damage variable, $D_{\vec{n}}$, is the ratio between the area of distributed micro-defects on the total area of the element in the plane normal to the direction, $\vec{n}$.

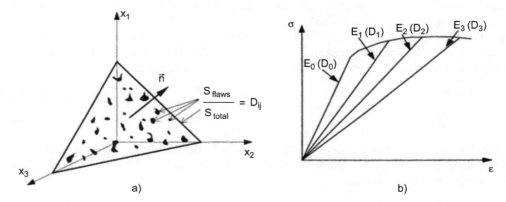

Figure 9 Damage in rocks: a) Definition of the damage parameter D_n; b) Elastic modulus for different states of damage. E_0 is the initial elastic modulus; the elastic modulus decreases ($E_0 > E_1 > E_2 > E$) as damage increases $D_0 = 0 < D_1 < D_2 < D_3$.

$D_{\vec{n}}$ allows the various states of damage to the material to be defined:

- $D_{\vec{n}} = 0$ corresponds to the undamaged material,
- $D_{\vec{n}} = 1$ corresponds to the material volume that has failed into two parts along the plane normal to, $\vec{n}$
- $0 < D_{\vec{n}} < 1$ characterizes the intermediate damaged state.

3.3.4 Isotropic and anisotropic damage

In the case where the orientation of the defects is assumed to be randomly distributed in all directions, the variable is independent of the orientation $\vec{n}$ and the scalar D completely characterizes the state of damage:

$$D_{\vec{n}} = D \qquad \forall \vec{n} \qquad (40)$$

The hypothesis of the equivalence principle in deformation directly leads to the definition of effective stress, $\tilde{\sigma}$:

$$\tilde{\sigma} = \frac{\sigma}{1-D} \qquad (41)$$

$\tilde{\sigma}$ is the stress related to the section that effectively resists forces and the law of transformation takes the particular form:

$$\overset{=4=}{\mathbf{M}_\sigma}(D) = \frac{1}{1-D} \overset{=4=}{\mathbf{I}} \qquad (42)$$

The most general form of the transformation law to describe isotropic damage was given by Ju (1990):

$$\overset{=4=}{\mathbf{M}_\sigma} = \alpha \overset{=4=}{\mathbf{I}} + \beta \mathbf{I} \otimes \mathbf{I} \qquad (43)$$

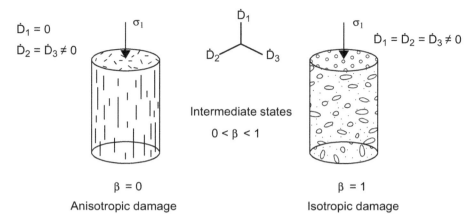

Figure 10 Isotropic damage β =1 (right hand side) and anisotropic damage β=0 (left hand side), after Pellet et al., 2005.

where α and β are two independent or dependent scalar variables, with $\alpha \geq 1$ and $-1/3 \leq \beta/\alpha \leq 0$ to ensure positivity of the stiffness tensor. The interpretation of isotropic damage does not necessarily imply a scalar variable.

The existing isotropic damage models with a scalar variable have the undeniable advantage of being simple to use. However, many experimental results (Tapponnier & Brace, 1976; Wong, 1982 for granite; Gatelier *et al.*, 2002 for sandstone; Lajtai *et al.*, 1994 for rock salt) have demonstrated that the mechanically-induced damage is anisotropic regardless of whether the intact rock is initially isotropic or anisotropic. In other words, material symmetries change during the loading process (Figure 10). In order to describe this phenomenon, the introduction of a tensor damage variable is needed. In the relevant literature we encounter second order variables (Kachanov, 1993; Murikami, 1988, Pellet *et al.*, 2005), or sometimes fourth order variables (Chaboche, 1979; Lubarda & Krajcinovic, 1993).

3.4 Discussion of Fracture Mechanics (FM) vs. Continuum Damage Mechanics (CDM)

Fracture Mechanics (FM) is an approach that aims to describe the rupture of solid bodies based on energy consideration at the crack scale. This method allows the analysis of how the development of micro-defects influences the overall behavior of the material. Despite the small number of assumptions, closed form solutions are often complex and difficult to establish for boundary values problems. Although the origin of linear fracture mechanics dates back to the early 20th century (Griffith, 1924), the development of this discipline has accelerated in recent years. We now speak about nonlinear fracture mechanics when the effects of plasticity or viscosity and dynamic fracture propagation are taken into account.

Continuum Damage Mechanics (CDM) relies on the concept of effective stress. It aims to describe the overall behavior of a Representative Elementary Volume

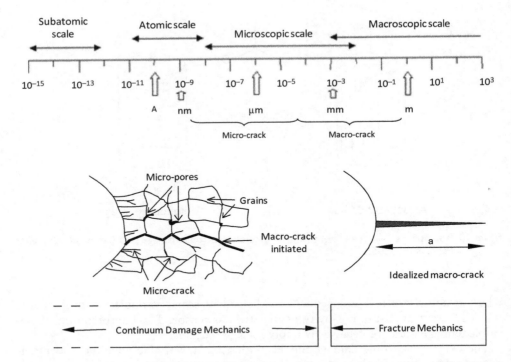

Figure 11 Schematic distinction between the representation of damage by Continuum Damage Mechanics and Fracture Mechanics (adapted from Chaboche, 1988).

(REV) by decreasing the material stiffness (elastic moduli) with the damage states. This is a phenomenological approach, which is easier to incorporate in computational codes.

Returning to phenomenological considerations, it may be said that FM is more appropriate for describing the propagation of a large single micro-crack. In contrast, CDM could be more suitable for describing the behavior of materials with multiple micro-cracks randomly spread in the body. Figure 11 illustrates with respect to scale the optimum applicability domain of both methods.

4 OTHER TYPES OF LOADING

4.1 Time-dependent damage

Studying time-dependent damage requires rate-dependent constitutive models (viscoplastic or viscoelastic) to be considered in Continuum Damage Mechanics. The first such attempts were proposed by Kachanov (1958) and Rabotnov (1969) for metallic alloys.

Based on Lemaitre's works, Pellet et al. (2005) developed a constitutive model to account for anisotropic damage and dilation of rock specimens. The strain rate is expressed by the following equation:

$$\dot{\varepsilon}^{vp} = \frac{\partial \Omega}{\partial \sigma_{ij}} = \frac{3}{2}\frac{1}{1-D}\left[\frac{\sigma_{eq}}{(1-D)\,K\,p^{1/M}}\right]^{N}\frac{S}{\sigma_{eq}} \qquad (44)$$

where Ω is the visco-plastic potential; K, N and M are the viscoplastic parameters of the model; σ_{eq} is the von Mises stress; S is the stress deviator and p is a variable that represents strain hardening.

It is also necessary to associate a law for the evolution with respect to time of the damage parameter. For uniaxial loading, the latter is expressed as follows:

$$\dot{D} = \left[\frac{\sigma}{A(1-D)}\right]^{r} \qquad (45)$$

where r is the damage exponent, q is the damage progression parameter and A is the tenacity coefficient.

From a physical point of view, this constitutive model is realistic as it accounts for volumetric strain (*i.e.* contraction and dilation) and damage-induced anisotropy through a second-rank tensor. Moreover, taking time into account allows numerical regularization and ensures the uniqueness of the solution.

Using this model, it is possible to predict delayed rupture by determining the time to failure, as observed in creep tests shown in Figure 12.

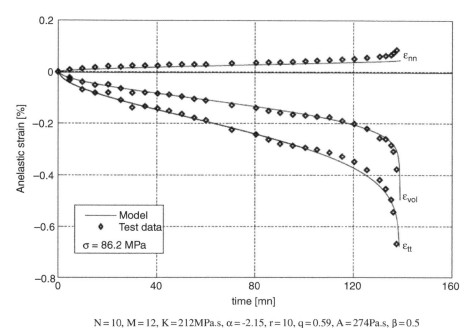

$N=10$, $M=12$, $K=212$ MPa.s, $\alpha=-2.15$, $r=10$, $q=0.59$, $A=274$ Pa.s, $\beta=0.5$

Figure 12 Comparison between the strain–time curves obtained from the model and experimental results from a creep test performed on marble, Singh (1975), after Pellet *et al.* (2005).

4.2 Damage under cyclic loading

The gradual weakening of rock properties can also be highlighted by performing cyclic loading tests. There are two types of cyclic tests: type 1, where the loading is cycled between two prescribed limits, and type 2, where the loading is increased from one cycle to the next. Most of the published data on type 1 tests are aimed at producing S–N curves, that relate the maximum stress, S, applied to a specimen to the number of cycles prior to specimen failure N. It has been shown (Costin & Holcomb, 1981) that cycling decreases rock strength, possibly by a combination of cyclic fatigue and stress corrosion. Over the last few decades several experimental programs have been performed to characterize rock behavior under static and cyclic behavior. The objective is to characterize the progressive development of damage in rocks under cyclic loading (Erarslan & Williams, 2012).

Gatelier *et al.* (2002) presented an extensive laboratory investigation of the mechanical properties of sandstone, which exhibits transversely isotropic behavior. Particular attention was paid to the influence of the structural anisotropy on the progressive development of pre-peak damage. Uniaxial and triaxial cyclic tests were performed for several orientations of the isotropy planes with respect to the principal stress directions in order to quantify the irreversible strains and the changes of oriented moduli with the cumulative damage. Two main mechanisms are involved throughout the loading process: compaction and micro-cracking.

Compaction is active at all stress levels. In uniaxial tests, both mechanisms were shown to be strongly influenced by the inclination of loading with respect to the planes of isotropy. However, with increased confining pressures, the influence of anisotropy is significantly reduced.

For an unconfined compression test carried out on sandstone specimens, Gatelier *et al.* (2002) showed that during the cycling, progressive damage is accompanied by a change in the volumetric strain. Figure 13 clearly shows that, at the early stage, the rock specimen

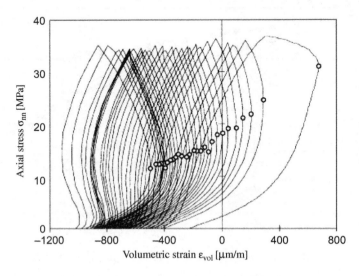

Figure 13 Axial stress as a function of volumetric strain for an unconfined cyclic compression test (from Gatelier *et al.* 2002, with permission from Elsevier).

tends to contract whereas later in the cycling process, it exhibits dilation. This observation is useful when developing an appropriate constitutive model.

4.3 Thermo-mechanical damage

In many situations, rock formations can be subjected to high temperatures that can lead to drastic changes in their mechanical properties. For example, in energy and environmental engineering, such as geothermal energy production, deep geologic disposal of high-level radioactive waste or tunnels that have experienced fire accidents, special attention has to be paid to thermal damage. In all these examples, the question that needs to be addressed is the same: how do changes in temperature influence the physical and mechanical properties of rocks?

Keshavarz *et al.* (2010) showed that thermal loading of rocks at high temperatures induces changes in their mechanical properties. In this study, a hard gabbro was tested in the laboratory. Specimens were slowly heated to a maximum temperature of 1,000° C. Following this thermal loading, the specimens were subjected to uniaxial compression. A drastic decrease of both unconfined compressive strength and elastic moduli was observed (Figure 14). The thermal damage to the rock was also highlighted by measuring elastic wave velocities and monitoring acoustic emissions during testing. The micro-mechanisms of rock degradation were investigated by analyzing thin sections after each stage of thermal loading. It was found that there is a critical temperature above which drastic changes in mechanical properties occur. Indeed, below a temperature of 600°C, micro-cracks start to develop due to a difference in the thermal expansion coefficients of the crystals. At higher temperatures (above 600°C), oxidation of Fe and Mg atoms, as well as bursting of fluid inclusions, are the principal causes of damage.

4.4 Rock joint damage

Thus far, we have focused on damage to the rock material (rock matrix). However, it is well known that discontinuities (joints, faults, etc.) are of the utmost importance in analyzing rock mass behavior (Vallier *et al.*, 2010). Traditionally, the mechanical behavior of rock discontinuities is analyzed with shear tests in different loading conditions: Constant Normal Load (CNL), Constant Normal Stiffness (CNS) or Constant Volume (CV). The two latter tests require advanced testing equipment to allow control of the displacements in the 3 spatial directions (Boulon, 1995) (see also Selvadurai & Boulon, 1995; Nguyen & Selvadurai, 1998).

Jafari *et al.* (2004) studied the variation of the shear strength of rock joints due to cyclic loadings. Artificial joint surfaces were prepared using a developed molding method that used special mortar; shear tests were then performed on these samples under both static and cyclic loading conditions. Different levels of shear displacement were applied to the samples to study joint behavior before and during considerable relative shear displacement. It was found that the shear strength of joints is related to the rate of displacement (shearing velocity), number of loading cycles and stress amplitude. Finally, based on the experimental results, mathematical models were developed for the evaluation of shear strength under cyclic loading conditions.

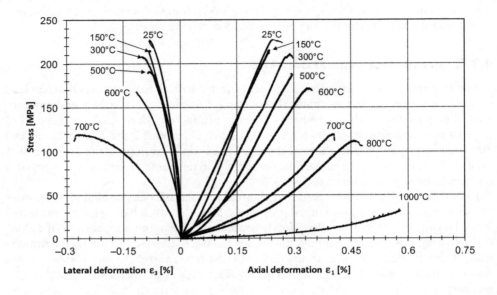

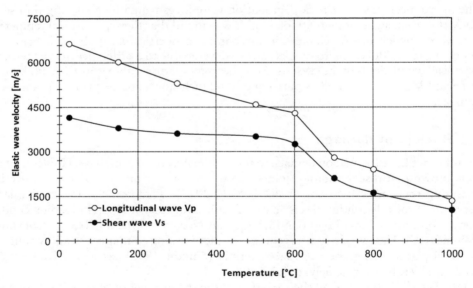

Figure 14 Damage of thermally altered gabbro specimens: top fig.) Stress-deformation curves from uniaxial compression tests. Axial and lateral deformations at failure increase with the loading temperature. A noticeable increase in both deformations occurs at temperature higher than 700°C. Lower fig.) Normalized elastic wave velocities *Vp*, *Vs* (After Keshavarz et al., 2010).

Figure 15a presents the evolution of rock joint dilation during cycling. It is shown that dilation progressively decreases as asperities damage is developed. In Figure 15b the associated decrease of shear strength is represented with respect of the number of cycles for both peak strength and residual strength.

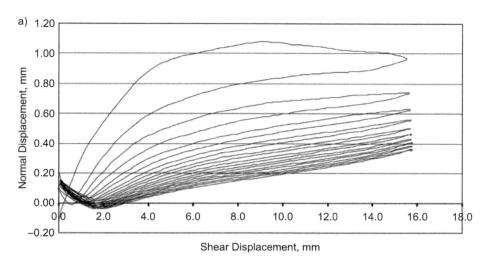

Figure 15a Cyclic shear test on rock discontinuity: Normal displacement versus shear displacement while cycling; (Jafari et al., 2004).

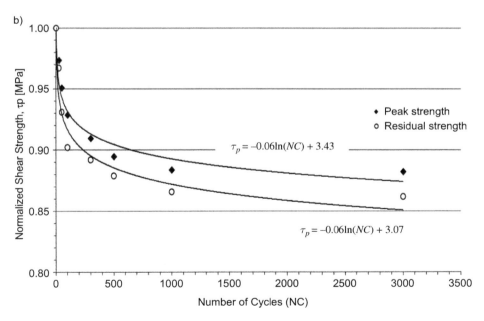

Figure 15b Cyclic shear test on rock discontinuity: Normalized shear strength as a function of the number of cycles (Jafari et al., 2004).

5 PERMEABILITY EVOLUTION WITH DAMAGE

During the development of damage in a rock, the permeability characteristics can be altered through either the development of discrete fractures or the development of continuum damage in the form of micro-cracks and micro-voids. The consequences of such damage development not only cause a reduction in the stiffness characteristics of the rock but also contribute to enhanced fluid flow through the accessible pore space of the rock. The development of enhanced fluid flow through rocks is of particular interest to geoenvironmental endeavors that focus on deep geologic disposal of hazardous nuclear wastes (Selvadurai & Nguyen, 1995, 1997; Nguyen *et al.*, 2005) and contaminants (Testa, 1994; Apps & Tsang, 1996; Selvadurai, 2006), geologic sequestration of greenhouse gases (Pijaudier-Cabot & Pereira, 2013; Selvadurai, 2013) and other energy extraction endeavors. In particular, excavation damaged zones in repositories used for storage of hazardous waste or in resources extraction activities can experience permeability alterations that will alter the fluid transport characteristics and consequently the potential for the enhanced migration of hazardous substances from repositories. The alteration of permeability of rocks during damage evolution is therefore of considerable importance to geoenvironmental applications. The influence of micro-crack generation and damage on the evolution of hydraulic conductivity of saturated geomaterials has been discussed by Zoback and Byerlee (1975), who present results of tests conducted on granite that indicate increases of up to a factor of four in the magnitude of the permeability. Results by Shiping *et al.* (1994) on sandstone indicate that for all combinations of stress states employed in their tests, the permeability increased by an order of magnitude. In triaxial tests on anisotropic granite Kiyama *et al.* (1996) presented results that also indicate increases in the permeability characteristics. Coste *et al.* (2001) present the results of experiments involving rocks and clay stone; their conclusions support the assumption of an increase in the permeability, of up to two-orders of magnitude, with an increase in deviator stresses. Focusing on the behavior of cementitious materials, the experimental results presented by Samaha and Hover (1992) indicate an increase in the permeability of concrete subjected to compression. Results by Gawin *et al.* (2002) deal with the thermo-mechanical damage of concrete at high temperatures. In their studies, empirical relationships have been proposed to describe the alterations in the fluid transport characteristics as a function of temperature and damage; here the dominant agency responsible for the alterations is identified as thermally-induced generation of micro-cracks and fissures (see also, Schneider & Herbst, 1989; Bary, 1996). Also, Bary *et al.* (2000) present experimental results concerning the evolution of permeability of concrete subjected to axial stresses, in connection with the modeling of concrete gravity dams that are subjected to fluid pressures. Gas permeability evolution in natural salt during deformation is given by Stormont and Daemen (1992), Schulze *et al.* (2001), and Popp *et al.* (2001). While the mechanical behavior of natural salt can deviate from a brittle response that is usually associated with brittle rocks, the studies support the general trend of permeability increase with damage evolution that leads to increases in the porosity. The article by Popp *et al.* (2001) also contains a comprehensive review of further experimental studies in the area of gas permeability evolution in salt during hydrostatic compaction and triaxial deformation, in the presence of creep. Time-dependent effects can also occur in brittle materials due to the sub-critical propagation of micro-cracks, a phenomenon known as *stress corrosion* (Shao *et al.*, 1997). In a recent study, Bossart *et al.* (2002) observed an alteration in permeability of an argillaceous material around deep excavations, a phenomenon attributed to the alteration of the properties of the material in the EDZ.

Souley *et al.* (2001) describe the excavation damage-induced alterations in the permeability of granite of the Canadian Shield; they observed an approximately four-orders of magnitude increase in the permeability in the EDZ. It is noted, however, that some of this data is applicable to stress states where there can be substantial deviations from the elastic response of the material as a result of the generation of localized shear zones and foliation-type extensive brittle fracture, which may be indicative of discrete fracture generation as opposed to the development of continuum damage. The data presented by Souley *et al.* (2001) has recently been re-examined by Massart and Selvadurai (2012, 2014), who used computational homogenization techniques to account for permeability evolution in rock experiencing dilatancy and damage.

It should also be noted that not all stress states contribute to increases in the permeability of geomaterials. The work of Brace *et al.* (1978) and Gangi (1978) indicates that the permeability of granitic material can indeed *decrease* with an increase in confining stresses. These conclusions are also supported by Patsouls and Gripps (1982) in connection with permeability reduction in chalk with an increase in the confining stresses. Wang and Park (2002) also discuss the process of fluid permeability reduction in sedimentary rocks and coal in relation to stress levels. A further set of experimental investigations, notably those by Li *et al.* (1994, 1997) and Zhu and Wong (1997) point to the increase in permeability with an increase in the deviator stress levels. These investigations, however, concentrate on the behavior of the geomaterial at post-peak and, more often in the strain softening range. Again, these experimental investigations, although of considerable interest in their own right, are not within the scope of the current chapter that primarily deals with permeability evolution during damage in the elastic range. The experimental studies by Selvadurai and Głowacki (2008) conducted on Indiana Limestone, similarly indicate the permeability reduction of a rock during isotropic compression up to 60 MPa. The studies also indicate the presence of permeability hysteresis during quasi-static load cycling. The internal fabric of the rock can also affect the enhancement of permeability of rocks even during isotropic compression. The experimental work conducted by Selvadurai *et al.* (2011) on the highly heterogeneous Cobourg Limestone indicates that the permeability can increase during the application of isotropic stress states. In this case the external isotropic stress state can result in locally anisotropic stress states that can lead to permeability enhancement through dilatancy at the boundaries of heterogeneities.

5.1 Modeling saturated porous media with continuum damage

The modeling of fluid-saturated porous media that experience continuum damage can be approached at various levels, the simplest of which is the modification of the equations of classical poroelasticity developed by Biot (1941) (see also Rice & Cleary, 1976; Detournay & Cheng, 1993; Selvadurai, 1996; Wang, 2000; Cheng, 2016; Selvadurai & Suvorov, 2016) to account for damage evolution in the porous fabric and the enhancement of the permeability characteristics of the fluid-saturated medium by damage evolution. In the case of an intact poroelastic material, considering Hookean isotropic elastic behavior of the porous skeleton, the principle of mass conservation and Darcy's law applicable to an isotropic porous medium, the coupled constitutive equations governing the displacement field $\mathbf{u}(\mathbf{x}, t)$ and the pore pressure field $p(\mathbf{x}, t)$ in a fluid-saturated body void of body or inertia forces can be expressed in the form

$$\mu \nabla^2 \mathbf{u} + \frac{\mu}{(1 - 2\nu)} \nabla(\nabla.\mathbf{u}) + \alpha \nabla p = 0 \tag{46}$$

and

$$\kappa \beta \nabla^2 p - \frac{\partial p}{\partial t} + \alpha \beta \frac{\partial}{\partial t}(\nabla.\mathbf{u}) = 0 \tag{47}$$

In (46) and (47)

$$\alpha = \frac{3(\nu_u - \nu)}{\widetilde{B}(1 - 2\nu)(1 + \nu_u)} \quad ; \quad \beta = \frac{2\mu(1 - 2\nu)(1 + \nu_u)^2}{9(\nu_u - \nu)(1 - 2\nu_u)} \tag{48}$$

ν and μ are the "drained values" of Poisson's ratio and the linear elastic shear modulus applicable to the porous fabric, ν_u is the undrained Poisson's ratio and $\widetilde{B}$ is the pore pressure parameter introduced by Skempton (1954), κ ($= k/\gamma_w$) is a permeability parameter, which is related to the hydraulic conductivity k and the unit weight of the pore fluid γ_w and ∇, ∇ and ∇^2 are, respectively, the gradient, divergence and Laplace's operators. When damage occurs in a general fashion the elasticity and fluid transport characteristics of the rock can result in the development of properties that can be anisotropic. If full anisotropy evolution with damage is considered, the elasticity parameters increase to 21 constants and the fluid transport parameters increase to six. Therefore the generalized treatment of damage-induced alterations to coupled poroelasticity effects can result in an unmanageable set of parameters and damage evolution laws. This has restricted the application of damage mechanics to the study of coupled poroelasticity problems. The prudent approach is to restrict attention to isotropic damage evolution that can be specified in terms of the effective stress defined by (41) and the strain equivalent hypothesis. The alterations in the elastic stress–strain relationships for the damaged isotropic elastic skeleton can be represented in the form

$$\boldsymbol{\sigma} = 2(1 - D)\mu \boldsymbol{\varepsilon} + \frac{2(1 - D)\mu\nu}{(1 - 2\nu)}(\nabla.\mathbf{u})\mathbf{I} + \alpha p \mathbf{I} \tag{49}$$

where D is the isotropic damage parameter and implicit in (49) is the assumption that Poisson's ratio remains unaltered during the damage process and alterations to the isotropic Lamé and other elastic constants and the permeability parameters maintain the energetic requirements

$$\mu(D) > 0; \quad 0 \leq \widetilde{B}(D) \leq 1; \quad -1 < \nu(D) < \nu_u(D) \leq 0.5; \quad \kappa(D) > 0 \tag{50}$$

This assumption makes the incorporation of influences of damage on the poroelastic response more manageable. In addition to the specification of the constitutive relations for the damaged geomaterial skeleton, it is also necessary to prescribe damage evolution criteria that can be postulated either by appeal to micro-mechanical considerations or determined by experiment. Examples of the evolution of the damage parameter with either stress or strain have been presented in the literature and references to these studies are given by Mahyari and Selvadurai (1998), Selvadurai (2004) and Selvadurai and Shirazi (2004, 2005); the experimental information that describes the variation of the poroelastic parameters with damage is generally scarce. Also, the extent of damage necessary to create substantial alterations to permeability evolution is generally small. By and large, the porous skeleton of the fluid-saturated rock can remain elastic with a small change in the elasticity parameters during which the permeability of the rock can change substantially. Based on this observation, Selvadurai (2004) introduced the concept of *"Stationary Damage"* in poroelastic solids where the skeletal elasticity properties remain at a constant value after the initiation of elastic damage and alterations in the hydraulic conductivity characteristics are determined at the damage level corresponding to the stationary damage or elastic estimates. Zoback and Byerlee (1975) have documented results of experiments conducted on granite and Shiping *et al.* (1994) give similar results for tests conducted on sandstone. They observed that the permeability characteristics of these materials can increase by an order of magnitude before the attainment of the peak values of stress and they can increase up to two-orders of magnitude in the strain softening regime where micro-cracks tend to localize in shear faults. Kiyama *et al.* (1996) also observed similar results for the permeability evolution of granites subjected to a tri-axial stress state. This would suggest that localization phenomena could result in significant changes in the permeability in the localization zones. It must be emphasized that in this study the process of localization is excluded from the analysis and all changes in permeability are at stress states well below those necessary to initiate localization or global failure of the material. Furthermore, in keeping with the approximation concerning scalar isotropic dependency of the elasticity properties on the damage parameter, we shall assume that the alterations in the permeability characteristics also follow an isotropic form. This is clearly an approximation with reference to the mechanical response of brittle geomaterials that tend to develop micro-cracking along the dominant direction of stressing, leading to higher permeabilities in orthogonal directions. The studies by Mahyari and Selvadurai (1998) suggest the following postulates for the evolution of hydraulic conductivity as a function of the parameter ξ_d:

$$k^d = (1 + \widetilde{\alpha}\xi_d)k \quad ; \quad k^d = (1 + \widetilde{\beta}\xi_d^2)k \tag{51}$$

where k^d is the hydraulic conductivity applicable to the damaged material, k is the hydraulic conductivity of the undamaged material, ξ_d is the equivalent shear strain, which is related to the second invariant of the strain deviator tensor, and $\widetilde{\alpha}$ and $\widetilde{\beta}$ are material constants. This approach has enabled the application of isotropic damage mechanics concepts to examine the time-dependent effects that can materialize in fluid-saturated poroelastic media where the porous fabric undergoes mechanical damage with an attendant increase in the hydraulic conductivity characteristics.

5.2 Application of the concept of "Stationary Damage"

As an illustration of the application of the concept of "Stationary Damage" to problems in poromechanics, we specifically consider the problem of the indentation of a poroelastic halfspace by a rigid circular indenter with a flat smooth base. This is a celebrated problem in contact mechanics; the elastic solutions were first presented in the classic studies by Boussinesq (1885) and Harding and Sneddon (1945). Computational techniques are applied to examine the influence of elastic damage-induced fluid transport characteristics on the time dependent indentational response of the Boussinesq indenter.

Indentation and contact problems occupy an important position in both engineering and applied mechanics. Solutions derived for classical elastostatic contact problems have been applied to examine the mechanics of indentors used for materials testing, mechanics of nano-indentors, tribology, mechanics of foundations used for structural support, biomechanical applications for prosthetic implants and, more recently, in the area of contact mechanics of electronic storage devices. We consider the problem of the *frictionless indentation* of a poroelastic material by a rigid circular punch with a flat base (Figure 16), which is subjected to a total load P_0, which is in the form of a Heaviside step function of time. The associated classical elasticity solution was first given by Boussinesq (1885), who examined the problem by considering the equivalence between the elastostatic problem and the associated problem in potential theory.

Harding and Sneddon (1945) subsequently examined the problem in their classic paper that uses Hankel transform techniques to reduce the problem to the solution of a system of dual integral equations. The procedure also resulted in the evaluation of the load-displacement relationship for the indentor in exact closed form. In a subsequent paper, Sneddon (1946) presented complete numerical results for the distribution of stresses within the halfspace region. The classical poroelasticity problem concerning the static indentation of a poroelastic halfspace and layer regions by a rigid circular indentor with a flat smooth base and related contact problems were considered by a number of authors including Agbezuge and Deresiewicz (1974), Chiarella and Booker (1975), Gaszynski (1976), Gaszynski and Szefer (1978), Selvadurai and Yue (1994), Yue and Selvadurai (1994, 1995a, b) and Lan and

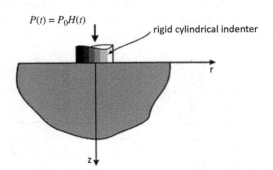

Figure 16 Indentation of a damage susceptible poroelastic halfspace.

Selvadurai (1996), using differing boundary conditions related to the pore pressure at the surface of the halfspace both within the contact zone and exterior to it. These authors also use different computational schemes for the numerical solution of the resulting integral equations and for the inversion of Laplace transforms. Of related interest are problems associated with the dynamic problem of a rigid foundation either in smooth contact or bonded to the surface of a halfspace (Halpern & Christiano, 1986; Kassir & Xu, 1988; Philippacopoulos, 1989; Senjuntichai & Rajapakse, 1996), where, in certain circumstances, the static transient poroelasticity solution can be recovered. The former studies will form a basis for a comparison with the modeling involving stationary damage; here we will consider *only* changes in the hydraulic conductivity characteristics, which will be altered corresponding to the initial elastic strains induced during the loading of the indenter. Also, the load applied is specified in the form of a Heaviside step function in time. In order to determine the stationary spatial variation of hydraulic conductivity properties within the halfspace region, it is first necessary to determine the distribution of the equivalent shear strain ξ_d in the halfspace region. Formally, the distribution $\xi_d(r, z)$, is determined by considering the stress state in the halfspace region associated with the elastic contact stress distribution at the indenter-elastic halfspace region, which is given by

$$\sigma_{zz}(r,0) = \frac{P_0}{2\pi a \sqrt{a^2 - r^2}} \qquad ; \qquad r \in (0,a) \tag{52}$$

and the classical solution by Boussinesq (1885) for the problem of the action of a concentrated normal load at the surface of a halfspace region (see also Selvadurai, 2000a, 2001). The displacement distribution at the surface of the halfspace region is given by

$$u_z(r,0) = \begin{cases} \Delta & ; \quad r \in (0,a) \\ \dfrac{2\Delta}{\pi} \sin^{-1}\left(\dfrac{a}{r}\right) & ; \quad r \in (a, \infty) \end{cases} . \tag{53}$$

The stress state in the halfspace region is given by

$$\sigma_{rr}(\rho,\zeta) = -\frac{P_0}{2\pi a^2}\left[J_1^0 + 2\tilde{\nu}\{J_1^0 - J_0^1\} - \zeta J_2^0 - \frac{1}{\rho}\{(1-2\tilde{\nu})J_0^1 - \zeta J_2^1\} \right]$$

$$\sigma_{\theta\theta}(\rho,\zeta) = -\frac{P_0}{2\pi a^2}\left[2\tilde{\nu} J_0^1 + \frac{1}{\rho}\{(1-2\tilde{\nu})J_0^1 - \zeta J_2^1\} \right]$$

$$\sigma_{zz}(\rho,\zeta) = -\frac{P_0}{2\pi a^2}[J_1^0 + \zeta J_2^0]$$

$$\sigma_{rz}(\rho,\zeta) = -\frac{P_0}{2\pi a^2}[\zeta J_2^1] \tag{54}$$

where $\tilde{\nu}$ is Poisson's ratio for the elastic solid and the infinite integrals $J_n^m(\rho,\zeta)$ are defined by

$$J_n^m(\rho, \zeta) = \int_0^\infty s^{n-1} \sin(s) \exp(-s\zeta) J_m(s\rho) ds. \tag{55}$$

As has been shown by Sneddon (1946), these infinite integrals can be evaluated in explicit closed form as follows:

$$J_1^0(\rho, \zeta) = \frac{1}{\sqrt{R}} \sin\left(\frac{\phi}{2}\right) \; ; \; J_0^1(\rho, \zeta) = \frac{1}{\rho}\left(1 - \sqrt{R}\sin\left(\frac{\phi}{2}\right)\right)$$

$$J_1^1(\rho, \zeta) = \frac{\Psi}{\rho\sqrt{R}} \sin\left(\theta - \frac{\phi}{2}\right) \; ; \; J_2^1(\rho, \zeta) = \frac{\rho}{R^{3/2}} \sin\left(\frac{3\phi}{2}\right)$$

$$J_2^0(\rho, \zeta) = \frac{\Psi}{R^{3/2}} \sin\frac{3}{2}(\phi - \theta) \tag{56}$$

where

$$\tan\theta = \frac{1}{\zeta} \; ; \; \tan\phi = \frac{2\zeta}{(\rho^2 + \zeta^2 - 1)} \; ; \; \Psi^2 = (1 + \zeta^2)$$

$$R^2 = [(\rho^2 + \zeta^2 - 1)^2 + 4\zeta^2] \; ; \; \rho = \frac{r}{a}; \; \zeta = \frac{z}{a}. \tag{57}$$

The principal stress components are determined from the relationships

$$\left.\begin{array}{c}\sigma_1 \\ \sigma_3\end{array}\right\} = \frac{1}{2}\left[(\sigma_{rr} + \sigma_{zz}) \pm \sqrt{(\sigma_{rr} - \sigma_{zz})^2 + 4\sigma_{rz}^2}\right] \; ; \; \sigma_2 = \sigma_{\theta\theta} \tag{58}$$

and the equivalent shear strain ξ_d can be expressed in the form

$$\xi_d = \frac{1}{2\sqrt{3}\tilde{\mu}}\left[(\sigma_1 - \sigma_3)^2 + (\sigma_3 - \sigma_2)^2 + (\sigma_2 - \sigma_1)^2\right]^{1/2} \tag{59}$$

where $\tilde{\mu}$ is the linear elastic shear modulus for the elastic solid. In these general elasticity solutions, the elastic constants $\tilde{\mu}$ and $\tilde{\nu}$ can be assigned values corresponding to their values at time $t = 0$, to reflect the undrained behavior of the poroelastic solid.

An examination of both (52) and (54) indicates that the elastic stress state is singular at the boundary of the rigid indenter. This places a restriction on the rigorous application of the stress state (54) for the determination of damaged regions. By definition, damaged regions are assumed to experience only finite levels of isotropic damage that would maintain the elastic character of the material. The singular stress state can result in either plastic failure of the rock (Ling, 1973; Johnson, 1985) or even brittle fracture extension in the halfspace region (Selvadurai, 2000b). Such developments are assumed to be restricted to a very limited zone of the halfspace region in the vicinity of the boundary of the indenter region. Also, in the computational modeling of the contact problem, no provision is made for the incorporation of special elements at the boundary of the contact zone to account for the singular stress state that can be identified from mathematical considerations of the contact problem. In the computational modeling, the mesh

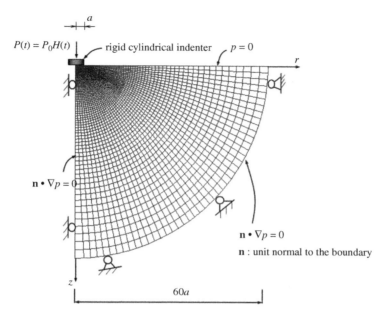

Figure 17 Finite element discretization of the damage susceptible spherical poroelastic region for the cylindrical indenter problem.

configuration is suitably refined to account for the sharp stress gradients that will result from the elastic stress state (Figure 17).

The distribution of equivalent shear strain is accounted for by assigning the values of the equivalent strains to the integration points within the elements. These in turn are converted to alterations in the hydraulic conductivity characteristics of the medium through the use of the expressions (51) that relate the hydraulic conductivity to the equivalent shear strain. In the study by Selvadurai (2004), the computational modeling was performed using the general purpose finite element code ABAQUS, although any computational code that is capable of examining poroelasticity problems for inhomogeneous media can be adopted for the purpose.

An 8-noded isoparametric finite element is used in the modeling and the integrations are performed at the nine Gaussian points. The displacements are specified at all nodes and the pore pressures are specified only at the corner nodes.

The application of the "Stationary Damage Concept" to the examination of the poroelastic contact problem where the porous skeleton can experience damage due to stress states that will not induce plastic collapse or extensive plastic failure in the rock involves the following steps:

– Apply the loadings to the elastic domain and determine the equivalent shear strain ξ_d in the domain.
– Determine the distribution of permeability alteration in the region by appeal to the prescribed variation of permeability with ξ_d. This will result in a poroelastic domain with a permeability heterogeneity.

- Solve the poroelasticity problem where the loads are applied at a prescribed time variation, where the elasticity and permeability characteristics are as implied by the previous steps.

We consider the computational modeling of the indentation problem where the pore fluid that saturates the porous elastic solid is assumed to be incompressible. The indenter is subjected to a load with a time variation in the form of a Heaviside step function *i.e.*:

$$P(t) = P_0 H(t); \ t > 0 \tag{60}$$

where P_0 is the magnitude of the total load acting on the indentor. Since the saturating fluid is assumed to be incompressible, the undrained behavior of the fluid-saturated poroelastic medium at time $t = 0$ corresponds to an elastic state with $\tilde{\nu} = 1/2$. The initial elastic strains that induce the spatial distribution of damage during the indentation are evaluated by setting $\tilde{\nu} = 1/2$ in the principal stresses computed, using the stress state (52), relevant to the smooth indentation of the halfspace with the porous rigid circular cylinder.

The finite element discretization of the halfspace domain used for the analysis of the indentation of the poroelastic halfspace by the porous rigid circular smooth indenter, of radius $a = 2m$, is shown in Figure 17. To exactly model a poroelastic halfspace region it is necessary to incorporate special infinite elements that capture the spatial decay in the pressure and displacement fields (see *e.g.* Simoni & Schrefler, 1987; Selvadurai & Gopal, 1989). These procedures were, however, unavailable in the computational modeling software. The poroelastic halfspace region is modeled as a hemispherical domain, where the outer boundary is located at a large distance ($60\ a$) from the origin (Figure 17). This external spherical boundary is considered to be rigid and all displacements on this boundary are constrained to be zero. The boundary is also considered to be impervious, thereby imposing Neumann boundary conditions for the pore pressure field at this spherical surface. (It is noted here that computations were also performed by prescribing Dirichlet boundary conditions on this surface. The consolidation responses computed were essentially independent of the far field pore pressure boundary condition.) The accuracy of the discretization procedures, particularly with reference to the location of the external spherical boundary at a distance $60\ a$, was first verified through comparisons with Boussinesq's solution for the indentation an elastic halfspace region by a cylindrical punch. The value of the elastic displacement of the rigid circular indenter can be determined to an accuracy of approximately three percent. The computational modeling of the poroelastic indentation problem is performed by specifying the following values for the material parameters generally applicable to a geomaterial such as sandstone (Selvadurai, 2004): hydraulic conductivity $k = 1 \times 10^{-6}\ m/s$; unit weight of pore fluid $\gamma_w = 1 \times 10^4 N/m^3$; Young's modulus. $E = 8.3\ GPa$; Poisson's ratio $\nu = 0.195$; the corresponding coefficient of consolidation, defined by

$$c = \frac{2\mu k}{\gamma_w} = 0.6946\ m^2/s. \tag{61}$$

Figure 18 illustrates the comparison between the analytical solution for the time-dependent variation in the rigid displacement of the circular indentor given by Chiarella and Booker (1975) and the computational results obtained for the

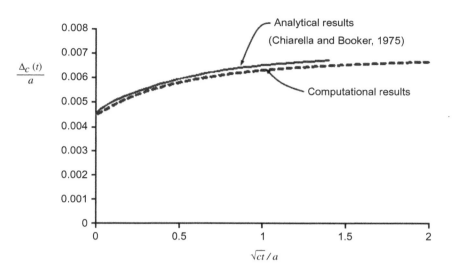

Figure 18 Comparison of analytical results and computational estimates for the cylindrical rigid indenter problem.

indentation of a poroelastic region with an external boundary in the shape of a hemisphere. The comparison is between an estimate for a halfspace region and a region of finite extent and there is reasonable correlation between the two sets of results. Since the influence of "Stationary Damage" will be assessed in relation to the computational results derived for the indentation of the hemispherical region, the accuracy of the computational scheme in modeling the indentation process is considered to be acceptable. The normal stress distribution at the contact zone can be determined to within a similar accuracy except in regions near the boundary of the indenter where, theoretically, the contact stresses are singular. In the computational treatments, the stationary damage-induced alteration in the hydraulic conductivity is evaluated according to the *linear dependency* in the hydraulic conductivity alteration relationship given by (51) and the parameter α is varied within the range $\alpha \in (0, 10^4)$. Figure 19 illustrates the results for the time-dependent displacement of the rigid cylindrical smooth indentor resting on a poroelastic hemispherical domain that displays either stationary damage-induced alteration in the hydraulic conductivity or is independent of such effects. These results are for a specific value of the hydraulic conductivity-altering parameter, $\alpha = 10^3$. Computations can also be performed to determine the influence of the parameter α and the stationary damage-induced alterations in the hydraulic conductivity on the settlement rate of the rigid indenter. The results can be best illustrated through the definition of a *"Degree of Consolidation"*, defined by

$$U_C = \frac{\Delta_C(t) - \Delta_C(0)}{\Delta_C(\infty) - \Delta_C(0)} \qquad (62)$$

The value $\Delta_C(t)$ corresponds to the time-dependent rigid displacement of the indenter. Both the initial and ultimate values of this displacement, for the purely poroelastic case, can be evaluated, independent of the considerations of the transient poroelastic

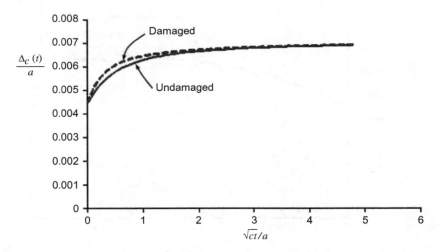

Figure 19 Influence of stationary damage on the displacement of the rigid cylindrical indentor [α= 1000].

responses since the poroelastic model allows for purely elastic behavior at $t = 0$ and as $t \to \infty$, with $\nu = 1/2$ and $\tilde{\nu} = \nu$ respectively: *i.e.*

$$\Delta_C(0) = \frac{P_0}{8\mu a} \; ; \; \Delta_C(\infty) = \frac{P_0(1-\nu)}{4\mu a}. \tag{63}$$

Figure 20 illustrates the variation of the degree of consolidation of the rigid indenter as a function of non-dimensional time and the parameter α that defined the alteration of the hydraulic conductivity with the equivalent shear strain ξ_d.

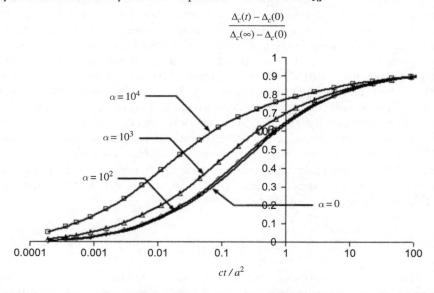

Figure 20 Influence of stationary damage and hydraulic conductivity alteration on the consolidation rate for the rigid cylindrical indenter.

6 NUMERICAL MODELING OF DAMAGE

We have seen in Section 3 that, from a theoretical point of view, two main approaches exist to model rock damage. These are based either on Continuum Mechanics, or on Fracture Mechanics.

In terms of numerical modeling, the continuum approach is classically handled using the Finite Element Method whereas fracturing is more difficult to deal with numerically. However, relatively recently, methods have been developed to incorporate the nucleation and propagation of cracks in an originally continuous medium. These methods belong to the family of Extended Finite Element Methods (XFEM). An alternative approach is the Finite Discrete Element Method (FDEM), which is also a recent computational tool that combines the Finite Element Method with the Discrete Element Method (DEM).

Currently, these methods are difficult to use in an engineering context because of the complexity of rock structures and the lack of knowledge of the boundary and initial conditions. For rock engineering problems, the computational time required is still cumbersome.

6.1 Modeling a loading test with CDM

Identification of model parameters is based on experiments such as creep tests, relaxation tests and quasi-static tests. Figure 21 shows the numerical modeling of a compression test performed with a low strain rate of loading (Fabre & Pellet, 2006). Both axial strain and lateral strain are well reproduced by the model. Additionally, the calculated change in volumetric strain (contraction-dilation) is consistent with the observed location.

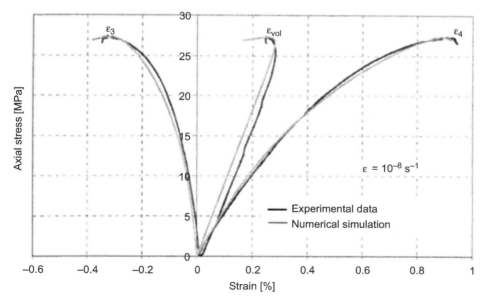

Figure 21 Modeling of a uniaxial compressive test performed on shale (Fabre & Pellet, 2006).

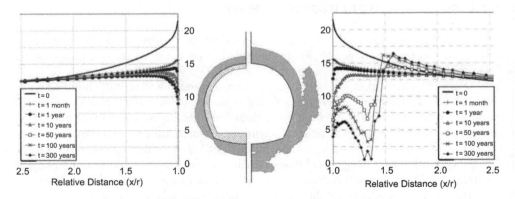

Figure 22 Excavation Damage Zone (EDZ) and stress distribution. In the centre diagram the greyish areas represent the extension of the EDZ, 300 years after the tunnel construction, for a supported tunnel (left hand side) and for an unsupported tunnel (right hand side). The tangential stresses in the rock mass at different times are also represented (adapted from Pellet et al., 2009).

6.2 Excavation Damage Zone (EDZ) around underground openings

When an underground opening is excavated in a stressed rock mass, short-term instabilities may occur during or right after operation. Collapses in the gallery may also take place many years or decades after the completion of the work. Indeed, the time-dependent behavior of rock has always been an issue for underground construction. This question has been recently studied, both experimentally and theoretically, by several authors for different types of rock (Blümling et al., 2007; Hudson et al., 2009).

A 3D numerical simulation of the mechanical behavior of deep underground galleries with a special emphasis on time-dependent development of the Excavation Damage Zone (EDZ) was presented by Pellet et al. (2009). In this study, the rock mass behavior is modeled by a damageable viscoplastic constitutive model (see section 4.1) in which both viscous and damage parameters are taken into account. Finite-element analysis investigates the evolution of near field stresses, progressive development of the damage zone as well as delayed displacements during the sequential construction process of the gallery. The influence of the orientation of in situ stresses with respect to the gallery axis is also highlighted. The effect of a support system is to reduce the damage zone and the displacements around the gallery (Figure 22).

7 CONCLUSION

Damage to rock depends largely on its geological nature and the state of stress it is subjected to. The physico-chemical identification of rocks is therefore a prerequisite that requires special attention. This involves a precise characterization of the mineralogical content and a detailed description of the rock fabric. Using these observations, the petrophysical properties of the rock (porosity, specific gravity, water content and degree of saturation ...) will be estimated. There have been considerable advances in the digital treatment of experimental data over the last two decades, associated with the development of new testing equipment (Scanning Electron Microscope, Magnetic Resonance Imaging, Acoustic Tomography, etc.) that has greatly facilitated the identification of the rock type and its in situ state.

Laboratory testing can then be used to determine the mechanical properties of the rock under investigation; these include the strength and deformability. Here, too, modern techniques allow complex mechanical tests involving high pressure-high temperature, to be combined with physical measures such as acoustic activity or the velocity of elastic waves. 3D visualization by X-ray computed tomography (CT) of the pore space helps to localize the areas of deformation and cracking.

Constitutive models can only be developed based on an understanding and knowledge of the detailed characterization of the state of the rock both in its intact state and after mechanical testing. We have seen that there are two main approaches employed in Rock Mechanics for damage characterization: Continuum Damage Mechanics and Fracture Mechanics. The first, formulated in the context of the thermo-dynamics of irreversible processes, lends itself to numerical modeling although certain hypotheses (shear failure mode, small deformations, etc.) must be invoked. For example, processing softening behavior inevitably requires regularizing the solution by introducing a time variable to establish a rate-dependent model. If the model is rate-independent, such an approach should be reserved for the damage suffered before the peak stress is achieved.

The second approach, Fracture Mechanics, is older and also more physical in that it addresses the failure mode by extension (Mode I: crack opening), which is frequently observed experimentally. However, this fails to meet the assumption of a continuum and is therefore inconvenient to implement, particularly if multiple crack generation, crack branching and discrete fragmentation is involved. However the introduction in the last ten years of the XFEM technique (eXtended Finite Element Method) allows the two approaches to be combined. The engineering applications of these advances are limited; in particular, it is difficult to account for the merging of several networks of cracks.

The use of a constitutive model for full-scale processes requires many adaptations that go beyond the simple consideration of the scale effect. The need to account for damage discontinuities is paramount and often the time effects are equally important. Under certain loading conditions the effects of repeated actions (cyclic and/or dynamic) can be the most important consideration: loadings of a thermo-hydro-mechanical origin are often related to damage through permeability changes.

Over the past two decades, great progress has been made on the experimental characterization of damage in rocks and rock masses and numerical methodologies have improved substantially. Future developments should particularly focus on the development of numerical tools for the identification of the parameters to be used in constitutive models by means of inverse analysis of measurements made on real structures.

REFERENCES

Agbezuge, L.K. and Deresiewicz, H. (1974) On the indentation of a consolidating halfspace. *Israel J. Tech.*, **12**: 322–338.

Apps, J.A. and Tsang, C.-F. (Eds.) (1996) *Deep Injection Disposal of Hazardous and Industrial Waste*, Academic Press, San Diego.

Atkinson, B.K. (1984) Sub-critical crack growth in geological materials. *J. Geophys. Res.*, **89**(B6): 4077–4114.

Atkinson, B.K. and Meredith, P.G. (1987) Experimental fracture mechanics data for rocks and minerals, In *Fracture Mechanics of Rock*, B.K. Atkinson Ed., Academic Press Geology Series, San Diego: 477–525.

Barber, J.R. (2010) *Elasticity, Solid Mechanics and Its Applications*, Vol. 172, Springer-Verlag, Berlin.

Bary, B. (1996) *Etude de couplage hydraulique-mecanique dans le beton endomage*, Lab. Meca. et Tech., CNRS, de Cachan, Univ. Paris XI, Publ. 11.

Bary, B., Bournazel, J.-P. and Bourdarot, E. (2000) Poro-damage approach to hydrofracture analysis of concrete. *J. Eng. Mech., ASCE*, **126**: 937–943.

Bažant, Z.P. (1991) Why continuum damage is nonlocal: Micromechanics arguments. *J. Eng. Mech.*, **117**(5): 1070–1087.

Bieniawski, Z.T. (1967) Mechanism of brittle fracture of rock, Parts I, II and III. *Int. J. Rock Mech. Min. Sci. Geomech. Abstr.*, **4**(4): 395–430.

Biot, M.A. (1941) General theory of three-dimensional consolidation. *J. Appl. Phys.*, **12**, 155–164.

Blümling, P., Bernier, F., Lebon, P. and Martin, C.D. (2007) The excavation damaged zone in clay formations time-dependent behaviour and influence on performance assessment. *Phy. Chem. Earth*, Parts A/B/C **32**(8–14): 588–599.

Bossart, P., Meier, P.M., Moeri, A., Trick, T. and Mayor, J.-C. (2002) Geological and hydraulic characterization of the excavation disturbed zone in the Opalinus Clay of the Mont Terri Rock Laboratory. *Eng. Geol.*, **66**: 19–38.

Boulon, M. (1995) A 3D direct shear device for testing the mechanical behaviour and the hydraulic conductivity of rock joints, in: Rossmanith, P. (ed.), *Proc. Mechanics of Jointed and Faulted Rock*: 407–413.

Boussinesq, J. (1885) *Applications des potentials a l'etude de l'equilibre et du mouvement des solides elastique*, Gauthier-Villars, Paris.

Brace, W.F., Paulding, B.W. and Scholz, C. (1966) Dilatancy in the fracture of crystalline rocks. *J. Geophys. Res.*, **71**(16): 3939–3953.

Brace, W.F., Silver, E., Hadley, K. and Goetze, C. (1972) *Cracks and Pores: A Closer Look*, New Series Published by: American Association for the Advancement of Science Vol. 178, No. 4057: 162–164,

Brace, W.F., Walsh, J.B. and Frangos, W.T. (1978) Permeability of granite under high pressure. *J. Geophys. Res.*, **73**: 2225–2236.

Chaboche, J.L. (1979) Le concept de contrainte effective appliqué à l'élasticité et à la viscoplasticité en présence d'un endommagement anisotrope, in J. P. Boehler ed., *Mechanical Behaviour of Anisotropic Solids*, Colloque Euromech 115, Villard de Lans, France : 737–759.

Chaboche, J. L. (1988) Continuum damage mechanics, Part I: General concepts, Part II: Damage growth, crack initiation, and crack growth. *J. Appl. Mech. Trans. ASME*, **55**: 55–72.

Cheng, A.H.-D. (2016) *Poroelasticity*, Springer-Verlag, Berlin.

Chiarella, C. and Booker, J.R. (1975) The time-settlement behaviour of a rigid die resting on a deep clay layer. *Q. J. Mech. Appl. Math.*, **28**: 317–328.

Cook, N.G.W. (1965) The failure of rocks. *Int. J. Rock Mech. Min. Sci.*, **2**: 389–403.

Coste, F., Bounenni, A., Chanchole, S. and Su, K. (2001) *A method for measuring hydraulic and hydromechanical properties during damage in materials with low permeability*, G.3S, Ecole Polytechnique, Palaiseau (Unpublished Report).

Costin, L.S. and Holcomb, D.J. (1981) Time dependent failure of rocks. *Tectonophysics*, **79**: 279–296.

Davis, G.H. and Reynolds, S.J. (1996) *Structural Geology of Rocks and Regions*, John Wiley & Sons, New York.

Detournay, E. and Cheng, A.H.-D. (1993) Fundamentals of poroelasticity, in: *Comprehensive Rock Engineering Principles, Practice and Projects Vol. II, Analysis and Design Methods* (Fairhurst, C. Ed.) Ch. 5, Pergamon Press, Oxford, pp. 113–171.

Dorlot, J.M., Bailon J.P. and Masounave J. (1986) *Des Matériaux*, Ecole Polytechnique de Montréal, Deuxième Edition.

Eftis, J. (1987) On the fracture stress for the inclined crack under biaxial load. *Eng. Fract. Mech.*, 26: 105–125.

Erarslan, N. and Williams, D.J. (2012) The damage mechanism of rock fatigue and its relationship to the fracture toughness of rocks. *Int. J. Rock Mech. Min. Sci.*, 56: 15–26.

Fabre, G. and Pellet, F.L. (2006) Creep and time dependent damage in argillaceous rocks. *Int. J. Rock Mech. Min. Sci.*, 43(6): 950–960.

Fortin, J., Gueguen, Y. and Schubnel, A. (2007) Effect of pore collapse and grain crushing on ultrasonic velocities and Vp/Vs. *J. Geophys. Res.*, 112: 1–16.

Gangi, A.F. (1978) Variation of whole and fractured porous rock permeability with confining pressure, *Int. J. Rock Mech. Min. Sci.*, 15: 249–257.

Gaszynski, J. (1976) On a certain solution of dual integral equations and its application to contact problems. *Arch. Mech. Stos.*, 28: 75–88.

Gaszynski, J. and Szefer, G. (1978) Axisymmetric problem of the indenter for the consolidating semi-space with mixed boundary conditions. *Arch. Mech. Stos.*, 30: 17–26.

Gatelier, N., Pellet, F.L. and Loret, B. (2002) Mechanical damage of an anisotropic rock under cyclic triaxial tests. *Int. J. Rock Mech. Min. Sci.*, 39(3): 335–354.

Gawin, D., Pesavento, F. and Schrefler, B.A. (2002) Simulation of damage-permeability coupling in hygro-thermo-mechanical analysis of concrete at high temperature. *Commun. Numer. Meth. Eng.*, 18: 113–119.

Griffith, A.A. (1924) Theory of rupture, *Proc. Int. Congr. Appl. Mech.*, Delft, pp. 55–63.

Halpern, M.R. and Christiano, P. (1986) Steady state harmonic response of a rigid plate bearing on a liquid-saturated poroelastic halfspace. *Earthquake Eng. Struct. Dyn.*, 14: 439–454.

Harding, J.W. and Sneddon, I.N. (1945) The elastic stresses produced by the indentation of the plane of a semi-infinite elastic solid by a rigid punch. *Proc. Cambridge Philos. Soc.*, 41: 16–26.

Hudson, J.A., Bäckström, A., Rutqvist, J., Jing, L., Backers, T., Chijimatsu, M., Christiansson, R., Feng, X.-T., Kobayashi, A., Koyama, T., Lee, H.-S., Neretnieks, I., Pan, P.-Z., Rinne, M. and Shen, B.-T., (2009) Characterising and modelling the excavation damaged zone in crystalline rock in the context of radioactive waste disposal. *Environ. Geol.*, 57(6): 1275–1297.

Irwin, G.H. (1957) Analyses of stresses and strains near the end of the crack traversing a plate. *J. Appl. Mech.*, 24: 361–364.

Jaeger, J.C., Cook, N.G. and Zimmerman, R.W. (2008) *Fundamentals of Rock Mechanics, Fourth Edition*, Blackwell Publishing, Oxford, UK.

Jafari, M.K, Pellet, F., Boulon, M. and Amini Hosseini, K. (2004) Experimental study of mechanical behaviour of rock joints under cyclic loadings. *Rock Mech. Rock Eng*, 37: 3–23.

Johnson, K.L. (1985) *Contact Mechanics*, Cambridge University Press, Cambridge.

Ju, J.W. (1990) Isotropic and anisotropic damage variables in continuum damage mechanics. *J. Eng. Mech.*, 116: 2764–2770.

Kachanov, M. (1993) Elastic solids with many cracks and related problems. *Adv. Appl. Mech.*, 30: 259–445.

Kachanov, L.M. (1958) Time of the rupture process under creep conditions. *Izv. Adad. Nauk. USSR Otd. Tekh.*, 8: 26–31.

Kassir, M.K. and Xu, J. (1988) Interaction functions of a rigid strip bonded to a saturated elastic halfspace. *Int. J. Solids Struct.*, 24: 915–936.

Keshavarz, M., Pellet, F.L. and Amini Hosseini, K. (2009), *Comparing the effectiveness of energy and hit rate parameters of acoustic emission for prediction of rock failure*, Proc. 2nd Sinorock Symp., Hong Kong, China, CDRom.

Keshavarz, M., Pellet, F. L. and Loret, B. (2010) Damage and changes in mechanical properties of a gabbro thermally loaded up to 1000 °C. *Pure Appl. Geophys.*, **167**: 1511–1523.

Kiyama, T., Kita, H., Ishijima, Y., Yanagidani, T., Akoi, K. and Sato, T. (1996) *Permeability in anisotropic granite under hydrostatic compression and tri-axial compression including post-failure region*, Proc. 2nd North Amer. Rock Mech. Symp., 1643–1650.

Kranz, R.L. (1983) Microcracks in rocks: A review, *Tectonophysics*, **100**: 449–480.

Lajtai, E.Z., Carter, B.J. and Duncan, E. J. S. (1994) En echelon crack-arrays in potash salt rock. *Rock. Mech. Rock. Eng.*, **27**(2): 89–111.

Lan, Q. and Selvadurai, A.P.S. (1996) Interacting indentors on a poroelastic half-space, *J. Appl. Math. Phys. (ZAMP)*, **47**: 695–716.

Lemaitre, J. (1996) *A Course on Damage Mechanics*. Berlin Heidelberg, Springer-Verlag.

Li, L.C., Tang, C.A., Li, G., Wang, S.Y., Liang, Z.Z. and Zhang, Y.B. (2012) Numerical Simulation of 3D Hydraulic Fracturing Based on an Improved Flow-Stress-Damage Model and a Parallel FEM Technique. *Rock Mech. Rock Eng.*, **45**:801–818.

Li, S.P., Li, Y.S. and Wu, Z.Y. (1994) Permeability-strain equations corresponding to the complete stress-strain path of Yinzhuang sandstone. *Int. J. Rock Mech. Min. Sci.*, **31**: 383–391.

Li, S.P., Li, Y.S. and Wu, Z.Y. (1997) Effect of confining pressure, pore pressure and specimen dimensions on permeability of Yinzhuang Sandstone. *Int. J. Rock Mech. Min. Sci.*, **34**: Paper No. 175.

Ling, F.F (1973) *Surface Mechanics*, John Wiley, New York

Little, R.W. (1973) *Elasticity*, Prentice-Hall, Upper Saddle River, NJ.

Lubarda, V.A. and Krajcinovic, D. (1993) Damage tensors and the cracks density distribution. *Int. J. Solids Struct.*, **30**: 2859–2877.

Mahyari, A.T. and Selvadurai, A.P.S. (1998) Enhanced consolidation in brittle geomaterials susceptible to damage. *Mech. Cohes.-Frict. Mater.*, **3**: 291–303.

Martin, C.D. and Chandler, N.A. (1994) The progressive fracture of Lac du Bonnet granite. *Int. J. Rock Mech. Min. Sci. Geomech. Abst.*, **31**: 643–659.

Martin, C.D. (1997) The effect of cohesion loss and stress path on brittle rock strength. *Seventeenth Canadian Geotechnical Colloquium: Can Geotech J.*, **34**(5): 698–725.

Massart, T.J. and Selvadurai, A.P.S. (2012) Stress-induced permeability evolution in quasi-brittle geomaterials. *J. Geophys. Res (Solid Earth)*, **117**(B1): doi 10.1029/2012JB009251.

Massart, T.J. and Selvadurai, A.P.S. (2014) Computational modelling of crack-induced permeability evolution in granite with dilatant cracks. *Int. J. Rock Mech. Min. Sci.*, **70**: 593–604.

McClintock, F.A. and Argon, A.S. (1966) *Mechanical Behavior of Materials*, Addison-Wesley Publishing Company, Reading, MA.

Murikami, S. (1988), Mechanical modeling of material damage, *J. Appl. Mech. Transaction of the ASME*, **55**: 280–286.

Muskhelishvili, N.I. (1953) *Some Basic Problems of the Mathematical Theory of Elasticity*, P. Noordhoff Ltd., Groningen, Holland.

Nguyen, T.S. and Selvadurai, A.P.S. (1998) A model for coupled mechanical and hydraulic behavior of a rock joint. *Int. J. Numer. Anal. Methods Geomech.*, **22**: 29–48.

Nguyen, T.S., Selvadurai, A.P.S. and Armand, G. (2005) modeling of the FEBEX THM experiment using a state surface approach. *Int. J. Rock Mech. Min. Sci.*, **42**: 639–651.

Patsouls, G. and Gripps, J.C. (1982) An investigation of the permeability of Yorkshire Chalk under differing pore water and confining pressure conditions. *Energy Sources*, **6**: 321–334.

Patterson, M.S. and Wong, T.F. (2005) *Experimental Rock Deformation – the Brittle Field*, Springer, Berlin Heidelberg.

Pellet, F.L., Hajdu, A., Deleruyelle, F. and Besnus, F. (2005) A viscoplastic constitutive model including anisotropic damage for the time dependent mechanical behaviour of rock. *Int. J. Numer. Anal. Methods Geomech.*, 29(9): 941–970.

Pellet, F.L. (2015) Micro-structural analysis of time-dependent cracking in shale. *Environ. Geotechnics*, 2(2): 78–86.

Pellet, F.L. and Fabre, G. (2007) Damage evaluation with P-wave velocity measurements during uniaxial compression tests on argillaceous rocks. *Int. J. Geomech. (ASCE)*, 7(6): 431–436.

Pellet, F.L., Keshavarz, M. and Amini-Hosseini, K. (2011) Mechanical damage of a crystalline rock having experienced ultra high deviatoric stress up to 1.7 GPa. *Int. J. Rock Mech. Min. Sci.*, 48: 1364–1368.

Pellet, F.L., Roosefid, M. and Deleruyelle, F. (2009) On the 3D numerical modelling of the time-dependent development of the Damage Zone around underground galleries during and after excavation. *Tunnelling Underground Space Technol.*, 24(6): 665–674.

Philippacopoulos, A.J. (1989) Axisymmetric vibration of disk resting on saturated layered half-space. *J. Eng. Mech. Div., Proc. ASCE*, 115: 1740–1759.

Pijaudier-Cabot, G. and Pereira, J.-M. (2013) *Geomechanics of CO_2 Storage Facilities*, John Wiley, Hoboken, NJ.

Popp, T., Kern, H. and Schulze, O. (2001) Evolution of dilatancy and permeability in rock salt during hydrostatic compaction and triaxial deformation. *J. Geophys. Res.*, 106: 4061–4078.

Rabotnov, Y.N. (1969) *Creep rupture, Proceedings 12th International Congress on Theoretical and Applied Mechanics (ICTAM)*, Springer, Stanford, CA: 342–349.

Rice, J.R. and Cleary, M.P. (1976) Some basic stress diffusion solution for fluid-saturated elastic porous media with compressible constituents. *Rev. Geophys. Space Phys.*, 14: 227–241.

Samaha, H.R. and Hover, K.C. (1992) Influence of micro-cracking on the mass transport properties of concrete. *ACI Mat. J.*, 89: 416–424.

Schneider, U. and Herbst, H.J. (1989) Permeabilitaet und porositaet von beton bei hohen temperature. *Deut. Aussch. Fuer Stahlbeton*, 403: 345–365.

Schulze, O., Popp, T. and Kern, H. (2001) Development of damage and permeability in deforming rock salt. *Eng. Geol.*, 61: 163–180.

Selvadurai, A.P.S. (Ed.) (1996) *Mechanics of Poroelastic Media*, Kluwer Acad. Publ., Dordrecht, The Netherlands.

Selvadurai, A.P.S. (2000a) *Partial Differential Equations in Mechanics, Vol. 2. The Bi-harmonic Equation, Poisson's Equation*, Springer-Verlag, Berlin.

Selvadurai, A.P.S. (2000b) Fracture evolution during indentation of a brittle elastic solid. *Mech. Cohes.-Frict. Mater.*, 5: 325–339.

Selvadurai, A.P.S. (2001) On Boussinesq's problem. *Int. J. Eng. Sci.*, 39: 317–322.

Selvadurai, A.P.S. (2004) Stationary damage modelling of poroelastic contact. *Int. J. Solids Struct.*, 41: 2043–2064

Selvadurai, A.P.S. (2006) Gravity-driven advective transport during deep geological disposal of contaminants. *Geophys. Res. Lett.*, 33: L08408, doi: 10.1029/2006GL025944.

Selvadurai, A.P.S. (2013). Caprock breach: A potential threat to secure geologic sequestration, Ch. 5 in *Geomechanics of CO_2 Storage Facilities* (G. Pijaudier-Cabot and J.-M. Pereira Eds.), John Wiley, Hoboken, NJ, pp.75–93.

Selvadurai, A.P.S. and Boulon, M.J. (eds.) (1995) *Mechanics of Geomaterial Interfaces, Studies in Applied Mechanics, Vol 42*, Elsevier Sciences, Amsterdam.

Selvadurai, A.P.S. and Głowacki, A. (2008) Evolution of permeability hysteresis of Indiana Limestone during isotropic compression. *Ground Water*, 46: 113–119.

Selvadurai, A.P.S. and Gopal, K.R. (1989) A composite infinite element for modelling un bounded saturated soil media. *J. Geotech. Eng., ASCE*, 115: 1633–1646.

Selvadurai, A.P.S. and Nguyen, T.S. (1995) Computational modelling of isothermal consolidation of fractured media. *Comput. Geotech.*, 17:39–73.

Selvadurai, A.P.S. and Nguyen, T.S. (1997) Scoping analyses of the coupled thermal-hydrological-mechanical behaviour of the rock mass around a nuclear fuel waste repository. *Eng. Geol.*, **47**: 379–400.

Selvadurai, A.P.S. and Shirazi, A. (2004) Mandel-Cryer effects in fluid inclusions in damage susceptible poroelastic media. *Comput. Geotech.*, **37**: 285–300.

Selvadurai, A.P.S. and Shirazi, A. (2005) An elliptical disc anchor in a damage-susceptible poroelastic medium. *Int. J. Numer. Methods Engng.*, **16**: 2017–2039.

Selvadurai, A.P.S. and Suvorov, A.P. (2016) *Thermo-Poroelasticity and Geomechanics*, Cambridge University Press, Cambridge.

Selvadurai, A.P.S. and Yue, Z.Q. (1994) On the indentation of a poroelastic layer. *Int. J. Numer. Anal. Meth. Geomech.*, **18**: 161–175.

Selvadurai, A.P.S., Letendre, A. and Hekimi, B. (2011) Axial flow hydraulicpulse testing of an argillaceous limestone. *Environ. Earth Sci.*, **64**(8): 2047–2058.

Senjuntichai, T. and Rajapakse, R.K.N.D. (1996) Dynamics of a rigid strip bonded to a multi-layered poroelastic medium, in: *Mechanics of Poroelastic Media* (A.P.S. Selvadurai, Ed.) Proceedings of the Specialty Conference on Recent Developments in Poroelasticity Kluwer Academic Publ., Dordrecht, 353–369.

Shao, J.F., Duveau, G., Hoteit, N., Sibai, M. and Bart, M. (1997) Time-dependent continuous damage model for deformation and failure of brittle rock. *Int. J. Rock Mech. Min. Sci.*, **34**: 385, Paper No. 285.

Shiping, L., Yushou, L., Yi, L., Zhenye, W. and Gang, Z. (1994) Permeability-strain equations corresponding to the complete stress-strain path of Yinzhuang Sandstone. *Int. J. Rock Mech. Min. Sci. and Geomech. Abstr.*, **31**: 383–391.

Sih, G.C. (1991) *Mechanics of Fracture Initiation and Propagation*, Springer-Verlag, Berlin.

Simoni, L. and Schrefler, B.A. (1987) Mapped infinite elements in soil consolidation. *Int. J. Num. Meth. Eng.*, **24**: 513–527.

Singh, D.P. (1975) A study of creep of rocks. *Int. J. Rock Mech. Min. Sci. Geomech. Abstr.*, **12**:271–276.

Skempton, A.W. (1954) The pore pressure coefficients A and B. *Geotechnique*, **4**: 143–147.

Sneddon, I.N. (1946) Boussinesq's problem for a flat-ended cylinder. *Proc. Camb. Phil. Soc.*, **42**: 29–39.

Souley, M., Homand, F., Pepa, S. and Hoxha, D. (2001) Damage-induced permeability changes in granite: a case example at the URL in Canada. *Int. J. Rock Mech. Min. Sci.*, **38**: 297–310.

Stanchits, S., Vinciguerra, S. and Dresen, G. (2006) Ultrasonic velocities, acoustic emission characteristics and crack damage of basalt and granite. *Pure Appl. Geophys.*, **163**(5–6): 974–993.

Stormont, J.C. and Daemen, J.J.K. (1992) Laboratory study of gas permeability of rock salt during deformation. *Int. J. Rock Mech. Min. Sci.*, **29**: 325–342.

Tapponnier, P. and Brace, W.F. (1976) Development of stress-induced microcracks in Westerly granite. *Int. J. Rock Mech. Min. Sci. & Geomech. Abstr.*, **13**: 103–112, 1976.

Testa, S.M. (1994) *Geological Aspects of Hazardous Waste Management*, CRC Press, Boca Raton, FL.

Timoshenko, S.P. and Goodier, J.N (1970) *Theory of Elasticity*, McGraw-Hill, New York.

Vallier, F., Mitani, Y., Boulon M., Esaki, T. and Pellet, F.L. (2010) A shear model accounting scale effect in rock joints behavior. *Rock Mech. Rock Eng.* **43** (5): 581–595.

Wang, H. (2000) *Theory of Linear Poroelasticity with Applications to Geomechanics and Hydrogeology*, Princeton University Press, Princeton.

Wang, J.-A. and Park, H.D. (2002) Fluid permeability of sedimentary rocks in a complete stress-strain process. *Eng. Geol.*, **63**: 291–300.

Wawersik, W.R, and Fairhurst, C. (1970) A study of brittle rock fracture in laboratory compression experiments. *Int. J. Rock Mech. Min. Sci. & Geomech. Abstr.*, **7**: 561–575.

Westergaard, H.M. (1939) Bearing pressures and cracks. *J. Appl. Mech.*, **6**: 49–53.

Wong, T.F. (1982) Micromechanics of faulting in Westerly granite. *Int. J. Rock Mech. Min. Sci. & Geomech. Abstr.*, **19**: 49–64.

Yue, Z.Q. and Selvadurai, A.P.S. (1994) On the asymmetric indentation of a poroelastic layer. *Appl. Math. Modelling*, **18**: 161–175.

Yue, Z.Q. and Selvadurai, A.P.S. (1995a) Contact problems for saturated poroelastic solid. *J. Eng. Mech., Proc. ASCE*, **121**:502–512.

Yue, Z.Q. and Selvadurai, A.P.S. (1995b) On the mechanics of a rigid disc inclusion embedded in a fluid saturated poroelastic medium. *Int. J. Eng. Sci.*, **33**: 1633–1662.

Zheng, Q.S. and Betten J., (1996) On damage effective stress and equivalence hypothesis. *Int. J. Damage Mech.* **5**: 219–240.

Zhu, W. and Wong, T.-F. (1997) Transition from brittle faulting to cataclastic flow: permeability evolution. *J. Geophys. Res.*, **102**: 3027–3041.

Zimmerman, R.W. (2010) Some rock mechanics issues in petroleum engineering, Rock Mechanics in Civil and Environmental Engineering, *Proceedings of the European Rock Mechanics Symposium, EUROCK 2010*, Lausanne; Switzerland, pp. 39–44.

Zoback, M.D. and Byerlee, J.D. (1975) The effect of micro-crack dilatancy on the permeability of Westerly Granite. *J. Geophys. Res.*, **80**: 752–755.

Chapter 4

Experimental and numerical anisotropic rock mechanics

Ki-Bok Min[1], Bona Park[1], Hanna Kim[1], Jung-Woo Cho[2] & Lanru Jing[3]

[1]Department of Energy Resources Engineering, Seoul National University, Seoul, Republic of Korea
[2]Technology Convergence R&D Group, Korea Institute of Industrial Technology, Daegu, Republic of Korea
[3]Engineering Geology and Geophysics Research Group, Royal Institute of Technology, Stockholm, Sweden

---

Abstract: Although characterization and numerical modeling of anisotropic rock is a longstanding difficulty in rock mechanics, development and advances are being made for anisotropic rock mechanics in spite of the hurdles associated with it. This chapter provides an overview of anisotropic rock mechanics issues and introduces a series of experimental and numerical anisotropic rock mechanics studies conducted in the past 15 years. Experimental investigations are made on elastic, thermal conductivity, seismic and permeability anisotropy of rock based on cores taken from directional coring system. The first part of numerical anisotropic rock mechanics introduces the numerical experiments to determine the compliance tensor of fractured rock mass with Discrete Fracture Network (DFN) modeled as equivalent continuum anisotropic rock. Blocky Discrete element method (DEM) is employed for this numerical experiment using three boundary conditions in two dimensions. The second part deals with representation of transversely isotropic rock using bonded-particle DEM model with smooth joint model as layers. Both deformation and strength behavior modeled by DEM showed a reasonable agreement with analytical solutions, and laboratory observations. Upscaled model applied to anisotropic foundation demonstrate that large scale application anisotropic DEM model is also feasible.

I INTRODUCTION

Rock anisotropy has been a long-standing issue in rock engineering, beginning in the early developmental stages of rock mechanics and one of the most distinct features that must be considered in rock engineering applied for civil, mining, geo-environmental, or petroleum engineering disciplines. Many rocks have anisotropic characteristics, *i.e.*, their mechanical, thermal, seismic, and hydraulic properties vary with direction, and engineering applications that do not consider the anisotropic behavior of rock produce errors of differing magnitudes, depending on the extent of rock anisotropy (*e.g.*, Amadei, 1996). Anisotropic characteristics generally originate from the mineral foliation in metamorphic rocks, stratification in sedimentary rocks, and discontinuities in

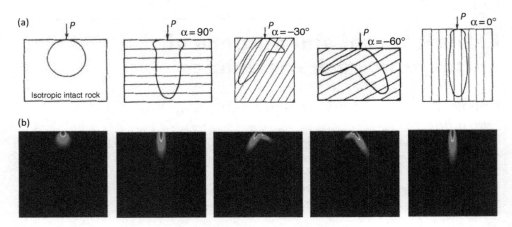

Figure 1 Radial stress distribution of transversely isotropic rock under concentrated line load. (a) cartoon presented in Goodman (1989), (b) results by FEM modeling (Park & Min, 2015a). The anisotropy ratio of two elastic moduli used in FEM is around 2, and red color denotes higher stress while cooler blue color denotes smaller stress.

the rock mass. Rock anisotropy is important for *in situ* stress measurements, especially for the overcoring method in which the constitutive relation of rock plays an important role (Amadei, 1996; Min *et al.*, 2003), displacement control in rock, and the development of excavation damage occurring during underground construction (*e.g.*, review in Cho *et al.*, 2012). When the equivalent continuum approach is used for regularly fractured rock masses, anisotropy must be considered because anisotropy can be pronounced due to the major deformations along the discontinuities. The extent of anisotropy is generally defined by the elastic modulus ratio, which decides whether it is necessary and relevant to consider anisotropy before certain operations begin depending upon anisotropy ratio (Cho *et al.*, 2012).

Figure 1 shows the radial stress distribution in a transversely isotropic rock subjected to concentrated line loads. Figure 1(a) is based on the analytical solution (Goodman, 1989), and Figure 1(b) shows results from finite element method (FEM) modeling (Park & Min, 2015a). By comparing the isotropic and anisotropic analysis in the figures, it is clear that stress distribution is much affected by the extent of anisotropy and that analysis not considering anisotropy can produce erroneous results.

The term 'anisotropy' is somewhat misleading in that the prefix 'an-' gives an impression that it is a *special* case of 'isotropy'. In fact, the opposite is true. The most general anisotropic elastic behavior of material is defined by 21 independent elastic constants. If the internal composition of a material possesses symmetry of any kind, then symmetry can be observed in its elastic properties. While there are many forms of elastic symmetry, the most relevant model for applications involving rock mechanics is the transversely isotropic model with five independent parameters. The stratification, foliation, and discontinuity planes, which are encountered frequently in rock mechanics applications, can be viewed as the transversely isotropic plane. Isotropy is then defined by only two independent elastic constants after the assumption of complete symmetry is made (Lekhnitskii, 1963). Therefore, complete anisotropy is the most *general* case and isotropy should be regarded as the most *special* one.

Although anisotropic rock mechanics has been considered as distinct characters of rock mechanics discipline, its application has been somewhat limited for a number of reasons. First, it is difficult, if not impossible, to determine the elastic properties when more independent parameters are needed, *e.g.*, five for transversely isotropic materials. Core samples in different directions are generally required for this purpose, which is not feasible in many industrial applications. Furthermore, there is no standard method established in the rock mechanics community yet, and this makes professionals more hesitant to consider rock anisotropy (ISRM, 2007; ISRM, 2015). Second, anisotropic rock mechanics gives the impression that it is difficult to master since analytical solutions are more complex with more associated parameters. Even if there are several analytical solutions available, their usage is not often straightforward due to its complexity in their equations (Lekhnitskii, 1963; Ting, 1996). Third, anisotropic rock mechanics research or industrial tools are still limited. Assumption of isotropy is in many cases the norm and there is still a dearth of tools applicable to anisotropy. For example, when rock mass deformation moduli are determined by empirical method such as Q, RMR or GSI, consideration of anisotropy is largely missing (Min & Jing, 2003).

We believe that the above-mentioned hurdles can be overcome with concerted research efforts, and this chapter intends to serve the purpose of introducing the most recent development in the area of anisotropic rock mechanics, mostly carried out by the authors. Focus is given to the elastic behavior of anisotropic rock with limited coverage of strength behavior since strength and damage of anisotropic rock will be covered elsewhere in this volume.

This chapter starts with a brief introduction of the anisotropic elasticity theory for completeness. Experimental results are presented to provide the insight into the overall anisotropic behavior of rocks in terms of elastic, thermal, seismic and permeability behaviors. Special focus was given to the method of determining anisotropic parameters and comparison with analytically predicted anisotropic behavior. Numerical results with their verification and validation to real cases are presented using discrete element method (DEM) in both bonded-particle and blocky systems (Jing & Stephansson, 2007). Blocky DEM is applied to determine the elastic properties of fractured rock masses by treating them as anisotropic equivalent continua. Bonded-Particle DEM models are being used to account for both elastic and strength behavior of anisotropic rock.

2 THEORETICAL BACKGROUND

2.1 Anisotropic constitutive equation

The constitutive relation for general linear elasticity can be expressed as

$$\varepsilon_{ij} = S_{ijkl}\sigma_{kl} \tag{1}$$

where ε_{ij} and σ_{kl} are stress and strain tensors of a second order rank and S_{ijkl} is the compliance tensor of a fourth order rank, involving 21 independent. By adopting a contracted matrix form of S_{ijkl}, Equation 1 can be expressed as

$$
\begin{pmatrix} \varepsilon_x \\ \varepsilon_y \\ \varepsilon_z \\ \gamma_{yz} \\ \gamma_{xz} \\ \gamma_{xy} \end{pmatrix} = \begin{pmatrix} S_{11} & S_{12} & S_{13} & S_{14} & S_{15} & S_{16} \\ S_{21} & S_{22} & S_{23} & S_{24} & S_{25} & S_{26} \\ S_{31} & S_{32} & S_{33} & S_{34} & S_{35} & S_{36} \\ S_{41} & S_{42} & S_{43} & S_{44} & S_{45} & S_{46} \\ S_{51} & S_{52} & S_{53} & S_{54} & S_{55} & S_{56} \\ S_{61} & S_{62} & S_{63} & S_{64} & S_{65} & S_{66} \end{pmatrix} \begin{pmatrix} \sigma_x \\ \sigma_y \\ \sigma_z \\ \tau_{yz} \\ \tau_{xz} \\ \tau_{xy} \end{pmatrix} \tag{2}
$$

where matrix S_{ij} is called the compliance matrix. The symbols of ε_i and γ_{ij} (i, j = x, y, z) denote the normal and shear strains, respectively, and symbols of σ_i and τ_{ij} (i, j = x, y, z) denote the normal and shear stresses, respectively. The compliance matrix can be described explicitly by giving the physical meaning of each element as functions of elastic moduli, Poisson ratios, shear moduli and other technical constants of the solids (Lekhnitskii, 1963).

Since S_{ijkl} is a fourth order tensor, its rotational transformation can be also defined by the following mapping operations

$$
S'_{ijkl} = \beta_{im}\beta_{jn}\beta_{kp}\beta_{lq}S_{mnpq} \tag{3}
$$

where S'_{ijkl} and S_{mnpq} are the compliance tensors in the transformed and the original axes, respectively, and β_{im}, β_{jn}, β_{kp}, and β_{lq} are the direction cosines representing rotational operations. Equation 3 is mathematically elegant but not convenient for practical calculations because it involves fourth order tensor operations. The following mapping operation with a 6 by 6 matrix for the transformation of compliance matrix is introduced to simplify the operations (Lekhnitskii, 1963).

$$
S'_{ij} = S_{mn}q_{mi}q_{nj} \tag{4}
$$

where S'_{ij} is the compliance matrix in the transformed axes and S_{mn} is the one in the original axes, respectively. The component of the q_{ij} matrix can be obtained purely from the direction cosine as available in Lekhnitskii (1963) and Min and Jing (2003). When only the rotation of axes is concerned, the transformation form becomes drastically simple. If the angle of anti-clockwise rotation of axes about z-axis is φ, the matrix of the direction cosines is then given by

$$
\beta_{ij} = \begin{pmatrix} \cos\varphi & \sin\varphi & 0 \\ -\sin\varphi & \cos\varphi & 0 \\ 0 & 0 & 1 \end{pmatrix} \tag{5}
$$

and the final matrix form for q_{ij} has the following form:

$$
q_{ij} = \begin{pmatrix} \cos^2\varphi & \sin^2\varphi & 0 & 0 & 0 & -2\sin\varphi\cos\varphi \\ \sin^2\varphi & \cos^2\varphi & 0 & 0 & 0 & 2\sin\varphi\cos\varphi \\ 0 & 0 & 1 & 0 & 0 & 0 \\ 0 & 0 & 0 & \cos\varphi & \sin\varphi & 0 \\ 0 & 0 & 0 & -\sin\varphi & \cos\varphi & 0 \\ \sin\varphi\cos\varphi & -\sin\varphi\cos\varphi & 0 & 0 & 0 & \cos^2\varphi - \sin^2\varphi \end{pmatrix} = Q \tag{6}
$$

Therefore, by substituting the matrix of Equation 6 into Equation 4, it is possible to express the elastic constants, *i.e.* elastic modulus or Poisson's ratio, in rotated axes in terms of direction cosines and components in the original axes.

The practical implication of this tensorial transformation for rock mechanics is that, once elastic properties in a given reference axis is determined, elastic properties in any arbitrary direction can be readily calculated. Comparison of theoretical tensorial transformation with actual laboratory measurement in core taken in various directions shows that a certain type of rock can be indeed modeled as transversely isotropic rock with moderate extent of discrepancy which can be more attributable to rock heterogeneity (Cho et al., 2012). Numerical study using blocky DEM also supports that fractured rock also follows this transformation rule especially when the size of the considered domain reaches the representative elementary volume (REV) (Min & Jing, 2003).

On the other hand, thermal conductivity and permeability obey the rotational transformation rules as a second-order tensor. The tensor of the anisotropic thermal conductivity and permeability is formulated follows with respect to the rotation of the axes (Carslaw & Jaeger, 1959; Bear, 1972).

$$k'_{pq} = \beta_{pi}\beta_{qj}k_{ij} \tag{7}$$

where k_{ij} and k_{pq} are the thermal conductivity or permeability tensors in the original and rotated axes, respectively, and β_{pi} and β_{qj} are the direction cosines. A study also confirms that measured thermal conductivity and permeability in core samples with different directions also match with theoretical predictions (Kim et al., 2012; Yang et al., 2013).

2.2 Constitutive equation of anisotropic rock

When there are three orthogonal planes of elastic symmetry with the axes of the coordinates perpendicular to these planes, the model is called orthogonal (or orthogonally isotropic) and the constitutive equation can be expressed as following (Lekhnitskii, 1963).

$$
\begin{pmatrix} \varepsilon_x \\ \varepsilon_y \\ \varepsilon_z \\ \gamma_{yz} \\ \gamma_{xz} \\ \gamma_{xy} \end{pmatrix} =
\begin{pmatrix}
\frac{1}{E_x} & -\frac{\nu_{yx}}{E_y} & -\frac{\nu_{zx}}{E_z} & 0 & 0 & 0 \\
-\frac{\nu_{xy}}{E_x} & \frac{1}{E_y} & -\frac{\nu_{zy}}{E_z} & 0 & 0 & 0 \\
-\frac{\nu_{xz}}{E_x} & -\frac{\nu_{yz}}{E_y} & \frac{1}{E_z} & 0 & 0 & 0 \\
0 & 0 & 0 & \frac{1}{G_{yz}} & 0 & 0 \\
0 & 0 & 0 & 0 & \frac{1}{G_{xz}} & 0 \\
0 & 0 & 0 & 0 & 0 & \frac{1}{G_{xy}}
\end{pmatrix}
\begin{pmatrix} \sigma_x \\ \sigma_y \\ \sigma_z \\ \tau_{yz} \\ \tau_{xz} \\ \tau_{xy} \end{pmatrix} \tag{8}
$$

where E_x, E_y, and E_z are elastic moduli in x, y, and z direction, respectively, and G_{yz}, G_{xz}, and G_{xy} are shear moduli defined in yz, xz and xy planes. ν_{xy}, ν_{yx}, ν_{zx}, ν_{xz}, ν_{yz} and ν_{zy} are Poisson's ratios and the property ν_{ij} determines the ratio of strain in the j-direction to the strain in the i-direction due to a stress acting in the i-direction).

When the symmetry planes are the z=0 plane and any plane that contains the z-axis, the z-axis is the axis of symmetry and this material is classified as transversely isotropic with the relationships of $E=E_x=E_y$, $E'=E_z$, $v=v_{xy}=v_{yx}$, $v'=v_{zx}=v_{zy}$ and $G=E/2/(1+v)$.

Thus, the constitutive equation of the transversely isotropic rock can be expressed as follows in a matrix form;

$$
\begin{bmatrix} \varepsilon_x \\ \varepsilon_y \\ \varepsilon_z \\ \gamma_{yz} \\ \gamma_{xz} \\ \gamma_{xy} \end{bmatrix} = \begin{bmatrix} \dfrac{1}{E} & -\dfrac{v'}{E'} & -\dfrac{v}{E} & 0 & 0 & 0 \\ -\dfrac{v'}{E'} & \dfrac{1}{E'} & -\dfrac{v'}{E'} & 0 & 0 & 0 \\ -\dfrac{v}{E} & -\dfrac{v'}{E'} & \dfrac{1}{E} & 0 & 0 & 0 \\ 0 & 0 & 0 & \dfrac{1}{G'} & 0 & 0 \\ 0 & 0 & 0 & 0 & \dfrac{2(1+v)}{E} & 0 \\ 0 & 0 & 0 & 0 & 0 & \dfrac{1}{G'} \end{bmatrix} \begin{bmatrix} \sigma_x \\ \sigma_y \\ \sigma_z \\ \tau_{yz} \\ \tau_{xz} \\ \tau_{xy} \end{bmatrix} \tag{9}
$$

In the above compliance matrix, there are five independent elastic constants. E and E' are the elastic moduli in the plane of transverse isotropy and in a direction normal to it, respectively. The terms v and v' are the Poisson's ratios that characterize the ratio of lateral strain to axial strain in the plane of transverse isotropy subjected to axial stress acting parallel and normal to it, respectively. The term G' is the shear modulus in the plane normal to the plane of transverse isotropy. This is one of the most popular models for rock mechanics applications, because the stratification, foliation, and discontinuity planes, which are encountered frequently in rock mechanics applications, can be viewed as the symmetry plane.

3 EXPERIMENTAL ANISOTROPIC ROCK MECHANICS

3.1 Introduction

It is generally accepted that three prismatic or cylindrical specimens that are inclined parallel with vertical to and at 45 degrees from the isotropic plane are required in order to determine the five independent elastic constants for a transversely isotropic model (Barla, 1974; Amadei, 1982; Worotnicki, 1993). Many researchers have sought ways to minimize the laborious task of preparing multiple specimens. One conventional approach is to reduce the five independent elastic constants to four by using Saint-Venant's empirical equations for shear modulus in the plane normal to the isotropic plane (Lekhnitskii, 1963). However, these empirical equations are contrary to the fundamental assumption of constitutive modeling in transverse isotropic rock, and they are not acceptable for use with rocks of high anisotropy (Worotnicki, 1993). Talesnick and Ringel (1999) used a single specimen of a thin walled hollow cylinder to determine the five independent elastic

Experimental and numerical anisotropic rock mechanics 115

constants, and this approach avoids the need to prepare more than one specimen and the complexity of applying the three different boundary conditions associated with axial compression, radial compression, and torsion. Nunes (2002) used the Council for Scientific Industrial Research's (CSIR's) triaxial cell, which is essentially a hollow cylinder under biaxial loading to determine the parameters, but the assumptions used, such as using isotropic solutions for stress distribution in the thick walled hollow cylinder, produced excessive approximations, as noted by Gonzaga *et al.* (2008). Gonzaga *et al.* (2008) presented a methodology of using two different laboratory tests composed of one hydrostatic compression test followed by one uniaxial and one triaxial compression test with a single cylindrical specimen to determine the five elastic constants, but the methodology still rely on the empirical Saint Venant's equation when the specimen axis is parallel with or normal to the isotropy plane in which no shear stresses are present. Also, the Brazilian test was used to determine the five elastic constants using an analytical, numerical and combination of two approaches (Chen *et al.*, 1998; Claesson & Bohloli, 2002; Exadaktylos & Kaklis, 2001; Nasseri *et al.*, 2003; Chou & Chen, 2008).

Despite significant efforts over the past 50 years to determine the elastic constants of anisotropic rocks, no standard method has yet been suggested (ISRM, 2007; ISRM, 2015).

The strength of anisotropic rock is significantly affected by the existence of weak planes, which are often the transversely isotropic planes. It is reported that failure occurs in the configuration along a weak plane in many cases of uniaxial, triaxial compression and Brazilian tensile tests (Vervoort, A. *et al.*, 2014). The strength anisotropy can be much greater than elastic anisotropy and the strength decreases significantly when the weak planes are inclined with the loading direction (*e.g.*, Cho *et al.*, 2012). Extensive list of study exists for compressive failure of anisotropic rock (Rodrigues, 1966; Barla, 1974; Ramamurthy, 1993; Nasseri *et al.*, 1997, 2003; Tien & Tsao, 2000; Hakala *et al.*, 2007) and tensile failure of anisotropic rock (Pinto, 1979; Claesson & Bohloli, 2002; Chen *et al.*, 1998; Exadaktylos & Kaklis, 2001; Chou & Chen, 2008).

Experimental study in this chapter presents a method of determining elastic constants of transversely isotropic rock using two rock samples. This study also fills the literature gap concerning the anisotropic characteristics of the mechanical, seismic and thermal properties of rocks by investigating the correlations between these properties and offering a more systematic examination of the applicability of tensorial properties.

3.2 Experimental setup

The laboratory directional coring system was established in order to extract the core sample in different directions (Cho *et al.*, 2012). Coring a block with angles of 0, 15, 30, 45, 60, 75, and 90 degrees with respect to the transverse isotropic plane by using our laboratory-scale directional coring system (Figure 2), cylindrical samples are obtained for anisotropic test for mechanical, thermal, seismic and permeability behavior (Cho *et al.*, 2012; Kim *et al.*, 2012; Yang *et al.*, 2013). Samples with 70 mm in length and 38 mm in diameter were used for the mechanical, seismic and permeability experiments and samples with 7 mm in length and 25.4 mm in diameter were used for thermal conductivity tests. The thicknesses of the Brazilian test specimens were approximately equal to their radii, in the size of about 18–22 mm.

Biaxial and triaxial strain gages were used for the strain measurements. Two biaxial strain gages were glued on the two perpendicular sides of each core specimen. Figure 3

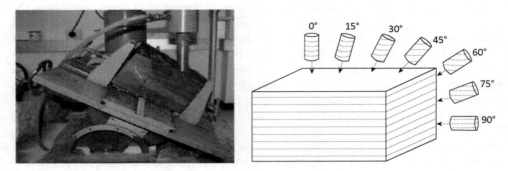

Figure 2 Directional coring system and schematic of anisotropic rock samples (Cho et al., 2012).

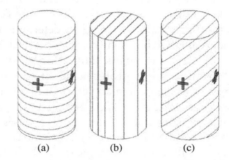

Figure 3 Three specimens for determination of anisotropic parameters (Cho et al., 2012).

(a–c) shows the schematic features of specimens having 0°, 90°, and an angle between 0° and 90° with respect to the isotropy plane. In the case of the specimens between 0° and 90° foliation (Figure 3 (c)), one biaxial strain gage was glued on the surface having the maximum apparent dip angle of isotropic plane, and the other strain gage was glued on the surface having the minimum apparent dip angle, *i.e.*, perpendicular to the first surface.

A seismic velocity measurement system was used to measure the seismic velocity of each sample. The divided-bar method was chosen for measuring thermal conductivity (Beardsmore & Cull, 2001) because this method directly measures the thermal properties following the Fourier's law.

3.3 Determination of anisotropic elastic constants

A method to determine the five elastic constants for transversely isotropic rocks were presented in the literature (Pinto, 1970; Barla, 1974; Worotnicki, 1993; Amadei, 1996; Hakala *et al.*, 2007). In this method, three or more specimens (*i.e.*, either cylindrical or prismatic) are generally used in uniaxial compression tests with $\theta = 0°$, $90°$, and an inclined angle θ between 0° and 90°, in order to determine the five independent constants. While five independent measurements of strain are sufficient to obtain the five elastic constants, there are more than five strain measurements. In this case, the least square method was used to obtain the best-fit elastic constants.

We can extract the independent equations from the uniaxial compression tests on the three specimens introduced in the literature (Amadei, 1996) as shown in Figure 3. From specimen (a) in which two strain gages measured identical strains, two equations were obtained as shown in Equation 10. Three equations were obtained from specimens (b) and (c), as shown in Equations 11 and 12, respectively, because the two lateral strain measurements were different in those specimens. In these equations, unknown elastic constants can be determined from the known values of stress (σ), strain (ε), and anisotropy angle (θ).

$$\frac{\varepsilon_x}{\sigma_y} = \frac{\varepsilon_z}{\sigma_y} = -\frac{\nu'}{E'}$$
$$\frac{\varepsilon_y}{\sigma_y} = \frac{1}{E'} \tag{10}$$

$$\frac{\varepsilon_y}{\sigma_y} = \frac{1}{E}$$
$$\frac{\varepsilon_x}{\sigma_x} = -\frac{\nu'}{E'} \tag{11}$$
$$\frac{\varepsilon_z}{\sigma_y} = -\frac{\nu}{E}$$

$$\varepsilon'_x/\sigma'_y = \frac{\sin^2 2\theta}{4}\left(\frac{1}{E} + \frac{1}{E'} - \frac{1}{G'}\right) - \frac{\nu'}{E'}\left(\cos^4\theta + \sin^4\theta\right)$$
$$\varepsilon'_y/\sigma'_y = \frac{\sin^4\theta}{E} + \frac{\cos^4\theta}{E'} + \frac{\sin^2 2\theta}{4}\left(-\frac{2\nu'}{E'} + \frac{1}{G'}\right) \tag{12}$$
$$\varepsilon'_z/\sigma'_y = -\sin^2\theta\frac{\nu}{E} - \cos^2\theta\frac{\nu'}{E'}$$

The minimum number of specimens required to determine the five elastic constants can be investigated in a matrix form composed of Equations 10, 11, and 12. When specimens (a) and (c) in Figure 3 are used, Equations 10 and 12 are obtained from the tests, and these equations can be summarized in a least square matrix form as follows:

$$
\begin{bmatrix} \varepsilon_{a(x1)}/\sigma_a \\ \varepsilon_{a(y1)}/\sigma_a \\ \varepsilon_{a(z1)}/\sigma_a \\ \varepsilon_{a(y2)}/\sigma_a \\ \varepsilon_{c(x1\theta)}/\sigma_c \\ \varepsilon_{c(y1\theta)}/\sigma_c \\ \varepsilon_{c(z1\theta)}/\sigma_c \\ \varepsilon_{c(y2\theta)}/\sigma_c \end{bmatrix}
=
\begin{bmatrix}
0 & 0 & 0 & -1 & 0 \\
0 & 1 & 0 & 0 & 0 \\
0 & 0 & 0 & -1 & 0 \\
0 & 1 & 0 & 0 & 0 \\
\sin^2 2\theta/4 & \sin^2 2\theta/4 & 0 & -\cos^4\theta/4 - \sin^4\theta/4 & \sin^2 2\theta/4 \\
\sin^4\theta & \cos^4\theta & 0 & -2\sin^2 2\theta/4 & \sin^2 2\theta/4 \\
0 & 0 & -\sin^2\theta & -\cos^2\theta & 0 \\
\sin^4\theta & \cos^4\theta & 0 & -2\sin^2 2\theta/4 & \sin^2 2\theta/4
\end{bmatrix}
\times
\begin{bmatrix} \dfrac{1}{E} \\ \dfrac{1}{E'} \\ \dfrac{\nu}{E} \\ \dfrac{\nu'}{E'} \\ \dfrac{1}{G'} \end{bmatrix}
$$
$$\tag{13}$$

The first, second, fifth, sixth and seventh rows in Equation 13 are independent of each other, and five independent equations were obtained. Thus, the five constants, *i.e.*, E, E', v, v', and G' can be determined from Equation 13.

In a similar way, when specimens (b) and (c) in Figure 3 are used, the five independent equations can be obtained from Equations 11 and 12, which again allow the determination of the five elastic constants. When two different specimens of (c) in Figure 3 are used with two different anisotropy angles (*i.e.*, $\theta_{C1} \neq \theta_{C2}$) five independent equations and the five constants also were obtained from the tests. Consequently, the minimum number of specimens to determine the five elastic constants is two, provided that one of the specimens was inclined with respect to the isotropic plane (Cho *et al.*, 2012).

3.4 Mechanical, thermal, seismic and hydraulic behavior of anisotropic rock

The experiments were conducted on three rock types, *i.e.*, Asan gneiss, Boryeong shale, and Yeoncheon schist for mechanical, seismic and thermal properties Berea sandstone was tested for permeability measurement. Except sandstone, samples showed a clear evidence of transverse isotropy due to the arrangements of some mineral particles as observed (Kim *et al.*, 2012). Asan gneiss is biotite gneiss consisting of plagioclase, hornblende, quartz, and biotite. Flat minerals are arrayed parallel to the foliation plane. Yeoncheon schist has schistosity, in which the platy minerals, such as feldspar and mica, are aligned with the schistose plane. The gneiss and the shale were unweathered, fresh rock with very few visible micro-cracks. However, the schist had a number of micro-cracks. Two or three sets of cylindrical samples for each type of rock were prepared to measure the mechanical, thermal, seismic and hydraulic properties.

Figure 4 shows the variations in the elastic moduli, P-wave velocity, and thermal conductivity with respect to the anisotropy angle in Asan gneiss, Boryeong shale, and Yeoncheon schist. Solid lines in each graph show the predicted value based on tensorial

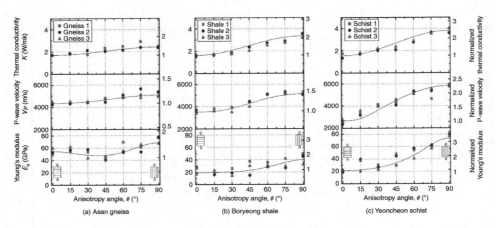

Figure 4 Young's modulus, P-wave velocity and thermal conductivity in different directions (Kim *et al.*, 2012).

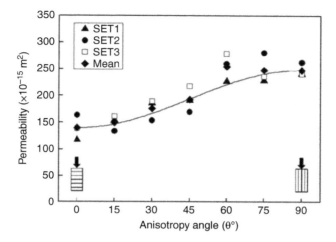

Figure 5 Permeability of Berea sandstone in different directions (Yang et al., 2013).

transformation for transversely isotropic model (elastic moduli, thermal conductivity) and approximated equation for transversely isotropic model (P-wave velocity). The right axis of each graph in Figure 4 indicates the normalized value with respect to the average minimum value. In general, variations of the mechanical, seismic, and thermal properties showed similar trends with anisotropy angle. The maximum values of all three properties occurred in the direction parallel to the isotropic plane, and the minimum value occurred in the direction perpendicular to the isotropic plane. The exception was the elastic moduli of Asan gneiss, which had a minimum value at an inclined angle of 45°. It is noted that minimum elastic modulus is often observed at around this angle due to the low shear stiffness of foliation planes or fracture planes in case of equivalent continuum model (Min & Jing, 2003).

Figure 5 shows the anisotropic permeability measured in Berea sandstone, in which degree of anisotropy can be larger than 1.5 even in sandstone which does not have clear visual anisotropy (Yang *et al.*, 2013). Characterization of anisotropy through X-ray CT scanning system turned out to be a useful option as discussed in Yun *et al.* (2013).

The anisotropy ratios of the elastic moduli parallel with and perpendicular to the isotropic planes (E/E') for Asan gneiss, Boryeong shale, and Yeoncheon schist were determined to be 1.3, 2.1, and 3.4, respectively. The anisotropy ratios of thermal conductivity parallel to and perpendicular to isotropic planes, ($K_{(90°)}/K_{(0°)}$), were 1.4 for Asan gneiss, 2.1 for Boryeong shale, and 2.5 for Yeoncheon schist. The P-wave velocity anisotropy ratio ($V_{P(90°)}/V_{P(0°)}$) was 1.2 for Asan gneiss, 1.5 for Boryeong shale, and 2.3 for Yeoncheon schist.

The mean prediction error (*MPE*) defined as the average relative difference between measured and predicted elastic moduli were lower than 17%, which indicates that transversely isotropic model is a reasonable assumption for the rock types chosen in this study (Cho *et al.*, 2012). The *MPE*s of seismic velocity for Asan gneiss, Boryeong shale, and Yeoncheon schist were 3.5%, 4.6%, and 8.9%, respectively. The *MPE*s of thermal conductivity for Asan gneiss, Boryeong shale, and Yeoncheon schist were 6.1%, 9.3%, and 8.6%, respectively. These low *MPE*s imply that seismic properties can be modeled effectively by the approximated equations for the transversely isotropic model and thermal conductivities follows the transformation rule of a second-order tensor for the rocks included in this study. This *MPE* can be a useful parameter as a criterion to determine whether the selected rock follows the transversely isotropic model (Kim *et al.*, 2012).

The correlations between elastic moduli, P-wave velocities, and thermal conductivities were evident, even though there were some outliers. The best correlations were observed between thermal conductivities and P-wave velocities. There were good correlations between the elastic moduli and thermal conductivities as well as between the elastic moduli and P-wave velocities for Boryeong shale and Yeoncheon schist. Asan gneiss showed poorest correlations due to more heterogeneous samples and the fact that the angle for minimum elastic moduli did not match with those for minimum P-wave velocities and thermal conductivities (Kim *et al.*, 2012).

The permeability anisotropy ratio ($K_{(90°)} / K_{(0°)}$) of Berea sandstone was 1.8 and this anisotropy trend corresponded with the porosity anisotropy due to the bedding plane.

The anisotropy ratios of maximum to minimum uniaxial compressive strength were 2.6, 2.6 and 18.6 for Asan gneiss, Boryeong shale and Yeoncheon schist, respectively. The anisotropy ratios of maximum to minimum tensile strength determined by Brazilian Tensile test were 3.2, 2.2 and 7.1 for Asan gneiss, Boryeong shale and Yeoncheon schist, respectively (Cho *et al.*, 2012).

4 NUMERICAL ANISOTROPIC ROCK MECHANICS – BLOCKY DISCRETE ELEMENT METHOD APPLICATION

4.1 Introduction

Fractures are common in rock masses. Any rock engineering structures on and in the fractured rock masses must take into account the existence of fractures. Figure 6 shows an outcrop rock mass observed in Forsmark, Sweden, in which numerous fractures were observed. The mechanical behavior of such fractured rock mass will be greatly affected by the existence of fractures, and the equivalent mechanical behavior in representative scale is of great importance in rock engineering in such fractured rock masses. Direct measurement by in-situ experiments with large or very large scale samples is technically possible, but is costly and often involves uncertainties related to the effects of hidden fractures, control of boundary conditions and interpretation of results (Min & Jing, 2003). Research needs for more accurate methods determining the rock mass properties have been widely recognized (Fairhurst, 1993) and considerable efforts have been made in the past on the methodology for determining the equivalent elastic mechanical properties (essentially elastic modulus and Poisson's ratio) of fractured rock mass employing empirical, analytical and numerical approaches.

The empirical methods 'infer' the rock mass properties from the rock mass classification or characterization results (Bieniawski, 1978; Hoek & Brown, 1997; Barton, 2002; Palmström, 1996). Although it has gained wide popularity for practical applications for design, especially in tunneling, it often gives too conservative estimates for property characterizations, largely because it makes use of categorized parameters based on case histories. The main shortcoming of this approach is that it lacks a proper mathematical platform to establish constitutive models suitable for rock anisotropy.

Figure 6 An example of fractured rock masses in the dimensions about 5 m x 5 m (Min, 2004). The picture was taken in Forsmark, Sweden, which is the site for geological repository of spent nuclear fuel in Sweden.

The efforts to find analytical solutions for estimating the equivalent properties of fractured rock masses have a rather long history and several analytical solutions were proposed for cases of simple fracture system geometry (Salamon, 1968; Singh, 1973; Amadei & Goodman, 1981; Gerrard, 1982; Oda, 1982; Fossum, 1985). The closed-form solutions have the advantage of being compact, clear and straightforward, but they work only for regular and often persistent and orthogonal fracture system geometries and simple constitutive behavior of fractures. It is difficult, often even impossible, to derive closed-form solutions with general irregular fracture systems. Analytical methods fail to consider the interaction between the fractures and the blocks divided by the fractures, which may have significant impacts on the overall behavior of rock masses because the intersections of the fractures are often the locations with the largest stress and deformation gradients, damage and failure.

With the rapid growth of computing capacity, numerical methods are attracting more attention to determine individual properties such as strength or deformability of the fractured rock mass. In comparison with the empirical and analytical approaches, the numerical approach has a certain advantage that the influence of irregular fracture system geometry and complex constitutive models of intact rock and fractures can be directly included in the derivation of the equivalent mechanical properties of rock masses. Although numerical experiments on realistic irregular fracture networks were large research subjects, they are now becoming applicable practicing tools thanks to the recent advances in the associated area of numerical methods, DFN treatment and improved computing power (Hart *et al.*, 1985; Pouya & Ghoreychi, 2001; Min & Jing, 2003; Min *et al.*, 2005; Fredriksson & Olofsson, 2007; Mas Ivars *et al.*, 2011). This section introduces some of the recent research development carried out by the authors with special focus on determining the rock mass properties considering full anisotropy.

4.2 Discrete Fracture Network – Discrete Element Method (DFN-DEM) approach

The DFN-DEM approach (Min & Jing, 2003) uses the fracture system realizations as the geometric models of the fractured rock masses and conducts numerical experiments using a DEM program, UDEC (Itasca, 2000), for the calculation of mechanical and hydraulic properties.

When six sets of boundary conditions are imposed and counter-part responses (stress when strain boundary condition is applied and vice versa) are measured, general constitutive relation shown in Equation 2 becomes as follows (Min, 2005; Mas Ivars *et al.*, 2011).

$$
\begin{pmatrix}
\varepsilon_{xx}^1 & \varepsilon_{xx}^2 & \varepsilon_{xx}^3 & \varepsilon_{xx}^4 & \varepsilon_{xx}^5 & \varepsilon_{xx}^6 \\
\varepsilon_{yy}^1 & \varepsilon_{yy}^2 & \varepsilon_{yy}^3 & \varepsilon_{yy}^4 & \varepsilon_{yy}^5 & \varepsilon_{yy}^6 \\
\varepsilon_{zz}^1 & \varepsilon_{zz}^2 & \varepsilon_{zz}^3 & \varepsilon_{zz}^4 & \varepsilon_{zz}^5 & \varepsilon_{zz}^6 \\
\gamma_{yz}^1 & \gamma_{yz}^2 & \gamma_{yz}^3 & \gamma_{yz}^4 & \gamma_{yz}^5 & \gamma_{yz}^6 \\
\gamma_{zx}^1 & \gamma_{zx}^2 & \gamma_{zx}^3 & \gamma_{zx}^4 & \gamma_{zx}^5 & \gamma_{zx}^6 \\
\gamma_{xy}^1 & \gamma_{xy}^2 & \gamma_{xy}^3 & \gamma_{xy}^4 & \gamma_{xy}^5 & \gamma_{xy}^6
\end{pmatrix}
$$

$$
=
\begin{pmatrix}
S_{11} & S_{12} & S_{13} & S_{14} & S_{15} & S_{16} \\
S_{21} & S_{22} & S_{23} & S_{24} & S_{25} & S_{26} \\
S_{31} & S_{32} & S_{33} & S_{34} & S_{35} & S_{36} \\
S_{41} & S_{42} & S_{43} & S_{44} & S_{45} & S_{46} \\
S_{51} & S_{52} & S_{53} & S_{54} & S_{55} & S_{56} \\
S_{61} & S_{62} & S_{63} & S_{64} & S_{65} & S_{66}
\end{pmatrix}
\begin{pmatrix}
\sigma_{xx}^1 & \sigma_{xx}^2 & \sigma_{xx}^3 & \sigma_{xx}^4 & \sigma_{xx}^5 & \sigma_{xx}^6 \\
\sigma_{yy}^1 & \sigma_{yy}^2 & \sigma_{yy}^3 & \sigma_{yy}^4 & \sigma_{yy}^5 & \sigma_{yy}^6 \\
\sigma_{zz}^1 & \sigma_{zz}^2 & \sigma_{zz}^3 & \sigma_{zz}^4 & \sigma_{zz}^5 & \sigma_{zz}^6 \\
\sigma_{yz}^1 & \sigma_{yz}^2 & \sigma_{yz}^3 & \sigma_{yz}^4 & \sigma_{yz}^5 & \sigma_{yz}^6 \\
\sigma_{zx}^1 & \sigma_{zx}^2 & \sigma_{zx}^3 & \sigma_{zx}^4 & \sigma_{zx}^5 & \sigma_{zx}^6 \\
\sigma_{xy}^1 & \sigma_{xy}^2 & \sigma_{xy}^3 & \sigma_{xy}^4 & \sigma_{xy}^5 & \sigma_{xy}^6
\end{pmatrix}
\tag{14}
$$

in other words,

$$
[\varepsilon] = [S][\sigma] \tag{15}
$$

Hence, compliance matrix can be obtained as follows.

$$
[\varepsilon][\sigma]^{-1} = [S] \tag{16}
$$

Obtained compliance matrix must be symmetric and the asymmetric parts are considered to be the numerical errors.

If we have more than six sets of boundary conditions, we have more equations than the unknowns, *i.e.* the problem needs to be solved through the least square method as follows

$$
\begin{aligned}
[\varepsilon] &= [S][\sigma] \\
[\varepsilon][\sigma]^T &= [S][\sigma][\sigma]^T \\
[\varepsilon][\sigma]^T [[\sigma][\sigma]^T]^{-1} &= [S]
\end{aligned}
\tag{17}
$$

For the case of a two-dimensional plane strain condition, the following equation holds (Min & Stephansson, 2011)

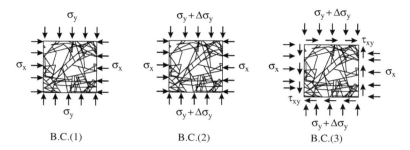

Figure 7 Three linearly independent boundary conditions for numerical experiments in UDEC (Min & Jing, 2003).

$$\begin{pmatrix} \varepsilon_{xx} \\ \varepsilon_{yy} \\ \varepsilon_{zz} \\ \gamma_{xy} \end{pmatrix} = \begin{pmatrix} S_{11} & S_{12} & S_{13} & S_{16} \\ S_{21} & S_{22} & S_{23} & S_{26} \\ S_{31} & S_{32} & S_{33} & S_{36} \\ S_{61} & S_{62} & S_{63} & S_{66} \end{pmatrix} \begin{pmatrix} \sigma_{xx} \\ \sigma_{yy} \\ \sigma_{zz} \\ \sigma_{xy} \end{pmatrix} \qquad (18)$$

Components S_{13}, S_{23} and S_{33} are pre-determined as those of intact rock, and S_{31} and S_{32} can be determined from the condition of symmetry. S_{63} and S_{36} are zero because fractures in 2D are always in parallel to the z-axis. When the pre-determined components are separated, the constitutive equation becomes as follows:

$$\begin{pmatrix} \varepsilon_{xx} \\ \varepsilon_{yy} \\ \gamma_{xy} \end{pmatrix} = \begin{pmatrix} S_{11} & S_{12} & S_{16} \\ S_{21} & S_{22} & S_{26} \\ S_{61} & S_{62} & S_{66} \end{pmatrix} \begin{pmatrix} \sigma_{xx} \\ \sigma_{yy} \\ \sigma_{xy} \end{pmatrix} + \begin{pmatrix} S_{13} \\ S_{23} \\ S_{63} \end{pmatrix} (\sigma_{zz}) \qquad (19)$$

With three independent boundary conditions as shown in Figure 7, Equation 19 becomes

$$\begin{pmatrix} \varepsilon_{xx}^{(1)} & \varepsilon_{xx}^{(2)} & \varepsilon_{xx}^{(3)} \\ \varepsilon_{yy}^{(1)} & \varepsilon_{yy}^{(2)} & \varepsilon_{yy}^{(3)} \\ \gamma_{xy}^{(1)} & \gamma_{xy}^{(2)} & \gamma_{xy}^{(3)} \end{pmatrix} - \begin{pmatrix} S_{13} \\ S_{23} \\ S_{63} \end{pmatrix} \begin{pmatrix} \sigma_{zz}^{(1)} & \sigma_{zz}^{(2)} & \sigma_{zz}^{(3)} \end{pmatrix}$$

$$= \begin{pmatrix} S_{11} & S_{12} & S_{16} \\ S_{21} & S_{22} & S_{26} \\ S_{61} & S_{62} & S_{66} \end{pmatrix} \begin{pmatrix} \sigma_{xx}^{(1)} & \sigma_{xx}^{(2)} & \sigma_{xx}^{(3)} \\ \sigma_{yy}^{(1)} & \sigma_{yy}^{(2)} & \sigma_{yy}^{(3)} \\ \sigma_{xy}^{(1)} & \sigma_{xy}^{(2)} & \sigma_{xy}^{(3)} \end{pmatrix} \qquad (20)$$

where $\sigma_{ij}^{(1)}$, $\sigma_{ij}^{(2)}$, and $\sigma_{ij}^{(3)}$ represent three linearly independent stress boundary conditions. These relationships can be expressed in terms of a matrix notation as follows.

$$[\varepsilon] - [S_z][\sigma_z] = [S][\sigma] \qquad (21)$$

The compliance matrix is solved as

$$[\varepsilon][\sigma]^{-1} - [S_z][\sigma_z][\sigma]^{-1} = [S] \qquad (22)$$

This numerical experiment can be an effective tool in determining rock mass properties considering the existence of numerous fractures and overall anisotropy. Constitutive behavior of fractures can be readily incorporated by considering e.g., non-linear behavior. Extension to three-dimensional application should be straight-forward. A comparison of two- and three-dimensional study showed that two-dimensional approach can underestimate the elastic modulus because three-dimensional DFN geometry is approximated in two dimensions (Min & Thoraval, 2012). Several studies exist that determined the strength properties of fractured rock masses (*e.g.*, Noorian-Bidgoli & Jing, 2015). This type of numerical experiment has been also used in understanding hydraulic and coupled hydromechanical behavior especially for stress-dependent permeability of fractured rock mass (Min *et al.*, 2004a, b)

4.3 Verification

In order to verify the DFN-DEM approach, a model with two orthogonal fracture sets is used for the comparison between the elastic properties produced by the numerical experiments and the closed-form solution (Amadei & Goodman, 1981). The computational models are rotated in intervals of 10 degrees to evaluate the variation of elastic moduli in rotated directions. Note also that the model boundary is located in the center of fracture spacing to avoid the need to adjust the equivalent spacing. Elastic modulus of intact rock was 84.6 GPa, Poisson's ratio was 0.24, normal stiffness of fracture was 434 GPa/m, shear stiffness of fracture varied from 43.4 GPa/m to 2170 GPa/m and the spacing between fracture was set to be 0.5 m (Min & Jing, 2003). Three boundary conditions described in Figure 7 are applied to obtain the compliance matrix. As shown in Figure 8, the two sets of results show an almost perfect agreement. Elastic moduli variation with respect to rotation angle was plotted as a function of the ratio of shear to normal stiffness of fractures. Importantly, each component of compliance tensor such as S_{11}, S_{12}, S_{16}, S_{22}, S_{26} and S_{66} also match perfectly with analytical solution as shown in Figure 8(b).

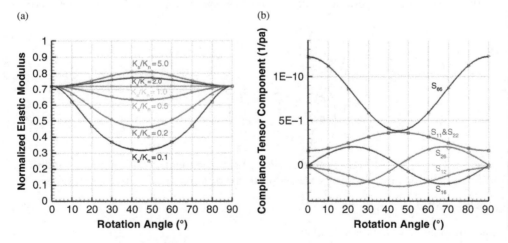

Figure 8 Verification of DEM modeling results against analytical solution for a fractured rock mass with two orthogonal sets of fractures. (a) Elastic modulus variation with respect to variation angles. Different ratios of shear stiffness and normal stiffness of fracture are indicated, (b) compliance tensor components variation with respect to rotation angle. In both figures, symbols correspond to numerical results while the lines are from analytical solution.

4.4 Demonstrating examples from Forsmark, Sweden

Case study was taken from Forsmark in Sweden which is now chosen as a candidate site for geological repository of spent nuclear fuels. In order to consider the stress dependency, different sets of boundary stresses were applied in accordance with the stress distribution, from depth 20 m to 1,000 m. In the modeling, stress-dependent normal behavior was implemented by the step-wise normal stiffness model, and different shear stiffness values were used as input data corresponding to the stress levels. Figure 9 shows the fractured rock geometry observed in the site and represented in the numerical experiment.

Implemented fracture behavior is shown in Figure 10(a), and the results of elastic moduli obtained by the stress-dependent fracture model are shown in Figure 10(b) where the calculated elastic moduli versus depth under increasing stress condition are presented. At low stress levels under 50 m of depth, the elastic moduli are about 50 percent of those of intact rock. However, at high stress levels over 400 m of depth, the moduli reach nearly the same values as that of intact rock. Stress induced anisotropy can be observed at shallow depths, mainly because of the low stiffness response of fractures under low stresses. The maximum anisotropy was about 20 percent at shallow depths. However, this effect was not significant at greater depths.

The Poisson's ratio of fractured rock mass is often greater than 0.5 because of the existence of fractures and their low shear stiffness. The discussion on this aspect can be found in Min and Jing (2004).

Figure 11 shows examples of horizontal displacements in the fractured rock mass after application of boundary stresses of different magnitudes. The displacement is highly influenced by the existence of the fractures. As the figures show, the influence of fracture is more evident at lower and middle stresses, *i.e.*, depth of 20~200 m (Figure 11(a) and (b)), while nearly uniform displacement is observed at high stress level, *i.e.*, depth of 1000 m (Figure 11(c)), where fractures are mainly closed with very high stiffness.

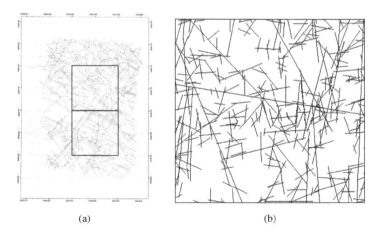

(a)　　　　　　　　　　(b)

Figure 9 Fractured rock geometry. (a) fracture trace map (SKB, 2004), (b) generated DFN. The two squares in the left and DFN in the right have dimension of 10 m × 10 m (Min et al., 2005; Min & Stephansson, 2011).

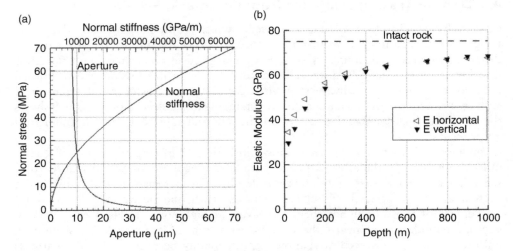

Figure 10 Elastic properties measured by DFN-DEM approach. (a) aperture and normal stiffness of a fracture with respect to normal stress implemented in the numerical experiment, (b) Elastic moduli of fractured rock mass versus depth. Because of stress dependent stiffness of fractures, elastic parameters are also stress dependent (Min et al., 2005; Min & Stephansson, 2011).

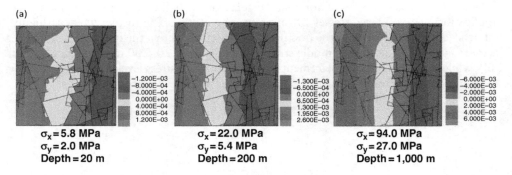

Figure 11 Horizontal displacement at different stress levels (modified from Min et al., 2005; Min & Stephansson, 2011).

5 NUMERICAL ANISOTROPIC ROCK MECHANICS – BONDED PARTICLE DISCRETE ELEMENT METHOD APPLICATION

5.1 Introduction

The bonded and particulate discrete element method (DEM) has been enjoying a wide variety of applications since its first introduction (Cundall & Strack, 1979; Potyondy & Cundall, 2004; Potyondy, 2015). It is now recognized as both a scientific tool to investigate the micro-mechanisms that produce complex macroscopic behaviors and an engineering tool to predict these macroscopic behaviors. Compared to DEM in block system in which deformable blocks are represented by polygon and polyhedral,

DEM in bonded-particle system can be more effective in simulating the progressive failure of rock. The theory and applications of DEM in bonded-particle system are introduced in Potyondy and Cundall (2004) and comprehensive review and theoretical background on explicit and implicit DEM are available in Jing and Stephansson (2007). Rock mechanics traditionally attempted to use the procedure developed in other branches of engineering such as FEM. DEM seems to be one of the unique contribution from rock mechanics which attract the interest from other areas such as powder technology, granular material, fluid mechanics and mineral processing as evidenced by a vast amount of citation being made on rock mechanics literature, *e.g.*, Cundall and Strack (1979). Despite significant advances in DEM modeling of rock to account for mechanical failure, most of these studies and their applications were applied entirely on the isotropic rock. The DEM modeling on anisotropic rock can be effective since it does not require pre-defined macroscopic failure criteria which can be very hard, if not impossible, to be obtained especially in anisotropic case. While some other approaches for modeling anisotropic rock have been suggested, systematic verification and direct comparison with anisotropic rock require more substantial research.

We present the DEM modeling of a transversely isotropic rock with systematic verification in both elastic and strength properties. The key conceptual idea of a transversely isotropic rock modeling is to include weak cohesive planes using a smooth joint model by assigning relatively larger cohesion compared to joints or fractures (Park & Min, 2015a, b). The smooth joint model simulates the behavior of a smooth interface by assigning new bonding models that have pre-defined orientations (Mas Ivars *et al.*, 2011). The developed model was validated against laboratory observations of three rock types (Cho *et al.*, 2012) and was extended to upscaled foundation problem under a surface line load that captured the stress distribution in the transversely isotropic rock formation.

5.2 Modeling methodology and verification

The results presented in this is obtained using PFC code (Itasca, 2008) which is a bonded-particle Discrete Element Method (DEM) defined as a dense packing of non-uniform sized circular or spherical particles joined at their contact points with parallel bond (Potyondy & Cundall, 2004). The calculation of particle movements is governed by Newton's second law of motion and a force-displacement law. The bonded particle model was adopted to construct isotropic rock without weak planes, which was calibrated based on elastic modulus and strengths that have the least effect of weak planes. Then, the smooth joint model (Mas Ivars *et al.*, 2011) was inserted to create the weak cohesive planes to simulate the behavior of the equivalent anisotropic continuum. Once the smooth joint, which consists of newly assigned properties, such as dip angle, normal and shear stiffness, friction coefficient, dilation angle, tensile strength, and cohesion, is created, pre-existing parallel bonds are deleted and replaced (Park & Min, 2015a, b). Figure 12 shows the rock samples made for transversely isotropic DEM modeling.

Figure 13(a) shows that the variations of the normalized elastic moduli of the transversely isotropic model with respect to various inclined angles from 0° to 90° in terms of different stiffness ratios of the weak planes (0.2, 0.5, 1.0, 2.0, and 5.0). The same definition of the stiffness ratio, K, is defined as the ratio of shear stiffness to

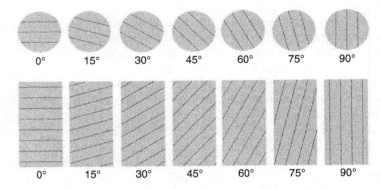

Figure 12 Transversely isotropic rock specimens made by bonded DEM model (Park & Min, 2015b).

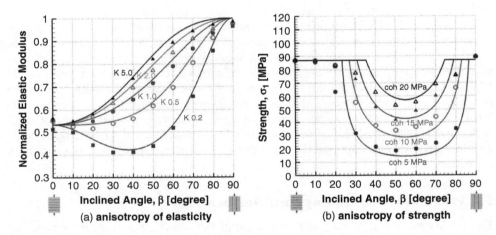

Figure 13 Verification of elastic and strength anisotropy using DEM. (a) Normalized elastic modulus in transversely isotropic rock model (symbol) with different stiffness ratios (K) is compared to that of the analytical solution (line). (b) Strength in the transversely isotropic rock model (symbol) with various cohesions of 5, 10, 15, and 20 MPa is compared to that of the analytical solution (line) (Park & Min, 2015a).

normal stiffness of the weak planes (k_s/k_n) and same analytical solution was used (Amadei & Goodman, 1981). The input normal stiffness ($k_{n,sj}$) was fixed as 3180 GPa/m, while the input shear stiffness ($k_{s,sj}$) was varied depending on the stiffness ratio (636, 1590, 3180, 6360, and 15900 GPa/m). The weak cohesive planes such as bedding planes normally have smooth interfaces without bumpiness of the grains, so that the dilation angle (ψ) of smooth joints was regarded as 0°. The friction coefficient (μ), tensile strength ($\sigma_{n,sj}$), and cohesion (C_{sj}) of smooth joints was determined as 57.29, 100 MPa, and 100 MPa, which were very high and thereby avoiding failure along the smooth joints when the elastic deformation occurs. There was a good match between the results of the numerical simulations and the analytical solutions, and a better match was achieved with higher resolutions with small particles. The particle size of DEM is one of the key parameters in reproducing the mechanical behavior of rock. The results acquired in this study show that reasonable resemblance was provided by about 88 particles across the width of 38 mm. These agreements demonstrate the validity of the smooth joint model as weak cohesive planes, so that the suggested bonded particle model with smooth joints model can be used to simulate the elastic modulus of a transversely isotropic rock (Park & Min, 2015b).

Furthermore, the DEM model can capture strength anisotropy to a reasonable extent as seen in Figure 13(b). The strength trends of the analytical solution indicated that a smaller friction angle or cohesion makes the curve wider and the minimum strength smaller (Jaeger *et al.*, 2007). Since the bonded particle model is not a continuum, a perfect match between the analytical and numerical models was not expected for strength variation. Some discrepancies are inevitably noticed as a form of transition between failures due to the creation of a new fracture, and the DEM model may be more realistic in this regard than the analytical model.

5.3 Validation against laboratory measurement

Three different types of a transversely isotropic rock (Asan gneiss, Boryeong shale and Yeoncheon schist) were reproduced as numerical models by assigning the microparameters of the bonded particle model and the smooth joint model listed in Table 1. In the numerical model, there are about 88 balls across 38 mm width of the specimen. As a result of the calibration process, macroproperties including elastic modulus, uniaxial compressive strength and Brazilian tensile strength of the numerical experiments were compared with those of the laboratory tests as presented in Figure 14. The squares indicate the laboratory test results, and the solid lines indicate the numerical test results. In the laboratory experiment, *UCS* (uniaxial compressive strength) of the specimen with the vertical weak planes was greater than the value of the specimen with horizontal weak planes although only identical *UCS*s are possible in the analytical solutions. It appears that rock matrix between layers also has a certain directional characteristics with stronger strength in the direction parallel to the layers. As this second-order anisotropy present in rock matrix was not considered in numerical model, current DEM model cannot adequately address this strength difference observed in direction parallel and perpendicular to the weak planes. Only a shoulder-type form of strength variation predicting the equal strength parallel and perpendicular to the loading axis was observed in the numerical experiments. It appears that the DEM model captured the elastic and

Table 1 Microparameters of smooth joint model for Asan gneiss, Boryeong shale, and Yeoncheon schist (Park & Min, 2015b).

Microparameter	Asan gneiss	Boryeong shale	Yeoncheon schist
Normal stiffness, $k_{n,sj}$ [GPa/m]	11,450	3,360	1,730
Shear stiffness, $k_{s,sj}$ [GPa/m]	2,540	960	1,440
Friction coefficient, μ (φ)	0.577 (30°)	0.364 (20°)	0.268 (15°)
Dilation angle, ψ [degree]	0°	0°	0°
Tensile strength, $\sigma_{n,sj}$ [MPa]	3	3	3
Cohesion, C_{sj} [MPa]	30	15	10

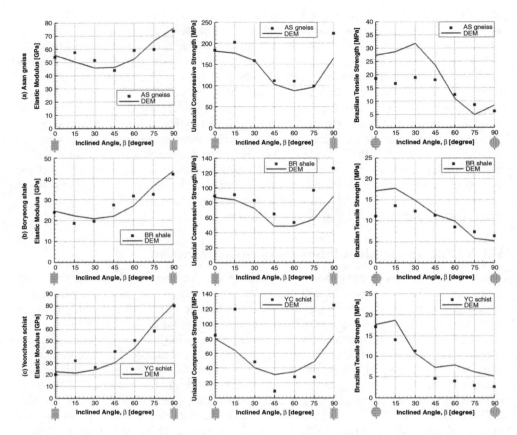

Figure 14 Comparison of elastic modulus, uniaxial compressive strength and Brazilian tensile strength from laboratory tests, and those from bonded-particle DEM modeling (AS gneiss: Asan gneiss, BR shale: Boryoeng shale, YC schist: Yeoncheon schist) (Park & Min, 2015b).

strength behaviors of laboratory observation to a reasonable extent, given that the variation of the elastic and strength properties in the laboratory measurements fluctuated and did not necessarily follow the theoretical trends.

Failure mechanisms during uniaxial and tensile tests were analyzed by comparing the failure patterns observed during numerical and laboratory experiments as shown in Figure 15. In the DEM model, the bond breakage of parallel bonds and smooth joints induced by either tensile or shear failure matched well with failure patterns on rock specimens (Vervoort et al., 2014; Park & Min, 2015b).

5.4 Numerical demonstration of a larger scale problem

The developed DEM model was extended to Boussinesq's problems subjected to concentrated line load that is often regarded as a foundation problem as depicted in Figure 16(a). When the line load (p) is applied normal to the surface, radial stress (σ_r) arising from the

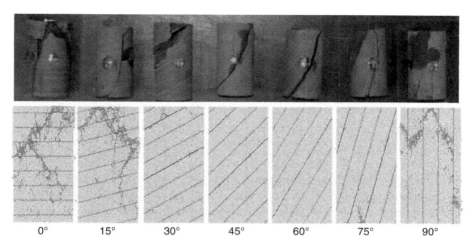

Figure 15 Comparison of post failure specimens obtained from compressive test on Boryeong shale (from Cho et al., 2012) and bonded-particle DEM modeling (Park & Min, 2015b).

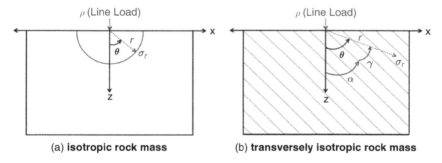

Figure 16 Conceptual figures of foundation problem: (a) isotropic rock mass model, and (b) transversely isotropic rock mass model (Park & Min, 2015b).

line load as a function of distance (r) and the angle (θ) rotated from the direction of applied load (vertical axis). At a certain point can be expressed as (Goodman, 1989):

$$\sigma_r = \frac{2\rho \cos \theta}{\pi r} \tag{23}$$

Equation 23 is the simplified two-dimensional solution of Boussinesq's problem which is often called Boussinesq-Flamant's Problem (Davis & Selvadurai, 1996). For transversely isotropic rock mass embedding a single set of weak planes (see Figure 16 (b)), following analytical solution can be used (Goodman, 1989):

$$\sigma_r = \frac{h}{\pi r} \left(\frac{X \cos \gamma + Yg \sin \gamma}{(\cos^2 \gamma - g \sin^2 \gamma)^2 + h^2 \sin^2 \gamma \cos^2 \gamma} \right) \tag{24}$$

where γ equals θ minus α as shown in Figure 16. X and Y are x-directional load and y-directional load, respectively. Both g and h are dimensionless quantities accounting for the properties of isotropic rock mass and are given by

$$g = \sqrt{1 + \frac{E}{(1 - \nu^2)k_n\delta}}$$

$$h = \sqrt{\frac{E}{(1 - \nu^2)}\left(\frac{2(1 + \nu)}{E} + \frac{1}{k_s\delta}\right) + 2\left(g - \frac{\nu}{1 - \nu}\right)}$$

(25)

Two material constants are composed of the mechanical properties of not only the intact rock elastic modulus (E) and Poisson's ratio (ν) but also the weak planes such as normal and shear stiffness (k_n, k_s) and mean spacing between weak planes (δ). By substituting Equation 25 into Equation 24, the radial stress distribution in transversely isotropic rock mass can be obtained.

For simulating foundation problems using bonded-particle DEM, a 2 x 2 m square isotropic rock mass model that consisting of 114,984 particles was used, and all boundaries were constrained except the upper boundary on which the surface line load was applied. Three models with 0°, 60°, and 90° inclined angle of smooth joints were taken into account. The stress distribution along the same distance line of isotropic rock mass model was compared to that of the numerical calculation, thereby obtaining the results shown in Figure 17(a). More detailed input microparameters of bonded particle model can be found in Park and Min (2015b). In this model, the radial stress was calculated at the distance (r) of 0.4, 0.6, and 0.8 m from the applied line load, which is 10 MN/m (Park & Min, 2015b). Contact force induced by the applied surface loading is shown in the left, and the pressure bulbs of radial stress (σ_r) are presented in the right with respect to the distance (r) from the point of applied line load for the angle (θ) ranging from 0° to 180°. The measured stress in the numerical model was in good agreement with the value from the analytical solution. As the pressure bulb becomes larger, the magnitude of measured stress increases, meaning that the measured point is closer to the applied load (ρ).

In transversely isotropic models, the radial stress was estimated under the same boundary and loading conditions as those employed in the isotropic rock mass model. As a result, the stress distribution in the numerical model and that in the analytical solution were compared, as illustrated in Figure 17(b–d). The observations in transversely isotropic rock also showed a close match between the results of the numerical and analytical solutions. In particular, the developed bonded-particle DEM successfully captured the effect of the bedding planes in horizontal, inclined, and vertical directions on the stress distribution. The closer the measured points are to the applied load, the more discrepancy is observed due to the higher magnitude of stress and gradient in the vicinity. This is because there were relatively fewer particles in the vicinity of the loading point, which resulted in a more heterogeneous stress distribution. The large-scale DEM modeling of transversely isotropic rock mass matched well in accordance with the analytical solutions, which demonstrated that the developed model can be used in anisotropic rock mass.

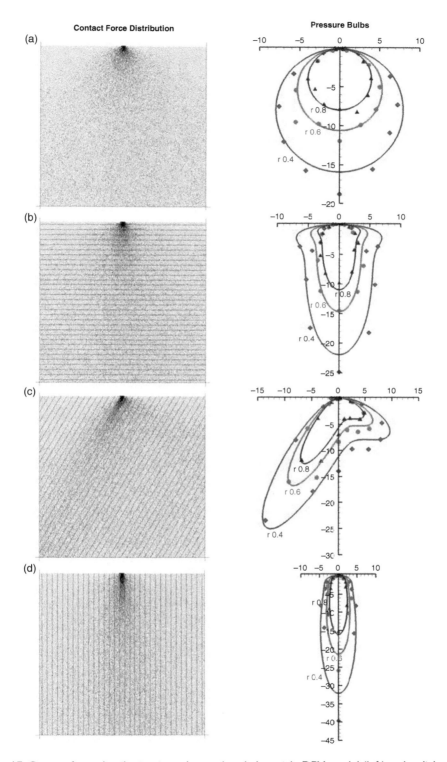

Figure 17 Contact force distribution in rock mass bonded-particle DEM model (left) and radial stress from both bonded-particle DEM model and analytical solutions (right): (a) isotropic case, and (b–d) transversely isotropic cases with the inclined angles of 0°, 60°, and 90°, respectively (Park & Min, 2015b).

6 CONCLUSION

This chapter provides a brief overview of theoretical anisotropic rock mechanics and introduces a series of experimental and numerical studies applied to anisotropic rock mechanics mostly conducted in the past 15 years by the authors.

Experimental investigations are made on elastic, thermal conductivity, seismic and permeability anisotropy of rock, based on cores taken from directional coring system. Methods are presented to obtain five independent parameters for transversely isotropic rocks using two rock samples cored in different directions. Experimental investigation of elastic, seismic, thermal conductivity and permeability anisotropy of some of rock types found in Korea showed a clear evidence of transverse isotropy. The anisotropy ratios of elastic modulus parallel to and perpendicular to isotropic planes (E/E') were determined to be 1.3, 2.1 and 3.4 for Asan gneiss, Boryeong shale and Yeoncheon schist, respectively, and this shows that extent of anisotropy can be great and cannot be ignored in practice. Currently, standard method of determining anisotropic parameters does not exist, and adoption of such standard will greatly help in considering anisotropy in practical rock engineering.

First part of numerical anisotropic rock mechanics in this chapter introduces the numerical experiments named as DFN-DEM (Discrete Fracture Network-Discrete Element Method) approach to determine the compliance tensor of fractured rock mass modeled as equivalent continuum anisotropic rock. Blocky Discrete element method (DEM) is employed for this numerical experiment using three boundary conditions in two dimensions. Systematic verification and application to Forsmark, Sweden, show that this numerical method is readily applicable in practice. The strength of the numerical approach over other empirical or analytical methods is that the mechanical properties of fractured rock mass with very irregular fracture system geometry can be directly determined and that complex constitutive models of fractures can be incorporated. Using this approach, we can investigate the effect of different factors in the determination of overall mechanical properties which are often anisotropic.

Second part of numerical anisotropic rock mechanics in this chapter deals with representation of transversely isotropic rock using bonded-particle DEM model. Bonded-particle DEM with embedded smooth joints was applied to model the mechanical behaviors of transversely isotropic rock with systematic verifications and extensions to laboratory and upscaled problems. Both deformation and strength behavior modeled by DEM showed a reasonable agreement with analytical solution, and laboratory observation. Upscaled model applied to anisotropic foundation demonstrate that large scale application anisotropic DEM model is also feasible.

ACKNOWLEDGMENTS

This work is a result of research conducted in the Royal Institute of Technology and Seoul National University. We thank Prof. Ove Stephansson for interaction and encouragement during the course of this work. Financial supports have been given from European Commission through BENCHPAR project (FIKW-CT-2000-00066), Swedish Nuclear Power Inspectorate (SKI) through DECOVALEX project, Swedish Radiation Safety Authority (SSM) and Seoul National University. Writing of this paper was supported by the New & Renewable Energy Technology Development Program of the Korea Institute of Energy Technology Evaluation and Planning (KETEP) through a grant that was funded by the Ministry of Trade, Industry & Energy, Republic of Korea (No. 20133030000240).

REFERENCES

Amadei, B. & Goodman, R.E. (1981) A 3-D constitutive relation for fractured rock masses. In: Selvadurai, A.P.S. (ed.) *Proceedings of the International Symposium on the Mechanical Behavior of Structured Media, Ottawa.* pp. 249–268.

Amadei, B. (1982) The influence of rock anisotropy on measurement of stresses in-situ. PhD Thesis. University of California, Berkeley.

Amadei, B. (1996) Importance of anisotropy when estimating and measuring in situ stresses in rock. *Int J Rock Mech Min Sci & Geomech Abstr*, 33 (3), 293–325.

Barla, G. (1974) *Rock anisotropy: Theory and laboratory testing.* New York, Springer-Verlag. pp. 131–169.

Barton, N. (2002) Some new Q-value correlations to assist in site characterization and tunnel design. *Int J Rock Mech Min Sci*, 39 (2), 185–216.

Bear, J. (1972) *Dynamics of fluids in porous media.* New York, Elsevier.

Beardsmore, G.R. & Cull, J.P. (2001) *Crustal heat flow: A guide to measurement and modelling.* New York, Cambridge University Press.

Bieniawski, Z.T. (1978) Determining rock mass deformability: Experience from case histories. *Int J Rock Mech Min Sci & Geomech Abstr*, 15 (5), 237–247.

Carslaw, H.S. & Jaeger, J.C. (1959) *Conduction of heat in solids* (2nd ed.). Oxford, Clarendon Press.

Chen, C.S., Pan, E. & Amadei, B. (1998) Determination of deformability and tensile strength of anisotropic rock using Brazilian tests. *Int J Rock Mech Min Sci*, 35 (1), 43–61.

Cho, J.W., Kim, H., Jeon, S. & Min, K.-B. (2012) Deformation and strength anisotropy of Asan gneiss, Boryeong shale, and Yeoncheon schist. *Int J Rock Mech Min Sci*, 50 (12), 158–169.

Chou, Y.C. & Chen, C.S. (2008) Determining elastic constants of transversely isotropic rocks using Brazilian test and iterative procedure. *Int J Numer Anal Meth Geomech*, 32 (3), 219–234.

Claesson, J. & Bohloli, B. (2002) Brazilian test: Stress field and tensile strength of anisotropic rocks using an analytical solution. *Int J Rock Mech Min Sci*, 39 (8), 991–1004.

Cundall, P.A. & Strack, O.D. (1979) A discrete numerical model for granular assemblies. *Géotechnique*, 29 (1), 47–65.

Davis, R.O. & Selvadurai, A.P.S. (1996) *Elasticity and geomechanics.* Cambridge, Cambridge University Press.

Exadaktylos, G.E. & Kaklis, K. N. (2001). Applications of an explicit solution for the transversely isotropic circular disc compressed diametrically. *Int J Rock Mech Min Sci*, 38 (2), 227–243.

Fairhurst, C. (1993). The strength and deformability of rock masses – an important research need. Proceedings of the ISRM International Symposium-EUROCK 93, 21–24 June 1993, Lisbon, Portugal. pp. 931–933.

Fossum, A.F. (1985) Effective elastic properties for a randomly jointed rock mass. *Int J Rock Mech Min Sci & Geomech Abstr*, 22 (6), 467–470.

Fredriksson, A. & Olofsson, I. (2007) *Rock mechanics modelling of rock mass properties – theoretical approach. Preliminary site description Laxemar subarea – version 1.2, SKB R-06-16.* Swedish Nuclear Fuel and Waste Management Co (SKB).

Gerrard, C.M. (1982) Equivalent elastic moduli of a rock mass consisting of orthorhombic layers. *Int J Rock Mech Min Sci & Geomech Abstr*, 19 (1), 9–14.

Gonzaga, G.G., Leite, M.H. & Corthésy, R. (2008) Determination of anisotropic deformability parameters from a single standard rock specimen. *Int J Rock Mech Min Sci*, 45 (8), 1420–1438.

Goodman, R.E. (1989) *Introduction to rock mechanics* (2nd ed.). New York, John Wiley & Sons.

Hakala, M., Kuula, H. & Hudson, J.A. (2007) Estimating the transversely isotropic elastic intact rock properties for in situ stress measurement data reduction: a case study of the Olkiluoto mica gneiss, Finland. *Int J Rock Mech Min Sci*, 44 (1), 14–46.

Hart, R.D., Cundall, P.A. & Cramer, M.L. (1985) Analysis of a loading test on a large basalt block. *Proceedings of the 26th U.S. Symposium on Rock Mechanics, 26–28 June 1985, Rapid City, USA*. pp. 759–768.

Hoek, E. & Brown, E.T. (1997) Practical estimates of rock mass strength. *Int J Rock Mech Min Sci & Geomech Abstr*, 34 (8), 1165–1186.

ISRM. (2007) *The complete ISRM suggested methods for rock characterization testing and monitoring: 1974–2006*. In: Ulusay, R. & Hudson, J.A. (eds.) Ankara, ISRM Turkish national group.

ISRM. (2015) *The ISRM suggested methods for Rock characterization, testing and monitoring: 2007–2014*. In: Ulusay, R. (ed), Berlin, Springer International Publishing.

Itasca Consulting Group Inc. (2000) *UDEC user's guide*. Minneapolis, ICG.

Itasca Consulting Group Inc. (2008) *PFC 2D (Particle Flow Code in 2 Dimensions) Version 4.0*. Minneapolis, ICG.

Jaeger, J.C., Cook, N.G.W. & Zimmerman, R.W. (2007) *Fundamentals of rock mechanics* (4th ed.). Oxford, Wiley-Blackwell.

Jing, L. & Stephansson, O. (2007) *Fundamentals of discrete element methods for rock engineering: Theory and applications*. Amsterdam: Elsevier Science.

Kim, H., Cho, J.W., Song, I. & Min, K.-B. (2012) Anisotropy of elastic moduli, P-wave velocities, and thermal conductivities of Asan Gneiss, Boryeong Shale, and Yeoncheon Schist in Korea. *Eng Geol*, 147–148, 68–77.

Lekhnitskii, S.G. (1963) *Theory of elasticity of an anisotropic body*. (City), Holden-Day.

Mas Ivars, D., Pierce, M.E., Darcel, C., Reyes-Montes, J., Potyondy, D.O., Young, P.R. & Cundall, P.A. (2011) The synthetic rock mass approach for jointed rock mass modelling. *Int J Rock Mech Min Sci*, 48 (2), 219–244.

Min, K.-B. & Jing, L. (2003) Numerical determination of the equivalent elastic compliance tensor for fractured rock masses using the distinct element method. *Int J Rock Mech Min Sci*, 40 (6), 795–816.

Min, K.-B., Lee, C.I. & Choi, H.M. (2003) An experimental and numerical study of the in-situ stress measurement on transversely isotropic rock by overcoring method. In: Sugawara, K., Obara, Y. & Sato, A. (eds.) *3rd International Symposium on Rock Stress – RS Kumamoto '03, 4–6 November 2003, Kumamoto, Japan*, pp. 189–195.

Min, K.-B. (2004) *Fractured rock masses as equivalent continua – A numerical study*. PhD Thesis. Royal Institute of Technology (KTH).

Min, K.-B. & Jing, L. (2004) Stress-dependent mechanical properties and bounds of Poisson's ratio for fractured rock masses investigated by a DFN-DEM technique. *Int J Rock Mech Min Sci*, 41 (3), 431–432.

Min, K.-B., Jing, L. & Stephansson, O. (2004a) Determining the equivalent permeability tensor for fractured rock masses using a Stochastic REV approach: Method and application to the field data from Sellafield, UK. *Hydrogeol J*, 12 (5), 497–510.

Min, K.-B., Rutqvist, J., Tsang, C.-F. & Jing, L. (2004b) Stress-dependent permeability of fractured rock masses: A numerical study. *Int J Rock Mech Min Sci*, 41 (7), 1191–1210.

Min, K.-B. (2005) Compliance tensor calculation. Technical Note, Itasca Geomekanik, Jul 2005.

Min, K.-B., Stephansson, O. & Jing, L. (2005) Effect of stress on mechanical and hydraulic rock mass properties – application of DFN-DEM approach on the date from site investigation at

Forsmark, Sweden. *Proceedings of the ISRM International Symposium-EUROCK 2005, 18–20 May 2005, Brno, Czech Republic*. pp. 389–395.

Min, K.-B. & Stephansson, O. (2011) The DFN-DEM approach applied to investigate the effects of stress on mechanical and hydraulic rock mass properties at Forsmark, Sweden. *J Korean Soc Rock Mech*, 21 (2), 117–127.

Min, K.-B. & Thoraval, A. (2012) Comparison of two- and three-dimensional approaches for the numerical determination of equivalent mechanical properties of fractured rock masses. *J Korean Soc Rock Mech*, 22 (2), 93–105.

Nasseri, M.H., Rao, K.S. & Ramamurthy, T. (1997) Failure mechanism in schistose rocks. *Int J Rock Mech Min Sci*, 34 (3–4), 219.e1–219.e15.

Nasseri, M.H., Rao, K.S. & Ramamurthy, T. (2003) Anisotropic strength and deformational behavior of Himalayan schists. *Int J Rock Mech Min Sci*, 40 (1), 3–23.

Noorian-Bidgoli, M. & Jing, L. (2015) Stochastic analysis of strength and deformability of fractured rocks using multi-fracture system realizations. *Int J Rock Mech Min Sci*, 78, 108–117.

Nunes, A.L. (2002) A new method for determination of transverse isotropic orientation and the associated elastic parameters for intact rock. *Int J Rock Mech Min Sci*, 39 (2), 257–273.

Oda, M. (1982) Fabric tensor for discontinuous geological materials. *Soils Fdns*, 22(4), 96–108.

Palmström, A. (1996) Characterizing rock masses by the RMi for use in practical rock engineering, Part 2: Some practical applications of the Rock Mass index (RMi). *Tunn Undergr Sp Tech*, 11 (3), 287–303.

Park, B. & Min, K.-B. (2015a) Discrete element modeling of transversely isotropic rock applied to foundation and borehole problems. *Proceedings of the 13th International ISRM Congress 2015, 10–13 May 2015, Vancouver, Canada, Paper No.: 843*.

Park, B. & Min, K.-B. (2015b) Bonded-particle discrete element modeling of mechanical behavior of transversely isotropic rock. *Int J Rock Mech Min Sci*, 76, 243–255.

Pinto, J.L. (1970) Deformability of schistous rocks. *Proceedings of the 2nd International ISRM Congress, September 1970, Belgrade, Yugoslavia*. pp. 1–19.

Pinto, J.L. (1979) Determination of the elastic constants of anisotropic bodies by diametral compression tests. *Proceedings of the 4th International ISRM Congress, September 1979, Montreux, Switzerland*. pp. 359–363.

Potyondy, D.O. & Cundall, P.A. (2004) A bonded-particle model for rock. *Int J Rock Mech Min Sci*, 41 (8), 1329–1364.

Potyondy, D.O. (2015) The bonded-particle model as a tool for rock mechanics research and application: current trends and future directions. *Geosys Eng*, 18 (1), 1–28.

Pouya, A. & Ghoreychi, M. (2001) Determination of rock mass strength properties by homogenization. *Int J Numer Anal Meth Geomech*, 25, 1285–1303.

Ramamurthy, T. (1993) Strength and modulus responses of anisotropic rocks. In: Hudson, J.A. (ed. in-chief) *Comprehensive rock engineering, Vol. 1, Chap. 3*. Oxford, Pergamon Press. pp. 313–329.

Rodrigues, F.P. (1966). Anisotropy of granites. Modulus of elasticity and ultimate strength ellipsoids, joint systems, slope attitudes, and their correlations. *Proceedings of the 4th International ISRM Congress, September 1966, Lisbon, Portugal*.

Salamon, M.D.G. (1968) Elastic Moduli of a stratified rock mass. *Int J Rock Mech Min Sci & Geomech Abstr*, 5, 519–527.

Singh, B. (1973) Continuum characterization of jointed rock masses. *Int J Rock Mech Min Sci & Geomech Abstr*, 10, 311–335.

SKB, (2004) *Preliminary site description Forsmark area – Version 1.1, SKB R 04-15*. Swedish Nuclear Fuel and Waste Management Co (SKB).

Talesnick, M.L. & Ringel, M. (1999) Completing the hollow cylinder methodology for testing of transversely isotropic rocks: torsion testing. *Int J Rock Mech Min Sci*, 36 (5), 627–639.

Tien, Y.M. & Tsao, P.F. (2000) Preparation and mechanical properties of artificial transversely isotropic rock. *Int J Rock Mech Min Sci*, 37 (6), 1001–1012.

Ting, T.C.T. (1996) *Anisotropic elasticity – Theory and applications*. Oxford, Oxford University Press.

Vervoort, A., Min, K.B., Konietzky, H., Cho, J.W., Debecker, B., Dinh, Q.D., Fruhwirt, T. & Tavalali, A. (2014) Failure of transversely isotropic rock under Brazilian test conditions. *Int J Rock Mech Min Sci*, 70, 343–352.

Worotnicki, G. (1993) CSIRO triaxial stress measurement cell. In: Hudson, J.A. (ed. in-chief) *Comprehensive rock engineering, Vol. 3, Chap. 13*. Oxford: Pergamon Press. pp. 329–394.

Yang, H.Y., Kim, H., Kim, K., Kim, K.Y. & Min, K.B. (2013) A study of locally changing pore characteristics and hydraulic anisotropy due to bedding of porous sandstone. *Tunnel & Underground Space: Journal of Korean Society for Rock Mechanics*, 23 (3), 228–240.

Yun, T.S., Jeong, Y.J., Kim, K.W. & Min, K.B. (2013) Evaluation of rock anisotropy using 3D X-ray computed tomography. *Eng Geol*, 163, 11–19.

Chapter 5

Characterization of rock masses based on geostatistical joint mapping and rock boring operations

M. Stavropoulou[1] & G. Exadaktylos[2]

[1] Faculty of Geology and Geoenvironment, National and Kapodistrian University of Athens, Greece

[2] School of Mineral Resources Engineering, Technical University of Crete, Greece

Abstract: In mechanized rock excavations using Tunnel Boring Machines (TBM's) the contact of the disc cutters with the rock precedes every other work like mucking and support. Penetration into the cohesive/frictional ground by the tip of a disc cutter subjected to a given thrust force can be achieved only a fraction of a cm at a time instant. In a backward analysis if cutting processes in rocks are properly registered and analyzed they give valuable information on the mechanical properties of the heterogeneous rock masses like strength, deformability, index properties such as abrasivity and hardness, or physical properties like content in abrasive or clay minerals. It is evident that a proper backward scheme may lead to a "smart" rock cutting process. On the other hand, the forward problem is the optimum design of cutting discs then of the cutting head and finally of the operational parameters of TBM's under a given heterogeneous and "opaque" geomechanical environment. In order to achieve both aims the logical step is to study first the mechanics of a single disc cutter penetrating and cutting the rock and then compose the whole response model of the cutting head of a boring machine. Before the outset of the excavation of a TBM several geotechnical and geological data are collected along the tunnel "corridor". This data should be interpolated along the planned tunnel axis. This interpolation should be done on a 3D grid of the geological model around and along tunnel axis. The Geostatistical approach accompanied with computer aided design tools provide the mathematical and the geometrical frames needed for the interpolation of spatially or temporally correlated data from sparse spatial sample data. Then it is possible to correlate the TBM performance with the geological-geotechnical conditions at regular intervals along the tunnel and extract valuable information regarding the former issue. Herein, we are going to display the above concepts and tools with an example case study.

1 INTRODUCTION

Drilling comprises a set of processes for breaking and removing rock to create drill-holes, boreholes, tunnels, and other types of excavations. It may be percussive-rotary or rotary with tri-cone bits or drag bits depending on hole diameter and ground conditions. On the other hand full-face boring machines like the Tunnel Boring Machine (TBM) are basically rotary cutting systems dedicated for the entire excavation of a

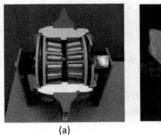 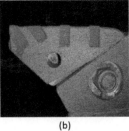

(a)   (b)

Figure 1 Main cutting tools of a TBM; (a) Sliced section of a disc cutter and its bearing with constant section for cutting medium to hard rocks, and (b) side section cutting knife (or scraper) with hard metal inserts for cutting soft rocks and soils (courtesy of Herrenknecht).

circular cross-section of an underground work (*i.e.* a tunnel, an incline etc) during their advancement. Such a mechanical excavation system includes the cutter head or cutting wheel with cutting tools attached on it (discs knives or chisel tools) as is shown in Figures 1a and b, mucking components, power supply systems for cutting wheel thrust and rotation, and a steering system. In hard rocks an "open" TBM has hydraulic grippers that exert to the sidewalls a certain thrust for the necessary resistance required for the provision of the thrust and torque to the cutting wheel, while in heavily fractured rock masses it is equipped with one or two shields for protection of a prone to fall (cave) roof like in longwall mining. Single shields move ahead by exerting thrust on the previous erected lining in the rear of the shield (*i.e.* reinforced concrete segments). Double shield or telescopic shield includes a sidewall gripping thrust system while the lining thrust system is an auxiliary component of the system. Additional back-up equipment includes tunnel support with concrete segments, muck transport, personnel transport, material supply and utilities and ventilation to the tunnel's face. In this chapter we are mainly concerned as of how the design and operational parameters of a TBM and the rock mass characteristics are affecting the performance of such an integrated mechanical excavation system.

    The cutting action of drag bits in rotary drilling, cutting picks in roadheaders and knifes (or scrapers) in soft rock TBM's differs slightly from that of the rolling disc cutters, although all types of tools act principally like indenters. A scraper, drag bit or chisel tool cuts the rock along a prescribed circular path under the combined action of the normal force and the drag or cutting force that is parallel with the cutting face and planning the rock ahead of it. Pick width, as well as, the tool, rake and back-rake angles among other bit design parameters greatly influence the forces necessary to cut the rock. On the other hand cutting of rocks with discs is performed by indenting them into the rock to the required depth with sufficiently high normal force and then to roll them along circular grooves with the aid of the rolling force. The geometry of the tip (*i.e.* V-shaped or constant cross-section) and disc's radius, among other parameters, depict the cutting forces transmitted to the rock. Cutter design has been evolved the past years both in regards to their geometry and materials. Regarding the bit shape the discs are now designed with a 'constant section' with flat tip rather than a 'wedge-shaped' tip, in order to maintain a stable profile after wear and a longer period of use. Special hardened steels or hard metal inserts are also employed exhibiting balanced properties of hardness, toughness and abrasion resistance. Disc cutters are more efficient than drag tools in hard and abrasive grounds and vice versa, that is in soft or heavily fractured and not abrasive geological formations the drag type tools perform in more efficient manner; disc cutters cannot freely rotate due to insufficient interfacial friction between the soft ground and the disc leading to excessive wear of the latter. Since the cutting tools are used to transfer the energy from the machine to the rock it may be realized that a central place in predicting the performance of a boring machine referring to the advance rate, and unit excavation cost, holds the model for the calculation of the cutting forces.

2 DESIGN CONCEPTS AND PERFORMANCE OF TUNNEL BORING MACHINES

2.1 Introduction

A central problem of considerable practical interest in rock mass boring operations is how to predict the spatial distribution of rock strength and possibly abrasivity over the tunnel length before the commencement of the excavation process. This is because strength and abrasivity are the principal parameters required for the prediction of the performance of the TBM like penetration and advance rates, capital and operational excavation costs, required thrust and torque, but also the stability of the tunnel itself and the disturbance of the ground surface (*i.e.* subsidence) in shallow excavations. For this purpose one may rely on the use of the conceptual 3D volume geological model (see Section 4), relevant geological interpretations (soft data) and limited number of samples obtained in the exploration phase (hard data). It is obvious that the initial "conceptual ground model" is incomplete due to rather limited size of sample compared to the scale of the excavation. So, the next challenging task is how to continuously upgrade this initial ground model by exploiting TBM registered data (hard data). Legitimate ways to characterize the rock mass may be based on RMR, Q, GSI, Fracture Frequency (FF) rock mass characterization schemes or through other appropriate scheme tailored to a certain project. This is because there are enough data from rock mass excavation projects worldwide in order to link these empirical rock mass indices with fundamental concepts like damage of rock mass. Herein, our concern on these aspects of TBM excavation is graphically depicted in Figure 2, and may be listed as follows:

- Knowing RMR, Q or GSI rock mass quality indices (that could be considered as a pre-existing damage inherited to the rock by the joints) and intact rock mechanical properties to appropriately design the TBM machine and predict its performance along the chainage of the tunnel. That is, predict penetration rate and torque based on input values of thrust and rotational speed of the cutting head.
- Alternatively, by collecting thrust, penetration rate and other TBM recorded operational data, to estimate RMR (or Q or GSI) and then to upgrade the model of the rock mass in front of the tunnel by combining with RMR (or Q or GSI) estimated from boreholes or drillholes in front of the tunnel.

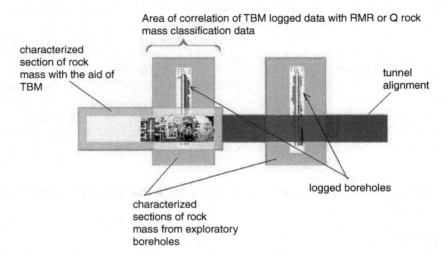

Figure 2 Sketch of the concept of using TBM as a tool for geotechnical characterization of the rock mass between logged boreholes.

2.2 Fundamentals of rock boring operations

The proper selection and design features of a TBM (*i.e.* installed power, number, type and location of cutting tools on the cutting wheel etc.) depends on the anticipated geological conditions, the tunnel cross-section diameter, the length of the tunnel and the permissible excavation cost of the project. Geological conditions refer to the strength and hardness of the rock masses, their content in abrasive and clay minerals, and the joint network. Ground conditions refer to the in situ stress field and the presence of water or temperature gradients. Absrasive minerals impose wear to the discs while clay minerals create sticky conditions for them (*i.e.* clay trapped in the housing of the disc cutter or at the interface of the tip of the cutter and the rock creating "sticky" conditions). Also, the presence of water and in situ stresses influence the stiffness and strength of the rock mass. Hence, it is quite probable that geological parameters are varying from face to the neighboring faces and inside the area of the tunnel's face itself. These parameters in turn influence the net Penetration Rate (PR [m/h]), the Advance Rate (AR [m/h]) and the unit excavation cost (*i.e.* cost per meter of drive). The latter refers to the cost of replacement of worn disc cutters or tools, capital and operational costs. The penetration rate, PR, in [m/h] is depicted by the cutting depth per revolution, denoted by the symbol p, and the rotational speed of the cutting wheel denoted with ω, in the following manner

$$PR = p[cm/rev] \cdot \omega[rev/\min] \cdot 60[\min/h] \cdot 10^{-2}[m/cm] \qquad (1)$$

It is noted that units appear in brackets. Also, it should be noticed that the maximum rotational speed of the cutting wheel ω [1/min] cannot be larger than a limiting value depending on the prescribed maximum linear velocity of the gauge (peripheral) cutters. That is the maximum rotational speed is depicted from the following inequality

$$v_{gauge} = \omega \frac{D}{2} = \Omega \frac{d}{2} \leq v_{\max} \tag{2}$$

wherein D is the diameter of the cutting wheel, and Ω, d denote the rotational speed and diameter of the cutting disc (the equality holds true since both expressions should give the rolling speed v_{gauge} of the disc). The reasons for the limitation of the rolling speed is the frictional heat produced in the cutter bearing - the cutters risk to block due to too high temperature and may thus hinder the boring process - the avoidance of machine vibrations and the reduction of the centrifugal forces on the rock chips. With the cutter discs currently available and their material characteristics, the number of revolutions per minute Ω range from 140 to 160 rpm (Maidl et al., 2012). Standard disc diameters are d = 394 mm and d = 432 mm, although smaller diameters are available like d = 280 mm and larger have been developed d = 483 mm. With this increase of diameter, the permissible rolling speed also increased up to about 190 m/min (Maidl et al., 2012), which considering Equation 2 also lead to the increase of cutting wheel revolution speed. It is noticed here that the as the disc cutter steel wears out, its diameter decreases. Hence the monitoring of the rotation speed of disc cutters and the comparison with the cutter head rotation may allow the estimation of the degree of disc wear.

The advance rate AR of the TBM is found if we know the degree of utilization or mechanical efficiency n_{util} of the TBM solely for cutting and PR. For TBM excavation Barton (2000) proposed the following empirical relation linking AR and PR

$$AR = PR \cdot n_{util} \tag{3}$$

with

$$n_{util} = t^m \tag{4}$$

where t is the time of duration of the tunnel excavation and the exponent m, is a negative real number with typical values ranging from –0.15 to –0.45. Alternatively, instead of the time that depends on other factors like shift hours etc., one may use the distance traveled by the TBM divided by the total planned length of the tunnel. For example for 50% utilization of the machine for rock cutting, p = 0.5 cm/rev, at a given time instant, and ω=12 rev/min we get the following value of the AR

$$\begin{aligned} AR &= n_{util} \cdot PR = n_{util} \cdot p[cm/rev] \cdot \omega[rev/\min] \cdot 60[\min/h] \cdot 10^{-2}[m/cm] = \\ &= 0.5 \cdot 0.5 \cdot 12 \cdot 60 \cdot 10^{-2} = 1.8 \, m/h \end{aligned} \tag{5}$$

According to the position of the disc cutters on the cutter head (e.g. Figure 3a), three sets of cutters can be distinguished, namely the centrals which perforate the center of the section, the frontals which attack the zone between the center and the periphery and the gauge cutters or 'of galibo' which are in the periphery of the head, supporting the diameter of the excavation and the needed form.

The number of discs mounted on the wheel depends on the diameter D of the cutting wheel and the prescribed spacing S of neighboring cuts, i.e.

$$N_t = N + N_{gauge} \tag{6}$$

wherein N_t is total number of cutting discs, N is the number of discs mounted on the forehead of the cutting wheel, and N_{gauge} denote the gauge (or peripheral) discs.

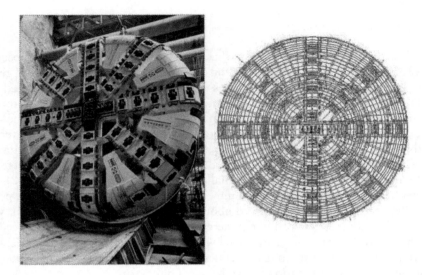

Figure 3 (a) Cutting head, and (b) traces of the disc cutters (right) of the shield convertible TBM used in the KCRC DB320 project for the construction of twin tunnels in Hong-Kong (courtesy of BOUYGUES TRAVAUX PUBLICS).

As the cutting wheel rotates each disc contacts the face along a different circular path as illustrated in Figure 3b. The average spacing between adjacent groves is denoted by the symbol S, and can be found from the circular traces of the disc cutters as is shown in Figure 3. The number of discs attached on the wheel may be calculated in the following approximate manner

$$N_t \approx \frac{D}{2S} \qquad (7)$$

Average spacing between neighboring cutter grooves is generally about 65–100 mm, whereas typical constant cross section cutter tip widths range from 12 to 19 mm. The increase of nominal contact force from 90 kN to about 312 kN for disc cutters increased in diameter from 280 mm (11") to 483 mm (19") enables not only a considerable increase of the average contact force, but also leads to a significant improvement in the lifetime of the disc cutters.

The cutting process involves initially indentation of the disc into the rock accompanied with pulverization (highly damaged powdered material) of the rock around the contact zone with it, formation during loading of radial cracks and microcracks beneath the tip of the cutter for pre-conditioning the indentation of the cutter in the next pass (the radial tensile crack directly below the tip called "median vent") and later after the passing of the rolling disc above the area, the unloading of the rock around the contact region of the cutting edge that is responsible for the closure of the median vent but on the other hand for imposing sufficiently large tensile stresses leading to the formation of inclined tensile radial cracks called "lateral vents" or "radial vents" (Swain & Lawn, 1976; Snowdon et al., 1982). It is worth noticing here that in another technique of rock excavation, namely that of blasting, apart from the much larger strain rates involved compared to disc cutting, the mechanism of rock fragmentation is principally with radial cracks extending around the blasthole which have been initiated by the tail of the traveling shock wave that produces tensile tangential stresses (Kutter & Fairhurst, 1971). This means that in rock destruction processes with cutting tools or blasting both tensile and compressive strengths of the rock should be considered. At a subsequent stage of the cutting process one or more mixed mode (*i.e.* mode-I and mode II) cracks propagate more or less parallel with the free surface toward the neighboring newly formed groove or free surface that finally lead to chip formation that resemble a slab. In this manner one expects that the chips have a width of the order of kerf spacing S, thickness of the order of disc penetration p, and length one to three times the chip width (Nelson, 1993). In some cases of large penetration depths p, the path of the mixed-mode crack is concave upwards leading to "undercutting" (Snowdon et al., 1982). Neighboring kerfs are not loaded at the same time, *i.e.* there is already a kerf at distance S from the neighboring rolling disc in order to facilitate crack propagation toward this kerf and chip formation.

A graph of forces versus time will have a characteristic form showing for a certain cycle a rise to peak, with a sudden drop as chipping occurs. In this manner the force-time diagram for both normal and rolling forces will exhibit a wave-form due to successive loading-unloading cycles like those illustrated in Figures 4a and b. It may be observed that the inherent heterogeneity of rocks is manifested with the fluctuation of the local average force values around the global average corresponding to a certain length traveled by the cutter.

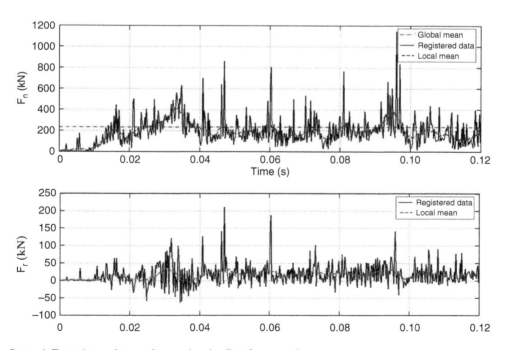

Figure 4 Typical waveforms of normal and rolling forces with.

A rather useful concept in the performance analysis of a TBM (or any other mechanical excavation technique) is the Specific Energy (SE) that has been proposed by Teale (1965) and initially has been used for the analysis of percussive and rotary drilling of rocks but later it has been also applied for boring machines. SE has units of stress, such as [MPa] or [MJ/m^3], has a direct effect on the efficiency of the boring operation, expressed by the PR, and the excavation time and hence on AR and on the cost of excavation.

SE is estimated by two different but practically equivalent methods. In the first method the estimation involves a single cutter and SE is expressed by the ratio of the rolling force F_r over the product of penetration depth per cutter head revolution p by the average spacing of neighboring cuts S, and is denoted here by the symbol SE_1, i.e.

$$SE_1 = \frac{F_r}{S \cdot p} \qquad (8)$$

The sketch of Figure 5 illustrates the basic force and length parameters entering the above equation. This formula is valid since the contribution of the normal force F_n to the specific energy is negligible (Teale, 1964) considering the fact that the cutting path of the discs tangential to the face is significantly larger than the distance perpendicular to the face.

In the second method the estimation of SE is done by dividing the consumed power of the cutting head multiplied by a coefficient of mechanical efficiency, by the Instantaneous Cutting Rate (ICR) which is the excavated volume of rock in the unit of time, and is denoted here by the symbol SE_2

$$SE_2 = \eta \cdot \frac{P}{V/t} \quad \Leftrightarrow \quad ICR = \frac{V}{t} = \frac{\eta \cdot P}{SE_2} \qquad (9)$$

where η is the coefficient of mechanical efficiency that is influenced by other types of actions apart from cutting like cutting on an incline, friction at cutter head and shield, turn it in idle position, lifting of buckets, worn discs, sticky rock etc., P is the consumed power in the time interval t, and V is the volume of rock that is excavated in the same time interval.

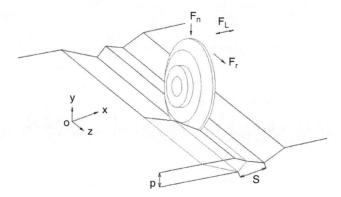

Figure 5 Forces exerted on the rolling disc in an isometric view of the cutting process.

The product of the angular velocity $\omega_{ang} = 2\Pi\omega$ [rad/min] with the torque T [kNm] of the cutting head gives the instantaneous power consumption,

$$P = T \cdot \omega_{ang} \tag{10}$$

The total thrust exerted on the face by the jacks, F_T, can be found by the sum of normal cutting forces of all disc cutters $F_{n,i}$, $i = 1..N_t$, the sliding resistance of the TBM cutterhead and shield casing on the tunnel wall, denoted by F_S, that is the sum of the weight of the machine and the normal force exerted on the shield by the tunnel boundary due to stress relaxation multiplied by a friction coefficient, the gradient resistance denoted by F_G, as well as the force for pulling the back-up F_P. The thrust is the force that should be applied on the face by the thrust cylinders through the cutting wheel in order the cutters to penetrate the rock mass at a depth equal to p and also surpass frictional forces. The total torque T exerted by the cutting head on the rock can be found by the summation of the products of the rolling cutting forcers $F_{r,i}$, $i = 1..N_t$ multiplied by the lever-arms R_i of the cutters with respect to the cutter head axis, the torque of the cutter head to overcome friction T_F, and the torque dedicated to overcome the lifting forces of the material buckets T_B. It is noted here that in EPB (Earth Pressure Balance) machines part of the torque is consumed for the stirring by the rotating cutting head of the grinded material plus additives inside the excavation chamber. This is yet an open problem remained to be analytically solved and is not considered here. Also, for shield tunneling with face support (*e.g.* EPB or slurry shields), part of the total thrust is expended for the provision of the pressure on the face. In summary, these relationships are given below,

$$F_T = \sum_{i=1}^{N_t} F_{n,i} + F_R + F_G + F_P \approx N_t \cdot F_n + F_S + F_G + F_P$$

$$T = \sum_{i=1}^{N_t} F_{r,i} \cdot R_i + T_F + T_B \approx \frac{1}{4\eta} \cdot N_t \cdot D \cdot F_r \tag{11}$$

It may be noted that we consider the torque is given by a single expression by including all the types of action other than cutting into the coefficient of mechanical efficiency η. The approximation in the second of relations of Equation 11 may be deduced by considering that SE_2 can be expressed in the following equivalent form

$$SE_2 = \eta \frac{P}{Vt} = \eta \frac{T\omega_{ang}}{\pi R^2 PR}, \qquad p = \frac{PR}{\omega} \tag{12}$$

Further elaboration on the above formula by setting the relation

$$T = \frac{1}{4\eta} \cdot N_t \cdot D \cdot F_r \tag{13}$$

gives the following result

$$SE_2 = \eta \frac{T\omega_{ang}}{\pi R^2 PR} = \frac{F_r}{pS} = SE_1, \qquad q.e.d. \tag{14}$$

For example by setting a mechanical efficiency of $\eta = 0.8$ the above Equation 13 gives

$$T \approx 0.3 \cdot N_t \cdot D \cdot F_r \tag{15}$$

Provided that we have a model for predicting the cutting forces, then Equations 15 and 10 could be used to predict the required operational parameters of the boring machine such as torque T and cutting head power P under given thrust force F_T and rotational speed ω of the wheel necessary for achieving a certain penetration depth per revolution p of the cutting wheel, and hence penetration rate PR by means of Equation 1.

2.3 Disc cutting models

Most of the proposed models for the prediction of cutting forces are empirical and database oriented. Therefore the predictions of such models are valid only for the range of conditions these databases refer to. In this section we first revisit two models, among many others, that have been proposed for wedge and constant profile cutters, respectively, and finally we create a new model based on the Limit Analysis theory of Perfect Plasticity Theory (Chen, 1975).

2.3.1 Roxborough's model

One of the cutting models that is interesting and is outlined here is that proposed by Roxborough & Phillips (1975) for wedge disc cutters. These investigators proposed that the normal force F_n that should be exerted on the cutter having a 'V-profile' (or wedge-shaped section) with a wedge angle $2a$ in order to penetrate into the rock at a depth p (*i.e.* Figure 6a), is the product of the UCS with the projected contact area A of the disc as is shown in Figure 6b.

The contact area may be found to be with reasonable approximation assuming that its shape is rectangular as follows

$$A \approx 2\,pl\tan\alpha, \quad l = 2\sqrt{2Rp - p^2} \tag{16}$$

where l denotes the chord length as shown in Figure 6b, and R is the radius of the disc. Based on the above argument the normal force required for the full penetration of the cutter at a depth p is given by the following expression

$$F_n = 4UCSp\sqrt{2Rp - p^2}\tan\alpha \tag{17}$$

Further, the same authors assume – based on experimental evidence – that this force remains constant during rolling of the disc as is shown in Figure 6b. It is worth noticing that in the frame of this model, the predicted indentation force on the disc is independent of the spacing of neighboring relieved cuts. In a next step they calculated the Cutting Coefficient (CC) that is defined as the ratio of the rolling force to the normal force. This is an important coefficient since it indicates the amount of torque required for a given amount of thrust according to the TBM model presented previously. For this purpose, they assumed that the resultant force passes is directed along the bisector of the angle 2ψ (*e.g.* Figure 6c) and then they apply the moment

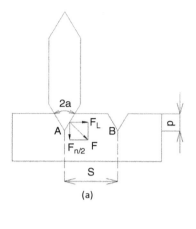

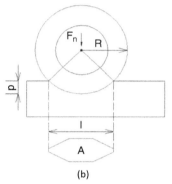

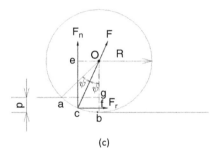

Figure 6 Indentation and rolling of wedge shaped disc cutter; (a) Front view of the wedge shaped disc with normal and lateral forces and geometry, (b) Side view of the cutter, and (c) side view of the disc with normal and rolling forces.

equilibrium equation for pure rolling conditions neglecting possible relative slide of the disc on the rock,

$$F_r \cdot (Of) = F_n \cdot (Oe) \tag{18}$$

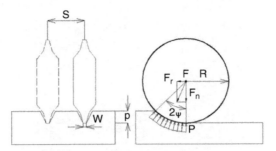

Figure 7 Front view (left) and lateral view (right) of disc cutter with constant section cutting edges.

The above relationship has been derived by inherently assuming that the load distribution along the disc-rock contact (ab) is uniform or symmetrical w.r.t. Oc-axis (*e.g.* Figure 6c). In this manner the rolling force is related with the normal force through the following relationship

$$F_r = F_n \cdot \tan \psi \qquad (19)$$

where 2ψ denotes the contact arc angle that may be easily found to be

$$2\psi = \cos^{-1}\left(\frac{R-p}{R}\right) \qquad (20)$$

2.3.2 Colorado School of Mines (CSM) model

The CSM model refers both to wedge and constant-section disc profiles. Figure 7 shows the front and side views of relieved cutting of constant profile discs.

In the frame of the Colorado School of Mines (CSM) model (Rostami & Ozdemir, 1993; Rostami *et al.*, 1996) the resultant cutting force exerted on the disc is given by the following relation (*e.g.* Figure 7)

$$F = \int dF = \int_0^{2\psi} P \cdot R \cdot w \cdot d\alpha = 2P \cdot R \cdot w \cdot \psi \qquad (21)$$

where w is the width of the tip of the disc, P is the uniform pressure applied along the contact area of the cutting disc, that depends on the Uniaxial Compressive Strength (UCS), the Uniaxial Tensile Strength (UTS) of the rock[1], the geometry of the cutting disc (radius R, and tip width w), and the geometry of the ledge cutting such as the distance from the neighboring cut S, and the cutting depth p through Equation 20. This equation refers to constant cross-section tip of cutters that replaced the discs with wedge tips, for uniform wear for a long period of time.

---

1 The authors who proposed this parameter do not clarify if the tensile strength is derived from uniaxial tensile tests or indirect tensile tests like Brazilian or beam bending.

According to this model the contact pressure is given by the following empirical relation derived from regression analysis of disc cutting data (Rostami *et al.*, 1996)

$$P = C \cdot \sqrt[3]{\frac{S \cdot UTS \cdot UCS^2}{2\psi \cdot \sqrt{R \cdot w}}} \tag{22}$$

where C is a dimensionless constant approximately equal to 2.12. Then for the uniform contact pressure that is most probably occurring in rock cutting, the normal and rolling forces can be determined in the following manner

$$F_n = \int_0^{2\psi} P \cdot R \cdot w \cdot \cos \alpha \cdot d\alpha = P \cdot R \cdot w \cdot \sin 2\psi$$

$$F_r = \int_0^{2\psi} P \cdot R \cdot w \cdot \sin \alpha \cdot d\alpha = P \cdot R \cdot w \cdot (1 - \cos 2\psi) \tag{23}$$

In contrast to the model of Roxborough and Phillips, the CSM model accounts for the effect of spacing S of neighboring cuts. Also the effect of the penetration depth is considered through the contact angle 2ψ as may be seen from Equation 20. CC is then found from Equation 23 as follows

$$CC = \frac{F_r}{F_n} = \tan \psi \tag{24}$$

It could be noted that the cutting coefficient is the same with that predicted by the Roxborough and Phillips model, *e.g.* Equation 19, since it was assumed that the contact pressure is uniformly distributed along the edge of the disc. Gertsch *et al.* (2007) have performed disc cutting experiments using a linear cutting rig and demonstrated that the resultant force exhibits the trend to bisect the contact angle as has been assumed by Roxborough and Phillips.

2.3.3 Limit analysis model of disc cutting

One way to predict the cutting forces exerted to the rock when a cutter acts solely or with a neighboring relief cut (due to action of a cutter before the one we study) is by using the theorems of limit analysis (Chen, 1975). One issue is the failure model of the rock. The simplest model is the linear Mohr-Coulomb but overestimates tensile strength of the rock mass. In this case one should use a modified Mohr-Coulomb failure model with arbitrarily small tensile strength.

Chen & Drucker (1969) solved analytically by applying the lower and the upper bound theorems of Limit Analysis the problem of the axial splitting of a block under the action of a punch. The punch with flat end penetrates the block creating a wedge and a median or tensile crack at the tip of the wedge that splits the block in two halves. Subsequently under the action of the wedge-shaped block, the two newly formed blocks displace in opposite directions along the horizontal frictionless floor. In the case of cutting we may assume that the block's lower horizontal boundary is formed by the propagation of two lateral horizontal mode I cracks in the unloading phase as was explained previously. If it is assumed that the rock is unable to undertake any tension and the tool-rock interface obeys a straight-line Mohr-Coulomb yield model with cohesion and internal friction angle c, ϕ, respectively, as is shown in Figure 8a. Further only rigid body discontinuous velocity conditions along cracks are considered. In this manner the rate of dissipation per unit area of discontinuity or crack surface is given by the formula (Drucker & Prager, 1952)

$$D_A = c\delta u \frac{\tan\left(45 + \dfrac{\phi}{2}\right)}{\tan\left(45 + \dfrac{\theta}{2}\right)} \tag{25}$$

where δu denotes the jump in the tangential velocity across the discontinuity, with the relative velocity vector δw forming an angle $\theta \geq \phi$ with the surface of the discontinuity, as is illustrated in Figure 8b. Considering the wedge penetrating the rock as is shown in Figure 8c under the action of load Q_u and equating the rate of the external and internal work dissipated along the two sides of the wedge it is found

$$q_u = c\frac{\cos\phi}{\cos(\phi + \alpha)\sin\alpha} \tag{26}$$

in which q_u denotes an upper bound of the average indentation pressure, and no energy is dissipated along the tensile cracks, *i.e.* along the median vent (vertical) and the two lateral horizontal cracks. Since

$$UCS = 2R_o = 2c\tan\left(\frac{\pi}{4} + \frac{\phi}{2}\right) \tag{27}$$

then Equation 26 may be written in the following form

$$q_u = \frac{UCS}{2\tan\left(\dfrac{\pi}{4} + \dfrac{\phi}{2}\right)} \frac{\cos\phi}{\cos(\phi + \alpha)\sin\alpha} \tag{28}$$

The application of lower bound theorem that satisfies loading and boundary conditions gives an approximate lower bound indentation pressure for obtuse wedge angles that is close to the UCS (*e.g.* Figure 8d)

$$q_l \approx UCS \tag{29}$$

This means that for the wedge shaped roller disc the indentation pressure is bracketed as follows

$$UCS \leq q \leq q_u \tag{30}$$

Figure 9 graphically presents the dependence of the lower and upper bound indentation pressures on various possible friction angles of the rock and wedge angles considered by Roxborough & Phillips (1975). It may be seen that most of the predictions of the

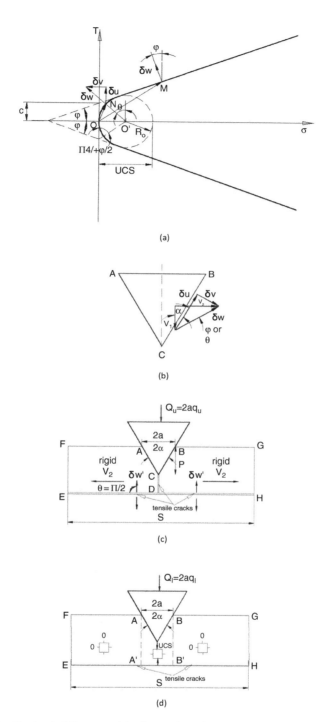

Figure 8 (a) Mohr-Coulomb failure model of a rock with zero tensile strength and obeying the normality condition (Chen, 1975), (b) velocity diagram, (c) upper bound pressure on rigid wedge shaped punch penetrating rock by axial splitting, and slabbing due to two horizontal mode I cracks, and (d) lower bound pressure on wedge penetrating the block.

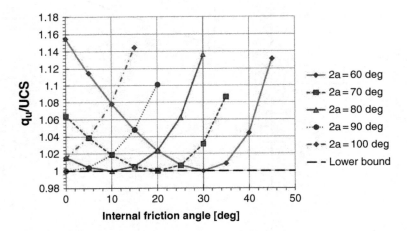

Figure 9 Dependence of the upper and lower bound disc indentation dimensionless pressures on the interfacial friction angle for various wedge subtended angles.

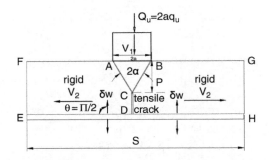

Figure 10 Flat punch penetrating a finite block with width S at depth p formed by the median crack and two lateral horizontal mode I (tensile) cracks dissipating no work.

Limit theorems presented in this figure justify the ad hoc assumption made by the authors, namely that the indentation pressure is equal to the UCS of the rock.

The punch with flat end originally considered by Chen & Drucker (1969) is shown in Figure 10.

In this case the angle subtended by the two shear cracks AB and BC formed under the flat punch is initially unknown. However, minimizing the previous Equation 28 with respect to the angle α is derived

$$\alpha = \frac{\pi}{4} - \frac{\phi}{2} \tag{31}$$

And finally one derives for the case of zero tensile strength rock

$$q_z^u = c \tan\left(\frac{\pi}{4} + \frac{\phi}{2}\right) = UCS \tag{32}$$

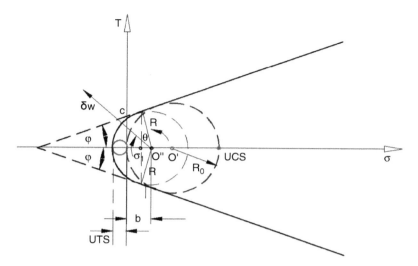

Figure 11 Mohr-Coulomb failure model of a rock possessing finite tensile strength UTS.

This value coincides with that predicted by the lower bound theorem, hence in the case of the flat punch the indentation pressure is the UCS of the rock. Along the way to compose a more general model we consider the more realistic case of rock possessing a tensile strength. In this case we follow the suggestion of Chen & Drucker (1969) of the modification of the Mohr-Coulomb model to account for the finite tensile strength of rock as is illustrated in Figure 11. It is remarked here that of great interest for efficient cutting of rock with discs, are the cracks which propagate in a lateral direction to meet the adjacent pre-existing groove or grooves. In the frame of this model the mathematical expression for the rock failure is discontinuous and is described by

$$\begin{aligned} \tau &= c + \sigma \tan\phi, \quad \sigma \geq \sigma_1 \\ (b-\sigma)^2 + \tau^2 &= R^2, \quad \sigma \leq \sigma_1 \end{aligned} \tag{33}$$

where we have set (tensile stresses are considered negative quantities),

$$R = \frac{1}{2}UCS + \frac{UTS\sin\phi}{1-\sin\phi},$$
$$b = R + UTS,$$
$$\sigma_1 = \frac{-(c\tan\phi - b) + \sqrt{(c\tan\phi - b)^2 - (1+\tan^2\phi)(b^2 + c^2 - R^2)}}{(1+\tan^2\phi)}, \tag{34}$$
$$c = \frac{UCS \cdot (1-\sin\phi)}{2 \cdot \cos\phi}$$

Also it can be shown that the energy dissipated along the cracks is given by the expression (Chen, 1975)

$$D_A = \delta w \left(UCS \frac{1-\sin\theta}{2} + UTS \frac{\sin\theta - \sin\phi}{1-\sin\phi} \right); \tag{35}$$

$$\tan\theta = \frac{\delta v}{\delta u} \geq \tan\phi$$

Figure 12 shows a two dimensional section of a penetrating wedge disc that is assumed here. The wedge moves downward into the rock mass as a rigid body and displaces the surrounding rock sideways that slides along gently inclined discontinuities CD and CG. The plane sliding along these inclined surfaces involves both shearing and separation so that Equation 35 can be used to calculate the rate of energy dissipation. For the penetration of the wedge the rock mass shear strength should be reached along the interfaces AC and BC under conditions of large mean stress. The total dissipation of energy in the block can then be found by adding the rates of dissipation at the discontinuity surfaces AB, BC, CD and CG. The free surfaces ED and FG represent pre-existing grooves (relieved cuts) or they end at the tips of neighboring radial cracks from adjacent discs. Equating the rate of work performed by the normal force on the it F_n it is found that an upper bound of this force is given by the expression

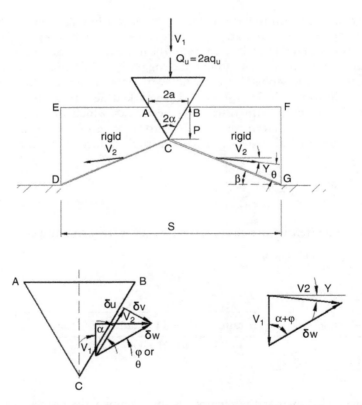

Figure 12 Mohr-Coulomb failure model of a rock possessing finite tensile strength UTS; (a) Penetration mechanism, and (b) hodographs.

Characterization of rock masses based on geostatistical joint mapping   157

$$F_n = Q_u = \frac{2 \cdot UCS \cdot p \cdot S \cdot \tan\alpha}{\cos (\alpha + \phi - \gamma)} \left\{ \left(\frac{1 - \sin \theta}{2} + \frac{UTS}{UCS} \frac{\sin \theta - \sin \phi}{1 - \sin \phi} \right) \times \right.$$

$$\left. \times \frac{\sin (\alpha + \phi)}{\cos \beta} + \frac{p}{S} \frac{\cos \gamma \cdot (1 - \sin \phi)}{\cos \alpha} \right\};$$

$$\theta = \beta - \gamma \tag{36}$$

For the derivation of the above expressions it was assumed that the relative velocity vector δw along the disc-rock interfaces AC and BC is inclined at an angle ϕ so the work dissipated there is given by Equation 36 for $\theta = \phi$. For the case of the constant section disc with tip width w the same Equation 36 is still valid provided that the relationship below is used

$$\alpha = \tan^{-1} \left(\frac{w}{2p} \right) \tag{37}$$

where w denotes the width of the cutting tip (e.g. Figure 7). In this case the above Equation 36 takes the form

$$F_n = Q_u = \frac{UCS \cdot w \cdot S}{\cos (\alpha + \phi - \gamma)} \left\{ \left(\frac{1 - \sin \theta}{2} + \frac{UTS}{UCS} \frac{\sin \theta - \sin \phi}{1 - \sin \phi} \right) \times \right.$$

$$\left. \times \frac{\sin (\alpha + \phi)}{\cos \beta} + \frac{p}{S} \frac{\cos \gamma \cdot (1 - \sin \phi)}{\cos \alpha} \right\} \tag{38}$$

Equation 36 has been checked against experimental linear disc cutting rig data on Plas Gwilym limestone that have been presented by Snowdon et al. (1982). In these tests a wedge disc with angle of $2\alpha = 80°$ and diameter of 200 mm has been employed for all tests. The reported mechanical properties of the limestone were UCS=155 MPa and UTS=13.72 MPa, while the internal friction angle has not been given. Here it is assumed to be $\phi = 35°$ for the limit analysis model that is reasonable for calcitic rocks. Then the only remaining parameters to be found are the angles θ and γ. The latter angle is found by the minimization of the value of the force, while the former is found by matching the minimum value of the force with that recorded during the experiment. We have used the disc cutting data with constant penetration depth p = 4 mm and S/p = 6.25, 12.5, 18.75 and 25, respectively. Calibrating the model to match experimental data for each S/p ratio, we have found that the angle θ reduces from $65°$ (indicating mixed tensile and shear crack propagation) to $35°$ (indicating pure shear crack propagation) as the ratio S/p increases, as is illustrated in Figure 13. In the same figure the prediction of the Roxborough and Phillips model (F_n = 40.8 kN) has been plotted as a continuous line since it does not depend neither on spacing, penetration depth or the inclination angle γ employed in the limit analysis model.

Hence, following first principles of Limit Analysis a simple model for relieved disc cutting has been composed, that contains fundamental rock parameters like the UCS, the ratio UTS./UCS that is a measure of rock brittleness, and the friction angle of rocks that manifests itself after the fracturing, as well as the basic parameters of the disc cutting process like the penetration depth, the width or angle of the tip of the cutter and

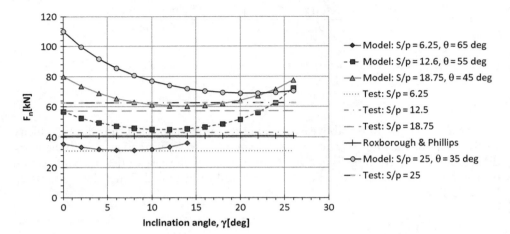

Figure 13 Comparison of model predictions with test results of linear cutting rig tests on Plas Gwilym limestone.

the spacing to penetration depth ratio. A more elaborate model could be found based on a more complicated fracture mechanism, but at this stage we want to retain the simplicity of a model that contains the basic features of the rock, as well as disc design and arrangement on the cutting wheel.

3 ESTIMATION OF JOINTED ROCK MASS STRENGTH PARAMETERS USING THE DAMAGE MECHANICS APPROACH

3.1 Introduction

It may be said that the rock mass characteristics probably affect much more the performance of a boring machine than the laboratory parameters of the intact rock (Nelson, 1993). Despite of this fact, much of the efforts are aiming towards characterizing the rock mass for rock support calculations rather than for the performance of tunnel boring driving or rotarry-percussive drilling in rocks.

The effects of joints on rock mechanics properties include increased deformability, decreased strength, increased permeability due to dilatancy and induced anisotropy, among other. The upscaling problem common to many aspects of rock mechanics is translating knowledge of microcracks to knowledge of fractured rock mass behavior. Currently, it is reasonably feasible to study microfracture orientation, frequency, and permeability characteristics in the laboratory, but this is not yet possible for the field scale, which is the scale of most interest.

Other possible approaches that one could employ to derive a model at the scale of the discretization element of the numerical model, called herein the "macroscale", are Linear Elastic Fracture Mechanics (LEFM) and Continuum Damage Mechanics (CDM). Although it may be shown that the two theories are equivalent, the most appropriate to start with for practical purposes, is the latter, since the LEFM approach, each crack should be considered explicitly, there is nonlinearity induced by the contact of joint lips, and so on.

In this paragraph we aim at a robust methodology to deteriorate the strength parameters of the intact rock due to the presence of cracks, assuming that the rock obeys the linear Mohr-Coulomb model with tension cut-off as has been presented in the previous paragraph. The strength rock parameters to be reduced are three for this model, *i.e.* the UTS, UCS and the internal friction angle ϕ.

3.2 Estimation of rock mass model parameters

A possible upscaling methodology may be founded on two basic hypotheses, namely:
Hypothesis A: In a first approximation upscaling may be based on the scalar damage parameter D or vector damage parameter $\mathbf{D} = D \cdot \mathbf{n}$ *for the anisotropic case of joint induced anisotropy of the rock mass (n is the unit normal vector of the plane of interest).*

Hypothesis B: A universal relationship between the damage parameter D and RMR (or equivalently with Q or GSI rock quality indices) exists for all rock masses.

Regarding Hypothesis A above, it is noted that joints may also alter the constitutive law of the rock during the transition from the lab scale to the scale of the discretization element of the numerical model, *i.e.* they may induce anisotropy or they may induce nonlinearity and/or creep. Let us assume for the sake of simplicity that there are more than three joint systems in the rock mass so that it behaves like an isotropic material (the case of anisotropic geomaterial may be easily considered through appropriate tensorial analysis). If the area δA with outward unit normal n_j of the Representative Elementary volume (REV) with position vector x_i of Figure 14 is loaded by a force δF_i the usual apparent traction vector $\sigma_i = \sigma_{ij} n_j$ is

$$\sigma_i = \lim_{\delta A \to A} \frac{\delta F_i}{\delta A}, \quad i = 1, 2, 3 \tag{39}$$

where A is the representative area of the REV.

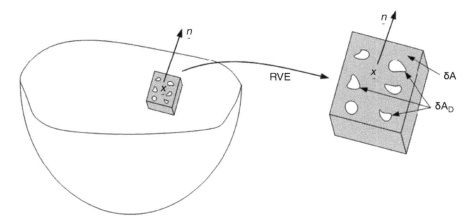

Figure 14 Representative Elementary Volume (REV) of damaged rock due to rock mass discontinuties (joints, fractures, cracks etc).

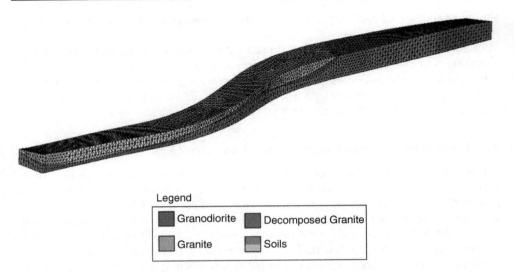

Figure 15 Discretized geological model on a mesh with tetrahedral elements and lithological data assigned to grid points.

The REV could be the finite element of the discretized geological model such as that shown in Figure 15. Each tetrahedral element in this 3D model in has assigned a lithological unit that has been previously characterized in the lab from core drilling logs and from the geological conceptual model based on a map in a 1:1000 or 1:500 scale created in the pre-design or design phase of the project. The interpolation of lithological units or stratigraphical surfaces between the boreholes could be performed by the Block Kriging method of Geostatistics as will be shown in the next Section 4.

The value of the dimensionless scalar damage parameter $D(n_i, x_i)$ (that is a function of orientation and position of the surface) may be defined as follows

$$D = \frac{\delta A_D}{\delta A} \qquad (40)$$

where δA_D is the total area occupied by the joints. In a similar fashion as it is displayed in Figure 11, the vector damage parameter may be referred to the directed area $\delta A\, n$. At this point we may introduce an 'effective traction vector' $\widetilde{\sigma}_i$ that is related to the surface that effectively resists the load, namely

$$\widetilde{\sigma}_i = \lim_{\delta A \to S} \frac{\delta F_i}{\delta A - \delta A_D}, \quad i = 1, 2, 3 \qquad (41)$$

From Equations 39–41 it follows that

$$\widetilde{\sigma}_i = \frac{\sigma_i}{1 - D}, \quad i = 1, 2, 3 \qquad (42)$$

According to the above definitions the elastic deformation of the intact rock can be described with the following relations:

Characterization of rock masses based on geostatistical joint mapping 161

- The relation $\widetilde{\sigma}_{ij} \to \varepsilon_{ij}^{(el)}$ which is obtained from elasticity, and
- the relation $\sigma_i \to \widetilde{\sigma}_i$ which is obtained by employing the concept of damage (Kachanov, 1986; Krajcinovic, 1989; Lemaitre, 1992), wherein the superscript (el) denotes elastic strains. It is remarked here that the damage parameter D should be multiplied with the joint closure factor $D_c\,(0 \le D_c \le 1)$ in order to take into account partial contact of joint lips in compressive loading normal to the plane of the joint.

Exadaktylos & Stavropoulou (2008) have shown how this closure factor is related to the initial normal stiffness of the joint, the relative joint closure w.r.t. the maximum joint closure and the Young's modulus of the intact rock. It was also demonstrated that D_c is close to unity for low compressive stress regimes or for low initial normal stiffness of the joints; therefore from hereafter we consider $D_c \cong 1$.

In thermodynamics, the axiom of the 'local state' assumes that the thermomechanical state at a point is completely defined by the time values of a set of continuous state variables depending upon the point considered. This postulate applied at the mesoscale imposes that the constitutive equations for the strain of a microvolume element are not modified by a neighboring microvolume element containing a microcrack or inclusion. Extrapolating to the macroscale, this means that the constitutive equations for the strain written for the surface $\delta A - \delta A_D$ are not modified by the damage or that the true stress loading on the rock is the effective stress $\widetilde{\sigma}_i$ and no longer the stress σ_i. The above considerations lead to the 'Strain Equivalence Principle' (Lemaitre, 1992), namely: *'Any strain constitutive equation for a damaged geomaterial may be derived in the same way for an intact geomaterial except that the usual stress is replaced by the effective stress'.*

The above principle may be applied directly on the isotropic rigid perfectly plastic geomaterial obeying the linear Mohr-Coulomb yield criterion with tension cut-off,

$$\begin{aligned}
\widetilde{\tau} &= c + \widetilde{\sigma} \tan\phi, \quad \widetilde{\sigma} \ge \sigma_1 \\
(\alpha - \widetilde{\sigma})^2 + \widetilde{\tau}^2 &= R^2, \quad \widetilde{\sigma} \le \sigma_1
\end{aligned} \tag{43}$$

In the presence of damage, assuming that it is constant, the coupling between the damage and the plastic strain is written in accordance with the above principle, that is to say the yield function is written in the same way as for the non-damaged material except that the stress is replaced by the effective stress according to Equation 42, that is to say

$$\left. \begin{aligned}
\frac{\tau}{1-D} &= c + \frac{\sigma}{1-D}\tan\phi, \quad \frac{\sigma}{1-D} \ge \sigma_1 \\
\left(b - \frac{\sigma}{1-D}\right)^2 + \frac{\tau^2}{(1-D)^2} &= R^2, \quad \frac{\sigma}{1-D} \le \sigma_1
\end{aligned} \right\} \tag{44}$$

Hence, the criterion for the rock mass takes the form

$$\left. \begin{aligned}
\tau &= (1-D)c + \sigma\tan\phi, \quad \sigma \ge (1-D)\sigma_1 \\
[(1-D)b - \sigma]^2 + \tau^2 &= [(1-D)R]^2, \quad \sigma \le (1-D)\sigma_1
\end{aligned} \right\} \tag{45}$$

Finally, in a more compact form the criterion takes the form

$$\left.\begin{array}{l}\tau = c_m + \sigma \tan \phi, \quad \sigma \geq \sigma_{1m} \\ [b_m - \sigma]^2 + \tau^2 = R_m^2, \quad \sigma \leq \sigma_{1m}\end{array}\right\} \quad (46)$$

wherein with the subscript 'm' are denoted the rock mass parameters, *i.e.*

$$R_m = \frac{1}{2} UCS_m + \frac{UTS_m \sin \phi}{1 - \sin \phi},$$

$$b_m = R_m + UTS_m,$$

$$\sigma_{1m} = \frac{-(c_m \tan\phi - b_m) + \sqrt{(c_m \tan\phi - b_m)^2 - (1 + \tan^2\phi)(b_m^2 + c_m^2 - R_m^2)}}{(1 + \tan^2\phi)},$$

$$c_m = \frac{UCS_m \cdot (1 - \sin \phi)}{2 \cdot \cos \phi}$$

(47)

and

$$\begin{array}{l}UCS_m = UCS \cdot (1 - D), \\ UTS_m = UTS \cdot (1 - D)\end{array} \quad (48)$$

It turns out that the internal friction angle of the equivalent rock mass is not affected by the presence of the joints. Provided that the joint walls are not weathered significantly, then this assumption is valid. Otherwise, the friction angle should be modified properly to account for weathering of joint lips.

Figure 16 shows how the failure line in the $\sigma - \tau$ plane is affected by the inherited (previous) damage of the rock mass. It may be observed that the above arguments from Damage Mechanics lead to a *self-similar* yield surface for the rock mass.

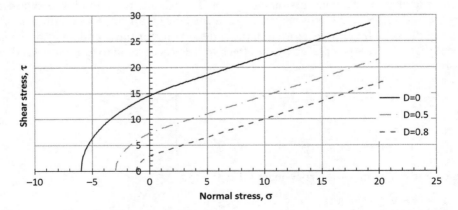

Figure 16 Schematic representation of the linear Mohr-Coulomb criterion with tension cut-off applied to damaged geomaterial in the $\sigma - \tau$ plane (compressive stresses are taken as positive quantities here) where the intact rock parameters are UCS = 28.8 MPa, UTS = −3 MPa, φ = 35°.

Hypothesis B mentioned at the outset, is based on the fact that RMR, Q or GSI indices do take into account explicitly the most important features of the rock mass joints that are responsible for the deterioration of its parameters compared to those the intact state; hence they may be linked to the damage parameter D with an appropriate relationship.

Such a function must have a sigmoidal shape resembling a cumulative probability density function giving D in the range of 0 to 1 for RMR or GSI varying between 100 to 0 or for Q varying from 1000 to 0.001, respectively. On the grounds of the above considerations, the Lorentzian cumulative density function (obtained directly from the Cauchy probability density function) has been proposed (Exadaktylos & Stavropoulou, 2008), namely

$$D = 1 - \left\{ \hat{a} + \frac{\hat{b}}{\pi} \left[\tan^{-1} \left(\frac{RMR - \hat{c}}{\hat{d}} \right) + \frac{\pi}{2} \right] \right\} \tag{49}$$

where $\hat{c}$ denotes the location parameter, specifying the location of the peak of the distribution, $\hat{d}$ is the scale parameter which specifies the half-width at half-maximum, respectively, $\hat{b}$ is a proportionality constant and $\hat{a}$ can be found in terms of the other three parameters by setting D = 1 for RMR = 100. It is noticed that Equation 49 is not used here as a statistical function but merely as a deterministic function. In this formula the RMR does not include the correction term due to unfavorable tunnel orientation with respect to joints and the ground water (hence RMR considers only the joints through the mean block size and joint shear strength, *i.e.* RMR_{89}). The three unknown parameters of this function $\hat{b}, \hat{c}, \hat{d}$ were calibrated on the empirical relationship proposed by Hoek & Brown (1997) for the dependence of $UCS_m/UCS = 1 - D$ on RMR, with UCS_m being the rock mass UCS, namely

$$\frac{UCS_m}{UCS} = \sqrt{e^{\frac{(RMR-5)-100}{9}}} = \sqrt{e^{\frac{GSI-100}{9}}} \tag{50}$$

Based on this best-fit procedure, performed for RMR<80, the following values for the unknown parameters were found, namely:

$$\hat{c} = 72.9968, \ \hat{d} = 12.9828, \ \hat{b} = 1.2377, \ \hat{a} = -0.06111 \tag{51}$$

The above empirical relationship of the damage parameter with GSI substituted by RMR_{89} is plotted in Figure 17a. Furthermore, D as it given by Equations 49 and 51 can be expressed as a function of Q index by virtue of the following empirical relationship linking the latter index with RMR

$$Q \approx 10^{\frac{RMR-50}{15}} \tag{52}$$

The graphical representation of the dependence of inherited rock damage on the value of Q is presented in Figure 17b. It may be noted that the above linking of damage with rock mass indices RMR and Q permits the use of geophysical measurements (*e.g.* seismic P- or S-waves) for the direct determination of damage based on the following damage mechanics formula

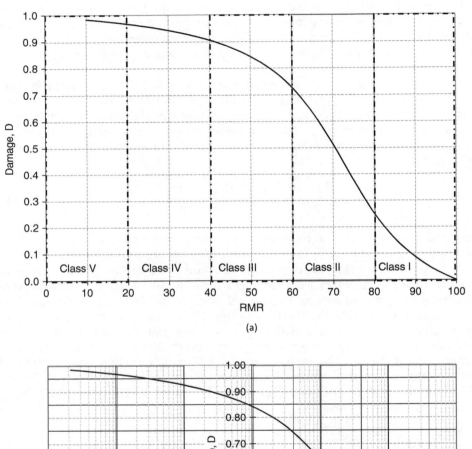

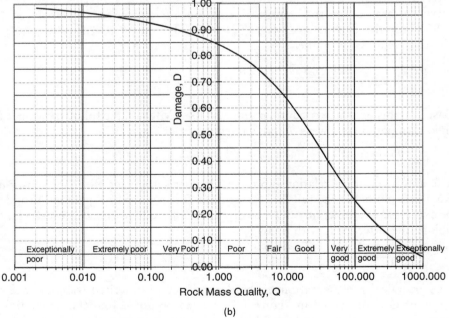

Figure 17 Dependence of damage (D) on (a) RMR and (b) Q rock quality index.

$$D = \frac{G_m}{G} = \frac{E_m}{E} = 1 - \frac{\tilde{V}_S^2}{V_S^2} = 1 - \frac{\tilde{V}_P^2}{V_P^2} \qquad (53)$$

where G_m denotes the shear modulus of the rock mass, G the shear modulus of the intact rock, V_S and V_P denote the shear and compressional wave velocities, respectively, measured on intact rock cores in the laboratory, while the respective symbols with curly overbars indicate the measured velocities in the field.

4 GEOSTATISTICAL CHARACTERIZATION OF ROCK MASSES FOR MINING AND TUNNELING DESIGN

Herein, in a first stage an attempt is made for the transformation of the conceptual qualitative geological model into a '3D ground model'. This model then serves as the input for subsequent analyses of TBM performance and numerical models of rock mass and support responses to altered ground conditions due to the excavation. This transformation is achieved by virtue of CAD techniques, cutting models for rocks and the RMR or Q or other appropriate rock mass classification scheme, as well as on the concept of damage mentioned previously. However, rock masses are heterogeneous continuous or discontinuous media. Heterogeneity is indirectly described by means of the Geostatistical approach that is employed for the interpolation of field parameters registered at certain locations to the grid of the discretized model instead of using average values. These concepts that form essential parts of an integrated design approach for a tunnel project are best described in Figure 18.

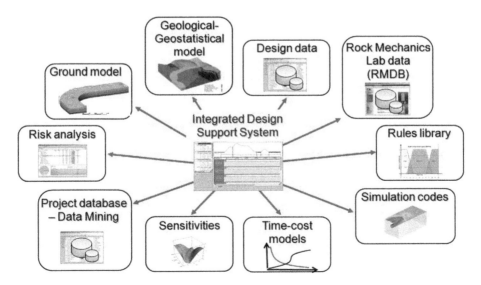

Figure 18 Main components or tools for the integrated design of underground excavations (Meschke et al., 2008).

4.1 Fundamental theoretical considerations of the Kriging approach

The objective of Geostatistics is the analysis of spatial data in contrast to classical descriptive Statistics that pertains to the extraction of pertinent information from large data sets regardless their spatial position. That is to say, each data value is associated with a location in space and there is at least an implied connection between the location and the data value. The main principle behind Geostatistics is that regions close together – *provided that they belong to the same lithological unit* – are more likely to have similar values than regions further apart. We can use geostatistical tools for the exploration and statistical characterization of sample point data. Hence the following two hypotheses that were assumed to be valid:

Hypothesis #1: RMR or Q classification index for rock masses and recorded Specific Energy (SE) during TBM advancement are spatial attributes that can be modeled as random functions. It may be also recalled here that a random field or otherwise a stochastic process denoted as $X(S)$, wherein S denotes the position vector in space, is the rule to correspond to every outcome ζ of an experiment a function $X(\underset{\sim}{S}, \underset{\sim}{\zeta})$.

Hypothesis #2: After the removal of the possible trend of data, the mean of RMR, Q, or SE random functions in the neighborhood of the estimation is constant for the same geological formation but unspecified, and the two-point mean square difference depends only on the distance between two locations and possibly on the orientation (anisotropy effect). This is the so-called 'intrinsic isotropic or anisotropic concept' of Geostatistics (Kitanidis, 1997).

Kriging is a linear interpolation method using fundamental geostatistical concepts that was developed by the South African mining engineer D. G. Krige (from whom the name Kriging was derived). Krige's main motivation was to develop optimal methods of interpolation for use in the mining industry. Kriging makes good use of geostatistical tools helping one to solve questions that didn't have a clear answer with more general interpolation methods. With Kriging, it is possible to define the best domain in which to interpolate (*i.e.* what is the extent to which we should consider data to get an optimal interpolation at a given point), it defines the shape and orientation for optimal interpolation, estimates the weight λ_i of each sampling point in a more thoughtful way than a mere function of Euclidean distance, and also makes possible to estimate the errors associated with each interpolated value.

Kriging is called an optimal interpolation method because the interpolation weights λ_i are chosen to provide for the value at a given point the Best Linear Unbiased Estimate (BLUE) based on the following two conditions:

a) Unbiased or null mean values of estimation error, *i.e.*

$$E[z^*(\mathbf{s}_o) - z(\mathbf{s}_o)] = 0 \tag{54}$$

b) Minimization of the mean squared estimation error, that is

$$\min E\left[\left(z^*(\mathbf{s}_o) - z(\mathbf{s}_o)\right)^2\right] = 0 \tag{55}$$

where $\mathbf{s}_0$ denotes the position vector of the estimation. The above conditions are known as the 'universality conditions' or 'unbiasedness conditions'.

Kriging rests on the following simple linear interpolation formula

$$z^*(x_p) = \sum_{i=1}^{n} \lambda_i z(x_i); \quad \sum_{i=1}^{n} \lambda_i = 1 \tag{56}$$

where $z(x_i)$ represents the known value of variable z at point x_i, and n represents the total number of measurements (hard data) used in the interpolation. The extra equation appearing in the right of Equation 56 is obtained from the satisfaction of the condition (a) or Equation 54.

Regionalized variable theory, in which Kriging has its basis, lies on the 'intrinsic hypothesis' (*i.e.* Hypothesis # 2), namely that the mean is constant and the two-point covariance function C depends only on the distance between two points, *i.e.*

$$E[z] = constant \quad \wedge \quad C[z(x), z(x+r)] = C[z(x+h), z(x+r+h)] \quad \forall \ x_a \tag{57}$$

which means that the spatial variation of any variable can be expressed as the sum of three major components, a deterministic variation, a spatially autocorrelated variation and an uncorrelated noise, respectively. So the value at a point is given by

$$z\left(\underset{\sim}{x}\right) = m\left(\underset{\sim}{x}\right) + \varepsilon'\left(\underset{\sim}{x}\right) + \varepsilon''\left(\underset{\sim}{x}\right) \tag{58}$$

where $\varepsilon'(x\sim) = \gamma(h)$, also called the semivariance or semivariogram function. The variance of differences depends only on the distance between measurements, h, so that we can calculate

$$E\left[\left\{z\left(\underset{\sim}{x}\right) - z\left(\underset{\sim}{x} + \underset{\sim}{h}\right)\right\}^2\right] = E\left[\left\{\varepsilon'\left(\underset{\sim}{x}\right) - \varepsilon'\left(\underset{\sim}{x} + \underset{\sim}{h}\right)\right\}^2\right] = 2\gamma\left(\underset{\sim}{h}\right) \tag{59}$$

Also, we can calculate the semivariance – or the experimental semivariogram - from the point data

$$\gamma\left(\underset{\sim}{h}\right) = \frac{1}{2m} \sum_{i=1}^{m} \left[z(x_i) - z\left(x_i + \underset{\sim}{h}\right)\right]^2 \tag{60}$$

where m is the number of pairs of sample points of observations of the values of attribute z separated by distance (or lag) h. When the nugget variance ε'' is too high and the experimental variogram does not diminish when $h \rightarrow 0$, then the data is too noisy and interpolation is not sensible. Possibly there aren't enough sample points. Kriging is an exact interpolator in the sense that the interpolated values, or best local average, will coincide with the values at the data points. Together with the interpolation at a certain location the associated Kriging error $z(x) - z^*(x)$ may be also computed. In this manner we obtain a sense of the amount of uncertainty associated with this prediction that may be useful for estimation of risks of some operation related with the variable at hand or for the estimation of additional effort (*i.e.* additional measurements) for the improvement of the final model.

The tool that is employed in a Kriging procedure for both the minimization of the estimation error and the interpolation is the theoretical semivariogram that is best-fitted to the experimental one that reflects the spatial correlation of a certain regionalized variable. For the posterior validation of the estimations there are used the statistical variables of the normalized errors Q_1-Q_2 or other procedures such as leave-one-out or cross-validation schemes (Journel & Huijbregts, 1978). In the case of Ordinary Kriging (OK), where the mean value of the field variable is unknown but remains constant in the neighborhood of the search, the weight values λ_i are found with the solution of the following linear system of n + 1 equations with n + 1 unknowns that was derived from condition (b) above (Chiles & Delfiner, 1999)

$$\sum_{j=1}^{n} \lambda_i \cdot \gamma_x(\| \, s_i - s_j \, \|) + L = \gamma_x(\| \, s_i - s_o \, \|) \tag{61}$$

where L denotes the Lagrange multiplier, $s_i, s_j, i, j = 1...n$ are the sampling locations and s_o the location where the estimation is being made, and the left-hand-side is the OK variance that is given by the relation

$$\sigma_{OK}^2(s_o) = E\left[\left(z^*(s_o) - z(s_o)\right)^2\right] \tag{62}$$

Such a geostatistical algorithm (Figure 19) has been developed and verified in several TBM tunnel applications (Exadaktylos *et al.*, 2008; Exadaktylos & Stavropoulou, 2008; Stavropoulou *et al.*, 2010).

4.2 Case study of TBM driving

The above concepts developed so far are subsequently applied for the analysis of TBM data gathered during a twin-tunnel project that was a part of the KCRC West Rail Project linking Mai Foo Station and Tsuen Wan West station in Hong Kong. Because of the **mixed ground conditions** that were encountered along the route of the excavation, as it may be seen in Table 1 and also shown with the geological model presented in Figure 15, the tunnel boring machine worked in EPB (earth pressure balanced) mode when operating in soft ground (highly fractured rock or soil). The pressure was applied in the sealed excavation chamber at the head of the machine to support the tunnel face, and spoil was carried away from the face using a screw conveyor. When operating in mixed face or, the TBM worked either in EPB mode, or in compressed air mode. TBM driving in competent rock mass was done in the open mode whereas in regions of low RMR it worked in the compressed air mode. Table 2 illustrates the operating modes of the TBM along the chainage of the tube 1.

The tunnel construction schedule was first the excavation of Tube 1 from April 2000 to December 2000, and then the excavation of Tube 2 from March 2001 to July 2001. When excavation of the first tunnel was completed, the machine had to be transported back to the starting shaft for excavation of the second tunnel. The shield was removed and broken down into smaller pieces for transportation by road, while the back-up cars were taken back through the tunnel.

The main features of the double shield TBM and a photo of the cutter head are shown in Table 3 and Figure 3, respectively. Pre-cast concrete tunnel lining segments of 1.8 m length and thickness of 400 mm support the excavated area behind the TBM. Erection of the segments follows immediately the tail of the TBM. When excavating, the thrust required for excavation is provided by jacks at the tail of the TBM which push against the lining segments. The chosen TBM was able to excavate on one hand rock with average UCS of 80 MPa up to 250 MPa, and on the other hand, soil materials composed of CDG *i.e.* clay with sand and marine alluvium.

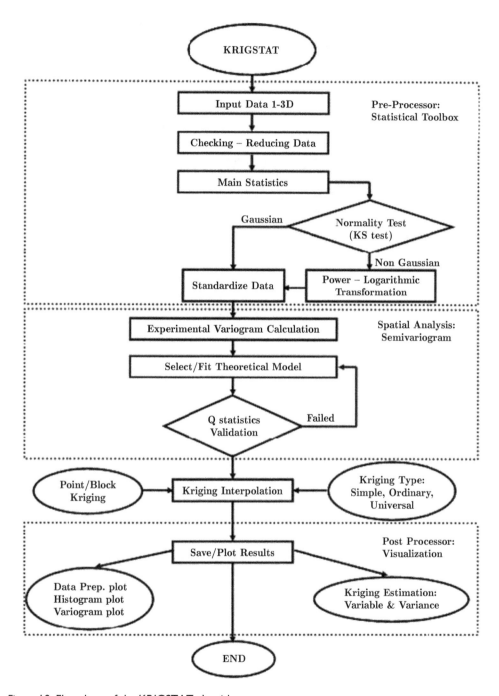

Figure 19 Flowchart of the KRIGSTAT algorithm.

Table 1 Main geological profile (courtesy BOUYGUES TRAVAUX PUBLICS).

Section length	Rock types	Cover
200 m	Hard granodiorite rock	Up to 30 m
400 m	Fair strength granite	Up to 50 m
700 m	Highly fractured and faulted granite rock	Up to 50 m
200 m	Mixed face (rock/soil)	15 to 25 m
300 m	Soft soil made up of decomposed granite, alluvium and marine deposit	15 m

Table 2 Operating modes of the TBM along the tunnel (courtesy of BOUYGUES TRAVAUX PUBLICS).

Chainage start	Chainage stop	Operating mode
0	600	Open
600	800	Compressed air
800	1260	Open
1260	1300	Compressed air
1300	1341	EPB
1341	1420	Compressed air
1420	1820	EPB

Table 3 TBM main features (courtesy of Michel de Broissia of BOUYGUES TRAVAUX PUBLICS).

General characteristics of the TBM	Excavation diameter: 8.75 m, Minimum tunnel radius: 400 m, Max tunnel slope: 4%, Total length: 108 m, Total weight: 1400 t, Spoil transport: 1 m wide conveyor belt
Shield in 3 parts:	Total weight: 860 t
Cutter head (CH)	CH drive 9 motorsx240 kW/motor, 0 to 3 rpm, 61 Disc Cutters (18 to 19"), Thrust: 0 to 5200 t, 13 pairs of thrust rams
	Excavation speed: 0 to 80 mm/min
Center shield Articulated tail-skin	
Backup	6 gantries, length is 90 m, Grouting system: 0 to 40 m^3/h

The geological information from the Hong Kong tunnel plus collected geotechnical data from existing boreholes and from tunnel face are combined for creation of the 3D geotechnical/ground model. For the creation of the Discretized Solid Geological Model (DSGM) around the tunnel alignment four separate geological formations are considered as is illustrated in Figure. 15 (a) initially in granodiorite (purple color), (b) then in granite (pink color), (c) fully decomposed granite (red color) and (d) finally in the soil formations (yellow color).

RMR data evaluated from the drill cores along the exploratory boreholes bored at distances 100–150 m apart are illustrated in Figure 20. As it may be seen from Figures 21a and b the histogram of the RMR values follows a normal probability density function (pdf). The above histogram indicates the following values of the first two moments of the RMR frequency distribution, *i.e.* a mean value and a variance of $m = 59.6$, $\sigma^2 = 65.2$, respectively.

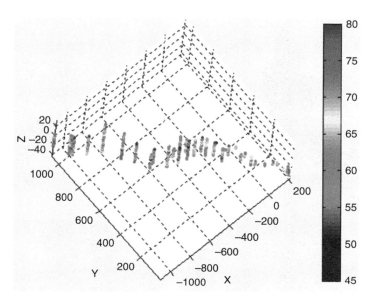

Figure 20 Screenshot of KRIGSTAT code showing the RMR values along the boreholes.

The omnidirectional (isotropic) experimental semivariogram of RMR from the boreholes was calculated by using a lag of 5m and it is shown in Figure 21. A theoretical model was fitted to the data. The general exponential model was used to fit the experimental data that has the following expression

$$\gamma(h) = (S - So) \cdot \left(1 - \exp\left(-(h/l)^n\right)\right) + So \tag{63}$$

It is shown from the semivariogram that the maximum correlation is approximately 15 m that is much lower than the average distance among the boreholes that is 120 m. It could be realized that the experimental points forming the semivariogram have been obtained solely from the downhole RMR measurements in each borehole, rather among neighboring boreholes. This means that the estimations of RMR at larger distances than this will be roughly equal to the average of the RMR values recorded between neighboring boreholes.

In order to gain a better knowledge of the RMR spatial distribution along the tunnel, successive faces after few meters of respective TBM advancements, were mapped and evaluated with respect to the rock mass RMR. In Figure 22 the RMR evaluated from the drill cores and at the successive faces are shown for comparison purposes. The much lower RMR exhibited by the boreholes at certain sections is attributed to weathered granite at higher elevations from the tunnel's crown. In this case the RMR exhibits a normal pdf as is shown in Figure 23a with a mean value and a variance of $m = 65$, $\sigma^2 = 90$, respectively, and a better experimental semivariogram following the Gaussian model (*e.g.* Figure 23b). The range of correlation is in this case 70 m (indicated by the arrow and found from the length L of the model with the approximate estimation of $\approx 7L/4$), that is the half of the distance separating the exploratory boreholes, which

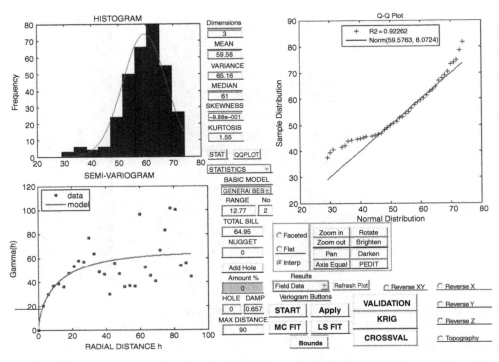

Figure 21 Screenshot of KRIGSTAT code showing the statistical and semivariography analysis; histogram of RMR data compared with normal distribution pdf (top left), normality check of sampled RMR values along the boreholes by examining the goodness of fit of quantiles of sampled distribution with the quantiles of normal distribution (Q-Q plot) (top right), and semivariogram (bottom left).

means that the model has no undefined areas. From Figure 22 it may be seen that the boreholes give valuable information regarding the expected RMR along the tunnel, but it cannot be used for the prediction of the performance of the TBM or rock mass response after excavation, along tunnel's sections between the boreholes.

Since it is impossible in the exploratory phase of a tunnel to drill boreholes at small distances apart in order to predict the rock mass quality that shall be encountered by the TBM between neighboring boreholes, then the TBM itself could be employed to upgrade the rough initial geostatistical model gained from the sparse boreholes according to the concept presented in Paragraph 2.1. In this frame, the best parameter that could be obtained from TBM data in a daily basis is the SE (*i.e.* specific energy) that may be estimated either from Equation 8, or 12. For the tunnel at hand, it was found from the boring results from the first few 10's of m's that SE correlates with RMR with the following hyperbolic function with two free parameters a, b (Exadaktylos et al., 2008)

$$RMR = 100 - \frac{a}{SE + b},$$
$$a = 1253.5, b = 10.4$$

(64)

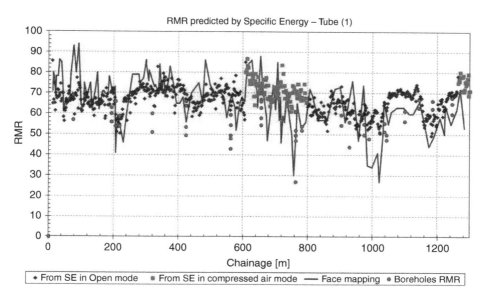

Figure 22 Distribution of RMR along the granodioritic and granitic rock mass of tube I estimated independently from the drill cores closest to the tunnel (circles), from joint mapping at successive faces of the tunnel (squares), and from the SE consumed by the TBM (continuous line).

where SE is expressed in units of MPa. SE exhibits also a Gaussian semivariogram model with a slightly larger correlation range of 90 m as is indicated in Figure 23c. Also, from Figure 22 it may be observed that the above indirect estimation of RMR compares very well with both RMR estimations independently from boreholes and from the mappings of joint conditions at the successive faces along the tunnel.

The significant upgrade of the RMR model along the tunnel with the aid of the SE estimated from the daily recordings of the TBM is illustrated in Figures 24 a-c. The initial geotechnical model created from borehole data is characterized by large Kriging error as is given by Equation 62 in regions between successive boreholes as is shown in Figure 24b. However, from inverse analysis of TBM Specific Energy (SE) data along the tunnel advance at a certain time using the phenomenological model linking RMR with SE as is shown in Figure 24a, the spatial distribution of the Kriging STD (error) in the tunnel (color bars indicate Kriging error) leads to the significant reduction of Kriging error along the sections between the boreholes that has been traversed by the TBM. Most importantly, the improvement of the model does not only happen behind the TBM but also in the front of the tunnel's face as may be realized by comparing Figures 24b and c.

It is recalled here that the RMR is linked with the damage of the rock due to joints through the empirical Equation 49. Then it is a straightforward task for someone to estimate the rock mass strength parameters using the Damage Mechanics approach presented in Section 3.

Another approach to the problem is the prediction of penetration depth p [mm/rev] based on input values of normal force F_n per cutter considering the rock mass strength

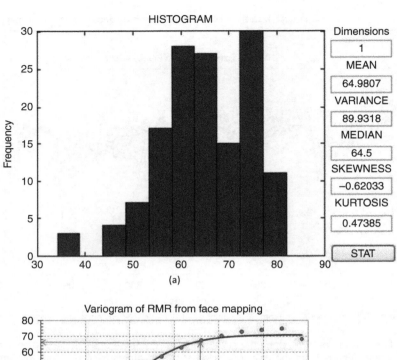
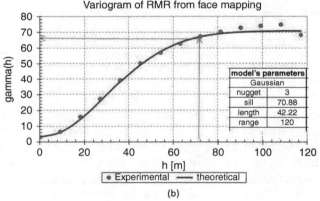
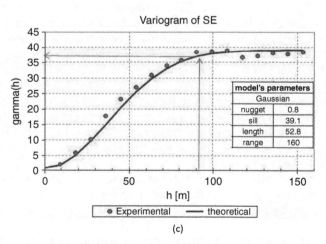

Figure 23 Statistical and variography analysis of RMR mapped on successive tunnel's faces; (a) Frequency histogram of composited RMR data mapped on the exposed sequential faces along tube 1, (b) Gaussian semi-variogram model of RMR, and (c) Gaussian model of SE.

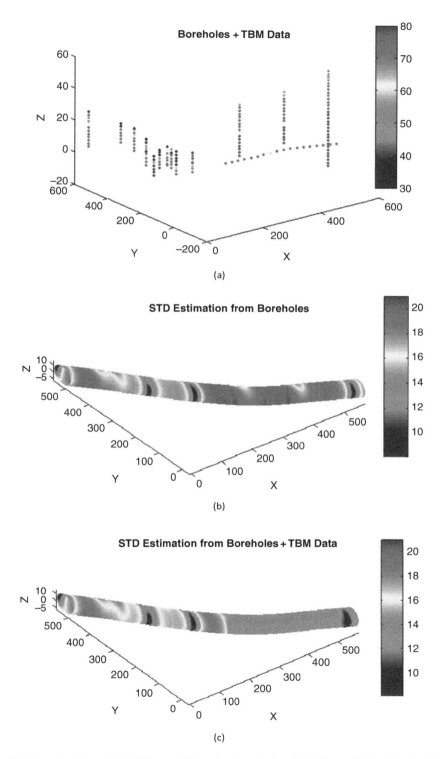

Figure 24 Upgrade of the initial RMR model based on boreholes; (a) RMR spatial distribution inferred from boreholes and TBM specific energy along a segment of tunnel, (b) spatial distribution of the Kriging Standard Deviation STD (error) along the tunnel (color bars indicate Kriging error), and (c) significant reduction of the Kriging error along the initial section by combining RMR from boreholes and from the records of specific energy.

parameters and TBM design and operational parameters (*i.e.* number of discs, rotational speed etc.). This could be based either in the empirical approach of linking SE with RMR presented before, or from the cutting model presented in Subparagraph 2.3.3. Indeed in the former method one could rely on the formula

$$p = \frac{CC \cdot F_n}{S \cdot SE} \tag{65}$$

where SE is obtained from the empirical Equation 64 after solving in terms of SE. Plots of the variation of the normal force per cutter and of the penetration depth along the chainage of tube #1, are shown in Figures 25a and b, respectively. From these figures the significant variation of both these parameters around the average values may be seen. It should be noticed that in these graphs the test data in the chainage range 600 m to 800 m have been omitted since they correspond to the compressed-air mode of TBM operation.

At this point we may avoid the empirical approach and follow the approach of the disc cutting model for rock chip formation shown in Figure 26 and elaborated in subparagraph 2.3.3. The two rock mass strength parameters UCS_m, UTS_m are obtained from the damage D that is linked with RMR according to the Damage mechanics approach, and the unconfined compression tests on intact rock cores sampled from boreholes and the tunnel, assuming a strength ratio $UCS/UTS = 8$ (since tensile strength tests have not be conducted in this project); the internal friction angle of both the granodioritic and granitic rock formations is assumed to be $\phi = 53^\circ$. Further the two extra angles entering the model have been assumed constant with values $\theta = 70^\circ$ and $\gamma = 8^\circ$. The other constant geometrical parameters are the spacing of neighboring kerfs $S = 75$ mm and the width of the tip of the disc w = 12.5 mm. The comparison of the limit analysis model predictions with the estimation from the TBM in the open mode recorded data referring to the normal force exerted on a disc cutter, may be seen in Figure 27. It may be seen from this graph that the predictions are in good agreement with estimations from TBM registered data considering that the former are based on a model, and the inherent uncertainty of the test data (intact rock strength, RMR, calculation of normal force per cutter from registered TBM data). The predictions made by means of the CSM model by virtue of Equations 23 & 24 are also shown in the same graph. It may be observed that always the latter predictions are smaller compared to the limit analysis model.

5 SUMMARY

A rock boring modeling method has been presented that is based on a simple disc cutting model of the Limit Analysis theory of Perfect Plasticity, as well as on damage mechanics and geostatistical approaches. These three main components of the proposed method are new compared with existing empirical models. This method could be used either for the prediction of the performance of hard rock gripper or shielded TBM or the back-analysis of registered TBM excavation data for evaluating the performance of the given TBM. The proposed method is best suited for mechanized excavation processes in heterogeneous rock formations, as has been demonstrated with the case study presented here. Another component of the method is the Digital

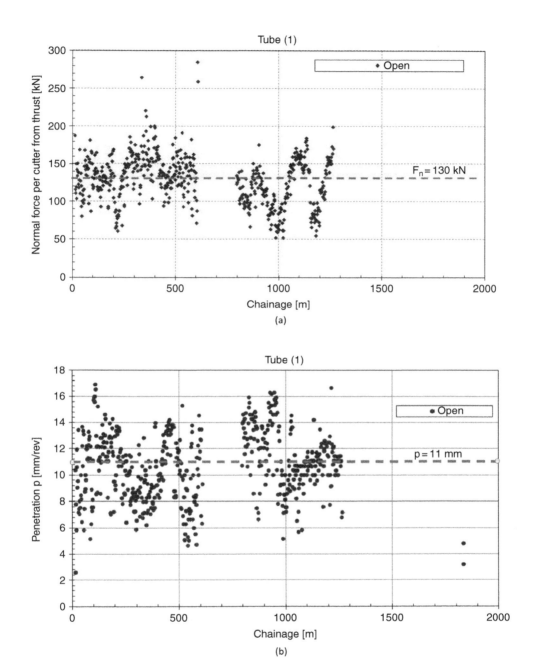

Figure 25 Variation of (a) normal force per disc cutter and (b) penetration depth along the chainage of tube 1.

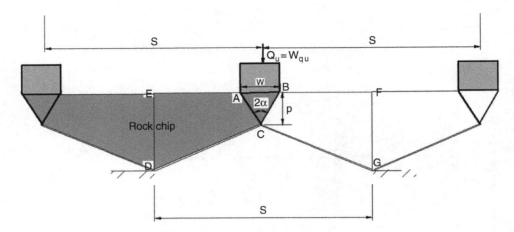

Figure 26 The model used for the prediction of the normal force on a disc.

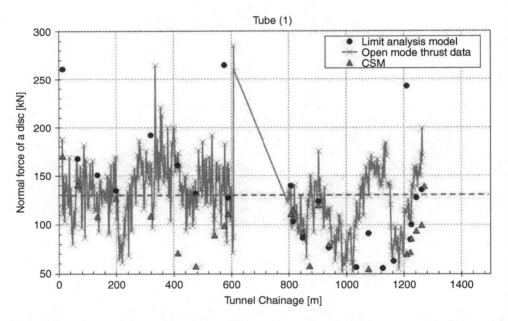

Figure 27 Comparison of normal force per cutter estimated from the TBM registered data, from the limit analysis model (bullets), and CSM model (triangles).

Discretized Ground Model that is continuously upgraded with TBM advance in front of the tunnel's face in order to make predictions of TBM performance and excavation behavior in rock mass regions that haven't been excavated yet. Such a method could be greatly supported with wireless sensors embedded in the cutting tools and transmitting real time to the operator, data like rotational speed and loads exerted on the discs.

REFERENCES

Barton, N. (2000) *TBM tunnelling in jointed and faulted rock*. A.A. Balkema, Rotterdam.

Chen, W-F. (1975) *Limit analysis and soil plasticity*. Elsevier Scientific Publishing Company, Amsterdam.

Chen, W.F. & Drucker, D.C. (1969) Bearing capacity of concrete blocks on rock. *Journal of Eng. Mech. Division*, Proceedings of the American Society of Civil Engineers, 955–978.

Chiles, J.P. & Delfiner, P. (1999) *Geostatistics – Modeling spatial uncertainty*, John Wiley & Sons, New York.

Drucker, D.C. & Prager, W. (1952) Soil mechanics and plastic analysis of limit design. *Q. Appl. Mech.*, 10, 157–165.

Exadaktylos, G. & Stavropoulou, M. (2008) A Specific Upscaling Theory of Rock Mass Parameters Exhibiting Spatial Variability: Analytical relations and computational scheme. *Int. J. of Rock Mech. & Min. Sci.*, 45, 1102–1125.

Exadaktylos, G., Stavropoulou, M., Xiroudakis, G., Broissia, M. & Schwarz, H. (2008) A spatial estimation model for continuous rock mass characterization from the specific energy of a TBM. *Rock Mech Rock Eng*, 41, 797–834.

Gertsch, R., Gertsch, L. & Rostami, J. (2007) Disc cutting tests in Colorado Red Granite: Implications for TBM performance prediction. *Int. J. of Rock Mech. & Min. Sci.*, 44, 238–246.

Hoek, E. & Brown, E.T. (1997) Practical estimates of rock mass strength. *Int J Rock Mech Min Sci. Geomech. Abstr.*, 34(8), 1165–1186.

Journel, A.G. & Huijbregts, C.J. (1978) *Mining geostatistics*. Academic Press, London.

Kachanov, L.M. (1986) *Introduction to continuum damage mechanics*. Kluwer Academic Publishers, Dordrecht.

Kitanidis, P.K. (1997) *Introduction to Geostatistics: Applications in hydrogeology*, Cambridge University Press, United Kingdom.

Krajcinovic, D. (1989) Damage mechanics. *Mech. Mater.*, 8, 117–197.

Kutter, H.K. & Fairhurst, C. (1971) On the fracture process in blasting. *Int. J. Rock Mech. Min. Sci.*, 8, 181–202.

Lemaitre J. (1992) *A course on damage mechanics*. Springer-Verlag, Berlin.

Maidl, B., Herrenknecht, M., Maidl, U. & Wehrmeyer, G. (2012) *Mechanised shield tunnelling*, 2nd Edition. Wilhelm Ernst & Sohn.

Meschke G., Nagel, F., Stascheit, J., Stavropoulou, M. & Exadaktylos, G. (2008) Numerical simulation of mechanized tunnelling as part of an integrated optimization platform for tunnelling design, In *12th International Conference of International Association for Computer Methods and Advances in Geomechanics (IACMAG)* 1–6 October, 2008, Goa, India.

Nelson, P.P. (1993) TBM performance analysis with reference to rock properties, In *Comprehensive Rock Engineering, Principles, Practice & Projects, Vol. 4 Excavation, Support and Monitoring*, Ed. J.A. Hudson, 261–292.

Rostami, J. & Ozdemir, L. (1993) A New Model for Performance Prediction of Hard Rock TBMS. In: *Proceedings of Rapid Excavation and Tunnelling Conference, USA*, pp. 794–809.

Rostami, J., Ozdemir, L. & Nilsen, B. (1996) Comparison Between CSM and NTH Hard Rock TBM Performance Prediction Models. In: *Proceedings of Annual Technical Meeting of the Institute of Shaft Drilling and Technology (ISDT), Las Vegas, NV*, p. 11.

Roxborough, F.F. & Phillips, H.R. (1975) Rock Excavation by Disc Cutter. *Int. J. Rock Mech. Min. Sci. & Geomech. Abstr.* 12, 361–366.

Snowdon, R.A., Ryley, M.D. & Temporal, J. (1982) A Study of Disc Cutting in Selected British Rocks. *Int. J. Rock Mech. Min. Sci. & Geomech. Abstr.*, 19, 107–112.

Swain, M.V. & Lawn, B.R. (1976) Indentation fracture in brittle rocks and glasses. *Int. J. Rock Mech. Min. Sci. & Geomech. Abstr.*, 13, 311–319.

Teale, R. (1965) The concept of specific energy in rock drilling. *Int. J. Rock Mech. Min. Sci.*, 2, 57–73.

Rock Stress

Chapter 6

Hydraulic fracturing stress measurements in deep holes

Douglas R. Schmitt[1] *& Bezalel Haimson*[2]

[1]*Department of Physics, University of Alberta, Edmonton, Canada*
[2]*Department of Materials Science and Engineering and Geological Engineering Program, University of Wisconsin, Madison, WI, USA*

Abstract: A comprehensive description is provided of the most common technique of quantitatively estimating the in situ state of stress at depth in the Earth's crust. Basic concepts central to crustal stress measurement are first provided as motivation. The equations of stress concentration around the borehole resulting from far-field in situ stresses and borehole fluid pressure provide the basis of the theory of hydraulic fracturing. The advantages and disadvantages of the various borehole testing setups are discussed with particular emphasis on straddle-packer assemblies that allow for the most reliable stress measurements. The recommended protocols for a multi-cycle pressurization stress measurement test are described and justified in light of expected behavior of fracture initiation and propagation. This is assisted by illustrative idealized pressurization curves that highlight the critical pressures that are interpreted in terms of the prevailing in situ principal stresses. The effect of the differing assumptions with regards to breakdown, fracture closure, and fracture re-opening on the determined principal in situ stresses is reviewed. This highlights the complexity associated with computing the magnitude of the greatest horizontal compression. Finally, the various directions of future research on hydraulic fracturing stress measurements are described.

I INTRODUCTION

The hydraulic fracturing (HF) method of in situ stress measurement is the best known method of quantitatively assessing the state of in situ stress at great depths. In the last 50 years since the historic first use of the method to determine the state of stress at Rangely, Colorado (Haimson, 1973, 1975, 1978; Raleigh *et al.*, 1976), HF has become an indispensable tool in the design of large underground projects such as hydroelectric powerhouses, tunnels, mines, waste disposal galleries, energy storage caverns, etc., is routinely employed in the design of oil and gas fields in conjunction with production stimulation in both vertical and horizontal boreholes, and has also become a requisite field measurement in plate tectonics research and earthquake prediction, mechanism and control studies.

Briefly, the method consists of sealing off a short segment (minimum recommended length: ten diameters) of a wellbore or borehole at the desired depth, injecting fluid (usually small quantities of clean water) into it at a rate sufficient to raise quite rapidly the hydraulic pressure (about 0.1–1 MPa s^{-1}), until a critical level is reached ('breakdown' pressure, P_b) at which a tensile crack ("hydraulic fracture") unstably develops at the borehole wall. At that point borehole fluid penetrates the fracture, and hence a drop in pressurization rate occurs. When pumping is stopped, the pressure will immediately decay, first very fast as the fluid chases the extending fracture tip, and eventually at a slower rate as the fracture closes and the only fluid loss is due to permeation of the injected fluid into the rock. The 'shut-in' pressure P_s is the transition level between fast and slow pressure decay and signifies the closure of the fracture. Several minutes after shut-in the pressure is bled off, completing the first pressurization cycle. Several additional pressure cycles are normally conducted. From these cycles supplementary shut-in values are obtained, as well as the pressure required to reopen the induced fracture, P_R. Pressure and flow rate of the injection-fluid are continuously recorded. The far-field stresses are calculated from the pivotal pressures (P_b, P_s, and P_R) recorded during the test. The directions of the in situ stresses are determined from the attitude of the induced hydraulic fracture trace on the borehole wall. When conditions permit, several tests are carried out in one borehole within the depth range of interest. This contribution seeks to convey the details necessary to interpret pressurization records in order to obtain values of the in situ stress.

1.1 Historical notes

The term 'hydraulic fracturing' comes from an oil-field stimulation method developed in the 1940s, by which a segment of a wellbore was injected with a mixture of water and various chemicals and propping agents, and the pressure raised until the surrounding rock fractured. The injected fluid penetrated the induced fracture and extended it, while the propping agents kept the fracture sufficiently open to provide a sink for the reservoir oil to flow into. The stimulation method proved very successful and its use grew tremendously over the years. Today it serves as the basis for the operation colloquially called 'fracking', used for oil and gas production stimulation from reservoirs in nearly impermeable shales and similar formations.

Hubbert and Willis (1957) attempted to understand the process of hydraulic fracturing. They used the theory of elasticity to conclude that the fluid pressures required to initiate and extend the hydraulic fracture, as well as fracture attitude, are directly related to the pre-existing in situ stress field. They submitted that the initiation of hydraulic fractures is the result of tensile failure and rupture at the borehole wall, and developed equations relating the pressure needed to induce hydraulic fracture to the state of stress at the borehole wall. With appropriate adjustments, their elastic model is still the basis of our understanding of hydraulic fracturing today.

Fairhurst (1964, 1965) was among the first to advocate the use of hydraulic fracturing for a diametrically opposed purpose: rather than use the presumed state of in situ stress to predict the induced-fracture direction and the pressure required to initiate that fracture, to employ the method as a tool for in situ stress determination. Haimson and Fairhurst (1967) and Haimson (1968) extended Biot's (1941) theory of poroelasticity to cover pressurized boreholes, and generalized the Hubbert and Willis (1957) model to include both nonpenetrating and penetrating injection-fluid cases. They also demonstrated in the laboratory the reliability of the suggested pressure-stress relations (Haimson, 1968; Haimson & Fairhurst, 1969a, 1970).

The breakthrough for hydraulic fracturing as an important method of in situ stress measurement came when the Menlo Park, California, branch of the U.S. Geological Survey engaged in studying the unprecedented series of earthquakes at Rangely,

Colorado. It was this group's decision in 1971 to financially support hydraulic fracturing stress measurements at Rangely, and the subsequent success of the experiment in explaining the conditions which had brought about local earthquakes (Haimson, 1973, 1975a; Raleigh *et al.*, 1972), that paved the way for other important field tests and for the general acceptance of the method.

During the 1970s there were several important milestones in the development of hydraulic fracturing, such as: the first measurements in an underground tunnel as part of the pre-excavation design of underground hydroelectric powerhouses (Haimson, 1975b), the first measurements outside the United States (Rummel & Jung, 1975; Haimson & Voight, 1977), the deepest stress measurements ever undertaken until then (Michigan 5km deep wellbore, Haimson, 1978), and the first measurements along the San Andreas fault (Zoback & Roller, 1979; Zoback *et al.*, 1980). Research groups dedicated to hydraulic fracturing stress measurements sprung up in Europe, Asia and Australia, and the proliferation of measurements brought to the forefront some lingering problems in the proper interpretation of test results in terms of the in situ stress.

Two international workshops held in the U.S. in 1981 and 1988, brought together the best known practitioners of the method for an exchange of experiences and a resolution of the different approaches to estimating the state of stress. The Proceedings of these workshops (Zoback & Haimson, 1983; Haimson, 1989a,b) constitute a must read for anyone interested in entering this field of endeavor.

Several important developments have evolved over the years in the practice of hydraulic fracturing for stress measurements. HF is developing along two distinct courses, one for the use in open holes drilled in hard rock, usually associated with the design of underground openings and with the study of the earth structure and tectonics, and the other for testing mainly cased holes in soft and permeable sedimentary formations typical of oil fields. Typically, hard rock HF testing is conducted in 'slim holes' (76–100 mm in diameter); oil field tests are carried out in large diameter (150 mm or larger) and usually much deeper wellbores. Testing procedures, data interpretation, and stress information obtained are quite different in the two approaches. The entire reason for carrying out hydraulic fracturing tests in hard rocks is to obtain quantitative information on the in situ stress magnitudes and directions, and techniques have developed to provide through multiple cycles some redundancy that provides additional confidence in the results. Conversely, in traditional petroleum production practice crustal stresses are usually only of secondary interest, with the primary goal of transient pressure testing being to determine pore pressures and bulk permeability; repeated pressurization cycles are rare and the complicated analyses of such data presume knowledge of fracture geometries and dimensions and of factors that retard fluid transfer to the formation. Here we focus on the original practice in order to highlight the relationships between in situ states of stress and the borehole pressure records. Given the public scrutiny with regard to earthquakes induced by long term fluid injection and massive hydraulic stimulations, we expect that the practice of repeated pressurization cycles will be increasingly applied to better understand stress states in petroleum reservoirs in the future.

Hydraulic fracturing equipment has evolved to where at least four different types are now in use: drillpipe, drillpipe and hose, wireline and hose, and multi-hose. Fracture tracing techniques presently in use also number four or so: oriented impression packer, borehole acoustic and optical televiewers, televiewer-impression packer, and electric resistivity. Stress calculations still mainly follow the elastic model first described by Hubbert and Willis (1957), but the poroelastic (Haimson & Fairhurst, 1969), and fracture mechanics approaches (Abou-Sayed *et al.*, 1978; Rummel & Winter, 1983) are also employed, albeit less frequently.

2 BASIC DEFINITIONS AND ASSUMPTIONS

A basic understanding of the definitions of stresses and the Hooke's Law relationships between stress and strain is presumed. This information is found in many texts today (*e.g.*, Jaeger *et al.*, 2007). As is normal in the geo-science and geo-engineering communities, compressive normal stresses, pore and borehole pressures, are by convention taken to have positive sign.

For purposes of illustration here we further make the assumption that within the rock mass of interest the three in situ principal stresses are vertical σ_V and horizontal σ_H and σ_h with magnitudes $\sigma_H > \sigma_h$ (Figure 1). We will refer to these as the far-field stresses with the meaning that these existed prior to the drilling of the borehole; these are the stresses that we seek to determine. Note that in the earth we expect that all three of these stresses to be compressive. The vertical stress σ_v is taken to be equal to the overburden weight per unit area at the depth of interest (Terzaghi & Richart, 1952):

$$\sigma_v = \sum_{i=1}^{n} \rho_i g D_i \qquad (1)$$

where ρ_I is the mean mass density of rock layer I; g is the local gravitational acceleration; D_i is the thickness of layer i; and n is the number of rock layers overlying the test

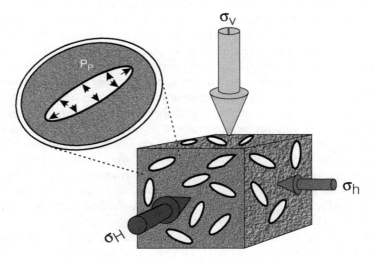

Figure 1 The rock mass is subject to the three principal stresses σ_V, σ_H, and σ_h and pressure P_P. The rock has porosity ϕ and permeability κ.

zone. This assumption is commonly employed in areas where there is little variation in the topography, but of course cannot hold near the surface in mountainous regions (Figueiredo *et al.*, 2014; Savage *et al.*, 1985), in the vicinity of underground workings, or if the structures and mechanical properties within the rock mass distort the stress field (Amadei *et al.*, 1987; Cornet, 1993). In this contribution, for the sake of clarity, we will assume however that σ_v is one of the principal stresses and that the borehole is nearly vertical. Of course, this will not always be the case particularly in inclined or horizontal drilling, but here we provide the reader with the essential background to independently further consider alternate cases.

Further, the fluid residing in the pore space of the rock has an *original* formation pore pressure P_o. However, if fluid infiltrates into the formation during the HF test the pore pressure in the vicinity of the borehole will vary both spatially and temporally and in this case we will employ P_P to represent the pore pressure that will be active in assisting creation of the fracture.

While it may be obvious, it is important to point out that the relative magnitudes of the three in situ principal stresses generally differ from one another. Structural geologists have long known (Anderson, 1951) that the relative magnitudes of these principal stresses control both the style of faulting and the opening and growth of tensile fractures. The three faulting regimes are:

i.   Thrust (or reverse) faulting regime with $\sigma_H > \sigma_h > \sigma_v$
ii. Strike-slip (or wrench) regime $\sigma_H > \sigma_V > \sigma_h$,
iii. Normal faulting regime $\sigma_V > \sigma_H > \sigma_h$

Away from the stress concentrations around the borehole, tensile fractures will most easily open in the direction of the minimum compressive principal stress σ_{min} in each case, so that the plane of such fractures lies vertically in normal and strike-slip faulting regimes but horizontally under thrust fault scenarios as was simply demonstrated by Hubbert and Willis (1957). While their demonstration illustrates nicely the growth of fractures in a stress field away from the borehole, great care must be taken in the interpretation of their tests to the fracture initiation at the borehole wall in a real situation. At the borehole wall in cases using inflatable packers, however, the fluid pressure in the wellbore induces a considerable azimuthal tension. Although the near borehole stress environment may in most, but not all (Bjarnason *et al.*, 1989) cases first influence the fracture initiation, once the fracture has propagated away from the borehole the undisturbed stresses will reassert themselves and control its orientation (*e.g.*, Warren & Smith, 1985). That means that the initial vertical fracture is aligned in the direction of σ_H (see Fig. 2c). Haimson and Lee (1980) observed this phenomenon directly during tests at Darlington, Ontario where, despite being in a thrust fault stress regime, they initiated vertical fractures in the boreholes which then later turned and traversed back into the boreholes horizontally.

The rock mass can also be characterized in a number of ways. It is first assumed that its physical properties are isotropic and that they are not dependent on the effective stress. The rock's elastic *'frame'* or *'drained'* properties may be described by its bulk K, Young's E, or shear μ moduli and its Poisson's ratio v, only two of these need to be known to fully describe the material's elastic properties. Normally the rock will have porosity ϕ and permeability κ. The frame of the rock and its

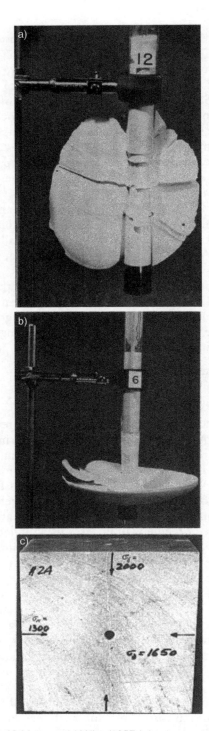

Figure 2 Fractures made by Hubbert and Willis (1957a) by pressurizing boreholes in thick gelatin subject to different states of stress. The pressurizing borehole fluid was a plaster-of-Paris slurry that was allowed to set. In a) the gelatin mass was subject to a horizontal compression and in b) a vertical compression. Vertical a) and horizontal b) fractures result. Images modified from Figures 23 and 24 of Hubbert and Willis (1957a), permission to use through expired copyright. c) Vertical hydraulic fracture propagating in the direction of σ_H made by Haimson and Fairhurst (1970).

constituent solid minerals will have bulk moduli $K < K_s$, respectively, and the Biot-Willis poroelastic parameter that is a measure of the volumetric strain induced by changes in P_P may be defined:

$$\alpha^P = 1 - \frac{K}{K_s}. \tag{2}$$

The limits $0 \leq \alpha^P \leq 1$ respectively correspond to extremely stiff ($K \rightarrow K_s$) or compliant ($K << K_s$) porous materials.

A useful shorthand used later is the simply named '*poroelastic co-efficient*' (Detournay *et al.*, 1989; Rice & Cleary, 1976)

$$\eta = \frac{\alpha^P(1 - 2\nu)}{2(1 - \nu)} \tag{3}$$

which for isotropic porous materials could range over $0 \leq \eta \leq 0.5$.

The rock, in its natural state in the earth, will also have a thermal conductivity k and a coefficient of thermal expansion α^T. Knowledge of these properties becomes important if corrections for pore-elastic or thermal effects are necessary.

Finally, Rankine's tensile strength criterion of $T > 0$ is assumed. Under this simple criterion the rock will fail in tension once

$$\sigma - P_P < -T \tag{4}$$

where $\sigma - P_P$ is the classic *Terzhagi* effective stress that applies generally for failure (Cornet & Fairhurst, 1974; Rice & Cleary, 1976; Robin, 1973).

One important point is that the analyses described below strictly only apply for the case in which the wellbore fluid transmitting P_w is in direct contact with the intact rock. Many of the equations presented, particularly those relating to fracture initiation or breakdown, cannot apply to perforated steel cased and cemented boreholes. Similarly, thick mudcakes may complicate interpretations (Raaen *et al.*, 2001). This contribution is focused on the quantitative determination of stress during relatively small injections of fluids in rock that could be considered isotropic. Several workers have studied the problem of pressure testing in naturally fractured or jointed rock masses and even usefully exploit such information to gain knowledge of the stress field (Baumgärtner & Rummel, 1989; Cornet & Valette, 1984).

3 CONCENTRATION OF STRESS AROUND THE BOREHOLE

In the discussions to follow the borehole is drilled in the vertical direction that is its axis is parallel to the vertical and subsequently also to σ_V. This simplifies the analysis and allows Kirsch's (1898) plane strain formulation to be applied to the calculation of the stresses concentrated in the vicinity of the borehole. This is done in a cylindrical coordinate system and for an arbitrary point A defined by its azimuth θ and radial distance r from the borehole axis. The concentrated tectonic stresses are for a uniaxial stress σ_{xx} applied along the azimuth $\theta = 0°$ (Figure 3).

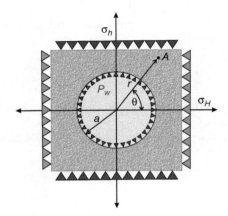

Figure 3 View down the vertical axis of the borehole of radius a drilled into a rock mass subject to the far-field horizontal principal stresses σ_H and σ_h. Point A is described by its *azimuth* θ and distance from the borehole *axis* r. The borehole is filled with fluid at borehole *fluid* pressure P_w.

$$\sigma_{\theta\theta} = \frac{\sigma_{xx}}{2}\left(1 + \frac{a^2}{r^2}\right) - \frac{\sigma_{xx}}{2}\left(1 + \frac{3a^4}{r^4}\right)\cos(2\theta)$$

$$\sigma_{rr} = \frac{\sigma_{xx}}{2}\left(1 - \frac{a^2}{r^2}\right) + \frac{\sigma_{xx}}{2}\left(1 + \frac{3a^4}{r^4} - \frac{4a^2}{r^2}\right)\cos(2\theta) \tag{5}$$

$$\tau_{\theta r} = -\frac{\sigma_{xx}}{2}\left(1 - \frac{3a^4}{r^4} + \frac{2a^2}{r^2}\right)\sin(2\theta)$$

with $\sigma_{\theta\theta}$, σ_{rr}, and $\tau_{\theta r}$ being the azimuthal (or hoop), the radial, and the shear stress at point (θ,r) within the cylindrical co-ordinate system centered on the borehole axis.

The formula describing the hydraulic fracturing process assume that the stresses in the vicinity of the borehole are controlled by the concentration of the greatest and least horizontal far-field compressive stresses σ_H and σ_h, respectively, and by the borehole fluid pressure P_w. The solution for the far-field stress concentrations is obtained from Kirsch's (1898) derivation for stress concentrations induced by application of a stress to a thin plate containing a circular hole. The effects of the wellbore pressure are obtained from Lamé's (1852) hollow cylinder expressions (see also Bickley, 1928 for more complex situations). Superposition of these gives (Haimson & Fairhurst, 1967; Haimson, 1968; Jaeger et al., 2007):

$$\sigma_{\theta\theta} = \frac{\sigma_H + \sigma_h}{2}\left(1 + \frac{a^2}{r^2}\right) - \frac{\sigma_H - \sigma_h}{2}\left(1 + \frac{3a^4}{r^4}\right)\cos(2\theta) - P_w\frac{a^2}{r^2}$$

$$\sigma_{rr} = \frac{\sigma_H + \sigma_h}{2}\left(1 - \frac{a^2}{r^2}\right) + \frac{\sigma_H - \sigma_h}{2}\left(1 + \frac{3a^4}{r^4} - \frac{4a^2}{r^2}\right)\cos(2\theta) + P_w\frac{a^2}{r^2} \tag{6}$$

$$\tau_{\theta r} = -\frac{\sigma_H - \sigma_h}{2}\left(1 - \frac{3a^4}{r^4} + \frac{2a^2}{r^2}\right)\sin(2\theta)$$

This result can also be obtained from Hiramatsu and Oka's (1962) 3D solution of a borehole arbitrarily oriented with respect to the stress field, which allows for the consideration of the vertical principal stress σ_V. Examination of these shows that σ_V does not influence the horizontal stress concentrations described in Equation 6. Conversely, if plain strain applies the horizontal stresses must induce a vertical stress (Aadnoy, 1987; Bjarnason *et al.*, 1989). No elastic properties appear in Equation 6 but the reader must keep in mind that this strictly only applies if the rock is isotropic and linearly elastic; introduction of anisotropy may complicate the situation.

Figure 4 illustrates graphically Equation 6 at the borehole wall where $r = a$ for two specific magnitudes of the far field stresses. As is well known, application of a uniaxial stress σ_H results in large variations in the azimuthal, or hoop, stress $\sigma_{\theta\theta}$ such that in the azimuths that point in the direction of σ_H (*i.e.* at $\theta = 0°$ and $180°$) a pure tension is generated with $\sigma_{\theta\theta} = -\sigma_H$ (Figure 4a). Indeed, the rock will fail in tension if $\sigma_H > T$. In contrast, σ_H's compression is amplified by a factor of 3 perpendicular to this (*i.e.* at $\theta = 90°$ and $270°$). Superposing increasing σ_h compressions attenuates the extremes and once σ_h exceeds a $\sigma_H/3$ then $\sigma_{\theta\theta}$ is compressive everywhere.

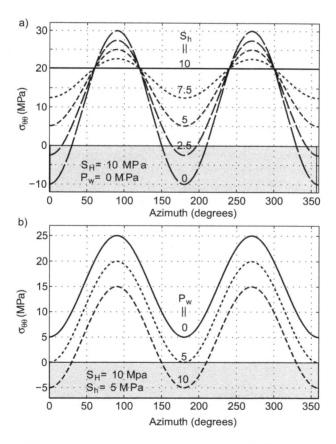

Figure 4 Illustration of the variations in stress concentration of $\sigma_{\theta\theta}$ with θ at the borehole wall $r = a$ due to a combination of far-field stresses σ_H and σ_h and borehole fluid pressure P_w. Tensile stresses indicated with gray background. a) Case with no fluid pressure $P_w = 0$ with constant $\sigma_H = 10$ MPa and with varying $0 \leq \sigma_h \leq \sigma_H$. b) Case with $\sigma_H = 10$ MPa and $\sigma_h = 5$ MPa with $0 \leq P_w \leq \sigma_H$.

The addition of any borehole pressure P_w superposes an additional tensile tangential stress with the same magnitude at the borehole wall (Figure 4b) and can easily push $\sigma_{\theta\theta}$ into tension at the azimuths surrounding $\theta = 0°$ and $180°$. Consequently, once $\sigma_{\theta\theta} \leq -T$ we will expect the rock to fail in tension and to initiate a hydraulic fracture. This is an important point, as the pressure at which the fracture initiates, which we provisionally associate here with a *breakdown pressure P_b*, is key in attempting to estimate the magnitude of σ_H. A further important implication is that we expect the hydraulic fractures created to lie at the azimuths that point in σ_H's direction.

It is also useful to consider the behavior of the stress concentrations that are assumed responsible for growth of the hydraulic fracture. This has implications for both fracture initiation and subsequent re-opening during later pressurization cycles (Hickman & Zoback, 1983; Ito *et al.*, 1999). The evolution of $\sigma_{\theta\theta}(r)$ at the borehole wall ($r = 1$) is plotted for cases of σ_H/σ_h from 1 to 4 in Figure 5a. As already mentioned, we expect the growing hydraulic fracture to propagate in a direction parallel to σ_H (Figure 2c) opening normal to the direction of $\sigma_{\theta\theta}(r)$; radial variations in these stresses will influence the growing fracture. Farther from the borehole as expected $\sigma_{\theta\theta}(r)$ asymptotically approaches σ_h for all cases. When $\sigma_H = \sigma_h$, the concentrated $\sigma_{\theta\theta}(r)$ monotonically decreases but generally remains strongly compressive. Conversely, when $\sigma_H >> \sigma_h$ zones near the borehole wall are in pure tension, a state that would assist fracture initiation. Indeed, one might expect tensile rupture to occur spontaneously for the extreme case of $\sigma_H/\sigma_h = 4$ without any borehole pressure. As this fracture grows into the formation, however, its propagation will be retarded by increasingly greater $\sigma_{\theta\theta}(r)$ compression.

This illustration is taken one step further (Figure 5b) by applying within the borehole the pressure P_w necessary to first bring the hoop stress $\sigma_{\theta\theta}(r = 1)$ to zero for a representative set of the σ_H/σ_h cases from Figure 5a. The purpose of this exercise is to show how P_w influences the hoop stress concentrations into the formation along the azimuth of σ_H. A small further increase in P_w would put the borehole wall into tension and closer to failure, and in later discussions this will be taken by some workers to be the *re-opening pressure. P_r*. Unexpectedly, the closer to tension the case in Figure 5a, the greater the corresponding compression encountered in Figure 5b once the counteracting wellbore pressure is applied. The results of Figure 4b provide insight into both hydraulic fracture initiation and subsequent propagation. For the case $\sigma_H/\sigma_h = 1, 5, P_w > \sigma_{\theta\theta}(r)$ but this situation evolves to $\sigma_H/\sigma_h = 3, P_w < \sigma_{\theta\theta}(r)$. This means that the borehole pressure required to initiate a tensile fracture is large for the former and small for the latter. However, any further tensile fracture propagation into the formation for this latter case is prevented as $\sigma_{\theta\theta}(r)$ rapidly becomes increasingly compressive. The opposite situation occurs for $\sigma_H/\sigma_h = 1.5$ as $\sigma_{\theta\theta}(r)$ decreases monotonically into the formation. Fracture growth is facilitated by the already high value of P_w relative to $\sigma_{\theta\theta}(r)$.

3.1 Thermo-elastic effects

The concentrated stresses of Equation 5 may often not be the only ones that need to be considered. Transient effects produced by the flow of heat or fluids both into and out of the borehole can also generate substantial stresses that will influence failure of the rock. Both of these processes are diffusive and in many ways have similar equations describing them.

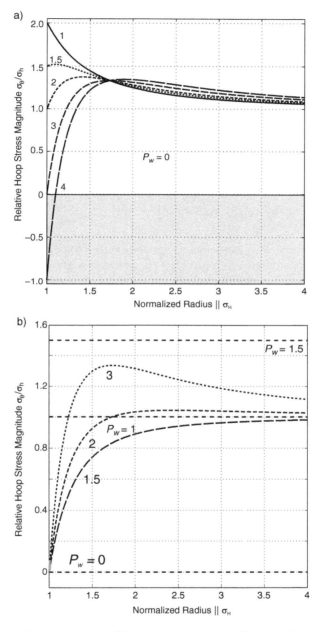

Figure 5 The value of the concentrated hoop stress σ_θ normalized by the minimum horizontal compression σ_h versus the normalized radius at the azimuth $\theta = 0$ parallel to σ_H for the cases of σ_H/σ_h of 1, 1.5, 2, 3, and 4. There is no P_w applied. Tensile zone indicated with gray background. b) The stresses in a) but modified by application of the wellbore pressure P_w (denoted by horizontal dashed lines) necessary to bring the concentrated hoop stress $\sigma_{\theta\theta}$ at the borehole wall to zero.

Thermo-elastic stress concentrations are generated once a temperature difference exists between the borehole fluid and the rock mass as first noted by Stephens and Voight (1982). This almost always exists during drilling itself where the drilling fluid temperature might often be expected to be less than the in situ temperature. In this case heat will flow from the hot rock to the cooler fluid filled borehole and generate additional stresses that superpose those described in Equation 5. The mathematics of this is rather involved. A rigorous treatment would require first time-dependent knowledge of the distribution of the temperature field $V(r,t)$ as it varies radially from the borehole, that in turn is controlled by the initial and boundary conditions on temperature. Once $V(r,t)$ is determined, the corresponding hoop stresses can be found using the radially symmetric thermo-elastic relationships

$$\sigma_{\theta\theta}^{T} = \frac{\alpha^{T} E}{1-\nu} \frac{1}{r^2} \left[\int_{a}^{r} V(r,t) r dr - V(r,t) r^2 \right] \tag{7}$$

Stephens and Voight (1982) employed an approximation for $V(r,t)$ provided by Ritchie and Sakadura (1956) based on results that are found in texts such as that by Carslaw and Jaeger (1959). They assumed a case in with the rock mass initially at uniform temperature V_o, and at time $t = 0$ the wall of a borehole drilled through it is subject to a temperature V_1 that thenceforth remains constant. The temperature difference is simply $\Delta V = V_1 - V_0$ and, surprisingly, at the borehole wall $r = a$ the induced stresses simplify to

$$\sigma_{\theta\theta}^{T}(r = a) = \frac{\alpha^{T} E}{1-\nu} \Delta V \tag{8}$$

that is independent of time.

As might be intuited, Equation 7 shows that increasing and decreasing temperatures will respectively superpose compression or tension to the hoop stresses at the borehole wall. These thermal stresses can be unexpectedly large. Broadly, the term $\alpha^{T}E/(1-v)$ can easily range from about 0.1 MPa/°C to 1 MPa/°C; an even modest temperature difference of only, say, −5 °C could generate hoop stresses of the same magnitude as the tensile strength.

3.2 Poroelastic effects

All rocks are to some degree porous and in the earth are nearly always saturated with fluids that will reside at some native formation pore pressure. The rock, too, will be permeable allowing fluid diffusion. The very geometry of the hydraulic fracturing experiment means that some such fluid diffusion from the borehole into the rock mass cannot be avoided in a realistic case. The infiltration of pressurized fluid into the rock mass increases local pore pressures and consequently induces additional stresses in a manner directly analogous to the thermoelastic situation. The differences are that instead the elastic properties of the rock must be used and that the stress distribution is controlled by the $P_P(r,t)$ instead of $V(r,t)$. Haimson and Fairhurst (1967) first examined this problem and by adapting Equation 6 using Biot's theory of poro-elasticity, obtained

$$\sigma_{\theta\theta}^P = \frac{\alpha^P(1-2\nu)}{1-\nu}\frac{1}{r^2}\left[\int_a^r P_P(r,t)rdr - P_P(r,t)r^2\right] \tag{9}$$

where the superscript P indicate stresses induced by poroelastic processes. If at the borehole wall $P_w = Pp$ and in an infinite medium the solution that considers in isolation only the stress generated due to fluid infiltration collapses simplifies to

$$\sigma_{\theta\theta}^P(r=a) = 2\eta\Delta P \tag{10}$$

where $\Delta P = P_w - P_o$. Fluid flow from the borehole into the formation generates a net compression the magnitude of which depends on η. Similarly to the thermo-elastic case, in a formation extending infinitely away from the borehole the stress immediately at the borehole wall is also independent of time (Aadnoy 1987; Detournay *et al.*, 1989; Rice & Cleary 1976), even though the pore pressures near the borehole must necessarily vary with time (See Schmitt & Zoback, 1992).

3.3 Additional effects

Sections 3.1 and 3.2 mention two possible corrections due to transport of fluid or thermal energy into the rock mass. There are also additional effects that could be important, and while we focus this contribution on simple elasticity, it is important to at least mention other factors that might need to be considered. These effects may be increasingly important at greater depths in the earth where the rock near the borehole is likely damaged by high stress concentrations or may in some cases become more ductile due to higher temperatures. This results in a redistribution of the concentrated stresses such that the validity of Eqns. 6 is no longer strictly valid, requiring modification of the break-down equations that are presented later. However, we expect that once past the borehole stress concentrations, the pressures required to propagate the fracture into the formation would behave approximately the same as for the perfectly elastic case and the values of σ_h determined would still be valid.

Even without damage, the response of most rocks to stress occurs nonlinearly. One major source of this nonlinearity arises from the existence of crack-like porosity in the rocks. Such pores are readily compressed and can with relatively modest confining pressures close (see Schmitt (2015)) with the consequence that that rock is nonlinearly elastic. This nonlinearity has been long known but is often ignored. Haimson and Tharp (1974) attempted a partial solution to this dilemma by assigning different values of Young's modulus depending upon whether the material was subject to either tension or compression, they called this a bilinear relationship. Their calculations showed that the magnitude of $\sigma_{\theta\theta}$ at the borehole wall was somewhat diminished relative to the linearly elastic case. Santarelli *et al.* (1986) and Brown *et al.* (1989) obtained pressure-dependent empirical curves for the Young's moduli of a set of sedimentary rocks. The models they developed, too, gave lower values of $\sigma_{\theta\theta}$ than the simple elastic case. Schmitt and Zoback (1992, 1993) also noted the effects of nonlinear behavior in failure of a series of internally pressurized hollow cylinders. This topic requires further study to determine how the deformation characteristics of such materials should be properly characterized. The great advantage of the linearly elastic case is that the stress distributions are independent of the material's elastic properties; this advantage is lost once nonlinear elasticity is invoked. Further, the studies above consider only axisymmetric geometries whereas it is long known that deviatoric stress states induce a state of azimuthal anisotropy around a borehole, a condition that is exploited to find stress directions in dipole sonic logging (see review in Schmitt et al., 2012). Solutions to this problem are only recently appearing (e.g., Ortiz et al. (2012)) but given the variety of material behaviors and situations encountered it is likely that one may need to examine particular cases by numerical modeling.

Time-dependent or plastic behavior of the rock could introduce additional complications (e.g., Detournay & Fairhurst (1987); Wang & Dusseault (1994)) that are in need of additional investigation also. Regardless of these additional complications, this contribution focuses on the classic analyses within the context of linear elastic or poroelastic models.

4 FIELD CONFIGURATIONS

As noted above, we are considering stress measurements in which the pressurized wellbore fluid acts directly on and can infiltrate into the borehole wall rock. There are two basic geometries that could be considered with either only the bottom-most section of the borehole being pressurized (Figure 6a–c) or a shorter interval along the

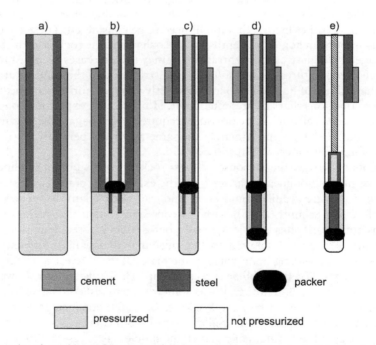

Figure 6 Examples of possible configurations of the hydraulic fracturing experiment. Cased and cemented configuration with pressurized zone consisting the bottom open hole interval and a) the entire casing string or b) the interval isolated by a packer within the casing string. Open hole configurations with c) a single packer within the open hole, d) a straddle packer isolating a smaller interval along the open borehole, and e) the same as d) but conveyed on a wireline with a downhole pump for pressurizing the interval.

borehole being isolated by a 'straddle-packer' arrangement (Figure 6d–e and Figure 7). In the first two, the borehole is entirely cased and cemented save for a small interval of open hole at the very bottom. Case c) essentially differs from those only in that the packer is activated within the open hole section. The most reliable stress determinations can be obtained, however, in cases d) and e) where a small interval along the borehole is isolated by packers; these cases too allow for multiple tests at different depths along the borehole.

There are advantages and disadvantages to the various geometries. Numerous factors come into consideration in the design of a stress measurement program including time available, costs, practicality, and safety. In the following, brief discussion of the different geometries is provided.

Case a) is encountered in petroleum drilling where 'leak-off' tests (LOT) are commonly carried out to ensure the competency of the casing and cementing, and the control of mud densities (*e.g.*, Addis *et al.*, 1998). This simply consists of a single cycle of pressurization in which breakdown may or may not be achieved with the pressures measured at the surface. The results from LOT's are commonly interpreted to estimate σ_h (*e.g.*, Breckels & van Eekelen, 1982) but such values are generally deemed unreliable and this motivated workers to develop what are now usually referred to as 'extended leak-off tests' (ELOT or XLOT). Kunze and Steiger (1992) suggested that ideally the XLOT include repeated cycles of pressurization, accurate downhole pressure measurement (or at least appropriate corrections for the pressure head), and sufficient time after shut-in for the pressure to decrease in order that the fracture closes. Practice usually does not achieve this regrettably, and Zoback (2007) provides an extensive critical discussion of the analysis of such data.

Some of the more important disadvantages (Li *et al.*, 2009) of these tests include the poor pressure sensitivity of the large volume of fluid within the casing and bottom-hole interval to the small changes due to creation or closure of an fracture (Ito *et al.*, 1999), the use of compressible (relative to water) and often non-Newtonian fluids, the lack (usually) or corresponding information on fracture orientations, and the complicated 3-D concentration of stresses in the vicinity of the bottom-hole that have resulted in horizontal fractures in the laboratory (Haimson & Fairhurst, 1969a). Raaen *et al.* (2006) warn that use of open intervals of more than a few meters length should not be used for stress determination. If the pressure is measured at the surface, viscous losses through the drill string may attenuate the pressure responses making determination of a breakdown pressure even less reliable. Despite these cautions, this geometry may be all that is available in some cases.

Cases b) and c) are variations of the XLOT geometry that include a packer deployed within either the casing or in the open hole. One advantage of these is that the pressure sensitivity may be increased due to the fact that a smaller mass of fluid need is used and hence the pressure changes are more substantial. With sufficient foresight, the packer system could be constructed to incorporate pressure transducers. A problem with single packer systems, however, is that substantial lifting forces push upwards during pressurization and if this exceeds the packers' frictional resistance then the assembly will have no choice but to move upwards. In some examples, this is known to result in damage to the equipment, blockage of the borehole, and danger for operational personnel at the surface.

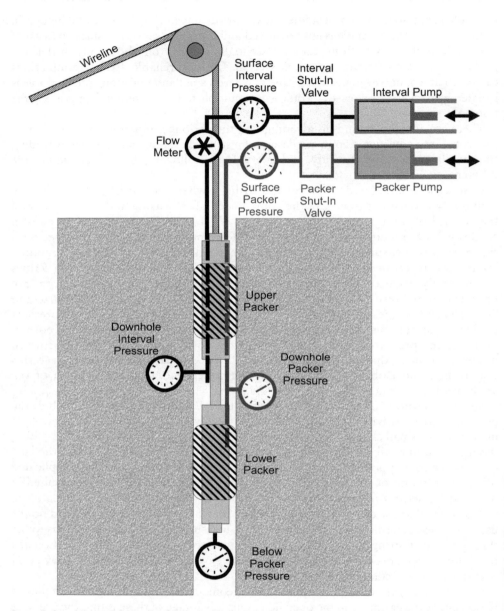

Figure 7 Straddle-packer configuration for hydraulic fracture stress measurement consisting of a pair of inflatable packers that seal the interval between them. Downhole pressure transducers continuously record the pressures in the packers, in the interval, and in the open borehole beneath the packers. Both the packer and interval pressures are also simultaneously measured at the surface. The amount of fluid pumped both into and returned from the interval is measured with a flowmeter. Valves are placed past the pumps in order to allow the packers and the interval hydraulic circuits to be shut-in or opened at the appropriate time during a test cycle.

In deep wellbores subject to significant borehole instability, packers in the open hole may not be able to provide a proper pressure-tight seal and Case b) may be the only option. This geometry was successfully used to estimate stresses at a depth of 9 km in the KTB deep borehole project in Germany (Brudy *et al.*, 1997). In contrast, safety considerations did not admit a similar test carried out at 6 km to reach sufficient pressures for an unambiguous determination of the fracture closure pressure. We are not aware of any publications in which stresses have been measured from the geometry of Case c) but this configuration has been popular in attempts to locally relieve stress in deep mines.

Cases d) and e) show straddle packer assemblies that are pressurized either by a pump at the surface or one built into the system itself, respectfully. Case d) may be deployed either by a steel casing or by wireline with the surface pump connected to the interval (and also independently to the packers) by the drill string itself, by drill string and high-pressure hoses, or, in the case of wireline deployment, by separate hoses. Straddle packers have a number of advantages. They allow for numerous tests along the open hole section of the borehole as opposed to only the single test that can be carried out for the XLOT cases. For example, Evans *et al.* (1989) obtained 75 measurements to depths as great as nearly 1040 m in three closely spaced boreholes in New York, Klee *et al.* (1999) nearly 150 in a 18 different boreholes to depths of 255 m, and Schmitt *et al.* (2012) were able to obtain 17 successful tests at depths of nearly 1500 m from the rig floor over a 24 hour period in order to constrain the state of stress in McMurdo Sound, Antarctica. Further, safety is increased substantially as the fluid pressure in the interval is both contained to a small volume and pushes equally against the two packers with no resultant net force that could push the packers upwards. In HF measurements conducted in slim boreholes drilled from the surface or from underground openings, in conjunction with civil or mining operations, case d) is exclusively used.

Case e) includes its own downhole pumping system. This capability has been available commercially for larger wellbores typical of petroleum drilling (*e.g.*, Desroches & Kurkjian, 1999; Thiercelin *et al.*, 1996) but smaller systems that might be applied in scientific or mining drilling applications have been designed but not yet constructed (Ito *et al.*, 2006). The great advantages of the downhole pumping system is that overall it is sufficiently stiffer, *i.e.*, a small abrupt change in the volume of the system due to breakdown or fracture reopening effect a larger relative change in the interval pressure. Conversely, pumping rates are much more restricted and this could be a problem if the formation is so permeable that P_w cannot build rapidly enough to create a fracture.

Hydraulic fracturing stress measurements go by many different names that are not necessarily consistently applied. The terms microfrac and minifrac have been associated generically with any of the above geometries. According to de Bree and Walters (1989) these are characterized on the volume and rate of fluid pumping in the test. In a microfrac a total of only 10 to 500 l are pumped at rates of 5 to 50 l/minute. A minifrac is significantly larger with 1.5 m^3 to 15 m^3 pumped at rates of 0.75 to 1.5 m^3/minute. These contrast with massive hydraulic fracture stimulations that in a single stage would typically pump 100 m^3 to 200 m^3 at rates of ~10 m^3/minute (Anonymous, 2009). Further, within the petroleum industry the acronym DFIT (diagnostic fracture-injection test) has evolved past the limits of its original definition as a transient pressure test primarily aimed at estimating in situ permeabilities in 'tight' formations (Craig & Brown, 1999) to also encompassing nearly any test carried out in which stresses are estimated regardless of procedure or geometry. Generally, however, determination of stresses from DFIT's is often complicated by the fact that they usually only consist of a single pressurization cycle that allows for only one measure of a fracture closure pressure P_{FC} and they are often carried out through perforated casing that disallows use of any of the stress concentration formulas presented here. Workers must take care to note carefully how 'DFIT's are actually carried out in order to minimize the risk that the results are over interpreted.

5 SYSTEM CONFIGURATION

The discussion here focuses on stress measurements made using the straddle packer configurations of Figure 6d and e. The interpretation of the pressurization records from the straddle-packer geometry is simplified by the fact that the experimental geometry is well constrained. In contrast the interpretation of records from system configurations 6a to 6c can be complicated by numerous factors such as the stress concentrations at the wellbore bottom (Rumzan & Schmitt, 2001) or irregular geometry and fractures of the open bottom-hole segment of the borehole (Haimson & Fairhurst, 1970; Zoback, 2007).

The design of the HF tool and the fluid pressure sensor location are key to a successful measurement. Ideally, in deep holes, like those encountered in oil-field wellbores, fluid pressure should be measured downhole in order to obtain the highest quality pressure records. As such, the tool should host pressure transducers that record in situ both the interval and the packer pressures. This would be done in real time but unless the tool is conveyed on a wireline doing this becomes difficult. At a minimum one can employ memory logging sensors many of which are now commercially available although care must be taken in selection to ensure that adequate sampling rates are achieved. Additional sensors that measure the pressure in the borehole both above but particularly below the packers can provide useful quality control information, a change in the pressure outside of the packed interval indicate improper sealing such that the pressurized interval fluid could leak past. Downhole sensors have the additional advantages in that any viscous losses through the pressurization system may be ignored and that no corrections for head pressures need be made.

In tests conducted in relatively short slim holes (say 0–500 m), and typically in hard rock of low permeability, surface recording of fluid flow rate and test-interval pressure is adequate.

Most configurations will include a surface pump. At the very minimum one must be able to track these surface pressures in real time in order to control the pressurization rate and to cease pumping once break-down is reached.

The pressure transducers need also be carefully selected to match the expected range of pressures encountered. Transducers whose range greatly exceeds the maximums may not be sufficiently accurate. Those whose range is too small may be damaged by overpressures. Other factors to consider are the transdicers' response times. In selecting pressure sensors, traditionally strain-gage-based transducers have been used. However, quartz resonator gages that measure pressure from fundamental changes in the resonance frequency of a pure quartz crystal have the advantages of fast response times, negligible drift, and high accuracy.

A flow meter to track both the amount and rate at which fluid is pumped in and subsequently allowed to flow back should also be a component of any system. Ito *et al.* (1999) assert that the position at which this flow meter is placed can have important consequences for stress measurement, if the flow meter is placed near the packers then the effects of the overall system compliance are minimized and better records of re-opening pressures are obtained.

Workers should also carefully consider the rate at which any data is digitally sampled. Regrettably, many of the systems employed were designed for other purposes and may not even record pressures at a uniform sampling rate or at a sampling rate that is insufficient to obtain a clear pressure record. A rough guideline is that one sample/second should be the absolute minimum, but faster rates of 10 or even 100 samples per second are recommended in order to better ensure that fine details of the pressurization curves, such as the peak at breakdown, are adequately captured. Such data sampling is easily achievable today with many off the shelf data loggers readily available. However, power consumption is also related to digitization rates and the energy that can be provided by small batteries in adverse downhole conditions remains problematic.

We focus our discussion on a design that has been particularly popular for quantitative stress measurements The major components of a system for stress determinations would include (Figure 7):

i.   A means to convey the packer system into the borehole. This is accomplished in Figure 7 by a wireline, which can also accommodate conductors, allowing the down-hole pressures to be monitored in real time. Alternatively, conveyance of the packer assembly at the bottom of a drill string is also popular. A system employed successfully in slim holes utilizes two high –pressure hoses, and, when deemed necessary, a data cable, attached on the outside of the drill string or wireline. The pressure hoses are for pressurizing the packers and the test interval; the data cable is for surface monitoring of fluid pressures in the packers and interval that are recorded at depth. There are also commercial systems for deep wellbores that allow switching between pressurization of the packers and of the interval using downhole valves that are controlled by set movements of the drill string up or down. This last configuration does not easily allow for the downhole pressures to be displayed at the surface during the test but they may still be continuously recorded with memory gages and retrieved later. An advantage of this configuration, however, is that the drill rig's pulling force is significantly greater than that pulling the wireline, and as such there is much better chance that the packer system can be retrieved should it become stuck.

ii.   A pair of packers connected via a steel pipe or mandrel. The mandrel must be sufficiently strong that it can safely hold the opposing pull of the two packers upon pressurization of the interval. A great deal of care must be taken in the selection of the packers, and workers will need to consider in situ temperatures, the peak differential pressure (*i.e.* the difference between the actual pressure in the packer and the ambient borehole fluid pressure outside of it), and the packer diameter. This is a particular concern in boreholes drilled from underground workings that may have only small head pressures or even be dry. Special quadruple packer designs have been constructed to the innermost packers sealing the interval to be supported (Ask *et al.*, 2009), thus greatly increasing the overall peak pressure that can be attained.

iii. A pumping system that can attain both sufficient pressures and flow rates. Conversely, higher pumping rates may increase viscous losses to the point that pressures measured at the surface will not properly track those within the interval and substantial errors in the interpretation of surface gage recordings may result. In Figure 7 two separate pumping systems are shown but use of an appropriate manifold can allow a single pump to be employed.

iv. Valves that will allow the packers and the interval to be 'shut in' or opened at the appropriate times during the test. Ideally these will be as close to the interval as possible in order to 'stiffen' the system (Klee *et al.*, 1999).

v. At least one flow meter that can measure the flow rate $q(t)$ to allow for determination of amount of fluid entering the interval's hydraulic circuit prior to shut in and returning after release. The amount of fluid returned may also be measured by collecting this volume directly.

vi. A series of pressure transducers to record the interval and packer pressures and, ideally, both at the surface and downhole. Downhole transducers are much preferred as they measure the changes in pressure directly at the point of the measurement. In contrast, the pressures sensed by the surface transducers are affected by viscous losses and system compliance and must be corrected for differences in head between the surface and the measurement point in the borehole. A fifth pressure transducer has been added below the lower packer in Figure 7 the purpose of which is to assess leakage of pressurized interval fluid past an improperly seated lower packer. A further pressure sensor above the upper packer would provide further confidence in the measurements.

This data should all be digitally recorded at a sufficiently high sampling rate in order that rapid changes in the pressures may be detected (Holzhausen *et al.*, 1989). This should be easily achieved with wireline deployed systems where live signal is readily accessible at the surface, but may be problematic for memory gauge systems that often sample at best once per second due to power and memory restrictions.

One example of the full set of pressure and flows recorded by a wireline hydraulic fracturing system designed for the ANDRILL project is provided in Figure 8. This configuration included a flow meter that recorded the rates of fluid injected and returned from the interval, and four pressure transducers recording the packer and interval pressures both downhole and at the surface. The liquid in the borehole and the system was seawater. It is particularly instructive to compare the responses between transducers recording at test interval at depth with that recorded at the surface after the appropriate correction for the head difference of 1412 m (Figure 8). The surface recorded pressures during period of rapid pressure increase are substantially greater than those measured in situ due to the high viscous losses through the 2 km of hose. Those pressures agree when pressures are only slowly changing. While this is to some degree an extreme case it does highlight the importance of being able to measure pressures as close to the zone being studied as possible.

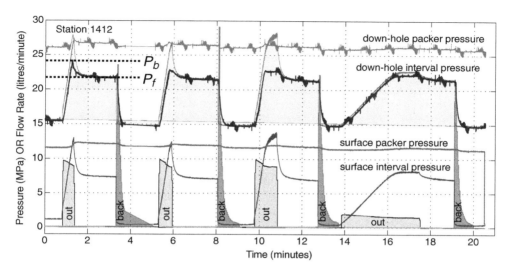

Figure 8 An example of actual set of recordings from a straddle-packer system at a depth of 1412 m from the surface. Filled light and dark gray areas represent flow into and out of the interval, respectively. Records include the pressure transducers recording the pressure within the packers both at the surface and downhole and the interval pressure Pw(t) both at the surface and downhole. The upper gray-filled area is the surface interval pressure after correction for the fluid column head pressure to allow for direct comparison to the interval pressure measured downhole. Note that both downhole transducers were affected by periodic electronic noise. Figure after Schmitt et al. (2012).

6 TEST PROTOCOL AND INTERPRETATION OF $P_w(T)$

The ultimate goal of hydraulic fracturing stress measurement is to obtain the in situ values of σ_H and σ_h in the earth. This is accomplished by pressurizing a section of the borehole in order to create a fracture that will propagate a sufficient distance into the formation such that it will be subject to the same virgin stresses. Protocols developed on the basis of experience have been developed (Haimson & Fairhurst, 1970; Haimson & Cornet, 2003; Hickman & Zoback, 1983) to carry this out, and it is essential that this first be described in order that the reader can understand the rationale for a given pressurization record. We take the perhaps unusual approach in this section of providing the conventional interpretations of fracture initiation and propagation and provide the basic theory used to explain this.

The protocol described here closely follows Haimson and Cornet (2003) recommendations:

i. Due diligence must be exercised in finding positions along the borehole suitable for the hydraulic fracture tests, using all the information available. Zones containing natural fractures or drilling induced borehole failure must be avoided primarily because these will compromise the packer's ability to seal the interval or even lead to packer rupture. Unless they are specifically sought for purposes of stress determination (*e.g.*, Cornet & Valette (1984)), their existence will complicate analysis of the pressurization records. The experiences of the drilling personnel can highlight both prospective zones and those to be avoided. Rock cores provide information on the rock quality and on the existence of pre-existing natural fractures. The core may allow workers to assess whether such natural fractures are presently open or sealed and impermeable. The core material, too, can be studied in the laboratory to provide measures of the essential physical properties above.

However, cores do not provide any direct information on the borehole geometry and geophysical logs should be acquired. The following requirements pertain especially to deep holes drilled in weak rock. At the very minimum, a simple caliper log can demonstrate whether or not the borehole is at the expected diameter for the drill bit used (this may change slightly with depth due to bit wear). In low permeability formations, single-point-resistance (SPR), spontaneous potential (SP), and other electrical resistance logs could expose the open natural fractures. Similarly, anomalies in fluid chemistry or the temperature gradient can provide supportive information. Tube waves seen in full waveform sonic logs and even vertical seismic profiles will also indicate open fractures.

The best methods, however, will rely on oriented borehole imaging (Luthi, 2005). Ultrasonic borehole televiewers first appeared in analog forms in the late 1960s (Zemanek *et al.*, 1969) with digitization appearing in the 1980s. They are popular in smaller diameter, water-filled boreholes because they provide high-resolution mappings of azimuth (typically from 0.5° to 2°) versus depth (typically less than 1 cm) of both the ultrasonic acoustic reflectivity and two-way transit time. The large number of transit times around the borehole azimuth allow for ready assessment of the circularity of the borehole. In the petroleum industry, micro-resistivity imaging techniques that employ multi-electrode pads pushed against the borehole wall to provide millimeter scale images of the electrical resistivity of the wall rock are usually preferred (Ekstrom *et al.*, 1987). These tools do provide oriented caliper measurements but cannot provide the same azimuthal resolution of borehole radius as the ultrasonic techniques. Such tools have been employed in boreholes drilled through igneous or metamorphic formations (Pezard & Luthi, 1988) but the high resistivities of these rocks is problematic for the processing of an interpretable image (*e.g.*, Chan, 2013). Taken together, workers should employ as much information as possible to best site the hydraulic fracturing measurement.

ii. Lower the packer assembly to the selected test zone and inflate the packers to pressures P_k sufficient to inflate them and to provide a suitable seal. These pressures typically be 2 to 4 MPa above the ambient borehole pressures. Some care must be taken to avoid having the packers inadvertently fracture the formation prior to the test. Workers may also need to make corrections for differences in depths arising between the core, log, and packer conveyance systems to ensure that the packers are correctly positioned.

iii. With the packers properly seated, workers may wish to attempt a slug test in order to estimate the in situ permeability and the system compliance. This test can be accomplished by rapidly decreasing or increasing the pressure within the system and then monitoring the time it takes for the borehole pressure to re-equilibrate (*e.g.*, Doan *et al.*, 2006). If downhole pressures can be monitored in real time this test could also reveal if the packers have properly sealed the interval. The pressure excursions should not be large, however, in order to avoid disturbing the pore pressure field within the borehole.

iv. Knowledge of the formation pore pressure P_p is also useful. This possibly could be found in detailed analysis of the slug test or by other means but as mentioned by Gaaremstroom *et al.*, (1993) the errors can be large. The key components of the test itself are described in this section and referenced to the highly idealized pressurization records $P_w(t)$ of Figure 9 and Figure 10, which would hypothetically be expected for a 'stiff' hydraulic system (*i.e.* P_w is highly sensitive to changes in the contained volume) not subject to any viscous losses and containing liquid whose compressibility for practical purposes remains constant during the test. Further, the hydraulic system of the interval is assumed to be perfectly sealed with no leakage due to faulty equipment or inadequately sealed packers. Of course, in reality some or all of these factors may influence the observed pressures $P_w(t)$. A typical test would consequently consist of the following $P_w(t)$ sequence of pressurization cycles.

First cycle: Following Figure 9 the interval pressure P_w is first increased from the ambient equilibrated pressure P_{EQ} within the wellbore at a constant rate sufficient such that 'breakdown' at pressure P_b occurs within a reasonably short period of time (<~3 minutes). As pumping continues, the newly created fracture continues to grow at a more or less uniform pressure P_{FP} until pumping ceases and the interval is isolated, or 'shut-in' whereupon the pressure drops abruptly to the '*instantaneous shut-in pressure*' P_{SI}. $P_w(t)$ continues to decline at first due to permeable infiltration into the surrounding rock mass via both the borehole and the faces of the fractures until the '*fracture closure*' pressure P_{FC} and from only the borehole subsequently. This first cycle terminates after a suitable period of time by venting the interval and allowing the interval pressure to re-equilibrate back to P_{EQ}.

Note that at all times the packer pressure must sufficiently exceed the interval pressure to prevent the interval fluid from escaping past the packers. Choice of the packers becomes important, sliding-end packers are preferred as their pressure will increase automatically. In slim holes and shallow depths separate pumps and lines to the packers are often used to independently control their pressure.

Initially, $P_w(t)$ increases linearly once uniform pumping begins after t_o until the first break in slope is observed at the incipient *fracture initiation pressure* P_{FI}, this is also often referred to as the *leak-off pressure LOP*. This change reflects the slight increase of the system's compliance upon formation of the incipient fracture.

It is important to investigate the various criteria that have been used to predict when this occurs particularly as these will in turn be used in later sections to provide constraints on σ_H. The simplest criteria assumes the rock to be completely impermeable to the wellbore fluid, the basis of this comes directly from Hubbert and Willis (1957) original contribution as modified by Scheidegger (1960) who added the strength T to arrive at the simple and classic formula

$$P_{FI} = 3\sigma_h - \sigma_H + T - P_o \tag{11}$$

Building on this, Haimson and Fairhurst (1967) further superposed the poroelastic stress at the borehole wall resulting from infiltration into the formation of Equation 11 and took the pore pressure P_P responsible for failure in Equation 4 to be equal P_{FI}

$$P_{FI} = \frac{3\sigma_h - \sigma_H + T - 2\eta P_o}{2(1-\eta)} \qquad (12)$$

which they referred to as the critical pressure. Recalling from above that $0 \le \eta \le 0.5$ the denominator in Equation 12 is always greater than or equal to unity and as such this means that if fluid infiltration is included we would expect P_{FI} to be smaller than expected for the classic case Equation 11. The differences between Equations 11 and 12 are described in more detail in a following section as these are important to the practical interpretations of the pressurization curves. It is worthwhile to point out that they may provide the upper and lower bounds for P_{FI} attained by 'fast' and 'slow' pressurization, respectively (Detournay & Carbonell, 1997; Garagash & Detournay, 1997).

$P_w(t)$ continues to rise until the peak 'breakdown' pressure P_b is reached (Figure 9). This is usually taken to be the point where a hydraulically induced tensile fracture extends unstably, increasing the volume of the interval's hydraulic system more rapidly than can be sustained by flow into it from the pump.

There are a few points worth noting here in regards to the interpretation of real $P_w(t)$ records. First, P_b will depend on the relative magnitudes of the horizontal stresses, particularly σ_h, and this may influence the shape of the pressurization curve (see

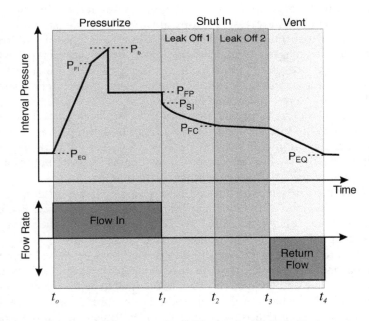

Figure 9 Idealized classic pressurization curve for the first cycle of a hydraulic fracturing measurement. Upper and lower graphs respectively represent the simultaneously recorded interval pressure PI and the flow rate either into or out of the interval. Four distinct time periods of 1) Pressurization (to to t1) with fluid entering the interval at a constant flow rate, 2) Leak off 1 (t1 to t2), 3) Leak off 3 (t2 to t3), and 4) Vent (t3 to t4) are delimited by differing background colors. The interval pressures begin (t0) and end (t4) at the equilibrium pressure P_{EQ}. The other interval pressures indicated are those at fracture initiation P_{FI}, break-down Pb, fracture propagation while pumping P_{FP}, instantaneous shut-in pressure P_{SI}, and fracture closure pressure P_{FC}.

Hickman & Zoback, 1983; Thiercelin *et al.*, 1996). Second, the actual volume change to the interval hydraulic system due to the incipient fracture at P_{FI} is small and actually detecting it may be difficult (see Morita *et al.*'s (1996) laboratory test records). Recent advancements in downhole sensor technologies that allow displacements of the borehole wall rock to be recorded within the interval during a test, however, may improve this situation (Cornet *et al.*, 2003; Guglielmi *et al.*, 2014, Holzhausen *et al.*, 1989, Ito & Hayashi, 1996). Consequently, P_b often provides an upper bound for P_{FI} and is often for practical purposes used in its place (see discussion by Detournay & Carbonell, 1997).

The wellbore fluid viscosity may also play a role. Morita *et al.* (1996) carried out a significant number of hydraulic fracturing tests from boreholes drilled into large (70 cm × 70 cm × 70 cm) cubes and showed that the initial fracture could extend several centimeters into the formation prior to breakdown if highly viscous drilling muds filled the borehole. It is important to measure this accurately as it provides the only means to estimate σ_H as will be discussed in later sections. It should be noted that in HF tests in slim holes the fluid used for pressurizing the test interval is typically water, in which case viscosity is not an important factor.

There has been considerable debate with regards to the use of Equations 11 and 12 as they can give quite different answers. Detournay and Carbonell (1997) for example point out that as $\eta \to 0$ with $K \to K_s$ in the limit of a nonporous solid we would expect the P_{FI} predicted by Equations 11 and 12 to match but in fact they will differ by a factor of 2. Paradoxically, the two formulas match when $\eta = 0.5$ where we might least expect them to agree. Schmitt and Zoback (1989) determined η from combinations of laboratory measurements on core and sonic log wave speeds from two boreholes drilled in crystalline rock in which hydraulic fracturing stress testing had been carried out. Using these data, they applied Equations 11 and 12 to calculate the σ_H and found, perhaps not surprisingly, that the values differed significantly and that some of the predicted values were nonphysical. To overcome this, they modified the Terzaghi effective stress law for failure of Equation 4 with a term $0 \leq \beta \leq 1$ that served to diminish the influence of the pore pressure on failure. Although at the time they had little theoretical justification for this, later laboratory experiments on rocks and glass provided some empirical support (Schmitt & Zoback, 1992, 1993).

While pumping continues the newly created fracture propagates into the formation at the *fracture propagation pressure* P_{FP}, sometimes also called the 'formation parting pressure'. Various models suggest this fracture growth is promoted if it contains pressurized fluid (*e.g.*, Shimizu *et al.*, 2011; Zoback & Pollard, 1978) that is of sufficiently low viscosity to fill the fracture. In order for the fracture to propagate, P_{FP} must act against the total stress normal its plane, overcome viscous losses due to fluid flow into and within it, and transmit sufficient energy to the crack tip to exceed the material's tensile toughness. Once this fracture has propagated outside of the zone where the borehole stress concentrations dominate (about 3 radii) the fracture will open parallel to the least principal compression σ_{min}, Consequently P_{FP} provides an upper bound to σ_{min}, whether it be σ_h or σ_V depending upon the faulting regime.

At t_1 the interval is *shut-in* and the pressure rapidly falls to the *instantaneous shut in pressure* P_{SI} where the fluid flow and fracture growth has been arrested or nearly so (Desroches & Thiercelin, 1993). In the literature there is no consensus on how long pumping should continue after breakdown is established. Some workers *shut-in* the system as immediately as possible upon breakdown while others prefer to allow the fracture to grow into the formation sufficiently that it is outside of the zone of stress concentrations (Figure 5). Presumably the pressure within the interval hydraulic system has equilibrated and P_{SI} just maintains the fracture open. $P_w(t)$ must continue to decrease, however due to percolation of the interval fluid into the formation until at t_2 it reaches the fracture closure pressure P_{FC} that is nearly equal to but just less than σ_{min} and the fracture closes. As such, P_{FC} should give the best quantitative measure of σ_{min} although one should take care in equating these (McLennan & Roegiers, 1982). In the context of small volume stress testing here, it may be difficult in actual practice to actually distinguish P_{SI} from P_{FC} and in much of the oil field-related literature they are taken to be one and the same; confusion in the use of the terms can arise. In deep wellbore HF testing P_{SI} is usually taken as the pressure that equals σ_{min}. Another pragmatic reason may be that P_{SI} is more easily determined from $P_w(t)$ than P_{FC} (Breckels & van Eekelen, 1982; White et al., 2002), This is not normally the case in slim hole testing, where the shut-in pressure P_{SI} is often equal to σ_{min}.

The first pressurization cycle ends with the venting of the pressurized interval liquid. The flow rate and amounts of fluid returned should be measured if possible in order to assess the volume lost to the formation. Sufficient time then needs to be taken for this 'flow-back' (Hickman & Zoback, 1983) to allow $P_w(t)$ to return to its pre-test equilibrium level P_{EQ} prior to commencing the second pressurization cycle. One may want to take advantage of this time by again shutting in the interval and monitoring a rise in pressure that would provide additional certainty that the packers were properly sealed (Haimson & Cornet, 2003).

It is important to note that the idealized pressurization curve of Figure 9 is not the only one that may be encountered (Fjaer et al., 2008, Hickman & Zoback, 1983). Returning to Figure 5b the reader will note that if the magnitude of σ_H is sufficiently larger that of σ_h, the borehole pressure $P_w(t)$ necessary to put the borehole wall into tension will be less than σ_h. Consequently P_{FI} will be less than P_{FP} and P_{FC} and instead of reaching a sharp breakdown $P_w(t)$, will more gradually approach peak value (see Hickman & Zoback (1983)). Appropriate care must be taken in the interpretation of the breakdown pressure in these situations.

Second (and subsequent) Pressurization Cycles: A proper hydraulic fracturing stress test will consist of a number of repeated pressurization cycles (Gronseth & Detournay, 1979; Hickman & Zoback, 1983). Haimson and Cornet (2003) recommend at least three of these cycles be carried out. The rationale for this are manifold and include ensuring that a stable fracture, P_{SI} and P_{SC} are achieved, providing additional information as to the relative magnitudes of the stresses in the formation from the character of the re-pressurization curves (e.g., Hickman & Zoback, 1983), and allowing for an estimation of the tensile strength T when compared to the first cycle (Bredehoeft et al., 1976). The elements between the first and the subsequent pressurization cycles are essentially the same but with the key difference that now a fracture already exists and it, presumably, has no tensile strength. Consequently, $P_w(t)$ increases to the *re-opening pressure* P_R at which point the concentrated hoop stresses are overcome and the fracture re-opens unstably as at P_b in the first cycle. Similarly, continued pumping into the interval will extend the fracture further into the formation. There is little difference between the remainder of this second cycle and the latter parts of the first cycle and all of the descriptions remain the same.

It is important to comment on the physical meaning of P_R as, like P_{FI} there are diverging views as to how it should be explained. Bredehoeft et al.'s (1976) classic interpretation assumes that the stress concentrations around the borehole return to their pre-fracture state but now the rock's tensile strength T need not be overcome. Their key assumption is that despite having no strength, the hydraulic fracture hydraulically seals upon closing and no pressurized fluid is admitted to it prior to re-opening with P_R occurring when:

$$P_R = 3\sigma_h - \sigma_H - P_P \tag{13}$$

that is simply Equation 11 with T omitted and with the corollary that $T = P_b - P_R$.

Various workers have questioned the validity of the sealed closed fracture assumption (Bunger et al., 2010, Hardy & Asgian, 1989; Ito et al., 1999; Ito & Hayashi, 1993; Rutqvist et al., 2000) and two simplified limiting cases may be considered with regards to fluid pressures within the already created fracture during the test. In the first the fluid pressure within the fracture is the same as in the wellbore and consequently the effect of P_w (t) on opening is essentially doubled by the combination of the Lamé tension and $P_w(t)$'s normal loading of the fracture surfaces, such that the re-opening criteria becomes

$$P^R = \frac{3\sigma_h - \sigma_H}{2} \tag{14}$$

The reader should note that P_P is omitted from Equation 14. Rutqvist et al. (2000) suggest that this is because $P_w(t)$ replaces P_P but alternatively one can arrive at this by more simply considering the total loading stresses on the borehole wall and fracture faces in which case the formation pore pressure may be neglected as no failure criterion (Equation 4) need be invoked.

In the second case the fracture is sufficiently large and it is so permeable that $P_w(t)$ uniformly applies a uniform total normal stress to the fracture faces that is just sufficient to overcome σ_{min}; consequently at best

$$P_R \approx \sigma_{min} \tag{15}$$

Lee and Haimson (1989), Sano et al. (2005), White et al. (2002) and Zoback et al. (2003) provide convincing evidence that this may often be the case and, indeed, Debree and Walters (1989) assume this in their analyses. Some care must be taken with this, however, and White et al. (2002) are careful to note that the P_{FI} or the P_R are similar to σ_{min} within experimental errors inherent to measurement in deep wells but that they are not strictly the same.

Finally, If poroelastic effects are included then following Detournay et al. (1989)

$$P_R = \frac{3\sigma_h - \sigma_H - 2\eta P_P}{2(1 - \eta)} \tag{16}$$

Step tests. The stress measurement experiment may sometimes conclude with a step-pressure or hydraulic jacking test (Lee & Haimson, 1989; Rutqvist & Stephansson, 1996) that can provide additional information on σ_h. Here, while carefully monitoring the flow rate, $P_w(t)$ is increased in a series of discrete pressure steps. At each level, $P_w(t)$ is held constant until the flow has stabilized. Once this is accomplished the $P_w(t)$ is increased to the next step. Ideally, the previously created hydraulic fracture remains closed until $P_w(t) > \sigma_h$ whereupon the fracture re-opens and the flow rate increases. This

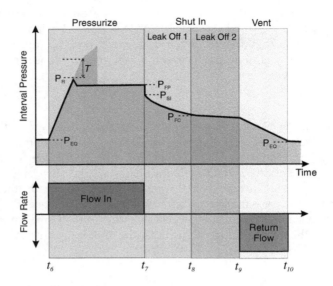

Figure 10 Idealized classic pressurization curve $P_w(t)$ for the second and subsequent cycles of a hydraulic fracturing stress measurement. Upper and lower graphs represent the simultaneously recorded interval pressure $P_w(t)$ and the flow rate into or out of the interval hydraulic system. Four distinct time periods are shown as in Figure 9. To allow for comparison, $P_w(t)$ for the first cycle is also delineated by the darker shaded area within the interval pressure plot.

re-opening pressure is then found by a break in the slope of the plot of $P_w(t)$ versus the flow rate $q(t)$. In the petroleum industry, step-rate tests in which the flow $q(t)$ is instead held constant at each step are popular. In more permeable formations such tests can also provide estimates of the formation permeability (*e.g.*, Weng *et al.*, 2002).

Debree and Walters (1989) introduced their 'tuned injection rate test' which also follows the main suite of cycles. In this test the operator raises $P_w(t)$ to a pressure just below the expected value of $\sigma_{min,}$ (assumed by them to be equal to P_R) and then 'tunes' the injected flow rate such that the pressure alternates from just above to just below σ_{min} such that the fracture cyclically re-opens and closes. Consequently the maximum and minimum pressure seen should match P_R and P_{FC} respectively. By varying the rate, however, one should be able to find the level at which $P_w(t)$ remains constant and this presumably gives σ_{min}.

Once these cycles are satisfactorily completed the packers are deflated and the assembly moved to the next test site in the borehole.

v. Upon completion of the pressure testing the packer assembly is removed from the borehole. The pressure records by themselves can only indicate stress magnitudes, but knowledge of the actual geographic direction of the horizontal stresses is also key. Given the 2θ azimuthal symmetry of the borehole (Figure 4), hydraulic fracturing would ideally have created two opposing vertical fractures radiating laterally into the formation. These fractures will lie at the azimuths parallel to σ_H; consequently determination of the azimuth of the fractures gives directly the direction of σ_H as can be inferred from Figure 4b.

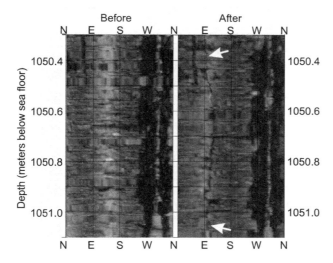

Figure 11 An example a 1-m section of the ultrasonic borehole televiewer images acquired before and after a hydraulic fracturing stress test at depths from 1050.2 to 1051.2 m below the sea floor. A single East-striking hydraulic fracture is seen in the right-hand panel. Texture of the image is caused by stick-slip motion of the televiewer through a high clay zone. After Schmitt et al. (2012).

This can be accomplished in numerous different ways with the various borehole imaging tools available today as already mentioned in Step 1. An example of before and after ultrasonic borehole televiewer image logs (Figure 11) shows a clear hydraulic fracture generally striking East. The existence of the fracture in the image is also further evidence that the test in question was successful. It is important to note, however, that in this case it is not clear whether an opposing fracture was created or not because of lost signal at the western azimuths.

The original technique that is still popular is to make an impression of the created fractures by pressurizing a single packer, wrapped with a thin deformable elastomer, in the fractured zone. When the packer is pressurized and allowed to set for a sufficient period of time, the elastomer molds to the small variations in the topography of the borehole wall associated with the fracture. This packer includes on-board orientation sensors, usually magnetometers, that allow its orientation with respect to magnetic North to be known. However, the fracture traces are not always perfect vertical lines and the variations in the orientations calls for statistical analysis. When the impression packer is returned to the surface, the impression of the fracture is readily mapped from the elastomer sheet. Lee and Haimson (1989) describe a technique that employs circular statistics to determine the azimuths of the induced vertical fractures as well as providing a rigorous measure of the uncertainty.

In hard rocks, the fracture created can often be subtle and it may not be readily visible in the image logs. Baumgärtner and Zoback (1989) overcame this difficulty by disrupting the borehole wall with an 'impression' packer to make the fractures more visible with the ultrasonic borehole televiewer.

This step completes the hydraulic fracturing stress test. Before the stresses may be determined, however, one must first determine the approximate key pressures of P_{FI}

and P_{FC} necessary for stress interpretation; and common strategies used to do this are described in the next section.

There is little discussion in the literature with regards to the determination of the pressures associated with fracture initiation, breakdown, and re-opening. This is in part due to the fact that often P_b and sometimes P_R are clearly recognized as the peak pressure after which the fracture propagates unstably. This is not necessarily always the case, however, because as noted earlier σ_H if significantly larger than σ_h then the $P_w(t)$ necessary to cancel out the resulting stress concentrations will be less than σ_{min} (Figure 5). Regardless of how unambiguous the breakdown pressure might be, it is actually the fracture initiation pressure that is actually desired but it, too, can be difficult to detect and will depend critically on the overall compliance of the hydraulic system.

One way to overcome this is to actually monitor the compliance of the system during pressurization. By plotting the pressure $P_w(t)$ versus the total volume $V(t)$ injected to the interval, Lee and Haimson (1989), and Ito *et al.* (1999) for example, were able to refine their determination of P_R from $P_w(t)$ records by locating the pressure at which this curve first deviated from a straight line indicating a change in the system compliance.

Lee and Haimson (1989) provide some statistical rigor in the determination of P_R. They note that P_R is often defined to be where the initial ascending portion of $P_w(t)$ in a re-opening cycle departs from that in the first cycle prior to breakdown. This of course assumes that the flow rate in both cycles remains the same. However, determination of this point can be subjective. They overcome this limitation by finding the minimum sum or squares errors between the $P_w(t)$ observed for the first cycle (Fig 9) and the subsequent re-opening cycle (Fig. 10).

More discussion has centered on finding P_{SI} or P_{FC} It is generally accepted that P_{FC} provides the best measure of σ_{min} provided the fracture is sufficiently large that its plane is outside of the borehole's stress concentration and that this fracture will 'close' once the pressure in the fracture falls just beneath σ_{min}. That said, and as discussed already, there are minor variations in the opinions as to where this might actually occur within a pressure record, and in most of the petroleum engineering literature the shut-in pressure P_{SI} is used. In actual practice, however, the differing preferences for naming of this are somewhat moot as nearly all of the procedures described below essentially rely on finding a point in $P_w(t)$ during the shut in period that reflects a change in the leak-off behavior. As such, and as re-iterated by Guo *et al.* (1993) this decline will depend on the permeability of the formation and this fact will be one factor to consider when attempting to determine P_{SI}. In civil or mining related slim holes, there are simple techniques for the determination of the actual P_{SC}, as well as for the correct magnitude of P_R (Lee & Haimson, 1989).

Table 1 gives an overview of a number of the different strategies that have been used to estimate σ_{min} from $P_w(t)$. These examples come from a wide range of situations, however, particularly with regards to whether they are employed in nearly impermeable crystalline rocks and shales or in more permeable sediments. As such, care should be taken in applying a given criterion to the case at hand. Permeability, for example, will be the primary factor controlling the length of the shut-in period and it is questionable whether those methods that rely on pressure decline due to infiltration into the formation will apply in low permeability formations. Conversely, in the same low permeability formations the time required to actually reach P_{FC} may be prohibitively long and this may need to be considered in the design of the experiment.

Table 1 Strategies for determining the shut-in pressure.

Method	Authors	Comments
$P_w(t)$ vs t	Gronseth and Kry (1981)	Draw tangent line to $P_w(t)$ vs t immediately after shut-in (t_I in Figure 9). σ_{min} is defined as the point at which $P_w(t)$ departs from this line. See also Baumgärtner and Zoback (1989).
$P_w(t)$ vs $log(t+\Delta t)/\Delta t$	McLellan and Roegiers (1981)	t is the time in the pressurization curve and $\Delta t = t - t_I$. σ_{min} is pressure at the point of inflection on this curve.
Asymptotic Method	Aamodt and Kuriyagawa (1981)	Plot $log(Pw(t) - p_a)$ vs Δt where p_a is a trial value for the pressure to which the pressure decline will approach asymptotically during a long shut-in period. P_a is varied until the best fit line is found, this line when extrapolated back to Dt = 0 will have intercept P_e. The, calculate $\sigma_{min} = P_e + P_a + $ 'hydrostatic pressure'. See also Cornet and Valette (1984) adaptation.
Variable Flow Rate	Aamodt and Kuriyagawa (1981)	During the test at one depth, carry out a number of pressurization cycles at different flow rates. Draw a curve through the peak pressure reached in each of these cycles (P_R) versus the flow rate. σ_{min} is taken to be the pressure at the 'knee' in this smooth curve.
Pressure-flow rate method	Lee and Haimson (1989)	This is carried out following the main test. The flowrate is incrementally increased in small steps allowing $P_w(t)$ to stabilize in each case. Essentially, at lower flow rates and pressures the fracture remains closed and the slope of the plot of $P_w(t)$ vs flow rate $q(t)$ remains high. At a higher pressure the fracture opens and the flow rate increases such that this slope is diminished. Under the assumption that $P_w(t)$ is linearly related to $q(t)$ and that the slope changes between the times that the fracture is closed or open allows a bilinear regression to find the best pressure at which the fracture reopens.
$P_w(t)$ vs $log(t)$	Doe et al. (1981)	Concept based on analogy to a two-stage pulse test for transient flow in fractures. σ_{min} is the point of inflection on this curve.
$log(P_w(t))$ vs $log(t)$	Haimson and Rummel (1982)	Different leak-off conditions will occur before and after closure of the fracture; in the $log(P_w(t))$ vs $log(t)$ space these appear as two lines of differing slopes. σ_{min} is taken to be the pressure at which these two lines cross.
$dP_w(t)/dt$ vs $P_w(t)$ or t	Baumgärtner and Zoback (1989), Guo et al. (1993)	Plot $dP_w(t)/dt$ vs $P_w(t)$ Again, different leak off conditions will exist before and after closure of the fracture resulting in different rates of change of $Pw(t)$ in each segment of the shut-in and that each segment can be approximated by a line. σ_{min} is taken to be the pressure at the point where these two lines intersect.
Bilinear Curve Fitting to $dP_w(t)/dt$ vs $P_w(t)$	Lee and Haimson (1989)	The pressure decay curves in the *Leak Off 1* and *Leak Off 2* time windows (Fig. 9) differ due to the variations in geometry. It is assumed that within each time window that the derivative of $P_w(t)$ will take the form $dP_w(t)/dt = aexp(at+b) \sim a(P_w(t) - P_{asym})$ where a and b are unknown constants and P_{asym} is an unknown asymptotic pressures, this is a line in $dP_w(t)/dt$ vs $P_w(t)$ space. The A nonlinear regression of this equation is individually applied to the time windows for the *Leak Off 1* and *Leak Off 2* (Fig. 9) periods using arbitrarily chosen values of the as yet unknown transition time t_2. The t_2 at which the lowest value of the sum of squared error for both curves is found is taken as the best measure of the time for P_{FC}.

Table 1 (Cont.)

Method	Authors	Comments
$P_w(t)$ vs $(\Delta t)^{\frac{1}{2}}$	Guo et al. (1993)	Plot $P_w(t)$ vs $(\Delta t)^{\frac{1}{2}}$. The initial time period of this curve will be nearly linear; σ_{min} is taken to be the pressure at the point the curve diverges from linearity.
$d^2 P_w(t)/dt^2$ vs $P_w(t)$	Guo et al. (1993)	This is also called the maximum curvature method as one would expect that the most rapid change in the decline behavior of $P_w(t)$ to be recognized. σ_{min} is taken to be the pressure at the maximum.
Decline Function	Debree and Walters (1989)	Plot $[P_w(t) - P_o]^\alpha$ vs $(1 + \delta)^{3/2} - \delta^{3/2}$ where $\delta = (t \text{-} t_1)/t$ is the dimensionless time since shut-in and $0 \leq \alpha \leq 1$ is an exponent they called the 'filtration power index' governing the pressure dependence of the leak-off. Plotting with this modified time base ideally forces even complex pressure decline curves to be linear. One must, however, determine two values of α that dictate behaviour before and after fracture closure either by special pressure tests in the well or by trial and error. This has many of the elements of 'G-time' analysis applied widely in the petroleum industry for the interpretation of pressure transient records.
System Stiffness	Raaen et al. (2006)	During a pump-in flow-back test, the volume returned is plotted versus $P_w(t)$.
LOP	White et al. (2002)	This suggestion comes from observations that leak off pressures (P_{FI}) and even fracture initiation pressures (P_R) are not meaningfully different in deep wells. σ_{min} is *approximated* by P_{FI} or P_R. See also discussion in Raaen et al. (2006). This is not recommended if more accurate understanding of the stress tensor is required.
Reopening Pressure	Ito and Hayashi (1993)	Here, as in Equation 15 $\sigma_{min} = P_R$ directly. This differs from the LOP in which P_R only suggests a value for σ_{min}.
Fracture Propagation Pressure	Zoback and Pollard (1978)	As discussed above, continued pumping past break-down or re-opening results in further propagation of the fracture into the formation. As indicated in Figures 9 and 10, during this time $P_w(t)$ remains relatively constant at P_{FP} and provides an upper bound to σ_{min}.
Nonlinear Regression of Post Closure Pressure-decay	Lee and Haimson (1989)	This iterative method seeks to delineate P_{FC} by assuming that the pressure decay after fracture closures is dominated by radial flow from interval only into the formation as defined by an exponential decay $P_w(t) = exp(d_1 t + d_2)$ for $t_f \geq t \geq t_2$ (i.e. the *Leak Off 2* period in Fig. 9). The essential idea is that a nonlinear regression of the pressures within zone t_2-t_3 is well described by the exponential decay formula; and that the fit is poorer should the data points within the *Leak Off 1* period from t_1 to t_2 be included. Hence, the method iteratively calculates the nonlinear regression residuals over progressively smaller time windows starting from t_1 to t_3 but successively removing the earliest data point in the series. The $P_w(t)$ at which the normalized residuals stabilize is declared equal to P_{FC}.

7 DETERMINATION OF THE STRESS TENSOR

Following the ISRM Suggested Method for Rock Stress Estimation using Hydraulic Fracturing (Haimson & Cornet, 2003), the calculations of the in situ principal stresses presented here are for HF tests conducted in vertical boreholes, which result in vertical to sub-vertical hydraulic fractures. This corresponds to the case in which the vertical in situ stress component acts along a principal direction in a reasonably isotropic rock.

To reiterate from the above:

1. σ_V can be estimated from the lithological load if knowledge of the densities are available as given by Equation 1.
2. As already discussed, in general the induced vertical hydraulic fractures will initially strike perpendicular to the direction of the minimum horizontal principal stress, σ_h. Consequently, the azimuth of the hydraulic fracture on the borehole wall indicates the direction of σ_H, which because the principal stresses are all orthogonal, is sufficient to orient all three stresses.
3. The P_{FC}, or in some cases P_{SI}, is taken to be just sufficient to counteract the principal stress component normal to the hydraulic fracture, Regardless of the faulting regime, these pressures give σ_{min}. In the normal and reverse faulting regime σ_h is unambiguously the least principal compression. However, in thrust faulting regimes σ_h is the intermediate principal stress and care need to be taken in the interpretation of P_{SI} as it may instead represent σ_V, one should compare the value so determined with that estimated using densities in Equation 1.
4. Unfortunately, σ_H cannot be so directly determined and it must be calculated by re-arranging the terms in Equations 11 through 16 into 'breakdown' equations using the already obtained value of σ_h. This maximum principal stress is calculated based on the assumption of linear elasticity and insignificant effect of fracturing fluid infiltration into the rock. Using Equation 11 the classic breakdown equation is

$$\sigma_H = 3\sigma_h - P_b + T - P_o \tag{17}$$

where P_{FI} has been replaced for historical reasons with P_b.

Solving Equation 17 requires that the rock tensile strength be known. The tensile strength can only be directly measured in the laboratory on core samples. The most common tensile test is the Brazilian test, which enables the testing of many disks cut directly from the extracted core. The Brazilian test configuration, however, does not simulate conditions under hydraulic fracturing, and the reliability of this test as representative of the tensile strength for hydraulic fracturing has not been established. Core is also used to prepare hollow cylinders, which are fractured by applying internal pressure, with no external confining stress. This test accurately simulates a hydraulic fracturing test in which there are no far-field stresses, and therefore the peak pressure is equal to the tensile strength T. The only unknown in such tests is the well-established scale effect between field and laboratory dimensions.

When extracted core is not available, or laboratory tests are not feasible, or when tension tests appear to yield an unreasonable value for use in Equation 17 an alternative relation has been used, invoking the fracture reopening pressure P_R. This pressure is assumed to represent the level at which the previously induced fracture reopens during a subsequent pressure cycle. The reopening pressure does not have to overcome the tensile strength T, since the rock has already been fractured. Thus, Equation 17 becomes:

$$\sigma_H = 3\sigma_b - P_R - P_o \tag{18}$$

This equation for calculating σ_H has been widely used in field measurement campaigns. There is, however, some controversy regarding its reliability in some circumstances as discussed earlier. These classic breakdown Equations 17 and 18 provide the upper estimate to σ_H.

Now, if fluid infiltration is allowed Equation 12 is organized as

$$\begin{aligned}\sigma_H &= 3\sigma_b - 2(1-\eta)P_b + T - 2\eta P_o \\ &= 3\sigma_b + 2\eta(P_b - P_o) - P_b + (T - P_b)\end{aligned} \tag{19}$$

The last expansion is done purposefully in order to explicitly highlight all of the factors influencing $\sigma_{\theta\theta}(r=a)$ including the amplified compression of σ_b, the compression induced by fluid flow into the formation, and Lamé tension of the wellbore pressure, and the tensile strength as modified by the pore pressure that acts at failure (Schmitt & Zoback, 1993) For the re-opening case:

$$\sigma_H = 3\sigma_b - 2(1-\eta)P_R - 2\eta P_o \tag{20}$$

where again $0 \leq \eta \leq 0.5$. In the absence of a formation pore pressure P_o, the lower bound to σ_H occurs when $\eta = 0$.

Even though these are relatively simple equations, it is useful to explore them in more detail to assess their influence on stress interpretations by varying both T and P_o and calculating the value of σ_H that would be obtained for a given P_b under differing scenarios (Figure 12). In the first case both T and P_o are omitted and the upper and lower limits to σ_H are then given by the classic Equation 17 and the infiltration Equation 19 when $\eta = 0$, respectively, and the range of possible values indicated by the dark fill. The second case indicated by light fill is the same as the first except that a tensile strength T = $0.2\sigma_b$ has been included. The difference between these two would ideally be the same as considering the initial cycle with breakdown and later cycles with re-opening. The largest range of possible σ_H values is encountered when $P_b = \sigma_b$ where ignoring T, $\sigma_b \leq \sigma_H \leq 2\sigma_b$. Further, the plots suggest that $P_R \leq 2\sigma_b$. The range of possible solutions decreases as $\sigma_H \rightarrow \sigma_b$.

A more interesting situation arises when the same T is used but now a formation pore pressure $P_o = 0.4\sigma_b$ is included and σ_H is calculated using only the infiltration Equation 19 with limiting values of $\eta = 0.5$ and $\eta = 0$. The range of possible solutions in this case is marked by the cross-hatched region. At about $P_b/\sigma_b \sim 0.4$ the lines cross and this represents the point bounding zones dominated by the additional compression due to fluid infiltration to the left and those controlled by the borehole fluid pressure $P_w(t)$ promoting failure to the right. According to this scenario, the maximum breakdown pressure can be no more than about $1.75\ \sigma_b$. A useful observation from this plot is that if P_o is included the allowable σ_H range is more restricted. One should carefully note, however, that Figure 12 only displays the range of physical solutions and that one can obtain nonsensical values (*i.e.* $\sigma_H < \sigma_b$) depending on the choice of η (Schmitt & Zoback, 1989).

Owing to the difficulties with both pore pressure effects and tensile strength estimation, the evaluation of the maximum horizontal principal stress magnitude involves a

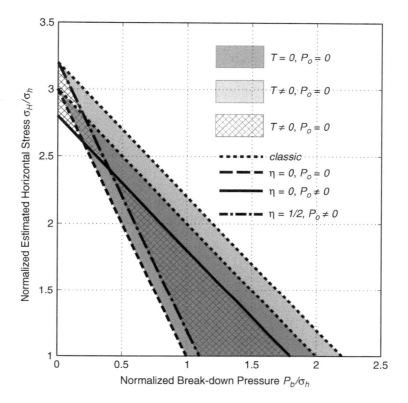

Figure 12 Normalized break-down pressure P_b/σ_h versus estimated greatest horizontal principal compressions σ_H/σ_h for Case 1 (dark gray transparent fill): omitting tensile strength T and formation pore pressure P_o, Case 2 (gray fill area): omitting pore pressure P_o but including $T = 0.2\sigma_h$; Case 3: Including $P_o = 0.4\sigma_h$ and $T = 0.2\sigma_h$. Heavy black lines represent the limiting estimates of σ_H. Dotted line: σ_H calculated using the classic breakdown Equation 17 for the cases with and without T; dashed line: σ_H calculated for $\eta = 0$ and with no pore pressure P_o in the formation; dot-dash line: two limiting cases coincide both including T but either σ_H estimated with $\eta = 0.5$ with Po = 0.4 σ_h under Equation 19 or with $\eta = 0$ omitting Po under Equation 18; solid line: σ_H estimated with $\eta = 0$ with Po = 0.4 σ_h. under Equation 19.

greater uncertainty than that of the minimum horizontal principal stress. This is where the qualification and experience of the test personnel is particularly important in order to ascertain whether the values determined are reasonable.

8 CONCLUDING REMARKS

A grand challenge in the Geosciences and Geological Engineering is to be able to obtain quantitatively the full stress tensor at a point in the earth from deep boreholes. The last quarter century has seen developments in the interpretation of borehole breakouts, drilling induced tensile fractures both in core and on the wellbore wall, and azimuthal acoustic anisotropy. These have been successful in determining stress directions. However, none of these techniques are able to provide quantitative stress magnitudes without making assumptions about rock properties. As such, at this point in history, we generally accept (but may need to still definitively show) that knowledge of the overlying density gives us σ_V. We also generally accept that if properly interpreted the hydraulic fracturing test gives σ_{min} that represents the magnitude of σ_h. However, there is still no way to directly measure σ_H.

The hydraulic fracturing technique is the only stress measurement method that explicitly applies forces to the rock mass transmitted via fluid pressure. This leads to the simplicity in the general interpretation of P_{SI}, or P_{FC} for σ_{min}. It also leads to the development of relatively simple equations to predict the pressures at which breakdown or fracture re-opening occur, and these are rearranged to conversely obtain estimates of σ_H. The discussions above give some of the flavor of the problems that can arise when attempting to do this, however, and there remain some problems to resolve.

One technical issue is the reliability of the determination of the various pressures. We rely almost exclusively on interpretation of the pressurization record $P_w(t)$, but this can be complicated by a variety of factors, such as viscous losses, inherent to the system. As many workers have recognized, being able to monitor the deformation of the borehole wall during the test can provide important complementary information and more developments along the lines of electrical imaging (*e.g.*, Cornet *et al.*, 2003) or strain recording (*e.g.*, Guglielmi *et al.*, 2014) show promise.

A more major issue relates to how these pressures are then used to estimate σ_H. There have been numerous theoretical analyses advanced over the last few years that would warrant an entire review on their own, many of which employ fracture mechanics concepts (*e.g.*, Bunger *et al.*, 2005; Mathias & van Reeuwijk, 2009; Sarvaramini & Garagash, 2015) or numerical models (*e.g.*, Jiao *et al.*, 2015; Sheng *et al.*, 2015, Shimizu *et al.*, 2011). Regardless of this progress, in many ways we still only have a rudimentary understanding of the physics of fracture initiation and propagation in materials even simpler than rock (Pook, 2010), and this calls for renewed laboratory and field testing against which these models can be tested. A corollary to this statement is that we also need renewed laboratory investigations leading to a better understanding of the nonlinear and time-dependent physical properties of the rocks under investigation.

Finally, while in many ways the hydraulic fracture technique will provide the best quantitative constraints on stress magnitudes, practitioners should avoid using it in isolation. Workers should consider all of the information available to them from geology, geophysics, and rock mechanics (Sano *et al.*, 2005; Schmitt *et al.*, 2012; Zang & Stephansson, 2010; Zoback *et al.*, 2003) when attempting to estimate the state of stress in the earth.

REFERENCES

Aadnoy, B. 1987, A complete elastic model for fluid-induced and in-situ generated stressses with the presence of a borehole. *Energy Sources*, 9:239–259.

Aamodt, R.L., and M. Kuriyagawa. 1981, Measurement of Instantaneous Shut-In Pressure in Crystalline Rock. In M.D. Zoback and B. Haimson, eds., *Hydraulic Fracturing Stress Measurements*. Monterey, CA: Natinoal Academy Press.

Abousayed A.S., C.E. Brechtel, and R.J. Clifton 1978. In situ stress determination by hydrofracturing – fracture mechanics approach. *Journal of Geophysical Research* 83:2851–2862. doi: 10.1029/JB083iB06p02851.

Addis, M.A., T.H. Hanssen, N. Yassir, D.R. Willoughby, and J. Enever. 1998, *A Comparison of Leak-Off Test and Extended Leak-Off Test Data for Stress Estimation*. Richardson, Texas. Society of Petroleum Engineers.

Amadei, B., W.Z. Savage, and H.S. Swolfs. 1987, Gravitational stresses in anisotropic rock masses. *International Journal of Rock Mechanics and Mining Sciences & Geomechanics Abstracts*, 24, no. 1,5–14. doi: 10.1016/0148-9062(87)91227-7.

Anderson, E.M. 1951, *The Dynamics of Faulting and Dyke Formation with Applications to Britain*. 2nd ed: Edinburgh, Scotland: Oliver and Boyd.

Anonymous. 2009, *Modern Shale Gas Development in the United States: A Primer*. National Energy Technology Laboratory.

Ask, D., F.H. Cornet, F. Fontbonne, T. Nilsson, L. Jonsson, and M.V.S. Ask. 2009, A quadruple packer tool for conducting hydraulic stress measurements in mines and other high stress settings. *International Journal of Rock Mechanics and Mining Sciences*, 46, no. 6,1097–1102. doi: 10.1016/j.ijrmms.2009.04.004.

Baumgärtner, J., and Rummel, F. 1989, Experience with fracture pressurization tests as a stress measuring technique in a jointed rock mass. *International Journal of Rock Mechanics and Mining Sciences and Geomechanics Abstracts*, 26:661–671.

Baumgärtner, J., and M.D. Zoback. 1989, Interpretation of hydraulic fracturing pressure-time records using interactive analysis methods. *International Journal of Rock Mechanics and Mining Sciences & Geomechanics Abstracts*, 26, no. 6,461–469.

Bickley, W.G. 1928, The distribution of stress around a circular hole in a plate. *Philosophical Transactions of the Royal Society of London: A: Mathematical, Physical, and Engineering Sciences*, 227: 647–658.

Biot, M.A. 1941, General theory of three-dimensional consolidation. *Journal of Applied Physics*, 12:155–164. doi: 10.1063/1.1712886.

Bjarnason, B., C. Ljunggren, and O. Stephansson. 1989, New developments in hydrofracturing stress measurements at Luleå University of technology. *International Journal of Rock Mechanics and Mining Sciences & Geomechanics Abstracts*, 26, no. 6,579–586. doi: http://dx.doi.org/10.1016/0148-9062(89)91438-1.

Breckels, I.M., and H.A.M. van Eekelen. 1982, *Relationship Between Horizontal Stress and Depth in Sedimentary Basins*. doi: 10.2118/10336-PA.

Bredehoeft, J.D., R.G. Wolff, W.S. Keys, and E. Shuter. 1976, Hydraulic fracturing to determine regional in situ stress field, Piceance Basin, Colorado. *Geological Society of America Bulletin*, 87:250–258. doi: 10.1130/0016-7606(1976)87<250:hftdtr>2.0.co;2

Brown, E.T., J.W. Bray, and F.J. Santarelli. 1989, Influence of stress-dependent elastic moduli on stresses and strains around axisymmetric boreholes, *Rock Mechanics and Rock Engineering*, 22:189–203. doi: 10.1007/bf01470986.

Brudy, M., M.D. Zoback, K. Fuchs, F. Rummel, and J. Baumgartner. 1997, Estimation of the complete stress tensor to 8 km depth in the KTB scientific drill holes: Implications for crustal strength. *Journal of Geophysical Research-Solid Earth*, 102, no. B8,18453–18475. doi: 10.1029/96jb02942.

Bunger, A.P., E. Detournay, and D.I. Garagash. 2005, Toughness-dominated hydraulic fracture with leak-off. *International Journal of Fracture*, 134, no. 2,175–190. doi: 10.1007/s10704-005-0154-0.

Bunger, A.P., A. Lakirouhani, and E. Detournay. 2010, Modelling the effect of injection system compressibility and viscous fluid flow on hydraulic fracture breakdown pressure. In X. Furen, ed., *Rock Stress and Earthquakes*. Boca Raton, Florida: CRC Press. 59–67.

Carslaw, H., and J. Jaeger. 1959, *Conduction of Heat in Solids*. 2nd. ed: London, UK: Oxford Univeristy Press.

Chan, J. 2013, *Subsurface Geophysical Characterization of the Crystalline Canadian Shield in Northeastern Alberta: Implications for Geothermal Development*. Edmonton, Alberta: University of Alberta.

Cornet, F.H. 1993, 12. Stresses in Rock and Rock Masses: Principles, Practice, and Projects. In J. A. Hudson, ed., *Comprehensive Rock Engineering*. Pergamon Press. 297–327.

Cornet, F.H., M.L. Doan, and F. Fontbonne. 2003, Electrical imaging and hydraulic testing for a complete stress determination. *International Journal of Rock Mechanics and Mining Sciences*, 40, no. 7–8,1225–1241. doi: 10.1016/s1365-1609(03)00109-6.

Cornet, F.H., and C. Fairhurst. 1974, Influence of pore pressure on the deformation behavior of saturated rocks. In Advances in Rock Mechanics: Proceedings of the Third Congress of the International Society for Rock Mechanics, Denver, Colorado, September 1–7.

Cornet, F.H., and B. Valette. 1984, In situ stress determination from hydraulic injection test data. *Journal of Geophysical Research*, 89:1527–1537. doi: 10.1029/JB089iB13p11527.

Craig, D.P., and T.D. Brown. 1999, Estimating pore Pressure and Permeability in Massively Stacked Lenticular Reservoirs Using Diagnostic Fracture-Injection Tests. In *SPE Annual Technical Conference and Exhibition*. Houston: Society of Petroleum Engineers.

Debree, P., and Walters, J.V. (1989) Micro minifrac test procedures and interpretation for in situ stress determination. International Journal of Rock Mechanics and Mining Sciences & Geomechanics Abstracts 26:515–521. doi: 10.1016/0148-9062(89)91429-0.

Desroches, J., and A.L. Kurkjian. 1999, Applications of Wireline Stress Measurements. *SPE Reservoir Evaluation & Engineering*, 2, no. 5. doi: 10.2118/58086-PA.

Desroches, J., and M. Thiercelin. 1993, Modelling the propagation and closure of micro-hydraulic fractures. *International Journal of Rock Mechanics and Mining Sciences & Geomechanics Abstracts*, 30, no. 7,1231–1234. doi: http://dx.doi.org/10.1016/0148-9062(93)90100-R.

Detournay, E., and R. Carbonell. 1997, Fracture-Mechanics Analysis of the Breakdown Process in Minifracture or Leakoff Test. *SPE Production & Facilities*, 12, no. 3,195–199. doi: 10.2118/28076-PA.

Detournay, E and C. Fairhurst. 1987, Two-dimensional elastoplastic analysis of a long, cylindrical cavity under nonhydrostatic loading. *International Journal of Rock Mechanics and Mining Sciences & Geomechanics Abstracts*, 24: 197–211.

Detournay, E., A.H.D. Cheng, J.C. Roegiers, and J.D. McLennan. 1989, Poroelasticity considerations in in situ stress determination by hydraulic fracturing. *International Journal of Rock Mechanics and Mining Sciences & Geomechanics Abstracts*, 26:507–513. doi: 10.1016/0148-9062(89)91428-9.

Doan, M.L., E.E. Brodsky, Y. Kano, and K.F. Ma. 2006, In situ measurement of the hydraulic diffusivity of the active Chelungpu Fault, Taiwan. *Geophysical Research Letters*, 33, no. 16,5. doi: 10.1029/2006gl026889.

Doe, W.T., W.A. Hustrulid, B. Liejon, K. Ingevald, and L. Strindell. 1981, Determination of the state of stress at the Stripa Mine, Sweden. In M.D. Zoback and B. Haimson, eds., *Hydraulic Fracturing Stress Measurements* Monterey, CA: Natinoal Academcy Press.

Ekstrom, M., C. Dahan, M.-Y. Chen, P. Lloyd, and D. Rossi. 1987, Formation imaging with microelectrical scanning arrays. *The Log Analyst*, 28,294–306.

Evans, K.F., T. Engelder, and R.A. Plumb. 1989, Appalacian stress study. 1. A detailed description of in situ stress variations in Devonian shales of the Appalacian Plateau. *Journal of Geophysical Research-Solid Earth and Planets*, 94, no. B6,7129–7154. doi: 10.1029/JB094iB06p07129.

Fairhurst, C. 1964, Measurement of in situ rock stresses with particular reference to hydraulic fracturing. *Felsmechanik und Ingenieurgeologie*, 2:129–147.

Fairhurst, C. 1965, On the determination of the state of stress in rock masses. In *Conference on Drilling and Rock Mechanics*. Austin: Society of Petroleum Engineers.

Figueiredo, B., F.H. Cornet, L. Lamas, and J. Muralha. 2014, Determination of the stress field in a mountainous granite rock mass. *International Journal of Rock Mechanics and Mining Sciences*, 72,37–48. doi: 10.1016/j.ijrmms.2014.07.017.

Fjaer, E., R.M. Holt, P. Horsrud, A.M. Raaen, and R. Risnes. 2008, Mechanics of hydraulic fracturing. *Petroleum Related Rock Mechanics.* 2nd ed: Elsevier Science Bv. 369–390.

Gaarenstroom, L, R.A.J. Tromp, M.C. de Jong, and A.M. Brandenberg, 1993 Overpressures in the central north sea: Implications for trap integrity and drilling safety. *Geological Society, London, Petroleum Geology Conference Series,* 4:1305–1313. doi: 10.1144/0041305.

Garagash, D., and E. Detournay. 1997, An analysis of the influence of the pressurization rate on the borehole breakdown pressure. *International Journal of Solids and Structures,* 34, no. 24,3099–3118. doi: http://dx.doi.org/10.1016/S0020-7683(96)00174-6.

Gronseth, J., and P. Kry. 1981, Instantaneous shut-in pressure and its relationship to the minimum in situ stress. In M.D. Zoback and B. Haimson, eds., *Hydraulic Fracturing Stress Measurements.* Monterey, CA: National Academcy Press.

Gronseth, J. M., and E. Detournay. 1979, Improved stress determination procedures by hydraulic fracturing: final report. Minneapolis, MN.

Guglielmi, Y., F. Cappa, H. Lancon, J. B. Janowczyk, J. Rutqvist, C. F. Tsang, and J. S. Y. Wang. 2014, ISRM Suggested Method for Step-Rate Injection Method for Fracture In-Situ Properties (SIMFIP): Using a 3-Components Borehole Deformation Sensor. Rock Mechanics and Rock Engineering, 47, no. 1,303–311. doi: 10.1007/s00603-013-0517-1.

Guo, F., N. R. Morgenstern, and J. D. Scott. 1993, Interpretation of hydraulic fracturing pressure: A comparison of eight methods used to identify shut-in pressure. International Journal of Rock Mechanics and Mining Sciences & Geomechanics Abstracts, 30, no. 6, 627–631. doi: http://dx.doi.org/10.1016/0148-9062(93)91222-5.

Haimson, B.C., and C. Fairhurst. 1967, Initiation and Extension of Hydraulic Fractures in Rocks. Journal of Petroleum Technology, 7, no. 3,310–318. doi: 10.2118/1710-pa.

Haimson B.C., 1968, Hydraulic fracturing in porous and nonporous rock and its potential for determining in situ stresses at great depth, PhD Thesis Univ. of Minnesota and Technical Report 4-68 Missouri River Division Corps of Engineers, 235 pp.

Haimson, B.C., and C. Fairhurst. 1969a, Hydraulic Fracturing in Porous-Permeable Materials. Journal of Petroleum Technology, 21, no. JUL,811–817.

Haimson, B.C., and C. Fairhurst, 1970, In-situ stress determination at great depth by means of hydraulic fracturing. In Somerton W.H., ed., *Rock Mechanics – Theory and Practice.* Am. Inst. Mining. Engrg. 559–584.

Haimson, B.C., (1973, Earthquake related stresses at Rangely, Colorado. In H.R. Hardy and R. Stefanko, eds., *New Horizons in Rock Mechanics.,* Am. Soc. of Civil Engr., 689–708.

Haimson, B.C., 1975a, Deep in-situ stress measurements by hydrofracturing. *Tectonophysics,* 29:41–47. doi: 10.1016/0040-1951(75)90131-6.

Haimson, B.C., 1975b, Design of underground powerhouses and the importance of pre-excavation stress measurements. In *16th U.S Symposium on Rock Mechanics.* Minneapolis: American Rock Mechanics Association.

Haimson, B.C., and B. Voight, 1977, Crustal stress in Iceland. *Pure and Applied Geophysics,* 115:153–190. doi: 10.1007/bf01637102.

Haimson, B.C., 1978a, Crustal stress in the Michigan Basin. *Journal of Geophysical Research* 83:5857–5863. doi: 10.1029/JB083iB12p05857

Haimson, B.C., 1978b, Hydrofracturing stress measuring method and recent field results. *International Journal of Rock Mechanics and Mining Sciences,* 15, no. 4,167–178. doi: 10.1016/0148-9062(78)91223-8.

Haimson, B.C., ed.,1989a, *Hydraulic Fracturing Stress Measurements* (Special Volume of the *International Journal of Rock Mechanics and Mining Sciences,* 26: 240).

Haimson, B.C. 1989b, Advances in test record interpretation: 1. Introduction. *International Journal of Rock Mechanics and Mining Sciences & Geomechanics Abstracts,* 26, no. 6, 445–445. doi: 10.1016/0148-9062(89)91419-8.

Haimson, B.C., and F.H. Cornet. 2003, ISRM suggested methods for rock stress estimation – Part 3: hydraulic fracturing (HF) and/or hydraulic testing of pre-existing fractures (HTPF). *International Journal of Rock Mechanics and Mining Sciences*, 40, no. 7–8, 1011–1020.

Haimson, B.C., and F. Rummel. 1982, Hydrofracturing stress measurements in the Iceland-research-drilling-project drill hole at Reydarfjordur, Iceland. *Journal of Geophysical Research*, 87:6631–6649. doi: 10.1029/JB087iB08p06631.

Hardy, M.P., and M.I. Asgian. 1989, Fracture reopening during hydraulic fracturing stress determinations. *International Journal of Rock Mechanics and Mining Sciences & Geomechanics Abstracts*, 26:489–497. doi: 10.1016/0148-9062(89)91426-5.

Hickman, S., and M.D. Zoback. 1983, The interpretation of hydraulic fracturing pressure-time data for in-situ stress determination. In M.D. Zoback and B.C. Haimson, *Hydraulic Fracturing Stress Measurements: Proceedings of a Workshop* by. Monterey, CA: National Academy Press.

Hiramatsu, Y., and Y. Oka. 1962, Stress around a shaft or level excavated in ground with a three-dimensional stress state. *Memoirs of the Faculty of Engineering, Kyoto University*, Part I., 24, 56–76.

Holzhausen, G., P. Branagan, H. Egan, and R. Wilmer. 1989, Fracture closure pressures from free-oscillation measurements during stress-testing in complex reservoirs. *International Journal of Rock Mechanics and Mining Sciences & Geomechanics Abstracts*, 26, no. 6, 533–540. doi: 10.1016/0148-9062(89)91431-9.

Hubbert, M.K., and D.G. Willis. 1957, Mechanics Of Hydraulic Fracturing. *Petroleum Transactions of the AIME*, 210:153–163.

Ito, H., and K. Hayashi. 1996, A study for in-situ stress measurements from circumferential deformation of a borehole wall due to pressurization. In M. Aubertin, F. Hassani and H. Mitri, eds., *NARMS'96, a Regional Conference of ISRM*. Montréal: Rotterdam; Brookfield: A.A. Balkema, 1996.

Ito, T., K. Evans, K. Kawai, and K. Hayashi. 1999, Hydraulic fracture reopening pressure and the estimation of maximum horizontal stress. International Journal of Rock Mechanics and Mining Sciences, 36, no. 6,811–825. doi: 10.1016/s0148-9062(99)00053-4.

Ito T., and K. Hayashi. 1993, Analysis of crack reopening behavior for hydrofrac stress measurement. *International Journal of Rock Mechanics and Mining Sciences & Geomechanics Abstracts*, 30:1235–1240. doi: 10.1016/0148-9062(93)90101-i.

Ito, T., H. Kato, and H. Tanaka. 2006, Innovative concept of hydrofracturing for deep stress measurement. In M. Lu, C.C. Li, H. Kjorholt and H. Dahle, eds., *In-Situ Rock Stress Measurement*. Interpretation and Application: Taylor & Francis Ltd.

Jaeger, J.C., N.G.W. Cook, and R.W. Zimmerman. 2007, *Fundamentals of Rock Mechanics*. 4th ed.: Blackwell.

Jiao, Y.Y., H.Q. Zhang, X.L. Zhang, H.B. Li, and Q.H. Jiang. 2015, A two-dimensional coupled hydromechanical discontinuum model for simulating rock hydraulic fracturing. *International Journal for Numerical and Analytical Methods in Geomechanics*, 39, no. 5,457–481. doi: 10.1002/nag.2314.

Kirsch, G. 1898, Die Theorie der Elastizitat und die Bedurfnisse der Festigkeitslehre. *Zeit Verein Deutsch Ing*, 42,797–807.

Klee, G., F. Rummel, and A. Williams. 1999, Hydraulic fracturing stress measurements in Hong Kong. *International Journal of Rock Mechanics and Mining Sciences*, 36, no. 6,731–741. doi: http://dx.doi.org/10.1016/S0148-9062(99)00036-4.

Kunze, K.R., and R.P. Steiger. 1992, *Extended Leakoff Tests to Measure In Situ Stress During Drilling*. In Rock Mechanics as a Multidisicplinary Science, J. Roegiers (ed), Proceedings of the 32nd U.S. Symposium on Rock Mechanics, Norman, Oklahoma, July 10–12, 1991, A.A. Balkema, Amsterdam.

Lamé, G. 1852, *Lecons Sur La Theorie Mathematique De L'Elasticite Des Corps Solides*. Gauthier-Villars.

Lee, M.Y., and B.C. Haimson. 1989, Statistical evaluation of hydraulic fracturing stress measurement parameters. *International Journal of Rock Mechanics and Mining Sciences & Geomechanics Abstracts*, **26**:447–456. doi: 10.1016/0148-9062(89)91420-4.

Li, G., A. Lorwongngam, and J.-C. Roegiers. 2009, *Critical Review of Leak-Off Test As A Practice for Determination of In-Situ Stresses*. American Rock Mechanics Association.

Luthi, S.M. 2005, Fractured reservoir analysis using modern geophysical well techniques: Application to basement reservoirs in Vietnam. In P.K. Harvey, T.S. Brewer, P.A. Pezard and V.A. Petrov, eds., *Petrophysical Properties of Crystalline Rocks: Geological Soc. Publishing House.* 95–106.

Mathias, S.A., and M. van Reeuwijk. 2009, Hydraulic Fracture Propagation with 3-D Leak-off. *Transport in Porous Media*, **80**, no. 3,499–518. doi: 10.1007/s11242-009-9375-4.

McLellan, J., and J.-C. Roegiers. 1981, Do instantaneous shut-in pressures accurately represent the minimum principal stress? In M.D. Zoback and B. Haimson, eds., *Hydraulic Fracturing Stress Measurements*. Monterey, CA: National Academy Press.

McLennan, J.D., and J.C. Roegiers. 1982, How Instantaneous are Instantaneous Shut-In Pressures? In *SPE Annual Technical Conference and Exhibition*. New Orleans: Society of Petroleum Engineers.

Morita, N., A.D. Black, and G.F. Fuh. 1996, Borehole breakdown pressure with drilling fluids.1. Empirical results. *International Journal of Rock Mechanics and Mining Sciences & Geomechanics Abstracts*, **33**, no. 1,39–51. doi: 10.1016/0148-9062(95)00028-3.

Pezard, P.A., and S.M. Luthi, 1988, Borehole electrical images in the basement of the Cajon Pass scientific drillhole, California – fracture identification and tectonic implications. *Geophysical Research Letters*, **15**:1017–1020. doi: 10.1029/GL015i009p01017.

Pook, L.P. 2010, Five decades of crack path research. *Engineering Fracture Mechanics*, **77**, no. 11, 1619–1630.

Raaen, A.M., P. Horsrud, H. Kjorholt, and D. Okland. 2006, Improved routine estimation of the minimum horizontal stress component from extended leak-off tests. *International Journal of Rock Mechanics and Mining Sciences*, **43**, no. 1,37–48. doi: 10.1016/j.ijrmms.2005.04.005.

Raaen, A.M., E. Skomedal, H. Kjørholt, P. Markestad, and D. Økland. 2001, Stress determination from hydraulic fracturing tests: the system stiffness approach. *International Journal of Rock Mechanics and Mining Sciences*, **38**, no. 4,529–541.

Raleigh, C.B., J.H. Healy, and J.D. Bredehoeft, 1972, Faulting and crustal stress at Rangely, Colorado. In *Flow and Fracture of Rocks*. Vol. 16, *Geophysical Monograph Series*, eds. H.C. Heard, L. Y. Borg, N. L. Carter, and C. B. Raleigh. American Geophysical Union, Washington, DC. 275–284.

Raleigh, C.B., J.H. Healy, and J.D. Bredehoeft, 1976, Experiment in earthquake control at Rangely, Colorado. *Science*, **191**:1230–1237. doi: 10.1126/science.191.4233.1230.

Rice, J.R., and M.P. Cleary. 1976, Some basic stress diffusion solutions for fluid-saturated elastic Porous-Media with compressible constituents. *Reviews of Geophysics*, **14**, no. 2, 227–241.

Ritchie, R.H., and A.Y. Sakakura. 1956, Asymptotic expansions of solutions of the heat conduction equation in internally bounded cylindrical geometry. *Journal of Applied Physics*, **27**, no. 12,1453–1459.

Robin, P.Y.F. 1973, Note on effective pressure. *Journal of Geophysical Research*, **78**, no. 14,2434–2437. doi: 10.1029/JB078i014p02434.

Rummel F, and R. Jung, 1975, Hydraulic fracturing stress measurements near Hohenzollern-graben-structure, SW Germany. *Pure and Applied Geophysics*, **113**:321–330. doi: 10.1007/bf01592921.

Rummel, F., and R.B. Winter, 1983, Fracture-mechanics as applied to hydraulic fracturing stress measurements. *Earthquake Prediction Research*, **2**:33–45.

Rumzan, I., and D.R. Schmitt. 2001, The influence of well bore fluid pressure on drilling penetration rates and stress dependent strength. In D. Elsworth, J.P. Tinucci, and K.A. Heasley, eds., *Rock Mechanics in the National Interest*: Proceedings of the 38th U.S. Rock

Mechanics Symposium: DC Rocks 2001, Waghington, D.C., July 7–10, A.A. Balkema, Amsterdam. Vols 1 and 2.

Rutqvist, J., and O. Stephansson. 1996, A cyclic hydraulic jacking test to determine the in situ stress normal to a fracture. *International Journal of Rock Mechanics and Mining Sciences & Geomechanics Abstracts*, 33, no. 7,695–711. doi: http://dx.doi.org/10.1016/0148-9062(96)00013-7.

Rutqvist, J., C.-F. Tsang, and O. Stephansson. 2000, Uncertainty in the maximum principal stress estimated from hydraulic fracturing measurements due to the presence of the induced fracture. *International Journal of Rock Mechanics and Mining Sciences*, 37, no. 1–2,107–120. doi: http://dx.doi.org/10.1016/S1365-1609(99)00097-0.

Sano, O., H. Ito, A. Hirata, and Y. Mizuta. 2005, Review of Methods of Measuring Stress and its Variations. *Bull. Earthquake Res. Inst., Univ. Tokyo*, 80,87–103.

Santarelli, F.J., E.T. Brown, and V. Maury. 1986, Analysis of borehole stresses using pressure-dependent, linear elasticity, *International Journal of Rock Mechanics and Mining Sciences & Geomechanics Abstracts*, 23, 445–449, 1986.

Sarvaramini, E., and D.I. Garagash. 2015, Breakdown of a Pressurized Fingerlike Crack in a Permeable Solid. *Journal of Applied Mechanics-Transactions of the ASME*, 82, no. 6,10. doi: 10.1115/1.4030172.

Savage, W.Z., Swolfs, H.S., and P.S. Powers, 1985, Gravitational stresses in long symmetric ridges and valleys. *International Journal of Rock Mechanics and Mining Sciences*, 22:291–302. doi: 10.1016/0148-9062(85)92061-3.

Scheidegger, A.E. 1960, On the connection between tectonic stresses and well fracturing data. *Pure and Applied Geophysics*, 46, no. 1,66–76. doi: 10.1007/bf02001098.

Schmitt, D.R., 2015, Seismic Properties.In L. Slater, Vol. 11: Geophysical Properties of the Near Surface Earth, in Schubert, G., *Treatise on Geophysics*. 2nd ed: 48-87, doi:10.1016/B978-0-444-53802-4.00190-1.

Schmitt, D.R., C.A. Currie, and L. Zhang. 2012, Crustal stress determination from boreholes and rock cores: Fundamental principles. *Tectonophysics*, 580,1–26. doi: 10.1016/j.tecto.2012.08.029.

Schmitt, D.R., T.J. Wilson, R.D. Jarrard, T.S. Paulsen, S. Pierdominici, D. Handwerger, and T. Wonik. 2012, Wireline hydraulic fracturing stress determinations In the ANDRILL South McMurdo Sound Drill Hole. In *46th U.S. Rock Mechanics/Geomechanics Symposium*. Chicago: American Rock Mechanics Association.

Schmitt, D.R., and M.D. Zoback, 1989, Poroelastic effects in the determination of the maximum horizontal principal stress in hydraulic fracturing tests – a proposed breakdown equation employing a modified effective stress relation for tensile failure. *International Journal of Rock Mechanics and Mining Sciences & Geomechanics Abstracts*, 26:499–506. doi: 10.1016/0148-9062(89)91427-7.

Schmitt, D.R., and M.D. Zoback. 1992, Diminished pore pressure in low-porosity crystalline rock under tensional failure – Apparent strengthening by dilatancy. *Journal of Geophysical Research-Solid Earth*, 97, no. B1,273–288.

Schmitt, D.R., and M.D. Zoback. 1993, Infiltration Effects in the Tensile Rupture of Thin-Walled Cylinders of Glass and Granite – Implications for the Hydraulic Fracturing Breakdown Equation. *International Journal of Rock Mechanics and Mining Sciences & Geomechanics Abstracts*, 30, no. 3,289–303.

Sheng, Y., M. Sousani, D. Ingham, and M. Pourkashanian. 2015, Recent developments in multiscale and multiphase modelling of the hydraulic fracturing process. *Mathematical Problems in Engineering*,15. doi: 10.1155/2015/729672.

Shimizu, H., S. Murata, and T. Ishida. 2011, The distinct element analysis for hydraulic fracturing in hard rock considering fluid viscosity and particle size distribution. *International Journal of Rock Mechanics and Mining Sciences*, 48, no. 5,712–727. doi: http://dx.doi.org/10.1016/j.ijrmms.2011.04.013.

Stephens, G., and B. Voight. 1982, Hydraulic fracturing theory for conditions of thermal-stress. *International Journal of Rock Mechanics and Mining Sciences*, 19, no. 6,279–284.

Terzaghi, K., and F. Richart. 1952, Stresses in Rock About Cavities. *Géotechnique*, 3, no. 2, 57–90. doi: doi:10.1680/geot.1952.3.2.57.

Thiercelin, M.J., R.A. Plumb, J. Desroches, P.W. Bixenman, J.K. Jonas, and W.A.R. Davie. 1996, A new wireline tool for In-Situ stress measurements. *SPE Formation Evaluation*, 11, no. 1. doi: 10.2118/25906-pa.

Wang, Y and M.B. Dusseault. 1994, Stresses around a circular opening in an elastoplastic porous-medium subjected to repeated hydraulic loading *International Journal of Rock Mechanics and Mining Sciences & Geomechanics Abstracts*, 31:597–616.

Warren, W.E., and C.W. Smith. 1985, In situ stress estimates from hydraulic fracturing and direct observation of crack orientation. *Journal of Geophysical Research-Solid Earth and Planets*, 90:6829–6839. doi: 10.1029/JB090iB08p06829.

Weng, X., V. Pandey, and K.G. Nolte. 2002, Equilibrium test – a method for closure pressure determination. In *SPE/ISRM Rock Mechanics Conference*. Irving: SPE.

White, A.J., M.O. Traugott, and R.E. Swarbrick. 2002, The use of leak-off tests as means of predicting minimum in-situ stress. *Petroleum Geoscience*, 8, no. 2,189–193. doi: 10.1144/petgeo.8.2.189.

Zang, A., and O. Stephansson. 2010, *Stress Field of the Earth's Crust*. Springer Netherlands.

Zemanek J, R.L. Caldwell, E.E. Glenn, S.V. Holcomb, L.J. Norton, and A.J.D. Straus. 1969, Borehole televiewer – a new logging concept for fracture location and other types of borehole inspection. *Journal of Petroleum Technology*, 21:762–774.

Zoback, M.D. 2007, *Reservoir Geomechanics*. Cambridge University Press.

Zoback, M.D., C.A. Barton, M. Brudy, D.A. Castillo, T. Finkbeiner, B.R. Grollimund, D.B. Moos, P. Peska, C.D. Ward, and D.J. Wiprut. 2003, Determination of stress orientation and magnitude in deep wells. *International Journal of Rock Mechanics and Mining Sciences*, 40, no. 7–8,1049–1076. doi: 10.1016/j.ijrmms.2003.07.001.

Zoback, M.D., and D.D. Pollard. 1978, Hydraulic Fracture Propagation and the Interpretation of Pressure-Time Records for In-Situ Stress Determinations. In *19th U.S. Symposium on Rock Mechanics (USRMS)*. Reno: American Rock Mechanics Association.

Zoback, M.D., and J.C. Roller, 1979, Magnitude of shear-stress on the San-Andreas fault – implications of a stress measurement profile at shallow depth. *Science*, 206:445–447. doi: 10.1126/science.206.4417.445.

Zoback, M.D., H. Tsukahara, and S. Hickman, 1980, Stress measurements at depth in the vicinity of the San-Andreas fault – implications for the magnitude of shear-stress at depth. *Journal of Geophysical Research*, 85:6157–6173. doi: 10.1029/JB085iB11p06157.

Zoback, M.D., and B. Haimson, eds., 1983, *Hydraulic Fracturing Stress Measurements*. Washington, DC: National Academy Press.

Chapter 7

Hydrofracturing

S.O. Choi
Department of Energy and Resources Engineering, Kangwon National University, Chuncheon, Kangwon, South Korea

Abstract: This chapter describes understanding of in-situ stresses in rock mass and various methods for estimating the state of stress, focusing on the hydrofracturing techniques. With a brief description of theory and procedure of the hydrofracturing test, various methods for determining the ambiguous shut-in pressure in hydrofracturing pressure history curves are introduced by numerical analysis. The hydrojacking test is also introduced as an example of application of hydrofracturing techniques.

I INTRODUCTION

In general, the order of magnitude of in-situ stresses and their directions are used for design parameters in the layout of complex underground works. An emergency spillway tunnel in a dam site for example, can be constructed and operated considering in-situ stresses normal to a fracture around an internal pressure tunnel. It is also dealt with essentially and directly in constructing an underground powerhouse or deep transportation tunnel, since the direction of excavation or the shape of a tunnel section can be altered according to in-situ stress regime at the construction site. Determination of in-situ stresses in rock mass and their spatial variation is technically difficult, since the current state of in-situ stresses is the final product of a series of past geological events. Further, since the physical quantity of in-situ stress is not tangible and rock masses are rarely homogeneous, no rigorous methods are available to predict in-situ stresses exactly. This chapter explains the problems of in-situ stresses in rock mass, the methods for measuring those stresses, and their importance in rock engineering as well as in conventional or unconventional energy engineering.

2 UNDERSTANDING OF IN-SITU STRESSES IN ROCK MASS

Rock stresses can be divided into in-situ stresses and induced stresses. In-situ stresses, which are also called virgin stresses, are the stresses that exist in the rock, generated naturally by a series of past geological events. On the other hand, induced stresses are relevant to artificial disturbance by excavation or are induced by changes in geological conditions. Stresses resulting from the weight of the overlying strata and from locked in stresses of tectonic origin denote in-situ stresses, for example, and the locally disrupted stresses in the rock surrounding the excavation opening denote induced stresses. Knowledge of the magnitudes and directions of these in-situ and induced stresses is an essential component of underground excavation design since the instability resulting from the stress concentration exceeding rock strength can have serious consequences for the behavior of the excavations.

2.1 Vertical stresses

The principal stress directions are horizontal and vertical in horizontal ground, in general, and are often assumed to be similar at depth. This simplifying assumption has been widely adopted in practice, but it is not adapted well at shallow depths beneath hilly terrain, because of no shear stresses in ground surface. Beneath a valley side, in other words, one principal stress is normal to the slope and equals zero, and the other two principal stresses lie in the plane of the slope. Nevertheless, the vertical stress can be regarded as one of the principal stresses since the effect of terrain decreases with depth. It is demonstrated from the field measurement experiences that this assumption is considerably reliable (Zoback, 1992).

2.1.1 Estimation of vertical stress

The state of stress at a point within the ground can be defined by the weight of its overlying rock. Calculate the vertical stress at a depth of Z below the surface, when the unit weight of the overlying rock mass is γ. Considering a cube from the surface to a certain depth, the downward force of the weight of overlying rock acting on this point at a depth of Z equals the product of the unit weight of the overlying rock and the volume of a cube.

$$F_v = \gamma Z A \tag{1}$$

where F_v is the downward force at a depth of Z and A is the unit area of base plane of a cube. Hence the vertical stress σ_v is estimated from dividing this force by the unit area.

$$\sigma_v = \gamma Z \tag{2}$$

Measurements of vertical stress at various sites around the world confirm that this relationship is valid although there is a significant amount of scatter in the collected data, as shown in Figure 1. Even though the increasing rate of vertical stress can be changed with the unit weight of rock mass, the vertical stress is predicted to increase as much as 0.027 MPa per 1 m depth for normal granitic rock mass.

In rock masses with complex strata of different density, the vertical stress can be estimated using this principle. When the unit weight of rock mass is assigned as a function $\gamma(Z)$, the vertical stress at a certain depth can be obtained by a definite integral from surface to Z.

$$\sigma_v = \int_0^Z \gamma(Z) dZ \tag{3}$$

When the rock density with depth is measured by density logging or the rock density changes with depth are excessive, for example, Equation 3 can be used for obtaining

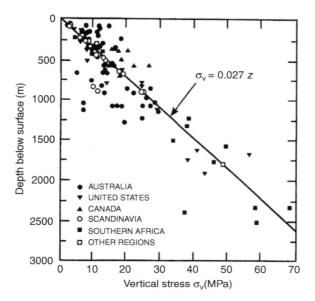

Figure 1 Vertical stress measurements from mining and civil engineering projects around the world (Brown & Hoek, 1978).

more accurate vertical stresses. However it is known that there is not a great discrepancy between the measurements by density logging data and the calculations by average unit weight of rock.

2.2 Horizontal stresses

The horizontal stresses acting on an element of rock at a depth of Z are much more difficult to estimate than the vertical stresses, because the magnitude and direction of horizontal stresses are influenced heavily by tectonic characteristics. Generally the ratio of the average horizontal stress to the vertical stress is denoted by K

$$K = \frac{\sigma_h}{\sigma_v} \qquad (4)$$

For a gravitationally loaded rock mass in which no lateral strain was permitted during formation of the overlying strata, the theory of elasticity can be invoked to predict that K will be equal to $v/(1-v)$, where v is the Poisson's ratio of the rock mass (Terzaghi & Richart, 1952). This expression derives from the symmetry of one-dimensional loading of an elastic material over a continuous plane surface, which infers a condition of no horizontal strain; such a formula has no validity in a rock mass that has experienced cycles of loading and unloading, and it proved to be inaccurate and seldom used today (Hoek, 2007). Measurements of horizontal stress with depth for different regions of the world discern that a hyperbolic relation for the limits of $K(Z)$, as in Equation 5 (Brown & Hoek, 1978).

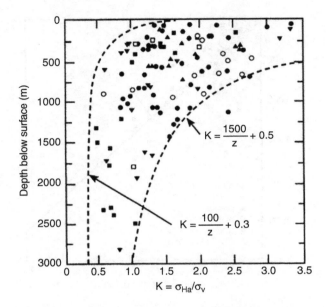

Figure 2 Variation of average horizontal to vertical stress ratio with depth Z, below surface (Brown & Hoek, 1978).

$$\frac{100}{Z} + 0.3 < K(Z) < \frac{1500}{Z} + 0.5 \qquad (5)$$

3 EFFECT OF GEOLOGICAL STRUCTURES ON THE STATE OF STRESS

Variations in rock mass geology and the existence of geological structures may affect the distribution and magnitude of in-situ stresses. Contrary to the vertical stress, the horizontal in-situ stress may vary substantially from one layer to the next in a stratified rock formation due to changes in rock stiffness. For instance, it is revealed that the major principal stress is diverted parallel to the discontinuity when the discontinuity is open, the principal stresses are not affected when the discontinuity is made of a material with similar properties as the surrounding rock, and the major principal stress is diverted perpendicular to the discontinuity when the material in the discontinuity is rigid (Hudson & Cooling, 1988). Many cases of field measurements for non-homogeneous stress regimes have been reported. Choi (2007) gives an example of hydrofracturing in-situ stress measurements conducted on both sides of the Yangsan fault in South Korea showing the effect of the fault on the in-situ stress field (Figure 3 & Table 1). Table 1 shows a completely different trend in the magnitudes and directions of horizontal principal stresses measured on both sides of the fault.

It is noteworthy that the magnitude of horizontal principal stresses in sedimentary rock formation is slightly larger than in andesite and granodiorite formation. It was also reported by Amadei and Pan (1992) regarding the horizontal principal stress at the

Table 1 Results of in-situ stress measurements by hydrofracturing tests on Yangsan fault in South Korea (Choi, 2007).

Depth (m)	P_c (MPa)	P_r (MPa)	P_s (MPa)	S_v (MPa)	S_h (MPa)	S_H (MPa)	K_h	K_H	S_H Dir. (°) Mea.	Ave.
Borehole No.: TB-35-4 (in andesite & granodiorite formation)									
26	2.32	1.934	1.431	0.697	1.431	2.239	2.05	3.21	30±5	
28	2.11	2.009	1.643	0.750	1.643	2.760	2.19	3.68	25±5	
34	2.23	1.672	1.353	0.911	1.353	2.167	1.49	2.38	25±5	
46	0.94	–	–	1.233	–	–	–	–	–	24±5
48	1.30	–	–	1.286	–	–	–	–	–	
64	3.03	2.675	2.398	1.715	2.398	3.999	1.40	2.33	20±5	
66	4.25	3.614	2.951	1.769	2.951	4.699	1.67	2.66	20±5	
Borehole No.: TB-37-6 (in sedimentary rock formation)									
20	4.54	3.455	2.264	0.536	2.264	3.157	4.22	5.89	0±5	
23	5.67	3.872	2.100	0.616	2.100	2.218	3.41	3.60	10±5	
30	5.34	3.881	2.389	0.804	2.389	3.006	2.97	3.74	10±5	
34	5.28	4.011	2.526	0.911	2.526	3.247	2.77	3.56	5±5	6±5
38	5.02	3.860	2.246	1.018	2.246	2.518	2.21	2.47	5±5	
41	5.10	4.056	2.639	1.099	2.639	3.471	2.40	3.16	5±5	
45	4.63	3.677	2.668	1.206	2.668	3.897	2.21	3.23	–	

P_c: initial breakdown pressure, P_r: reopening pressure, P_s: the shut-in pressure, S_v: vertical stress, S_h: minimum horizontal principal stress, S_H: maximum horizontal principal stress, $K_h=S_h/S_v$, $K_H=S_H/S_v$.

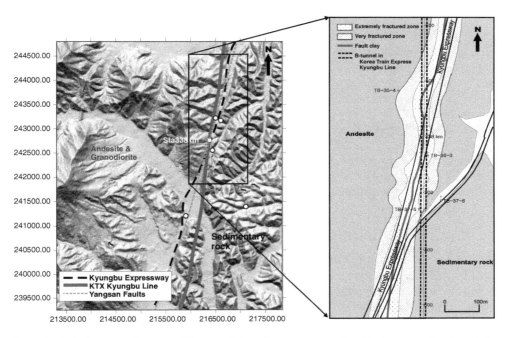

Figure 3 Shaded topographic map of Yangsan fault making an acute angle with the routes for the existing expressway and the proposed high speed railroad. Andesite and granodiorite are dominant on the west side of the fault, and the east side is covered mostly by sedimentary rocks (Choi, 2007).

site of sedimentary rocks larger than at igneous rocks. Difference in directions of maximum horizontal principal stress in different rock formations is also remarkable; the average direction of maximum horizontal principal stress in andesite and granodiorite formation of Yangsan faults is N24E, but it is N6E in sedimentary rock formation. It can be explained by not only the differences of geology at both sides of the fault but also the changes of the fault running which is generally NNE in the northern part and is nearly NS in the southern part. The results of the horizontal stress direction depending on the existence of fault zone coincide with Moos and Zoback (1993) and Sugawara and Obara (1993).

Amadei and Stephansson (1997) revealed that the K-value in the fault zone can be estimated to be less than or equal to the measured one observed in the surrounding rock mass.

4 METHODS OF IN-SITU STRESS MEASUREMENT

The techniques for measurement of in-situ stresses were developed in diverse methods with a new attempt to quantify the stability of underground structures. Among them the most used techniques in field measurements are overcoring, the flat jack method, and hydraulic fracturing. Rock cores can also be used for the AE (acoustic emission) method, the DRA (deformation rate analysis) method and the ASR (anelastic strain recovery) method in laboratory test. Each of these methods is complementary. In the overcoring test, the rock is unloaded by drilling out a large core sample, while radial displacements are monitored in a central, parallel borehole. In the flat jack test, the rock is unloaded by cutting a slot and reloaded by jacking a flat jack installed in the slot. In the hydraulic fracturing method, the rock is cracked by high pressure water pumped into a borehole, and the tensile strength of the rock and the inferred concentration of stress at the borehole wall are processed to yield the in-situ stress in the plane perpendicular to the borehole. In the AE method, acoustic emission on core samples is monitored while the rock is loaded cyclically in uniaxial compression in the laboratory, since it is known that there is a significant increase in the rate of acoustic emissions as the stress exceeds its previous higher value, known as the Kaiser effect (Holcomb, 1993). Similar to the AE method, the DRA method uses inelastic strains instead of acoustic emissions in the AE method (Yamamoto *et al.*, 1990). In the ASR method, strain response is monitored while an oriented core sample is removed from a borehole, and a viscoelastic model for the rock response to unloading is required for determining the in-situ stress magnitudes (Teufel, 1982).

4.1 Overcoring method

Overcoring tests use a large-diameter hole drilled to the required depth in the volume of rock in which in-situ stresses have to be determined. An instrumented device for measuring strains or displacements is inserted into the pilot hole, and then changes of strain or displacement within the instrumented device are recorded while a large-diameter hole is drilled. There is a variety of instrumented devices, such as the South African CSIR triaxial strain cell (Leeman & Hayes, 1966), the Australian CSIRO Hollow Inclusion (HI) Cell (Worotnick & Walton, 1976), and the US Bureau of

Table 2 Comparison of common methods in overcoring test.

Method	Principles	Advantage	Disadvantage	Devices
Borehole deformation gauge	• In-situ stress is estimated by elastic modulus of rock mass and changes of borehole diameter measured during overcoring • 6 gauges (60 intervals) are used for detecting the changes of borehole diameter in 3 directions	• Process is easy and incidence of test failure is low	• 2 dimensional state of stress in the plane perpendicular to borehole is measured • 3 tests in different directions are needed for the 3 dimensional state of stress	USBM gage
Triaxial strain cell	• Several strain rosettes are attached on borehole wall to measure the strain during overcoring • Elastic modulus of rock mass is needed for estimating in-situ stress	• The 3 dimensional state of stress is determined completely by single test borehole	• Process is difficult • Full adhesion of rosette gauge is hard in wet borehole	CSIR cell, CSIRO HI cell

Mines (USBM) gage (Merrill, 1967). Depending on the instrument used to monitor the rock during overcoring, the complete state of stress can be determined in one, two or three non-parallel boreholes. No assumption needs to be made regarding the in-situ stress field as with the hydraulic fracturing method. However, the overcoring method is limited by the magnitude of the in-situ stresses themselves; namely, it can only be used at depths for which the strength of the rock in the wall and bottom of the borehole is not exceeded (Herget, 1993). Table 2 compares the basic principle and features of two most common methods in the overcoring test.

4.2 Flat jack method

The flat hydraulic jacks, consisting of two plates of steel welded around their edges, are used for in-situ stress measurement in a rock face, such as the wall of an underground gallery. The equilibrium of a rock mass is disturbed by cutting slots on rock surfaces, and this will create deformations that are measured with reference pins placed in the near vicinity of the slots. Equilibrium is recovered when the inserted flat jacks are pressurized until all deformations vanish. Therefore, the flat jack method is sometimes called the stress compensating method. A total of six tests need to be carried out to obtain the complete three dimensional state of stress since one component of the in-situ stress field is obtained from each flat jack test. The flat jack method has an advantage in that it does not require knowledge of the elastic modulus of the rock to determine the tangential stress at points in the wall of an excavation, and the stresses are measured directly. However, the flat jack method also has disadvantages that limit its application ranges. Standards for determining in-situ stresses with flat jacks have been suggested by the American Society for Testing of Materials (ASTM D 4729-87, 1993) and the International Society for Rock Mechanics (Kim & Franklin, 1987); according to

234 Choi

ASTM D 4729-87, the measurement points should be installed within a distance, $L/2$ of the flat jack slot, where L is the flat jack width.

5 HYDROFRACTURING IN-SITU STRESS MEASUREMENT

5.1 Theory and background

The hydrofracturing technique used first for the well stimulation in the petroleum industry is currently applied for determining the in-situ stresses in the Earth's crust (Hubbert & Willis, 1957). The basic concept for the determination of the in-situ stresses starts from several assumptions that the rock mass is an impermeable, homogenous, and isotropic elastic material and one of the principal stresses is parallel to the direction of wellbore. Consequently the other principal stresses are assumed to be on the plane perpendicular to the wellbore and calculated from the pressure-time history curves obtained during hydrofracturing (Haimson & Fairhurst, 1970). While calculating the principal stresses from the pressure-time history curve, however, the accurate determination of shut-in pressure, which shows a pressure decline gradually after stopping the fluid injection, is very important because the shut-in pressure is equal to the normal stress acting on the induced hydraulic fractures and indicates the minimum horizontal principal stress directly (Aamodt & Kuriyagawa, 1983; Rummel, 1987). However, in most field cases the pressure-time history after shut-in shows an ambiguous curve rather than a sharp break, so a lot of methods for determining the shut-in pressure have been proposed. Some compare the several shut-in pressures obtained from several methods for more accurate determination of the minimum horizontal principal stress. A review of these various methods can be found in Kim and Franklin (1987) and Lee and Haimson (1989), and in more details in Amadei and Stephansson (1997). They described the strengths and weaknesses of each method by comparing the shut-in pressures, but were not able to propose which method is best for determining the minimum horizontal principal stress. Because the absolute minimum horizontal principal stress in the Earth's crust cannot be determined, that is, the error between the real value and the calculated value cannot be defined.

5.2 Equipment and procedure

In the hydrofracturing test, a certain interval in borehole is completely sealed off by straddle packer (Figure 4), and then this interval is pressurized until tensile crack starts to generate. When the rock mass is isotropic, the hydrofracturing tensile cracks will be generated to the plane perpendicular to the horizontal minimum principal stress in which resistance to the tensile failure is lowest. When the pumping into the interval continues after hydrofractures are generated, the pressure of the interval does not increase and reaches to a breakdown pressure. An idealized hydrofracturing pressure history is shown in Figure 5, and the horizontal maximum and minimum principal stresses are obtained from the equations for the state of stress around a circular opening (Figure 6).

From the hydrofracturing pressure history curves, it is possible to determine the fracture initiation (breakdown) pressure P_C, the shut-in pressure P_S and the fracture reopening pressure P_r. The shut-in pressure is the pressure at which a hydrofracture stops propagating and closes following pump shut-off, so it is the pressure for

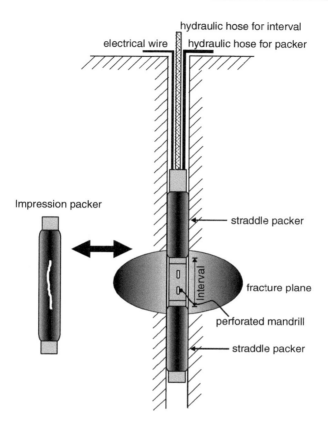

Figure 4 Schematic diagram of hydrofracturing test.

maintaining the balance with the minimum principal stress acting perpendicularly on the fracture plane. That is, the shut-in pressure equals the horizontal minimum principal stress in the test site.

$$\sigma_h = P_S \tag{6}$$

In Figure 6, tangential stress σ_t at a point in a distance of r from the origin of a circular opening of radius a is given by Kirsch solution (Jaeger et al., 2007) and described below in Equation 7.

$$\sigma_t = \frac{1}{2}(\sigma_H + \sigma_h)\left(1 + \frac{a^2}{r^2}\right) - \frac{1}{2}(\sigma_H - \sigma_h)\left(1 + 3\frac{a^4}{r^4}\right)\cos(2\theta) \tag{7}$$

where

σ_H; maximum horizontal principal stress,
σ_h; minimum horizontal principal stress,
σ_t; tangential compressive stress at a point in a distance of r from borehole axis,
a; radius of borehole.

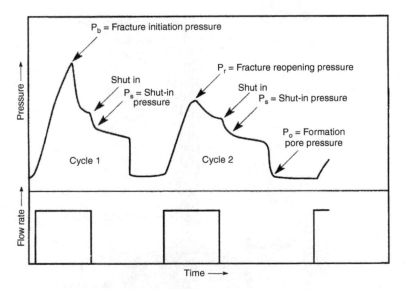

Figure 5 Idealized hydrofracturing pressure history curves (Amadei & Stephansson, 1997).

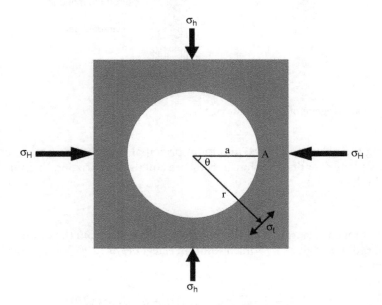

Figure 6 State of stresses around a circular opening (Jaeger et al., 2007).

Increasing the pressure of the interval, a crack starts to propagate at point A and the tangential compressive stress at this point becomes to be a minimum. At point A, $r = a$ and $\theta = 0$, so Equation 7 can be expressed as Equation 8.

$$\sigma_t = 3 \cdot \sigma_h - \sigma_H \tag{8}$$

The conditions for a new, vertical tensile crack are that the tensile stress at point A should become equal to the tensile strength $-T_0$. Applying this to the hydrofracturing experiment yields as a condition for creation of hydrofractures

$$3 \cdot \sigma_h - \sigma_H - P_C = -T_0 \tag{9}$$

$$T_0 = P_C - P_r \tag{10}$$

Equations 9 and 10 allow the maximum horizontal principal stress to be determined as Equation 11.

$$\sigma_H = 3 \cdot \sigma_h - P_r \tag{11}$$

Also the pore pressure into the rock mass around the borehole should be considered. Therefore, Equation 11 can be expressed as Equation 12.

$$3 \cdot \sigma_h - \sigma_H - P_C = -T_0 + P_0$$

$$\sigma_H = 3 \cdot \sigma_h - P_r - P_0 \tag{12}$$

5.3 Data analysis and interpretation

The shut-in pressure obtained from the hydrofracturing pressure history curves is essential for determining the in-situ stresses. However, in many situations the pressure decay is gradual with no obvious breaks and the shut-in pressure cannot be readily defined. Several methods for determination of the shut-in pressure in hydrofracturing have been investigated using the numerical code UDEC (Choi, 2012). The (P vs. t) method, the (P vs. log(t)) method and the (log(P) vs. log(t)) method have been introduced as some of graphical intersection methods, and the (dP/dt vs. P) method has been adopted for a statistical method. Through a series of numerical analyses with the different physical properties and the different remote stress regimes on the randomly sized polygonal joint model, it is revealed that numerical analysis using the discrete element method is probably suitable for simulating the hydraulic fracturing. And also from the pressure-time history curves, the shut-in pressures obtained from various methods are known to be usually higher than the applied minor horizontal principal stress. It can be explained by the classical Kirsch solution (Jaeger *et al.*, 2007), which defines the stress distributions around a cylindrical hole in an infinite isotropic elastic medium under plane strain conditions. In other words, the stress in the direction of x-axis at the wellbore wall is same to the applied far-field stress but increases with radial distance from the wellbore. This phenomenon could be exaggerated when the differential stress is high, so care should be taken in using the various methods for shut-in pressure. Figure 7 shows procedures for determining the shut-in pressure in various methods (Choi, 2012).

5.4 Numerical examples of in-situ stress determination

When there is no obvious discontinuity in rock mass and it is covered with the same base rock, the in-situ stress measurement can be performed in several locations and then extrapolated to cover the whole area. Figure 8 shows the example of a hydrofracturing test in this manner. The testing boreholes are located at the beginning and

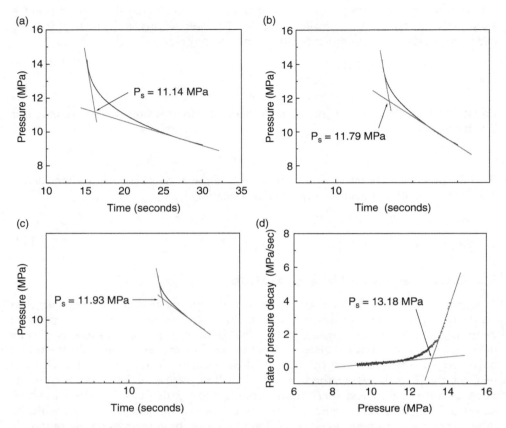

Figure 7 Determination of shut-in pressure P_S from hydrofracturing pressure versus time records: (a) tangent intersection method, (b) pressure versus log(time) method, (c) log(pressure) versus log(time) method, and (d) bilinear pressure decay rate method (Choi, 2012).

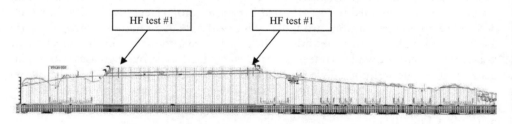

Figure 8 Location of hydrofracturing test boreholes at the beginning and end part of tunnel route.

end part of the proposed tunnel route. From the numerical analysis as shown in Figure 9, the K-value for the whole tunnel route is suggested in Figure 10.

Figure 11 shows another example of the hydrofracturing in-situ stress measurement under and next to the Han River in South Korea. One hydrofracturing test was carried out on the terrace land on the river, and the other was performed in the middle of the

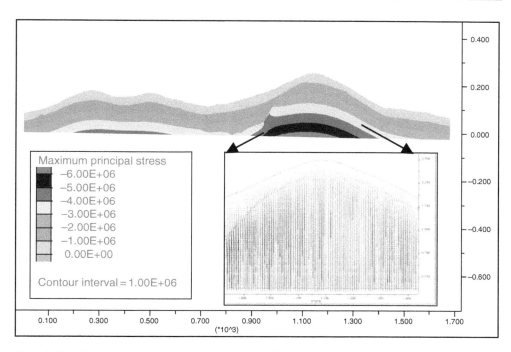

Figure 9 Distribution of principal stress in rock mass obtained from numerical analysis.

river. Comparing the results from two hydrofracturing tests and numerical analysis, the principal stresses were obtained for the whole tunnel route by extrapolation technique.

6 APPLICATION OF HYDROFRACTURING TECHNIQUE

6.1 Hydrojacking test

When designing an internally pressurized tunnel, the normal stress required to reopen the natural fractures should be a major parameter for its stability. Also the relationship between the fracture opening and the fluid pressure should be pre-interpreted for the stability of structures, such as the spillway tunnel of a dam. Apart from the general hydrofracturing in-situ stress measurements, the stress normal to the natural fracture plane should be identified with respect to the dip direction and the dip angle of the pre-existing fracture. A hydrojacking test can be adopted for this purpose, and is used to determine the mechanical and hydraulic behaviors of pre-existing fractures around an internal pressure tunnel site. Figure 12 presents a plan of the project area geology and a longitudinal section along the spillway tunnel. Precambrian foliated leucocratic granitic gneiss composition occupies the project area. For in-situ stress measurements via hydrofracturing, the fracture-free intervals in borehole NSP-6 were chosen with the aid of borehole televiewer images (Figure 13). Figure 14 is an example curve obtained from the hydrojacking test with the hydrofracturing test equipment. As shown in this figure, the hydrojacking pressure and the stress normal to the fracture plane (σ_n) have been

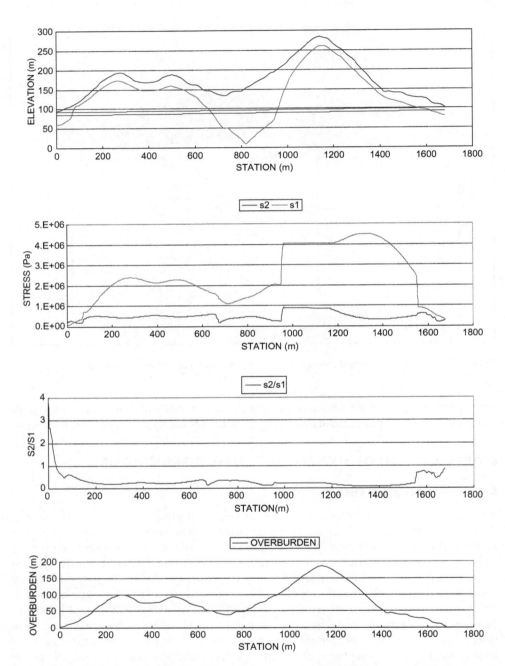

Figure 10 Extrapolation of K-value after comparing results from field measurement and numerical analysis (S1 and S2 are principal stresses obtained from numerical analysis).

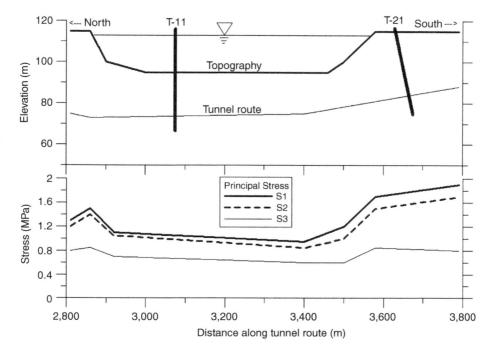

Figure 11 Extrapolation of principal stress for the whole tunnel route after comparing results from field measurement and numerical analysis.

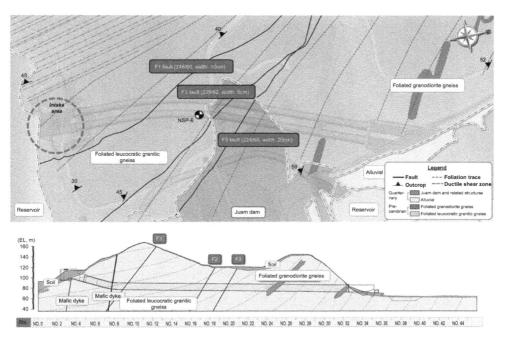

Figure 12 Plan and tunnel section of dam spillway project in Juam, South Korea (Choi, 2014).

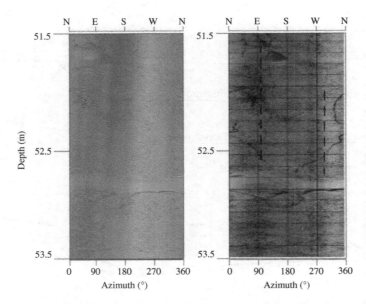

Figure 13 Examples of borehole televiewer images taken before and after the hydrofracturing test in Korea (Choi, 2014). (Dotted lines in (b) denote the estimated average direction of hydrofractures.)

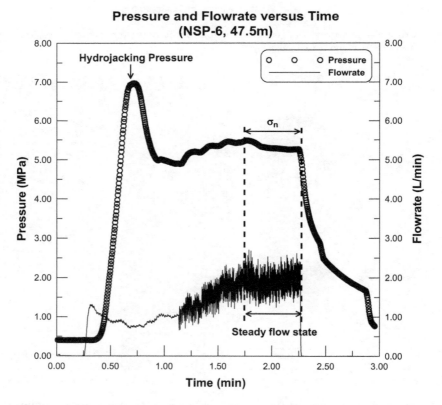

Figure 14 Example of well pressure and flowrate vs. time curve obtained from the hydrojacking test at 47.5 m depth (Choi, 2014).

estimated through the hydrojacking interval and the constant flowrate interval, respectively. As in Rutqvist and Stephansson's elastic model (1996), the hydrojacking pressure can be defined as the pressure needed to reopen the natural fracture as well as overcome the stiffness between the two fracture planes, so it could be shown as the peak pressure.

REFERENCES

Aamodt, R.L. & Kuriyagawa, M. (1983) Measurements of instantaneous shut-in pressure in crystalline rock. In: Zoback, M. and Haimson, B. (eds.) *Hydraulic fracturing stress measurements*. Washington, DC, National Academy Press. pp. 139–142.

Aggson, J.R. & Kim, K. (1987) Analysis of hydraulic fracturing pressure history: a comparison of five methods used to identify shut-in pressure. *Int. J. Rock Mech. Min. Sci. Geomech. Abstr.*, 24, 75–80.

Amadei, B. & Pan, E. (1992) Gravitational stresses in anisotropic rock masses with inclined strata. *Int. J. Rock Mech. Min. Sci. Geomech. Abstr.*, 29, 225–236.

Amadei, B. & Stephansson, O. (1997) *Rock stress and its measurement*. London, Chapman & Hall.

ASTM D 4729-87 (1993) Standard test method for in-situ stress and modulus of deformation using the flatjack method. In: *1993 Annual Book of ASTM Standards*, Vol. 04–08.

Brown, E.T. & Hoek, E. (1978) Trends in relationships between measured in situ stresses and depth. *Int. J. Rock Mech. Min. Sci. Geomech. Abstr.*, 15, 211–215.

Choi, S.O. (2007) A decade's hydrofracturing experiences of in-situ stress measurements for tunnel construction in Korea. *Chinese J. Rock Mech. Eng.*, 26(11), 2200–2206.

Choi, S.O. (2012) Interpretation of shut-in pressure in hydrofracturing pressure-time records using numerical modeling. *Int. J. Rock Mech. Min. Sci.*, 50, 29–37.

Choi, S.O. (2014) Determination of in situ stresses normal to a fracture around an internal pressure tunnel by hydrojacking testing. *Tunnelling Underground Space Technol.*, 40, 228–236.

Haimson, B.C. & Fairhurst, C. (1970) In-situ stress determination at great depth by means of hydraulic fracturing. *Proc. 11th US Symp. Rock Mech.*, Berkeley, SME/AIME. pp. 559–584.

Herget, G. (1993) Rock stresses and rock stress monitoring in Canada. In: Hudson, J.A. (ed.) *Comprehensive rock engineering. Vol. 3. Rock Testing and Site Characterization*, Pergamon. pp. 473–496.

Hoek, E. (2007) *Practical rock engineering*. Available from: http://www.rocscience.com/hoek/corner/Practical_Rock_Engineering.pdf

Holcomb, D.J. (1993) Observations of the Kaiser effect under multiaxial stress states: implications for its use in the determining in-situ stress. *Geophys. Res. Lett.*, 20, 2119–2122.

Hubbert, M.K. & Willis, D.G. (1957) Mechanics of hydraulic fracturing. In: *Petroleum Transactions*, AIME, 210, 153–166.

Hudson, J.A. & Cooling C.M. (1988) In situ rock stresses and their measurement in the UK – Part I. The current state of knowledge. *Int. J. Rock Mech. Min. Sci. Geomech. Abstr.*, 25, 363–370.

Jaeger, J.C., Cook, N.G.W. & Zimmerman, R.W. (2007) *Fundamentals of rock mechanics*. 4th edition, Wiley-Blackwell.

Kim, K. & Franklin, K.A. (1987) Suggested methods for rock stress determination. *Int. J. Rock Mech. Min. Sci. Geomech. Abstr.*, 24, 53–73.

Lee, M.Y. & Haimson, B.C. (1989) Statistical evaluation of hydraulic fracturing stress measurement parameters. *Int. J. Rock Mech. Min. Sci. Geomech. Abstr.*, 26(6), 447–456.

Leeman, E.R. & Hayes, D.J. (1966) A technique for determining the complete state of stress in rock using a single borehole. *Proc. 1st Cong. Int. Soc. Rock Mech.* Lisbon, Vol. II. pp. 17–24.

Merrill, R.H. (1967) *Three component borehole deformation gage for determining the stress in rock.* US Bureau of Mines Report of Investigation RI 7015.

Moos, D. & Zoback, M.D. (1993) Near-surface 'thin skin' reverse faulting stresses in the southeastern United States. *Int. J. Rock Mech. Min. Sci. Geomech. Abstr.*, 30(7), 965–971.

Rummel, F. (1987) Fracture mechanics approach to hydraulic fracturing stress measurements. In: Atkinson (ed.). *Fracture mechanics of rock.* London, Academic press. pp. 217–239.

Rutqvist, J. & Stephansson, O. (1996) A cyclic hydraulic jacking test to determine the in situ stress normal to a fracture. *Int. J. Rock Mech. Min. Sci. Geomech. Abstr.*, 33(7), 695–711.

Sugawara, K. & Obara, Y. (1993) Measuring rock stress: case examples of rock engineering in Japan. In: Hudson, J.A. (ed.) *Comprehensive rock engineering. Vol.3. Rock Testing and Site Characterization*, Pergamon. pp. 533–552.

Terzaghi, K. & Richart, F.E. (1952) Stresses in rock about cavities. *Geotechnique*, 3, 57–90.

Teufel, L.W. (1982) Prediction of hydraulic fracture azimuth from anelastic strain recovery measurements of oriented core. *Proc. 23rd US Symp. Rock Mech.*, Berkeley, SME/AIME. pp. 238–245.

Worontnicki, G. & Walton, R.J. (1976) Triaxial hollow inclusion gauges for determination of rock stresses in-situ. *Proc. ISRM Symp. on Investigation of Stress in Rock, Advances in Stress Measurement*, Sydney, pp. 1–8.

Yamamoto, K.Y., Kuwahara, N., Kato, N. & Hirasawa, T. (1990) Deformation rate analysis: a new method for in situ stress examination from inelastic deformation of rock samples under uniaxial compressions. *Tohoku Geophys. J.*, 33, 127–147.

Zoback, M.L. (1992) First and second order patterns of stress in the lithosphere: the world stress map project. *J. Geophys. Res.*, 97, 11703–11728.

Chapter 8

Methodology for determination of the complete stress tensor and its variation versus depth based on overcoring rock stress data

D. Ask [1,2]

[1] Department of Civil, Environmental and Natural Resources, Division of Geosciences and Environmental Engineering, Research group Exploration Geophysics, Luleå University of Technology, Luleå, Sweden
[2] FracSinus Rock Stress Measurements AB, Luleå, Sweden

Abstract: This paper presents a methodology for evaluating the complete stress field and its variation versus depth. The validity of the protocol is visualized in the case study at the Äspö Hard Rock Laboratory (HRL), south-eastern Sweden, a site predominantly sampled using the overcoring method. Overcoring data involve explicit (measurement-related) as well as implicit uncertainties. The former include for example uncertainties regarding determination of the location of the test sections in physical space and of the value of elastic parameters, as well as uncertainties in strain-/displacement measurements, etc. The explicit types of uncertainties are fairly straightforward to analyze and correct for during the stress calculation procedure. The implicit uncertainties, on the other hand, such as the assumption of homogeneity and linear-elasticity, are much more difficult to appreciate and correct for, if possible at all. Yet, as for explicit errors, they may render an individual test or a series of tests completely meaningless, and it is therefore crucial that both categories of uncertainties are identified, understood, and properly considered within the process of stress field determinations.

The proposed methodology follows the directions outlined by ISRM for rock stress estimation using overcoring methods (Sjöberg *et al.*, 2003; Sjöberg & Klasson, 2003). In addition, we pay particular attention on avoiding, identifying, and correcting for various potential sources of error, the sampling strategy, and considerations of the continuity hypothesis.

1 INTRODUCTION

Knowledge of the prevailing stress field is important for rock mechanical studies because it provides means to analyze the mechanical behavior of bodies of rock and serves as boundary conditions in rock engineering applications. One of the most commonly applied methods for estimating the *in situ* stress is overcoring. Several cells of more or less different design exist. However, they are collectively often referred to as borehole relief methods, and the theory, field application, and analysis of data are quite similar for the various cells (Amadei & Stephansson, 1997).

Overcoring cells sample a partial or a complete local stress tensor in each individual measurement attempt. The regional stress tensor for larger rock volumes can be determined from a number of local stress tensors with proper attention to stress field continuity and stress gradients. Cornet (1993) defines the regional stress tensor with six functions of spatial coordinates from a number of local stress tensors. The full parameterization involves 22 unknowns, but for most cases realistic assumptions can be made that significantly reduce the number of unknown parameters. The requirement of a suitable parameterization may be regarded as a scale issue, and this is common for all *in situ* stress measurement methods as they all sample stress at a scale significantly smaller than that of the mechanical problem in question. At larger scales, proper attention must be paid to fractures and faults as they are incompatible with the continuum concept.

The nature of the overcoring method, which involves fairly sophisticated drilling operations in conjunction with the measurement, renders the method sensitive to explicit uncertainties; some of which will be highlighted in this paper. Additionally, also implicit uncertainties are entailed due to the small scale; a strain gauge length is generally about 1 cm. Because the vast majority of bodies of different rock types are neither perfectly linear elastic, nor completely homogeneous, the small scale introduces the question to what degree the small sample is representative for the larger rock volume investigated. This leads to the Representative Elementary Volume (REV) concept. The size of the REV is defined as the smallest volume of rock for which there is equivalence between the idealized continuum material and the real rock. The REV is a physical description of a volume, and the various functions defined by homogenizing the body within the REV cannot be used to understand, or model, phenomena that occur at scales smaller than the REV.

Besides explicit violations of the inherent assumptions in the overcoring method, examples of implicit data uncertainties will also be addressed in this paper which will take us through the different phases during a complete stress field determination using overcoring data. Focus will be on data uncertainties in the overcoring measurement technique. Finally, the highlights of a complete stress field determination at the Äspö Hard Rock Laboratory (HRL), south-eastern Sweden is given.

2 OVERCORING DATA ANALYSIS

In the following section, the testing principle, theory, and interpretation of overcoring data using the Borre Probe (a CSIR-type cell) are described. The case study (Section 3) also involves the CSIRO HI cell and some specific issues for this cell have also been described in the section. These two cells include a strain gauge configuration that enables capturing of the full stress tensor at each measurement attempt. The cells are not described in detail here and the reader is referred to Sjöberg & Klasson (2003), Worotnicki (1993), and ES&S (2015).

2.1 The overcoring measurement principle

The overcoring method implies measurement of the relaxation of a pre-stressed, instrumented rock cylinder. However, before reaching the measurement phase, *i.e.* the overcoring, a specialized drilling protocol is followed (Figure 1). These drilling operations are important for the success rate of any overcoring stress measurement campaign and a well-experienced drilling crew is essential. Key objectives of the

Methodology for determination of the complete stress tensor 247

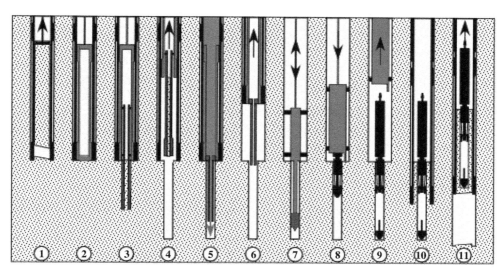

Figure 1 Measurement procedure for the Borre Probe: (1) advance of main borehole to measurement depth; (2) Grinding of the borehole bottom; (3) Drilling of pilot hole; (4) Recovery of core for appraisal; (5) Flushing of pilot hole to remove drill cuttings (and optionally compressed air afterwards); (6) Hoisting of flushing tool and entire drill string; (7) Cleaning wall with brush/cloth; (8) Lowering of cell with installation tool; (9) Cell at position, mechanical release of cell from installation tool; (10) Hoisting of installation tool; (11) Overcoring; and (11) Core break and recovery of core sample for data reading and subsequent biaxial testing.

overcoring measurement principle are outlined below, based on the current version of the Borre Probe, version III.

2.1.1 Coring of main borehole

The first drilling operation involves coring using NQ-size or larger down to chosen borehole testing length. It is important that the borehole is drilled as straight as possible to avoid problems to descent with the installation tool, but also to prevent drill string vibrations during the subsequent overcoring, which may damage the overcore sample. Secondly, and the most important step, the borehole must be flushed thoroughly to remove all drill cuttings. Failure to achieve this will unquestionably hamper the glue bonding between the strain rosettes and the pilot hole walls.

If the borehole penetrates a fracture zone, indicated by a water loss during drilling, drill water and cuttings will be injected into the fractured section, and as soon as the coring and flushing stops, water and cuttings will re-enter into the borehole. This will effectively prevent successful testing in the borehole, why significantly fractured sections must be grouted and borehole re-drilled.

2.1.2 Grinding of borehole bottom

It is important that the entry of the pilot hole is centralized and coaxial with the full sized borehole. To maximize the chances of achieving this outcome, the bottom of the full sized borehole is grinded using a full face drill bit.

2.1.3 Pilot hole drilling, flushing, and cleaning

The pilot hole drilling (diameter ca 36 and 38 mm for Borre and CSIRO HI cells, respectively) uses centralizers to ensure that the hole is drilled coaxially with the full sized borehole. The retrieved core is examined in detail with respect to fractures, inhomogeneities, such as large grains and features that may lead to anisotropy (*e.g.* schistosity/foliation).

If the location is accepted for testing, a flushing tool is used to clean the entire borehole from drill cuttings. For CSIRO HI cells, requiring dry boreholes, the borehole is subsequently dried using a cloth or using compressed air.

2.1.4 Installation of cell

The installation of the Borre Probe is undertaken with an installation tool that centers the probe of the main borehole, thereby ensuring that the cell enters into the pilot hole. Once the cell is in correct position, it is mechanically released from the installation tool. The installation can be made using rods or a wireline to substantial borehole lengths. Prior to the installation, the probe is programmed and started.

Upon start-up, strain readings are undertaken every 15 minutes and hence allows post-examination of the installation and the glue hardening process (Figure 2). At a pre-set time, generally 8–10 h after installation, glue hardening is complete and sampling rate is increased to a pre-determined frequency, generally one reading per second. The overcoring phase can then be initiated.

2.1.5 Overcoring

The overcoring sequence involves several steps: (i) hoisting of installation tool; (ii) lowering of drill string; (iii) initiation of flushing water; (iv) rotation start of drill string; (v) initiation of drilling; (vi) drilling stop; (vii) flush water stop; (viii) core break; and (ix) hoisting to surface for data dump. The timing of each of these operations is noted and provides useful information when judging the reliability of each strain rosette.

It is important that the drill water circulation commences at least 10–15 min prior to the overcoring starts and continues for at least equally long time after completed overcoring, in order to let the strain gauges stabilize and also to minimize temperature effects.

2.1.6 Biaxial testing

After completed overcoring, the recovered overcore sample is usually placed in a biaxial test chamber to determine the elastic parameters Young's modulus, E, and the Poisson's ratio, v. During biaxial testing, the overcore sample is first subjected to a step-wise increase of applied pressure, followed by a step-wise decrease to zero pressure while the resulting strains are measured (*e.g.* Amadei & Stephansson, 1997).

Methodology for determination of the complete stress tensor 249

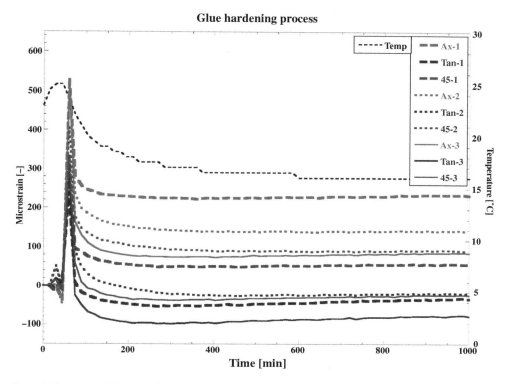

Figure 2 Example of glue bonding period including from start-up of cell (at surface at time = 0 min), installation in the borehole (peak strains at t = 30 min), to start of dense sampling (at ends of file). Sampling rate is one reading per 15 minutes.

For the thin-walled overcore sample when using main hole dimension NQ, pressurization is not made higher than 10 MPa to avoid introduction of microfractures in the core. The elastic parameters are based on secant, unloading values.

2.1.7 Core sample analysis

Before and after the biaxial test, the core sample is evaluated with respect to inhomogeneities, potential anisotropy, grain sizes, fractures or other effects that may affect the results. In addition, the core inner and outer diameter and the rosette positions are measured, and potential decentralization and/or inclination of the pilot hole is documented. These values are required for the calculation of the elastic parameters and for potential data corrections.

Particular focus is placed on the adhesion of the strain rosettes on the rock sample. In ideal conditions, the rosettes cannot be removed from the core sample without breaking into pieces upon removal, leaving the adhesive and the strain rosette left inside the overcore sample. The force required to remove the gauges is noted as well as any traces of drill cuttings in the pilot hole and on the glue surface for potential requirement of adjustment of flow rate and length of flushing period during subsequent overcoring tests.

2.2 Theory for CSIR- and CSIRO HI-type cells

The overcoring theory is based on that the rock can be described as continuous, homogeneous, isotropic (although also anisotropic solutions exist), and linear-elastic. It is further assumed that the relief during overcoring is assumed to be identical in magnitude to that produced by the *in situ* stress field, but of opposite sign, and that the measuring probe is mounted far enough from the end of the borehole. With these assumptions, the displacement fields in cylindrical components are given by *e.g.* Hirashima & Koga (1977):

$$u_r = \frac{1+\nu}{2E} r \left[\frac{R^2}{r^2} (\sigma_x + \sigma_y) + \left\{ 1 + 4(1-\nu)\frac{R^2}{r^2} - \frac{R^4}{r^4} \right\} \left\{ \begin{array}{c} (\sigma_x - \sigma_y)\cos 2\theta \\ +2\tau_{xy}\sin 2\theta \end{array} \right\} \right.$$

$$\left. + \left\{ \frac{1-\nu}{1+\nu} (\sigma_x + \sigma_y) - 2\frac{\nu}{1+\nu}\sigma_z \right\} \right] \tag{1}$$

$$u_\theta = -\frac{1+\nu}{2E} r \left[\left\{ 1 + 2(1-2\nu)\frac{R^2}{r^2} + \frac{R^4}{r^4} \right\} \left\{ (\sigma_x - \sigma_y)\sin 2\theta - 2\tau_{xy}\cos 2\theta \right\} \right] \tag{2}$$

$$u_z = \frac{1+\nu}{E} \left[2r\left(1 + \frac{R^2}{r^2}\right)(\tau_{yz}\sin\theta + \tau_{yz}\cos\theta) + \frac{z}{1+\nu}\{\sigma_z - \nu(\sigma_x + \sigma_y)\} \right] \tag{3}$$

where R is the borehole radius, r the radial distance to the measurement point, E and ν are Young's modulus and Poisson's ratio, respectively, σ_i is the far field stress component i, and θ the circumferential angle describing the location of the strain gauge.

For CSIR- and CSIRO HI-type of measuring devices including axial gauges, tangential gauges, and gauges inclined ±45° from the axial direction, the following relationships are valid:

$$\varepsilon_\theta = \frac{1}{R} \left\{ (u_r)_{r=R} + \left(\frac{\partial u_\theta}{\partial \theta}\right)_{r=R} \right\} \tag{4}$$

$$\varepsilon_z = \left(\frac{\partial u_z}{\partial z}\right)_{r=R} \tag{5}$$

$$\gamma_{\theta z} = \frac{1}{R} \left(\frac{\partial u_z}{\partial \theta}\right)_{r=R} \tag{6}$$

$$\varepsilon_{45°} = \frac{1}{2}\left(\varepsilon_\theta + \varepsilon_z + \gamma_{\theta z}\right) \tag{7}$$

where

$$\left(\frac{\partial u_\theta}{\partial \theta}\right)_{r=R} = -\frac{4(1-\nu^2)}{E} R\left[(\sigma_x - \sigma_y)\cos 2\theta + 2\tau_{xy}\sin 2\theta\right]. \tag{8}$$

$$\left(\frac{\partial u_z}{\partial z}\right)_{r=R} = \frac{1}{E}\left[\sigma_z - \nu(\sigma_x + \sigma_y)\right] \tag{9}$$

$$\left(\frac{\partial u_z}{\partial \theta}\right)_{r=R} = \frac{R}{E}\left[4(1+\nu)\left(\tau_{yz}\cos\theta - \tau_{zx}\sin\theta\right)\right] \tag{10}$$

Combining Equations 12 to 18 and using $r = R$, gives the final solution:

$$\varepsilon_\theta = \left[(\sigma_x + \sigma_y)K_1 - 2(1-\nu^2)\left\{(\sigma_x - \sigma_y)\cos 2\theta + 2\tau_{xy}\sin 2\theta\right\}K_2 - \nu\sigma_z K_4\right]/E \tag{11}$$

$$\varepsilon_z = \left[\sigma_z - \nu(\sigma_x + \sigma_y)\right]/E \tag{12}$$

$$\varepsilon_{\pm45^\circ} = 0.5\left(\varepsilon_\theta + \varepsilon_z \pm 4(1+\nu)\left(\tau_{yz}\cos\theta - \tau_{zx}\sin\theta\right)K_3/E\right) \tag{13}$$

where K_1-K_4 are correction factors for the CSIRO HI cells and represent the effect of locating the strain gauges at some distance from the rock surface and of the resistance of the HI cell to deformation (Worotnicki, 1993; Duncan, Fama and Pender, 1980). For the Borre Probe, K_1-K_4 are equal to 1.

The elastic properties of the overcore sample are derived using the theory for an infinitely long, thick-walled circular cylinder subject to uniform external pressure, and the assumption that plane stress applies (*e.g.* Worotnicki, 1993; Amadei & Stephansson, 1997):

$$E = K_1 \frac{p}{\varepsilon_\theta}\frac{2}{1-\left(\dfrac{D_i}{D_o}\right)^2} \tag{14}$$

$$\nu = -K_1 \frac{\varepsilon_z}{\varepsilon_\theta} \tag{15}$$

where p is the applied load, ε_θ and ε_z are the tangential and axial strains, respectively, and D_i and D_o are the inner and outer diameters of the cylinder, respectively. Ask (2006) suggested that also the $\pm45^\circ$ inclined gauges are used in cases where few data are available:

$$E = K_1 \frac{p}{(2\varepsilon_{\pm45} - \varepsilon_z)\varepsilon_\theta}\frac{2}{1-\left(\dfrac{D_i}{D_o}\right)^2} \tag{16}$$

$$\nu = -K_1 \frac{(2\varepsilon_{\pm45} - \varepsilon_\theta)}{(2\varepsilon_{\pm45} - \varepsilon_z)} \tag{17}$$

2.3 Explicit data uncertainties

Ideally, the differences between observed strains before and after completed overcoring yield sufficient information for stress determination. However, in reality, the testing curves incorporate also a whole series of factors that affect the results, which in the worst of cases may lead to erroneous interpretation. In this section, the explicit (measurements-related) uncertainties are highlighted, *i.e.* errors associated with the measurement operation itself. Common for measurement-related uncertainties is that they most often can be appreciated and compensated for in a straight-forward fashion.

2.3.1 Drilling operations and installation of cell

The drilling operations and installation procedure of the cell may for different reasons deviate from planned outcome. This may involve a slightly decentralized or inclined pilot hole, which are measurable deviations that may be corrected for using numerical methods. In general, however, the deviations are generally very small when using the Borre Probe drilling equipment, and the effect on calculated stresses are as a result negligible (*e.g.* decentralization is commonly less than 1 mm).

However, due to the construction of the Borre Probe, with three strain rosettes mounted 120° apart and at the end of thin cantilever arms, dislocations from the stipulated 120° may take place. Sensitivity studies have shown that the effect on calculated stresses are quite moderate (less than 5%) for the commonly observed dislocations (Ask *et al.*, 2002).

2.3.2 Diagnostic strain curve response of Borre Probe data

The strain value for each gauge used for stress calculation involves a stable plateau before and after overcoring. Generally, the pre-readings are taken before drill circulation water has been turned on, prior to the overcoring, whereas post-readings are taken after drilling water has been turned off, but before the core break.

The obtained strain versus time plots from an overcoring test include not only information of the strain/stress field, but also clues that help estimating if the data collection has been successful, or if it is hampered with difficulties. Firstly, the glue hardening process should show a gradual strain increase for all strain gauges, which eventually level out and become linear toward the ends of the period (Figure 2). Secondly, if the cell is properly installed, it is coaxial with the borehole and neither removal of the installation tool, nor the descent with and rotation of the drill string over the cell is visible on the strain curve (Figure 3). If there is a reaction, the bonding between the strain gauges and the rock may have been damaged. Thirdly, the onset and turnoff of the drilling water circulation implies some turbulence downhole, and if a strong reaction can be observed, one or more strain rosettes are not glued properly.

Finally, looking at the resulting strain values, a few controls can be made regarding the validity of results. The Borre Probe configuration involves three axial gauges which should be equal in theory. Furthermore, the configuration entails that the following relationship is valid:

$$\varepsilon_{45,1} + \varepsilon_{45,2} + \varepsilon_{45,3} = \frac{1}{2}\left(\varepsilon_{z,1} + \varepsilon_{z,2} + \varepsilon_{z,3} + \varepsilon_{\theta,1} + \varepsilon_{\theta,2} + \varepsilon_{\theta,3}\right) \tag{18}$$

or reformulated to rosettes i

$$\varepsilon_{45,i} = \frac{1}{2}\left(\varepsilon_{Z,i} + \varepsilon_{\theta,i}\right) \tag{19}$$

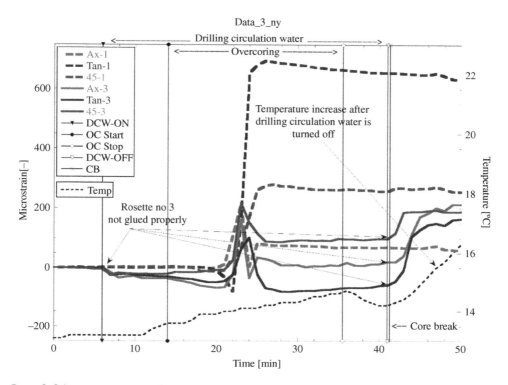

Figure 3 Schematic response of a strain gauge versus time during overcoring. Ax denotes axial strain gauges, Tan denotes tangential gauges, and 45 denots strain gauges inclined 45° from the axial direction, and the numbers 1 and 3 denote rosette number (rosette 2 not displayed to avoid overcrowding the picture). The strongest strain gauge response occurs when the drill bit is at the position of the strain gauge (about t = 22 m). The drilling circulation and overcoring intervals are given as well at the core break (vertical bars). The strains in rosette 3 are not stable when drilling circulation water (DCW) is turned on/off, indicating a bonding problem. In addition, a temperature increase is observed after DCW has been turned off, indicating a too short post-cooling period and thus temperature induced strains. Modified after Ask (2003) with permission from Elsevier.

2.3.3 Temperature effects

If there is a temperature difference between the initial and final strain reading, the stress field may not be fully relieved and may require correction. The temperature correction is a function of the thermal expansion of the rock and the inherent thermal expansion of the strain gauges. The former is seldom known or investigated, but for the Äspö HRL main rock types, the thermal correction was determined to 8 microstrain/°C (Ask et al., 2004).

It is fairly common that a fairly large strain drop after peak strains have been reached are observed in overcoring data. This is to some extent expected as the drilling continues for some time/distance after having passed the strain gauges. However, after completed drilling, and if the conditions are ideal, only temperature effects are responsible for strain variations. Hence, if temperature effects cannot explain the strain variation after completed drilling, implicit errors are evident.

2.3.4 Boundary yielding

Boundary yielding is a phenomenon that relates to CSIRO HI cells when the drill-induced heat leads to softening of the adhesive grout and the cell is exposed to expansion. Radially, the expansion is limited to the relaxation associated with the relief of the overcore sample, whereas it is free to expand in the borehole direction. As a result, the axial gauges and also the 45°/135°-inclined gauges show abnormally large strains. In terms of stress, this implies that the major stress component becomes aligned with the borehole axis. Data can be corrected with respect to boundary yielding, provided that a realistic axial strain can be estimated (Ask, 2003).

2.3.5 Biaxial testing

The biaxial testing serves, apart from a method for determination of the elastic characteristics of the overcore sample, as a quality control method for the overcoring test. Firstly, the three groups of strain orientations should respond identically for each loading step (Figure 4), which is a test of the hypothesis of isotropy. Again, Equation 19

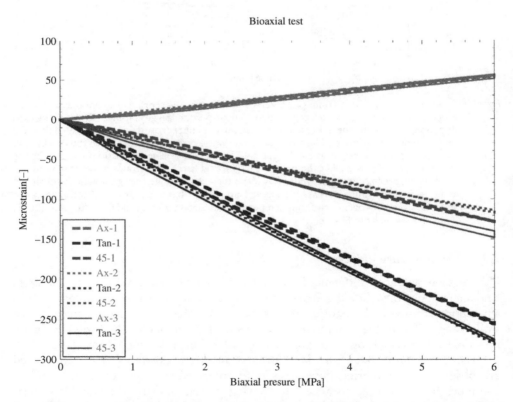

Figure 4 Hypothetical biaxial test result with slight non-linearity and hysteresis. Ax denotes axial strain gauges, Tan denotes tangential gauges, and 45 denotes strain gauges inclined 45° from the axial direction and the numbers 1, 2 and 3 denote rosette number.

may be used for verification of results, although one cannot expect perfect fit with real data. Deviation should preferably be within some 15%. Secondly, the biaxial tests gives an indication of the amount of microfractures in the core, expressed as the non-linearity at low pressures and hysteresis effects. If deviations are pronounced, it implies an obvious violation of the inherent assumptions of the overcoring method, and data must be regarded with skepticism, or even be discarded.

In some cases, especially in strong hard and brittle rocks such as quartzite, the core break can be violent and damage the glue bonding. Hence, overcoring strains may be satisfactory but biaxial test results meaningless due to damaged bonding. In such cases, it is recommended that lab tests on full size cores retrieved as close to the testing location as possible are conducted.

2.3.6 Validation of local stress tensor

For each local stress tensor the validity of the measurement can often be evaluated. For example, in the absence of topographical effects or secondary stress fields, two components should have a very small dip, whereas the third should be nearly vertical. In addition, under named circumstances, the measured vertical stress should be of the same order of magnitude as the theoretical weight of the overburden rock mass.

2.4 Implicit data uncertainties

Implicit data uncertainties refer to errors associated with the interpretative model and can be divided into two categories; those associated with violations of inherent assumptions in the overcoring methodology and those related to the attempts to describe the stress field at a large scale based on a set of local stress tensors. It is not as straightforward to associate implicit errors with *e.g.* an expected value, variance and possibly covariances as for the explicit uncertainties.

2.4.1 Violations of theoretical assumptions

As mentioned, the overcoring method rests upon the assumption that the rock mass is continuous, homogeneous, isotropic, and linear-elastic. Yet, we know that rocks do not completely fulfill these assumptions. For example, deviations from linear-elasticity are displayed in most biaxial tests. In fact, because the method relies on complete relaxation of the overcore sample, the most pronounced non-linear phase is incorporated in the data. The displayed non-linearity in the overcore sample is attributed to the overcoring phase, *i.e.* to microfractures introduced as a result of stress relaxation and as a result of the drilling process. However, in many cases, the deviation from linear-elasticity has negligible effect on calculated stresses but, regrettably, there is no generally accepted guideline/rule of thumb stating when to trust data and when to discard them.

Anisotropic behavior of the rock can be adjusted for (Amadei, 1996) but requires knowledge of all elastic parameters of the rock as the principal strain direction differs from the principal stress directions. Studies have shown that the degree of anisotropy must be fairly large before it seriously affects stress magnitudes and orientations (larger than 1.3–1.5, based on Young's modulus; Worotnicki, 1993).

Sometimes overcoring cells produce seemingly good individual test data, but when considering a group of tests at a site, a different picture may arise. This is for example the case for the Olkiluoto site of the Finnish Nuclear Fuel and Waste Management Co. (Posiva), where both the Borre Probe and the CSIRO HI cells have been employed. The site is dominated by a migmatitic, foliated gneiss, but the foliation direction is not persistent, but instead highly variable. Hence, at this site, the overcoring testing scale (a few dm^3) is smaller than the representative elementary volume, and data therefore indicate just about anything. The same problems is observed in minifrac tests, which are made at about the same scale as the overcoring tests, whereas conventional hydraulic methods have produced consistent results in this heterogeneous, or small scale anisotropic, rock (Posiva, 2012).

As described in Section 2.3.3, significant strain drop after peak strains have been reached are signs of implicit errors, often related to bonding problems or microfracturing of the overcore sample. Bonding problems are often visible during onset/turn-off of drilling water, but may indicate seemingly realistic overcoring curves in the initial phase of the overcoring process. If the gauges are more or less de-bonded from the rock sample, the strain values often drop to near sub-equality of the three strain gauges in the rosette (rosette 3 in Figure 5). Bonding problems are obvious during the subsequent biaxial tests and can also be verified during the core sample analysis when the rosettes are removed from the core sample. Microfracturing of the core sample can, if the stress is high relative to the strength of the rock sample, be severe and disqualify the results. Again, the subsequent biaxial test and tests of the adhesion of the strain rosettes can be used for verification. If the core yields, it is hard to keep the biaxial pressures steady during the pressurization steps; excessive pumping and draining must be undertaken during the loading and unloading phases, respectively, in order to maintain stability during the biaxial pressure steps. If this occurs, while the adhesion of the rosettes are still acceptable, the strain drop can be correlated to yielding of the overcore sample due to significant microfracturing (rosette 1 in Figure 5). In such cases, coring dimension must be increased in order to reach reliable results, although some improvements can also be reached by optimizing the drilling process.

A fairly common observation of yielding of the overcore sample is related to ring or core disking (*e.g.* Li & Schmitt, 1998). The failure of the rock specimen is related to the strength of rock relative to the stress field, but also on a number of drilling parameters. The phenomenon may in overcoring data be displayed in different ways, ranging from non-visible fractures for the naked eye but with abnormal axial strains (primarily), to fully developed disks that effectively prevent recording of data. Core disking involves a serious violation of the theoretical assumptions associated with the overcoring technique, and data are not useful for stress determination.

If disking appears, it may be advisable to use the core disking phenomenon itself for stress field determination. The methodology is still somewhat premature, but significant improvements have been made in recent years (*e.g.* Corthésy & Leite, 2008).

2.4.2 The continuity hypothesis and choice of integration procedure

A stress measurement campaign using the Borre Probe typically involves three successful tests within a few meters of borehole (*i.e.* a very small volume of rock), which are clustered together to define a single local tensor. However, sometimes, several such

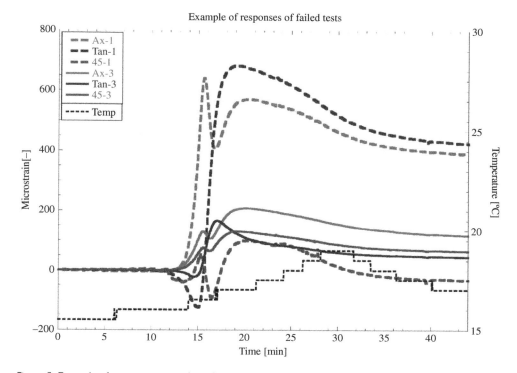

Figure 5 Example of excessive strain drop during overcoring resulting from yielding of the overcore sample (rosette 1) and debonding of a strain rosette (rosette 3). Microfracturing of the overcore sample will be manifested in non-linearity during the subsequent biaxial test, and if proper bonding of the rosette can be verified, the strain drop can be correlated to yielding of the overcore sample due to significant microfracturing. Bonding problems are often visible during onset/turn-off of drilling water but may indicate seemingly realistic overcoring curves in the initial phase of the overcoring process. When strains increase as the drill bit approaches the strain rosettes, bonding is further damaged and resulting strain levels are often unexpectedly low and declining significantly during the drilling process. If the rosette more or less de-bonds from the rock sample, the strain values of the three strain gauges in the rosette drop to near sub-equality.

clusters are made in a borehole with several tenths of meters apart, why stress gradients need to be considered. For larger campaigns, it is strongly recommended that choice of test locations is based on consideration of the continuity hypothesis to ascertain that the data indeed sample the same continuum. This requires detailed analysis of site geology, drill cores, and geophysical logs.

Stress field distributions world-wide predominantly indicate a sub-linear depth dependency and a first approximation of the distribution can be described as (Cornet, 1993, 2015):

$$\sigma(X^m) = \sigma(X) + (x^m - x)\alpha^{[x]} + (y^m - y)\alpha^{[y]} + (z^m - z)\alpha^{[z]} \qquad (20)$$

where $\sigma(X^m)$ and $\sigma(X)$ are the stress tensors in points X^m and X, respectively, and $\alpha^{[x]}$, $\alpha^{[y]}$, and $\alpha^{[z]}$ are second-order symmetrical tensors characterizing the stress gradient in the x-, y- and z-directions. The stress field satisfies the following equilibrium condition:

$$div\left(\sigma(X^m)\right) - \rho(X)b_i = 0 \tag{21}$$

where $\rho(X)$ is the density of the rock mass and b_i is the gravitational acceleration $(b_i = g\delta_{i3}; \delta_{i3} = 0$ for $i \neq 3; \delta_{i3} = 1$ for $i = 3)$. A major drawback with this representation is that it requires some 30 tests for acceptable determination of the 22 model parameters. However, a series of assumptions can be made that significantly reduce the number of unknowns. In practice, a tentative distribution of the stress field is first assumed and the subsequent calculations will outline if this model will hold, if it need adjustments, or if the data set requires grouping into subsets as a result of $e.g.$ discontinuities.

In the following, a linear stress distribution with respect to the vertical direction is chosen to complete the theory. This is a common description of the stress field distribution as stress measurements are often made along a single borehole:

$$\sigma(X^m) = \sigma(X) + (z^m - z)\alpha^{[z]} \tag{22}$$

The assumed parameterization involves the full stress tensor and its variation with depth ($i.e.$ 12 unknown parameters). Hereafter, we express $\sigma(X)$ and $\alpha^{[z]}$ with three Euler angles and three principal values. For $\sigma(X)$, the eigenvalues are S_1 to S_3 and the three Euler angles are E_1 to E_3, which are expressed in the geographical frame of reference. Corresponding eigenvalues for $\alpha^{[z]}$ are α_1 to α_3, and the three Euler angles E_4 to E_6 are expressed in the $\sigma(X)$ frame of reference. Thus, the gradients α_1 to α_3 correspond to the vertical gradient of S_1 to S_3, only if E_1 to E_3 are equal to E_4 to E_6.

For overcoring data, the general equations for the measured strains related to the stresses in a local xyz frame of reference are given in Equations 11 to 13. The equation for σ_x^n of the nth measurement in matrix form is thus:

$$\sigma_x^n = \left(\left[SB \cdot \left(S^o + (z^m - z) \cdot AB \cdot A^o \cdot AB^T\right) \cdot SB^T\right]\vec{n}_x^n\right)\vec{n}_x^n \tag{23}$$

where n_x^n is the direction of the local x-axis with respect to the geographical frame of reference. The expressions for the remaining stress components (σ_y^n, σ_z^n, σ_{xy}^n, σ_{xz}^n, σ_{yz}^n) are analogous. A vector function $f(\pi)$ may be introduced in which the mth component is defined by:

$$f^m(\pi) = \sigma_x^n - \left(\left[SB \cdot \left(S^o + (z^m - z) \cdot AB \cdot A^o \cdot AB^T\right) \cdot SB^T\right]\vec{n}_x^n\right)\vec{n}_x^n \tag{24}$$

Borre Probe data involve three different expressions for $f^n(\pi)$; for axial, tangential and 45°-inclined strain gauge, and the inversion may be solved using the iterative algorithm based on the fixed point method (Tarantola & Valette, 1982):

$$\pi_{n+1} = \pi_o + C_o F_n^T \left(F_n C_o F_n^T\right)^{-1} [F_n(\pi_n - \pi_o) - f(\pi_n)] \tag{25}$$

where F is a matrix of partial derivatives of $f(\pi)$ valued at point π.

3 INTERPRETATION OF THE STRESS FIELD: CASE STUDY ÄSPÖ HRL

So far, we have discussed how to avoid, identify, and correct for various potential sources of error. In the following, the analysis protocol is applied at the Äspö HRL where several of the previously discussed problems were present. At this site, a majority of the data involved overcoring but also a number of hydraulic stress measurements existed that proved useful for verifications. The section is a shorted version of Ask (2003), Ask (2000a), and Ask (2000b) and the interested reader is referred to these papers for more details.

3.1 The Äspö Hard rock laboratory

The Äspö HRL of the Swedish Nuclear Fuel and Waste Management Co (SKB) is located at the Baltic sea shore line, about 300 km south of Stockholm (Figure 6). The island of Äspö is characterized be a mildly undulating topography and composed of ca 1.8 Ga intrusive rocks associated with the 1.86–1.65 Ga Transscandinavian Igneous Belt (TIB, Wahlgren, 2010). The TIB rocks in the area show variable composition and are affected by magma-mixing processes and consequently display close genetic relationship. The main bedrocks at Äspö are grouped into Småland (Ävrö) granite, Äspö diorite (a more mafic variety of Småland granite), greenstone and fine-grained granites. Overall, this may be described as a rock mass subjected to local inclusions and dikes, *i.e.* inhomogeneities. The Småland granites exhibit a general foliation trend about N70°E to E-W and commonly steeply dipping. Intrusions of fine-grained granites and aplites are common as well as greenstones in larger massifs and as smaller lenses.

The intrusive rocks at Äspö are well preserved and hardly affected by ductile deformation and metamorphism. As a result, the bedrock is generally considered more or less isotropic. The majority of the major deformations zones are characterized by polyphase brittle deformations, although with ductile precursors. Hence, the fracture zone network was formed when the bedrock still responded to deformation in the ductile regime with discrete brittle-ductile to ductile shear zones forming the most prominent ductile structures in the area. These are sub-vertical and strike N-S, NE-SW, and E-W.

The brittle structures have formed as a result of multiple reactivations of fracture and fault sets related to orogenic episodes affecting the region, starting from approximately 1.5 Ga. The structures have been grouped according to strike and inclination; (1) NE-SW, moderate to steep dip; (2) N-S, moderate to steep dip; (3) E-W to NW-SE, steep to moderate dip toward south; (4) E-W to NW-SE, moderate dip toward north; and (5) gently dipping.

3.2 Collected data at the Äspö HRL

At the Äspö HRL and its surroundings, about 100 hydraulic measurements and 140 overcoring measurements had been conducted (Bjarnason *et al.*, 1989; Lee & Stillborg, 1993; Lee *et al.*, 1994; Litterbach *et al.*, 1994; Ljunggren & Klasson, 1996, 1997; Nilsson *et al.*, 1997; Ljunggren & Bergsten, 1998; Klee & Rummel, 2002; Klasson *et al.*, 2002; Klasson & Andersson, 2002). In this paper, focus is on the stress data

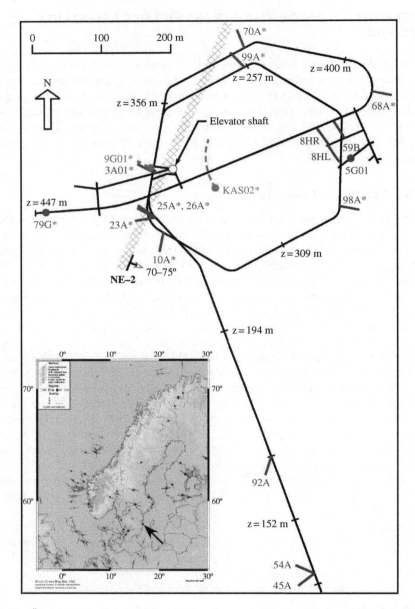

Figure 6 The Äspö Hard Rock Laboratory, SE Sweden, marked in the European Map in the lower left corner. The detailed map of the Äspö HRL displays boreholes included in the study together with the NE-2 fracture zone at the tunnel intersection depth (green dashed line) that divides the HRL into a NW (red) and a SE (blue) domain. Stress measurement boreholes are named and displayed with black lines (sub-horizontal boreholes), black circles (sub-vertical boreholes), or a combination of these (inclined boreholes). Modified after Ask (2003) with permission from Elsevier.

collected in the immediate surroundings of the HRL and only on data sampling of the *in situ* stress field, *i.e.* tests sampling the secondary stress field are discarded. This reduces the amount of data to 34 hydraulic and 63 overcoring measurement points.

The overcoring data were collected using four different overcoring cells: (1) the 9-gauge CSIRO HI cell; (2) the thick-walled 12-gauge CSIRO HI cell, (3) the thin-walled 12-gauge CSIRO HI cell; and (4) the 9-gauge Borre Probe (Worotnicki, 1993; Sjöberg & Klasson, 2003). The CSIRO HI cells require dry boreholes and the tests were conducted in short boreholes oriented slightly upwards from the ramp between 140 and 420 m vertical depth (mvd), *cf.* Figure 6. The Borre Probe measurements were conducted in both sub-horizontal and sub-vertical boreholes in the vertical depth range 410–480 mvd. The hydraulic data were collected using standard hydraulic fracturing equipment based on impression packer technique to obtain fracture orientation data.

Despite the number of stress measurements within a fairly limited volume of rock, the stress field was poorly understood due to the pronounced scatter in data and because of large differences in results between methods.

3.3 Results from data re-evaluation

The entire data set at Äspö HRL was subjected to a thorough, uniform and diagnostic data analysis based on ISRM directions (Sjöberg *et al.*, 2003; Haimson & Cornet, 2003). Firstly, several strains from both cells were discarded as a result of non-expected behavior resulting from fracturing of the overcore sample and bonding problems. Secondly, 25% of the overcoring data were corrected for induced temperature, as cooling was not undertaken long enough after completed overcoring, leading to a temperature difference between the pre- and post-reading of strains. Thirdly, it was found that the CSIRO HI data suffered from boundary yielding, and corrections were made to almost 60% of the CSIRO HI data set. Finally, the biaxial testing of the CSIRO HI overcore samples were of very poor quality as a result of too high loading pressure, leading to fracturing of the core sample. Conclusively, a large number of difficulties were observed in the data, but after applied corrections, the stress distribution versus depth appeared markedly different with a much improved agreement between methods (Ask, 2003) compared to the picture prior to the corrections. Yet, it was not yet clear why tests at similar depths seemed to indicate two different stress fields.

3.4 Integration of stress data

Detailed analysis of site geology with special attention to larger structures and possible sub-domains was then undertaken. Stress field determinations involving smaller rock volumes were made, *e.g.* at the Prototype Repository (Ask *et al.*, 2001) and at the Cedex Test Site (Ask *et al.*, 2003), to validate stress field continuity.

The final step of the evaluation process, integration of the entire data set, proved to be a difficult task, but it became clear that a North-East striking fracture zone, denominated NE-2, decoupled the stress field at Äspö into a NE and SE domain. The NE-2 fracture zone is 0.7 to 5 m wide, strikes 21°N and dips about 77° toward south-east.

The fracture zone forced division of the data set, which was somewhat unfortunate, because the available data in the SE domain were limited (125 CSIRO HI strains). However, the NW domain included 19 hydraulic data, 27 strains from the Borre Probe and 224 strains from the CSIRO HI cells.

Four different calculations using the data in the NE domain were conducted: (I) using only hydraulic data; (II) using only overcoring data; (III) joint inversion based on elastic parameters from biaxial testing; and (IV) joint inversion with elastic parameters chosen as unknown (Figure 7). The differences between the solutions are small, less than 10° for stress directions and about 2 MPa for σ_2 and σ_3. The main difference concerns the magnitude of σ_1 (σ_H), but this parameter is poorly resolved in the hydraulic solution and the results is somewhat expected. In the joint inversion using known elastic parameters (E=61.6 GPa and v=0.26), the effect of the hydraulic data is hardly noticed as the overcoring strains are so much more numerous. However, when elastic parameters are chosen as unknown (found to be equal to 50.8 GPa and 0.33, respectively), the hydraulic fracturing data help constrain the overcoring set with respect to magnitudes. The difference in stress state between the known and *in situ* elastic parameters is overall fairly small, which is to be expected as the two data sets became quite comparable after the re-evaluation.

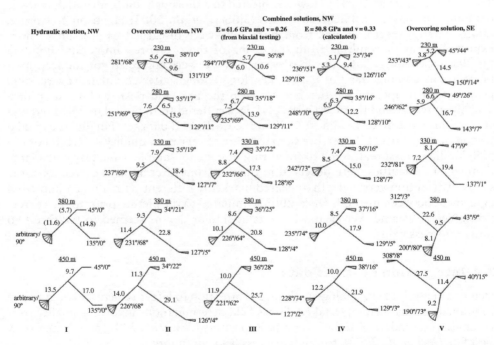

Figure 7 Three-dimensional state of stress based on hydraulic data in the NW domain (I), overcoring data in the NW and SE domains (II and V), and combined hydraulic and overcoring data sets in the NW domain between 230 and 450 m vertical depth (solutions III and IV). The combined solutions for the NW domain are based on calculations with known average elastic parameters (IV; E = 61.6 GPa and v = 0.26) and with *in situ* elastic parameters (V, E = 50.8 GPa and v = 0.33). Modified after Ask (2006) with permission from Elsevier.

Methodology for determination of the complete stress tensor 263

The joint solutions in the NE domain indicate that σ_1 and σ_3 are sub-horizontal and σ_2 is sub-vertical. In the SE domain, a pronounced rotation with depth was observed, where *e.g.* σ_1 rotates from 150°N to 128°N between 230 and 450 mvd. The observed stress rotation in the SE domain can be correlated to the physical location of the various measurement points in this domain; shallow data in the SE-domain are located much further away from the NE-2 zone compared to data in the NW domain and deeper data in the SE domain (see Figure 6). Hence, the rotation is suggested to be attributed to the radius of influence of the NE-2 zone.

At depths, both domains have the same direction of σ_1, suggesting that σ_1 is nearly perpendicular to the NE-2 zone, whereas the σ_2 and σ_3 components nearly coincide with the plane defined by the NE-2 zone. Hence, data indicate that the NE-2 fracture zone is a principal plane.

4 CONCLUDING REMARKS

A protocol for complete stress determination using the overcoring stress measurement method has been outlined. It is shown that the methodology, independent upon overcoring device, requires experienced personnel, well-established routines during all operational steps of the overcoring process, careful documentation, and a skilled drilling crew. Only with these pre-requisites, can uncertainties be kept an acceptable level.

In the case study, the effectiveness of the protocol is visualized; the stress field can be resolved even in cases where data are significantly hampered with difficulties. Indeed, the interpreted stress field at Äspö HRL have been verified recently in Hakala *et al.* (2013). Yet, it is unfortunate that errors are introduced as significant time must be allocated to understand and interpret the turn of events leading to the problem. In addition, a corrected datum is always regarded as somewhat ambiguous as the correction itself is associated with uncertainties; the case study was remedied by the existing hydraulic data that helped constrain and verify the overcoring solutions.

The major deficit with the overcoring methodology is the difficulty to comply with the many assumptions inherent in the theory. Regrettably, many of the implicit uncertainties in the overcoring methodology cannot be corrected for. As a result, it is absolutely critical that they are identified and as far as possible corrected for, as they otherwise may result in erratic stress field interpretation or completely disqualify the results. If deviations are identified, it is highly recommended that other stress measurement techniques are employed for data comparison, as this will be the only means to verify or reject the stress field determination.

Rock stress measuring methods always produce a certain degree of data scatter. Common for all methods is that explicit errors can be taken into consideration and properly be adjusted for. However, the smaller scale methods, such as overcoring, are more sensitive to implicit errors; manifested as a more pronounced data scatter compared to larger scale methods. Given this sensitivity, it is important that each individual test is analyzed carefully, taking into consideration overcoring strain data, biaxial strain data, overcore sample and glue bonding. This is the only approach that enables derivation of the source of the problem, weather it is explicit or implicit, and if it can be corrected for, or if the data are more suitable for the trash bin.

REFERENCES

Amadei, B. and Stephansson, O. (1997). *Rock Stress and Its Measurements*, Chapman and Hall Publ., London.

Ask, D., Stephansson, O., and Cornet, F.H. (2001). Analysis of overcoring stress data in the Äspö region. In: Elsworth, D., Tinucci, J.P., and Heasley, K.A. (eds.) *Rock Mechanics in the National Interest: Proceedings of the 38th US Rock Mechanics Symposium*, 7–10 July 2001, Washington, USA. AA. Balkema Publ. pp. 1401–1405.

Ask, D. (2003) Evaluation of measurement-related uncertainties in the analysis of overcoring rock stress data from Äspö HRL, Sweden: a case study. *Int J Rock Mech Min*, 40, 1173–1187.

Ask, D., Cornet, F.H., and Stephansson, O. (2003) Integration of CSIR- and CSIRO-type of overcoring rock stress data at the Zedex Test Site, Äspö HRL, Sweden. In: Handley, M. and Stacey, D. (Eds.) *Technology Roadmap for Rock Mechanics: Proceedings of the 10th International Congress on Rock Mechanics*, 8–12 September 2003, Johannesburg, South-Africa. SAIMM. pp. 63–68.

Ask, D., Cornet, F.H., and Stephansson, O. (2004) Analysis of overcoring rock stress data at the Äspö HRL, Sweden. Analysis of overcoring rock stress data from the Borre Probe. SKB International Progress Report, IPR-04-05, Stockholm.

Ask, D. (2006a) Measurement-related uncertainties in overcoring rock stress data at the Äspö HRL, Sweden, Part 2: Biaxial tests of CSIRO HI overcores. *Int J Rock Mech Min*, 43, 127–138.

Ask D (2006b). New developments of the Integrated Stress Determination Method and application to rock stress data at the Äspö HRL. *Int J Rock Mech Min*, 43, 107–126.

Bjarnason, B., Klasson, H., Leijon, B., Strindell, L. and Öhman, T. (1989) Rock stress measurements in boreholes KAS02, KAS03 and KAS05 on Äspö. SKB Progress Report 25-89-17, Stockholm.

Cornet, F.H. (2015) *Elements of Crustal Geomechanics*, Cambridge University Press.

Corthésy, R. and Leite, M.H. (2008) A strain-softening numerical model of core discing and damage. *Int J Rock Mech Min*, 45 (3), 329–350.

Duncan Fama, M.E. and Pender, M.J. (1980) Analysis of the hollow inclusion gauges with implications for stress monitoring. *Int J Rock Mech Min*, 17, 137–146.

ES&S (2015). 3D Rock stress – Digital HI cell. [Online] Available from: http://www.essearth.com/product/3d-rock-stress-digital-hi-cell/ [Accessed 3rd June 2015].

Haimson, B.C. and Cornet, F.H. (2003) ISRM Suggested Methods for rock stress estimation – Part 3: hydraulic fracturing (HF) and/or hydraulic testing of pre-existing fractures (HTPF). *Int J Rock Mech Min*, 40, 1011–1020.

Hakala, M., Sirén, T., Kemppainen, K., Christiansson, R., and Martin, D. 2012. In Situ stress measurement with the new LVDT-cell – method description and verification. Posiva Report 2012-43.

Hirashima, K. and Koga, A. (1977). Determination of stresses in anisotropic elastic medium unaffected by boreholes from measured strains or deformations. In: Kovári (Ed.) *Field Measurements in Rock Mechanics: Proceedings of the International Symposium*, 4–6 April 1977, Zurich, Switzerland. AA Balkema Publ, Rotterdam. pp. 173–182.

Klasson, H. and Andersson, S. (2002) 3D overcoring rock stress measurements in borehole KF0093A01 at the Äspö HRL. For comparison to the 2D AECL Deep Doorstopper Method. SKB Report, Stockholm.

Klee, G. and Rummel, F. (2002) Rock stress measurements at the Äspö HRL. Hydraulic fracturing in boreholes KA2599G01 and KF0093A01. SKB International Progress Report IPR-02-02, Stockholm.

Lee, M. and Stillborg, B. (1993) Äspö virgin stress measurement results in sections 1050, 1190 and 1620 m of the access ramp. SKB Progress Report 25-93-02, Stockholm.

Lee, M., Hewitt, T. and Stillborg, B. (1994) Äspö virgin stress measurement results. Measurements in boreholes KA1899A, KA2198A and KA2510A. In SKB Progress Report 25-94-02, Stockholm.

Litterbach, N., Lee, M., Struthers, M. and Stillborg, B. (1994) Virgin stress measurement results in boreholes KA2870A and KA3068A. SKB Progress Report 25-94-32, Stockholm.

Ljunggren, C. and Klasson, H. (1997) Deep hydraulic fracturing rock stress measurements in borehole KLX02, Laxemar, Drilling KLX02-Phase 2, Lilla Laxemar, Oskarshamn. SKB Project Report U-97-27, Stockholm.

Ljunggren, C. and Klasson, H. (1996) Rock stress measurements at Zedex test area, Äspö HRL. Äspö Hard Rock Laboratory, Technical Note TN-96-08z, Stockholm.

Ljunggren, C. and Bergsten, K.-Å. (1998) Rock stress measurements in KA3579G, Prototype repository. SKB Progress Report HRL-98-09, Stockholm.

Nilsson, G., Litterbach, N., Lee, M. and Stillborg, B. (1997) Virgin stress measurement results, borehole KZ0059B. Äspö Hard Rock Laboratory, Technical Note TN-97-25g, Stockholm.

Posiva (2012) Olkiluoto Site Description 2011. Posiva Report 2011–02.

Sjöberg, J. and Klasson, H. (2003) Stress measurement in deep boreholes using the Borre (SSPB) probe. *Int J Rock Mech Min*, 40, 1205–1223.

Sjöberg, J., Christiansson, R. and Hudson, J.A. (2003). ISRM Suggested Methods for rock stress estimation – Part 2: overcoring methods. *Int J Rock Mech Min*, 40, 999–1010.

Tarantola, A. and Valette, B. (1982) Generalized non-linear inverse problem solved using the least square criterion. *Rev Geophys Space Phys*, 20, 219–232.

Wahlgren, C.H. (2010). Bedrock geology – overview and excursion guide. SKB Report R-10-05, Stockholm.

Worotnicki, G. (1993) CSIRO triaxial stress measurement cell. In: Hudson J. (ed.) *Comprehensive Rock Engineering*, vol. 3. Pergamon Press, Oxford. pp. 329–394.

Chapter 9

Measurement of induced stress and estimation of rock mass strength in the near-field around an opening

Y. Obara[1] & K. Sakaguchi[2]

[1]Graduate School of Science and Technology, Kumamoto University, Kumamoto, Japan
[2]Graduate School of Environmental Studies, Tohoku University, Sendai, Japan

Abstract: In order to estimate rock mass strength and sliding criterion of discontinuity from the stress state around an opening, methods of measuring induced stress and monitoring stress change are described. Those methods are the Compact Conical-ended Borehole Overcoring (CCBO) technique and the Cross-sectional Borehole Deformation Method (CBDM). Firstly the concept of estimation of rock mass strength from the stress state measured by those methods is explained. Then applying the CCBO to measure around underground opening, rock mass strength and sliding criterion of discontinuity from the obtained stress distributions are estimated. Furthermore a case example of the CBDM such that stress change around an opening is monitored under construction is demonstrated. It is shown that the CCBO and CBDM are convenient for estimating not only induced stress and stress change but also rock mass strength and sliding criterion of discontinuity.

I INTRODUCTION

The estimation of rock mass strength is important for designing rock structure. For this purpose, a series of in situ shear test is performed at construction site of rock structure. The in situ shear test was developed at the dawn of rock mechanics. In 1961, Serafim & Lopes introduced the technique of in situ shear test, using rock blocks attached base rock and concrete blocks molded against the rock surface (Jeager, 1979). The suggested methods for determining shear strength were published by Franklin *et al.* in 1974. In Japan, testing method for shear strength was published as suggested methods for in-situ test from the Japan Society of Civil Engineers in 1983 and revised it in 2000. Performing a series of in situ shear test, we can estimate the rock mass strength.

On the other hand, rock stress also is one of important parameters for designing rock structures. The behavior of rock mass is dependent on initial stress, as well as mechanical properties of rock. Therefore the initial stress is measured for designing rock structure. Especially, in the use of overcoring method, the initial stress is measured at points far from the wall of a gallery excavated at construction site of it. Because that the immediate wall is damaged due to excavation. However, it is considered that the damaged area is under post failure and the stress states in the area satisfy a failure criterion. Therefore, if the stress sates at several points in the damaged area are measured, a failure criterion of rock mass can be estimated from the measured results.

Accordingly, measuring stress distribution, namely induced stress, around an underground opening, we can estimate not only initial stress in far field but also failure criterion of rock mass in near filed at the same time. Furthermore, monitoring stress change around an underground opening under construction, we can also estimate failure criterion of rock mass. It is concluded that the measurement of rock stress distribution and stress change is one of profitable in situ tests.

In this chapter, firstly the concept of estimating a failure criterion from stress distribution around an underground opening is described. Then methods of measuring stress distribution and monitoring stress change are introduced. The former is Compact Conical-ended Borehole Overcoring (CCBO) technique and the latter is Cross-sectional Borehole Deformation Method (CBDM). Demonstrating case examples by CCBO and CBDM, it is shown that the CCBO and CBDM are convenient for estimating not only induced stress and stress change but also rock mass strength.

2 METHOD FOR MEASURING INDUCED STRESS AND MONITORING STRESS CHANGE

2.1 The stress around a circular opening under hydrostatic initial stress field

The stress distribution around a circular opening, which is excavated in a Mohr-Coulomb material under hydrostatic initial stress field, is investigated. The Mohr-Coulomb failure criterion is represented as:

$$\tau = c + \sigma \tan\phi \tag{1}$$

Using principal stresses, the criterion is written as:

$$\sigma_1 = S_c + q\sigma_3 \tag{2}$$

where c is the cohesion, ϕ is the internal friction angle. S_c is the uniaxial compressive strength and q is a constant as follows:

$$S_c = 2c \tan\left(\frac{\pi}{4} + \frac{\phi}{2}\right) \tag{3}$$

$$q = \tan^2\left(\frac{\pi}{4} + \frac{\phi}{2}\right) \tag{4}$$

Assuming that the circular opening with a radius of R is excavated under hydrostatic initial stress field of $\sigma_0 \geq S_c/2$. In this case, the plastic region which material satisfies the failure criterion is produced in the region of $R \leq r \leq R^*$. The outside region of $R^* \leq r$ behaves elastically. The radius R^* of the boundary surface between the two regions has to be found, and continuity conditions have to be satisfied on it.

In the plastic region, the equation of stress equilibrium holds and is (Jeager & Cook, 1979):

$$\frac{d\sigma_r}{dr} = \frac{\sigma_r - \sigma_\theta}{r} \tag{5}$$

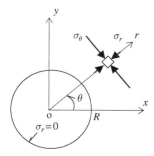

Figure 1 A circular opening and coordinates.

in the polar coordinate as shown in Figure 1. Assuming that $\sigma_r > \sigma_\theta$, Equation 2 becomes:

$$\sigma_\theta = S_c + q\sigma_r \tag{6}$$

Then using Equations 5 and 6 under the condition of $\sigma_r = 0$ at $r = R$,

$$\sigma_r = \frac{S_c}{1-q}\left\{1 - \left(\frac{r}{R}\right)^{q-1}\right\} \tag{7}$$

On the other hand, the stress state in the elastic region is represented as:

$$\sigma_\theta = 2\sigma_0 - \sigma_r \tag{9}$$

$$\sigma_r = \sigma_0 - \frac{(q-1)\sigma_0 + S_c}{q+1}\left(\frac{R^*}{r}\right)^2 \tag{10}$$

The radius R^* gives (Aoki et al., 1988):

$$\left(\frac{R^*}{R}\right)^{q-1} = \frac{2\{(q-1)\sigma_0 + S_c\}}{(q+1)S_c} \tag{11}$$

The stress distribution around the circular opening under the condition of hydrostatic initial stress field of $\sigma_0/S_c = 2$ and $q = 4.6$ ($\phi = 40$ degrees) is shown in Figure 2 (Obara et al., 1992).

2.2 Estimation of failure criterion of rock mass from stress measurement

Consider the stress distribution around an opening under the condition of $S_c = 20$MPa, $R = 3$m. In this time, cohesion becomes $c = 4.7$MPa. In this case, the Mohr's stress circles and the failure criterion are drawn in Figure 3. The dotted stress circles in the elastic region lie below failure criterion of rock mass as the solid line. That means that the failure will not take place. On the other hand, the solid stress circles in the plastic region touch the failure criterion. The stress states in the plastic

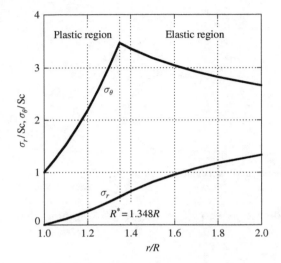

Figure 2 Stress distribution around an opening under the condition of hydrostatic initial stress field of $\sigma_0/S_c = 2$ and $q = 4.6$.

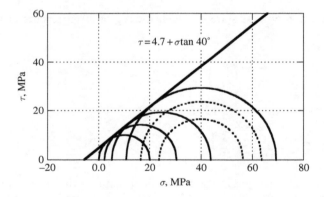

Figure 3 Mohr's stress circles in the plastic region (solid) and elastic region (dotted), and Mohr-Coulomb's failure criterion.

region satisfy the failure criterion. The depth of the plastic region around the opening is not so large and the rock condition of the region is not so good. Because of that the failure occurs and many fracture surfaces are induced in the plastic region and joint spacing becomes short. However, a rock block surrounded by the fracture surfaces in the plastic region maintains elasticity. If the stress state of the rock block at several points in the plastic region can be measured, the stress distribution in the plastic region and the failure criterion are estimated. For this purpose, it is indispensable that stress state is measured under the condition of short joint spacing and that stress measurement method which can be performed in short distance between measurement points is adopted. The Compact Conical-ended Borehole Overcoring (CCBO) technique is suitable.

2.3 Compact Conical-ended Borehole Overcoring (CCBO) technique for stress measurement

(Sugawara & Obara, 1999; Obara & Sugawara, 2003a, 2003b)

2.3.1 Field measurement system

The field measurement procedure is illustrated in Figure 4. Firstly, a pilot borehole having a diameter of 76mm is drilled to the stress measurement station. Then, the bottom of the pilot borehole is formed into a conical shape with a bortz crown bit (Figure 5(a)), the surface of which is ground smooth with an impregnated bit (Figure 5(b)). After bottom

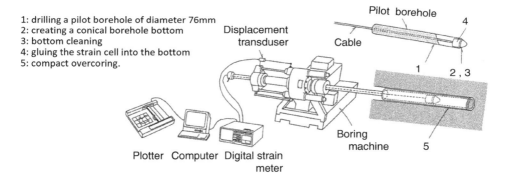

Figure 4 Field measurement system and procedure.

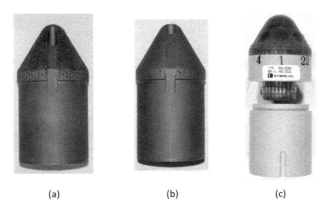

Figure 5 Special bits and strain cell, (a) bortz crown bit, (b) impregnated bit, (c) 24 element conical strain cell.

cleaning with water and acetone, the existence of cracks on the borehole surface is confirmed by a borehole camera. Then, the 24 element conical strain cell shown in Figure 5(c) is directly bonded to the conical borehole bottom surface with glue. Finally, the stress around the bottom of the pilot borehole is relieved by the compact overcoring, that is a thin-walled core boring having a diameter of 76mm which coincides with that of the pilot borehole as shown by broken line in Figure 4. During this operation, the changes in strain are continuously measured and recorded by a strain meter and a computer. For this purpose, the cable is linked to the strain meter through the boring rods and the water swivel.

2.3.2 Theory

For calculation of the initial stress from the measured strains, the spherical coordinates (ρ, θ, ϕ) and the cylindrical coordinates (r, θ, z) are defined as well as the Cartesian coordinates (x, y, z) with the z-axis coincident with the borehole axis, as illustrated in Figure 6(a). The initial stress tensor $\{\sigma\}$ can be expressed as follows:

$$\{\sigma\} = \{\sigma_x, \sigma_y, \sigma_z, \tau_{yz}, \tau_{zx}, \tau_{xy}\}^T \tag{12}$$

where σ_x, σ_y, σ_z, τ_{yz}, τ_{zx} and τ_{yx} are the stress components in the Cartesian coordinates.

The strains are required to be measured at eight specified points on the conical borehole socket of radius 38mm, as shown in Figure 6(b). The strain measuring

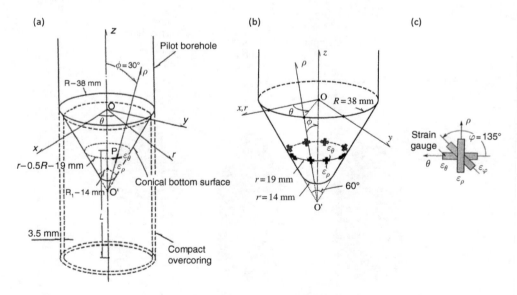

Figure 6 Strains to be measured on the conical borehole socket and compact overcoring, and arrangement of strains and definition of coordinates, (a) definition of strains and overcoring, (b) arrangement of strains for 16 element method, (c) strain gauge arrangement at measuring point for 24 element method from the inside.

points are axisymmetrically arranged along a measuring circle of radius 19mm, by rotating 45 degrees at a step. The specification of strain measuring points has been optimized through theory and experiment. In the 16 element method, the tangential strain ε_θ and the radial strain ε_ρ are measured at each strain measuring point, using a 16 element conical strain cell. The 24 element method requires the additional strain at each point, that is, the oblique strain ε_φ, as shown in Figure 6(c). Thus, the strains measured on a conical borehole socket can be denoted by:

$$\{\beta\} = \{\beta_1, \beta_2, \ldots\ldots, \beta_n\}^T \tag{13}$$

where n is the number of strains; $i.e.$ $n = 16$ for the 16 element method; $n = 24$ for the 24 element method.

The strains $\{\varepsilon_\theta, \varepsilon_\rho, \varepsilon_\varphi\}$ at a strain measuring point of a tangential angle θ are given, in the isotropic case, as follows:

$$\left\{\begin{array}{c} \varepsilon_\theta \\ \varepsilon_\rho \\ \varepsilon_\varphi \end{array}\right\} = \left[\begin{array}{ccc} A_{11} + A_{12}\cos 2\theta, & A_{11} - A_{12}\cos 2\theta, & C_{11}, \\ A_{21} + A_{22}\cos 2\theta, & A_{21} - A_{22}\cos 2\theta, & C_{21}, \\ A_{31} + A_{32}\cos 2\theta + A_{33}\sin 2\theta, & A_{31} - A_{32}\cos 2\theta - A_{33}\sin 2\theta, & C_{31}, \end{array}\right.$$

$$\left.\begin{array}{ccc} D_{11}\sin\theta, & D_{11}\cos\theta, & 2A_{12}\sin 2\theta \\ D_{21}\sin\theta, & D_{21}\cos\theta, & 2A_{22}\sin 2\theta \\ D_{31}\sin\theta - D_{32}\cos\theta, & D_{31}\cos\theta + D_{32}\sin\theta, & 2A_{32}\sin 2\theta - 2A_{33}\cos 2\theta \end{array}\right] \cdot \frac{\{\sigma\}}{E}$$

$$\tag{14}$$

where E is the Young's modulus of rock and $A_{11}, A_{12}, \ldots, D_{32}$ are the strain coefficients.

The values of the strain coefficients are dependent upon Poisson's ratio of the rock. They have to be evaluated by numerical analysis, since there is no analytical solution. The strain coefficients of the isotropic case computed by the BEM analysis are summarized in Table 1.

Observation equation of the initial stress tensor $\{\sigma\}$ is expressed by the following matrix equation:

Table 1 Strain coefficient for 24 element method.

Poisson's ratio	A_{11}	A_{12}	A_{21}	A_{22}	A_{31}	A_{32}	A_{33}
0.10	1.002	−1.762	0.109	0.343	0.562	−0.724	−0.802
0.20	1.000	−1.752	0.022	0.365	0.519	−0.707	−0.818
0.25	0.999	−1.733	−0.021	0.373	0.496	−0.693	−0.821
0.30	0.997	−1.704	−0.065	0.380	0.474	−0.679	−0.822
0.40	0.989	−1.611	−0.154	0.386	0.426	−0.625	−0.823

Poisson's ratio	C_{11}	C_{21}	C_{31}	D_{11}	D_{21}	D_{31}	D_{32}
0.10	−0.155	0.655	0.246	0.082	1.542	0.802	−1.725
0.20	−0.263	0.641	0.185	0.095	1.627	0.860	−1.860
0.25	−0.317	0.636	0.155	0.101	1.673	0.886	−1.923
0.30	−0.371	0.632	0.126	0.108	1.716	0.911	−1.983
0.40	−0.481	0.630	0.071	0.123	1.787	0.953	−2.091

$$[A]\{\sigma\} = E\{\beta\} \tag{15}$$

where $[A]$ is an n by 6 elastic compliance matrix. The elements of $[A]$ are computed by substituting the tangential angle θ of each strain measuring point in Equation 14.

The most probable values of the initial stress components are determined by the least square method, providing the normalized expression of Equation 15 as follows:

$$[B]\{\sigma\} = E\{\beta^*\} \tag{16}$$

where $[B] = [A]^T[A]$ and $\{\beta^*\} = [A]^T\{\beta\}$. The most probable values of the initial stress $\{\sigma^*\}$ can be expressed as:

$$\{\sigma^*\} = E[C]\{\beta^*\} \tag{17}$$

where $[C]$ is the inverse matrix $[B]$. Detail explanation of the CCBO is found in the suggested method of ISRM.

2.4 Cross-sectional Borehole Deformation Method (CBDM) for monitoring of stress change

The stress state around an underground opening is changed with progress of excavation. That is, the surrounding rock behaves elastically at the beginning of excavation. The stress in the immediate wall of the opening increases with progress of excavation. Then the stress reaches a failure criterion of rock mass, and the plastic region is produced as shown in Figure 2. In this region, magnitude of the stress decreases. If the stress change due to excavation is monitored under construction, the failure criterion may be estimated from the measured results.

There are some methods for stress change around an opening under construction. For example, the stress change of an underground power house has been measured by a vibrating wire strain gauge in Japan (Kudo *et al.*, 1998). However, using this gauge which is buried in the borehole, the stress in only one direction in a plane perpendicular to a borehole axis is measured. Since this instrument is contact type and has any rigidity, the measured stress change may be influenced by it. It is desired that the instrument is non-contact type without rigidity.

The Cross-sectional Borehole Deformation Method (CBDM) , which instrument is of non-contact type, developed by Taniguchi *et al.* (2003) and Obara *et al.* (2004a, b, 2010, 2011a, b, 2012 a, b, 2014) is a by which the two-dimensional state of stress change within a rock mass in a plane perpendicular to a borehole axis can be measured. This method is convenient to monitor stress change around the underground opening under construction.

2.4.1 Measurement of displacement on borehole wall and Instrument

The rock mass around a borehole is elastically deformed corresponding to subjected rock stress. Based on this principle, a method was developed for easily and accurately measuring two-dimensional stress change in a plane perpendicular to the borehole axis. This method is the Cross-sectional Borehole Deformation Method (CBDM). The

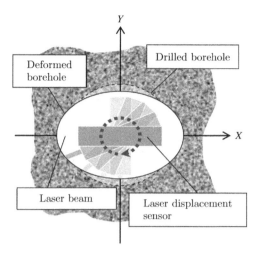

Figure 7 Principle of measuring displacement of a borehole wall.

displacement of the borehole wall is measured by a non-contact typed sensor, namely a laser displacement sensor, which is inserted and rotated in the borehole as shown in Figure 7. Accordingly, the rigidity of the instrument becomes zero for the measurement, because the displacement of the borehole wall can be measured without touching the wall. Then the rock mass around the borehole is undisturbed due to measurement, such as the hydraulic fracturing method.

In order to measure radial displacement of the wall in a cross section of the borehole, a compact and accurate laser displacement sensor is used. The dimensions are 43mm×40mm×18mm, and the resolution is 0.1 μm. A small stepping motor is adopted for rotation of the laser displacement sensor. The minimum angle of rotation step of the stepping motor is 0.1 degrees.

The prototype instrument for measurement and schematic view are shown in Figure 8. The tube of the instrument, 70 mm in diameter and 670 mm in length, is aluminum. The instrument is fixed in a borehole using two air pistons. The laser displacement sensor is located near small windows which are covered by acrylic plates for waterproof, and rotated by the stepping motor set in a head of the instrument. The motor is controlled by a computer through a controller and a driver. On the other hand, the output from the laser displacement sensor is stored in a computer through an amplifier unit and a data logger. These are assembled into the control box as shown in Figure 8(b). The sensor and motor are linked by the cables of about 30m length.

2.4.2 *Principle of measurement*

The schematic view of a cross section in a plane perpendicular to the borehole axis is shown in Figure 9. The borehole having a cross section of perfectly circular is drilled within a rock mass. Its radius is defined by R. The homogeneous and isotropic rock mass is assumed to be infinite and elastic. The principal stress subjected at infinity is

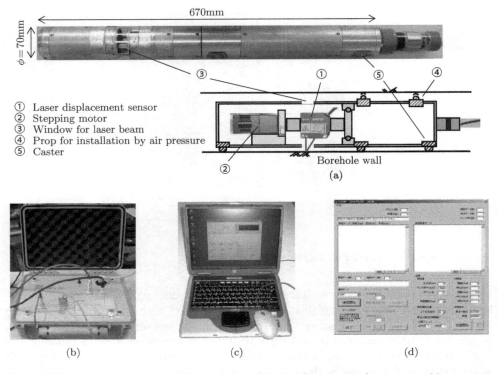

① Laser displacement sensor
② Stepping motor
③ Window for laser beam
④ Prop for installation by air pressure
⑤ Caster

Figure 8 Schematic view of prototype instrument and devices for control of instrument; (a) prototype instrument, (b) control box, (c) PC and display, (d) example of display of program.

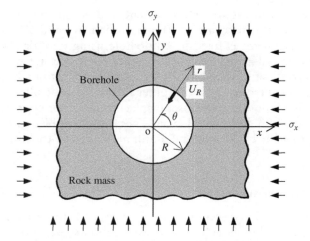

Figure 9 Schematic view of cross section of a bore-hole drilled within rock mass, which is assumed to be infinite and elastic.

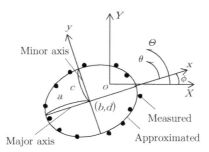

Figure 10 Schematic diagram of measured results and approximated ellipse by a non-linear least square method. X and Y axes are defined on an instrument, then x and y axes defined on a borehole coincide with principal direction.

defined in the x-y coordinate system which is set on the borehole with the origin at its axis in Figure 10:

$$\{\sigma\} = \{\sigma_x, \sigma_y\} \tag{18}$$

The axes in the coordinate system coincide with the principal directions.

The radial displacements due to each principal stress are described as:

$$U_R^x = -\frac{\sigma_x R}{E}(1-v^2)(1+2\cos 2\theta),$$
$$U_R^y = -\frac{\sigma_y R}{E}(1-v^2)(1-2\cos 2\theta) \tag{19}$$

In general, the radial displacement U_R is the vector sum of displacement U_R^x and U_R^y, which are generated corresponding to each principal stress (Jaegar & Cook, 1979):

$$U_R = U_R^x + U_R^y = H\{(\sigma_x + \sigma_y) + 2(\sigma_x - \sigma_y)\cos 2\theta\} \tag{20}$$

where $H = -R(1-v^2)/E$, E is Young's modulus and v is Poisson's ratio, then θ is rotation angle with the positive x axis. The radius R_R after deformation is represented:

$$R_R = R + U_R \tag{21}$$

In a measurement, the displacements and measured radii, number of n, are denoted by:

$$\{U_R\} = \{U_{R1}, U_{R2}, ..., U_{Ri}, ..., U_{Rn}\}$$
$$\{R_R\} = \{R_{R1}, R_{R2}, ..., R_{Ri}, ..., R_{Rn}\} \tag{22}$$

The coordinates of the measuring point i on the borehole wall are written in the X-Y coordinate system defined on the instrument with the origin at its axis in Figure 10 as follows:

$$Xi = R_{Ri}\cos\Theta \quad Yi = R_{Ri}\sin\Theta \tag{23}$$

where Θ is the rotation angle with the positive X axis in Figure 10. The measured results schematically are shown in this figure. The plots represent measurement values, and the solid curve is approximately expressed by an ellipse with a center of $(0, 0)$ in x-y and (b, d) in an X-Y coordinate system. The length of major and minor axes of the ellipse is $2a$ and $2c$, respectively. In general, the center of the ellipse does not coincide with that of the borehole as shown in the figure. In the cases that the distance between origins of each center is very short, the equation of the ellipse in the x-y coordinate system may be written as:

$$\frac{x^2}{a^2} + \frac{y^2}{c^2} = 1 \tag{24}$$

Using the coordinate transformation law from the X-Y to x-y coordinate system represented by Equation 25, the observation equations are obtained at each measurement point as Equation 26.

$$\begin{aligned} x &= (X - b)\cos\phi + (Y - d)\sin\phi, \\ y &= -(X - b)\sin\phi + (Y - d)\cos\phi \end{aligned} \tag{25}$$

$$\frac{(X\cos\phi + Y\sin\phi - b\cos\phi - d\sin\phi)^2}{a^2} + \frac{(-X\sin\phi + Y\cos\phi + b\sin\phi - d\cos\phi)^2}{c^2} = 1 \tag{26}$$

The most probable parameters of an ellipse, a, c, b, d, ϕ, are determined by applying a non-linear least square method to observation equations for measured values. When the axis of the instrument coincides with that of the borehole, the parameters b and d are equal to zero.

The displacements on major and minor axes of the determined ellipse are:

$$a = R + H(3\sigma_x - \sigma_y), \quad c = R + H(3\sigma_y - \sigma_x) \tag{27}$$

Accordingly, most probable principal stresses can be obtained in the x-y coordinate system as follows:

$$\sigma_x = \frac{3a + c - 4R}{8H}, \quad \sigma_y = \frac{a + 3c - 4R}{8H} \tag{28}$$

Then the stress components in the X-Y coordinate system are calculated by the stress transformation law.

The stress estimated from Equation 28 is not correct, because it is impossible to measure the radius of the borehole precisely just after boring. Therefore, the absolute stress state cannot be estimated. The stress determined by Equation 28 is considered to be a temporal stress.

The stress change can be estimated, using the temporal stress at more than two stages as follows. For example, the state of stress is changed with progress of construction of the underground opening. At the first stage, a borehole is drilled within a rock mass, and the cross-sectional shape of the borehole is measured at an early stage of excavation of an opening. Using the displacement, the temporal stress at the first stage can be determined. At the second stage, the shape at the same section of the borehole is measured again and the temporal stress is determined at an arbitrary stage during excavation. Then, the stress change is determined by the difference of temporal rock stress states determined at the first and second stages of excavation, assuming that rock is elasticity. Thus, the stress change due to elapsed time or excavation can be estimated by measuring the cross-sectional shape at the same cross section of one borehole repeatedly. Consequently, the temporal stress state $\{\sigma^I\} = \{\sigma_X{}^I, \sigma_Y{}^I, \sigma_{XY}{}^I\}$ is assumed at the first stage. The temporal stress $\{\sigma^{II}\} = \{\sigma_X{}^{II}, \sigma_Y{}^{II}, \sigma_{XY}{}^{II}\}$ is also assumed at the second stage. The stress change $\{\Delta\sigma\}$ can be estimated by the following equation, using the estimated temporal stress state at two stages:

$$\{\Delta\sigma\} = \{\Delta\sigma_X, \Delta\sigma_Y, \Delta\tau_{XY}\} = \{\sigma^{II}\} - \{\sigma^I\} \qquad (29)$$

3 STRESS STATE AROUND LARGE ROCK CAVERN AND ESTIMATION OF ROCK STRENGTH

3.1 Site description

The measurements of induced stress were performed at Kannagawa hydroelectric power plant during excavation of power plant cavern (Maejima et al., 2001). A pumped-storage power plant has the maximum output of 2700MW. The dimension of the cavern is 33m in width, 52m in height and 216m in length, as shown in Figure 11. The thickness of the overburden of the cavern is about 500m.

The geology of the Kannagawa site consists of mudstone-based rock that is irregularly mixed with olistoliths such as sandstone, chert, basic volcanic rock and limestone. The stratum at the site of the cavern is classified into 6 regions, considering the kind of gravel, coarse sandstone as the matrix and rate of mixture.

In order to predict the progression of the loosened zone around cavern during its excavation, the rock mass strength and initial rock stress were evaluated, then two-dimensional finite element analyses were performed using their values, and the model in which the variation of geology and the constitutive low of surrounding rock, namely strain softening, were introduced. Furthermore, the filed measurements of stress and displacement within rock, stress in shotcrete, load of PS anchor and AE were made during excavation of the cavern.

The initial stress was measured by the borehole deformation method with the eight-element strain gauge (Kanagawa et al., 1986), which is one of the overcoring methods. The results are shown in Figure 12. In the vertical cross section, the maximum principal stress has an inclination of 26 degrees to the penstock side from the vertical. Then the

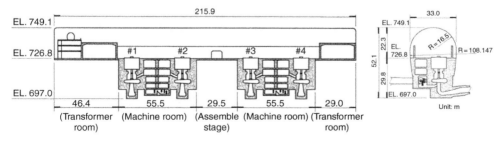

Figure 11 Dimension of underground power plant cavern (Obara and Sugawara, 2002).

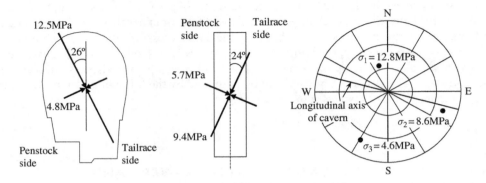

Figure 12 State of initial stress in vertical and horizontal cross section, and lower stereographic projection of principal directions (Obara & Sugawara, 2002).

longitudinal axis of the cavern is rotated at 24 degrees from the direction of the maximum principal stress in horizontal plane. Consequently the ratio of horizontal stress to vertical stress was 0.57 in vertical cross section.

3.2 Stress distribution

The induced stress was measured by the CCBO device at three stages of the excavation, namely after excavation of ceiling, seventh bench excavation and completion of final excavation. Then the results of the measurement were compared with those of the finite element analyses at the same stages in the vertical cross section as shown in Figure 13.

At the stage of excavation of ceiling, the five measurement stations rang from 2m to10m from the wall in each borehole at both shoulders of the cavern ceiling in Figure 13(a). In the tailrace side, the maximum principal stress is large near wall and decreases toward the deep zone. The directions of maximum principal stress at each measurement station are unchanged. On the other hand, the maximum principal stress at each measurement station on the penstock side represents almost the same value, then its direction inclines to the vertical with increasing depth. It was estimated that the depth of loosened zone is about 2m on both sides of the cavern ceiling.

In the second stage of seventh bench excavation, the stress distribution was measured at the tailrace side in Figure 13(b). The four measurement stations range from 3m to 10m from the side wall. The maximum principal stresses within the range of 7m from the wall were smaller and the state of stress is uniaxial. The stresses perpendicular to the side wall within the rage were almost zero, because of existence of the discontinuity with large persistence parallel to the side wall. In the range deeper than 7m, the values of principal stresses became large. From the results, the depth of loosened zone can be estimated at about 7m.

The stress distribution at the level of fourth bench was measured after completion of the excavation in Figure 13(c). The maximum principal stresses within the range of 6m were small, but those in the range deeper than 9m became large and were concentrated in the circumferential direction of the cavern. The depth of loosened zone is evaluated at 7.5m, based on the measured results.

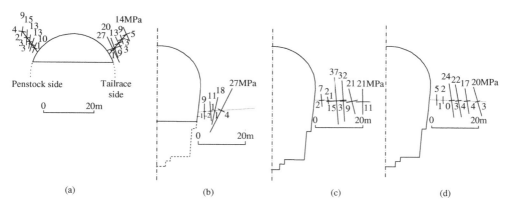

Figure 13 Induced stress measured at three excavation stages and comparison with numerical result of final excavation, (a) stress distribution after excavation of ceiling, (b) stress distribution after seventh bench excavation, (c) stress distribution of final excavation, (d) numerical result by FEM (Obara & Sugawara, 2002).

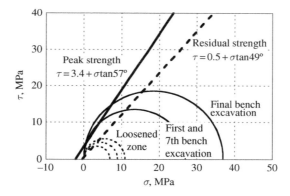

Figure 14 Mohr's stress circles at critical state with the failure criterion of rock mass estimated by in situ shear test (Obara & Sugawara, 2003).

Finally, the numerical result analyzed by FEM after completion of the excavation is shown in Figure 13(d). These results are good agreement with measured results in Figure 13(c). It is concluded that the induce stress measurements were successfully applied to estimate the loosened zone around the cavern, and that the numerical analysis by FEM is effective for prediction of the mechanical behavior of the surrounding rock of the cavern due to excavation.

3.3 Estimation of rock mass strength

The stress at the depth of loosened zone is considered to be under critical state of strength. On the measured stress distributions, the radial stress acting from the cavern is considered to be small and nearly equal to zero. Therefore, the Mohr's circles can be drawn by the stress state at a depth of 2m in Figure 13(a), 7m in (b) and 7.5m in (c), assuming that the minimum principal stress is zero. These circles are shown in Figure 14 with peak strength criterion and residual strength criterion estimated by a series of in situ rock shear tests. The estimated strength from stress distributions are good agreement with those of in situ rock shear tests. The dimension of block in rock shear test is 60 × 60 × 30cm. Therefore, it is concluded that the volume of rock block of in situ rock shear test was enough to evaluate rock strength in this case.

4 STRESS STATE OF AROUND TUNNEL UNDER HIGH STRESS FIELD AND ESTIMATION OF ROCK STRENGTH

4.1 Site description

The approach tunnel to the underground powerhouse is excavated at 36 m below the floor of the gallery, as shown in Figure 15 (Obara & Ishiguro, 2004). The dimensions of the cross section of the tunnel are 6.6min width and 6.0min height. A series of measurements of induced stress in the tunnel near-field was performed in four boreholes drilled in its roof. The distances between the boreholes, termed s-1, s-2, s-3, s-4, and the tunnel face are 14, 15 and 16 m. The elevation angles of the boreholes are 76 for s-1, s-2, s-4 and 87 for s-3. The measurement stations, 13 in number, are located within a range of 1.2 m in depth from the tunnel roof.

4.2 Strain change during overcoring

Figure 16 shows an example of a series of strain changes on the conical bottom surface during overcoring monitored by the 16-element conical strain cell. The ε_ρ and ε_θ are radial and tangential strains, respectively. This is the result from the fourth measurement station in borehole s-1, called s-14. The depth of the measurement station is 1.14m from the tunnel wall. The lateral axis is the distance between overcoring advance and the section of the strain measuring circle. The numerals in the figure represent the number of strain gauges pasted on the surface of the conical borehole bottom.

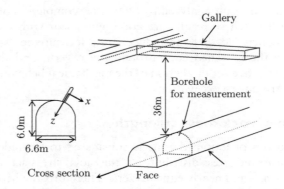

Figure 15 Schematic view of the approach tunnel and gallery (Obara & Ishiguro, 2004).

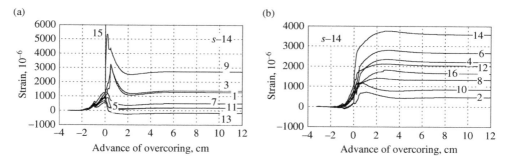

Figure 16 Changes in strain during overcoring at measurement station s-14, (a) radial strains ε_ρ, (b) tangential strains ε_θ (Obara & Ishiguro, 2004).

The changes in strain are rapid in all cases after the overcoring passed through the section of the strain measuring circle. In the change of ε_θ, the strain is relieved smoothly with the advance of overcoring, and then converges to a constant value in each gauge. On the other hand, in the change of ε_ρ, a large tensile strain is experienced in the section of the strain measuring circle and it finally converges to a constant value. However, the change of strains during the overcoring is not smooth and some undergone strains are large. Furthermore, the data from gauge no.15 diverges. The fractures due to core disking can be observed on the surface. Therefore, it is considered that the fractures penetrate to the surface of borehole bottom by the core disking.

4.3 State of induced stress

All of the cross sections including strain gauges were taken by the X-ray CT scanner nondestructively, and the existence of fractures near the gauges was investigated, and then the strain data to be used in the estimation of rock stress were selected. As a result, the tangential strain data were adopted, because that most radial strain gauges were intersected by fractures. Consequently, the state of stress was estimated as two dimensional stress in the plane perpendicular to the borehole axis for measurement, assuming plane stress state.

The results of induced stresses in a range of 0.23m to 1.17m from the tunnel wall are summarized in Table 2. The Young's modulus and Poisson's ratio used in the estimation are 30GPa and 0.2 respectively. These values are determined by the conventional multi-stage uniaxial compression test (Sugawara & Obara, 1999). The normal stress σ_x is large at all stations. The direction of horizontal projection of the x-axis is northeast, and the y-axis is defined in the direction parallel to its axis. The direction of maximum principal stress of the induced stress is East-West; that of the initial stress is also East-West. Both directions are similar. However, the magnitude of maximum principal stress of the induced stress is larger than that of the initial stress. The mean values of both principal stresses are the 37MPa and 28MPa respectively. This value is shown 1.3 times as much as that of initial stress.

The state of stress at each measurement station is shown in Figure 17 in three dimensional. All maximum principal stresses act in the direction of x-axis and minimum ones are directed parallel to the tunnel axis. The difference between the maximum

Table 2 Induced stress in the plane perpendicular to the borehole axis (Obara & Ishiguro, 2004).

Measurement station		s–13	s–14	s–21	s–31	s–32	s–33	s–42	s–43	s–44
Depth from tunnel wall [m]		0.83	1.14	0.54	0.23	0.54	0.90	0.64	0.90	1.17
Elevation angle of borehole [deg]		76	76	76	87	87	87	76	76	76
Stress component [MPa]	σ_x	40.7	37.9	42.1	33.1	47.8	40.6	37.9	33.4	19.8
	σ_y	17.9	17.0	18.8	17.0	26.0	17.8	17.0	15.9	10.3
	τ_{xy}	2.1	−2.7	7.3	−1.2	3.2	−0.3	−0.9	3.3	0.3
Principal stress [MPa] and	σ_1	40.9	38.2	44.2	33.2	48.3	40.6	37.9	34.0	19.8
Maximum principal direction [deg]	σ_2	17.7	16.6	16.7	16.9	25.6	17.8	17.0	15.3	10.3
		−5.3	7.5	−16.1	4.4	−8.1	1.9	2.6	−10.4	−1.8

Remarks: Anti-clockwise from x-axis is positive in principal direction.

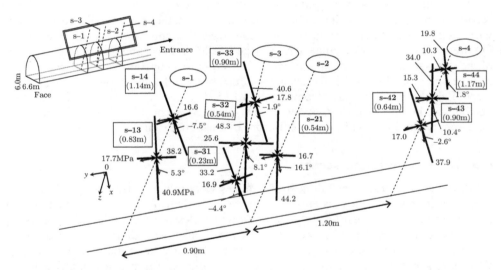

Figure 17 State of induced stress in immediate roof of the approach tunnel. The elevation angle of borehole s-1, s-2, s-4 for measurement is 76°, and that of s-3 is 87°(Obara & Ishiguro, 2004).

and minimum principal stress is large. This tendency is the same in the state of initial stress.

The stress distribution is shown in Figure 18. In regard of maximum principal stress, the magnitude is smaller near roof, and becomes large in a region of 0.5–0.8m from the tunnel wall and small in the deeper region. Therefore, it is considered that a shallow depth of roof is damaged by blasting and so on, and that the region of 0.5–0.8m from the tunnel wall is under critical state of rock mass strength. On the other hand, the magnitude of minimum principal stress parallel to the tunnel axis is almost constant.

In each region, the mean stress state is shown in Figure 19. These sketches represent the state of stress in the plane perpendicular to the borehole axis. Although the direction is slightly varied, the states are similar each other. The magnitude of maximum principal stress is largest in a region of 0.5–0.8m.

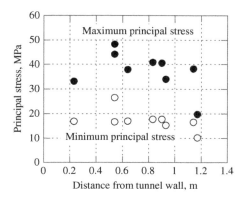

Figure 18 Distribution of principal stresses in the plane perpendicular to the borehole axis in immediate roof (Obara & Ishiguro, 2004).

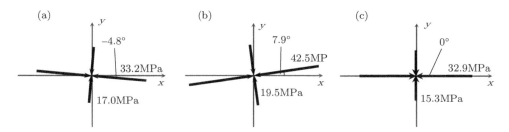

Figure 19 Direction and magnitude of mean principal stresses in immediate roof to the borehole axis, (a) at 0.23m from tunnel wall, (b) in a range of 0.5 to 0.8m, (c) in the deeper region than 0.8m (Obara & Ishiguro, 2004).

4.4 Estimation of strength of rock mass

A series of uniaxial compression tests using various specimen scales and in situ rock shear tests were conducted to determine the strength of rock mass. In the uniaxial compression tests, the specimen scale is 5.0, 19.8 and 29.3cm in diameter and 10, 40 and 60cm in length, respectively. The test was performed under frictionless edge condition. The mean value of uniaxial compressive strength in each scale was 81.2, 68.5 and 74.1MPa with increasing scale of specimen. Therefore, it is concluded that the strength of specimens more than 19.8cm in diameter is not likely to change.

Also, in situ rock shear tests were performed at the gallery in Figure 15. The scale of block is 50 × 50 × 20cm and the number of blocks is four. These results are shown in Figure 20 with the stress estimated under critical state in a region of 0.5 – 0.8m on the induced stress distribution. The results of in situ rock shear test are represented by plots; and those of uniaxial compression test and stress measurement by Mohr's circles. In drawing Mohr's circles, the minimum principal stress is assumed to be zero. These results are reasonably compatible. Consequently, putting these results together, the

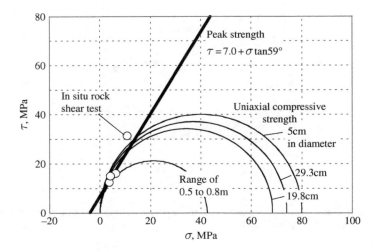

Figure 20 Comparison of Mohr's stress circle of critical stress state in the stress distribution and uniaxial compressive strength with shear strength of in situ rock shear test, and estimation of the strength criterion for the rock mass (Obara & Ishiguro, 2004).

peak strength criterion of Mohr-Coulomb type can be estimated with parameters of cohesion 7.0MPa and internal friction angle of 59 degrees.

As a result, it is noted that in situ rock shear test is effective for the estimation of rock strength used in design of a cavern, as well as a series of rock stress measurements under and after construction.

5 ESTIMATION OF SLIDING CRITERION OF DISCONTINUITY FROM MEASURED ROCK STRESS

The multiple times stress measurements were performed by the CCBO within diorite and granodiorite (Sakaguchi *et al.*, 1995). The borehole for the measurement was drilled horizontally from the wall of gallery at a depth of 520 m and three faults I, II and III are investigated in the measurement field, as shown in Figure 21. The eighteen measurement stations are ranging from 0.6 m to 29.5 m from the wall of gallery. The borehole passed through the geological boundary between the hard diorite and the comparatively soft granodiorite at 8.7 m from the wall of gallery. On this geological boundary, fault II is largest and 0.25 m in width dipping about 80 degrees, an immediate skarn of 1.5 m in width.

The magnitudes and directions of the principal stresses at each measurement station in the xz-plane and the xy-plane are shown in Figure 22. The measured results indicate clearly that the stress varies with the region bounded by the fault III. The stress distribution changes in two side of the fault III. In Figure 22(a), the maximum stresses act along the fault in the Region I and parallel to the x-axis in the Region II. Then, the direction of them in the Region II rotates clockwise with increasing the distance from gallery. In Figure 22(b), the maximum stresses act from the direction of y-axis due to the excavation of the gallery in the Region I. On the other hand, the direction of them is

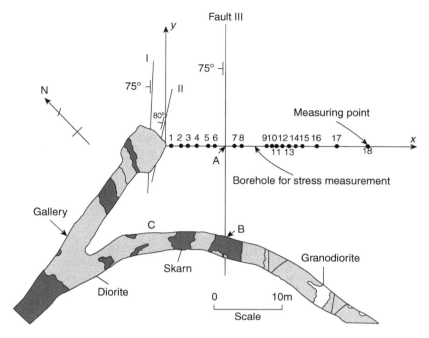

Figure 21 Plan view of the site for stress measurement; numerals in the figure represent the number of the stress measurement.

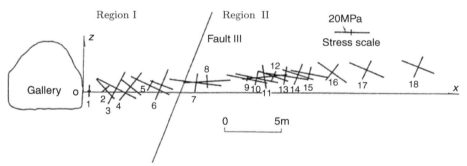

(a) Maximum and minimum normal stresses in the vertical section.

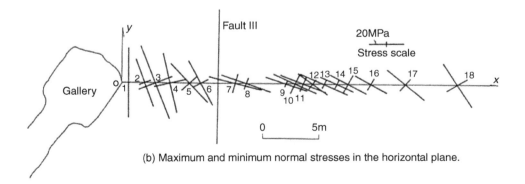

(b) Maximum and minimum normal stresses in the horizontal plane.

Figure 22 Two-dimensional expression of the magnitudes and directions of the principal stresses.

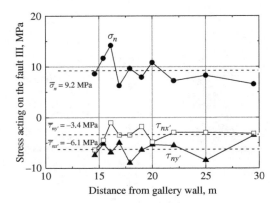

Figure 23 The normal stress and the shear stress acting on the fault III, which are evaluated by the measured stress in the Region III according to the coordinate transformation.

almost that of the x-axis and rotates clockwise with increasing the distance from gallery in the Region II. This result indicates clearly that the stress in this field varies with the region bounded by the fault and the skarn.

It is considered that the discontinuity of the stress distribution exists in between the Region I and Region II, and that this cause is the fault III. Here, let us further consider about the stress disturbance caused by the fault III. If the thickness of the fault III is constant and the direction of the fault III is constant, too, the traction acting on the fault plane becomes uniformity, and it is thought that the persistence of the traction must be satisfied. However, it is suggested that the traction is not uniformity according to the above measurement results. In other words, it is suggested that the stress field around the fault is considerably disturbed because the thickness of the fault is not constant and the fault plane is not flat surface.

Let us estimate the mean stress to act on the fault III using the results (measurement station 10 to 18) of the measurement of the region with some distance from the fault III. The value at each measuring station is calculated by the measured stress tensor according to the stress transformation law. Figure 23 shows the results of the distribution of the normal stress σ_n and the shear stresses ($\tau_{nx'}$, $\tau_{ny'}$) acting on the fault III, which were calculated by the results of from the measurement result 10 to 18. Where, n-axis is the outward normal vector on the fault III plane, x'-axis is horizontal axis on the fault plane and y'-axis is the perpendicular to the x'-axis. The shear direction of resultant shear stress estimated by the mean value of the shear stresses in Figure 23 is about 30 degrees from the horizontal. Because that the mean values ($\tau_{nx'}$, $\tau_{ny'}$) are equal to (-6.1MPa, -3.4MPa). Therefore, it is estimated that this fault is strike-slip fault type.

Using the results of the mean value of the shear stresses, the resultant shear stress (τ_s) acting on the fault plane is calculated as shown in Figure 24. The filled circle is the result calculated from the measurement station 10 to 18 and the open circles show the stresses calculated by the result of the measurement station 5 and 6 and that of the measurement station 7 and 8. If this fault expands in a planer manner, two results shown by the open circle should be plotted on the same position in the figure, because the traction should be uniformly and continuously. However, the actual results are

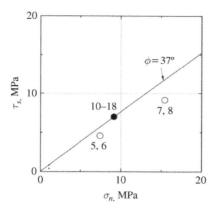

Figure 24 Mean stress acting on the fault III (filled circle) and stresses acting on the fault estimated by the results of the measurement in the vicinity of the fault (open circle).

plotted in difference positions and those lie below the line which is drawn as sliding criterion assuming that its cohesion is zero. This means that stress state in the vicinity of the fault is disturbed.

The existence of discontinuities such as fault affects on a stress distribution. The influence of discontinuity on the stress distribution is dependent on the stress acting on the discontinuity. That is, it may be large in case that the scale of the discontinuity is large and that the stress acting on the discontinuity approaches the sliding criterion of it. Accordingly, the more the absolute value of the ratio of shear stress to normal stress acting on the discontinuity increases, the more the stress distribution is influenced. Such discontinuity is termed active discontinuity.

The shear stress and normal stress acting on the active discontinuity, which are calculated by the measured stress near the discontinuities according to the stress transformation law, are shown in Figure 25 (Obara *et al.*, 2003b). Assuming the active discontinuity is under the static sliding condition, the plots can be approximately expressed by linear equation, namely siding criterion. The cohesion of each discontinuity is almost zero, and then the frictional angle of it ranges from 25 to 60 degrees. It is considered that these values are dependent on the type of rock mass, the cause of discontinuity generation, the state of discontinuity surface and so on. It is concluded that such analysis based on induced stress measurements is effective for not only the specification of active discontinuity but also the prediction of failure pattern of rock mass around opening.

6 MONITORING OF STRESS CHANGE UNDER CONSTRUCTION OF CAVERN

6.1 Site description

The plan view of the measurement site in Kamioka Mine is shown in Figure 26 (Obara *et al.*, 2014). A cavern was excavated at a depth of 900m within gneiss. The Young's

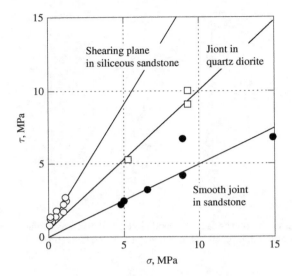

Figure 25 State of rock stress acting on active discontinuities, calculated by the measured stress near the discontinuities (Obara et al., 2003b).

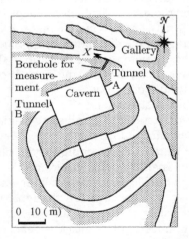

Figure 26 Location of borehole for measurement and cavern in the plan view of measurement site in Kamioka Mine (Obara et al., 2014).

modulus and Poisson's ratio are 30GPa and 0.2, respectively. The dimension of the cavern is 15m by 21m and 15m in height. A borehole with a length of 5m for measurement of stress change was drilled horizontally from the gallery to the cavern before the start of its excavation. The width of the rock between the gallery and the cavern is about 7m. The borehole for measurement was drilled in the wall. The four measuring points are located at depths of 1.0m, 1.8m, 4.0m and 4.5m, as

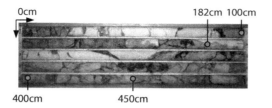

Figure 27 Core of borehole for measurement and measurement points (Obara et al., 2014).

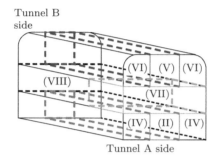

Stage I: Before excavation
Stage II: Excavation of center of lower part
Stage III: During excavation of lower part
Stage IV: After excavation of lower part
Stage V: Excavation of center of upper part
Stage VI: After excavation of upper part
Stage VII: During excavation of middle part
Stage VIII: Just after excavation of middle part
Stage IX: Three months after completion of excavation

Figure 28 Measurement stage for the excavation of the cavern: access tunnel A is linked to (II) of lower part and tunnel B is to (V) of upper part (Obara et al., 2014).

shown in Figure 27. The measuring points are determined from the condition of the recovered core.

The excavation process is shown in Figure 28 in three dimensions. Firstly, the lower part of the cavern was excavated from access tunnel A. Then the upper part was excavated from access tunnel B. Finally, the middle part was excavated. The excavation method was blasting and its period was about six months for the whole of stages I to IX. The measurements were performed at nine stages before, during, and after excavation.

6.2 Measurement results

The results at a depth of 4.0m are shown as an example. The measured and corrected displacements of the borehole wall at Stages V of the excavation are shown in Figure 29, assuming that the borehole radius is 37.85mm. The radius is not a real value but an expedient assumed one for calculating temporal stress. The solid line is approximation by the equation of the ellipse. The distributions of displacement in the measured results have a period of 2π. However, applying both the non-linear least square method and non-linear programming for optimization to the measured results, the corrected data for eccentric positioning of the instrument changes to have a period of π. The corrected data vary slightly, but represent a fairly good approximation. In Figure29(c), the cross-sectional shape in the plane perpendicular to the borehole axis is shown, adding 50 times displacement to the radius. The shape is represented by an

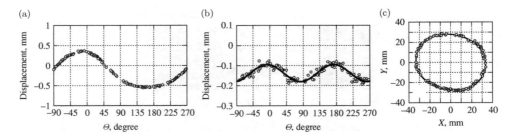

Figure 29 Measured and analyzed results: (a) measured data, (b) corrected data, (c) cross-sectional shape in the plane perpendicular to the borehole axis at Stage V; solid lines in (b) and (c) are approximations; deformation in (c) is described, adding 50 times displacement to radius (Obara et al., 2014).

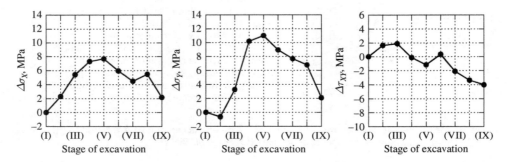

Figure 30 Changes of all components of stress in the X-Y coordinate system: X-axis is defined in the horizontal direction (Obara et al., 2014).

ellipse. The principal direction of the absolute stress state can be confirmed from this shape, although its value cannot be estimated from the CBDM.

6.3 Change of stress distribution

The stress change at the depth of 4.0m during the excavation is shown in Figure 30. Since the initial stress state is not measured in this location, the stress change is calculated to subtract each component of the temporal stress at Stage I from that at any stage respectively. The vertical stress change $\Delta\sigma_Y$ is zero until stage II, but increases at stage III, then reaches the maximum value at stage V. After that, the stress decreases gradually with the progress of excavation. The tendency of the horizontal stress change $\Delta\sigma_X$ is almost the same as that of $\Delta\sigma_Y$. It is considered that the rock near the measuring point was damaged and became the loosened zone. However, as the change of all stress components is continuous, it is also considered that that damage did not happen suddenly.

The distributions of stress change along the borehole axis at some stages are shown in Figure 31. The vertical cross section along the borehole axis is in Figure 31 (d). The width of the pillar between the gallery and the cavern is 7.0m. In shear stress change $\Delta\tau_{XY}$ of Figure 31 (c), the stress change is relatively small along the borehole axis during excavation. This means that there is not very much change in principal direction with elapsed time and geometry of cavern.

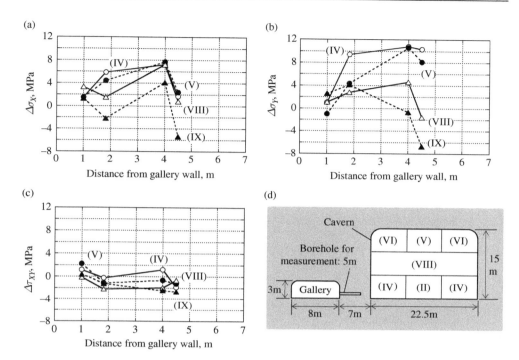

Figure 31 Distribution of stress change along borehole axis: (a) $\Delta\sigma_X$, (b) $\Delta\sigma_Y$, (c) $\Delta\tau_{XY}$, (d) vertical cross section (Obara et al., 2014).

The vertical stress change $\Delta\sigma_Y$ in Figure 31 (b) is comparatively large. The stress change $\Delta\sigma_Y$ at a depth of 1.0m is small. As this point is near the gallery wall, the rock mass in this area is considered to be damaged. On the other hand, the stress change at a depth of 1.8m, 4.0m and 4.5m is large in early stages of the excavation. At a depth of 1.8m, the stress represents the maximum value at Stage IV, then it decreases to a half of the maximum value at Stage V. This value is maintained until Stage IX, which is completion of excavation. This means that the rock mass near a depth of 1.8m was not damaged. On the other hand, the stresses at depths of 4.0m and 4.5m also represent the maximum value at Stage IV, then they decreases gradually with the advance of the excavation. At Stage IX, the stresses decrease to the stress level lower than that before excavation. It is considered that the rock mass near depth of 4.0–4.5m was damaged due to excavation. These trends can be seen in horizontal stress change $\Delta\sigma_X$ shown in Figure 31(a). However, the state of the damaged zone is not clear in the field. Therefore, that state should be confirmed by other methods such as numerical methods.

Finally, the stress change from initial stress state which was not measured in the field was estimated in this case example. Therefore if absolute stress change would be known, the initial stress state should be measured before excavation of underground opening. As a result we can estimate rock mass strength from the absolute stress change measured during excavation of it.

7 CONCLUSION

For estimating rock mass strength, the method from rock stress distribution around an underground opening and stress change under construction of it was explained instead of in situ shear test. Then Compact Conical-ended Borehole Overcoring (CCBO) technique and Cross-sectional Borehole Deformation Method (CBDM) were described to measure rock stress, and the case examples by them. it was concluded that CCBO and CBDM are convenient for estimating not only stress distribution and stress change but also rock mass strength.

REFERENCES

Aoki, T., Sugawara, K., Obara, Y. & Suzuki, Y. (1988) Elasto-plastic deformation of a circular opening under biaxial stress condition. *J. Min. Metall. Inst. Japan*, 104, 489–494.

Franklin, J. A. & 25 researches (1974) Suggested methods for determining shear strength. *Int. J. Rock Mech. Min. Sci. Geomech. Abstr.*, 12(3), A35.

Jaeger, C. (1979) *Rock mechanics and engineering*, 2nd edition, Cambridge University Press, pp. 118–120, 134–137, 414–424.

Jaeger, J. C. & Cook, N. G. W. (1979) *Fundamentals of rock mechanics*, 3rd edition, Chapman & Hall, London, Chapter 10.

Japan Society of Civil Engineers (2000) Suggested methods for in-situ test, Tokyo.

Kanagawa, T., Hibino, S., Ishida, T., Hayashi, M. & Kitahara, Y. (1986) In situ stress measurements in the Japanese Islands. *Int. J. Rock Mech. Min. Sci. Geomech. Abstr.*, 23, 29–39.

Kudo, K., Koyama, T. & Suzuki, Y. (1998) Application of numerical analysis to design for supporting large-scale underground caverns. *J. Constr. Manage. Eng.*, JSCE, 588(VI-38), 37–49.

Maejima, T., Morioka, H., Mori, T. & Aoki, K. (2001) Application of the observational construction technique for large underground cavern excavation. *Proc. EUROCK 2001*, 443–448.

Obara, Y., Nakayama, T., Sugawara, K., Aoki, T. & Jang, H. K. (1992) Determination of plastic regions around underground openings by a coupled boundary element – characteristics method. *Int. J. Num. Anal. Methods Geomech.*, 16, 401–716.

Obara, Y. & Sugawara, K. (2003a) Updating the use of the CCBO cell in Japan: overcoring case studies. *Int. J. Rock Mech. Min. Sci.*, 40, 1189–1203.

Obara, Y. & Sugawara, K. (2003b) Estimation of rock strength by means of rock stress measurement. *Proc. 3rd Int. Symp. on Rock Stress*, 35–45.

Obara, Y. & Ishiguro, Y. (2004a) Measurement of induced stress and strength in the near-filed around a tunnel and associated estimation of the Mohr-Coulomb parameters for rock mass strength. *Int. J. Rock Mech. Min. Sci.*, 41, 761–769.

Obara, Y., Matsuyama, T., Taniguchi, D. & Kang, S. S. (2004b) Cross-sectional borehole deformation method (CBDM) for rock stress measurement. *Proc. 3rd ARMS*, 2, 1141–1146.

Obara, Y., Shin, T., Yoshinaga, T., Sugawara K. & Kang, S. S. (2010) Cross-sectional borehole deformation method (CBDM) for measurement of rock stress change. *Proc. 5th ISRS*, CD.

Obara, Y., Shin, T. & Yoshinaga, T. (2011a) Development of cross-sectional borehole deformation method (CBDM) for measurement of rock stress change. *J. MMIJ*, 127, 20–25, in Japanese.

Obara, Y., Fukushima, S., Yoshinaga, T., Shin, T., Ujihara, T., Kimura, M. & Yokoyama, T. (2011b) Measurement of rock stress change by cross-sectional borehole deformation method (CBDM). *Proc. ISRM 12th Int. Cong.*, 1077–1080.

Obara, Y., Yoshinaga, T., Shin, T., Kataoka, M. & Yokoyama T. (2012a) Applicability of cross-sectional borehole deformation method (CBDM) to measure rock stress changes through laboratory and in-situ experiments. *J. MMIJ*, 128, 134–139, in Japanese.

Obara, Y., Kataoka, M. & Yoshinaga, T. (2012b) Cross-sectional borehole deformation method (CBDM) for measurement of rock stress change and its application. *Proc. 4th Traditional International Colloquium on Geomechanics and Geophysics*, 13–15.

Obara, Y., Yoshinaga, T., Kataoka, M. & Yokoyama, T. (2014) A method for measurement of rock stress change, – cross-sectional borehole deformation method. *Int. J. JCRM*, 10(1), 5–10.

Sakaguchi, K., Huang, X., Noguchi, Y., & Sugawara, K. (1995) Application of conical-ended borehole technique to discontinuities rock and consideration. *J. of MMIJ*, 110, 283–288, in Japanese.

Serafim, J. L. & Lopes, J. J. B. (1961) In situ shear tests and triaxial tests of foundation rocks of concrete dam. *Proc. 5th Int. Conf. on Soil Mech. Found. Eng.*, Paris 1.

Sugawara, K. & Obara, Y. (1999) Draft ISRM suggested method for in situ stress measurement using the compact conical-ended borehole overcoring (CCBO) technique. *Int. J. Rock Mech. Min. Sci.*, 36, 307–322.

Taniguchi, D., Yoshinaga, T. & Obara, Y. (2003) Method of rock stress measurement based on cross-sectional borehole deformation scanned by a laser displacement sensor. *Proc. 3rd ARMS*, 283–288.

Geophysics

Chapter 10

Compressive strength–seismic velocity relationship for sedimentary rocks

T. Takahashi & S. Tanaka
Fukada Geological Institute, Tokyo, Japan

Abstract: Estimating rock strength from seismic velocity is a very effective way for building a strength model of a large rock mass. This chapter first reviews the literature on empirical equations for estimating unconfined compressive strength of a rock from seismic P-wave velocity. It then focuses on the relationships between compressive strength and seismic velocity for soft rock, because there are few studies on the relationship for soft rock. To enable more accurate and reliable estimates, physical models of the compressive strength–seismic velocity relationship are proposed and applied to real data for demonstrating the applicability of the models. These demonstrations proved that physical models can be effectively used to estimate rock strength from seismic velocity.

1 INTRODUCTION

Knowledge of the mechanical properties, especially the compressive strength of a rock, is indispensable for designing and constructing tunnels, dams, and underground caverns in civil engineering; for evaluating the stability of wellbores, and the effective fracturing of rocks in oil and gas development; and for safe and effective tunnel excavation and underground caving in mining operations.

The compressive strength of a rock is usually measured in laboratory tests of rock specimens sampled at outcrops and/or in boreholes drilled in a rock mass. A strength model of the entire rock mass is then built by extrapolating these measurements based on a rock mass classification obtained from geological site characterization. However, model-building using data obtained from a limited number of boreholes and tunnels in the rock mass, especially in a complex rock mass, may cause problems in the accuracy and reliability of the model.

Geophysical methods can delineate a wide range of subsurface structures and properties effectively. If geophysical properties, like seismic velocity, can be used for estimating rock strength, then geophysical methods can be efficiently employed for profiling strength throughout the entire rock mass. Therefore, rock strength is often estimated from seismic velocity using correlations with data measured in the specific rock mass. However, since such correlations are generally made with a small number of measured data, they may occasionally be inaccurate and unreliable.

Rock strength is also estimated from seismic velocity using empirical relationships between seismic velocity and compressive strength. There are many empirical equations proposed and utilized for relating P-wave velocity and unconfined compressive strength for various types of rocks (*e.g.* Zhang, 2005; Zoback, 2007; Schön, 2011). As empirical equations are generally derived from many real data, they are more applicable and reliable than the correlation-based techniques described above. Therefore, this chapter firstly summarizes the empirical equations relating unconfined compressive strength and P-wave velocity proposed in the literature. These equations are compared with real data collected for different rock types in several dam sites in Japan.

The empirical equations for relating unconfined compressive strength and P-wave velocity described above are mostly derived for rocks with relatively large strength (hereafter referred to as hard rock). There are fewer empirical equations available for soft rock with an unconfined compressive strength of less than 25 MPa. A soft rock has many different characteristics from a hard rock. This chapter, therefore, describes the general characteristics of a soft rock and shows the relationship between compressive strength and seismic velocity.

Empirical equations are only effective if they are used for the rock type from which the equation is derived. However, they cannot usually be extrapolated to other rock types. For more general, accurate, and reliable estimation of rock strength from seismic velocity, a physical model is needed that can represent the relationship between these two properties of the rock. Physical models have been in widespread use in recent years for the interpretation of seismic data for oil and gas exploration and exploitation (*e.g.*, Dvorkin *et al.*, 2014). The effective medium model such as the Hashin-Shtrikman model (Hashin & Shtrikman, 1963) and the Kuster-Toksoz model (Kuster & Toksoz, 1974) are used for representing the mechanical properties of rocks. The authors have also been employing physical models for interpreting shallow geophysical data in geotechnical applications (Takahashi & Tanaka, 2009; Takahashi & Tanaka, 2013). One of the important applications of physical models is to estimate the compressive strength of a rock from seismic velocity. This chapter describes in detail how to model the compressive strength and seismic velocity relationship and then shows several examples of applications using real data to demonstrate the effectiveness of the physical model.

2 EMPIRICAL RELATIONSHIPS

Many empirical equations relating unconfined compressive strength and P-wave velocity have been proposed in civil and mining engineering, and in the oil and gas development fields. Table 1 summarizes empirical equations from the books of Schön (2011), Zhang (2005) and Zoback (2007), and from the authors' collection of recently published work (*e.g.*, Sharma & Singh, 2008; Karakul & Ulusay, 2015). The first two books mainly cover the empirical relationships proposed in civil and mining engineering, and the last book covers empirical relationships in the oil and gas development field. New additions have been made in both fields. The compressive strength is measured by a laboratory compression test for all empirical equations. However, the P-wave velocity is obtained from well log data or ultrasonic velocity measurements in the laboratory. Many of the P-wave velocities in the equations from the oil and gas development field are obtained from the well log data. Table 1 shows the empirical equations proposed in the literature and the rock type from which the equation is derived.

Figure 1 shows crossplots of the actual data of unconfined compressive strength and P-wave velocity measured in the laboratory tests for rock core samples from several

Table 1 Empirical equations between unconfined compressive strength and P-wave velocity with other basic properties such as density and porosity of rocks. The table shows the equation, rock type from which the equation is derived, and reference in which the equation is proposed.

No.	Equation	Rock type	Reference
1	$q_u=35.0V_p-31.5$	Sandstone	Freyburg (1972)
2	$q_u=1277e^{(-11.2/Vp)}$	Sandstone	McNally (1987)
3	$q_u=50.0V_p-114.5$	Sandstone	Schön (2011)
4	$q_u=3\times10^{-0.65}V_p^{3.45}$	Sandstone	Schön (2011)
5	$q_u=18V_p-16.26$	Sandstone and marble	Howarth et al. (1989)
6	$q_u=0.499V_p^3$	Weak and unconsolidated sandstones	Zoback (2007)
7	$q_u=3.3\times10^{-20}\rho^2V_p^2\{(1+v)/(1-v)\}^2(1-2v)(1+0.78V_{clay})$	Sandstone with $q_u>30MPa$	Fjaer et al. (1992)
8	$q_u=1.745\times10^{-9}\rho V_p^2-21$	Coarse grained sands and conglomerates	Moos et al. (1999)
9	$q_u=42.1\exp(1.9\times10^{-11}\rho V_p^2)$	Consolidated sandstones with $0.05<\phi<0.12$ and $q_u>80MPa$	Zoback (2007)
10	$q_u=3.87\exp(1.14\times10^{-10}\rho V_p^2)$	Sandstone	Zoback (2007)
11	$q_u=-0.98V_p+0.68Vp^2+0.98$	Sandy and shaly rocks	Gorjainov and Ljachovickij (1979)
12	$q_u=0.77Vp^{2.93}$	Mostly high porosity Tertiary shale	Horsrud (2001)
13	$q_u=0.43Vp^{3.2}$	Pliocene or younger shale	Zoback (2007)
14	$q_u=1.35Vp^{2.6}$	Shale	Zoback (2007)
15	$q_u=0.5Vp^3$	Shale	Zoback (2007)
16	$q_u=10(Vp-1)$	Mostly high porosity Tertiary shale	Lal (1999)
17	$q_u=\{(Vp-1.4)/0.2\}^{1.43}$	Tertiary tuffaceous mudstone with $q_u<10MPa$	Aydan et al. (1992)
18	$\log q_u=0.444V_p+0.003$	Schist	Golubev and Rabinovich (1976)
19	$q_u=36.0V_p-31.2$	Coal measure rocks	Göktan (1988)
20	$q_u=2.45V_p^{1.82}$	Limestone	Militzer and Stoll (1973)
21	$\log q_u=0.358V_p+0.283$	Limestone	Golubev and Rabinovich (1976)
22	$q_u=10^{(2.44+0.358Vp)}/145$	Limestone and dolomite	Golubev and Rabinovich (1976)
23	$q_u=31.5V_p-63.7$ ($r^2=0.80$)	Dolomite, marble, limestone	Yasar and Erdogan (2004)
24	$q_u=43V_p^{2.23}$	Mainly volcanic rocks	Inoue and Ohmi (1981)
25	$q_u=10.79\{V_p^2+4(\rho-1)\}^{1.447}$	Mainly volcanic rocks	Inoue and Ohmi (1981)
26	$q_u=22.03V_p^{1.247}$ ($r^2=0.72$)	Granites	Sousa et al. (2005)
27	$q_u=40.7Vp-36.31$	Granites	Vasconcelos et al. (2008)
28	$q_u=35.54V_p-55$ ($r^2=0.64$)	Granitic rocks	Tugul and Zarif (1999)
29	$q_u=1.02V_p^3$	Granites, andesite, sandstone	Ohkubo and Terasaki (1971)
30	$q_u=9.95V_p^{1.21}$ ($r^2=0.69$)	Dolomite, sandstone, limestone, marl, diabase, serpentine, hematite	Kahraman (2001)
31	$q_u=64.2Vp-117.99$	Sandstone, weathered basalt, phyllite, schist, coal, shaly rocks	Sharma and Singh (2008)
32	$\log q_u=1.368+0.794\log(1+0.001Vp)-0.201S_r-5.6\phi Vclay$	Marl, sandstone, andesite, limestone, tuff, ignimbrite, claystone	Karakul and Ulusay (2015)

where q_u: unconfined compressive strength in MPa, V_p: P-wave velocity in km/s, ρ : density in g/cm^3, V_{clay}: clay content in fraction, ϕ: porosity in fraction, Sr: degree of saturation in fraction. r^2 is the determination coefficient.

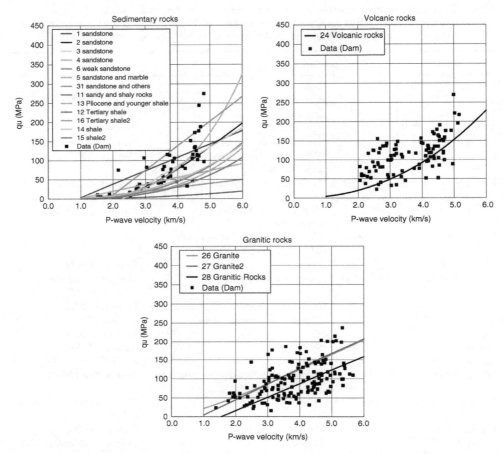

Figure 1 Empirical equations between unconfined compressive strength (q_u) and P-wave velocity with real data for sedimentary (uppermost), volcanic (middle) and granitic rocks (lowermost). Each curve is the empirical equation listed in Table 1. The number indicated in the legend is that shown in Table 1. Black dots are real data obtained in laboratory tests of rock cores sampled in several dam sites in Japan.

dam sites in Japan (Takahashi & Inazaki, 2010). The data are plotted for three different rock types: sedimentary, volcanic (mainly andesite and basalt), and granitic rocks. In these plots empirical equations proposed for each rock type shown in Table 1 are overlain for comparison. It can be seen from these plots that although data are scattered, the proposed empirical equations represent average features of the data except for the data with higher velocity in sedimentary rocks.

3 RELATIONSHIPS FOR SOFT ROCKS

Soft rocks, especially soft sedimentary rocks, are the main bedrocks of many large cities, such as Bangkok and Tokyo. Estimation of their mechanical properties,

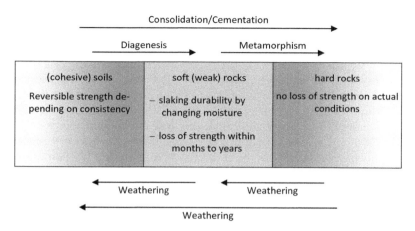

Figure 2 Position of soft (weak) rocks between soils and hard rocks (modified from Figure 1 in Nickmann et al., 2006).

including strength, is often essential in civil engineering and other applications. However, most of the empirical relationships described above have been proposed mainly for hard rocks, with few proposed for soft rocks. Soft rocks, intermediate in nature between soil and hard rocks (Nickmann et al., 2006) (Figure 2), have different compressive strength and seismic velocity relationships than hard rocks. This section, therefore, focuses on the relationship for soft rocks.

3.1 Main features of soft rocks

3.1.1 Geological features

Soft rock is geologically classified into three types of rocks based on their geological age, generation process, and mineral composition: sedimentary, weathered, and volcanic soft rocks (The Society of Material Sciences, Japan, 1993). Sedimentary soft rock is generally a Neogene or younger rock that is considered to be progressing toward consolidation and cementation in its sedimentary process. It is mainly soft mudstone, siltstone, and sandstone. A fine-grained soft sedimentary rock often contains clay minerals which may cause slaking and swelling. The weathered soft rock is generally a weathered or altered Paleozoic or Mesozoic sedimentary rock and granitic rock. Weathered granite is generated by the development of joints in the rock mass and groundwater intrusion into them caused by stress release due to uplift and erosion of the earth's crust. Weathered soft rock is also generated by hydrothermal processes in the rock mass. Volcanic soft rock is weakly welded volcanic and volcanic clastic rock generated by pyroclastic flow. Welded tuff is a typical volcanic soft rock which is generated in such a way that volcanic clastic materials are welded and consolidated. It is generally very heterogeneous in composition and physical properties.

3.1.2 Mechanical properties

The unconfined compressive strength is a basic mechanical property of a rock. For many engineering applications, the rock is classified based on the unconfined compressive strength as shown in Table 2 (The Japanese Geotechnical Society, 2007). This table shows the classifications by references to four major global organizations. The soft (weak) rock is defined as a rock with an unconfined compressive strength of less than 25 MPa in three of the references; while BS 5930-1999 defines it is as less than 12.5 MPa. There are some soft rocks with a very low unconfined compressive strength of less than 10MPa. The porosity and water content of soft rocks are generally high. Some soft rocks have a porosity of more than 50% and a water content of up to 50%. The failure strength of soft rocks is also large, and in some cases reaches 1%.

The mechanical properties of the soft rock can be significantly changed by the conditions under which the rock is placed. Figure 3 shows an example that demonstrates changes of P- and S-wave velocities and the unconfined compressive strength of the soft rock with its water content (The Society of Material Sciences, Japan, 1993). Figure 4 shows P- and S-wave velocities obtained with the suspension PS logging, and confined compressive strength measured on rock cores sampled in the same boreholes logged in soft sedimentary rocks. These profiles clearly show the dependency of these properties on depth and thus on confining stress.

3.2 Compressive strength and seismic velocity relationship for soft rock

3.2.1 Unconfined compressive strength–seismic velocity relationship

Figure 5 shows crossplots for P- and S-wave velocities vs. unconfined compressive strengths (The Japanese Geotechnical Society, 2004). P- and S-wave velocities were obtained with ultrasonic velocity measurements in the laboratory using a variety of rock core samples collected in Japan. The unconfined compressive strength was measured with the uniaxial compression test in the laboratory using the same rock samples as those for velocity measurements. In these plots, tuff breccia, lapilli tuff, tuff, sandy mudstone, and mudstone were the soft rocks defined geologically. These crossplots show that their compressive strengths are less than 20–30 MPa and P- and S-wave velocities are less than 3000 m/s and 1500 m/s, respectively.

3.2.2 Confined compressive strength-seismic velocity relationship

Figure 6 shows the relationship between the confined compressive strength and P- and S-wave velocities for three kinds of soft Pliocene sedimentary rocks from Japan: mudstone, sandy mudstone, and sandstone. The confined compressive strength was measured with the tri-axial compression test of the rock cores sampled in 350 m long boreholes in these rocks. The confinements corresponding to the core sampling depths were used in the tri-axial compression tests. Seismic velocities were measured with the suspension PS logging run in the same boreholes in which the rock cores were sampled. Although there is some scatter in the sandy mudstone data, the relationships for three types of rock appear to be on a single regression curve. Since these rocks have very large

Table 2 Rock mass classification and unconfined compressive strength defined in various references (modified from Table-1.2.2 in The Japanese Geotechnical Society (2007)).

Reference	Unconfined Compressive Strength (MPa)						
	0.25–1	1–5	5–25	25–50	50–100	100–250	250<
ISRM suggested methods for the quantitative description of discontinuities in rock masses, 1977.	Extremely weak	Very weak	weak	Medium strong	strong	Very strong	Extremely strong
Reference	<1.25	1.25–5	5–12.5	12.5–50	50–100	100–200	200<
BS 5930–1999 Code of practice for site investigation, 1999.	Very weak	weak	Moderately weak	Moderately strong	strong	Very strong	Extremely strong
Reference	<1	1–5	5–25	25–50	50–100	100–250	250<
ISO 14689–1 2003(E) Geotechnical investigation and testing – Identification and classification of rock-Part 1: Identification and description, 2003.	extremely weak	Very weak	weak	Medium strong	strong	Very strong	Extremely strong
Reference	<1	1–5	5–10	10–25	25–50	50–100	100<
JGS 3811–2004 Engineering classification of rock mass	G	F	E Soft rock	D	C	B Hard rock	A

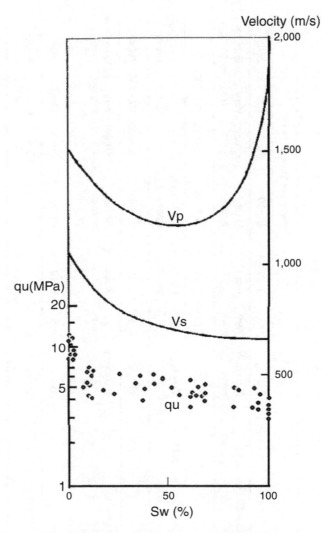

Figure 3 Changes of P- (Vp) and S-wave (Vs) velocities and unconfined compressive strength (q_u) of mudstones with water saturation of the rock (modified from Fig. 12.5 in The Society of Material Sciences, Japan, 1993).

porosities ranging from 30%–60%, the compressive strength and both P- and S-wave velocities are extremely small.

4 INTERPRETATION WITH PHYSICAL MODELS

To more generally, accurately, and reliably estimate rock strength from seismic velocity, it is effective to utilize physical models which can represent the relationship

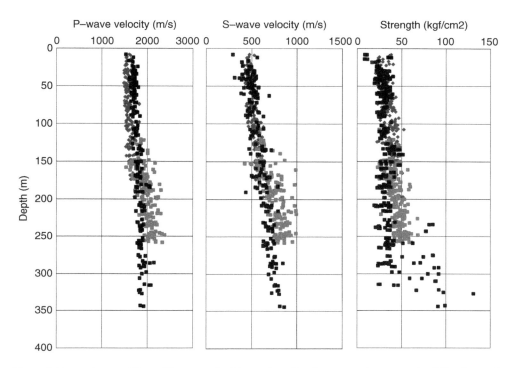

Figure 4 Dependency of P- and S-wave velocities and confined compressive strength of Tertiary soft sedimentary rocks (sandstone (brown), sandy mudstone (green) and mudstone (blue)) on the depth and thus confining stress.

between these two properties of a rock. In rock engineering, models have been widely proposed and utilized to interpret mechanical and hydraulic properties of a discontinuous rock mass. In the interpretation of seismic data for oil and gas exploration and exploitation, physical models have also been very extensively applied to estimate mechanical and hydraulic properties of oil and gas reservoirs and their changes in time for 4D seismic data. Recently, more attention has been given to the estimation of geomechanical properties such as deformability and strength of rocks from seismic data. In this section, the physical models which have been proposed by the authors are described; first regarding the model derivation, and then regarding actual applications to the relationships of unconfined and confined compressive strengths for seismic velocities, respectively.

4.1 Modeling of the compressive strength–seismic velocity relationship

The shaly sand model as an effective-medium model for sedimentary rocks (Avseth et al., 2005) is employed to represent the relationships between compressive strength and porosity, and between seismic velocity and porosity. Assuming that a rock is composed of two constituent rocks with a higher and lower stiffness, this model

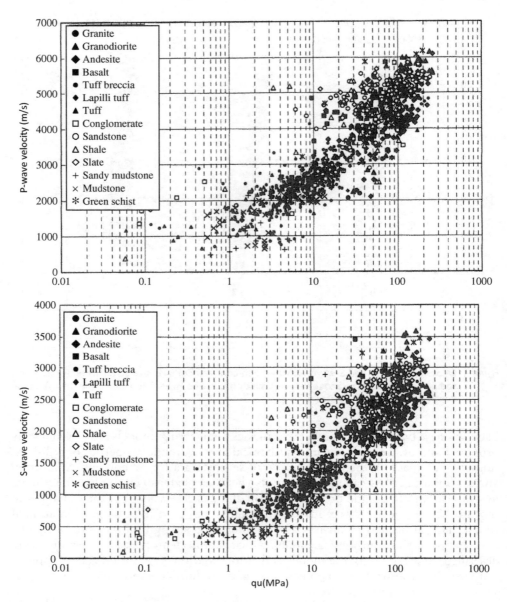

Figure 5 Relationship between the unconfined compressive strength (q_u) and ultrasonic P- (upper) and S-(lower) wave velocities for various types of rocks collected in Japan (modified from Figure 1.2.2 in Japanese Geotechnical Society, 2004).

represents a rock with a variety of mechanical properties by changing the ratios of the two constituent rocks. In this modeling, sandstone and shale are employed as the constituent rocks with higher and lower stiffness, respectively. This model is schematically drawn in Figure 7.

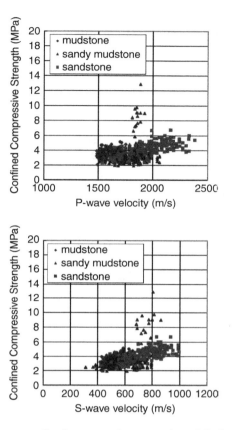

Figure 6 Relationship between confined compressive strength and P- (upper) and S-(lower) wave velocities for three kinds of soft sedimentary rocks (mudstone, sandy mudstone and sandstone) in Japan. Seismic wave velocity is measured with the suspension PS logging and confined compressive strength is measured in the tri-axial compressive test with the confining pressure corresponding to the core sampling depth.

4.1.1 Compressive strength–porosity model

The compressive strength of a rock under a confining stress is given in terms of the cohesion and angle of internal friction as:

$$\sigma = q_u + \sigma_0 \cdot \tan^2(45 + \frac{\varphi}{2})$$
$$q_u = 2c \cdot \tan(45 + \frac{\varphi}{2}) \quad (1)$$

where σ is the confined compressive strength, q_u is the unconfined compressive strength, σ_0 is the confining stress, and c and φ are the cohesion and angle of internal friction of the rock, respectively (Goodman, 1989).

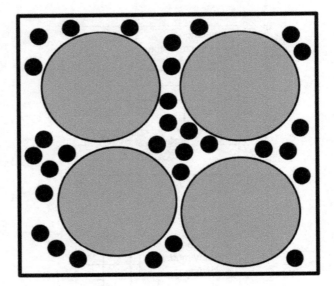

Figure 7 Schematic diagram of the shaly sand model.

Using the shaly sand model, the compressive strength of a rock for an arbitrary porosity can be calculated by averaging the strengths of sandstone and shale with two extreme porosities: zero and maximum (critical) porosities, respectively. In this modeling, the following Voigt-Reuss-Hill average (Mavko *et al.*, 2009), modified by the authors, is employed for this calculation:

$$\sigma_\phi = \omega\{(1-f)\sigma_s) + f\sigma_c\} + (1-\omega)\left(\frac{1-f}{\sigma_s} + \frac{f}{\sigma_c}\right)^{-1} \qquad \phi = \phi_0 \cdot f \qquad (2)$$

where σ_ϕ is the compressive strength of a rock with a porosity ϕ. The compressive strengths of the constituent rocks are σ_s for sandstone with zero porosity and σ_c for shale with the critical porosity ϕ_0, respectively, which can be calculated by Equation 1. The volume fraction of shale with the critical porosity is f ($0 < f < 1$). ω is added in this modification to represent the proportion of Voigt and Reuss averages of the model, which is also a fraction as $0 < \omega < 1$. Given c and φ of the constituent rocks and the confining stress σ_0, the compressive strength of a rock at an arbitrary porosity under a confining stress can be calculated with Equations 1 and 2.

4.1.2 Seismic velocity–porosity model

The seismic velocity–porosity relationship for a sedimentary rock can also be modeled with the shaly sand model. In this modeling, the Hashin-Shtrikman Bounds (Mavko *et al.*, 2009), which has been widely used for modeling sedimentary rocks, is employed for calculating the bulk and shear moduli necessary for estimating the seismic velocity of the rock as given below:

$$K_{dry} = \alpha \left\{ \left(\dfrac{\dfrac{\phi}{\phi_0}}{K_{HM} + \dfrac{4}{3}G_s} + \dfrac{1 - \dfrac{\phi}{\phi_0}}{K_s + \dfrac{4}{3}G_s} \right)^{-1} - \dfrac{4}{3}G_s \right\}$$

$$+ (1 - \alpha) \left\{ \left(\dfrac{\dfrac{\phi}{\phi_0}}{K_{HM} + \dfrac{4}{3}G_{HM}} + \dfrac{1 - \dfrac{\phi}{\phi_0}}{K_s + \dfrac{4}{3}G_{HM}} \right)^{-1} - \dfrac{4}{3}G_{HM} \right\}$$

$$G_{dry} = \alpha \left\{ \left(\dfrac{\dfrac{\phi}{\phi_0}}{G_{HM} + Z_s} + \dfrac{1 - \dfrac{\phi}{\phi_0}}{G_s + Z_s} \right)^{-1} - Z_s \right\}$$

$$+ (1 - \alpha) \left\{ \left(\dfrac{\dfrac{\phi}{\phi_0}}{G_{HM} + Z_{HM}} + \dfrac{1 - \dfrac{\phi}{\phi_0}}{G_s + Z_{HM}} \right)^{-1} - Z_{HM} \right\}$$

$$Z_s = \dfrac{G_s}{6} \left(\dfrac{9K_s + 8G_s}{K_s + 2G_s} \right)$$

$$Z_{HM} = \dfrac{G_{HM}}{6} \left(\dfrac{9K_{HM} + 8G_{HM}}{K_{HM} + 2G_{HM}} \right) \tag{3}$$

where K_{dry} and G_{dry} are the bulk and shear moduli of the dry rock, respectively. ϕ ($0 < \phi < \phi_0$) is its porosity. K_s and G_s are the bulk and shear moduli of sandstone with zero porosity, respectively. K_{HM} and G_{HM} are the bulk and shear moduli of shale at the critical porosity ϕ_0 and confining stress P, calculated with the Hertz-Mindlin model as given below (Mindlin, 1949; Mavko et al., 2009):

$$K_{HM} = \left\{ \dfrac{n^2(1 - \phi_0)^2 G_c^2}{18\pi^2(1 - \nu_c)^2} P \right\}^{\frac{1}{3}}$$

$$G_{HM} = \left\{ \dfrac{5 - 4\nu_c}{5(2 - \nu_c)} \right\} \left\{ \dfrac{3n^2(1 - \phi_0)^2 G_c^2}{2\pi^2(1 - \nu_c)^2} P \right\}^{\frac{1}{3}} \tag{4}$$

where G_c and ν_c is the shear modulus and Poisson's ratio of shale itself, respectively. n is the coordination number that indicates the number of contact points of two rock grains contacting each other. α is a parameter representing the proportion of the upper and lower bounds of the Hashin-Shtrikman Bounds, which is a fraction as $0 < \alpha < 1$.

Elastic moduli for the saturated rocks are calculated by applying Gassmann's equation (Gassmann, 1951; Mavko et al., 2009) as given below:

$$K_{sat} = K_{dry} + \frac{(1 - K_{dry}/K_s)^2}{\phi/K_f + (1 - \phi)/K_s - K_{dry}/K_s^2}$$

$$G_{sat} = G_{dry} \tag{5}$$

where K_{sat} and G_{sat} are the bulk and shear moduli of the saturated rock and K_f is the bulk modulus of the pore fluid. The seismic velocities can be calculated by:

$$V_P = \sqrt{\frac{K_{sat} + (4/3)G_{sat}}{\rho}}$$

$$V_S = \sqrt{\frac{G_{sat}}{\rho}} \tag{6}$$

where V_p and V_s are P- and S-wave velocities, respectively, and ρ is the density of the saturated rock.

4.1.3 Compressive strength–seismic velocity model

By combining the two models for the compressive strength–porosity and seismic velocity–porosity relationships derived above, a model for the compressive strength–seismic velocity relationship for a rock is easily obtained.

4.2 Application to unconfined compressive strength

The model was applied to sedimentary rock data sampled from several dam sites in Japan (Takahashi & Inazaki, 2010). Unconfined compressive strength and ultrasonic P-wave velocity data measured in rock cores in the laboratory are used in the modeling. Figure 8 shows the crossplots of the unconfined compressive strength and P-wave velocity against porosity. In these plots, the calculated model curves are overlain on the measured data. The parameters used in these calculations are listed in Table 3. Although there are large scatters in the data, it can be recognized that the calculated model curves closely matches the average features of the data.

By combining these two models, the unconfined compressive strength–P-wave velocity relationship can be modeled. The model curve is shown on the data in Figure 9. This model satisfactorily represents the features of the relationship where the unconfined compressive strength rapidly increases as P-wave velocity becomes larger than 4 km/s, which cannot be achieved by the empirical equations described in section 2.

4.3 Application to confined compressive strength

4.3.1 Example for soft sedimentary rock

The first example is an application to S-wave velocity and confined compressive strength data obtained in three different Pliocene soft sedimentary rocks in Japan: sandstone, sandy mudstone, and mudstone (Takahashi & Tanaka, 2010). S-wave velocity data were

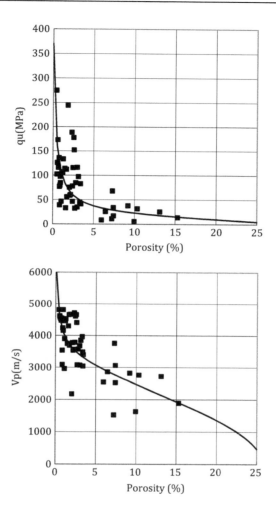

Figure 8 Relationship between unconfined compressive strength (q_u) vs. porosity (upper) and P-wave velocity (v_p) vs. porosity (lower) for sedimentary rocks in dam sites in Japan. These data were measured with the laboratory tests.

obtained by the Suspension PS logging in 350 m-deep boreholes. The confined compressive strength data were measured by the tri-axial compression test of rock cores sampled in the same boreholes as those that had been logged. Confining stresses corresponding to the core sampling depths were used in the tri-axial test.

A crossplot of measured data for confined compressive strength vs. porosity for the three soft sedimentary rocks is shown in Figure 10. Although the data for sandy mudstone are scattered, most of the measured confined compressive strengths are around 2 to 6 MPa. The modeled curves are overlain on the data. The modeled curves were calculated using Equations 1 and 2 for three different confining stresses (0.5, 1.0,

Table 3 Physical properties and parameters used in the calculation for modeling the relationships between unconfined compressive strength and P-wave velocity against porosity.

Physical property/Parameter	Value
Cohesion of sandstone	70.6 MPa
Angle of internal friction of sandstone	48 degree
Cohesion of shale	2.1 MPa
Angle of internal friction of shale	7.5 degree
Bulk modulus of sandstone	36.6 GPa
Shear modulus of sandstone	45.0 GPa
Bulk modulus of shale	21.0 GPa
Shear modulus of shale	7.0 GPa
Critical porosity of shale	0.25
Coordination number in Equation (4)	12

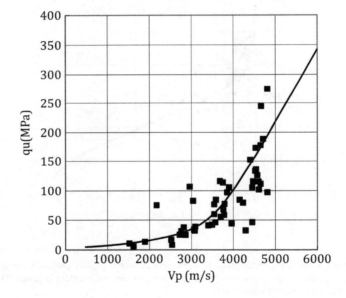

Figure 9 Relationship between unconfined compressive strength (q_u) and P-wave velocity (v_p) for sedimentary rocks in dam sites in Japan.

and 2.0 MPa), corresponding to log measurement and core sampling depths. In these calculations, the two constituent rocks with zero and critical porosities are assumed to be quartzite and shale, respectively. For these rocks, the values for c and φ (Goodman, 1989), and the critical porosity of shale (Takahashi & Tanaka, 2010), used in this calculation are shown in Table 4. The calculated curves are in good agreement with the measured data on depth dependency, and thus on confining stress.

Table 4 Physical properties and parameters used in the calculation for modeling the relationship between confined compressive strength and porosity.

Physical property/Parameter	Value
Cohesion of quartzite	70.6 MPa
Angle of internal friction of quartzite	30 degree
Cohesion of shale	0.3 MPa
Angle of internal friction of shale	7.5 degree

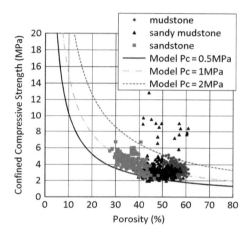

Figure 10 Relationship between confined compressive strength and porosity for three kinds of soft sedimentary rocks: sandstone (brown), sandy mudstone (green) and mudstone (blue). Model-calculated curves for three different confining stresses (0.5, 1.0 and 2.0MPa) are overlain on the real data.

The S-wave velocity vs. porosity crossplot of the measured data for the same three rocks is shown in Figure 11. The porosities used in this crossplot are also laboratory measurements on rock cores sampled at the same depth as the corresponding log measurement. S-wave velocity ranged from 400 to 800 m/s for mudstone and sandy mudstone, and 600 to 1000 m/s for sandstone. Three curves, calculated using Equations 3 to 6 with the model rock composed of quartzite and shale, are overlain on the measured data. The bulk and shear moduli of these rocks (Mavko et al., 2009), and the coordination number and critical porosity employed in this calculation, are shown in Table 5. The model calculations agree well with the measured data.

These two models were combined to model the confined compressive strength–S-wave velocity relationship, as shown in Figure 12. The model-predicted curves for three different confining stresses correspond well with the measured data, except for the higher velocity data. This is because of the poor fit of the model calculations in the S-wave velocity vs. porosity crossplot for the higher velocity at lower porosities for sandstone.

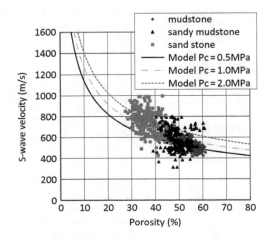

Figure 11 Relationship between S-wave velocity and porosity for three kinds of soft sedimentary rocks: sandstone (brown), sandy mudstone (green) and mudstone (blue). Model-calculated curves for three different confining stresses (0.5, 1.0 and 2.0MPa) are overlain on the real data.

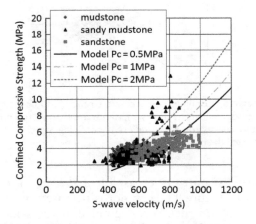

Figure 12 Relationship between confined compressive strength and S-wave velocity for three kinds of soft sedimentary rocks: sandstone (brown), sandy mudstone (green) and mudstone (blue). Modelcalculated curves for three different confining stresses (0.5, 1.0 and 2.0MPa) are overlain on the real data.

4.3.2 Example for hard sedimentary rock

The second example is an application to ultrasonic S-wave velocity and confined compressive strength data measured in laboratory tests on rock core samples. These core samples were collected from sedimentary rocks in different locations in Japan (Hoshino *et al.*, 2001). The confined compressive strength data were measured with the tri-axial compression test for four different confining stresses: 25, 50, 100, and 150 MPa. Ultrasonic P- and S-wave velocity data were measured under natural

Table 5 Physical properties and parameters used in the calculation for modeling the relationship between S-wave velocity and porosity.

Physical property/Parameter	Value
Bulk modulus of quartzite (quartz)	36.6 GPa
Shear modulus of quartzite (quartz)	45.0 GPa
Bulk modulus of shale	21.0 GPa
Shear modulus of shale	7.0 GPa
Critical porosity of shale	0.8
Coordination number in Equation (4)	21

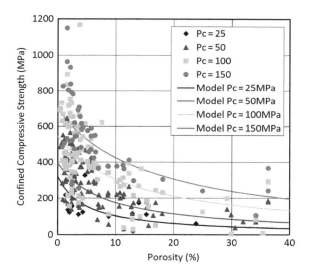

Figure 13 Relationship between confined compressive strength and porosity for sedimentary rocks collected in Japan. Confined compressive strength is measured for four different confining stresses (25, 50, 100 and 150MPa). Model-calculated curves for four different confining stresses are overlain on the real data.

conditions, without confinement, using the same rock core samples as those used in the tri-axial compression test. Only S-wave velocity data are presented in this section so as to be able to make comparisons with the relationships for the soft rock described above.

The confined compressive strength vs. porosity crossplots of the measured data (point symbols) and model calculations (solid curves) are shown in Figure 13 for the four confining stresses. This calculation also employs quartzite and shale as two constituent rocks, using the same physical properties as those in the first application, shown in Table 5. This figure clearly shows that the model-predicted curve for each different confinement agrees well with the measured data.

Combining this model with the shaly sand model for the seismic velocity vs. porosity relationship used in the first application, but with a different coordination

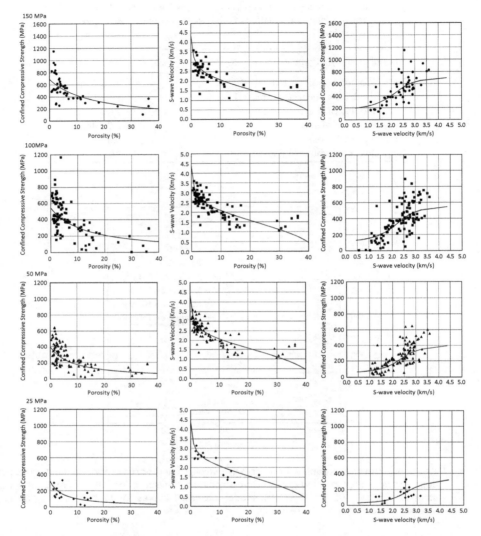

Figure 14 Relationships between confined compressive strength – porosity (left), S-wave velocity - porosity (middle) and confined compressive strength – S-wave velocity (right) for sedimentary rocks collected in Japan. Confined compressive strength is measured for four different confining stresses (25, 50, 100 and 150MPa from bottom to top rows). A model-calculated curve for each confining stress is overlain on the real data.

number (6) and critical porosity (0.4), the S-wave velocity–confined compressive strength relationship is modeled and shown in Figure 14, together with the other two relationships for four different confining stresses. Although there is a large scatter in the data for all crossplots, it can be seen that the model predicted curves represent the average features in the relationship between confined compressive strength and S-wave velocity.

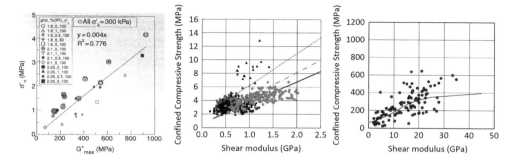

Figure 15 Confined compressive strength and dynamic shear modulus relationship for soils (left: after Sharma et al., 2011), soft (middle) and hard (right) sedimentary rocks. Note that each property is in a wide range. Horizontal axis is dynamic shear modulus determined from S-wave velocity measurement and vertical axis is confined compressive stress.

4.3.3 Comparison with the relationship for soils

Sharma *et al.* (2011) measured ultrasonic S-wave velocity during the tri-axial compression test for artificial soils to study the relationship between compressive strength and dynamic shear modulus determined by S-wave velocity and density measurements. They found a simple linear relationship between these two properties. Figure 15 shows this relationship compared with those for soft and hard sedimentary rocks used in the previous sections. These comparisons reveal that the model proposed here can be applicable to materials (soils and rocks) with a wide range of elasticity (Takahashi, 2015).

SUMMARY

The compressive strength of rocks or rock masses is an important mechanical property required in many applications in civil engineering, oil and gas development, and mining operations. Seismic profiling can be effectively used for building a strength model of a large rock mass if seismic data can be converted to rock strength. This chapter, therefore, reviews empirical equations to estimate the unconfined compressive strength of a rock from seismic P-wave velocity. For more accurate and reliable estimation than empirical equations, physical models of compressive strength–seismic velocity relationship are proposed and applied to real data for soft and hard rocks to verify the models. These applications of the models to real data revealed that the compressive strength of a rock can be estimated from seismic velocity using physical models with adequate parameters.

REFERENCES

Avseth, P., Mukerji, T. and Mavko, G. (2005) Quantitative seismic interpretation, Cambridge University Press, Cambridge.

Aydan, O., Akagi, T., Ito, T. and Kawamoto, T. (1992) Prediction of behaviour of tunnels in squeezing ground, J. JSCE, 448/III-19, 73–82 (*in Japanese*).

Dvorkin, J., Gutierrez, M.A. and Grana, D. (2014) Seismic reflections of rock properties, Cambridge University Press, Cambridge.

Fjaer, E., Holt, R.M., Horsrud, P., Raaen, A.M. and Risnes, R. (1992) Petroleum related rock mechanics, Elsevier, Amsterdam.

Freyburg, D. (1972) Der Untere und mittlere Buntsandstein SW-Thurringen in seinen gestein-stechnicshen Eigenscften, Ber.Dte.Ges.Geol.Wiss.A; Berlin, 17(6), 911–919.

Gassmann, F. (1951) Uber die Elastizitat poroser Medien, Vier. der Natur. Gesellschaft Zurich, 96, 1–23.

Göktan, R.M. (1988) Theoretical and practical analysis of rock rippability, Ph.D. Thesis, Istanbul Technical University.

Golubev, A.A. and Rabinovich, G.Y. (1976) Resultay primeneia appartury akusticeskogo karotasa dija predeleina procontych svoistv gornych porod na mestorosdeniaach tverdych isjopaemych, Prikladnaja GeofizikaMoskva, 73, 109–116.

Goodman, R.E. (1989) Introduction to rock mechanics, 2nd Edition, John Wiley & Sons, New York.

Gorjainov, N.N. and Ljachovickij, F.M. (1979) Seismiceskie Metody v Insenernoi Geologii, Izdat. Nedra, Moskva.

Hashin, Z. and Shtrikman, S. (1963) A variational approach to the elastic behavior of multiphase materials, J. Mech. Phys. Solids., 11, 127–140.

Horsrud, P. (2001) Estimating mechanical properties of shale from empirical correlations, SPE Drill. Complet., 16(2), 68–73.

Hoshino, K., Kato, S. and The Committee of Database in GSJ (2001) Summary of physical properties of rocks in deep formations in Japan, Geological Survey of Japan (*in Japanese*).

Howarth, D.F., Adamson, W.R. and Berndt, J.R. (1989) Correlation of model tunnel boring and drilling machine performances with rock properties, Int. Rock Mech. Sci. Geomech., 23, 171–175.

Inoue, M. and Ohmi, M. (1981) Relation between uniaxial compressive strength and elastic wave velocity of soft rock, Int. Symp. on Weak rock, Tokyo, 9–13.

Japanese Geotechnical Society (2007) A guideline for determination of design parameters for the subsurface model – For a rock mass, Japanese Geotechnical Society.

Japanese Material Sciences, Japan (1993) Chapter 12 Soft Rocks in Rock Mechanics, Maruzen, Tokyo.

Kahraman, S. (2001) Evaluation of simple methods for assessing the uniaxial compressive strength of rock, Int. Rock Mech. Min. Sci., 38, 981–994.

Karakul, H. and Ulusay, R. (2015) Multivariate predictions of geomechanical properties of rocks by ultrasonic velocities, physical and mineralogical properties under degrees of saturation, Proceedings of ISRM Congress 2015, Int'l Symposium on Rock Mechanics.

Kuster, G.T. and Toksoz, M.N. (1974) Velocity and attenuation of seismic waves in two phase media, Geophysics, 39, 587–618.

Lal, M. (1999) Shale stability: drilling fluid interaction and shale strength, SPE 54356, SPE Latin American and Caribbean Petroleum Engineering Conference, Caracas, Venezuela, Society of Petroleum Engineering.

Mavko, G., Mukerji, T. and Dvorkin, J. (2009) The rock physics handbook, 2nd Edition, Cambridge University Press, Cambridge.

McNally, G.H.N. (1987) Estimation of coal measures rock strength using sonic and neutron logs, Geoexploration, 24, 381–395.

Militzer, H. and Stoll, R. (1973) Einige Beiträge der geophysik zur primärdatenerfassung im Bergbau, Neue Bergautechnik, Leipzig, 3, 21–25.

Mindlin, R.D. (1949) Compliance of elastic bodies in contact, J. Appl. Mech., 16, 259–268.

Moos, N., Zoback, M.D. and Bailey, L. (1999) Feasibility study of the stability of openhole multilaterals, cook inlet, Alaska, 1999 SPE mid-continent operations symposium, Oklahoma City, OK, Society of Petroleum Engineers.

Nickmann, M., Spaun, G. and Thuro, K. (2006) Engineering geological classification of weak rocks, IAEG2006 Paper number 492.

Ohkubo, T. and Terasaki, A. (1971) Physical properties and seismic velocity of rocks, Soil Mech. Found. Eng., 19(7), 31–37 (*in Japanese*).

Sharma, R., Baxter, C. and Jander, M. (2011) Relationship between shear wave velocity and stresses at failure for weakly cemented sands during drained triaxial compression, Soils Found., 51, 761–771.

Schön, J.H. (2011) Physical properties of rocks, Elsevier, Amsterdam.

Sousa, L.M.O., del Rio, L.M.S., Calleja, I., de Argandona, V.G. and Rey, A.R. (2005) Influence of microfractures and porosity on the physico-mechanical properties and weathering of ornamental granites, Eng. Geol., 77, 153–168.

Standardization Committee for Engineering Classification of Rock Mass, JGS (2004) Method for engineering classification of rock mass, Japanese Geotechnical Society.

Takahashi, T. (2015) Rock physical interpretation of the compressive strength - dynamic shear modulus relationship for sedimentary rocks, Proceedings of Near Surface Asia Pacific Conference 2015.

Takahashi, T. and Tanaka, S. (2009) Rock physics model for interpreting elastic properties of soft sedimentary rocks, Int. J. JCRM, 4, 53–59.

Takahashi, T. and Inazaki, T. (2010) Collection and analysis of physical properties of rocks for enhancing the geotechnical database KuniJiban, Proceedings of the 123 SEGJ Conference, 9–12 (*in Japanese*).

Takahashi, T. and Tanaka, S. (2010) Rock physical interpretation of the compressive strength – seismic velocity relationship for sedimentary rocks, Explor. Geophys., 44, 31–35. Available from: http://www.publish.csiro.au/journals/eg.

Takahashi, T. and Tanaka, S. (2013) Rock physical interpretation of the compressive strength – seismic velocity relationship for sedimentary rocks, Explor. Geophys., 44, 31–35.

The Japanese Society Geotechnical Society (2007) A guideline for determination of mechanical and hydraulic properties of a rock mass for design, The Japanese Geotechnical Society, Tokyo (*in Japanese*).

The Society of Material Sciences, Japan (1993) Rock mechanics, Maruzen, Tokyo (*in Japanese*).

Tugul, A. and Zarif, I.H. (1999) Correlation of mineralogical and textural characteristics with engineering properties of selected granitic rocks from Turkey, Eng. Geol. 51, 303–317.

Vasconcelos, G., Lourenc, P.B., Alves, C.A.S. and Pamplona, J. (2008) Ultrasonic evaluation of the physical and mechanical properties of granites, Ultrasonics, 48, 453–466.

Yasar, E. and Erdogan, Y. (2004) Correlating sound velocity with the density, compressive strength and Young's modulus in gypsum from Sivas (Turkey), Eng. Geol., 66, 211–219.

Zhang, L. (2005) Engineering properties of rocks, Elsevier, Amsterdam.

Zoback, M. (2007) Reservoir geomechanics, Cambridge University Press, Cambridge.

Chapter 11

Elastic waves in fractured isotropic and anisotropic media

Laura J. Pyrak-Nolte[1,2,3], Siyi Shao[1] & Bradley C. Abell[4]
[1]*Department of Physics & Astronomy, Purdue University, West Lafayette, IN, USA*
[2]*Department of Earth, Atmospheric and Planetary Sciences, Purdue University, West Lafayette, IN, USA*
[3]*Lyle School of Civil Engineering, Purdue University, West Lafayette, IN, USA*
[4]*W.D. VonGonten Laboratories LLC, Houston, TX, USA*

Abstract: Elastic waves are an essential tool for characterizing rock on the laboratory and field scale. In this chapter, methods for interpreting the elastic constants for isotropic and anisotropic fractured rock are given. An overview of the discrete effects of fractures on elastic wave propagation is presented to illustrate behavior not captured by effective medium approaches. Maintaining the discreteness of fractures yields guided modes and energy partitioning that depend on the stiffness of fractures and the signal frequency. Given the complexity of rock and field sites, understanding both effective medium and discrete approaches is required to achieve the best interpretation of the properties of isotropic or anisotropic fractured rock.

1 INTRODUCTION

To quote Barton and Quadros (2015): "Anisotropy is everywhere. Isotropy is rare". An anisotropic rock is defined by material and physical properties that vary with direction. The source of anisotropy is intimately linked to the depositional, diagenetic and tectonic history of a rock that results in preferred orientation of minerals, micro-cracks, laminae, layering, fractures and other mechanical discontinuities. A rock may exhibit anisotropic mechanical properties because of one or more of these fabric and/or structural features. For example, anisotropy in shale has been attributed to the distribution of platy clay minerals (Hornby *et al.*, 1994; Johnston & Christensen, 1995; Sondergeld & Rai, 2011), compliant organic materials (Vernik & Nur, 1992; Vernik & Liu, 1997; Sondergeld *et al.*, 2000; Vernik & Milovac, 2011), microcracks (Hornby *et al.*, 1994; Vernik & Nur, 1992), as well as depositional features such as laminae and layering (Schoenberg *et al.*, 1996). These textural and structural features of shale lead to mechanical anisotropy (Sone & Zoback, 2013) as well as seismic anisotropy in elastic wave velocity that is sometimes on the order of 20-50% (Sondergeld & Rai, 2011).

Anisotropy in mechanical properties can also be induced in a rock from non-hydrostatic loading (for example see: Nur & Simmons, 1969). Sayers (1988) showed theoretically that an isotropic rock, with an isotropic distribution of microcracks, will exhibit mechanical and seismic anisotropy under non-hydrostatic stress conditions because of preferentially oriented cracks that remain open under loading. Similarly, an isotropic rock with a set of parallel fractures can also exhibit anisotropic properties that are stress dependent. Recently, Shao & Pyrak-Nolte (2013) demonstrated experimentally and theoretically that an anisotropic medium can appear nearly isotropic under low loads when a fracture is oriented perpendicular to the layering. Their work demonstrated the effect of competing anisotropies (matrix layering versus fracture orientation) on the interpretation of the isotropic or anisotropic nature of a fractured anisotropic medium under stress.

Knowledge of the isotropic or anisotropic properties of rock is important for many engineering activities in the surface and subsurface, for example the stability of underground excavations, foundations, and boreholes (Amadei, 1996). Stress determined from strain measurements can be misinterpreted if the anisotropy of a medium is not taken into account. For example, Amadei & Goodman (1982) showed for overcoring techniques that the magnitude and direction of stress for a transversely isotropic medium would be misinterpreted by as much as 50% and 100 degrees, respectively, if isotropy was assumed.

Measurements of compressional and shear wave velocities are commonly used to characterize the mechanical isotropy or anisotropy of rock because elastic wave velocities depend on the elastic moduli of a sample. In this chapter, we present the theoretical formulations for interpreting elastic constants from measurements of elastic wave velocities for anisotropy caused by the matrix and by fractures using effective medium approaches for transversely isotropic rock. We also provide an overview of the discrete effects of fractures on elastic waves that yield guided modes and energy partitioning that depends directly on the stiffness of fractures and does not arise when effective medium approaches are used.

2 CHARACTERIZATION OF INTACT ROCK

Using elastic waves to characterize rock requires a link between elastic wave velocities and the mechanical properties of a rock. When a compressional wave propagates through rock, the medium is alternately compressed and dilated in the direction of wave propagation, while transverse displacements occur perpendicular to the direction of propagation for shear waves. These normal and shear displacements are often described using linear elasticity to link the elastic moduli of rock to measured wave velocities. In this section, a brief summary of linear elasticity is provided for the discussion of the relationships between elastic wave velocities and the elastic constants that define a rock.

2.1 Elastic constants

Linear elasticity theory describes the deformation of an elastic medium in response to applied external forces. The amount of elastic deformation of a solid relative to its undeformed original size is referred to as the strain, which has both normal and shear components. Linear elasticity assumes a linear relationship between stress and strain that is given by a generalized form of Hooke's law: $s_{ij} = C_{ijkl} e_{kl}$ where C_{ijkl} is a fourth rank stiffness tensor with 81 components, and s_{ij} and e_{kl} are second rank tensors of stress and strain, respectively. For convenience, the strain (or stress) can be converted into a vector with six components (three normal components and three shear components) because stress and strain tensors are diagonally symmetric ($s_{ij} = s_{ji}$ and $e_{ij} = e_{ji}$).

Voigt's notation is often used to simplify the tensor notations from C_{ijkl} to C_{ab} using the conversion $11\rightarrow1$, $22\rightarrow2$, $33\rightarrow3$, $23=32\rightarrow4$, $13=31\rightarrow5$, and $12=21\rightarrow6$ (Thomsen, 1986). For example, C_{2312} becomes C_{46}, and s_{23} becomes s_4 (see Figure 1a). Hooke's law can be rewritten with Voigt's notation as

$$
\begin{pmatrix} \sigma_1 \\ \sigma_2 \\ \sigma_3 \\ \sigma_4 \\ \sigma_5 \\ \sigma_6 \end{pmatrix} = \begin{pmatrix} C_{11} & C_{12} & C_{13} & C_{14} & C_{15} & C_{16} \\ C_{21} & C_{22} & C_{23} & C_{24} & C_{25} & C_{26} \\ C_{31} & C_{32} & C_{33} & C_{34} & C_{35} & C_{36} \\ C_{41} & C_{42} & C_{43} & C_{44} & C_{45} & C_{46} \\ C_{51} & C_{52} & C_{53} & C_{54} & C_{55} & C_{56} \\ C_{61} & C_{62} & C_{63} & C_{64} & C_{65} & C_{66} \end{pmatrix} \begin{pmatrix} \varepsilon_1 \\ \varepsilon_2 \\ \varepsilon_3 \\ \varepsilon_4 \\ \varepsilon_5 \\ \varepsilon_6 \end{pmatrix}
\tag{1}
$$

For an *isotropic* medium, an external force causes the same displacements no matter which direction the force is applied, *i.e.*, the stiffness tensor has no preferred direction. The elastic constants for an *isotropic* material only depend on two independent parameters. The elastic constants for an *isotropic* medium are shown below in terms of the Young's modulus, E, and Poisson's ratio, ν, and also in terms of Lame's constants, λ and μ. μ is also the shear modulus.

$$
\begin{aligned}
C_{\alpha\beta} &= \begin{pmatrix} \frac{(1-\nu)E}{(1+\nu)(1-2\nu)} & \frac{\nu E}{(1+\nu)(1-2\nu)} & \frac{\nu E}{(1+\nu)(1-2\nu)} & 0 & 0 & 0 \\ \frac{\nu E}{(1+\nu)(1-2\nu)} & \frac{(1-\nu)E}{(1+\nu)(1-2\nu)} & \frac{\nu E}{(1+\nu)(1-2\nu)} & 0 & 0 & 0 \\ \frac{\nu E}{(1+\nu)(1-2\nu)} & \frac{\nu E}{(1+\nu)(1-2\nu)} & \frac{(1-\nu)E}{(1+\nu)(1-2\nu)} & 0 & 0 & 0 \\ 0 & 0 & 0 & \frac{E}{2(1+\nu)} & 0 & 0 \\ 0 & 0 & 0 & 0 & \frac{E}{2(1+\nu)} & 0 \\ 0 & 0 & 0 & 0 & 0 & \frac{E}{2(1+\nu)} \end{pmatrix} \\
&= \begin{pmatrix} \lambda+2\mu & \lambda & \lambda & 0 & 0 & 0 \\ \lambda & \lambda+2\mu & \lambda & 0 & 0 & 0 \\ \lambda & \lambda & \lambda+2\mu & 0 & 0 & 0 \\ 0 & 0 & 0 & \mu & 0 & 0 \\ 0 & 0 & 0 & 0 & \mu & 0 \\ 0 & 0 & 0 & 0 & 0 & \mu \end{pmatrix},
\end{aligned}
\tag{2}
$$

where a and b range from 1 to 6.

The mechanical properties of rock are often anisotropic, exhibiting elastic properties that differ in orthogonal directions. Rock is typically represented by a medium with either *cubic, transverse isotropy* (also known as hexagonal symmetry), or *orthorhombic symmetry*. *Transverse isotropy* is common in geologic formations because of layering and the stresses in the Earth's crust (Thomsen, 1986; Djikpesse, 2015; Urosevic & Juhlin, 1999; Li *et al.*, 2014; Kerner *et al.*, 1989; Brittan & Warner, 1996) and is the focus of this presentation. In a vertically *transversely isotropic* medium (VTI), the unique symmetry axis is perpendicular to the layering (x_3 in Figure 1a). Properties parallel to the layering are the same while those along the unique axis are different. For *transversely isotropic* media, five independent components are needed to characterize the material, and the stiffness tensor becomes:

$$C_{\alpha\beta} = \begin{pmatrix} C_{11} & C_{11} - 2C_{66} & C_{13} & 0 & 0 & 0 \\ C_{11} - 2C_{66} & C_{11} & C_{13} & 0 & 0 & 0 \\ C_{13} & C_{13} & C_{33} & 0 & 0 & 0 \\ 0 & 0 & 0 & C_{44} & 0 & 0 \\ 0 & 0 & 0 & 0 & C_{44} & 0 \\ 0 & 0 & 0 & 0 & 0 & C_{66} \end{pmatrix}$$

$$= \begin{pmatrix} \frac{1-\nu_{31}^2}{E_1 E_3 \Delta_{TI}} & \frac{\nu_{21}-\nu_{31}^2}{E_1 E_3 \Delta_{TI}} & \frac{\nu_{31}-\nu_{21}\nu_{31}}{E_1 E_3 \Delta_{TI}} & 0 & 0 & 0 \\ \frac{\nu_{21}-\nu_{31}^2}{E_1 E_3 \Delta_{TI}} & \frac{1-\nu_{31}^2}{E_1 E_3 \Delta_{TI}} & \frac{\nu_{31}-\nu_{21}\nu_{31}}{E_1 E_3 \Delta_{TI}} & 0 & 0 & 0 \\ \frac{\nu_{31}-\nu_{21}\nu_{31}}{E_1 E_3 \Delta_{TI}} & \frac{\nu_{31}-\nu_{21}\nu_{31}}{E_1 E_3 \Delta_{TI}} & \frac{1-\nu_{21}^2}{E_1^2 \Delta_{TI}} & 0 & 0 & 0 \\ 0 & 0 & 0 & \mu_{13} & 0 & 0 \\ 0 & 0 & 0 & 0 & \mu_{13} & 0 \\ 0 & 0 & 0 & 0 & 0 & \frac{E_1}{2(1+\nu_{21})} \end{pmatrix} \tag{3}$$

where $\Delta_{TI} = \left(1 - \nu_{21}^2 - 2\nu_{31}^2 - 2\nu_{21}\nu_{31}^2\right)/E_1^2 E_3$, and for the Poisson's ratio, ν_{ab}, the first subscript refers to the direction of the normal to a surface and the second subscript refers to the direction parallel to the surface (see Figure 1b); and for the Young's modulus, E_a, the subscript also refers to the direction of the normal to a surface.

2.2 Wave velocity in isotropic and anisotropic elastic media

Interpretation of rock properties from measurements of compressional and shear wave velocities depends on the relationship between wave velocity and the elastic constants. For this chapter, the assumed geometry and coordinate axes are shown in Figure 1. The three unique symmetry directions are assumed to align with the coordinate axes (x_1, x_2, x_3). Compressional waves measured along the x_1, x_2, and x_3 symmetry axes are labeled V_{P1}, V_{P2}, and V_{P3}, respectively (Figure 1b). Note that the phase and group velocities are identical for these velocities because they are measured along symmetry axes (Bucur, 2006; Mavko et al., 1998; Carcione, 1996). The shear wave velocities are labeled V_{S12}, $V_{S13}, V_{S21}, V_{S31}, V_{S23}$, and V_{S32}, where the first subscript represents the direction of propagation and the second subscript is the direction of polarization of the shear wave (Figure 1b). These definitions will be used throughout this section.

For *isotropic* rock, the compressional wave, V_P, and shear wave, V_S, velocities are

$$\rho V_P^2 = \frac{(1-\nu)E}{(1+\nu)(1-2\nu)} = C_{11} \quad and \quad \rho V_S^2 = \frac{E}{2(1+\nu)} = \mu = C_{66} = \frac{C_{11} - C_{12}}{2}, \tag{4}$$

where r is the rock density. $V_{P1} = V_{P2} = V_{P3} = V_P$, and $V_S = V_{S12} = V_{S13} = V_{S21} = V_{S31} = V_{S23} = V_{S32}$ (Figure 1b). The shear wave velocity is the same for all polarizations and only two unique constants, C_{11} and C_{66}, are needed to characterize an isotropic rock. From measurements of V_P and V_S, the Poisson's ratio, ν, of an *isotropic* medium can also be obtained by

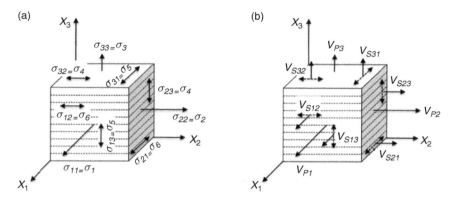

Figure 1 (a) Definition of stresses with Voigt notation also shown. (b) Compressional and shear wave velocities, V_P and V_S, respectively, measured along symmetry axes aligned with the Cartesian axes. The dashed arrows in (b) represent the direction of polarization for the shear waves. The labeled solid arrows next to the velocities represent the propagation direction.

$$\frac{V_P^2}{V_S^2} = \frac{2(1-\nu)}{1-2\nu} \quad \text{or} \quad \nu = \frac{V_S^2 - \frac{V_P^2}{2}}{V_S^2 - V_P^2} \tag{5}$$

For *transversely isotropic* rock, in addition to the compressional wave, there are two shear waves that travel with different speeds depending on the polarization of the shear wave relative to the symmetry planes (*i.e.*, layering). The compressional wave speed depends on the direction of propagation relative to the layering. V_{SH} refers to the velocity of a shear wave polarized parallel to the layering and $V_{SH} = V_{S12} = V_{S21}$, (Figure 1b). V_{SV} refers to the velocity of a shear wave polarized perpendicular to the layering and $V_{SV} = V_{S13} = V_{S23}$, (Figure 1b). For shear waves propagated along the x_3 axis, the symmetry axis, (perpendicular to the layering) $V_{SH} = V_{SV} = V_{S31} = V_{S32}$. For waves propagated with an angle θ relative to the symmetry axis, V_P, V_{SV} and V_{SH} depend on the angle of incident and are expressed as

$$\rho V_P^2(\theta) = \frac{1}{2}\left[C_{33} + C_{44} + (C_{11} - C_{33})\sin^2\theta + D(\theta)\right] \tag{6}$$

$$\rho V_{SV}^2(\theta) = \frac{1}{2}\left[C_{33} + C_{44} + (C_{11} - C_{33})\sin^2\theta - D(\theta)\right] \tag{7}$$

$$\rho V_{SH}^2(\theta) = C_{66}\sin^2\theta + C_{44}\cos^2\theta \tag{8}$$

where

$$D(\theta) = \sqrt{\left[(C_{33} - C_{44})\cos^2\theta - (C_{11} - C_{44})\sin^2\theta\right]^2 + (C_{13} + C_{44})^2\sin^2(2\theta)} \tag{9}$$

and ρ is the rock density. The diagonal components of the stiffness tensor for a transversely isotropic medium, C_{11}, C_{33}, C_{44}, and C_{66} can be directly obtained by measuring body waves along the coordinate axes and knowing the density of the rock using

$$C_{22} = C_{11} = \rho V_{P1}^2 = \rho V_{P2}^2, \qquad C_{33} = \rho V_{P3}^2, \qquad C_{66} = \rho V_{S12}^2 = \rho V_{S21}^2$$
$$C_{55} = C_{44} = \rho V_{S23}^2 = \rho V_{S32}^2 = \rho V_{S13}^2 = \rho V_{S31}^2,$$
(10)

where the first numerical subscripts refer to the normal to the surface and the second is the direction of polarization of the shear wave. C_{12} is determined from $C_{12} = C_{11} - 2C_{66}$. However, the determination of C_{13}, which shows up explicitly in the expression for $D(q)$ (Equation 9) requires the wave speed from an off-angle measurement, *i.e.*, off of the symmetric axis (see section 2.3). For V_{P45}, 45° off-angle measurement, C_{13} can be obtained using

$$C_{13} = \sqrt{\left[2\rho V_{P45}^2 - \frac{(C_{11} + C_{33})}{2} - C_{44}\right]^2 - \frac{(C_{11} - C_{33})^2}{4}} - C_{44}$$
(11)

Based on Equations 10–11, a minimum of 5 independent elastic wave measurements is needed to determine the elastic constants of a transversely isotropic rock.

2.3 Laboratory methods for characterizing anisotropy

While there are many methods to measure the elastic constants of rock, this section focuses on elastic wave transmission methods commonly used in rock engineering and geophysics. For a review of other techniques, the reader is referred to Every (1994) and Wolfe (1998). In the laboratory, ultrasonic transducers (typical frequency range 0.25 – 2 MHz) are used to send and receive compressional and shear waves propagated through rock. A rule of thumb for the size of sample is that it is of sufficient length to contain a propagation path length > 10–50 wavelengths to obtain group velocity measurements. Typical wavelengths at 1 MHz for compressional and shear waves are on the order of several mm. As mentioned earlier, group and phase velocities are only equal along symmetry axes.

For an *isotropic* rock, in addition to knowing the dimensions and density of a rock sample, only two elastic waves measurements are required to fully characterize the elastic constants: the arrival times of compressional and shear waves that are needed to calculate V_P and V_S. Using Equation (4), the Young's modulus and shear modulus of the material can be obtained as well as the Poisson's ratio.

Anisotropic rock requires additional measurements to determine the elastic constants because the velocity is no longer the same in every direction. In addition to measurements along the symmetry axis, off-axis velocity measurements are needed to determine the off-diagonal elastic constants such as C_{12}, C_{13}, and C_{23}. One approach is to prepare samples taken at different angles relative to the symmetry axes (Figure 2a). For example, Christensen & Ramananantoandro (1971) cut 7 samples of Dunite at 0°, 15°, 30°, 45°, 60°, 75°, and 90° to the symmetry axis to acquire sufficient data to determine the elastic constants. A drawback of this technique arises when the structure, composition, symmetry or density of a rock varies among the oriented core samples, especially when taken from rock obtained from different locations. Two methods that enable characterization of anisotropy on a single sample are wavefront imaging (Figure 2b) and surface wave (Figure 2c) techniques, both are described in the next two sections.

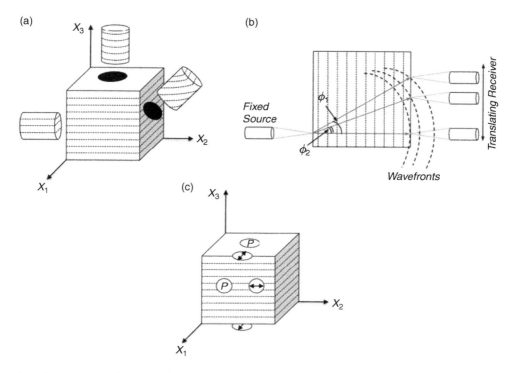

Figure 2 Techniques for determining the elastic constants of anisotropic rock. (a) Oriented core, (b) wavefront imaging, and (c) Rayleigh wave method. In (b) the source and receiver are water-coupled spherically-focused piezoelectric transducers. In (c), the circles represent contact transducers where P represents compressional wave transducers and the arrows represent the polarization of shear waves transducers.

2.3.1 Wavefront imaging techniques for determining elastic constants

Wavefront imaging methods (Figure 2b) are often used to visualize and quantify the effect of material/fabric (matrix) and structural (layering, fractures, etc.) properties on the energy distribution in a propagating wavefront (Hauser et al., 1992; Nagy et al., 1995; Mullenbach, 1996; Roy & Pyrak-Nolte, 1997; Xian et al., 2001; Oliger et al., 2003; Shao et al., 2015). For a typical experiment, a source is maintained at a fixed location while a receiver is translated over a two-dimensional region to record the spatial distribution of energy with time. Figure 3 shows examples of portions of three-dimensional data sets acquired over a 60 mm by 60 mm area in 1 mm increments for an isotropic solid (acrylic), a sample with an isotropic matrix that contains an orthogonal fracture network (aluminum with a fracture spacing of 20 mm), and for an anisotropic sample with a set of parallel fractures (Garolite with a fracture spacing of 10 mm) (Shao, 2015). The data can be viewed as snapshots in time to examine the spatial distribution of energy or as a spatial slice to examine the scattering or confinement of energy by structural features or the individual signals can be analyzed in the time-frequency domain. At a fixed time, the isotropy or anisotropy of a sample can be

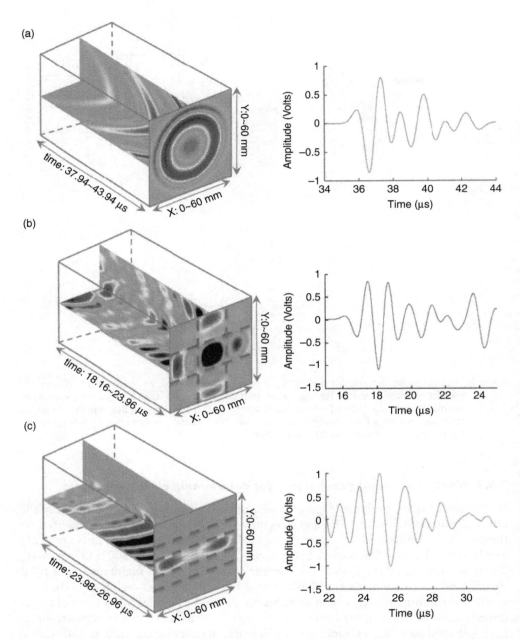

Figure 3 Examples of wavefront imaging datasets from (a) intact sample, (b) isotropic sample with an orthogonal fracture network, and (c) anisotropic sample with parallel fractures. The color corresponds to amplitude of the signal. For each sample, the central signal is shown to the right of the wavefront.

directly visualized. For example, a circular wavefront (Figure 3a) demonstrates that the energy spreads out with uniform amplitude and propagates with the same speed in all directions as expected for an isotropic medium. When fractures are present, energy confinement between fractures is observed (Figure 3b&c).

Recently, Abell *et al.* (2014) used a wavefront imaging method to extract the off-diagonal elastic constant, C_{13}, for a transversely isotropic material (Figure 2b). For a transversely isotropic medium, this approach requires the measurement of compressional and shear waves along symmetry axes to obtain C_{11}, C_{33}, C_{44} and C_{66} (or V_{P1}, V_{P3}, V_{S12}, and V_{S23} see Equation 10) with contact transducers. Then, wavefront imaging is performed on the sample with the source and receiver focused on the faces of the sample that are perpendicular to the symmetry axis (Figure 2b). While the source is held fixed near the lower corner of a sample, signals are recorded along a line to obtain data at angles off of the symmetry axis (ϕ in Figure 2b).

Equations 6–9 for determining the velocity from elastic constants or the elastic constants from velocity in section 2.2 are written in terms of phase velocity. For an isotropic medium, where the wavefronts are spherical, the group and phase velocities are equal. However, for an anisotropic medium the group and phase velocities are not equal (Bucur, 2006; Abell *et al.*, 2014; Mavko *et al.*, 1998). The conversion between group and phase velocities requires an understanding of the distinction between the phase angle, θ, and the group angle, ϕ (Figure 4a), as well as additional equations. The phase angle is measured between the symmetry axis and the normal to the wavefront, *i.e.*, the wave number vector, k. The group angle is the angle between the symmetry axis and a ray drawn from the source directly to the receiver (Figure 4a).

Wavefronts in an isotropic medium are spherical and hence $\theta = \phi$. For an anisotropic medium, the relationship between phase and group angle (Thomsen, 1986) is

$$\tan\phi = \frac{V\sin\theta + \frac{dV}{d\theta}\cos\theta}{V\cos\theta - \frac{dV}{d\theta}\sin\theta}, \tag{12}$$

where V is the phase velocity. From Thomsen (1986), the group velocity, U is given by

$$U(\phi) = \sqrt{V(\theta)^2 + \left(\frac{dV}{d\theta}\right)^2}. \tag{13}$$

The difference between phase and group angles is difficult to delineate in experimental measurements without taking measurements at several angles, because of the derivative in phase velocity (Equation 13) with respect to the phase angle (Berryman, 1979; Thomsen, 1986; Tsvankin, 1996, 1997). Though this is often attempted with measurements on angled core (Figure 2a) or with angled transducers (Sharf-Aldin *et al.*, 2013), uncertainty exists in whether group or phase velocity is measured. The use of the wavefront imaging technique reduces this uncertainty by acquiring information on the same sample at multiple group angles, *i.e.*, providing a more continuous $U - \phi$ curve for determining the derivative. A drawback of wavefront imaging is that it cannot, currently, be performed on samples under true-triaxial or triaxial loading conditions. However, Shao *et al.* (2015) and Shao & Pyrak-Nolte (2015) have performed wavefront imaging under uniaxial and biaxial conditions on isotropic and anisotropic fractured media.

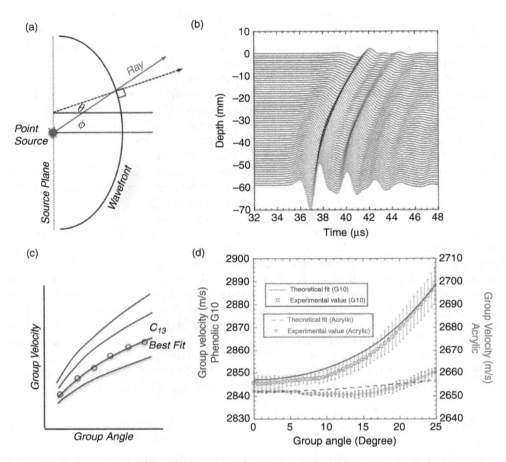

Figure 4 (a) Illustration of the difference between group, ϕ, and phase angle, θ, for an elliptical wavefront in an anisotropic medium; (b) an example of signals collected off of the symmetry axis using a wave-front imaging technique; (c) illustration of the theoretical fitting-approach, and (d) an example of a match to data from an acrylic sample and a sample of G_{10} phenolic, a transversely isotropic medium, for obtaining the off-diagonal constant C_{13} from the wavefront imaging data based on the methods in Abell et al. (2014).

The wavefront imaging method provides measurements of group velocity, U, and group angle, ϕ. The group angle equals zero when the source and receiver are aligned. After acquiring signals as a function of position (corresponding to group angles of $\phi = tan^{-1}(translated\ distance/central\ path\ length)$, Figure 4b), the value of C_{13} is determined by finding the best fit to theoretical predictions (Figure 4c). A step-by-step procedure for finding C_{13} can be found in Abell et al. (2014).

The wavefront imaging technique is a robust approach for determining off-diagonal elastic constants for a transversely isotropic medium. By using the full propagating wavefront, this method yields small uncertainties in the elastic constants. Additional theoretical work is required to extend this technique to orthorhombic symmetries.

2.3.2 Surface waves methods for determining elastic constant

Surface and interface wave techniques have been used to determine the elastic constants of different anisotropic material (Bucur & Rocaboy, 1988; Dahmen *et al.*, 2010; Shao & Pyrak-Nolte, 2013). In many of these approaches, first compressional and shear waves are measured along the symmetry axes of a sample using compressional and shear wave piezoelectric transducers. Then surface waves are propagated along the surface of the sample in the direction of the symmetry axis where the group and phase velocity are equal. Surface waves measured in this manner only depend on one off-diagonal elastic constant for *transversely isotropic* materials. The surface wave velocity and other bulk-wave measurements are then used to invert for the off-diagonal elastic constant. The reader is referred to Dahmen *et al.* (2010) for a method based on Lamb waves, to Deresiewicz & Mindlin (1957) or Bucur & Rocaboy (1988) for surface acoustic wave techniques and Shao & Pyrak-Nolte (2013) for a method based on fracture interface waves. In this section, the Rayleigh-wave technique developed by Abell *et al.* (2014) is presented as an example of a surface wave technique.

2.3.2.1 RAYLEIGH WAVE TECHNIQUE FOR DETERMINING C_{13}

For a Rayleigh wave propagated along the symmetry axis of a *transversely isotropic* medium, the Rayleigh wave velocity is given by (Vinh & Ogden, 2004):

$$V_{Rayleigh} = \sqrt{\frac{C_{66}bd\sqrt{a}}{\rho} \left[\frac{\sqrt{a}(bd+2)}{3} + \sqrt[3]{R+\sqrt{W}} + \sqrt[3]{R-\sqrt{W}}\right]^{-1}}, \tag{14}$$

where

$$a = \frac{C_{33}}{C_{11}}, \ b = \frac{C_{11}}{C_{66}}, \ d = 1 - \frac{C_{13}^2}{C_{11}C_{33}}, \ W = R^2 + \left[-\left(\frac{1}{3}\sqrt{a_2^2+3}\right)^2\right]^3, \tag{15}$$

$$R = -\frac{1}{54}\left(2a_2^3 + 9a_2 + 27a_o\right), \ a_o = -\sqrt{a}(1-d), a_2 = \sqrt{a}(1-bd). \tag{16}$$

Based on Equations 14–16, the Rayleigh wave velocity in a *transversely isotropic* medium depends on C_{11}, C_{33}, C_{66} and C_{13}. C_{11} and C_{33} are determined from measurements of V_{P1} and V_{P3}, *i.e.*, from compressional waves propagated parallel to the layers and along the symmetry axis using contact transducers (see Equation 10 and Figure 2c). C_{66} is obtained from measurements of the shear wave velocity, V_{S12}, also made with a contact transducer. The shear wave source and receiver are polarized parallel to the layers with a propagation direction along the layers (Figure 2c). Using shear wave transducers, a Rayleigh wave is propagated along the surface of the sample in the direction of the symmetry axis (Figure 2c) by placing half of the transducer face off the edge, polarized normal to the surface of propagation. By propagating the Rayleigh wave along a symmetry axis, the group, $U_{Rayleigh}$, and phase, $V_{Rayleigh}$, velocity of the Rayleigh are equal. $U_{Rayleigh}$ measured with the contact transducers is used directly in Equation 14 along with C_{11}, C_{33}, and C_{66} to solve for C_{13}. Thus, only four measurements on the sample are required to determine C_{13} for a transversely isotropic medium by using a cube-shaped sample aligned with the symmetry axis.

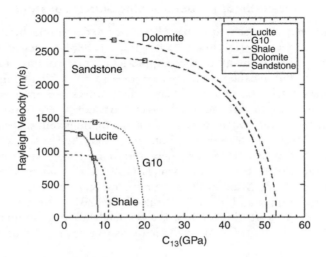

Figure 5 Theoretical curves demonstrating the effect of C_{13} on the Rayleigh wave velocity for several materials. The square symbols represent the measured values of C_{13} by Abell *et al.* (2014) for Lucite and G_{10} using the Rayleigh wave technique, and the Rayleigh wave velocity for shale, dolomite and sandstone, from Thomsen (1986) and Martinez and Schmitt (2013), calculated based on C_{13} values.

The effect of C_{13} on the Rayleigh velocity in a transversely isotropic medium is shown in Figure 5. As C_{13} increases the Rayleigh wave velocity decreases. For an isotropic material, C_{13} is given by $C_{13} = C_{11} - 2C_{66}$ (Helbig, 1994; Mavko *et al.*, 1998; Bucur, 2006). From the Rayleigh wave technique, Abell *et al.* (2014) determined for isotropic Lucite that $C_{13,Lucite} = 4.06 \pm 0.19$ GPa and for phenolic G10 that $C_{13,G10} = 7.7 \pm 0.7$ GPa which were within 0.03 ± 0.20 GPa and 0.3 ± 0.8 GP, respectively, of the values of C_{13} obtained using the wavefront imaging technique (see Section 2.3.1). An in depth discussion of the uncertainty of the Rayleigh wave technique is given in Abell *et al.* (2014).

While the Rayleigh wave technique was tested on synthetic material, Abell *et al.* (2014) used values of elastic constants for rock from the literature to determine if this method would theoretically apply to rock. Figure 5 contains theoretical curves of the Rayleigh wave velocity for shale, dolomite, and sandstone based on values from the literature (Thomsen, 1986; Martinez & Schmitt, 2013). For many rock types this approach works well. However, when the measured Rayleigh wave velocity is within 1% of the asymptotic value (low values of C_{13} in Figure 5), then an accurate measure of C13 cannot be obtained by this technique. The rock data from the literature that yielded high uncertainty in C_{13} tended to be from stress-sensitive rock under high pressures. Additional research is needed to determine the effect of micro-cracks and fractures on the Rayleigh wave technique because mechanical discontinuities give rise to frequency dispersion resulting in group and phase velocities that are not equal even along a symmetry axis.

Surface wave techniques, such as the Rayleigh wave technique, have several advantages over wavefront imaging approaches (see section 2.3.1) that include (1) measurements times typically under 10 minutes, (2) the theory exists for Rayleigh wave velocities for both transversely isotropic and orthorhombic materials, and (3) only

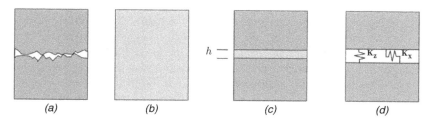

Figure 6 (a) Fracture topology. A fracture can be conceptually represented as (b) an effective medium with a reduced modulus, (c) a three-layer model where the middle layer has reduced moduli or (d) a displacement discontinuity. Light gray represents an effective moduli (b) or low moduli (c) compared to dark gray moduli of matrix.

four measurements are required with no need to calculate the shape of wavefronts or group versus phase angles. Additional research and development is required to adapt the Rayleigh wave technique to measure C_{13} for rocks under confining pressure.

3 CHARACTERIZATION OF FRACTURED ROCK

In this section, the effect of single fractures, fracture sets and fracture intersections on elastic wave propagation are briefly presented to demonstrate the discrete effects of fractures on energy partitioning that enables characterization of fracture properties.

Conceptually, a fracture is composed of two rough surfaces in contact. Voids of variable shape are created between points or regions of contacts (Figure 6a). As stress is applied to a fracture, the contact area between the two surfaces increases and the voids deform. Theoretical models for wave propagation across a fracture capture this complicated fracture topology with effective parameters. The three simplest ways to represent a fracture are shown in Figure 6b–d. First, fractures are known to reduce the elastic moduli of a medium (*e.g.*, Nur, 1971; Hudson, 1981; Crampin, 1984; Zimmerman, 1991; Kemeny & Cook, 1986; Moreland, 1974; White, 1983; Schoenberg & Douma, 1988) and thus the fracture behavior is often incorporated into an effective or reduced Young's and shear moduli (Figure 6b, also see section 3.2.1). The second approach is to represent the fracture with a three layer model (Fehler, 1982; Liu *et al.*, 2000) where a fracture is represented by a medium of thickness, h, with different seismic impedance (*i.e.*, density and moduli) than the surrounding rock matrix (Figure 6c). The third approach is to represent a fracture as a displacement discontinuity, also referred to as a non-welded contact or linear slip interface, (Mindlin, 1960; Kendall & Tabor, 1972; Schoenberg, 1980, 1983; Kitsunezaki, 1983; Pyrak-Nolte & Cook, 1987; Pyrak-Nolte *et al.*, 1990a&b) where the complicated fracture topology is represented by normal and shear fracture specific stiffnesses (Figure 6d). This non-welded contact approach arises from the three layer model by allowing $h \to 0$ (Schoenberg, 1983).

Here, the focus is on the non-welded contact approach or displacement discontinuity theory, for examining the effect of a single fracture on wave propagation across a fracture. The benefits of this approach are (1) that it maintains the discreteness of a fracture such that waves are not delayed nor attenuated until crossing the fracture, (2) that it uses an effective parameter, fracture specific stiffness, to capture the complicated void topology that is linked to the mechanical and hydraulic response of a fracture (Petrovitch *et al.*, 2013, 2014; Pyrak-Nolte & Nolte, 2016), (3) that it is a purely elastic representation but yields frequency dependent group time delays, transmission and reflection coefficients, and (4) that it produces energy partitioning of waves into body waves as well as guided modes that depend on fracture specific stiffness. A disadvantage of this theoretical approach is the assumption that fractures are infinite in extent and, as such, does not yield scattered modes from the edges or fracture tips. However, numerical methods that incorporate displacement discontinuity representations of finite-size fractures (de Basabe *et al.*, 2010) or that model explicitly the physical geometry of the fracture (Petrovitch, 2013; Shao *et al.*, 2015) overcome this difficulty.

3.1 Single fractures

3.1.1 Single fractures in isotropic and anisotropic media

3.1.1.1 DISPLACEMENT DISCONTINUITY THEORETICAL APPROACH

Fractures have regions of contacts between which are voids of variable geometry that can contain gas, liquid, solid, or some combination of the three. For example, cleats in coal beds may be partially open with voids containing methane and water along with partial infilling with minerals such as calcite or pyrite. The mathematical representation of a fracture will vary depending on field or laboratory conditions. Displacement and velocity discontinuity boundary conditions have been used to represent a fracture with different rheological responses, such as (i) a spring, (ii) a dashpot, (iii) a Kelvin material with a spring and dashpot in parallel, or (iv) a Maxwell material with a spring and dashpot in series (Schoenberg, 1980; Pyrak-Nolte *et al.*, 1990a&b; Pyrak-Nolte, 1996; Suarez-Rivera, 1992; Choi, 2013). Nakagawa & Schoenberg (2007) and Nakagawa & Myer (2009) extended this approach to a poroelastic linear slip interface that includes fluid transport between a fracture and the matrix of the host rock.

Here we present the theory for a fracture based on a non-welded contact with the rheological response of a Kelvin material because with appropriate choice of fracture parameters, one can obtain the solution for a fracture represented by only a spring, by only a dashpot, or by a spring and dashpot in parallel. The derivation of the transmission and reflection coefficients for body waves propagated across a fracture starts by representing the fracture as an interface between two elastic half-spaces. The coordinate system used for this derivation is shown in Figure 7. For an incident compressional wave (P-wave) impinging on a fracture represented by a Kelvin interface, the boundary conditions are:

$$\begin{aligned} \kappa_z(u_{z1} - u_{z2}) + \eta(\dot{u}_{z1} - \dot{u}_{z2}) = \tau_{zz}, \qquad \tau_{zz2} = \tau_{zz1}, \\ \kappa_x(u_{x1} - u_{x2}) + \eta(\dot{u}_{x1} - \dot{u}_{x2}) = \tau_{zx}, \qquad \tau_{zx2} = \tau_{zx1}, \end{aligned} \tag{17}$$

where a dot denotes a derivative with respect to time, k_z and k_x are the normal and shear specific stiffness (units of force per volume), η_z and η_x are the normal and shear specific viscosity (units of viscosity per length), and u is the displacement. The reader is referred to Pyrak-Nolte *et al.* (1990a) or Pyrak-Nolte (1996) for derivations related to incident shear waves (for both polarizations) for any angle of incidence.

The velocity discontinuity (Equation 17) provides a dissipative mechanism. The exact meaning of a specific viscosity for a fracture still needs further investigation.

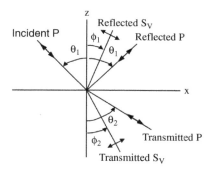

Figure 7 Incident, reflected, and transmitted components of compressional (P) wave incident on a fracture represented by a non-welded contact. (Based on Pyrak-Nolte et al., 1990a).

Suarez-Rivera (1992) investigated the influence of fluid viscosity on the specific viscosity but determined that specific viscosity was related to the adhesion or cohesion of a fluid to a surface. Pyrak-Nolte *et al.* (1990a) found that shear-waves propagated across a dry fracture were better simulated by assuming a Kelvin model for the tangential components of particle displacement.

For a *P*-wave incident on a Kelvin non-welded contact in an *isotropic* medium (Pyrak-Nolte, 1996):

$$\begin{bmatrix} -(\kappa_z - i\omega\eta_z)\cos\theta_1 & (\kappa_z - i\omega\eta_z)\cos\phi_1 & -(\kappa_z - i\omega\eta_z)\cos\theta_2 + i\omega Z_{P2}\cos 2\phi_2 \\ -(\kappa_x - i\omega\eta_x)\sin\theta_1 & -(\kappa_x - i\omega\eta_x)\cos\phi_1 & (\kappa_x - i\omega\eta_x)\sin\theta_2 - i\omega\frac{Z_{S2}^2}{Z_{P2}}\sin 2\theta_1 \\ -Z_{P1}\cos 2\phi_1 & Z_{S1}\sin 2\phi_1 & Z_{P2}\cos 2\phi_2 \\ \frac{Z_{S1}^2}{Z_{P1}}\sin 2\theta_1 & Z_{S1}\cos 2\phi_1 & \frac{Z_{S2}^2}{Z_{P2}}\sin 2\theta_2 \end{bmatrix}$$

$$\begin{bmatrix} (\kappa_z - i\omega\eta_z)\sin\phi_2 - i\omega Z_{S2}\sin 2\phi_2 \\ (\kappa_x - i\omega\eta_x)\cos\phi_2 - i\omega Z_{S2}\cos 2\phi_2 \\ -Z_{S2}\sin 2\phi_2 \\ Z_{S2}\cos 2\phi_2 \end{bmatrix} \begin{bmatrix} R_P \\ R_s \\ T_P \\ T_s \end{bmatrix} = \begin{bmatrix} -\kappa_z\cos\theta_1 \\ \kappa_x\sin\theta_1 \\ Z_{P1}\cos 2\phi_1 \\ \frac{Z_{S1}^2}{Z_{P1}}\sin 2\theta_1 \end{bmatrix} \quad (18)$$

where ω is the angular frequency, θ and φ are the angles for the compressional, *P*, and shear, *S*, waves, respectively. This is the complete solution for all angles of incidences and for half-spaces with difference seismic impedances, (Z = density * phase velocity) for a *P*-wave, incident on and propagated across a Kelvin non-welded contact. The effect of non-welded contacts on transmitted and reflected waves is examined for compressional waves propagated at normal incidence ($\theta_1 = 0°$) to the fracture plane to illustrate the dependence on frequency and fracture specific stiffness.

When the material is the same for the media on either side of a fracture (*i.e.*, $Z_1 = Z_2 = Z$) and the fracture is represented by κ_z and κ_x, the transmission and reflection coefficients for compressional waves reduce to

$$|R(\omega)| = \frac{\frac{\omega}{\omega_c}}{\sqrt{1 + \left(\frac{\omega}{\omega_c}\right)^2}}, \quad \text{and} \quad |T(\omega)| = \frac{1}{\sqrt{1 + \left(\frac{\omega}{\omega_c}\right)^2}}, \quad (19)$$

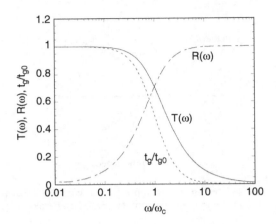

Figure 8 Transmission, T(w), and reflection coefficients, R(w), and normalized group time delay, t_g/t_{go}, as a function of normalized frequency ($\omega/\omega_c = \omega Z/2\kappa$).

where ω_c is the characteristic frequency and is equal to $\omega_c = 2\kappa/Z$. For this purely elastic model, the amount of energy transmitted or reflected from a fracture depends on the frequency of the signal, specific stiffness of the interface and the seismic impedance of the half-spaces (rock matrix). Though the solution is frequency-dependent, the energy is conserved at a displacement discontinuity interface, as demonstrated $|R(w)|^2 + |T(w)|^2 = 1$.

The fracture behaves as a low pass filter with a cutoff frequency of w_c. The low-pass filter behavior of the transmission coefficient is illustrated in Figure 8 where T(w) is shown as a function of the frequency normalized by ω_c. Conversely, the reflection coefficient, R(w), is similar to a high pass filter with a characteristic frequency that differs from that for transmission. As ω/ω_c goes to zero (or $\kappa \to 8$), a fracture behaves as a welded contact and all of the energy is transmitted across the fracture (T(w)→1 and R(w) → 0). Conversely, as $\kappa \to 0$ (or $w/w_c \to \infty$), the fracture behaves as a free surface where all of the energy is reflected from the fracture (T(w)→0 and R(w) → 1). Also shown in Figure 8 is the normalized group time delay. The group time delay for a P-wave propagated at normal incidence to a displacement discontinuity is derived from the change in the phase shift with frequency, $t_{gTP} = d\theta/d\omega = \omega_c/(\omega_c^2 + \omega^2)$. When t_{gTP} is normalized by the group time delay for $\omega = 0$, the dimensionless form of t_{gTP} is

$$\frac{t_{gTP}}{t_{gTP0}} = \frac{1}{1 + \left(\frac{\omega}{\omega_c}\right)^2}. \tag{20}$$

For very high stiffness or very low frequency, $t_g/t_{go} \to 1$ (Figure 8). For mid-range values, as the frequency increases or the stiffness decreases, t_g/t_{go} decrease monotonically. Normalized frequency, ω/ω_c, is a built-in scaling parameter that enables determination of the magnitude of facture specific stiffnesses that can be detected for a given frequency based on the time delay and/or the reflection/transmission coefficients and the matrix properties (Z). The normalized group time delay is a maximum at zero frequency and decreases with increasing frequency. Laboratory studies on wave

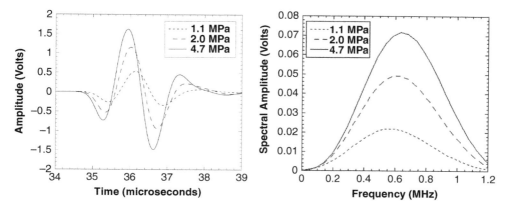

Figure 9 (*Left*) Compressional wave signals propagated at normal incidence to a fracture in gypsum subjected to normal stresses of 1.1, 2.0, and 4.7 MPa. (*Right*) Fourier spectra of the signals shown on the *left*.

propagation across single fractures in the ultrasonic frequency range often use wave attenuation rather than velocities to interpret changes in fracture properties as a function of stress. At ultrasonic frequencies, changes in time delays are relatively small compared to changes in transmission/reflection coefficients. This is illustrated by the data shown in Figure 9 (*on the left*) for compressional waves transmitted through 100 mm of gypsum containing a single fracture that was subjected to normal loading. Ultrasonic contact transducers with a central frequency of 1 MHz were used to send and receive the signal. As the stress increases from 1.1 to 4.7 MPa, the transmitted wave amplitude increases by 160% while the change in arrival time (first break) is on the order of 1%.

The spectral content of a transmitted wave is also affected by the stiffness of the fracture. In Figure 9 (*right*), spectra of the signals, shown in Figure 9 (*left*), were obtained by tapering the signals and applying a Fast Fourier transform (FFT). The transmitted wave's spectral amplitude is observed to increase and the dominant (or most probable) frequency shifts from 0.57 to 0.63 MHz as the stress increases from 1.1 MPa to 4.7 MPa. An increase in stiffness increases ω_c and enables the transmission of high frequency components.

As mentioned, the ability to detect a fracture is a function of ω/ω_c. In Figure 8, the transmission/reflection coefficients and the group time delay provide significantly measureable responses to the existence of a fracture for $0.03 < \omega/\omega_c < 30$. Implicit in this statement is that the stiffness of fractures detected at laboratory frequencies is much higher than that for field frequencies. For example, Figure 10 provides a comparison of the transmission/reflection coefficients and the normalized group time delay for signals at 10 Hz (field) and 1 MHz (laboratory) propagated at normal incidence to a fracture. The time delay produced by the fracture is normalized by the time delay that results from propagating a wave over λ, one wavelength, in the intact portion of the rock with an intact velocity, c. The maximum group time delay from the fracture occurs when $\omega/\omega_c = 1$ (Figure 10 *left*). Below $\omega/\omega_c = 1$, the group time delay caused by the fracture increases while above $\omega/\omega_c = 1$, the group time decreases. For a rock

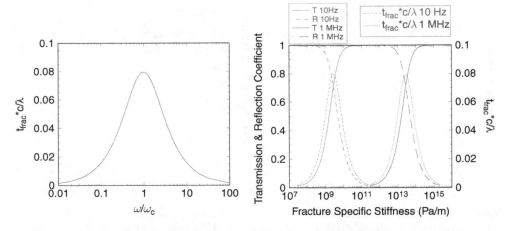

Figure 10 (Left) Normalized group time delay as a function of normalized frequency. (Right) Transmission and reflection coefficients and normalized group time delay as a function of fracture specific stiffness. The time delay is normalized by the ratio of the wavelength, λ, to the group velocity in the intact material.

containing a single fracture, the total time delay is $t_{gTotal} = t_{intact} + t_{fracture} = t_{intact} + t_{gTP}$, where $t_{intact} = path\ length/c = \lambda/c$ is the time for the signal to propagate through the intact portion of a rock, and $t_{fracture} = t_{gTP}$ is the delay caused by the fracture. The total group time delay is

$$t_{gTotal} = \frac{\ell}{c} + \frac{\omega_c}{\omega_c^2 + \omega^2} \qquad (21)$$

As the path length increases (λ where n is the number of wavelengths), the fraction of the time delayed contributed by the fracture decreases. In Figure 10, $n = 1$ and the maximum time delay caused by the fracture is ~8% of the delay measured over 1 in intact rock. Laboratory measurements are often made using signals with a central frequency of 1 MHz on samples of length 10λ to 50λ. For a granitic rock (c_P ~ 5000 m/s), this would yield an intact time delay of λ/c ~ 10.0 to 50.0 microseconds and a maximum ($\omega/\omega_c = 1$) delay from the fracture of $t_{fracture}$ ~ 0.15 to 0.03 microseconds or 0.8% to 0.15% of $n\lambda/c$. However, the transmission and reflection coefficients are both equal to ~ 0.7 at $\omega/\omega_c = 1$. The additional delay from the fracture at laboratory frequencies is often on the resolution of the measurement, yet the changes in amplitude are easily measured. Fracture specific stiffness can be estimated based on changes in the reflected or transmitted amplitude.

An important note is that in the laboratory, measurements are made at a fixed ω while ω_c is altered by increasing the stress on the fractured sample, thereby changing the fracture specific stiffness. In the field, it is more likely that a fracture will be at a fixed stiffness, κ. However, if broadband methods are used, the delay as a function of frequency, or velocity dispersion, can be used to identify fractures and interpret fracture specific stiffness (e.g., see Nolte et al. (2000)). Improved interpretation of fracture specific stiffness on the laboratory and field scales requires using measurements of both time delays and transmission/reflection coefficients to constrain predictions (Pyrak-Nolte *et al.*, 1987). Theoretically, the group time delay increases and then decreases as the fracture specific stiffness increases (Figure 10 *right*). If only a single value of velocity is used instead of the velocity dispersion, then two values of κ are possible. However, if the reflection and transmission coefficients are also measured, the prediction of stiffness can be constrained.

To summarize, a wave incident on a fracture is partitioned into transmitted and reflected body waves (both P and S) that depend on the specific stiffness (κ_z, κ_x, and κ_y) of the fracture, the frequency content of the signal and the seismic impedance of the rock matrix. For the effect of non-uniform fracture stiffness on elastic wave propagation, the reader is referred to Oliger *et al.* (2004) for radial distributions of fracture stiffness, Pyrak-Nolte & Nolte (1992) for uniform, bi-modal and multifractal probability distributions of fracture specific stiffness and to Acosta-Colon *et al.* (2009) for spatial distributions of stiffness. The reader is also referred to Nakagawa *et al.* (2000) for a discussion on shear-induced cross-coupling stiffnesses that give rise to P to S conversions at normal incidence that are sensitive to the state of shear stress on a fracture. In this section, fractures in *isotropic* media were examined. In the next section, the effect of a single fracture in an *anisotropic* matrix is presented.

3.1.1.2 SINGLE FRACTURES IN ANISOTROPIC MEDIA

In the previous section, the displacement discontinuity theory was presented for wave propagation across a single fracture in an *isotropic* medium. When a fracture occurs in an *anisotropic* medium, the seismic impedance depends on the angle of incidence relative to the unique symmetry axis of the medium. To illustrate this effect, we present the solution for a P-wave incidence on a fracture in a transversely isotropic medium with subwavelength layers, *i.e.*, layer thickness is << λ (where λ is the wavelength). Here, we only present the result for a fracture oriented parallel to the layering (parallel to x_1-x_2 plane in Figure 7) or a horizontal fracture in a vertical *transversely isotropic* medium (VTI). The reader is referred to Shao (2015) for a fracture oriented perpendicular to the layering.

For a horizontal fracture in a VTI medium, the displacement for a plane wave propagating in the x_1-x_3 plane (*x-z* plane in Figure 7) is $u = U_o\, exp[iw(t-s_1x_1-s_3x_3)]$ where s_1 and s_3 are components of the slowness vectors in the x_1 and x_3 directions (slowness is the inverse of the velocity, *e.g.*, $s_1 = 1/V_1$), U_o is a complex vector representing the amplitudes, w is the angular frequency of the plane wave, and t is the time. The elastic stiffness tensor is the same as that for a VTI medium (see Equation 3):

$$\sigma_{11} = C_{11}\frac{\partial u_1}{\partial x_1} + C_{13}\frac{\partial u_3}{\partial x_3}, \quad \sigma_{33} = C_{13}\frac{\partial u_1}{\partial x_1} + C_{33}\frac{\partial u_3}{\partial x_3},$$

$$\sigma_{13} = C_{44}\left(\frac{\partial u_3}{\partial x_1} + \frac{\partial u_1}{\partial x_3}\right), \tag{22}$$

where σ_{11} and σ_{33} are normal stresses along the x_1 and x_3 directions (x_1-x_3 plane in Figure 7), s_{13} is the shear stress (Figure 1), u_1 and u_3 are the horizontal and vertical components of the displacement field, and C_{ij} is the elastic stiffness constants. The slownesses, s_1 and s_3, in a VTI medium must satisfy the dispersion relation:

$$\left(C_{11}s_1^2 + C_{44}s_3^2 - \rho\right)\left(C_{33}s_3^2 + C_{44}s_1^2 - \rho\right) - \left(C_{13} + C_{44}\right)^2 s_1^2 s_3^2 = 0 \tag{23}$$

where ρ is the density of the medium. Slowness is the reciprocal of velocity. The solutions for Equation 23 correspond to the quasi P- and quasi S-waves. s_3 can be expressed in terms of s_1 and the elastic components as:

$$s_3 = \pm \frac{1}{\sqrt{2}}\sqrt{K_1 \mp \sqrt{K_1^2 - 4K_2 K_3}} \tag{24}$$

where

$$K_1 = \rho\left(\frac{1}{C_{44}} + \frac{1}{C_{33}}\right) + \frac{1}{C_{44}}\left[\frac{C_{13}}{C_{33}}\left(C_{13} + 2C_{44}\right) - C_{11}\right]s_1^2,$$

$$K_2 = \frac{1}{C_{33}}\left(C_{11}s_1^2 - \rho\right), K_3 = s_1^2 - \frac{\rho}{C_{44}}, \tag{25}$$

where the different combinations of + and – in the expression of s_3 (Equation 24) correspond to P- and S-waves with different propagation directions: (+,–) corresponds to downward propagating P-wave, (+,+) corresponds to downward propagating S-wave, and (–,–) corresponds to upward propagating P-wave, (–,+) corresponds to upward propagating S-wave. The components in the complex amplitude vector U are obtained from the Kelvin-Christoffel equation:

$$\beta = \sqrt{\frac{C_{44}s_1^2 + C_{33}s_3^2 - \rho}{C_{11}s_1^2 + C_{33}s_3^2 + C_{44}\left(s_1^2 + s_3^2\right) - 2\rho}},$$

$$\zeta = \pm\sqrt{\frac{C_{11}s_1^2 + C_{44}s_3^2 - \rho}{C_{11}s_1^2 + C_{33}s_3^2 + C_{44}\left(s_1^2 + s_3^2\right) - 2\rho}}, \tag{26}$$

where the signs are "+" for P-waves and "–" for S-waves. According to the Snell's law, the horizontal slowness, s_1, is continuous, for all reflected and transmitted waves, while the vertical slownesses, s_{P3} and s_{S3}, for the P- and S-waves can be calculated from Equation 24. Assuming U_o to be unity, and applying the displacement discontinuity boundary condition (Equation 17 with $\eta = 0$), the following matrix equation is generated for the reflection and transmission coefficients:

$$\begin{pmatrix} \beta_{P1} - \frac{W_{P1}}{\kappa_1} & \beta_{S1} - \frac{W_{S1}}{\kappa_1} & -\beta_{P2} & -\beta_{S2} \\ \zeta_{P1} - \frac{Z_{P1}}{\kappa_3} & \zeta_{S1} - \frac{Z_{S1}}{\kappa_3} & \zeta_{P2} & \zeta_{S2} \\ Z_{P1} & Z_{S1} & -Z_{P2} & -Z_{S2} \\ W_{P1} & W_{S1} & W_{P2} & W_{S2} \end{pmatrix} \begin{pmatrix} R_{PP} \\ R_{PS} \\ T_{PP} \\ T_{Ps} \end{pmatrix} = \begin{pmatrix} -\beta_{P1} - \frac{W_{P1}}{\kappa_1} \\ \zeta_{P1} + \frac{Z_{P1}}{\kappa_3} \\ -Z_{P1} \\ W_{P1} \end{pmatrix} \tag{27}$$

where κ_z and κ_x are the shear and normal fracture specific stiffnesses, respectively, W and Z represent the shear and normal stress (see Shao, 2015 or Carcione & Picotti, 2012). Note that the absolute values of the displacement coefficients are not equal to the energy partitions of the reflected/transmitted waves. The method to calculate the

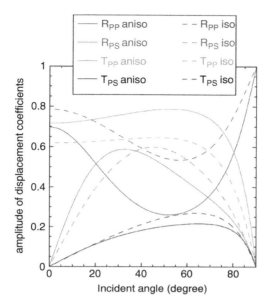

Figure 11 Comparison of the displacement coefficients for waves reflected and transmitted across a single fracture in an isotropic medium (iso) and in an anisotropic medium (aniso) with the fracture plane oriented parallel to the layering (FH).

energy partition for the reflected/transmitted waves, as well as the energy loss for the transmitted wave, is given in Carcione & Picotti (2012). For off-angle transmission, the summation of energy is complicated by the dependence of velocity on the angles of reflection and transmission (refraction).

Figure 11 provides a comparison of the displacement coefficients for the reflected and transmitted waves as a function of angle of incidence on a single fracture in an *isotropic* medium and in an FH medium (*i.e.*, *transversely isotropic* medium with a single fracture parallel to the layering). For this analysis, ρ = 1364 m/s kg/m^3, and C_{11}, C_{33}, C_{13}, and C_{55} are 12, 12, 6.48 and 2.78 GPa, respectively for an isotropic medium; and 12, 7.04, 3.74, and 2.76 GPa for an anisotropic medium; and $\kappa_z = \kappa_x = 10^{13}$ Pa/m. If the anisotropy of a material is not taken into account, interpretation of reflection and transmission coefficients (as well as velocities) would yield incorrect estimates of the orientation of the fracture and fracture specific stiffness.

3.1.2 Wave-guiding along single fractures

By retaining the discreteness of a fracture, additional waves are available for characterizing a single fracture that do not arise from an effective medium approach. Fractures represented as a non-welded interface (Equation 17) give rise to coupled Rayleigh waves that are also known as fracture interface waves (Murty, 1975; Pyrak-Nolte & Cook, 1987; Murty & Kumar, 1991; Pyrak-Nolte *et al.*, 1992; Gu, 1994; Gu *et al.*,

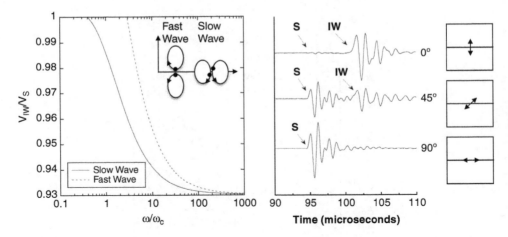

Figure 12 (*Left*) Interface wave group velocity as a function of normalized frequency, w/w_c. (*Right*) Experimental data from Pyrak-Nolte et al. (1992) showing the dependence of the existence of an interface wave on shear wave polarization (arrow for waves propagated into the page.)

1996; Shao & Pyrak-Nolte, 2013). Conceptually, each fracture surface can support a Rayleigh wave. As two fracture surfaces are brought into contact, the waves couple through the points of contact and propagate as coupled Rayleigh waves. As stress on a fracture increases, κ increases because of the increase in contact area between the two fracture surfaces from the deformations of the voids. For increasing κ (decreasing w/w_c in Figure 12 left), the interface wave velocity increases from the Rayleigh wave velocity for weakly coupled fracture surfaces to the bulk shear wave velocity when a fracture becomes completely welded. Two fracture interface waves exist: (1) a symmetric (fast wave) mode that depends only on the normal stiffness of the fracture and (2) an anti-symmetric (slow wave) mode that depends only on the shear stiffness of the fracture. The particle motion is in phase parallel to the fracture plane for the fast wave and perpendicular to the fracture plane for the slow wave (see inset of Figure 12 left). The generation of interface waves requires a component of the shear wave that is perpendicular to the fracture plane (Figure 12 right). The velocities of the symmetric and antisymmetric fracture interface waves are determined from the following secular equations for an *isotropic* medium (Gu et al., 1996)

$$(1 - 2\xi^2)^2 - 4\xi^2\sqrt{\xi^2 - \eta^2}\sqrt{\xi^2 - 1} - 2K_3\sqrt{\xi^2 - \eta^2} = 0 \quad (symmetric),$$
$$(1 - 2\xi^2)^2 - 4\xi^2\sqrt{\xi^2 - \eta^2}\sqrt{\xi^2 - 1} - 2K_1\sqrt{\xi^2 - 1} = 0 \quad (anti\text{-}symmetric).$$
(28)

where $x = V_S/V$, $h = V_S/V_P$, V is the interface wave velocity, $K_1 = \kappa_1/\omega_\rho V_S$ and $K_3 = \kappa_3/\omega_\rho V_S$ are the normalized normal and shear stiffnesses.

The secular equations for the symmetric and antisymmetric interface waves for a *vertical transversely isotropic* medium (VTI) with a *fracture oriented parallel to the layering* (FH) are

$$\left(\frac{\eta_2^2 - \eta_3^2}{2\eta_1^2}\right)\left[(2\xi^2 - 1)\left(2\xi^2 - \frac{2\eta_1^2\eta_2^2}{\eta_2^2 - \eta_3^2}\right) - 4\xi^2\sqrt{\xi^2 - 1}\sqrt{\xi^2 - \eta_1^2}\right]$$

$$-2K_3\sqrt{\xi^2 - \eta_1^2} = 0 \quad (symmetric)$$

$$\left(\frac{\eta_2^2 - \eta_3^2}{2\eta_1^2\eta_2^2}\right)\left[(2\xi^2 - 1)\left(2\xi^2 - \frac{2\eta_1^2\eta_2^2}{\eta_2^2 - \eta_3^2}\right) - 4\xi^2\sqrt{\xi^2 - 1}\sqrt{\xi^2 - \eta_1^2}\right]$$

$$-2K_1\sqrt{\xi^2 - 1} = 0 \quad (anti\text{-}symmetric), \tag{29}$$

where V is the interface wave velocity, $x = V_S/V$, $\eta_1 = V_S/V_P$, $\eta_2 = V^*{}_P/V_P$, $\eta_3 = \zeta/V_P$, $V^*{}_P$ is the compressional wave velocity perpendicular to the layering, and z is the velocity related to the off-diagonal component (see Shao, 2015 or Shao & Pyrak-Nolte, 2014).

For a VTI medium with a *fracture oriented perpendicular to the layering* (FV), the secular equations are very similar to Equation 29 when the interface waves propagate in an isotropic medium:

$$(1 - 2\xi^2)^2 - 4\xi^2\sqrt{\xi^2 - \eta^2}\sqrt{\xi^2 - 1} - 2K_2\sqrt{\xi^2 - \eta^2} = 0 \quad (symmetric),$$

$$(1 - 2\xi^2)^2 - 4\xi^2\sqrt{\xi^2 - \eta^2}\sqrt{\xi^2 - 1} - 2K_1\sqrt{\xi^2 - 1} = 0 \quad (anti\text{-}symmetric). \tag{30}$$

Only the definition of the parameters has changed where $x = V_{SH}/V$, $\eta = V_{SH}/V_P$, V is the interface wave velocity, and where $K_1 = \kappa_1/\omega_\rho V_{SH}$ and $K_2 = \kappa_2/\omega_\rho V_{SH}$ are the normalized normal and shear stiffnesses. If the interface wave is propagated along the fracture but perpendicular to the layering for a FV medium, the secular equations become

$$\left(\frac{\eta_2^2 - \eta_3^2}{2\eta_1^2}\right)\left[(2\xi^2 - 1)\left(2\xi^2 - \frac{2\eta_1^2\eta_2^2}{\eta_2^2 - \eta_3^2}\right) - 4\xi^2\sqrt{\xi^2 - 1}\sqrt{\xi^2 - \eta_1^2}\right]$$

$$-2K_2\sqrt{\xi^2 - \eta_1^2} = 0 \quad (symmetric),$$

$$\left(\frac{\eta_2^2 - \eta_3^2}{2\eta_1^2\eta_2^2}\right)\left[(2\xi^2 - 1)\left(2\xi^2 - \frac{2\eta_1^2\eta_2^2}{\eta_2^2 - \eta_3^2}\right) - 4\xi^2\sqrt{\xi^2 - 1}\sqrt{\xi^2 - \eta_1^2}\right]$$

$$-2K_3\sqrt{\xi^2 - 1} = 0 \quad (anti\text{-}symmetric) \tag{31}$$

where $x = V_{SH}/V$, $\eta_1 = V_S/V^*{}_P$, $\eta_2 = V_P/V^*{}_P$, $\eta_3 = \zeta/V^*{}_P$, and where $K_2 = \kappa_2/\omega_\rho V_{SH}$ and $K_3 = \kappa_3/\omega_\rho V_{SH}$ are the normalized normal and shear stiffnesses. Equations 28–31 enable calculation of the interface wave velocity for a fracture in an isotropic medium and for a fracture oriented either perpendicular or parallel to the layering in a VTI medium. Additional research is needed to examine wave speeds for orientations other than those presented here.

Facture interface waves are a potential method for characterizing both the normal and shear stiffnesses of a fracture. However, it is also important to keep in mind the

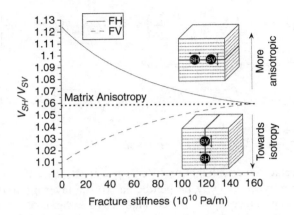

Figure 13 Shear wave group velocity ratio for waves polarized parallel to the layering (V_{SH}) to those polarized perpendicular to the layering (V_{SV}) as a function of fracture specific stiffness for FH and FV, i.e., for transversely isotropic media with fractures parallel (FH) and perpendicular (FV) to the layering.

existence of interface waves when interpreting shear wave anisotropy for a fractured anisotropic medium. Shao and Pyrak-Nolte (2013) demonstrated theoretically and experimentally that a medium can appear to exhibit more or less shear wave anisotropy depending on the orientation of a fracture relative to the layering and the stiffness of the fracture (Figure 13). For a *transversely isotropic* medium, a shear wave polarized parallel to the layering but perpendicular to a fracture propagates as a fracture interface wave for low values of fracture specific stiffness. Shear waves usually travel faster when polarized parallel to the layering than for perpendicular polarizations. As shown in Figure 13 for the FV case, for a low stiffness fracture, the ratio of V_{SH}/V_{SV} is close to the isotropic condition value of 1 because V_{SH} is traveling at the Rayleigh speed as a fracture interface wave. However, as the stiffness of a fracture increases, as the sample is loaded, the apparent anisotropy of the medium increases until the fracture is closed and the matrix anisotropy is recovered. The converse occurs when a fracture is oriented parallel to the layers, i.e., the FH case. For the FH case, the fractured anisotropic medium exhibits more shear wave anisotropy at low stress than the matrix because V_{SV} is traveling at the Rayleigh speed. Once again, the original matrix or background anisotropy is recovered by closing the fracture (high stiffness). This demonstrates the importance of considering the relevant orientation between the symmetry axes of the fractures and the matrix, as well as the existence of fracture interface waves, when the interpretation of material properties is based on measurements of the shear wave velocity.

Another type of guided-mode is the coupled wedge wave that occurs along the intersection of a fracture with a free surface (inset in Figure 14) like at an outcrop. A wedge wave travels along the corner of a single block with speeds that are slower than the Rayleigh wave (Moss et al., 1973; Maradudin et al., 1972; Lagasse et al., 1972). Abell & Pyrak-Nolte (2013) coupled two of these wedge waves (corner of two blocks) using the displacement discontinuity boundary conditions (Equation 17 for η=0). The full derivation of coupled wedge waves in isotropic and anisotropic media is given in

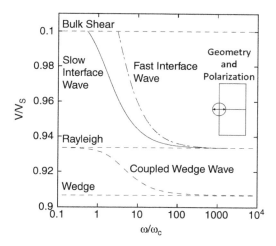

Figure 14 Group velocity dispersion of the coupled wedge wave as a function of w/w_c. Inset: Geometry and shear wave polarization for a coupled wedge wave propagated into the page.

Abell & Pyrak-Nolte (2013) and Abell (2015). Coupling between the corners is related to the stiffness of the fracture. At low stiffness, the coupled wedge wave travels at the wedge wave velocity because the corners of the block are uncoupled and each supports a wedge wave (w/w_c > 1000 Figure 14). Increasing the coupling (fracture stiffness) results in an increase in the coupled wedge wave velocity. When the fracture is fully closed, the coupled wedge wave travels at the Rayleigh wave velocity along the free surface (ω/ω_c < 1 Figure 14). Measurements of coupled-wedge waves provide a method for characterizing the specific stiffness of fractures at the surface of outcrops.

When using surface waves to characterize surface fractures care must be taken to discriminate between Rayleigh waves and coupled wedge waves. Coupled wedge waves are dispersive, while in general, Rayleigh waves are non-dispersive in isotropic and anisotropic rock. However, if the material properties (density and wave speeds) vary with depth in a rock mass then Rayleigh waves can become dispersive. Thus in a field setting, Rayleigh wave measurements should be made along fractures and adjacent to fractures to characterize both the fracture and the matrix.

3.2 Parallel fractures

3.2.1 Parallel fractures in anisotropic media

One approach for treating sets of parallel fractures or fracture networks in isotropic and anisotropic media is to use a compliance tensor (inverse of the stiffness tensor) that represents the excess compliance from the fractures, S_f, that is added to the rock matrix or background compliance, S_b (*e.g.*, Schoenberg & Douma, 1988; Hood & Schoenberg, 1989; Pyrak-Nolte *et al.*, 1990b; Diner, 2013):

$$C^{-1} \equiv S = S_b + S_f, \tag{32}$$

where $S_b = C_b^{-1}$ where C_b contains the elastic constants that describe the rock matrix. The excess compliance from the parallel fractures depends on the normal, b_N, and shear, b_V and b_H, fracture compliances (*i.e.*, inverse of normal and shear fracture specific stiffnesses). This approach assumes that the equivalent anisotropic representation of a fractured medium is equivalent to the static properties of the fractured medium, *i.e.* the stress distribution is nearly uniform with very small spatial variations in stress and the fracture response does not depend on frequency (Nakagawa, 1998).

Here we present an example of this approach for a VTI medium with vertical parallel fractures based on Schoenberg & Helbig (1997) and Bakulin *et al.* (2000). The excess compliance from a set of vertical fractures with the normal to the fractures in the x1 direction is

$$
S_f = \begin{pmatrix} \frac{\Delta_N}{C_{11b}(1-\Delta_N)} & 0 & 0 & 0 & 0 & 0 \\ 0 & 0 & 0 & 0 & 0 & 0 \\ 0 & 0 & 0 & 0 & 0 & 0 \\ 0 & 0 & 0 & 0 & 0 & 0 \\ 0 & 0 & 0 & 0 & \frac{\Delta_V}{C_{44b}(1-\Delta_V)} & 0 \\ 0 & 0 & 0 & 0 & 0 & \frac{\Delta_H}{C_{66b}(1-\Delta_H)} \end{pmatrix} \tag{33}
$$

where the weaknesses Δ_N, Δ_V, and Δ_H are

$$
\Delta_N = \frac{\beta_N C_{11b}}{1 + \beta_N C_{11b}}, \quad \Delta_V = \frac{\beta_V C_{44b}}{1 + \beta_V C_{44b}} \quad \text{and} \quad \Delta_H = \frac{\beta_H C_{66b}}{1 + \beta_H C_{6b}} \tag{34}
$$

The weaknesses range in value from 0 to 1, that is from an unfractured medium to a heavily fractured medium, respectively (Bakulin *et al.*, 2000). Using Equations 32 and 33 along with the elastic constants for a VTI medium given by Equation 3, the effective stiffness matrix for a VTI medium with a set of vertical parallel fractures is

$$
C = \begin{pmatrix} C_{11b}(1-\Delta_N) & C_{12b}(1-\Delta_N) & C_{13b}(1-\Delta_N) & 0 & 0 & 0 \\ C_{12b}(1-\Delta_N) & C_{11b} - \Delta_N \frac{C_{12b}^2}{C_{11b}} & C_{13b}\left(1 - \Delta_N \frac{C_{12b}}{C_{11b}}\right) & 0 & 0 & 0 \\ C_{13b}(1-\Delta_N) & C_{13b}\left(1 - \Delta_N \frac{C_{12b}}{C_{11b}}\right) & C_{33b} - \Delta_N \frac{C_{13b}^2}{C_{11b}} & 0 & 0 & 0 \\ 0 & 0 & 0 & C_{44b} & 0 & 0 \\ 0 & 0 & 0 & 0 & C_{44b}(1-\Delta_V) & 0 \\ 0 & 0 & 0 & 0 & 0 & C_{66b}(1-\Delta_H) \end{pmatrix}
$$
$$\tag{35}$$

with ($C_{12b} = C_{11b} - 2C_{66b}$). Schoenberg & Douma (1988) showed that Equation 34 describes a special type of orthorhombic media with the stiffnesses also constrained by $C_{13}(C_{22}+C_{12}) = C_{23}(C_{11}+C_{12})$ and characterized by only eight independent parameters: five stiffness coefficients (C_{11b}, C_{13b}, C_{33b}, C_{44b}, C_{66b}) of the VTI medium and three fracture weaknesses (Δ_N, Δ_V, Δ_H). These effective medium approaches enable calculation of compressional and shear wave velocities for long wavelength approximations but the discreteness of the fractures is lost. It is important to note that this approach can break down in the long wavelength approximation when fractures exhibit frequency dependent behavior (*e.g.*, see section 3.1.1), which is not captured by effective medium

Elastic waves in fractured isotropic and anisotropic media   349

static approximations. In the next section, the discreteness of the fractures is maintained to demonstrate the existence of guided-modes that do not exist in effective medium approximations but enable characterization of fracture specific stiffness.

3.2.2 Wave-guiding between parallel fractures in isotropic and anisotropic media

Traditionally, wave guiding is viewed as a result of impedance contrasts between rock units such as that observed in laboratory studies on Austin Chalk (Li *et al.*, 2009), in field-scale cross-hole studies in oil bearing sand-shale reservoirs (Leary *et al.*, 2005), in coal seams (Buchanan *et al.*, 1983), and sandstone – shale formations (Parra *et al.*, 2002). However, theoretical and laboratory studies have shown that parallel fractures in isotropic media produce wave guiding (Nihei *et al.*, 1999; Xian *et al.*, 2001) because energy partitioning by the fractures confines energy to a central wave guide. Shao *et al.* (2015) also demonstrated this behavior for fractured anisotropic media with matrix layering $<< \lambda$ but with a fracture spacing comparable to λ (Figure 15) using a wavefront imaging method (section 2.3.1). The amount of energy confined by a waveguide is a function of fracture specific stiffness. When fracture specific stiffness is low (Figure 15 top), very little energy is transmitted across a fracture. The reflected waves remained confined to the central layer but lose energy upon every reflection. As stress is increased on the sample, the transmission across the fractures increases (Figure 15 middle) because fracture specific stiffness increases (see Figure 8). The variation in stiffness among the fractures within a set is observed from the differences in the amplitude and arrival times of the guided-modes (vertical axis y-positions 12-22 and positions 32-42 in Figure 15 middle) on either side of the central wave-guide (vertical axis y-positions 22-32). At very high stress (Figure 15 bottom), the fractures are sufficiently closed to enable almost full transmission across the fracture planes which results in a wavefront that is dominated by the matrix anisotropy. Energy confinement by parallel fractures generates leaky modes because some energy is lost upon reflection/transmission at the fracture. However, laboratory studies have measured guided modes over distances of 25 wavelengths for frequencies between 0.5 and 1 MHz which would be equivalent to 300–3000 m for field frequencies of 50–350 Hz (Shao *et al.*, 2015).

Guided modes can be used to characterize fracture specific stiffness from the time delay of guided-modes relative to the direct modes through unfractured material. A short summary of the theory for guided modes is presented here to highlight the significant factors that affect wave guiding between fractures. Wave guiding between parallel fractures occurs when the direct wave interferes constructively with a twice-reflected wave. For example, in Figure 16, a monochromatic plane wave is reflected between the upper and lower fracture planes. The phase change between the original and twice reflected wave must equal an integer multiple of 2ϕ (Saleh & Teich, 2007) such that

$$\frac{2\pi(AC - AB)}{\lambda} + 2\phi_{R_{PP}} = 2m\pi, \tag{36}$$

where m is a non-negative integer that corresponds to different guided modes (1st guided mode $m = 0$, 2nd guided mode $m = 1$, etc.) and λ is the wavelength. AC-AB is the

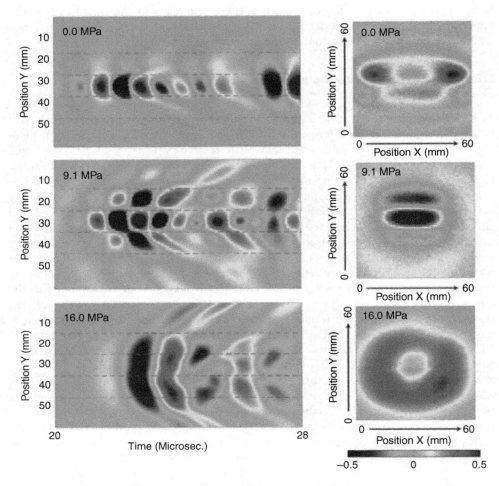

Figure 15 (*Left column*) Arriving wavefront as a function of time and position. (*Right column*) A snap shot of the wave front in time showing the spatial distribution of energy for increasing stress on a fractured anisotropic sample (top – middle – bottom) with FH orientation (see Figure 13).

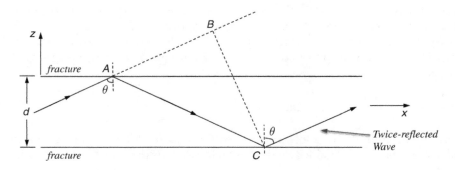

Figure 16 Wave-guiding geometry between two fractures.

Elastic waves in fractured isotropic and anisotropic media 351

difference in the travel path between the direct wave (A to B) and the twice reflected wave (A to C). ϕ_{Rpp} is the phase shift upon reflection from the fracture, which is a function of frequency, fracture specific stiffness, and the impedance of the matrix. For an anisotropic medium, the value of the impedance is directionally dependent. The relative time delay, ΔT, of a guided mode has two components (1) a geometric component from traveling a longer distance than the direct wave, and (2) a dynamic component that occurs because of the phase shift upon reflection from the fractures. For compressional wave guiding in an anisotropic medium, the total relative delay is

$$\Delta T = t_{geo} + t_{dyn} = \frac{L}{V_P(\theta)} \left(\frac{1}{\sin \theta} - 1 \right) + N \frac{d\phi_{R_{PP}}}{d\omega}, \tag{37}$$

where N is the number of reflections within a sample of length L, and $V_P(\theta)$ is the directionally dependent compressional wave velocity in anisotropic media. For an isotropic medium, V_P is independent of direction ($V_P(\theta) = V_P$). The number of reflections for a given sample length determines the magnitude of dynamic contribution to the ΔT. Both group time delay components, as well as the total time delay, are a function of frequency, because the existence of guided modes (Equation 36) depends on the wavelength of the signal λ and the fracture spacing d. However, an additional frequency dependence arises from the fracture that does not occur for waves guided only by layering. A fracture that is represented as a non-welded contact produces a group time delay for both transmitted and reflected waves (Equation 36). For an unfractured layered medium, the phase shift upon reflection is frequency independent, when the layer thickness is much smaller than a wavelength (the effective medium approximation). For low fracture stiffness, a rapid change in the phase shift occurs as a function of the signal frequency, which increases the magnitude of dynamic delay of a single reflection. Similar to the isotropic case of wave guiding by Xian et al. (2001), higher-order guided waves experience larger time delays than lower-order ones, because the number of reflections from the fractures increases as m increases. The number of reflection increases because the reflection angle approaches normal incident (as $\theta \rightarrow 0$) as the frequency decreases.

Estimates of fracture specific stiffness can be obtained from analysis of the arrival times of the guided modes (Figure 17). An example is presented from Shao (2015) and Shao et al. (2015) for measured guided modes in Garolite samples (cloth-epoxy laminate with sub-wavelength layering) that contained a set of parallel fractures. The FH sample contained fractures parallel to the layering, while the FV sample contained a fracture perpendicular to the layering (Figures 13). The analysis was performed by subtracting the central waveform (when the source and receiver were aligned) obtained from the fractured samples from that obtained from the intact reference sample (Figure 17 left & center). Before subtraction, the central waveforms from each fractured sample and the intact sample were processed using a wavelet transformation (Nolte et al., 2000) to achieve a single spectral component (e.g., 0.6 MHz in this analysis). Figure 17 (left) shows the first guided modes in the FV sample for a single frequency of 0.6 MHz for several different normal stresses. Shao (2015) observed that when the stress increased, the first guided mode arrived earlier. This observation is in an agreement with the theoretical calculation based on the theory described above, namely, an increase in fracture stiffness results in a decrease in the guided-mode time

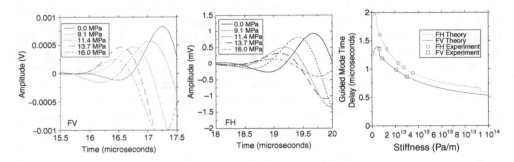

Figure 17 Central signals as a function of stress for sample (*left*), FV (*center*) and FH (*right*). Estimated fracture specific stiffness based on wave-guiding theory. For sample descriptions see Figure 13.

delay (*i.e.*, an earlier arrival time). A similar trend was observed for the FH sample (Figure 17 (center)). Fracture specific stiffness was estimated (Figure 17 (right)) by comparing the measured guided-mode time delays and theoretical predictions based on the material properties of the matrix. As fracture stiffness increases, the guided-mode time delay decreases because the dynamic and geometric time delays decrease. The amplitude of the first guided mode also decreases for both the FV and FH samples because of weaker energy confinement between the fractures.

Shao *et al.* (2015) also examined wave guiding in fractured anisotropic media composed of isotropic layers with thicknesses > λ and fracture spacing greater > λ. In such a medium, two potential wave-guiding mechanisms exist and compete: (1) wave-guiding in low velocity layers sandwiched between high velocity layers; and (2) wave-guiding between fractures. From numerical simulation, Shao *et al.* (2015) showed that wave guiding produced by impedance contrast among the layers could be either suppressed or enhanced by the presence of fractures and depended on the stiffness of the fractures.

In general, the key ideas for wave-guiding in fractured anisotropic media is that guided-mode behavior depends on contributions from the mechanical properties of the fractures and the matrix, the layer and fracture spacing, and the signal frequency or wavelength. The time delay of guided modes provides a measure of fracture specific stiffness for the fractures within a set.

3.3 Orthogonal intersecting fractures

Orthogonal fracture networks also occur in rock (for example see Bai *et al.*, 2002 or Gross, 1993). Effective medium approaches have been developed to obtain effective stiffness matrices for two orthogonal fracture networks in an isotropic medium (Bakulin *et al.*, 2000; Fuck & Tsvankin, 2006). Similar to the approach used for a set of parallel fractures (see section 3.2.1), an effective stiffness matrix can be determined theoretically that represents the combine contributions from the compliance of the fracture sets and the compliance of an isotropic (or anisotropic) rock matrix. In this approach, each fracture set can have a different fracture specific

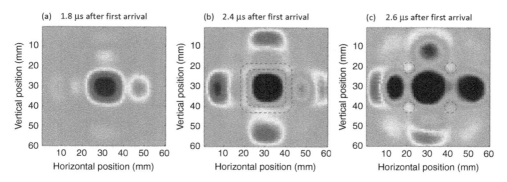

Figure 18 Snapshots in time of the arriving wave front in an aluminum sample with two orthogonal sets of fractures. (a) 1.8 µs, (b) 2.4 µs, and (c) 2.6 µs after the first arrival.

stiffness but the fractures within the set are assume to have uniform stiffness. In addition, the stiffness or compliance of fracture intersections are not taken into account and are often assumed to have negligible effects on propagating waves (*e.g.* Grechka & Kachanov, 2006).

Shao (2015) and Pyrak-Nolte & Shao (2016) used a wavefront imaging technique (Figures 2b and 3 in section 2.3.1) to visualize and quantify the effects of two orthogonal fracture sets on a propagating wave front (Figure 18). In his experiments, the fracture spacing, d, was 20 mm for each fracture set or approximately 3.1λ for a frequency of 1 MHz. Samples were subjected to bi-axial loading with the loads applied perpendicular to each fracture set. From the data (Figure 18), it is observed that (1) the stiffness of the fractures within a set and between fracture sets varied as indicated by the non-symmetric distribution of energy in the arriving wavefront (Figure 18a); (2) a compressional guided-mode appears along the fracture planes (Figure 18b); and (3) fracture intersections delay and attenuate the wave front, and appear to support a guided-mode (Figure 18c). These observations raise some challenges for remote characterization of orthogonal fracture networks. Current effective medium approaches require uniform fracture stiffness within a fracture set. Given the state of stress in the subsurface (Zoback & Zoback, 1980; Zoback, 1992) or in surface mines and outcrops, it is unlikely that fracture specific stiffness will be uniform within a fracture set, let alone between sets. Another challenge is the incorporation of the effect of fracture intersections on the interpretation of propagating elastic waves. Though intersections are long essentially 1D structural features in a 3D system of mechanical discontinuities, they significantly affect a propagating wavefront. No model or theory for an orthorhombic medium (or for any other form of anisotropy) currently accounts for fracture intersections. As shown in sections 3.1.2 and 3.2.2, single fractures and a set of parallel fractures support guided modes. Figure 18c shows the appearance of a guided-mode along each of the intersections that arrives shortly after the direct compressional wave, but the existence of compressional-mode intersection waves has not yet been studied. In the next section, we introduce recently discovered guided-modes that travel along intersections with speeds that range between the wedge wave and shear wave speeds.

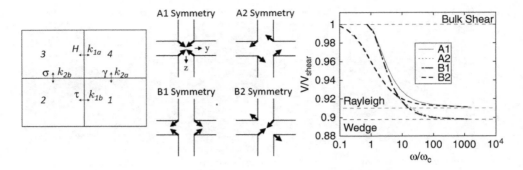

Figure 19 (*left*) Sketch of an intersection between two orthogonal fractures defined by fracture specific stiffnesses k_1 and k_2. (*center*) Displacements along an intersection for four intersection modes (after Abell, 2015). (*right*) Intersection wave velocity normalized by the bulk shear wave velocity as a function of w/w_c for A_1, A_2, B_1 and B_2 modes for $\kappa_1 = \kappa_2$.

3.3.1 Orthogonal fracture networks: Intersection waves

Recently, Abell (2015) derived the theoretical description of elastic *"intersection"* waves that propagate along a fracture intersection between two orthogonal fractures. The displacement field along a fracture intersection is modeled as a coupling of four wedges in partial contact (Figure 19). Each wedge (corner) supports a propagating wedge wave (Moss *et al.*, 1973; Lagasse *et al.*, 1972; Maradudin *et al.*, 1972). The wedge wave is modeled by expanding the displacement in a linear series of Laguerre polynomial functions to characterize the wedge geometry of each block. Displacement discontinuity boundary conditions were used to couple the four wedges across each fracture plane. Abell's solution is sufficiently general that the stiffness along each fracture plane can differ as well as the elastic constants for each block. Here we present the four vibrational modes that exist along an intersection for the case of equal fracture specific stiffness on the two fracture planes and the same material properties in each medium. The reader is referred to Abell (2015) for the case of unequal fracture specific stiffness.

For an intersection composed from two orthogonal fractures with equal fracture specific stiffness, four singlet vibrational modes exist and propagate with the particle motions illustrated in Figure 19 (center). The displacements are similar to that of a wedge wave (Moss *et al.*, 1973). Wedge waves can support two vibrational modes: breathing and wagging, depending on the material properties and wedge angle (Zavorokhin & Nazarov, 2011). In the breathing mode, both free surfaces of a corner expand and contract, similar to a bellows (modes A1 & B2 in Figure 19 center). For the wagging mode, both free surfaces twist/wag in the same direction like a dog's tail (modes A2 & B1 in Figure 19 center). For a further discussion of the difference between wagging and breathing modes, see Moss *et al.* (1973). For an intersection, the predicted motion is highly localized to the corners of each media (*i.e.*, within a few wavelengths) and oscillates in one of these two wedge modes that propagate along an intersection when $\kappa_1 = \kappa_2$. When $\kappa_1 \neq \kappa_2$, Abell (2015) showed that all four modes exhibit either breathing or wagging motion.

When $\omega/\omega_c > 1000$, the elastic waves either propagate at the Rayleigh wave or wedge wave velocity. The velocity of the four modes for media with positive Poisson's ratios, is shown in Figure 19 (right), note that modes A2 and B1 start at the wedge wave velocity. A1 and B2 modes start at the Rayleigh wave velocity (Figure 19). As the fracture plane closes and the intersection becomes welded (*i.e.*, $\omega/\omega_c < 1$), the four media weld and the intersection supports a bulk wave traveling at the bulk shear wave velocity for all four modes (Figure 19). Between these extremes, the four inter-section modes increase monotonically to the shear wave velocity. The sensitivity of intersection waves to the condition of the intersection (*i.e.*, opened, partially opened or closed) provides a potential characterization tool for assessing the connectivity of fracture networks.

4 CONCLUSIONS

As noted in the Introduction to this chapter, rock is rarely isotropic, and fractures lend an additional stress-dependent anisotropy to the rock. Mechanical properties of frac-tured rock are typically determined from measurements of elastic wave velocities. Whether the data are interpreted based on effective medium approaches or approaches that retain the effects of discrete fractures, the measured response is controlled by the length scales in the fractured rock system relative to the wavelength of the probing signal and by the specific stiffness of the fractures. Effective medium approaches for fractured rock are appropriate when (1) the wavelength of the probe signal is much larger than the scale of the source of anisotropy in the matrix (*i.e.*, greater than the layer thickness) and much larger than the fracture spacing, (2) stress is uniform or nearly uniform such that the fractures in a set have the same stiffness, and (3) the seismic response of the fractures is frequency-independent because fracture specific stiffness is high. Effective medium approaches breakdown when (1) fracture specific stiffness is low which results in frequency dependent behavior even though the fracture spacing is much smaller than a wavelength; (2) non-uniform stress distributions that cause gradients in fracture specific stiffness or variation in fracture specific stiffness among fractures in a set; and/or (3) the signal wavelength is smaller than, or comparable to, the fracture spacing, causing a wave to be multiply scattered, which affects time delays and attenuation. These breakdowns occur because effective medium approaches are essentially static approx-imations that do not account for the frequency-dependent effects that arise from single and multiple fractures, the variation in fracture stiffness that arises from the geometry of the fractures and state of stress, nor the existence of scattered-modes.

Discrete approaches take advantage of mode-conversions and guided-modes that occur at and along fractures to constrain the interpretation of matrix properties and stiffness of fractures. As shown in this chapter, all of these fracture-generated modes are frequency dependent where the characteristic frequency depends on fracture specific stiffness, on the impedance of the matrix, and on the frequency of the probe wave. The frequency-dependent mode conversions are affected by the matrix anisotropy because of the directionally dependent seismic impedance.

In the laboratory and field, overlapping behavior can occur from competing length scales (*e.g.*, layer thickness versus fracture spacing) and a transition from discrete behavior to effective medium behavior can occur from changes in stress. Stress can drive a fractured rock system into or out of different scattering and analysis regimes as fractures are opened or closed. Transitional and overlapping behaviors are strongly associated with length scales in the system (*e.g.*, fracture spacing, asperity spacing, stiffness distributions) and can be identified through spectral analysis. Many elastic wave measurements in the laboratory and field are made using broadband sources that produce signals with a range of frequencies. Broadband signals capture the transitions between different scattering regimes by using the data to measure velocity dispersions that arise from fracture-generated modes and that help define length scales associated with a fracture or fractured systems (Nolte *et al.*, 2000).

Elastic waves are an essential tool for characterizing rock on the laboratory and field scale. Given the complexity of rock and field sites, understanding both effective medium and discrete approaches is required to achieve the best interpretation of the properties of isotropic or anisotropic fractured rock.

ACKNOWLEDGMENTS

This material is based upon work supported by the U.S. Department of Energy, Office of Science, Office of Basic Energy Sciences, Geosciences Research Program under Award Number (DE-FG02-09ER16022) and by the Geo-mathematical Imaging Group at Purdue University.

REFERENCES

Abell, B.C. (2015) *Elastic Waves along a Fracture Intersection*. Ph.D. Thesis. Purdue University, West Lafayette, IN. 329 pages.

Abell, B.C. & Pyrak-Nolte, L.J. (2013) Coupled Wedge Waves. *Journal of the Acoustical Society of America*, 3551–3560, doi: http://dx.doi.org/10.1121/1.4821987.

Abell, B.C., Shao, S. & Pyrak-Nolte, L.J. (2014) Measurements of elastic constants in anisotropic media. *Geophysics*, 79(5), 1–14.

Acosta-Colon, A., Nolte, D.D. & Pyrak-Nolte, L.J. (2009) Laboratory-scale study of field of view and the seismic interpretation of fracture specific stiffness. *Geophysical Prospecting*, 57, 209–224.

Amadei, B. (1996) Importance of anisotropy when estimating and measuring in situ stresses in rock. *International Journal of Rock Mechanics, Mining Sciences and Geomechanical Abstracts*, 33(3), 293–325.

Amadei, B. & Goodman, R.E. (1982) The influence of rock anisotropy on stress measurements by overcoring techniques. *Rock Mechanics*, 15, 167–180.

Bai, T., Maerten, L., Gross, M.R. & Aydin, A. (2002) Orthogonal cross joints: Do they imply a regional stress rotation? *Journal of Structural Geology*, 24, 77–88.

Bakulin, A., Grecjka, V. & Tsvankin, I. (2000) Estimation of fracture parameters from reflection seismic data – P II: Fractured models with orthorhombic symmetry. *Geophysics*, 65(6), 1803–1817.

Barton, N. & Quadros, E. (2015) Anisotropy is everywhere, to see, to measure, and to model. *Rock Mechanics and Rock Engineering*, 48(4), 1323–1339, doi: 10.1007/s00603-014-0632-7.

Berryman, J. (1979) Long-wave elastic anisotropy in transversely isotropic media. *Geophysics*. 44(5), 869–917.

Brittan, J. & Warner, M. (1996) Seismic velocity, heterogeneity, and the composition of the lower crust. *Tectonophysics*, 264, 249–259.

Buchanan, D.J., Jackson, P.J. & Davis, R. (1983) Attenuation and anisotropy of channel waves in coal seams. *Geophysics*, 48(2), 133–147.

Bucur, V. (2006) *Acoustics of Wood*, 119. Springer, Berlin-Heidelberg.

Bucur, V. & Rocaboy, F. (1988) Surface wave propagation in wood: Prospective method for the determination of wood off-diagonal terms of stiffness matrix. *Ultrasonics*, 26, 344–348.

Carcione, J.M. (1996) Elastodynamics of a non-ideal interface: Application to crack and fracture scattering. *Journal of Geophysical Research: Solid Earth*, 101(B12), 28177–28188.

Caricione, J.M. & Picotti, S. (2012) Reflection and transmission coefficients of a fracture in transversely isotropic media. *Studia Geophysica et Geodaetica*, 56(2), 307–322.

Choi, M-K. (2013) *Characterization of Fracture Specific Stiffness under Normal and Shear Stress*, Ph.D. Thesis. Purdue University, West Lafayette, IN, 338 pages.

Christensen, N. & Ramananantoandro, R. (1971) Elastic moduli and anisotropy of Dunite to 10 kilobars. *Journal of Geophysical Research*, 76, 4003–4010, doi: 10.1029/JB076i017p04003.

Crampin, S. (1984) Effective anisotropic elastic constants for wave propagation through cracked solids. *Geophysical Journal of the Royal Astronomical Society*, 76, 133–145.

Dahmen, S., Ketata, H., Ghozlen, M. & Hosten, B. (2010) Elastic constants measurement of anisotropic Olivier wood plates using air-coupled transducers generated Lamb wave and ultrasonic bulk wave. *Ultrasonics*, 50, 502–507.

De Basabe, J.D., Sen, M.K. & Wheeler, M.F. (2010) Seismic wave propagation in fractured media. *SEG Technical Program Expanded Abstracts 2011*, SEG San Antonio 2011 Meeting, 2920–2924.

Deresiewicz, H. & Mindlin, R. (1957) Waves on the surface of a crystal. *Journal of Applied Physics*, 28(669), 669–671.

Diner, C. (2013) Decomposition of a compliance tensor for fractures and transversely isotropic medium. *Geophysical Prospecting*, 61, 409

Djikpesse, H.A. (2015) C_{13} and Thomsen anisotropic parameter distributions for hydraulic fracture monitoring. *Interpretation*, Special section: Recent advances with well whisperers. 3(3), SW1–SW10.

Every, A.G. (1994) Determination of the elastic constants of anisotropic solids. *NDT & E International*, 27(1), 3–10, doi: 0963-8695/94/01/0003-08.

Fehler, M. (1982) Interaction of seismic waves with a viscous liquid layer. *Bulletin of the Seismological Society of America*, 72(1), 55–72.

Fuck, R.F., & Tsvankin, I. (2006) Seismic signatures of two orthogonal sets of vertical microcorrugated fractures. *Journal of Seismic Exploration*, 15, 183–208.

Grechka, V. & Kachanov, M. (2006) Effective elasticity of fractured rocks: A snapshot of the work in progress. *Geophysics*, 71(6), W45–W58.

Gross, M.R. (1993) The origin and spacing of cross joints: examples from the Monterey formation, Santa Barbara coastline. *California, Journal of Structural Geology*, 15(6), 737–751

Gu, B. (1994) *Interface Waves on a Fracture in Rock*. Ph.D. Thesis, University of California, Berkeley.

Gu, B., Nihei, K., Myer, L.R. & Pyrak-Nolte, L.J. (1996) Fracture interface waves. *Journal of Geophysical Research*, 101(1), 827.

Hauser, M.R., Weaver, R.L. & Wolfe, J.P. (1992) Internal diffraction of ultrasound in crystals: phonon focusing at long wavelengths. *Physical Review Letters*, 68, 2605–2607.

Helbig, K. (1994) Foundations of anisotropy for exploration seismics, in Handbook of Geophysical exploration, 22, Elsevier Science, Inc., Great Britain.

Hood, J.A. & Schoenberg, M. (1989) Estimation of vertical fracturing from measured elastic moduli. *Journal of Geophysical Research*, 94, 15,611–15,618.

Hornby, B.E., Schwartz, L.M. & Hudson, J.A. (1994) Anisotropic effective medium modeling of the elastic properties of shales. *Geophysics*, 59, 1570–1583, doi: 10.1190/1.1443546.

Hudson, J.A. (1981) Wave speeds and attenuation of elastic waves in material containing cracks. *Geophysical Journal of the Royal Astronomical Society*, 64, 133–150.

Johnston, J.E. & Christensen, N.I. (1995) Seismic anisotropy of shales. *Journal of Geophysical Research*, 100, 5991–6003, doi: 10.1029/95JB00031.

Kemeny, J. & Cook, N.G.W. (1986) Effective moduli, non-linear deformation and strength of a cracked elastic solid. *International Journal of Rock Mechanics, Mining Sciences & Geomechanic Abstracts*, 23(2), 107–118.

Kendall, K. & Tabor, D. (1971) An ultrasonic study of the area of contact between stationary and sliding surfaces. *Proceedings of the Royal Society of London. Series A, Mathematical and Phys-ical Sciences*, 323, 321–340.

Kerner, K., Dyer, B. & Worthington, M. (1989) Wave propagation in a vertical transversely isotropic medium: Field experiment and model study. *Geophysical Journal*, 97, 295–309.

Kitsunezaki, C. (1983) Behavior of plane waves across a plane crack. *Journal of the mining college, akita university, series A: Mining Geology*, 3(6), 173–187.

Lagasse, P.C. (1972) Analysis of a dispersion free guide for elastic waves. *Electronics Letters*, 8, 372–373.

Leary, P.C., Ayres, W., Yang, W.J. & Chang, X.F. (2005) Crosswell seismic waveguide phenomenology of reservoir sands and shales at offsets >600 m, Liaohe oil field, NE China. *Geophysical Journal International*. 163, 285–307.

Li, J., Li, C., Morton, S.A., Dohmen, T., Katahara, K. & Toksoz, M.N. (2014) Micro-seismic joint location and anisotropic velocity inversion for hydraulic fracturing in a tight Bakken reservoir. *Geophysics*, 79(5), C111–C122.

Li, W., Petrovitch, C. & Pyrak-Nolte, L.J. (2009) The effect of fabric-controlled layering on compressional and shear wave propagation in carbonate rock. *International Journal of the JCRM*, 4(2), 79–85.

Liu, E., Hudson, J.A. & Popinter, T. (2000) Equivalent medium representation of fractured rock. *Journal of Geophysical Research*, 105(B2), 2981–3000.

Maradudin, A.A., Wallis, R.F., Mills, D.L. & Ballard, R.L. (1972) Vibrational edge modes in finite crystals. *Physical Review B*, 6(4), 1106–1111.

Martinez, J. & Schmitt, D. (2013) Anisotropic elastic moduli of carbonates and evaporites from the Weyburn-Midale reservoir and seal rocks. *Geophysical Prospecting*, 61, 363–379.

Mavko, G., Mukerji, T. & Dvorkin, J. (1998) *The rock physics handbook: Tools for seismic analysis in porous media*. Cambridge University Press.

Mindlin, R.D. (1960) Waves and vibrations in isotropic planes. *Structural Mechanics*. eds. J.W. Goodier & W.J. Hoff, Pergamon.

Moreland, L.W. (1974) Elastic response of regularly jointed media. *Geophysical Journal of the Royal Astronomical Society*, 37, 435–446.

Moss, S.L., Maradudin, A.A. & Cunningham, S.L. (1973) Vibrational edge modes for wedges with arbitrary interior angles. *Physical Review B*, 8(6), 2999–3008, doi: 10.1103/PhysRevB.8.2999

Mullenbach, B.L. (1996) *Acoustic imaging of sediments*. Master thesis. University of Notre Dame, South Bend, Indiana.

Murty, G.S. (1975) A theoretical model for the attenuation and dispersion of Stoneley waves at the loosely bonded interface of elastic half spaces. *Physics of the Earth and Planetary Interiors*, 11: 65–79.

Murty, G.S. & Kumar, V. (1991) Elastic wave propagation with kinematic discontinuity along a non-ideal interface between two isotropic elastic half-spaces. *Journal of Nondestructive Evaluation*, 10(2), 39–53.

Nagy, P.B., Bonner, B.P. & Adler, L. (1995) Slow wave imaging of permeable rocks. *Geophysical Research Letters*, 22(9), 1053–1056.

Nakagawa, S. (1998) Acoustic Resonance Characteristics of Rock and Concrete Containing Fractures. Ph.D. Thesis, University of California, Berkeley, 412 pages.

Nakagawa, S., Nihei, K.T. & Myer, L.R. (2000) Shear-induced conversion of seismic waves across single fractures, *International Journal of Rock Mechanics and Mining Sciences*, 37, 203–218.

Nakagawa, S. & Schoenberg, M.A. (2007) Poroelastic modeling of seismic boundary conditions across a fracture. *Journal of the Acoustical Society of America*, 122(2), 831–847.

Nakagawa, S. & Myer, L.R. (2009) Fracture permeability and seismic wave scattering – Poroelastic linear-slip interface model for heterogeneous fractures. *SEG Technical Program Expanded Abstracts* 2009, 3461–3465, doi: 10.1190/1.3255581.

Nihei, K.T., Yi, W., Myer, L.R. & Cook, N.G.W. (1999) Fracture channel waves. *Journal of Geophysical Research: Solid Earth*, 10(B3), 4749–4781.

Nolte, D.D., Pyrak-Nolte, L.J., Beachy, J. & Ziegler, C. (2000) Transition from the displacement discontinuity limit to the resonant scattering regime for fracture interface waves. *International Journal of Rock Mechanics and Mining Sciences*, 37, 219–230.

Nur, A. (1971) Effects of stress on velocity anisotropy in rocks with cracks. *Journal of Geophysical Research*, 76(8), 2022–2034.

Nur, A. & Simmons, G. (1969) Stress-induced velocity anisotropy in rock: An experimental study. *Journal of Geophysical Research*, 74(27), 6667–6674.

Oliger, A., Nolte, D.D. & Pyrak-Nolte, L.J. (2003) Seismic focusing by a single planar fracture. *Geophysical Research Letters*, 30(5), 1203, doi:10.1029/2002GL016264.

Parra, J.O., Hacker, C.L., Gorody, A.W. & Korneev, V. (2002) Detection of guided waves between gas wells for reservoir characterization. *Geophysics*, 67(1), 38–49.

Petrovitch, C.L. (2013) Universal Scaling of the Flow-Stiffness Relationship in Weakly Correlated Single Fractures. Ph.D. Purdue University, West Lafayette, IN.

Petrovitch, C.L., Pyrak-Nolte, L.J. & Nolte, D.D. (2013) Scaling of fluid flow versus fracture specific stiffness. *Geophysical Research Letters*, 40, 1–5, doi:10.1002/grl.50479.

Petrovitch, C.L., Pyrak-Nolte, L.J. & Nolte, D.D. (2014) Combined scaling of fluid flow and seismic stiffness in single fractures. *Rock Mechanics and Rock Engineering*, 47(5), 1613–1623, doi: 10.1007/s00603-014-0591-z.

Pyrak-Nolte, L.J. (1996) The seismic response of fractures and the interrelations among fracture properties. *International Journal of Rock Mechanics and Mining Sciences and Geomechanics Abstracts*, 33(8), 787–802.

Pyrak-Nolte, L.J., & Cook, N.G.W. (1987) Elastic interface waves along a fracture. *Geophysical Research Letters*, 11(14), 1107–1110.

Pyrak-Nolte, L.J., Myer, L.R. & Cook, N.G.W. (1987) Seismic Visibility of Fractures, *Rock Mechanics: Proceedings of the 28th U.S. Symposium*, ed. I.W. Farmer, J.J.K. Daemen, C.S. Desai, C.E. Glass and S.P. Neuman, University of Arizona, Tucson, June 1987, Pubs. A.A. Balkema (Boston). 47–56.

Pyrak-Nolte, L.J., Myer, L.R. & Cook, N.G.W. (1990a) Transmission of seismic waves across single natural fractures. *Journal of Geophysical Research*, 95(B6), 8617–8638.

Pyrak-Nolte, L.J., Myer, L.R. & Cook, N.G.W. (1990b) Anisotropy in seismic velocities and amplitudes from multiple parallel fractures. *Journal of Geophysical Research*, 95(B7), 11345–11358.

Pyrak-Nolte, L.J. & Nolte, D.D. (1992) Frequency Dependence of Fracture Stiffness. *Geophysical Research Letters*, 19(3), 325–328.

Pyrak-Nolte, L.J., Xu, J. & Haley, G.M. (1992) Elastic Interface Waves Propagating in a Fracture. *Physical Review Letters*, 68(24), p3650–3653.

Pyrak-Nolte, L.J., Roy, S. & Mullenbach, B.L. (1996) Interface waves propagated along a fracture. *Journal of Applied Geophysics*, 35, 79–87.

Pyrak-Nolte, L.J. and Nolte, D.D. (2016) Approaching a universal scaling relationship between fracture stiffness and fluid flow. *Nat Commun.* 7:10663. doi:10.1038/ncomms10663.

Roy, S. & Pyrak-Nolte, L.J. (1997) Observation of a Distinct Compressional-Mode Interface Waves on a Single Fracture. *Geophysical Research Letters*, 24(2), 173–176.

Saleh, B.E. & Teich, M.C. (2007) Fundamentals of Photonics, Wiley-Interscience, 2nd edition.

Sayers, C.M. (1988) Inversion of ultrasonic wave velocity measurements to obtain the micro-crack orientation distribution function in rocks. *Ultrasonics*, 26, 311–317.

Shao, S. (2015) Fractures in anisotropic media. PhD thesis. Purdue University, West Lafayette, IN.

Shao, S. & Pyrak-Nolte, L.J. (2013) Interface waves along fractures in anisotropic media. *Geophysics*, 78(4), 99–112.

Shao, S. & Pyrak-Nolte, L.J. (2015) Acoustic wavefront imaging of orthogonal fracture networks subjected to biaxial loading. In the *Proceedings from 49th US Rock Mechanics / Geomechanics Symposium* held in San Francisco, CA, USA, June 28–July 1, 2014, paper number 15-382.

Shao, S. and Pyrak-Nolte, L.J. (2016) Elastic wave propagation in isotropic media with orthogonal fracture sets, Rock Mechanics and Rock Engineering, DOI 10.1007/s00603-016-1084-z, 16 pages.

Shao, S., Petrovitch, C.L. & Pyrak-Nolte, L.J. (2015) Wave guiding in fractured layered media. in *Fundamental Controls on Fluid Flow in Carbonates, Geological Society, London, Special Publications*, 406, 375–400.

Sharf-Aldin, M., Rosen, R., Narasimhan, S. & PaiAngle, M. (2013) Experience using a novel 45 degree transducer to develop a general unconventional shale geomechanical model. In the *Proceedings of 47th US Rock Mechanics / Geomechanics Symposium*. San Francisco, CA, USA, 23–26 June 2013, paper number 2013–296.

Schoenberg, M. (1980) Elastic wave behavior across linear slip interfaces. *Journal of the Acoustical Society of America*, 5, 68, 1516–1521.

Schoenberg, M. (1983) Reflection of elastic waves from periodically stratified media with interfacial slip. *Geophysical Prospecting*, 31, 265–292.

Schoenberg, M. & Douma, J. (1988) Elastic wave propagation in media with parallel fractures and aligned cracks. *Geophysical Prospecting*, 36, 571–590.

Schoenberg, M., Muir, F. & Sayers, C.M. (1996) Introducing ANNIE: a simple three-parameter anisotropic velocity model for shales. *Journal of Seismic Exploration*, 5, 35–49.

Schoenberg, M. & Helbig, K. (1997) Orthorhombic media: Modeling elastic wave behavior in a vertically fractured earth. *Geophysics*, 62, 1954–1974.

Sondergeld, C.H. & Rai, C.S. (2011) Elastic anisotropy of shales. *The Leading Edge*, 30, 324–331, doi: 10.1190/1.3567264.

Sondergeld, C.H., Rai, C.S., Margesson, R.W. & Whidden, K.J. (2000) Ultrasonic measurement of anisotropy on the Kimmeridge shale. *70th International Meeting of the Society of Exploration Geophysicists, Expanded Abstracts*, 1858–1861.

Sone, H. & Zoback, M.D. (2013) *Mechanical properties of shale gas reservoir rocks and its relation to the in-situ stress variation observed in shale gas reservoirs*. Ph.D. thesis. Stanford University.

Suarez-Rivera, R. (1992) *The Influence of Thin Clay Layers Containing Liquids on the Propagation of Shear Waves*. Ph.D. Thesis. University of California, Berkeley.

Thomsen, L. (1986) Weak elastic anisotropy. *Geophysics*, 51(10), 1954–1966, 1986.

Tsvankin, I. (1996) P-wave signatures and notation for transversely isotropic media: an over-view. *Geophysics*, 61, 467–483.

Tsvankin, I. (1997) Anisotropic parameters and p-wave velocity for orthorhombic media. *Geophysics*, 62(4), 1292–1308.

Urosevic, M. & Juhlin, C. (1999) Seismic anisotropy in the upper 500 m of the southern Sydney Basin. *Geophysics*, 64(6), 1901–1911.

Vernik, L. & Liu, X. (1997) Velocity anisotropy in shales: A petrophysical study. *Geophysics*, 62, 521–532, doi: 10.1190/1.1444162.

Vernik, L. & Milovac, J. (2011) Rock physics of organic shales. *The Leading Edge*, 30, 318–323, doi: 10.1190/1.3567263.

Vernik, L. & Nur, A. (1992) Ultrasonic velocity and anisotropy of hydrocarbon source rocks. *Geophysics*, 57, 727–735, doi: 10.1190/1.1443286.

Vinh, P. & Ogden, R.W. (2004) Formulas for the Rayleigh wave speed in orthotropic elastic solids. *Archives in Mechanics*, 56(3), 247–265.

White, J. E. (1983) *Underground Sound*. Elsevier, New York.

Wolfe, J. (1998) *Imaging phonons: Acoustic wave propagation in solids*. Cambridge University Press.

Xian, C.J., Nolte, D.D. & Pyrak-Nolte, L.J. (2001) Compressional waves guided between parallel fractures. *International Journal of Rock Mechanics and Mining Sciences*, 38(6), 765–776.

Zavorokhin, G.L. & Nazarov, A.I. (2011) On elastic waves in a wedge. *Journal of Mathematical Sciences*, 175 (6), 646–650, doi: 1072-3374/11/1756-0646.

Zimmerman, R.W. (1991) Compressibility of Sandstones. *Developments in Petroleum Science* 29, Elsevier, New York.

Zoback, M.L. & Zoback, M.D. (1980), State of stress in the conterminous United States. *Journal of Geophysical Research*, 85(NB11), 6113–6156, doi: 10.1029/JB085iB11p06113.

Zoback, M.L. (1992), First- and second-order patterns of stress in the lithosphere: The world stress map project. *Journal of Geophysical Research*, 97(B8), 11703–11728, doi: 10.1029/92JB00132.

Strength Criteria

Chapter 12

On yielding, failure, and softening response of rock

J.F. Labuz[1], R. Makhnenko[2], S-T. Dai[3] & L. Biolzi[4]
[1]University of Minnesota, Minneapolis, USA
[2]University of Illinois at Urbana-Champaign, USA
[3]Minnesota Department of Transportation, USA
[4]Politecnico di Milano, Italy

Abstract: Mechanical descriptors of inelastic behavior of rock are reviewed and approaches to measure material and system response are highlighted. A non-traditional testing method that involves biaxial deformation developed through a Vardoulakis-Goldscheider plane strain apparatus is used to assess dilatancy and friction, as well as dilatancy of the shear band after localization. The Paul-Mohr-Coulomb failure criterion, which includes the intermediate stress, is applied to interpret the experiments at peak stress. Post-peak response, often referred to as softening, is explained as a size effect within the context of class I-class II stability.

This chapter is dedicated to I. Vardoulakis (1949–2009), a giant of modern geomechanics

1 INTRODUCTION

"In discussions with sensible professional men, I have not infrequently encountered the opinion expressed that it would be wasting vain efforts to develop a theory on which the strength of materials could be based scientifically. Homogeneous bodies of materials – I was told – do not exist, homogeneous states of stress are not encountered. It seems, therefore, utterly impossible to deduce a law of nature from experience. Since the existing irregularities furthermore are of such a nature that they nearly completely obscure any orderly behavior, it has little interest to track down the half-blurred traces of such laws. Under these circumstances nothing else remains than to make special tests in every case and to pay no heed to a physical interpretation of the results. I had to admit in each case that nothing could be said against this reasoning; and yet for more than one hundred years, there have been attempts again and again to establish order within the confusing abundance of the experiences. If one should succeed in finding a few rules under which many experiences could be subordinated – of course rules in which some confidence can be placed – no law of nature would have been derived, but some means found for judging the probability of new results of experience" (Mohr, 1901; from Nadai, 1950).

Rock is a complex material exhibiting, to various degrees, heterogeneity, anisotropy, pressure dependence, and irreversible damage. Nevertheless, extraordinary structures in (or of) rock have been designed and built, and general rules regarding their mechanical response have been identified (Vardoulakis & Sulem, 1995). In fact, when numerous observations from experiments on different rock types are compared, it is soon discovered that rock behaves in a similar manner from certain points of view. If the correct variables are used to measure material properties, then many phenomena can be explained by identical relations, *e.g.* Mohr-Coulomb failure criterion.

Concepts associated with the measurement of mechanical properties of rock are reviewed. As in many branches of engineering, specific tests have evolved for evaluation of response under conditions of interest; examples of tests in geomechanics are conventional triaxial compression or extension and plane strain compression. Furthermore, general behavior of a dilatant, frictional rock is highlighted and experimental results of interest are discussed.

2 PLANE STRAIN APPARATUS

An apparatus for determining the constitutive and failure response of rock, named the University of Minnesota Plane Strain Apparatus (Labuz *et al.*, 1996), was designed and built based on a passive stiff-frame concept of Vardoulakis & Goldscheider (1981). The biaxial device, in the sense of plane strain deformation, is unique because it allows the failure plane to develop and propagate in an unrestricted manner (Drescher *et al.*, 1990). By placing the upper platen on a low friction linear bearing, the prismatic specimen has the freedom to translate in the lateral direction if the deformation has localized. A schematic of the Apparatus is shown in Figure 1. A prismatic specimen 75–100 mm in height h, 27–40 mm in width w, 100 mm in length b, is placed within a stiff biaxial frame; the thick-walled cylinder limits the deformation of the specimen to very small values, approximating the plane strain condition (Makhnenko & Labuz, 2014). The specimen is subjected to lateral (confining) pressure by placing the entire assembly within a pressure vessel, and to axial load applied through displacing rigid platens by a servo-hydraulic actuator. In contrast to other plane strain devices with similar loading conditions, the upper loading platen is attached to a linear bearing. The low friction bearing allows free displacement of the upper part of the specimen upon formation of a failure mechanism

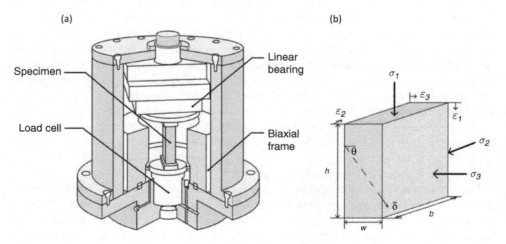

Figure 1 (a) University of Minnesota Plane Strain Apparatus; (b) Specimen geometry with principal stresses and corresponding strains.

On yielding, failure, and softening response of rock   367

with no restriction. The apparatus is internally instrumented with five pressure resistant LVDTs for measuring axial and lateral displacements, two load cells for measuring axial force, and four strain gages for measuring strains on the specimen and the deformation of the biaxial frame. The two surfaces of the specimen exposed to confining pressure are sealed by a polyurethane coating; metal targets glued to the specimen provide firm contact points for the lateral LVDTs. The four surfaces in contact with polished-steel platens are covered with a stearic acid lubricant to reduce frictional constraint (Labuz & Bridell, 1993).

2.1 Experiments

Plane strain experiments were performed on Red Wildmoor sandstone with a porosity of 25.8% and uniaxial compressive strength C_o = 10.0 MPa; the uniaxial test gave Young's modulus E = 2.0 GPa and Poisson's ratio v = 0.34. The grains of the sandstone are subrounded with a mean grain diameter of 0.1 mm. Closed-loop, servo-controlled tests with lateral displacement as the feedback signal were conducted at confining pressures (σ_3) of 3.5 and 10.0 MPa. For convenience, force and displacement are taken positive for compression and shortening (*cf.* Jaeger *et al.*, 2007).

Figure 2 shows the axial load and lateral displacement responses as a function of axial displacement. The brittle nature of the rock is apparent for 3.5 MPa test; in post-peak, the axial force dropped sharply with little change in axial displacement (Figure 2a). In fact, the response of the axial displacement was to decrease or snap-back, which was possible to observe because of lateral displacement control. Note that this class II snapback behavior is not constitutive response but is due to the "system" containing the displacements along the failure plane and from the elastic unloading of the specimen, and these involve modulus, fracture parameters, and size (explained in Section 4). As confinement increased to 10 MPa, the global response was no longer class II; load reduced with a small increase in axial displacement to a residual level (Figure 2b). The orientations of the failure plane relative to the lateral (minimum stress) direction were 63° and 58°, an indication that dilatancy and friction are pressure dependent.

2.2 Dilatancy and friction

The major principal strains are calculated from the global displacements ($\varepsilon_1 = \varepsilon_{yy} = \Delta h/h$, $\varepsilon_2 = \varepsilon_{zz} = 0$, $\varepsilon_3 = \varepsilon_{xx} = \Delta w/w$) such that volume strain ε and shear strain γ are simply

$$\varepsilon = \varepsilon_1 + \varepsilon_3, \ \gamma = \varepsilon_1 - \varepsilon_3 \tag{1}$$

which can be decomposed into incremental elastic and plastic (inelastic) components:

$$\Delta\varepsilon = \Delta\varepsilon^e + \Delta\varepsilon^p, \ \Delta\gamma = \Delta\gamma^e + \Delta\gamma^p \tag{2}$$

The plastic strains were determined from the measured total strains and the calculated elastic strains (constant E, v) at the start of nonlinear volume strain behavior (Riedel & Labuz, 2007), and the plastic volume response is shown in Figure 3a. The dilatancy angle ψ (Hansen, 1958) is determined by

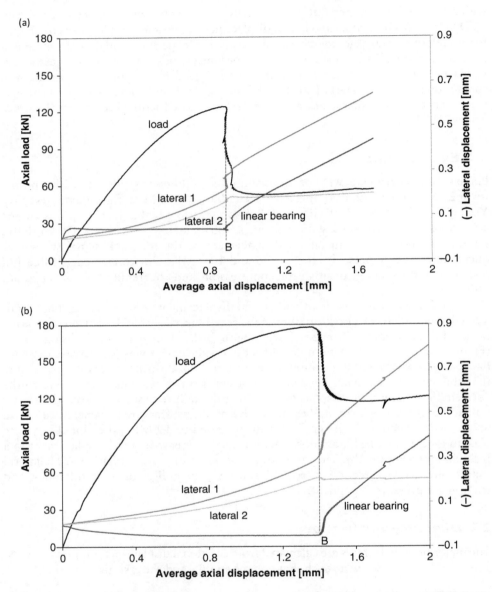

Figure 2 Load – displacement response of sandstone at (a) 3.5 MPa confinement and (b) 10.0 MPa confinement; the shear band traversed the specimen and slip displacement started at B.

$$\sin \psi = -\frac{\Delta \varepsilon^p}{\Delta \gamma^p} \qquad (3)$$

and ψ as a function of plastic shear strain is illustrated in Figure 3b. At low confinement ($\sigma_3 = 0.35 C_o$), the rock compacted very little before dilating and reached a dilation

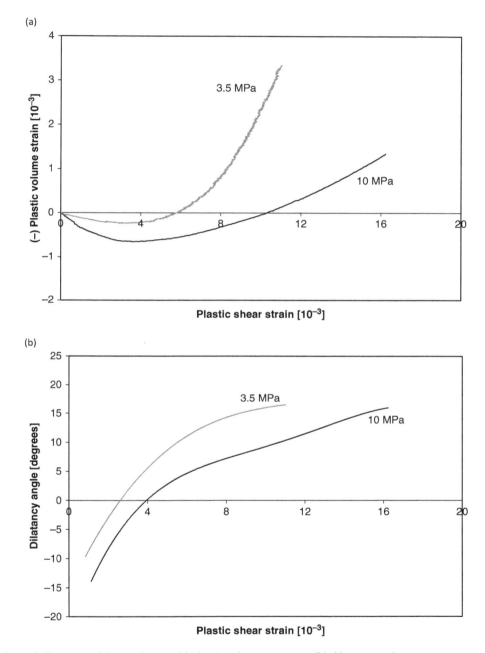

Figure 3 Behavior of the sandstone: (a) plastic volume response, (b) dilatancy angle.

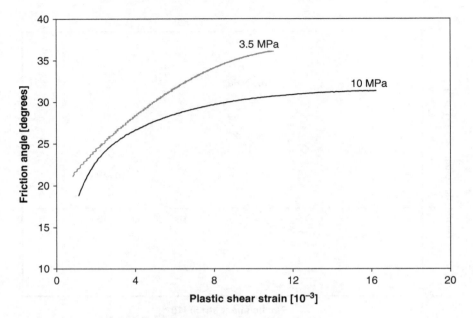

Figure 4 Mobilized friction angle at 3.5 MPa and 10.0 MPa confinement.

angle at peak stress $\psi = 17°$. At high confinement ($\sigma_3 = C_o$), the sandstone displayed compaction followed by peak dilation $\psi = 14°$.

The Mohr-Coulomb (MC) failure criterion can be written as

$$\sin \phi_m = \frac{t}{s - V_o} \quad (4)$$

where (2D) mean stress $s = (\sigma_1 + \sigma_3)/2$, shear stress $t = (\sigma_1 - \sigma_3)/2$, the parameter $V_o = S_o/tan\phi_p$ is the intercept at $t = 0$, S_o is the shear stress intercept also known as cohesion c, and ϕ_p is the peak friction angle. The sandstone is assumed to have a linear yield envelope with constant V_o and increasing friction (and cohesion). The relationship between mobilized friction angle ϕ_m and plastic shear strain showed sensitivity to pressure (Figure 4): as pressure increased, ϕ_m decreased. Friction angles were different at the two confining pressures, suggesting a nonlinear failure envelope.

3 FAILURE CRITERIA

The MC criterion of Equation 4, which is a reasonable approximation in the brittle regime over a limited range of mean stress, can be written (a, c are material parameters):

$$a\sigma_1 + c\sigma_3 = 1 \quad (5)$$

which represents a plane in principal stress space σ_1, σ_2, σ_3, but it is natural to consider

$$A\sigma_1 + B\sigma_2 + C\sigma_3 = 1 \tag{6}$$

which is called Paul-Mohr-Coulomb (PMC) by Meyer & Labuz (2013). Note that PMC includes σ_2. PMC can be evaluated by performing conventional triaxial testing on a right circular cylinder, where axial (vertical) stress $\sigma_a = \sigma_V$ is applied independent of radial (horizontal) stress $\sigma_r = \sigma_H = \sigma_h$ so that either compression ($\sigma_r = \sigma_2 = \sigma_3$) or extension failure ($\sigma_r = \sigma_1 = \sigma_2$) can be achieved. The generalized pyramidal failure surface, Equation 6, was proposed by Paul (1968), a simple version by Haythornthwaite (1962), and it can be written as

$$\frac{\sigma_1}{V_o}\left[\frac{1-\sin\phi_c}{2\sin\phi_c}\right] + \frac{\sigma_2}{V_o}\left[\frac{\sin\phi_c - \sin\phi_e}{2\sin\phi_c\sin\phi_e}\right] - \frac{\sigma_3}{V_o}\left[\frac{1+\sin\phi_e}{2\sin\phi_e}\right] = 1 \tag{7}$$

where ϕ_c is the internal friction angle for compression ($\sigma_2 = \sigma_3$), ϕ_e is the internal friction angle for extension ($\sigma_2 = \sigma_1$), and V_o is the intersection of the failure surface with the hydrostatic axis ($\sigma_1 = \sigma_2 = \sigma_3$); V_o represents all-around equal tension and it is not measured in experiments.

For isotropic rock, the orientation of the principal stresses does not matter and they can be relabeled as σ_V, σ_H, σ_h and interchanged as major, intermediate, and minor principal stresses. Because of the six orderings of the principal stresses, six planes can be constructed in σ_V, σ_H, σ_h space giving an irregular hexagonal pyramid (Figure 5a). The plane normal to the hydrostatic axis is called the π-plane and the projections of the coordinate axes are labeled σ'_V, σ'_H, σ'_h (Figure 5b); stress paths are shown for conventional triaxial compression and extension, and plane strain compression.

As reported in Table 1, triaxial compression and extension data from Papamichos et al. (2000), along with the plane strain data, were used to find the best-fit plane (Figure 6a) and the coefficients A, B, C (Makhnenko et al., 2015); note that because of isotropy, the extension data can be "moved" to the plane containing the compression

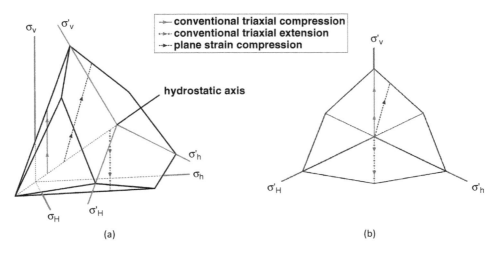

Figure 5 Linear failure surfaces with stress paths in (a) principal stress space, (b) π-plane view.

Table 1 Principal stresses at failure for plane strain compression (BXC) and conventional triaxial compression (TXCO) and extension (TXEX) experiments.

Test name	σ_1 [MPa]	σ_2 [MPa]	σ_3 [MPa]
BXC-1	17.0	7.8	0.4
BXC-2	33.4	12.3	3.5
BXC-3	56.0	22.5	10.0
TXCO-1	15.1	0.4	0.4
TXCO-2	21.3	1.4	1.4
TXCO-3	30.0	3.5	3.5
TXCO-4	41.7	6.9	6.9
TXCO-5	49.8	10.3	10.3
TXEX-6	60.0	60.0	8.1
TXEX-7	52.5	52.5	5.9
TXEX-8	48.0	48.0	4.4
TXEX-9	45.0	45.0	3.3

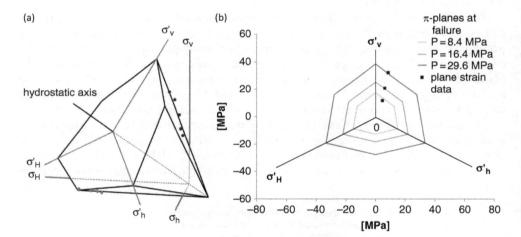

Figure 6 PMC failure surface and data for sandstone. (a) principal stress space with diamonds and circles from triaxial extension and compression. (b) π-plane view with plane strain data.

data. The friction angles in compression and extension are $\phi_c = 30.6°$ and $\phi_e = 37.8°$, demonstrating an intermediate stress effect for the rock. From the vertex $V_o = 8.7$ MPa, the cohesions in compression and extension are $c_c = 5.3$ MPa and $c_e = 7.0$ MPa. The view in the π-plane showing the plane strain data is given in Figure 6b.

4 SOFTENING

Figure 7 shows a close-up of the mechanical response around peak load, including the decrease in force with increase in lateral displacement known as softening, although this is not material response because of the failure process. From observations of

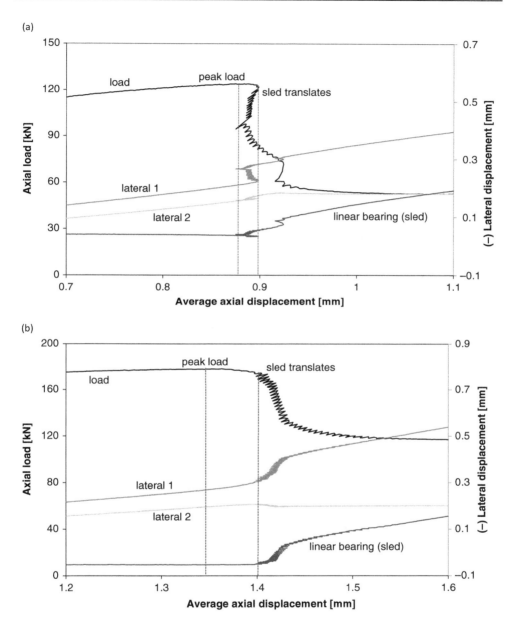

Figure 7 Load – displacement response of the sandstone around peak load at (a) 3.5 MPa confinement showing class II behavior and (b) 10.0 MPa confinement showing class I.

acoustic emission locations and optical microscopy, it is clear that strain localization – the shear band – is formed at or slightly before peak load but it does not traverse the specimen (Carvalho & Labuz, 2002; Labuz et al., 2006); propagation of the shear band (unstable crack growth) is responsible for initial softening after peak. Once a mechanism

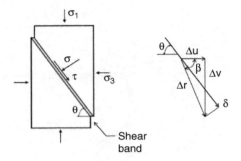

Figure 8 Loading and kinematics of the shear band.

is formed, which means that the shear band is fully developed, the upper block slides with a normal component associated with either compaction or dilation (Figure 8).

4.1 Shear band characteristics

From the plane strain experiments, the softening response of the specimen (Figure 8) with the shear band angle θ, measured from σ_3, can be determined from the shear stress $\tau = t(sin2\theta)$ and slip displacement δ:

$$\delta = [(\Delta u)^2 + (\Delta v)^2]^{1/2} \quad (8)$$

where Δu is the lateral displacement of the upper block sliding along the shear band and Δv is the axial component of sliding (Figure 9a). The angle β is the orientation of the resultant displacement $\Delta r = [(\Delta u)^2 + (\Delta v)^2]^{1/2}$:

$$\cos \beta = \frac{\Delta u}{\Delta r} \quad (9)$$

For both experiments, the shear band compacted (negative dilation) with slip before reaching the residual value of zero dilation (Figure 9b). For the 10 MPa experiment, Figure 10 describes how the shear stress degraded from a value τ_p corresponding to the onset of slip (not at peak force) to a constant residual level τ_r, when δ exceeds a critical amount of slip $\delta_c = 0.35$ mm for this sandstone. The energy per unit area dissipated along the shear band is (Palmer & Rice, 1973; Rice, 1980; Wong, 1982)

$$G_c = \int_0^{\delta_c} (\tau - \tau_r) d\delta \quad (10)$$

given by $G_c = 1.14$ kJ/m^2. A summary of the material and shear band characteristics is contained in Table 2, along with data from an additional experiment performed at $\sigma_3 = 15.0$ MPa.

4.2 Stability analysis

To explain the softening response for the plane strain compression test, we modify the plane stress analysis presented by Labuz & Biolzi (1991). Once the shear band develops

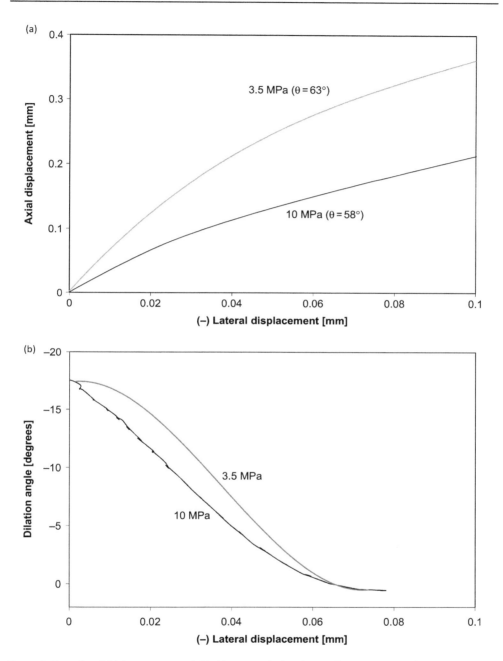

Figure 9 Shear band (a) kinematics and (b) dilation angle for the sandstone.

across the entire specimen, further displacement involves the rock outside the band unloading elastically, while rock inside the band deforms such as to give a net slip δ with decreasing stress τ. The relation between δ and τ follows the slip-weakening model of Palmer & Rice (1973), where the shear stress decreases from a value τ_p at the start of

Table 2 Summary of the sandstone parameters from plane strain experiments.

Parameter	3.5 MPa	10 MPa	15 MPa
Dilatancy angle (peak)	17	14	10
Friction angle (peak)	36	31	29
Shear band angle	63	58	52
Residual /peak shear stress	0.43	0.66	0.88
Critical slip displacement [mm]	0.40	0.35	0.27
Dissipated energy [kJ/m^2]	0.83	1.14	0.22
Brittleness number	0.96	1.46	1.56

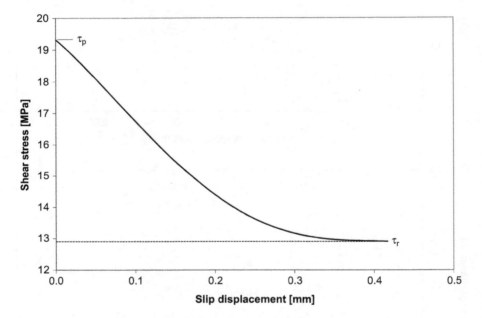

Figure 10 Slip weakening law of the sandstone at 10 MPa confinement.

slip to the residual value τ_r at a critical amount of slip δ_c. The behavior shown in Figure 10 is approximated by a linear function:

$$\delta = \delta_c \left(1 - \frac{\Delta \tau}{\Delta \tau_p}\right) \tag{11}$$

where $\Delta\tau = \tau - \tau_r$ and $\Delta\tau_p = \tau_p - \tau_r$.

Consider the plane strain specimen of height h and width w that behaves linearly up to peak load, at which point a shear band of angle θ forms with the slip-weakening constitutive relation. The total displacement v is due to the rock, testing machine, and shear band:

$$v = \frac{h}{2\mu}[\Delta\sigma(1-\nu) + \sigma_3(1-2\nu)] + \frac{w}{k'}(\Delta\sigma + \sigma_3) + \delta_c\left(1 - \frac{\Delta\tau}{\Delta\tau_p}\right)\sin\theta \tag{12}$$

where $\Delta\sigma = \sigma_1 - \sigma_3$, μ = shear modulus, $2\mu = E/(1 + v)$, and k' is the testing machine (load frame) stiffness per unit thickness b.

Apply the incremental stability condition for critical softening $dv/d\Delta\sigma = 0$ to Equation 12:

$$b\frac{(1-v)}{2\mu} + \frac{w}{k'} = \frac{\delta_c \sin^2\theta \cos\theta}{\tau_p - \tau_r} \tag{13}$$

where $\tau = \Delta\sigma \sin\theta \cos\theta$. Because fracture energy and Young's modulus are more readily identified than critical slip displacement and shear modulus, it is convenient to substitute G_f for δ_c and E for μ, with $f(\theta) = \sin2\theta \sin\theta$, $\lambda = b/w$, $E' = E/(1 - v^2)$ and the final result is

$$\frac{1}{f(\theta)}\left(\lambda + \frac{E'}{k'}\right) = \frac{G_f E'}{w(\tau_p - \tau_r)^2} \tag{14}$$

A number of interesting features in Equation 14 are noted:

- Material properties, specimen size and shape, as well as confinement (mean stress), which influences both G_f and $(\tau_p - \tau_r)$, are important for determining stability.
- Machine stiffness influences the post-peak response, but it is not the sole factor. Even for an infinitely stiff machine, a class II behavior can still be observed.
- A brittleness number $B_n = G_f E'/[w(\tau_p - \tau_r)^2]$ appears with a size dependence, such that for specimens smaller or larger than some width w and the same material properties E', G_f, and $(\tau_p - \tau_r)$, the response is stable for small w and unstable for large.
- The LHS > RHS in Equation 14 predicts class II response, and the smallest B_n is associated with the 3.5 MPa experiment (Table 2), which exhibited class II behavior.

5 SUMMARY

The material and softening behavior of rock can be studied using a Vardoulakis-Goldscheider plane strain apparatus, which combines the positive features of a constitutive (plane strain) compression test, such that the two-dimensional material behavior, including dilatancy and friction, can be evaluated, and a direct shear test, such that the shear stress-slip displacement and dilatancy characteristics of the shear band can be measured. Although the nature of the inelasticity (microcracking, intergranular sliding, etc.) was not identified, the inelastic (plastic) response can be determined by removing the elastic response, in an incremental approach.

The softening behavior of a rock is not an essential material property but simply a typical global response, as failure occurs in a manner described by the slip-weakening model of fracture. Furthermore, the stability is dependent not only on machine stiffness, but also on geometry and size of the specimen.

REFERENCES

Carvalho, F. & Labuz, J.F. 2002. Moment tensors of acoustic emission in shear faulting under plane-strain compression. *Tectonophysics*, 356, 199–211.

Drescher, A., Vardoulakis, I. & Han, C. 1990. A biaxial apparatus for testing soils. *Geotech. Testing J. ASTM* 13, 226–234.

Hansen, B. 1958. Line ruptures regarded as narrow rupture zones. Basic equations based on kinematic considerations. *Proc. Conf. Earth Pressure Problems*. Brussels, 1, pp. 39–48.

Haythornthwaite, R.M. 1962. Range of yield condition in ideal plasticity. *Trans. ASCE*, 127(I), 1252–1267.

Jaeger, J.C., Cook, N.G.W. & Zimmerman, R.W. 2007. *Fundamentals of Rock Mechanics, 4th edn*. London, Blackwell.

Labuz, J.F. & Biolzi, L. 1991. Class I vs. class II stability: a demonstration of size effect. *Int. J. Rock Mech. Min. Sci. Geomech. Abstr.*, 28, 199–205.

Labuz, J.F. & Bridell, J.M. 1993. Reducing frictional constraint in compression testing through lubrication. *Int. J. Rock Mech. Min. Sci. Geomech. Abstr.*, 30, 451–455.

Labuz, J.F., Dai, S.-T. & Papamichos, E. 1996. Plane-strain compression of rock-like materials. *Int. J. Rock Mech. Min. Sci. Geomech. Abstr.* 33, 573–584.

Labuz, J.F., Riedel, J.J. & Dai, S.-T. 2006. Shear fracture in sandstone under plane strain compression. *Eng. Frac. Mech.* 73, 820–828.

Makhnenko, R. & Labuz, J. 2014. Plane strain testing with passive restraint. *Rock Mech. Rock Eng.*, 47(6), 2021–2029.

Makhnenko, R.Y., Harvieux, J. & Labuz, J.F. 2015. Paul-Mohr-Coulomb failure surface of rock in the brittle regime. *Geophys. Res. Lett.*, 42, 6975–6981.

Meyer, J.P. & Labuz, J.F. 2013. Linear failure criteria with three principal stresses. *Int. J. Rock Mech. Min. Sci.*, 60, 180–187.

Nadai, A. 1950. *Theory of Flow and Fracture of Solids*. New York, McGraw-Hill.

Palmer, A.C. & Rice, J.R. 1973. The growth of slip surfaces in the progressive failure of over-consolidated clay. *Proc. Roy. Soc. Lond. A*, 332, 527–548.

Papamichos, E., Tronvoll, J. Skjvstein, A., Unander, T.E., Labuz, J.F., Vardoulakis, I. & Sulem, J. 2000. Constitutive testing of Red Wildmoor sandstone. *Mech. Cohes. Frict. Mater.*, 5, 1–40.

Paul, B. 1968. Generalized pyramidal fracture and yield criteria. *Int. J. Solids Struct.*, 4, 175–196.

Rice, J.R. 1980. The mechanics of earthquake rupture. In: Dziewonski, A.M. & Boschi, E. (eds.) *Physics of the Earth's Interior, Italian Physical Society, Bologna*. Amsterdam, North-Holland. pp. 555–649.

Riedel, J.J. & Labuz, J.F., 2007. Propagation of a shear band in sandstone. *Int J. Num. Anal. Meth. Geomech.*, 31, 1281–1299.

Vardoulakis, I. & Goldscheider, M. 1981. Biaxial apparatus for testing shear bands in soils. *Proc. 10th Int. Conf. Soil Mech. Found. Eng.* Rotterdam, Balkema. pp. 819–824

Vardoulakis, I. & Sulem, J. 1995. *Bifurcation Analysis in Geomechanics*. Glasgow, Blackie Academic & Professional.

Wong, T.F. 1982. Shear fracture energy of Westerly granite from post-failure behavior. *J. Geophys. Res.* 87, 990–1000.

Chapter 13

True triaxial testing of rocks and the effect of the intermediate principal stress on failure characteristics

Bezalel Haimson[1], Chandong Chang[2] & Xiaodong Ma[1,3]
[1]*Department of Materials Science and Engineering, University of Wisconsin, Madison, WI, USA*
[2]*Department of Geology, Chungnam National University, Daejeon, Korea*
[3]*Currently at Department of Geophysics, Stanford University, Stanford, CA, USA*

Abstract: We review the research conducted at the University of Wisconsin in the last two decades in the area of true triaxial testing of rocks. We designed and fabricated equipment (the UW true triaxial testing system) capable of applying three different compressive loads on mutually perpendicular faces of cuboidal rock specimens. The equipment was used to carry out extensive series of true triaxial experiments in an array of rocks, from strong crystalline to weak porous clastic. The results revealed the substantial effect of the intermediate principal stress on failure characteristics, in terms of the failure stress, failure-plane angle, and failure mode. True triaxial failure criteria incorporate the intermediate principal stress, and expose the inadequacies of the commonly employed criteria, such as the Mohr and Mohr-Coulomb theories, which consider only the two extreme principal stresses.

1 INTRODUCTION

1.1 Background

One of the main objectives of laboratory testing of rock is to determine its deformation and failure characteristics when subjected to stress conditions comparable to those encountered in situ. Most rock mechanics experiments to date are conducted on cylindrical specimens subjected to uniform lateral confining pressure. Such 'conventional triaxial tests' replicate unique crustal stress conditions, in which the intermediate and the minor principal stresses, σ_2 and σ_3, are equal (triaxial compression), or in which the intermediate and the major principal stresses σ_2 and σ_1 are equal (triaxial extension). Conventional triaxial tests have been widely used for the study of mechanical characteristics of rocks because of equipment simplicity and convenient specimen preparation and testing procedures. The justification for the use of conventional triaxial tests is the assumption that σ_2 has a negligible effect on rock failure characteristics, as expressed in the Coulomb, Mohr, and Griffith failure criteria (Jaeger & Cook, 1979, p. 106).

However, a growing number of in situ stress measurements have shown that the state of stress at varying depths is typically anisotropic, *i.e.* $\sigma_1 \neq \sigma_2 \neq \sigma_3$ (McGarr & Gay, 1978; Haimson, 1978; Brace & Kohlstedt, 1980). Moreover, increasingly researchers have faced situations in which the above-mentioned failure criteria were found lacking.

For example, Vernik & Zoback (1992) found that use of the linearized Mohr failure criterion (also called 'Mohr-Coulomb') in relating borehole breakout dimensions to the prevailing in situ stress conditions in crystalline rocks did not provide realistic results. They suggested the use of a more general criterion that accounts for the effect of the intermediate principal stress on failure. Also, Ewy (1998) reported that the Mohr-Coulomb criterion is significantly too conservative for the purpose of calculating the critical mud weight necessary to maintain wellbore stability. He suggested that the reason was the limitation of the Mohr-Coulomb, which neglects the perceived rock strengthening effect of the intermediate principal stress.

The potential effect of σ_2 on rock failure was studied as early as the 1960's by Murrell (1963) and Handin *et al.* (1967). Murrell compared the failure stress levels from two different series of conventional triaxial tests on Carrara marble: triaxial compression ($\sigma_1 > \sigma_2 = \sigma_3$) and triaxial extension ($\sigma_1 = \sigma_2 > \sigma_3$). He noted that the ultimate level of σ_1 ($\sigma_{1,\text{peak}}$) required to bring about failure for any known σ_3 was larger in triaxial extension than in triaxial compression, implying that the intermediate principal stress does affect the failure process. Handin *et al.* carried out similar tests in Solnhofen limestone, Blair dolomite, and Pyrex glass. Their test results were similar to those of Murrell's, reinforcing the need to further investigate the effect of σ_2 on failure of rock.

Wiebols & Cook (1968) took a different path to study the effect of σ_2 on rock failure. They employed the strain energy stored by the solid rock, and the additional strain energy around Griffith cracks resulting from the sliding of crack surfaces, in order to derive a failure criterion. They found that under true triaxial compressive stresses (*i.e.* $\sigma_1 \geq \sigma_2 \geq \sigma_3$) the intermediate principal stress has a definite effect on failure, which can be predicted if the coefficient of sliding friction between crack surfaces is known. Wiebols & Cook (1968) determined from their model that if σ_3 is held constant and σ_2 is increased from $\sigma_2 = \sigma_3$ to $\sigma_2 = \sigma_1$ the failure stress first rises, reaches a maximum at some value of σ_2 and then continuously decreases to a level slightly higher than that obtained in a conventional triaxial test, *i.e.* when $\sigma_2 = \sigma_3$.

Mogi (1971) took the investigation of the σ_2 effect on rock failure to a new level, with the fabrication of a novel apparatus that enabled the nearly friction-free application of three independent, mutually perpendicular, uniform loads to the faces of a cuboidal specimen. He subjected Dunham dolomite, followed by other rocks, to several levels of σ_3, and for each σ_3, to different intermediate principal compressive stresses, σ_2. For each applied stress configuration, he raised σ_1 to failure. With these first true triaxial experiments, Mogi demonstrated that rock failure is a function of σ_2 in a manner quite similar to that predicted by Wiebols & Cook (1968) theoretically.

At the University of Wisconsin (UW), we have designed and fabricated a new true triaxial testing system suitable for even the strongest rocks, which emulates Mogi's (1971) original design, with significant simplifications (Haimson & Chang, 2000). The tests described in this paper were all conducted in the UW true triaxial system.

1.2 The UW true triaxial testing system

The UW true triaxial testing system enables the application of high, uniform compressive loads in three principal directions to a cuboidal specimen of size $19 \times 19 \times 38$ mm^3 (up to: σ_1, $\sigma_2 = 1600$ MPa, $\sigma_3 = 400$ MPa). The system consists of two main parts, a biaxial loading apparatus and a true triaxial pressure vessel. The biaxial apparatus

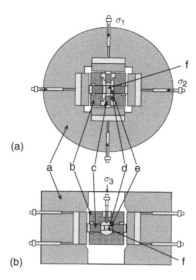

Figure 1 Schematic diagram of the UW true triaxial testing apparatus: (a) cross section and (b) profile; a. biaxial loading apparatus; b. true triaxial pressure vessel; c. loading pistons; d. confining fluid; e. metal spacers; f. rock specimen.

facilitates the application of two independent perpendicular stresses, σ_1 (along the long axis of the specimen) and σ_2 (along one of the lateral directions) (Figure 1). These two stresses are transmitted to the rock specimen via orthogonal pairs of pistons located in the true triaxial pressure vessel. Thin copper shims coated with a mixture of stearic acid and Vaseline at a 1:1 weight ratio, are placed between the pistons and the specimen faces in order to minimize friction. Loading in the third principal direction (σ_3) is applied directly by hydraulic fluid inside the pressure vessel. Thin polyurethane coating on the specimen open faces prevents fluid from infiltrating into the specimen. The entire system is compact, portable, and self-contained.

The pair of specimen lateral faces subjected directly to hydraulic pressure is also used for mounting strain gages, thermocouples, and acoustic emission transducers, as needed. Strains in the major and intermediate directions (ε_1 and ε_2) are measured by strain gages bonded directly to these faces and oriented in the respective principal directions. For the measurement of ε_3, we designed and fabricated a beryllium-copper strain-gaged beam, the ends of which are fixed, and whose center is forced to make contact with a pin bonded to the exposed face of the specimen (Figure 2). As the rock specimen expands in the σ_3 direction during compressive loading, the beam flexes, allowing the strain gages mounted near its ends to monitor the calibrated least principal strain.

The two loads applied mechanically to the specimen are monitored by load cells mounted on respective pistons inside the true triaxial pressure vessel and, as a backup, by pressure transducers installed in the corresponding hydraulic fluid lines. The load in the third direction is monitored directly by a pressure transducer in line with the hydraulic fluid pressure. The true triaxial pressure vessel has a total of sixteen insulated electrical feedthroughs, which allow data from strain gages, acoustic emission sensors,

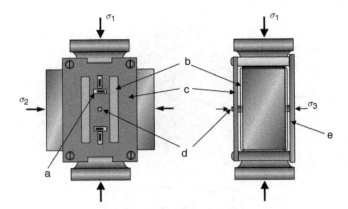

Figure 2 Schematic diagram of the specimen assembly with the installed strain-gaged beam: a. strain gages; b. polyurethane coating; c. beam; d. steel pins; e. base plate.

or thermocouples mounted on the specimen to be recorded using a data acquisition program.

A standard aluminum sample of same dimensions as those of the rock specimens (19 × 19 × 38 mm) was used to verify the uniformity of the stresses applied to the specimen faces by the two pairs of loading pistons, and to calibrate the respective load cells.

1.3 Calibration

Calibration of the axial load cell was carried out in the true triaxial pressure vessel by first applying a hydrostatic state of stress to the aluminum specimen, and then monotonically increasing σ_1 while maintaining uniform lateral pressure ($\sigma_2 = \sigma_3$), and recording the axial load-cell and strain outputs. The same procedure was followed for the calibration of the lateral load cell. A consistent ratio of load-cell voltage to stress was recorded both during the initial hydrostatic loading and through the axial or lateral stress increase, regardless of the amount of confining pressure.

The strain-gaged beam designed for measuring ε_3, *i.e.* the strain in the σ_3 direction, was calibrated by subjecting the aluminum specimen to uniaxial loading and recording simultaneously the voltage output from the beam strain gages and the strain measured in the same direction by strain gages mounted directly on the specimen.

1.4 True triaxial testing procedure

Testing procedure begins with the insertion of a cuboidal rock specimen (with or without strain gages bonded to the faces aligned the minimum principal stress, depending on the purpose of the test) into the true triaxial cell and the application of monotonically rising hydrostatic loading. As σ_3 and then σ_2 reach their predetermined magnitudes, they are kept constant for the remainder of the test. Loading in the σ_1 direction continues either by monotonically increasing the stress at a constant rate

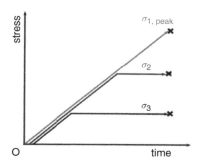

Figure 3 True triaxial loading path under stress-control mode.

of 1 MPa/sec, or by raising the least principal strain (ε_3) at a rate of 5×10^{-6}/sec, until failure occurs (Figure 3). The failure stress, or $\sigma_{1,peak}$ does not appear to be affected by the loading mode, but the ε_3-controlled tests allow one to observe post failure behavior.

2 TRUE TRIAXIAL FAILURE IN CRYSTALLINE ROCKS

2.1 Failure stress ($\sigma_{1,peak}$)

An extensive series of true triaxial tests was carried out in three crystalline rocks using the UW true triaxial testing system. We present herein the results obtained in Westerly granite (Haimson & Chang, 2000), KTB amphibolite (Chang & Haimson, 2000) and SAFOD granodiorite (Lee & Haimson, 2011).

All three rocks have a very low porosity (under 1%). Westerly granite, quarried in Rhode Island, USA, is informally considered a standard rock type for rock mechanics testing, because it is nearly isotropic, approximately linear elastic, and homogeneous (Brace, 1964; Wawersik & Brace, 1971; Wong, 1982; Lockner et al., 1991). KTB amphibolite, extracted from a depth of 6.4 km in the super-deep scientific borehole drilled by the German Continental Deep Drilling Program (KTB) in Bavaria, Germany, is also nearly isotropic and linear elastic (Vernik & Zoback, 1992; Röckel & Natau, 1995). The SAFOD granodiorite, recovered from a depth of 1.4 km in the San Andreas Fault Observatory at Depth (SAFOD) main drillhole, California, USA, although linearly elastic, exhibits some degree of inhomogeneity with respect to compressive failure (Lee & Haimson, 2011).

Testing procedure for all three crystalline rocks consisted of first simultaneously raising the three compressive principal stresses from zero to a preset level, then holding σ_3 constant while raising the other two principal stresses together to a higher preset level, followed by continuing loading in the σ_1 direction while holding both σ_2 and σ_3 at their predetermined values, until failure occurred (Figure 3).

Several levels of σ_3 were employed for each rock between 0 MPa and 100 MPa (in Westerly granite), 0–150 MPa (in KTB amphibolite), and 0–160 MPa (in SAFOD granodiorite), and for every σ_3 the magnitude of σ_2 was varied from test to test between $\sigma_2 = \sigma_3$ and $\sigma_2 = k\sigma_3$, where k was either 4 or 5. Figure 4 shows the results

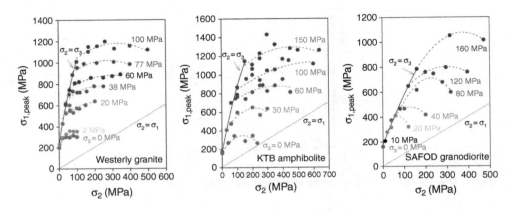

Figure 4 Maximum principal stress at failure ($\sigma_{1,peak}$) as a function of σ_2 for different constant σ_3 in three crystalline rocks: Westerly granite, KTB amphibolite, and SAFOD granodiorite.

of these tests for each rock in the form of the failure stress (peak σ_1 at failure, $\sigma_{1,peak}$) as a function of the applied σ_2 for different σ_3 levels. For each level of σ_3, the lowest σ_2 tested ($\sigma_2 = \sigma_3$) yielded a data point that can be used for establishing the conventional triaxial failure criterion, and can be fitted by a Mohr-type failure envelope (depicted by a solid curve in Figure 4). Mohr failure criteria, also called two-dimensional criteria (because of their definition in terms of only σ_1 and σ_3 at failure, neglecting σ_2), are commonly utilized in rock engineering industry under the assumption that failure stress is affected only the least principal stress, σ_3.

The true triaxial failure stress, $\sigma_{1,peak}$, plotted as a function of σ_2 for given constant σ_3, and fitted by second order polynomial functions (dashed lines) first rises significantly with increasing σ_2, but reaches a plateau as σ_2 attains magnitudes several times larger than σ_3. At the highest magnitudes of applied σ_2, an initial failure stress degradation is typically observed (Figure 4). The increase in $\sigma_{1,peak}$ with σ_2 is remarkable. For example, in Westerly granite at $\sigma_3 = 20$ MPa, $\sigma_{1,peak}$ at $\sigma_2 = \sigma_3$ is 430 MPa, but as σ_2 is elevated to 200 MPa, $\sigma_{1,peak}$ rises to 640 MPa, an increase of 49%. Even at $\sigma_3 = 100$ MPa, when the applied σ_2 is 260 MPa, failure stress rises by 18% over its base value. Similarly in KTB amphibolite, failure stress rises by about 50% when $\sigma_3 = 30$ MPa and the applied σ_2 is at least 200 MPa, and by 30% at $\sigma_3 = 100$ MPa, when σ_2 rises to 300 MPa. It should be noted that the general true triaxial failure characteristics of the tested crystalline rocks is qualitatively similar to the theoretical prediction of Wiebols & Cook (1968) and to experimental results obtained by Mogi (1971) in Dunham dolomite.

True triaxial testing clearly demonstrates that failure of crystalline rocks is significantly affected by the magnitude of the intermediate principal stress, and that the traditional two-dimensional failure criteria, such as Mohr-Coulomb, Mohr, and Griffith criteria, characterize only the special case in which $\sigma_2 = \sigma_3$, and not the general stress conditions. In fact the two dimensional failure criteria only predict the lowest failure stress for a given σ_3, and thus yield merely a conservative estimate of rock failure.

2.2 Failure plane angle (θ)

Failure of the three crystalline rocks in all tested ranges of σ_2 and σ_3 occurs unambiguously in brittle mode along a high-angle failure plane (Figure 5). The failure plane (or 'shear band'; also called 'fault') consistently strikes in the σ_2 direction and dips steeply in the σ_3 direction. The failure-plane angle (θ), defined as the angle between the normal to the failure plane and σ_1 direction, varies as a function of both σ_3 and σ_2 (Figure 6). The angle tends to decrease with increasing σ_3. However, for a given σ_3, failure-plane angle increases as the level of σ_2 is raised. The increase can be well fitted by a linear function (dashed line in Figure 6). The angle increase is qualitatively similar in all three rocks. For each constant σ_3, and within the σ_2 range tested, failure-plane angles increase by as much as 20° from their base values ($\sigma_2 = \sigma_3$). This demonstrates a significant σ_2 effect on rock failure characteristics.

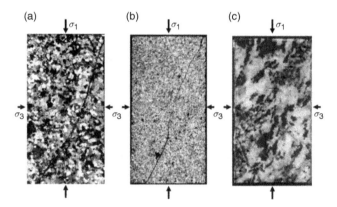

Figure 5 Typical examples of failed crystalline rock specimens after true triaxial tests, showing high-angle fault planes: (a) Westerly granite, (b) KTB amphibolite and (c) SAFOD granodiorite.

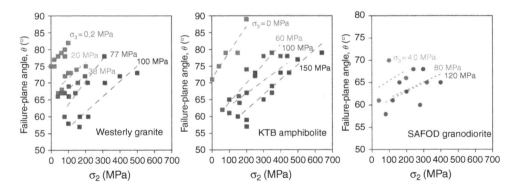

Figure 6 Variation of failure-plane angle with σ_2 for different magnitudes of σ_3 in the tested crystalline rocks.

2.3 True triaxial failure criterion for crystalline rocks

The true triaxial test results in crystalline rocks demonstrate that three-dimension criteria in terms of all the three principal stresses are necessary to fully characterize failure under general stress conditions. We propose a failure criterion based on Nadai (1950) that is entirely experiment-based, utilizing the results of a comprehensive series of true triaxial tests conducted at realistic stress levels of σ_2 and σ_3. The strategy is analogous to the case for establishing the empirical two-dimensional Mohr criterion.

The Nadai criterion is expressed in terms of the two stress invariants, $\tau_{oct,f}$ (octahedral shear stress at failure) and $\sigma_{oct,f}$ (octahedral normal stress, or mean stress, at failure) in the general form of

$$\tau_{oct,f} = f(\sigma_{oct,f}) \tag{1}$$

where $\tau_{oct,f} = 1/3[(\sigma_{1,peak} - \sigma_2)^2 + (\sigma_2 - \sigma_3)^2 + (\sigma_3 - \sigma_{1,peak})^2]^{0.5}$, $\sigma_{oct,f} = (\sigma_{1,peak} + \sigma_2 + \sigma_3)/3$, and f is a function that varies from rock to rock.

Figure 7 presents for each of the tested crystalline rocks, the variation of $\tau_{oct,f}$ with $\sigma_{oct,f}$, where both variables are calculated from the major principal stress at failure ($\sigma_{1,peak}$) and the respective applied σ_2 and σ_3. In all three rock types, the average $\tau_{oct,f}$ rises monotonically with $\sigma_{oct,f}$, although data points appear to scatter especially as $\sigma_{oct,f}$ increases. In Figure 7, we use power law functions to best fit the experimental data, in the form of

$$\tau_{oct,f} = m(\sigma_{oct,f})^n \tag{2}$$

where m and n are empirical constants determined from the experimental results.

Further study of the test results plotted in Figure 7 indicates that the observed scatter is not random. Rather, the wide range of $\sigma_{oct,f}$ that fit the same magnitude of $\tau_{oct,f}$ is a function of the different σ_2 in each of the data points. The relative magnitude of σ_2 between $\sigma_2 = \sigma_3$ and $\sigma_2 = \sigma_1$ can be expressed by a stress ratio parameter, b (Lade & Duncan, 1973; Zhang et al., 2010), defined as $b = (\sigma_2 - \sigma_3)/(\sigma_1 - \sigma_3)$. The parameter b is 0 when $\sigma_2 = \sigma_3$, and 1 when $\sigma_2 = \sigma_1$. In Figure 8 we plotted the true triaxial failure

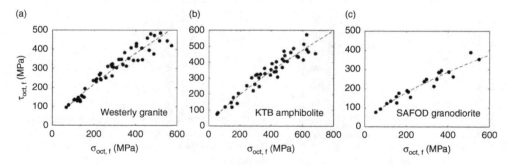

Figure 7 Variation of $\tau_{oct,f}$ with $\sigma_{oct,f}$ in (a) Westerly granite, (b) KTB amphibolite, and (c) SAFOD granodiorite. Dashed lines are the best-fit curves representing failure criteria of the form $\tau_{oct,f} = m(\sigma_{oct,f})^n$.

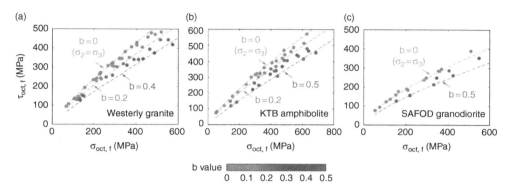

Figure 8 Variation of $\tau_{oct,f}$ with $\sigma_{oct,f}$ in (a) Westerly granite, (b) KTB amphibolite, and (c) SAFOD granodiorite, plotted with different symbol shades of gray for different stress ratios *b* indicated by the shades of gray bar. Failure criteria can be constructed for individual stress ratios with minimum scatter.

stresses in terms of $\tau_{oct,f}$ and $\sigma_{oct,f}$ using specific shades of gray matching individual *b* value. We observe that there is a systematic pattern of $\sigma_{oct,f}$ distribution for each magnitude of $\tau_{oct,f}$, depending on the *b* value. For each level of $\tau_{oct,f}$, the smallest $\sigma_{oct,f}$ is the one corresponding to *b*=0, and the magnitude of $\sigma_{oct,f}$ increases with the *b* value.

Thus, we propose an 'Extended Nadai Criterion' that presents the relationship $\tau_{oct,f} = f(\sigma_{oct,f})$ for each *b* value employed during testing (Figure 8). Since the number of *b* parameters tested for each $\tau_{oct,f}$ magnitude is finite, interpolation between adjacent tested *b* values should yield fairly accurate failure criteria even in cases where the respective *b* value has not been tested.

Within the range of σ_2 used in the true triaxial tests of the three crystalline rocks, the two extreme cases of the respective Extended Nadai Criterion are for *b* = 0 and for *b* = 0.4 or *b* = 0.5, depending on the tested rock. Specifically, for Westerly granite:

$$\tau_{oct,f} = 2.71(\sigma_{oct,f})^{0.85} \text{ (for } b = 0\text{)} \tag{3a}$$

$$\tau_{oct,f} = 1.57(\sigma_{oct,f})^{0.89} \text{ (for } b = 0.4\text{)} \tag{3b}$$

for KTB amphibolite:

$$\tau_{oct,f} = 2.85(\sigma_{oct,f})^{0.82} \text{ (for } b = 0\text{)} \tag{4a}$$

$$\tau_{oct,f} = 1.27(\sigma_{oct,f})^{0.91} \text{ (for } b = 0.5\text{)} \tag{4b}$$

for SAFOD granodiorite:

$$\tau_{oct,f} = 4.58(\sigma_{oct,f})^{0.70} \text{ (for } b = 0\text{)} \tag{5a}$$

$$\tau_{oct,f} = 2.88(\sigma_{oct,f})^{0.74} \text{ (for } b = 0.5\text{)} \tag{5b}$$

Similar best fitting curves, representing Extended Nadai Criteria for specific *b* values between the two extreme values, can be obtained experimentally.

3 TRUE TRIAXIAL FAILURE IN CLASTIC ROCKS

3.1 Failure stress ($\sigma_{1,peak}$)

True triaxial tests leading to compressive failure in clastic rocks were conducted on TCDP (Taiwan Chelungpu-fault Drilling Project) siltstone (Oku *et al.*, 2007), Coconino sandstone (quarried in northern Arizona; Ma, 2014) and Bentheim sandstone (quarried in Germany, Ma, 2014). The porosities of the three rocks vary from 7% (TCDP siltstone), to 17% (Coconino sandstone), to 24% (Bentheim sandstone).

Tests were carried out for six levels of constant σ_3 between 0 and 100 MPa in TCDP siltstone, and eight levels of constant σ_3 between 0 and 150 MPa in Coconino and Bentheim sandstones. Within each series of constant σ_3, the magnitude of σ_2 was varied from test to test, covering the range from $\sigma_2 = \sigma_3$ to $\sigma_2 = \sigma_1$.

We first ran a series of tests in which the applied intermediate and minimum principal stresses we kept equal ($\sigma_2 = \sigma_3$). These 'conventional triaxial' tests yielded the classic Mohr failure criterion (Figure 9), *i.e.* the experiment-based variation of $\sigma_{1,peak}$ with σ_3 (= σ_2). A comparison between the three clastic rocks shows that $\sigma_{1,peak}$ is slightly larger in Coconino sandstone than in TCDP siltstone while $\sigma_{1,peak}$ is substantially smaller in Bentheim sandstone, regardless of σ_3 (= σ_2) magnitude. In all three rocks, failure stress, $\sigma_{1,peak}$, rises with σ_3, albeit at a declining rate. This phenomenon is most apparent as porosity increases, to the point where in Bentheim sandstone, the rise in $\sigma_{1,peak}$ appears to level off when σ_3 is about 100 MPa.

The true triaxial stress conditions at failure were plotted in Figure 10 in the form of $\sigma_{1,peak}$ as a function of σ_2 for each constant σ_3. In all three rocks, $\sigma_{1,peak}$ increases with σ_2 for a given σ_3, until a plateau is reached at some σ_2 value, beyond which $\sigma_{1,peak}$ gradually declines so that when σ_2 approaches the magnitude of σ_1, $\sigma_{1,peak}$ is approximately equal to its base value when $\sigma_2 = \sigma_3$. This typical ascending-then-descending trend of $\sigma_{1,peak}$ can be well-fitted by a polynomial equation of the second order (as represented by dashed curves for each constant σ_3 series in Figure 10). Figure 10 supports observations in crystalline rocks that $\sigma_{1,peak}$ is a function of not only σ_3, but

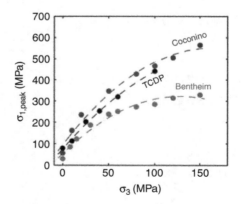

Figure 9 Variation of $\sigma_{1,peak}$ with σ_3 when $\sigma_2 = \sigma_3$ in TCDP siltstone, Coconino sandstone, and Bentheim sandstone.

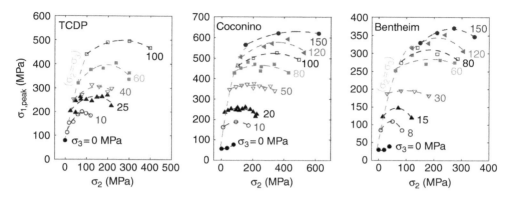

Figure 10 Variation of $\sigma_{1,peak}$ with σ_2 for each constant σ_3 in TCDP siltstone, Coconino sandstone, and Bentheim sandstone. The Mohr failure criterion is the special case when $\sigma_2 = \sigma_3$.

also of σ_2, as shown in Figure 4 (see also Haimson, 2006; Haimson & Chang, 2000; Chang & Haimson, 2000; Lee & Haimson, 2011; Mogi, 1971).

The σ_2 effect on failure was evaluated for each constant σ_3 series, by comparing the maximum $\sigma_{1,peak}$ with its base magnitude when $\sigma_2 = \sigma_3$. It was found that the maximum increase of $\sigma_{1,peak}$ beyond its value when $\sigma_2 = \sigma_3$ (conventional triaxial tests) is considerably lower than in crystalline rocks.

3.2 Failure mode and failure plane angle (θ)

In conventional triaxial compression tests ($\sigma_2 = \sigma_3$), failure-plane angle (θ) direction in the three clastic rocks was random. However, under true triaxial stress conditions, failure-plane angle was aligned with the σ_3 direction, in accord with the findings in crystalline rocks. Failure-plane angle was steep at low σ_3, but gradually became gentler with the rise in σ_3 (Figure 11). As σ_2 was raised beyond constant σ_3, θ generally increased monotonically (Figure 12). This trend can be approximated by a straight line. For example, in TCDP siltstone at $\sigma_3 = 100$ MPa, θ increased with σ_2 by as much as 10° (from 60° at $\sigma_2 = \sigma_3 = 100$ MPa to 70° at $\sigma_2 = 400$ MPa). In Coconino sandstone, θ in specimens subjected to $\sigma_3 = 20$ MPa increased with σ_2 by 14° (from 62° at $\sigma_2 = 40$ MPa to 76° at $\sigma_2 = 215$ MPa). The angle increase for higher σ_3 (= 100 MPa) was only 11°, from 52° at $\sigma_2 = 100$ MPa to 63° at $\sigma_2 = 494$ MPa (Figure 12). For the $\sigma_3 = 100$ MPa series, the number of parallel and conjugate failure planes observed at $\sigma_2 = \sigma_3$ (brittle-ductile transition threshold) diminished as σ_2 increased, while θ steepened. This suggests that by raising σ_2 beyond $\sigma_2 = \sigma_3 = 100$ MPa, Coconino sandstone became more dilatant and returned to its brittle condition.

The increase of θ in Bentheim sandstone was less pronounced than that observed in TCDP siltstone and Coconino sandstone, and was limited to less than 10°. For low σ_3 (= 30 MPa), the increase was 9° (from 57° at $\sigma_2 = 30$ MPa to 66° at $\sigma_2 = 180$ MPa) (Figure 13); at $\sigma_3 = 80$ MPa, the increase in θ was slightly less, from 45° at $\sigma_2 = 80$ MPa to 53° at $\sigma_2 = 285$ MPa. The extreme case was the test series in which $\sigma_3 = 150$ MPa, where all specimens developed compaction bands (θ ≈ 0°), indicating no σ_2

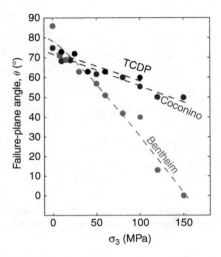

Figure 11 Failure-plane angle θ variation with σ_3 when $\sigma_2 = \sigma_3$ in TCDP siltstone, Coconino sandstone, and Bentheim sandstone.

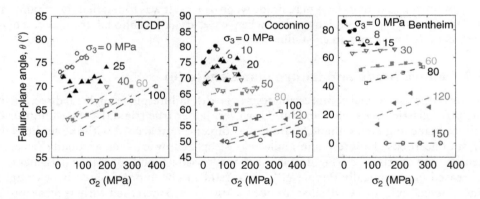

Figure 12 Failure-plane angle θ variation with σ_2 for constant σ_3 levels in TCDP siltstone, Coconino sandstone, and Bentheim sandstone.

dependence. Compaction bands result from localized compaction along a plane normal to σ_1 (θ ≈ 0°), reflecting compressive failure in highly porous rock subjected to high confining stresses, σ_2 and σ_3, and is exhibited by pore collapse and grain shattering and even crushing (Olsson, 1999).

As depicted in Figure 12, the increase in failure-plane angle in all three clastic rocks as a function of σ_2 was generally consistent with but was not as prominent as that observed in crystalline rocks (up to about 20°, see Figure 5). The failure-plane angle variation with σ_2 for a constant σ_3 revealed by the true triaxial tests contradicts the Coulomb criterion assumption that failure-plane angle is a unique rock property. The results are also in disagreement with Mohr failure criterion, which neglects the effect of the intermediate principal stress.

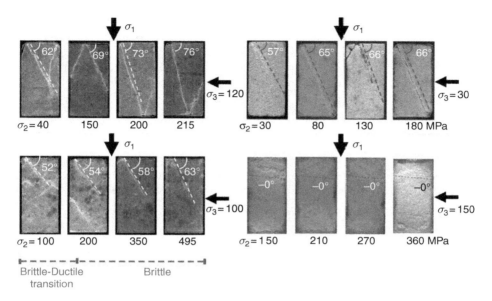

Figure 13 Photographs of specimen faces subjected to σ_2 in Coconino (left) and Bentheim (right) sandstones, showing the failure-plane angle increase with σ_2 in true triaxial tests under constant σ_3.

3.3 True triaxial failure criterion for clastic rocks

Following Nadai (1950), all failure related experimental results can be represented by a single relationship (Equation 1). A summary of all stress conditions at failure in the clastic rocks tested is displayed in Figure 14 in terms of $\tau_{oct,f} = f(\sigma_{oct,f})$. In TCDP siltstone and Coconino sandstone, as $\sigma_{oct,f}$ increases, the average $\tau_{oct,f}$ continuously rises, at a decreasing rate, up to the maximum applied $\sigma_{oct,f}$. In Bentheim sandstone, the rising $\tau_{oct,f}$ reaches a peak when $\sigma_{oct,f} \approx 200$ MPa, beyond which it continuously drops with further rise in $\sigma_{oct,f}$. The best-fitting curve of the $\tau_{oct,f} = f(\sigma_{oct,f})$ relationship forms a 'cap', the threshold of $\sigma_{oct,f}$ beyond which less three-dimensional shear stress ($\tau_{oct,f}$) is required to bring about true triaxial failure. Such 'caps' have been typically found to exist in high-porosity rocks within high $\sigma_{oct,f}$ range, where rocks primarily fail due to compaction rather than shearing (expansion, or dilatancy) (Wong et al., 1997; Olsson, 1999).

All three data sets in Figure 14 show considerable dispersion in the $\tau_{oct,f}$–$\sigma_{oct,f}$ domain. To further identify the effect of σ_2 on the $\tau_{oct,f}$–$\sigma_{oct,f}$ relationship, we introduced the above-mentioned stress ratio b parameter (Lade & Duncan, 1973; Zhang et al., 2010). In Figure 15, we re-plotted the same data points as shown in Figure 14, with unique shades of gray corresponding to their b values (shades of gray bar shown in figure). It is evident that for the same octahedral shear stress ($\tau_{oct,f}$), rock subjected to $\sigma_2 = \sigma_3$ ($b = 0$) requires the lowest mean stress ($\sigma_{oct,f}$) to induce failure, and that the higher the b value, the larger the mean stress required. The maximum mean stress needed for failure is when $b = 1$. For any b between 0 and 1, a $\tau_{oct,f} = f(\sigma_{oct,f})$ relationship

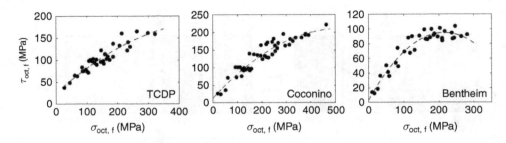

Figure 14 Variation of the octahedral shear stress at failure ($\tau_{oct,f}$) with the octahedral normal stress at failure ($\sigma_{oct,f}$) in the three tested clastic rocks. All data points were correlated with a second-order polynomial equation, showing a monotonically rising trend (albeit with considerable dispersion).

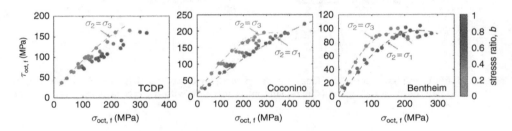

Figure 15 Variation of the octahedral shear stress at failure ($\tau_{oct,f}$) with the octahedral normal stress at failure ($\sigma_{oct,f}$) in the three tested clastic rocks. All data points are represented by a color corresponding to the stress ratio b at failure (= $(\sigma_2-\sigma_3)/(\sigma_{1,peak}-\sigma_3)$), as indicated by the color sidebar. Best fitting curves are plotted for b = 0 and b = 1.

can be fitted by a *b*-specific second-order polynomial equation, carrying a very small standard deviation.

For the two extreme *b* values the following true triaxial failure criteria were obtained and plotted in Figure 15.

TCDP siltstone:

$$\tau_{oct,f} = 24.98 + 0.63\sigma_{oct,f} - 0.00062\sigma_{oct,f}^2 \quad (\text{for } b=0) \tag{6}$$

(*b* = 1 tests were too few to produce a best-fitting curve in this rock.)
Coconino sandstone:

$$\tau_{oct,f} = 8.36 + 1.132\sigma_{oct,f} - 0.002\sigma_{oct,f}^2 \quad (\text{for } b=0) \tag{7a}$$

$$\tau_{oct,f} = 7.87 + 0.61\sigma_{oct,f} - 0.000332\sigma_{oct,f}^2 \quad (\text{for } b=1) \tag{7b}$$

Bentheim sandstone:

$$\tau_{oct,f} = 2.98 + 1.139\sigma_{oct,f} - 0.0036\sigma_{oct,f}^2 \quad (\text{for } b=0) \tag{8a}$$

$$\tau_{oct,f} = -5.82 + 0.836\sigma_{oct,f} - 0.0017\sigma_{oct,f}^2 \quad (\text{for } b=1) \tag{8b}$$

As in the case of crystalline rocks, we propose that the true triaxial failure of clastic rocks is best represented by the Extended Nadai Criterion in the form of $\tau_{oct,f} = f(\sigma_{oct,f})$ for each b value employed during testing. Since the number of b parameters that can be tested for each $\tau_{oct,f}$ magnitude is limited, interpolation between adjacent tested b values should yield fairly accurate failure criteria even in cases where the respective b value has not been tested directly.

4 CONCLUSIONS

- We have designed, fabricated, and successfully calibrated and tested a new true triaxial apparatus capable of applying three unequal loads to the faces of $19 \times 19 \times 38$ mm rock cuboids, with minimal friction.
- We completed an extensive series of tests in three crystalline rocks, Westerly granite, KTB amphibolite, and SAFOD granodiorite, and obtained a true triaxial failure criterion for each.
- We also completed similar series of tests in three clastic rocks, TCDP siltstone, Coconino sandstone, and Bentheim sandstone, of porosities 7%, 17%, and 24%, respectively, leading to a true triaxial failure criterion for each.
- We found that because of the effect of the intermediate principal stress on the mechanical behavior of all tested rocks, not one single equation can correctly represent the relationship between all the critical stresses leading to failure. Based on our laboratory results, we propose an experiment-based Extended Nadai Criterion, in which failure is defined by a Nadai-type relationship for the specific relative magnitude of the intermediate principal stress between the minimum and the maximum principal stresses.
- The true triaxial failure criterion for each of the crystalline and clastic rock-stested, is superior to the commonly used Mohr, or Mohr-Coulomb criteria in that it accounts for the significant effect of the intermediate principal stress. For the same least horizontal stress, the maximum principal stress at failure may increase by as much as 50% or more over that determined in a conventional triaxial test, depending on the intermediate principal stress magnitude. An important consequence is that when the difference between least and intermediate principal stresses is significant, use of a Mohr-type criterion may lead to overly conservative predictions of instability. Conversely, back-calculating the critical principal stresses active during an earthquake inducing fault initiation or a wellbore breakout, may yield erroneous results unless a true triaxial strength criterion is used.
- True triaxial tests reveal that for the same least principal stress, failure-plane angle increases with the magnitude of the intermediate principal stress. This increase is more prevalent in the crystalline rocks.

ACKNOWLEDGMENTS

The authors are indebted to the National Science Foundation for supporting this research under grants EAR-9418738 and EAR-0940323.

REFERENCES

Brace, W.F. & Kohlstedt, D.L. (1980) Limits on lithospheric stress imposed by laboratory experiments. *Journal of Geophysical Research*, 85, 6248–6252.

Chang, C. & Haimson, B. (2000) True triaxial strength and deformability of the German Continental Deep Drilling Program (KTB) deep hole amphibolite. *Journal of Geophysical Research*, 105, 18999–19013.

Ewy, R.T. (1998) Wellbore stability predictions using a modified Lade criterion: *Proceedings of Eurock 98: SPE/ISRM Rock Mechanics in Petroleum Engineering, Published by Society of Petroleum Engineers*, vol. 1, SPE/ISRM paper no. 47251, 247–254.

Haimson, B.C. (1978) The hydrofracturing stress measurement technique-method and recent field results. *International Journal of Rock Mechanics and Mining Sciences and Geomechanics Abstracts*, 15, 167–178.

Haimson, B. (2006) True triaxial stresses and the brittle fracture of rock. *PAGEOPH*, 163, 1101–1130.

Haimson, B. & Chang, C. (2000) A new true triaxial cell for testing mechanical properties of rock, and its use to determine rock strength and deformability of Westerly granite. *International Journal of Rock Mechanics and Mining Sciences*, 37, 285–296.

Handin, J., Heard, H.C., & Magouirk, J.N. (1967) Effect of the intermediate principal stress on the failure of limestone, dolomite, and glass at different temperature and strain rate. *Journal of Geophysical Research*, 72, 611–640.

Jaeger, J.C. & Cook, N.G.W. (1979) *Fundamentals of Rock Mechanics*, 3rd ed., Chapman and Hall, London, Blackwell Publishing.

Judd, W.R. (1963) *State of stress in the Earth's crust*. In: Brace, W.F. (ed.) *Brittle Fracture of Rocks*, New York, Elsevier, pp. 111–178.

Lade, P.V. & Duncan, J.M. (1973) Cubical triaxial tests on cohesionless soil. *Journal of the Soil Mechanics and Foundations Division, ASCE*, 99, 793–812.

Lee, H. & Haimson, B.C. (2011) True triaxial strength, deformability, and brittle failure of granodiorite from the San Andreas Fault Observatory at Depth. *International Journal of Rock Mechanics and Mining Sciences*, 48, 1199–1207.

Lockner, D.A., Byerlee, J.D., Kuksenko, V., Ponomarev, A., & Sidorin, A. (1991) Quasi-static fault growth and shear fracture energy in granite. *Nature*, 350, 39–42.

Ma, X. (2014) *Failure characteristics of compactive, porous sandstones subjected to true triaxial stresses*, PhD thesis, University of Wisconsin-Madison.

McGarr, A. & Gay, N.C. (1978) State of stress in the earth's crust. *Annual Review of Earth and Planetary Sciences*, 6, 405–436.

Mogi, K. (1971) Fracture and flow of rocks under high triaxial compression. *Journal of Geophysical Research*, 76, 1255–1269.

Murrell, S.A.F. (1963) A criterion for brittle fracture of rocks and concrete under triaxial stress and the effect of pore pressure on the criterion. *Proceedings of Fifth US Rock Mechanics Symposium of the ARMA*, Pergamon Press, Oxford, 563–577.

Nadai, A. (1950) Theory of Flow and Fracture of Solids, v. 1, New York, McGraw-Hill Publishing.

Oku, H., Haimson, B. & Song, S.R. (2007) True triaxial strength and deformability of the siltstone overlying the Chelungpu fault (Chi-Chi earthquake), Taiwan. *Geophysical Research Letters*, 34(9), L09306.

Olsson, W.A. (1999) Theoretical and experimental investigation of compaction bands in porous rock. *Journal of Geophysical Research*, 104(B4), 7219–7228.

Röckel, T. & Natau, O. (1995) *Rock mechanics, Niedersächsisches Landesamt für Bodenforschung*, Hannover, Germany, KTB Report number: 95–2.

Vernik, L. & Zoback, M.D. (1992) Estimation of maximum horizontal principal stress magnitude from stress-induced well bore breakouts in the Cajon Pass scientific research borehole. *Journal of Geophysical Research*, 97, 5109–5119.

Wawersik, W.R. & Brace, W.F. (1971) Post-failure behavior of a granite and diabase. *Rock Mechanics*, 3, 61–85.

Wiebols, G.A. & Cook, N.G.W. (1968) An energy criterion for the strength of rock in polyaxial compression. *International Journal of Rock Mechanics and Mining Sciences and Geomechanics Abstracts*, 5, 529–549.

Wong, T.-f. (1982) Micromechanics of faulting in Westerly granite. *International Journal of Rock Mechanics and Mining Sciences and Geomechanics Abstracts*, 19, 49–64.

Wong, T., David, C. & Zhu, W. (1997) The transition from brittle faulting to cataclastic flow in porous sandstones: Mechanical deformation. *Journal of Geophysical Research*, 102(B2), 3009–3025.

Zhang, C, Zhou, H., Feng X. & Huang S. (2010) A new interpretation for the polyaxial strength effect of rock. *International Journal of Rock Mechanics and Mining Sciences*, 47(3), 496–501.

Chapter 14

The $MSDP_u$ multiaxial criterion for the strength of rocks and rock masses

L. Li, M. Aubertin & R. Simon
Department of Civil, Geological & Mining Engineering, École Polytechnique, Montréal, Québec, Canada

Abstract: The natural and induced stresses near openings in rock media usually have different magnitudes along different directions. These have to be taken into account for stability analysis. A general multiaxial criterion was developed to assess the strength of rocks and rock masses. The $MSDP_u$ criterion was formulated in a modular manner, so it can be adapted to a variety of rock characteristics and structural calculations. The main equations are presented in this chapter. Specific conditions are described, including short term failure, initiation of damage, time effect, and the influence of scale and defects. In the case of rock mass, the criterion includes a continuity parameter that can be linked to the RMR geomechanical classification. The $MSDP_u$ criterion is compared with other formulations to highlight some of the similarities and differences. Validation and application of the criterion are illustrated using laboratory tests results; additional studies also present the analysis of borehole stability and failure around large scale excavations.

1 INTRODUCTION

The multiaxial stress state that exists in the rock around openings must be taken into account to assess their behavior and stability. Many criteria have been proposed to evaluate the possible occurrence of failure in rock media (*e.g.*, Jaeger & Cook, 1979; Franklin & Dusseault, 1989; Lade, 1993; Andreev, 1995; Sheorey, 1997) and other similar materials (Wastiels, 1979; Meredith, 1990; Theocaris, 1995; Yu, 2002; Papanikolaou & Kappos, 2007; Du *et al.*, 2010; Lu *et al.*, 2016). A variety of factors may affect rock failure, including scale, time, and characteristics of the defect population (from microcracks and pores to joints and faults).

The most popular expressions used in practice, such as the Coulomb (Lama, 1974; Goodman, 1980) and Hoek and Brown (1980a, 1980b, 1988, 1997) formulations, rely on equations that involve only the major σ_1 and minor σ_3 principal stresses, hence neglecting (or simplifying) the effect of the intermediate principal stress σ_2. Commonly used criteria also omit the influence of time and use simplified scaling parameters to go from laboratory specimen to rock mass size. When combined with idealized constitutive laws (such as linear elasticity), such simplifications may greatly help users faced with actual stability calculations, while reducing the effort needed to obtain the required material parameters. However, these simplifications can reduce significantly the accuracy of the calculations.

Detailed studies required for some of the most challenging rock engineering projects, such as large scale caverns, transportation tunnels, underground storage reservoirs, and toxic waste disposal facilities, typically require the use of more elaborate models developed over the years (*e.g.*, Desai & Salami, 1987; Aubertin *et al.*, 1994, 1998; Shao *et al.*, 1996; Cristescu & Hunsche, 1997). Such models must rely on representative expressions to define specific states, such as onset of cracking and ultimate strength. These are usually expressed in stress space using an appropriately formulated criterion.

Many relatively brittle materials (rocks, concrete, cast iron, ceramics) used by engineers share common features. For instance, their uniaxial compression strength C_0 largely exceeds their axial tensile strength T_0. Also, the maximum deviatoric stress that can be supported largely depends on the loading geometry, with all three principal stresses influencing their strength. The maximum load that can be supported is generally reached at small strain ($\varepsilon < 1\%$), and the stress-strain curve beyond the peak shows a significant drop (under a low confining pressure).

Failure of rock (and other brittle materials) is associated with the evolution of micro-cracks and appearance of macro-cracks associated with localized deformations. Failure thus constitutes a limit condition that engineers want to avoid. The failure condition is usually defined in stress space with a surface represented by a mathematical expression known as the failure criterion. This expression and the related parameters are generally obtained from laboratory tests performed under well-defined stress path. The maximum load supported by the tested specimen represents the failure strength. Performing several tests under different loading conditions gives different points on the failure surface. The selected mathematical expression is often (partially) empirical, based on a good adjustment of the formulation to experimental results. Many criteria have been developed for rocks and somewhat similar (brittle) materials (see reviews by Andreev, 1995; Sheorey, 1997; see also Aubertin *et al.*, 1999, 2000). Each of these criteria has its advantages and limitations, and none is universal or unanimously accepted.

The $MSDP_u$ criterion will be presented below, following a description of the main features that are deemed required for a rock failure criterion.

General description of rock failure

Like many other brittle materials, rocks often have a low porosity (usually less than a few percent). Their failure surface in the principal stress space, which represents the locus of the peak strength, is closed along the axes of negative (tensile) stresses and it remains open along the positive (compressive) stresses axes.

Before reaching failure, different stages are encountered (Paterson, 1978; Aubertin & Simon, 1997). For instance, a hard rock specimen submitted to an unconfined uniaxial compression test first shows a stage of tightening (or elastic contraction) due to the closure of micro cracks, followed by a stage of linearly elasticity, then a stage of inelasticity when the applied stress exceeds the threshold of crack propagation. This crack propagation can eventually drive the sample to failure at peak strength.

Such failure usually results from the propagation and eventual coalescence of micro-cracks (Li & Nordlund, 1993; Germanovitch *et al.*, 1996). The onset of crack propagation and their subsequent interaction leading to the failure depend on the path followed in the stress space. The application of compressive stresses, which tend to close some of the micro cracks, results in the mobilization of frictional resistance on the contact faces (McClintock & Walsh, 1962); this effect is related to the influence of the hydrostatic component of the stress tensor on the propagation threshold and on the ultimate strength of rocks.

Description of the failure envelope

The failure surface can be represented in the tridimensional space of principal stresses $\sigma_1, \sigma_2, \sigma_3$ (with $\sigma_1 \geq \sigma_2 \geq \sigma_3$). It can also be visualized in the octahedral (π) plane and in the $I_1 - J_2^{1/2}$ plane (where I_1 is the first invariant of the stress tensor σ_{ij}; J_2 is the second invariant of the deviatoric stress tensor S_{ij}).

Fig. 1 shows this surface, for the MSDP$_u$ criterion, in the conventional triaxial compression (CTC) test stress plane, $\sqrt{2}\sigma_x - \sigma_z$ (where $\sigma_x = \sigma_y$, represents the stress applied along the horizontal (radial) axes, and σ_z is the vertical stress). The shape of this criterion is based on physical and phenomenological considerations, which can be summarized by using specific points and segments identified on the curve of ultimate strength shown in Fig. 1.

When the specimen is submitted to uniaxial loading, its resistance is C_0 in compression and T_0 in tension ($T_0 < 0$); these two conditions are respectively represented by points A and B on Fig. 1. Point D represents a state of biaxial loading in tension, with $\sigma_x = \sigma_y < 0$ and $\sigma_z = 0$. In this case, the σ_x value should exceed T_0, i.e. $|\sigma_x| < |T_0|$ (e.g., Theocaris, 1995; Aubertin & Simon, 1998). A state of spherical loading in tension ($\sigma_x = \sigma_y = \sigma_z < 0$) should also leads to a lower strength in absolute value than axial tensile loading, i.e. $|\sigma_x| < |T_0|$ (see point C). It is to be noted that several existing criteria, often inspired from the two dimensional criterion of Griffith (1921, 1924), consider that failure in tension can only be produced when one of the three principal stresses becomes equal to T_0. However, this vision is not supported by the physics of the problem or by a theoretical analysis of the failure conditions under multiaxial loading; when the stress component perpendicular to the critical failure plane is equal to T_0, the

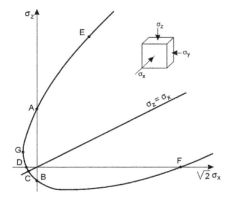

Figure 1 A schematic representation of the failure surface of rocks in the conventional triaxial plane (taken from Aubertin & Simon, 1996, 1998).

individual value of the principal stresses will be higher than T_0 (*i.e.* smaller than $|T_0|$ in absolute value) in the case of biaxial or spherical tensile loading. Therefore, such a Griffith type approach is not conservative, and it can lead to an overestimation of the strength of brittle materials submitted to these types of tensile loading. It also creates an apex (with a singularity) on the negative side of the stresses axes, which is problematic from a numerical point of view.

When one applies simultaneously some compressive and tensile stresses, as is the case between points D and A or between points B and F in Fig. 1, some criteria consider that the strength is only controlled by the highest tensile stress (in absolute value), with failure if $\sigma_z = T_0$ or $\sigma_x = T_0$, or by the maximum compressive stress that must be lower or equal to C_0. However, some laboratory test results, such as those reported by Andreev (1995, Figs. 6.175b and 6.176b) and Hunsche (1994, Fig. 3.19) indicate that the application of a relatively small compressive stress perpendicularly to the axis of tension can increase the material strength due to a tightening (increased stiffness) associated with closure of micro-cracks and mobilization of friction along the contact faces of cracks; this is happening around point G in Fig. 1. When the compressive stress becomes large enough, it also participates to (wing) crack propagation, leading to failure (Aubertin & Simon, 1998). Point F on Fig. 1 represents the strength of a material submitted to a biaxial compression, with $\sigma_x = \sigma_y > C_0$ (also shown by Ottosen, 1977; Maso & Lereau, 1980; Lade, 1982, 1993). Beyond point A ($= C_0$), all stresses are in compression, the propagation of closed cracks implies that the frictional resistance along contact faces must be overcome (McClintock & Walsh, 1962). The value of the available friction coefficient along the contact faces may change with the normal stress, due to the shearing of asperities as is the case with geological discontinuities (Patton, 1966; Ladanyi & Archambault, 1970). One thus expects that the slope of the failure criterion, which reflects the mobilization of friction, progressively decreases with the increased confining stress. This slope reduction ends when all asperities are sheared and the frictional sliding takes place over flattened surfaces. The contribution of this friction to the strength of the material becomes proportional to the residual friction angle ϕ_r (which is often close to the base friction angle ϕ_b). The slope of the failure criterion in the $\sqrt{2}\sigma_x - \sigma_z$ plane is then constant beyond point E in Fig. 1. The failure locus then corresponds to a linear (Coulomb type) criterion under higher mean stress, while for a lower means stress, the apparent friction angle tends to decrease progressively, as observed on rocks (Singh *et al.*, 1989; Charlez, 1991) and concrete (Chen & Chen, 1975).

The $MSDP_u$ failure criterion in the $I_1 - J_2^{1/2}$ plane is defined by a curve on which the conventional triaxial compression (CTC) test strength (where $\sigma_z = \sigma_1 \geq \sigma_x = \sigma_y = \sigma_2 = \sigma_3$) is located above that of reduced triaxial extension (RTE) test (where $\sigma_z = \sigma_3 \leq \sigma_x = \sigma_y = \sigma_2 = \sigma_1$). This difference reflects the effect of the intermediate principal stress σ_2 on the ultimate strength of rock and similar materials, as demonstrated by various experimental observations (*e.g.*, Mills & Zimmerman, 1970; Akai & Mori, 1970). In this $I_1 - J_2^{1/2}$ plane, the criterion is defined by a curve up to the point where it becomes a straight line of slope α in CTC (as shown in Fig. 2). In the π plane, the criterion forms a rounded triangle.

Many of the features described above are captured by the multiaxial criterion presented in the following sections.

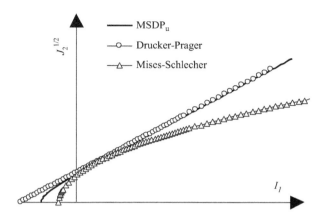

Figure 2 Schematical comparison between the MSDP$_u$ criterion and related criteria (taken from Li et al., 2005).

2 THE MSDP$_U$ CRITERION FOR INTACT ROCK

The proposed 3D criterion, named MSDP$_u$ (which stands for Mises-Schleicher & Drucker-Prager unified), was developed to define a general unified locus in stress space. It has been formulated in a modular manner, so it can be adapted to a variety of rock and rock masses (and other media) for structural calculations. This criterion was initially developed to describe the short term strength of intact rock (Aubertin & Simon, 1996) and other low porosity, brittle materials (Aubertin & Simon, 1998). It has later been extended to describe the short term (Aubertin et al., 1999) and long term strength of rocks (Aubertin & Simon, 1997) and rock masses (Aubertin et al., 2000), considering various influence factors such as porosity and the presence of discontinuities at different scales. The main components of the criterion are presented below; other features are also mentioned in relation with additional applications described elsewhere.

The formulation of the MSDP$_u$ criterion for intact rocks is presented in Table 1 (left column). The basic equation for the failure strength can be written as:

$$\sqrt{J_2} = F_0 F_\pi \qquad (1)$$

In this equation, F_0 is a function of the first stress invariant I_1. This function includes parameters α, a_1 and a_2 defined from basic material properties, *i.e.* σ_c and σ_t, the uniaxial compressive and tensile strength (in absolute values) respectively, and ϕ, the friction angle on plane surfaces ($\phi \cong \phi_r$, the residual friction angle). Parameter b reflects the ratio of the locus size at a Lode angle $\theta = -30°$ and $30°$ in the π plane (see Fig. 3); the value of b can range from about 0.7 to 1.

The MSDP$_u$ criterion adopts the shape of a rounded triangle in the octahedral (π) plane (Aubertin et al., 1994), to represent the higher strength under triaxial compression than under reduced extension (Fig. 3b). An alternative expression has also been proposed for the formulation of F_π, so it can be reduced to the shape of an isosceles triangle in the π plane (Aubertin & Li, 2004; Li et al., 2005); this formulation is better suited for granular materials and is not deemed required for rock media.

Table 1 Formulation of the MSDP$_u$ criterion.

For brittle intact rock	For porous rock[†]	For damaged rock and rock mass[†]

$$\sqrt{J_2} - F_0 F_\pi = 0$$

$$\sqrt{J_2} - F_0 F_\pi = 0$$

$$\sqrt{J_2} - F_0 F_\pi = 0$$

$$F_0 = \left[\alpha^2\left(I_1^2 - 2a_1 I_1\right) + a_2^2\right]^{1/2}$$

$$F_0 = \left\{\alpha^2\left(I_1^2 - 2a_{1n} I_1\right) + a_{2n}^2 - a_{3n}\langle I_1 - I_{cn}\rangle^2\right\}^{1/2}$$

$$F_0 = \left[\alpha^2\left(I_1^2 - 2\tilde{a}_1 I_1\right) + \tilde{a}_2^2 - a_3'\langle I_1 - I_c\rangle^2\right]^{1/2}$$

$$F_\pi = b[b^2 + (1 - b^2)\sin^2(45° - 1.5\theta)]^{-1/2}$$

$$a_{1n} = \left(\frac{\sigma_{cn} - \sigma_{tn}}{2}\right) - \left(\frac{\sigma_{cn}^2 - (\sigma_{tn}/b)^2}{6\alpha^2(\sigma_{cn} + \sigma_{tn})}\right)$$

$$\tilde{a}_1 = \Gamma a_1 = \left(\frac{\tilde{\sigma}_c - \tilde{\sigma}_t}{2}\right) - \left(\frac{\tilde{\sigma}_c^2 - (\tilde{\sigma}_t/b)^2}{6\alpha^2(\tilde{\sigma}_c + \tilde{\sigma}_t)}\right)$$

$$\theta = \frac{1}{3}\sin^{-1}\frac{3\sqrt{3}}{2}\frac{J_3}{\sqrt{J_2^3}}; -30° \leq \theta \leq 30°$$

$$a_{2n} = \left\{\left(\frac{\sigma_{cn} + (\sigma_{tn}/b^2)}{3(\sigma_{cn} + \sigma_{tn})} - \alpha^2\right)\sigma_{cn}\sigma_{tn}\right\}^{1/2}$$

$$\tilde{a}_2 = \Gamma a_2 = \left\{\left(\frac{\tilde{\sigma}_c + (\tilde{\sigma}_t/b^2)}{3(\tilde{\sigma}_c + \tilde{\sigma}_t)} - \alpha^2\right)\tilde{\sigma}_c\tilde{\sigma}_t\right\}^{1/2}$$

$$\alpha = \frac{2\sin\phi}{\sqrt{3}(3 - \sin\phi)}$$

$$a_{3n} = \frac{\alpha^2\left(I_{1n}^2 - 2a_{1n}I_{1n}\right) + a_{2n}^2}{(I_{1n} - I_{cn})^2}$$

$$a_3' = a_3(1 - \Gamma)$$

$$a_1 = \left(\frac{\sigma_c - \sigma_t}{2}\right) - \left(\frac{\sigma_c^2 - (\sigma_t/b)^2}{6\alpha^2(\sigma_c + \sigma_t)}\right)$$

$$\tilde{\sigma}_c = \Gamma\sigma_c \text{ and } \tilde{\sigma}_t = \Gamma\sigma_t$$

$$a_2 = \left\{\left(\frac{\sigma_c + (\sigma_t/b^2)}{3(\sigma_c + \sigma_t)} - \alpha^2\right)\sigma_c\sigma_t\right\}^{1/2}$$

[†]F_π, θ, and α have the same expressions as those for brittle intact rock

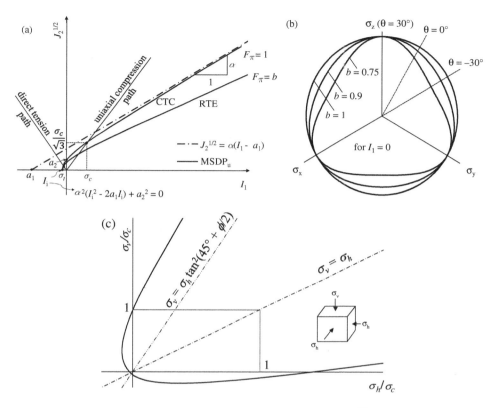

Figure 3 Schematic representation of the MSDP$_u$ criterion: a) in the $I_1 - J_2^{1/2}$ plane, CTC: conventional triaxial compression ($\theta = 30°$), RTE: reduced triaxial extension ($\theta = -30°$); b) in the octahedral plane; c) in the biaxial stress plane. Figures adapted from Aubertin et al. (2000).

Fig. 3c shows that, in the plane of the intermediate σ_2 and major σ_1 principal stresses (normalized by $C_0 = \sigma_c$), the shape of the criterion follows the conditions described above for Fig. 1.

The criterion can also be expressed in terms of the q and p invariants that are commonly used in soil mechanics (Li et al., 2005).

In the $I_1 - J_2^{1/2}$ plane (Fig 3a), the failure envelope corresponding to the CTC condition is higher than that corresponding to the RTE condition. In the octahedral (π) plane (Fig. 3b), the shape of the MSDP$_u$ criterion can change from a circle (with $b = 1$) to a rounded triangle (for $1 > b \geq 0.70$). For most brittle rocks, the value of b is typically close to 0.75. In Fig. 3c, it can be seen that the biaxial compressive strength is higher than the uniaxial compressive strength (i.e. $\sigma_1 = \sigma_2 > \sigma_c$).

Fig. 3 indicates that for isotropic rocks submitted to conventional triaxial compression (CTC) tests, the MSDP$_u$ formulation practically reduces to the Mises-Schleicher criterion (Lubliner, 1990) at low mean stress and approaches the Drucker-Prager equation (Drucker & Prager, 1952; Desai & Siriwardane, 1984) at higher mean stress, as shown in Fig. 3a.

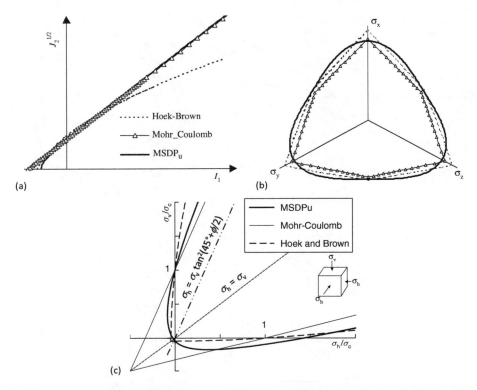

Figure 4 Schematic comparisons between the MSDP$_u$ criterion and two commonly used criteria for geomaterials (shown for normalized parameters): a) in the $I_1 - J_2^{1/2}$ plane; b) in the octahedral plane; c) in the biaxial stress plane.

Figure 4 shows a schematic representation of the MSDP$_u$ criterion (in the same three planes) plotted with the well-known Mohr-Coulomb and Hoek–Brown criteria. The MSDP$_u$ criterion is fairly close to these two criteria (Li *et al.*, 2005), but there are some important differences. For instance, the proposed criterion avoids the presence of a singularity on the tensile (negative stress) side, contrary to the Mohr-Coulomb and Hoek-Brown criteria; this apex may tend to overestimate the tensile strength of the materials because of the linear shape in the negative stresses quadrant (Fig. 4c). In the octahedral (π) plane (Fig. 4b) the MSDP$_u$ criterion takes a rounded triangle shape while the 3D Mohr-Coulomb and Hoek–Brown envelopes include singularities at $\theta = 30°$ and $-30°$. In the biaxial stress plane (Fig. 4c), the commonly used Mohr-Coulomb and Hoek–Brown formulations predict an equal strength under uniaxial compression and biaxial compression, contrary to the MSDP$_u$ criterion.

Also, the latter is based on the basic assumption that the uniaxial compression σ_c and tensile σ_t strengths are two distinct properties that are not directly related to each other, and which must be determined specifically (*e.g.* You 2015); the two other criteria shown in Fig. 4 consider that these two strength parameters are linked (*i.e.* one can thus be predicted using the other).

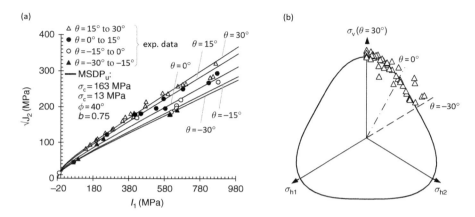

Figure 5 The MSDP$_u$ criterion applied to peak strength of Bowral trachyte: a) in $I_1 - J_2^{1/2}$ plane; b) in π plane. Tests results taken from Hoskins (1969); adapted from Aubertin et al. (1999).

The MSDP$_u$ criterion has been compared with various other criteria developed for geomaterials to illustrate the similarities and differences (Li et al., 2005); these specific features will not be repeated here. It is nonetheless interesting to note that its shape is quite similar to the one described by Lundborg (1974) for the strength of intact rock submitted to multiaxial loading. The MSDP$_u$ characteristics are also quite close to those of other criteria recently developed for different materials (e.g., Du et al., 2010; Lu et al., 2016).

The application of the MSDP$_u$ criterion to low porosity rock samples submitted to laboratory tests is straightforward and was illustrated in Aubertin et al. (1999, 2000), Li et al. (2000) and Li et al. (2005). Only the values of σ_c, σ_t, ϕ ($= \phi_r$) and b are required to apply the criterion. These values are obtained from independent tests that can include uniaxial, diametric, and triaxial compression tests and shear tests on sheared surfaces. For homogeneous rocks, the full failure surface in stress space can be defined using a fairly small number of representative test results (depending on data scattering). It has previously been shown by the authors that the multiaxial formulation of MSDP$_u$ represents well different types of test results. For instance, Fig. 5 shows results obtained on Bowral trachyte.

3 TIME EFFECTS

All rocks may show a mechanical response that is time (or rate) dependent (Cristescu & Hunsche, 1997; Aubertin et al., 1998). The proposed MSDP$_u$ criterion can be applied to define various stages of material failure.

Damage initiation and long term strength

Several studies have shown that a damage initiation threshold (DIT) exists in rock. It can be associated with the onset of micro-cracking, detected through volumetric strain measurements or acoustic emission activities (Paterson, 1978; Meredith, 1990; Martin & Chandler, 1994; Aubertin & Simon, 1997; Aubertin et al., 1998). This threshold, which can also be seen as the long term strength of rocks, can be defined using the

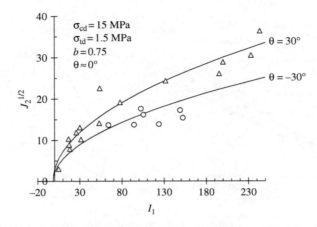

Figure 6 The MSDP$_u$ criterion applied to the damage initiation threshold (DIT) of rocksalt (data taken from Thorel, 1994). Figure adapted from Aubertin *et al.* (2000).

MSDP$_u$ criterion. For this type of application, parameters σ_c and σ_t are replaced by the corresponding values for the DIT (*i.e.* σ_{cd} and σ_{td}), which can be identified on the stress-strain curves. It has been observed that for many low porosity rocks, $\sigma_{cd} = 0.3$ to $0.7\,\sigma_c$ and $\sigma_{td} = 0.5$ to $0.9\,\sigma_t$. As for parameters ϕ and b, their value does not seem to differ much when going from the usual short term to the long term (DIT) conditions.

The MSDP$_u$ criterion has been used to describe the long term failure surface of hard and brittle rocks as well as for softer and more ductile materials. For instance, Fig. 6 shows that the criterion matches fairly well (considering the data scattering) the results obtained on rocksalt samples tested by Thorel (1994), using a very low value of ϕ ($\phi_r \cong 0$) because of the viscous and plastic nature of rocksalt response at high mean stresses. For such semi-brittle behavior, MSDP$_u$ reduces to the Mises-Schleicher criterion for CTC tests, an expression frequently used for metals (Skrzypek & Hetnarski, 1993; Hjelm, 1994). When the mean stress is large enough, the surface becomes almost parallel to the I_1 axis, thus resembling the von Mises criterion (for $b = 1$).

Delayed failure

The strength of rock specimens submitted to a sustained deviatoric loading is expected to decrease over time. For a given stress state, the time to failure can be expressed with an simple equation based on an extension of Charles (1958) law for subcritical crack growth (Aubertin *et al.*, 2000). The equation can be formulated as:

$$t_f = \alpha_1 \left(\frac{\delta_1 + \delta_2}{\langle \delta_1 \rangle} \right)^\beta \tag{2}$$

where δ_1 is the difference between the applied deviatoric stress σ_{app} and the DIT and δ_2 is the difference between the short term strength STF (standard test) and σ_{app}; α_1, β are material parameters. Equation 2 can be used to evaluate strength as a function of time. For instance, Fig. 7a shows the application of this equation to Lac du Bonnet granite

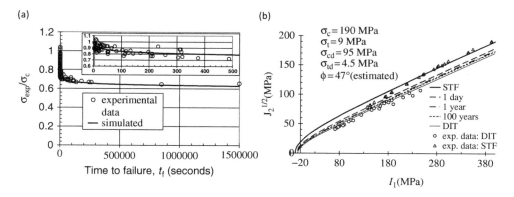

Figure 7 Time dependent failure strength of Lac du Bonnet granite cylinders: (a) uniaxial compression; STF tests performed on cylinders using a loading rate of about 1 MPa per second up to failure; other samples subjected to long term loading at constant loads (data taken from Schmidtke & Lajtai, 1985); (b) uniaxial and triaxial compression tests; application of MSDP$_u$ in $I_1 - J_2^{1/2}$ plane (data taken from Lau & Gorski, 1991). Figures adapted from Aubertin et al. (2000).

samples submitted to uniaxial compression tests, using data from Schmidtke & Lajtai (1985) with $\alpha_1 = 2.7$ s, $\beta = 9.73$, and a long term uniaxial strength (DIT) taken as half the short term strength (STF). On this figure, it can be seen that the strength drop is initially more rapid and then progresses rather slowly toward the long term strength (DIT), which would be attained after a very long time. It is assumed here that any deviatoric stress below the DIT could be supported indefinitely (Aubertin & Simon, 1997; Aubertin et al., 1998). Fig. 7b shows the corresponding isochronous curves for the strength of Lac du Bonnet granite for various time intervals; these curves are obtained by reducing σ_c and σ_t proportionally to the strength given by Equation 2 at a given time (data taken from Lau & Gorski, 1991).

4 SIZE EFFECT FOR INTACT ROCK

The strength of rock is influenced by scale, with the measured peak stress usually decreasing with sample size (Hoek & Brown, 1980b; Bieniawski, 1984; Cunha, 1993a, 1993b). This phenomenon has been linked mainly to statistical effects due to random strength and defect distribution (Jaeger & Cook, 1979) and to energy allocation and dissipation around cracks (Bazant & Planas, 1998). Size effect analysis is however complicated because it depends on the deformation processes, which in turn may vary with the loading state and testing method (Jaeger & Cook, 1979; Hudson & Harrison, 1997). Scale effects are usually more pronounced in very brittle materials, and they progressively decrease when going from a brittle to a semi-brittle behavior, altogether disappearing in the ductile (fully plastic) regime of inelastic flow. The influence of scale is also more pronounced in uniaxial tension than in uniaxial compression. It can be reduced substantially by applying a large confining pressure in triaxial compression tests.

Rock strength is decreased by larger defect size and defect density (Ramamurthy & Arora, 1994; Wong et al., 1996). As the initial size of rock defects is often related to grain size, it can be expected that an increase in mean grain dimension also reduces failure strength (Wong et al., 1996; Hatzor & Palchik, 1997).

Rock strength diminishes until the specimen size becomes equal to that of the large scale reference size d_L. Beyond this sufficiently large size, the effect of scale practically disappears. The strength then remains unchanged beyond d_L unless new types of defects are introduced (such as joint sets in a rock mass—see next section). The material strength σ_L on the scale of d_L can be much lower (< 25%) than the strength σ_S measured at the representative small scale d_S (typical of laboratory tested specimens), depending on the defect characteristics and loading state.

Various investigations have shown that the progressive decrease of strength can be related to the increasing size of the tested specimen using a power law function applied to the representative dimension (side, surface, volume; *e.g.* Jaeger & Cook, 1979; Cunha, 1993a, 1993b; Bazant & Chen, 1997; Hudson & Harrison, 1997). Such a scale effect function was proposed to define the strength from the smallest scale of the representative volume element d_S with a maximum strength σ_S, to the scale d_L where increasing the size does not affect strength any more (*i.e.* σ_L is the nominal large scale strength); this expression can be written as (Aubertin *et al.*, 2000, 2002):

$$\sigma_N = \sigma_S - x_1(\sigma_S - \sigma_L) \left\langle \frac{d_N - d_S}{d_L - d_S} \right\rangle^{m_1} \tag{3}$$

The first term on the right hand side is the strength σ_S at small scale d_S, and the second term represents the decreasing value as size increases until σ_L is reached at d_L. The rate at which the decrease takes place depends on two material parameters x_1 and m_1. In this equation, d_L is the reference size (length L, area L^2 or volume L^3) which has the minimum asymptotic strength σ_L, and d_S is the corresponding size when strength is considered maximum for a homogeneous representative volume element of the material. For rocks, the authors have proposed using $d_S \cong 10^y d_g$ and $d_L \cong 10^{2y} d_S$, where d_g is the mean grain size; here $y \cong 1$ for measures of length, $y \cong 2$ for area, and $y \cong 3$ for volume. In many practical cases, one finds that $d_S \cong 0.5^y$ to 5^y (cm, cm^2, cm^3), and typically $d_L \geq 10^2$ cm, 10^4 cm^2, 10^6 cm^3. $\langle \, \rangle$ are Macaulay brackets ($\langle x \rangle = (x + |x|)/2$), which limits the decrease of strength for $d \geq d_L$.

It can be noted here that an alternate equation (not presented here) has also been proposed by Aubertin *et al.* (2002) to represent a more progressive (and somewhat more representative) reduction of the strength with size.

A practical application procedure has been developed for Equation 3, based on statistical analysis of standard laboratory tests results. This has led to the following simple predictive equation to estimate the large scale strength of intact rocks (Aubertin *et al.*, 2001; Li *et al.*, 2001):

$$\sigma_L = z_1(\bar{\sigma}_{c50} + z_2 \times S_0) \tag{4}$$

where $\bar{\sigma}_{c50}$ is the average observed mean value of the uniaxial compressive strength on standard size specimens (50 mm), S_0 is the corresponding standard deviation of the test results (when at least 10 tests results are available), and z_1 and z_2 are two statistically obtained parameters (see details in Aubertin *et al.*, 2001). This equation was applied by Li *et al.* (2001) for the analyses of the URL tunnel (in Manitoba, Canada), with $z_1 \cong 0.08$ and $z_2 \cong 5$ to 6.

Li *et al.* (2007) later proposed a statistical approach to estimate the value of σ_S from standard laboratory tests results on relatively hard rocks. The results from this

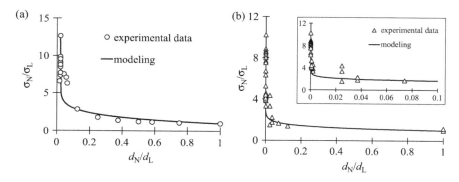

Figure 8 Influence of size on the uniaxial compressive strength: a) cubic coal specimens; $x_1 = 1$, $m_1 = 0.075$, $d_L = 121.92$ cm, $\sigma_L = 4.48$ MPa, $d_S = 2.54$ cm and $\sigma_S = 56.54$ MPa (data taken from Bieniawski, 1968); b) Cedar City quartz diorite with prismatic and cylindrical specimens; $x_1 = 1$, $m_1 = 0.025$, $d_L = 2.67 \times 10^5$ cm^3, $\sigma_L = 6.83$ MPa, $d_S = 28.4$ cm^3 and $\sigma_S = 86.7$ MPa (data taken from Pratt et al., 1972). Taken from Aubertin et al. (2000).

investigation indicate that the standard unconfined compressive strength σ_c is often close to about two thirds of the small scale unconfined compressive strength σ_S. This analysis suggests that the large scale strength of low porosity rocks can be as low as 20 % (or even less) of this standard strength.

Fig. 8 shows Equation 3 applied to test results presented by Bieniawski (1968) on coal (Fig. 8a) and by Pratt *et al.* (1972) on quartz diorite (Fig. 8b).

It is important to recall here that size effects are related to the presence and influence of defects (at various scales) on the behavior of rock. Increasing the confining pressure tends to reduce the influence of existing flaws, so it can also reduce size effects. This phenomenon has been illustrated, for instance, by the experimental results from Gerogiannopoulos & Brown (1978) on intact and granulated marble, and by measurements made for elastic properties and failure strength by Michelis (1987) and Medhurst & Brown (1998). This is also in accordance with the strength envelope of joints and intact rock which tends to converge at high normal stresses (Ladanyi & Archambault, 1970; Gerard, 1986). This factor however is not easily taken into account, and it has been largely neglected in previous scale effect investigations.

An approach was proposed to address this aspect with MSDP$_u$. To do so, parameters σ_t and σ_c are taken as variables whose values are corrected for scale and for stress state. The ensuing values of σ_{ts} and σ_{cs} are expressed according to Equation 3, with x_1 given by the following phenomenological function:

$$x_1 = \exp(x_0 \sigma_3 / T_0) \qquad (5)$$

where T_0 is the uniaxial tensile strength σ_t (with a negative value) of standardized size specimens; T_0 is used here as normalizing parameter because it corresponds to the stress state where scale effects are near their maximum. Fig. 9a shows a schematic representation of the MSDP$_u$ criterion with Equation 3 used for σ_{ts} and σ_{cs}, with x_1 taken as a constant ($x_1 = 1$ or $x_0 = 0$, *i.e.* no effect of the stress state) or given by Equation 5 with $x_0 > 1$.

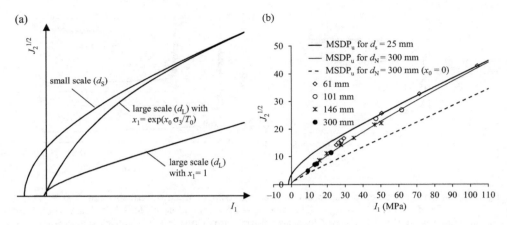

Figure 9 MSDP$_u$ criterion for different specimen sizes of intact rock: a) schematic representation of the MSDP$_u$ criterion for different specimen sizes of rock (x_0 = 0.5); b) application of MSDP$_u$ criterion to triaxial compression tests on large size cylindrical samples; description with x_0 = 0.065, m_1 = 0.015, d_L = 120 cm, d_S = 2.5 cm, σ_{cS} = 35.0 MPa, σ_{cL} = 3.5 MPa, σ_{tS} = 1.5 MPa, σ_{tL} = 0.15 MPa (s_{cS} and s_{cL} are the uniaxial compression strengths obtained on small and large size specimens, respectively; s_{tS} and s_{tS} are the uniaxial tensile strengths of small and large specimens, respectively; data taken from Medhurst & Brown, 1998). Figures taken from Aubertin et al. (2000).

The effect of size theoretically disappears only when a fully plastic behavior is encountered (often at very high mean stress). For practical calculations with low porosity rocks, it can be considered, as a first estimate, that this effect becomes negligible when the shear strength of a closed defect (microcrack) surface becomes equal to the cohesion of the surrounding rock material. Based on the Coulomb criterion, this condition can be approximated by the following expression (Aubertin et al., 2000):

$$\sigma_3 = \frac{\sigma_c}{2\tan\phi \, \tan(45° + \phi/2)} \tag{6}$$

which gives $\sigma_3 \cong 0.5 \, \sigma_c$ for $\phi \cong 30°$. Above this value of the confining pressure, scale effect becomes much less important, and the failure envelope at small scale d_S and large scale d_L tend to converge (see Fig. 9a). Fig. 9b shows how this concept applies to actual test results on rock samples of different sizes (data from Medhurst & Brown, 1998). Fig. 10 illustrates the effect of scale and of loading state on material strength according to Equations 3 and 5.

The complex influence of scale and stress state may explain why rock strength close to the walls of large underground openings may appear to be much lower than the value deduced for locations deeper in the rock mass, where the confining stresses are more significant.

5 YIELDING AND FAILURE OF POROUS ROCKS

It has long been known that rocks with a relatively high porosity typically show some inelastic yielding under a high mean stress, even with little or no deviatoric stress (Nova, 1986; Brown & Yu, 1988; Charlez, 1991; Shao & Henry, 1991; Abdulraheem

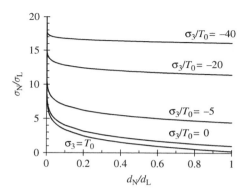

Figure 10 Representation of size effect with different confining pressures according to Equations 3 and 5 with $T_0 = -\sigma_t$, using $x_0 = 0.04$, $m_1 = 0.075$, $d_L = 1000$, $\sigma_L = 10$, $d_S = 1$, $\sigma_S = 200$ (taken from Aubertin et al. 2000).

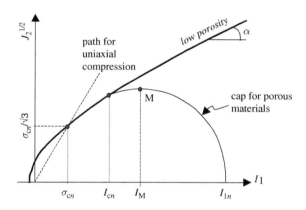

Figure 11 Schematic representation of the MSDP$_u$ criterion for dense and porous materials under CTC conditions ($\theta = 30°$); with key parameters. The maximum value of $J_2^{1/2}$ corresponds to point M (taken from Li et al., 2005).

et al., 1992). The corresponding yield surface (which is more or less equivalent to a DIT) can then curve downward and eventually close on itself on the compressive side of the stresses. A "cap" can be used to capture the curvature under a large hydrostatic stress component (as is commonly done in soil mechanics; *e.g.* Desai & Siriwardane, 1984).

This approach has been applied to the MSDP$_u$ criterion by adding the last term on the right hand side of the formulation given in the central column in Table 1. The corresponding shape with (and without) this cap is shown in Fig. 11.

Modifications were later introduced in the MSDP$_u$ criterion to describe the yield or failure conditions in terms of porosity (Aubertin & Li, 2004; Li et al., 2005). In this

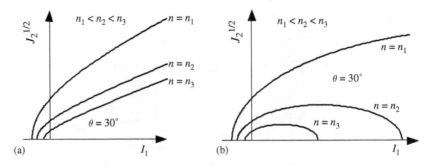

Figure 12 Schematic representation of the MSDP$_u$ criterion (a) for low porosity materials with $a_{3n} = 0$ and (b) for porous materials with $I_c \cong 0$, $a_{3n} \neq 0$, $b = 0.75$ (taken from Li et al., 2005).

case, material parameters are expressed explicitly as a function of initial porosity using the following (Li & Aubertin, 2003):

$$\sigma_{un} = \left\{ \sigma_{u0}\left(1 - \sin^{x'_1}\left(\frac{\pi}{2}\frac{n}{n_C}\right)\right) + \langle\sigma_{u0}\rangle\cos^{x'_2}\left(\frac{\pi}{2}\frac{n}{n_C}\right)\right\}\left\{1 - \frac{\langle\sigma_{u0}\rangle}{2\sigma_{u0}}\right\} \quad (7)$$

where σ_{un} may be used for compression ($\sigma_{un} = \sigma_{cn}$) or tension ($\sigma_{un} = \sigma_{tn}$); n_C is the critical porosity for which σ_{un} tends toward zero, in tension ($n_C = n_{Ct}$) and in compression ($n_C = n_{Cc}$); parameter σ_{u0} represents the theoretical value of σ_{un} for $n = 0$; x'_1 and x'_2 control the non-linearity of the σ_{un}–n relationship.

With the cap, the criterion closes down toward the I_1 axis; parameter I_{cn} represents the I_1 value where the locus departs from the "low porosity" condition (see Fig. 11), while I_{1n} corresponds to the intersection of the criterion with the positive I_1 axis (also shown on Fig. 11).

The values of I_{cn} and I_{1n}, which may be obtained experimentally, become very large for dense materials; the Cap portion can then be neglected.

Specific functions have been developed to define I_{1n} and I_{cn} as a function of porosity n (Li et al., 2005).

The effect of porosity on the criterion is illustrated in Fig. 12a for a typical high strength rock at low mean stress conditions (i.e. $I_1 < I_{cn}$), while Fig. 12b shows its shape for a relatively low strength material.

Fig. 13 show the locus with $b = 0.75$ defined for I_1 that extends beyond I_{cn}, in the case of rocks. The presence of the Cap is required in such cases to describe the elastic limit and the failure strength.

These results highlight the great flexibility of the proposed set of equations, which allow a good description of the inelastic loci and failure strength of rocks and various other geomaterials. Additional illustrations are presented in Aubertin et al. (2003), Aubertin & Li (2004), and Li et al. (2005).

It is also possible to predict the failure strength of rock for different porosities, when data is available for a given value of n, as demonstrated by Li et al. (2005) for sandstone.

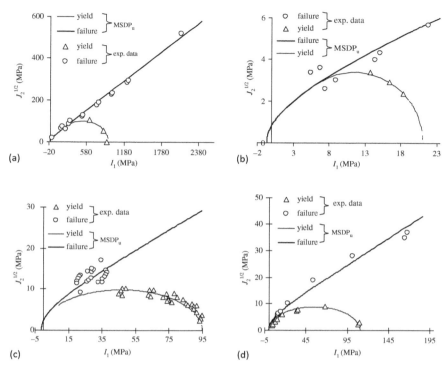

Figure 13 Failure strength and elastic limit (in CTC) of (a) Kayenta sandstone (data from Wong et al., 1992), (b) a tuff (data from Pellegrino, 1970; figure adapted from Aubertin et al. 2000), (c) Bath stone (data from Elliott & Brown, 1985), and (d) Castlegate sandstone (data from Coop & Willson, 2003). For Kayenta sandstone, the MSDP$_u$ criterion was applied with ϕ = 30° (estimated), σ_{cn} = 30 MPa (measured), σ_{tn} = 2 MPa (estimated), and a_{3n} = 0 (or I_{cn} >>) for failure and with ϕ = 30° (estimated), σ_{cn} = 30 MPa (measured), σ_{tn} = 2 MPa (estimated), a_{3n} = 0.115 (estimated), and I_{cn} = 250 MPa (estimated) for yield. For the tuff, the MSDP$_u$ criterion was applied with ϕ = 20° (estimated), σ_{cn} = 3.8 MPa (measured), σ_{tn} = 0.5 MPa (estimated), and a_{3n} = 0 (or I_{cn} >>) for failure and with ϕ = 20° (estimated), σ_{cn} = 3.8 MPa (measured), σ_{tn} = 0.5 MPa (estimated), a_{3n} = 0.115 (estimated), and I_{cn} = 6.5 MPa (estimated) for yield. For Bath stone, the MSDP$_u$ criterion was applied with ϕ = 30° (estimated), σ_{cn} = 15 MPa (measured), σ_{tn} = 1 MPa (estimated), and a_{3n} = 0 (or I_{cn} >>) for failure and with ϕ = 30° (estimated), σ_{cn} = 15 MPa (measured), σ_{tn} = 1 MPa (estimated), a_{3n} = 0.095 (estimated), and I_{cn} = 0 MPa (estimated) for yield. For Castlegate sandstone, the MSDP$_u$ criterion was applied with ϕ = 26° (estimated), σ_{cn} = 9 MPa (estimated), σ_{tn} = 0.1 MPa (estimated), and a_{3n} = 0 (or I_{cn} >>) for failure and with ϕ = 26° (estimated), σ_{cn} = 9 MPa, σ_{tn} = 0.1 MPa (estimated), a_{3n} = 0.064 (estimated), and I_{cn} = 1 MPa (estimated) for yield.

6 APPLICATION TO DAMAGED ROCK AND FRACTURED ROCK MASS

The strength of rocks and rock masses depends on the initial structural state that can be represented by a continuity parameter Γ (described below). The introduction of parameter Γ into the formulation of the MSDP$_u$ criterion is shown in the right hand side

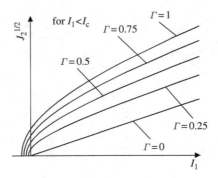

Figure 14 Schematic representation of the influence of the continuity parameter Γ on the failure surface of a tight undisturbed rock media (with $I_1 < I_c$); $\Gamma = 1$ corresponds to undamaged material, and $\Gamma = 0$ refers to a cohesionless media (taken from Aubertin et al. 2000).

column of Table 1 (Aubertin *et al.*, 2000). This continuity parameter can be seen as an equivalent damage parameter D (= $1 - \Gamma$), as defined in the Kachanov-Rabotnov approach forming the basis for Continuum Damage Mechanics (Lemaitre, 1992). It is treated here as a scalar, although it could be extended to deal with anisotropy under a tensorial format (Aubertin *et al.*, 1998).

When the rock has few defects (*i.e.*, a very small population of cracks and pores), $\Gamma \cong 1$. When defects (flaws of various sizes) become more abundant and their influence more important, the value of Γ is reduced. For a highly fractured but relatively dense medium that behaves as a cohesionless soil, the value of Γ becomes nil; this means that $\tilde{a}_1 = \tilde{a}_2 = 0$ and $a'_3 = a_3$ in the F_0 equation (Table 1, right column). The proposed criterion then becomes equivalent to the Coulomb criterion without cohesion (for $I_1 \leq I_c$), as shown in Fig. 14. This figure also illustrates how the value of Γ influences the position of the surface in the $I_1 - J_2^{1/2}$ plane; it shows that the strength is reduced as Γ decreases. This effect of the continuity parameter can be combined with that of porosity (Aubertin *et al.*, 2000), as is shown in the following illustration.

Applications of MSDP$_u$ to results obtained on porous rocks and rock-like materials are shown in Fig. 15 (using data from Nguyen, 1972 and Wong *et al.*, 1992). As can be seen, the MSDP$_u$ is able to properly represent the behavior of these porous rocks, when considering also the effect of the continuity parameter.

This approach can also be applied to rock masses, although going from the behavior of rock to that of the large scale rock mass is quite a challenge.

The continuous scale effect described above for rocks applies when there is no new type of flaws introduced in the media. However, going from intact rock to in situ rock mass implies not only the usual statistical and energy release size effects, but also the addition of other types of larger scale defects such as joint sets. Thus, the relationship used for scaling up properties may become more or less discontinuous. This phenomenon is schematically illustrated in Fig. 16, which shows that the scale effect function can be considered continuous until new types of defects are introduced; this is the case with grain boundaries (grains to rock) and joint sets (rock to rock mass). In the transition zone (shade areas), the strength–size function becomes ill defined

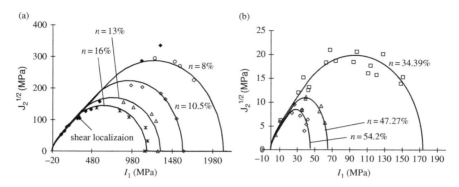

Figure 15 Applications of MSDP$_u$ criterion to porous materials: a) saturated Berea sandstone cylinders submitted to CTC loading condition (data taken from Wong et al., 1992); description of test results using: $b = 0.75$, $\phi = 35°$, $\sigma_c = 110$ MPa, $\sigma_t = 9$ MPa, $a_3 = 0.75$; for $n = 8\%$, $\Gamma = 0.65$ and $I_c = 710$ MPa; for $n = 10.5\%$, $\Gamma = 0.75$ and $I_c = 515$ MPa; for $n = 13\%$, $\Gamma = 0.79$ and $I_c = 325$ MPa; for $n = 16\%$, $\Gamma = 0.8$ and $I_c = 250$ MPa; b) application to Paris plaster cylinders (with water to plaster ratio of 70%), representation with $b = 0.75$, $\phi = 33°$, $\sigma_c = 10$ MPa, $\sigma_t = 0.5$ MPa, $a_3 = 0.75$ (data taken from Nguyen, 1972); $\Gamma = 0.825$ and $I_c = 40$ MPa for $n = 34.39\%$; $\Gamma = 0.675$ and $I_c = 25$ MPa for $n = 47.27\%$; $\Gamma = 0.575$ and $I_c = 20$ MPa for $n = 54.2\%$. Figure adapted from Aubertin et al. (2000).

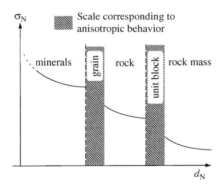

Figure 16 Schematic representation of scale effect on rock properties; the shaded areas represent scales at which strength is not isotropic. Figure taken from Aubertin et al. (2000).

because the material cannot be considered homogeneous and anisotropy needs to be considered.

Anisotropy is also a key aspect when only one or two defect families or few individual members are added. In this case, one must either use an anisotropic criterion or combine an isotropic expression (such as MSDP$_u$) with a shear strength criterion for the weakness planes (Li & Aubertin, 2000).

Rock masses with more than two distinct joint sets usually behave almost isotropically on a large scale. It is thus often considered that an isotropic criterion is appropriate for such cases (Hoek & Brown, 1980a, 1980b, 1988).

One of the challenges associated with defining the strength of a rock mass stems from the difficulty to perform appropriate tests and obtain adequate in situ rock mass properties (strength and deformability) at a scale corresponding to engineering structures. Consequently, the commonly used approach in rock mechanics has been to rely on laboratory properties "corrected" for scale and discontinuity conditions. For that purpose, several techniques have been proposed, with each technique suffering from some limitations and fairly large uncertainties.

The approach applied here is based on the use of the continuity parameter, introduced above. The value of Γ can be related to the reduction of strength parameters when compared to "undamaged" materials.

For strength and stability calculations, the continuity parameter Γ can be used as a correction factor acting on "undamaged" material properties (Aubertin et $al.$, 2000).

Parameter Γ of an isotropic rock mass can be defined from either the deformability or the strength. Its effect is to alter the stress state in the bearing portion areas, which in turn decreases proportionally all mechanical properties. For this application of the $MSDP_u$ criterion to rock mass strength, the authors have proposed the following expression:

$$ \Gamma = \Gamma_{100} \left[0.5 \left(1 - \cos \frac{\pi RMR}{100} \right) \right]^{p'} \tag{8} $$

with

$$ \Gamma_{100} = \sigma_{cL}/\sigma_c \tag{9} $$

Here, σ_{cL} is taken as the uniaxial compressive strength of the rock at size d_L (see Equations 3 and 4), while σ_c is the standard size specimen strength of intact rock. This equation was based on an expression proposed by Mitri et $al.$ (1994) for the deformability of rock masses; exponent p' was added to better represent strength parameters. In Equation 9, the value of σ_{cL} is usually found to be 0.2 to 0.3 times σ_c; σ_{cL} can be seen as the rock mass strength when RMR is 100 (n.b. the Bieniawski, 1989 RMR version is used).

Fig. 17 shows this relationship with $\Gamma_{100} = 0.3$ and $p' = 1, 2$ and 3 (a value of 3 is favored for practical calculations). Also shown in this figure is the $s^{1/2}$ parameter ($= \sigma_{cmass}/\sigma_c$) expressed from the relationship proposed by Hoek and Brown (1997). The two functions are fairly close to each other at RMR values below about 80, but differ at larger RMR. Here, Γ at high RMR values is bounded by Γ_{100} corresponding to the large scale strength of the rock σ_{cL}.

The value of parameter Γ given by Equation 8 can be introduced into the general $MSDP_u$ criterion, with values of σ_c and σ_t given for standard size samples. Alternately, one could use $\Gamma_{100} = 1$ in Equation 8 and use σ_c and σ_t values in the F_0 equation (Table 1, right column) corresponding directly to large scale conditions (from Equations 3 and 5).

Figure 18 shows the failure strength envelope (for $I_1 < I_c$) using Γ obtained from Equation 8 for different RMR values; note that the influence of loading mode (Equation 4) is neglected in this representation. As expected, reducing the RMR decreases the rock mass strength; a highly fractured mass may even behaves as a purely frictional (cohesionless) material.

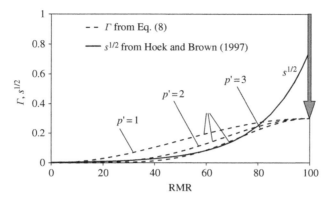

Figure 17 Continuity parameter Γ vs. rock mass rating (RMR)1989; also shown is the $s^{1/2}$ value as described by the relationship proposed by Hoek and Brown (1997). The vertical arrow on the right hand side shows the effect of scale on intact rock strength (going from the specimen size to the large scale unit volume). Figure taken from Aubertin et al. (2000).

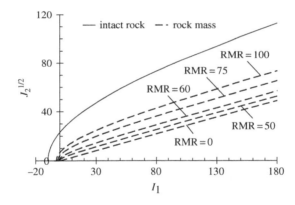

Figure 18 The MSDP$_u$ criterion for rock and rock masses (for $I_1 < I_c$) according to Equation 8 with $b = 0.75$, $\phi = 35°$, $\sigma_c = 220$ MPa, $\sigma_t = 9$ MPa, $\Gamma_{100} = 0.3$, $p' = 3$ for various RMR values (taken from Aubertin et al. 2000).

7 DISCUSSION

The application of the MSDP$_u$ criterion for the strength analysis of rock and other somewhat similar materials has been documented in many of the publications mentioned above. These include specific conditions not explicitly addressed here, such as the response of cohesionless or weakly cemented porous materials, and rocks with planar anisotropy.

The criterion has also been used to conduct analysis of engineered openings in rock media, as illustrated in several publications. These cannot be presented here due to space limitation; some key examples are nonetheless recalled in the followings.

The stability analysis of boreholes in isotropic rock can be performed directly using laboratory properties. This was shown using experimental results of Mazanti &

Sowers (1965) obtained from hollow cylinder tests on granite (Aubertin *et al.*, 2000). The $MSDP_u$ criterion has also been used to back-calculate in situ stresses around boreholes using breakout geometry (Li & Aubertin, 1999; Li *et al.*, 1999, 2000; Kikiessa-Kisaka, 2015).

Simulations of the short term failure and delayed failure around a large scale tunnel at the AECL underground facilities (Manitoba, Canada) has been described by Li *et al.* (2000) and Aubertin *et al.* (2000).

The $MSDP_u$ criterion equations have also been used to develop analytical solutions for the stresses around circular openings in an elasto-plastic media, with and without the Cap component (Li *et al.*, 2005, 2010; Li & Aubertin, 2009). These original solutions have been successfully compared with other specific formulations (encompass by $MSDP_u$) for the Coulomb and von Mises criteria; validation was also conducted using plane strain numerical simulations.

The implementation of the $MSDP_u$ criterion in FLAC (Itasca) as part of a new elasto-plastic constitutive model has been presented by Li *et al.* (2010); the latter includes the application of the ensuing $MSDP_u$–EP model to determine the stresses in backfilled mine stopes.

These applications have shown that the multiaxial criterion presented in the chapter is simple to use, and can be applied for a wide variety of rock media characteristics and loading conditions. Like any other criterion, it also has some limitations, some of which are briefly recalled in the following.

i) The criterion equations presented above are only applicable to isotropic media, so it cannot be applied to the shaded areas shown schematically in Fig. 16. This is also the case for most existing criteria. As mentioned however, it has also been adapted for planar anisotropy (Li & Aubertin, 2000), but this aspect needs to be investigated further.

ii) Size effects involve complex physical phenomena, and the strength magnitude at various scales depends on the controlling inelastic processes leading to failure (or yielding). With low porosity rocks, scale effects tends to be reduced when the mean stress is increased because the added confinement diminishes the influence of existing flaws by closing the opened crack faces (see Equation 4). However, not all flaws (microcracks to joints) will be perfectly matched upon closure, so it can be expected that scale effects cannot be fully eliminated simply by increasing the mean stress, especially when natural porosity increases. More work needs to be done on this aspect, especially to assess scale effects and other influence factors in the damaged rock zone around openings where the stress distribution is highly non-uniform.

iii) It is assumed here that rock strength (from DIT to STF) is related to the initial defect state. However, additional provisions are required to include the evolution of the damage state. An internal state variable approach can be used to treat complex load paths and history, including the effect of progressive damage growth (Aubertin *et al.*, 1994, 1998). For most practical calculations however, the simplified procedure presented in the referenced papers, commonly in rock engineering, provides a good estimate of failure occurrence.

This presentation has not taken into account the intrinsic variability of rock properties. As with any other criteria, this aspect should be treated adequately with $MSDP_u$, by using proper statistical tools.

Despite these limitations, the sound physical basis from which it has been formulated, its unified and modular nature, and its adaptability to treat hard and soft materials with little ($\Gamma \cong 1$) or many flaws ($0 < \Gamma < 1$) make the proposed criterion a practical engineering tool.

8 FINAL REMARKS

In this chapter, the authors have presented a summary of the main features for the general multiaxial criterion $MSDP_u$, initially developed for intact rock samples and extended for different types of rock media. The proposed criterion can be applied to describe the short term strength and the damage initiation threshold (DIT) of rocks. It can also address the effect of time to obtain isochronous failure surfaces. The effect of scale is also treated, taking into account the size of the element and the loading conditions. A simple continuity parameter, Γ, is used to define the influence of large size defects and extrapolate laboratory tests results to in situ conditions. The use of the $MSDP_u$ criterion is illustrated with a number of experimental results. Application of this criterion to engineered structures was summarized, based on earlier publications and ongoing work.

ACKNOWLEDGMENTS

The authors also acknowledge the financial support from the Natural Sciences and Engineering Research Council of Canada (NSERC), Institut de recherche Robert–Sauvé en santé et en sécurité du travail (IRSST), Fonds de recherche du Québec – Nature et technologies (FRQNT), the Industrial NSERC Polytechnique–UQAT Chair in Environment and Mine Wastes Management (2001–2012) and the industrial partners of Research Institute on Mines and Environment (RIME UQAT–Polytechnique; http://rime-irme.ca). Elsevier and NRC Research Press are acknowledged for the permission of reuse of some published figures in this chapter of the book.

REFERENCES

Abdulraheem, A., Roegiers, J.C. & Zamen, M. (1992) *Mechanics of pore collapse and compaction in weak porous rocks*. In: Tillerson, J.R. & Wawersik, W.R. (eds.) *Rock Mechanics*. Rotterdam, Balkema. pp. 233–242.

Akai, K. & Mori, H. (1970) Ein versuch über Bruchmecanismus von Sandstein unter mehrachsigen Spannungszustand. In: *Proceedings of the 2nd International Congress on Rock Mechanics, Belgrade*. Vol. 2, pp. 207–213.

Andreev, G.E. (1995) *Brittle Failure of Rock Materials – Test Results and Constitutive Models*. Rotterdam, Balkema.

Aubertin, M. & Li, L. (2004) A porosity-dependent inelastic criterion for engineering materials. *Int J Plast*, 20(12), 2179–2208.

Aubertin, M. & Simon, R. (1996) *A multiaxial failure criterion that combines two quadric surfaces*. In: Aubertin, M., Hassani, F. & Mitri, H. (eds.) *Rock Mechanics: Tools and Techniques*. Rotterdam, Balkema. pp. 1729–1736.

Aubertin, M. & Simon, R. (1997) A damage initiation criterion for low porosity rocks. *Int J Rock Mech & Min Sci*, 34(3–4), #017, CD-ROM.

Aubertin, M. & Simon, R. (1998) Un critère de rupture multiaxial pour matériaux fragiles. *Can J Civil Engineering*, 25(2), 277–290.

Aubertin, M., Gill, D.E. & Ladanyi, B. (1994) Constitutive equations with internal state variables for the inelastic behavior of soft rocks. *Appl Mech Rev*, 47(6–2), S97–S101.

Aubertin, M., Julien, M.R. & Li, L. (1998) Keynote Lecture: The semi-brittle behavior of low porosity rocks. In: *NARMS'98, Proc 3rd North Am Rock Mech Symp*, Cancun. Rotterdam, Balkema. Vol. 2, pp. 65–90.

Aubertin, M., Li, L., Simon, R. & Khalfi, S. (1999) Formulation and application of a short-term strength criterion for isotropic rocks. *Can Geotech J*, 36(5), 947–960.

Aubertin, M., Li, L. & Simon, R. (2000) A multiaxial criterion for short term and long term strength of rock media. Int J Rock Mech Min *Sci*, 37, 1169–1193.

Aubertin, M., Li, L. & Simon, R. (2001) *Evaluating the large scale strength of rock mass with the* $MSDP_u$ *criterion*. In: Elsworth, D., Tinucci, J.P. & Heasley, K.A. (eds.) *Rock Mechanics in the National Interest: Proceedings of the 38th U.S. Rock Mechanics Symposium, DC Rocks 2001*, Washington DC, 7–10 July 2001. A. A. Balkema. Vol. 2, pp. 1209–1216.

Aubertin, M., Li, L. & Simon, R. (2002) *Effet de l'endommagement sur la stabilité des excavations souterraines en roche dure*. Études et Recherches R-312, Institut de recherche Robert-Sauvé en santé et en sécurité du travail (IRSST), Montreal, Quebec, Canada.

Aubertin, M., Li, L., Simon, R. & Bussière, B. (2003) *A general plasticity and failure criterion for materials of variable porosity*. Technical Report EPM-RT-2003-11, École Polytechnique de Montréal.

Bazant, Z.P. & Chen, E.P. (1997) Scaling of structural failure. *Appl Mech Rev*, 50(10), 593–627.

Bazant, Z.P. & Planas, J. (1998) *Fracture and Size Effect in Concrete and Other Quasi-brittle Materials*. Boca Raton, CRC Press.

Bieniawski, Z.T. (1968) The effect of specimen size on compressive strength of coal. *Int J Rock Mech Min Sci*, 5, 325–335.

Bieniawski, Z.T. (1984) *Rock Mechanics Design in Mining and Tunneling*. Rotterdam, Balkema.

Bieniawski, Z.T. (1989) *Engineering Rock Mass Classification*. New York, John Wiley & Sons.

Brown, E.T. & Yu, H.S. (1988) A model for the ductile yield of porous rock. *Int J Num & Analy Methods Geomech*, 12, 679–688.

Charlez, Ph.A. (1991) Rock Mechanics, Volume 1 – Theoretical Fundamentals. Paris, Editions Technip.

Chen, A.C.T. & Chen, W.F. (1975) Constitutive relations for concrete. *ASCE Journal of the Engineering Mechanics Division*, 101, 465–481.

Coop, M.R. & Willson, S.M. (2003) Behavior of hydrocarbon reservoir sands and sandstones. *J Geotech* Geoenviron *Eng*, 129(11), 1010–1019.

Cristescu, N.D. & Hunsche, U. (1997) *Time Effects in Rock Mechanics*. New York, John Wiley & Sons.

Cunha, A.P. (1993a) *Research on scale effects in the determination of rock mass mechanical properties – The Portuguese experience*. In: Cunha, A.P. (ed.) *Scale Effects in Rock Masses 93*. Rotterdam, Balkema. pp. 285–292.

Cunha, A.P. (1993b) *Scale effects in rock engineering - An overview of the Loen Workshop and other recent papers concerning scale effects*. In: Cunha, A.P. (ed.) *Scale Effects in Rock Masses 93*. Rotterdam, Balkema. pp. 3–14.

Desai, C.S. & Salami, M.R. (1987) Constitutive model for rocks. *ASCE J Geotech Engng*, 113, 407–423.

Desai, C.S. & Siriwardane, H.J. (1984) *Constitutive Laws for Engineering Materials with Emphasis on Geologic Materials*. London, Prentice-Hall.

Drucker, D.C. & Prager, W. (1952) Soil mechanics and plastic analysis on limit design. *Quat Appl Meth*, 10(2), 157–165.

Du, X., Lu, D., Gong, Q. & Zhao, M. (2010) Nonlinear unified strength criterion for concrete under three-dimensional stress states. *J Eng Mech*, 136(1), 51–59.

Elliott, G.M. & Brown, E.T. (1985) Yield of a soft, high porosity rock. *Géotechnique*, 35(4), 413–423.

Franklin, J.A. & Dusseault, M.B. (1989) *Rock Engineering*, New York, McGraw-Hill.

Gerard, C. (1986) Shear failure of rock joints: appropriate constraints for empirical relations. *Int J Rock Mech Min Sci & Geomech Abstr*, 23(6), 421–429.

Germanovitch, L.N., Carter, B.J., Ingraffea, A.R., Dyskin, A.V. & Lee, K.K. (1996) Mechanics of 3-D crack growth under compressive loads. In: Aubertin, M., Hassani, F. & Mitri, H. (eds.) *Rock Mechanics, Tools and Techniques: Proceedings of the 2nd North American Rock Mechanics Symposium*. Balkema, Rotterdam. pp. 1151–1160.

Gerogiannopoulos, N.G. & Brown, E.T. (1978) The critical state concept applied to rock. *Int J Rock Mech Min Sci & Geomech Abstr*, 15(1), 1–10.

Goodman, R.E. (1980) *Introduction to Rock Mechanics*. New York, John Wiley & Sons.

Griffith, A.A. (1921) The phenomena of rupture and flow in solids. *Philos Trans R Soc London*, 221A, 163–198.

Griffith, A.A. (1924) The theory of rupture. *Proceedings of the First International Congress Applied Mechanics*. Delft. Partie 1. pp. 55–63.

Hatzor, Y.H. & Palchik, V. (1997) The influence of grain size and porosity on crack initiation stress and critical flaw length in dolomites. *Int J Rock Mech Min Sci*, 34(5), 805–816.

Hjelm, H.E. (1994) Yield surface for grey cast iron under biaxial stress. *ASME J Engng Materials & Technol*, 116, 148–154.

Hoek, E. & Brown, E.T. (1980a) Empirical strength criterion for rock masses. *J Geotech Engng Div*, 106, 1013–1035.

Hoek, E. & Brown, E.T. (1980b) *Underground Excavations in Rock*. London, Institution of Min & Metall.

Hoek, E. & Brown, E.T. (1988) *The Hoek-Brown failure criterion—a 1988 update*. In: *Rock Engineering for Underground Excavations: Proc 15th Can Rock Mech Symp*, Toronto. pp. 31–38.

Hoek, E. & Brown, E.T. (1997) Practical estimates of rock mass strength. *Int J Rock Mech Min Sci*, 34(8), 1165–1186.

Hoskins, E.R. (1969) The failure of thick-walled hollow cylinders of isotropic rock. *Int J Rock Mech Min Sci*, 6, 99–125.

Hudson, J.A. & Harrison, J.P. (1997) *Engineering Rock Mechanics – An Introduction to the Principles*. Oxford, Pergamon Press.

Hunsche, U. (1994) *Uniaxial and triaxial creep and failure tests on rock; experimental technique and interpretation*. In: Cristescu, N.D., & Gioda, G. (eds.) *Visco-plastic Behaviour of Geomaterials, Courses and Lectures*. New York, Springer Verlag, International Center for Mechanical Sciences. pp. 1–51.

Jaeger, J.C. & Cook, N.G.W. (1979) *Fundamentals of Rock Mechanics*. 3rd edition. New York, Chapman and Hall.

Kikiessa-Kisaka, P. 2015. *Évaluation des contraintes en place en profondeur à partir de l'écaillage des trous de forage*. M.Sc.A. thesis, École Polytechnique de Montréal.

Ladanyi, B. & Archambault, G. (1970) *Simulation of shear behavior of a jointed rock mass*. In: *Proc 11th US Symp on Rock Mech*, Berkeley. pp. 105–125.

Lade, P.V. (1993) Rock strength criteria - The theories and evidence. In: Hudson, J.A. (ed.) *Comprehensive Rock Engineering – Principles, Practice and Projects*. Oxford, Pergamon Press. Vol. 1, pp. 255–284.

Lade, P.V. (1982) Three parameter failure criterion for concrete. *ASCE J Eng Mech*, 108, 850–863.

Lama, R.D. (1974) *The uniaxial compressive strength of jointed rock mass*. Prof. L. Muller Festschrift, Univ Karlsruhe. pp. 67–77.

Lau, J.S.O. & Gorski, B. (1991) *The Post-Failure Behaviour of the Lac du Bonnet Grey Granite*. CANMET: Divisional Report MRL 91-079 (TR).

Lemaitre, J.A. (1992) *Course on Damage Mechanics*. New York, Springer-Verlag.

Li, C. & Nordlund, E. (1993) Deformation of brittle rocks under compression – with particular reference to microcracks. *Mech Mater*, 15, 223–239.

Li, L. & Aubertin, M. (1999) Estimation des contraintes dans les roches à partir de l'écaillage autour de trous de forage. *Rev Franç Geotech*, 89, 3–11.

Li, L. & Aubertin, M. (2000) Un critère de rupture multiaxial pour les roches avec une anisotropie planaire. Proceedings, 52nd Canadian Geotechnical Conference, 15–18 October 2000, Montreal, Quebec. CGS. Vol. 1, pp. 357–364.

Li, L. & Aubertin, M. (2003) A general relationship between porosity and uniaxial strength of engineering materials. *Can. J. Civ. Eng*, 30(4), 644–658.

Li, L. & Aubertin, M. (2009) An elastoplastic evaluation of the stress state around cylindrical openings based on a closed multiaxial yield surface. *Int. J. Numer. Anal. Methods Geomech*, 33(2), 193–213.

Li, L., Aubertin, M. & Simon, R. (1999) Stability analyses of underground openings using a multiaxial failure criterion. In: Fernandez, G. & Bauer, R.A. (eds.) Geo-Engineering for Underground Facilities: *Proceeding of the 3rd National Conf Geo-Institute ASCE*, Urbana-Champaign. pp. 471–482.

Li, L., Aubertin, M. & Simon, R. (2000) Analyses de la stabilité d'un tunnel non supporté avec un critère de rupture multiaxial. *J Can Tunnelling*, 2000, 119–130.

Li, L., Aubertin, M. & Simon, R. (2001) *Stability analyses of underground openings using a multiaxial failure criterion with scale effects*. In: Wang, S., Fu, B. & Li, Z. (eds.) *Frontiers of Rock Mechanics and Sustainable Development in the 21st Century: Proceedings of the 2nd Asian Rock Mechanics Symposium*, 11–14 September 2001, Beijing, China. Lisse, A.A. Balkema. pp. 251–256.

Li, L., Aubertin, M., Simon, R. & Bussière, B. (2005) Formulation and application of a general inelastic locus for geomaterials with variable porosity. *Can Geotech J*, 42(2), 601–623.

Li, L., Aubertin, M, Simon, R., Deng, D. & Labrie, D. (2007) *Chapter 97, Influence of scale on the uniaxial compressive strength of brittle rock. Rock Mechanics: Meeting Society's Challenges and Demands*. London, Taylor & Francis Group. pp. 785–79.

Li, L., Aubertin, M. & Shirazi, A. (2010) Implementation and application of a new elasto-plastic model based on a multiaxial criterion to assess the stress state near underground openings. *ASCE Int J Geomech*, 10(1), 13–21.

Lu, D., Du, X., Wang, G., Zhou, A. & Li, A. (2016). A three-dimensional elastoplastic constitutive model for concrete. *Comput Struct*, 163, 41–55.

Lubliner, J. (1990) *Plasticity Theory*. New York, McMillan Publishing.

Lundborg, N. (1974) *A statistical theory of the polyaxial strength of materials*. In: Advances in Rock Mechanics: Reports of Current Research; Proceedings of the Third Congress of the International Society for Rock Mechanics, Themes 1-2, Denver, Colorado, September 1–7, 1974. National Academy of Sciences, Washington, DC. Vol. II, Part A.

Martin, C.D. & Chandler, N.A. (1994) The progressive fracture of Lac du Bonnet granite. *Int J Rock Mech Min Sci & Geomech Abstr*, 31, 643–659.

Maso, J.C. & Lereau, J. (1980) Mechanical behavior of Darney sandstone in biaxial compression. *Int J Rock Mech Min Sci & Geomech Abstr*, 17, 109–115.

Mazanti, B.B. & Sowers, G.F. (1965) Laboratory testing of rock strength. In: Proc Symp Testing Techniques for Rock Mech, Seattle. pp. 207–227.

McClintock, F.A. & Walsh, J.B. (1962) Friction on Griffith cracks under pressure. In: Proceeding of the 4th US National Congr Appl Mech, Berkeley. Vol. 2, pp. 1015–1021.

Medhurst, T.P. & Brown, E.T. (1998) A study of the mechanical behaviour of coal for pillar design. *Int J Rock Mech Min Sci*, 35(8), 1087–1105.

Meredith, P.G. (1990) *Fracture and failure of brittle polycrystals: an overview*. In: Barber, D.J. & Meredith, P.G. (eds.) Deformation Processes in Minerals, Ceramics and Rocks. London, Unwyn Hyman. pp. 5–47.

Michelis, P. (1987) True triaxial yielding and hardening of rock. *J Geotech Engng*, 113(6), 616–635.

Mills, L.L. & Zimmerman, R.M. (1970) Compressive strength of plain concrete under multiaxial loading conditions. ACI Proc - J Am Concrete Inst, 67(10), 802–807.

Mitri, H.S., Edrissi, R. & Henning, J. (1994) *Finite element modelling of cable-bolted stopes in hard rock underground mines*. Reprint No. 94-116, Society for Mining, Metallurgy and Exploration Inc.

Nguyen, D. (1972) *Un concept de rupture unifié pour les matériaux rocheux denses et poreux*. Ph.D. thesis, École Polytechnique de Montréal – Université de Montreal.

Nova, R. (1986) An extended Cam-Clay model for soft anisotropic rocks. *Comput Geotech*, 2, 69–88.

Ottosen, N.S. (1977) A failure criterion for concrete. *ASCE J Eng Mech*, 103, 527–535.

Papanikolaou, V.K. & Kappos, A.J. (2007) Confinement-sensitive plasticity constitutive model for concrete in triaxial compression. Int J *Solids Struct*, 44(21), 7021–7048.

Paterson, M.S. (1978) *Experimental Rock Deformation – The Brittle Field*. Wien, Springer-Verlag.

Patton, F.D. (1966) Multiple modes of shear failure in rock joints in theory and practice. Proceedings of the 1st Congress of the International Society of Rock Mechanics, Lisbon. Vol. 1, pp. 521–529.

Pellegrino, A. (1970) Mechanical behaviour of soft rock under high stresses. In: Proceedings of the 2nd International Conference on Rock Mechanics, Beograd. Vol. 2, pp. 173–180.

Pratt, H.R., Black, A.D., Brown, W.S. & Brace, W.F. (1972) The effect of specimen size on the mechanical properties of unjointed diorite. *Int J Rock Mech Min Sci*, 9, 513–529.

Ramamurthy, T. & Arora, V.K. (1994) Strength predictions for jointed rocks in confined and unconfined states. *Int J Rock Mech Min Sci & Geomech Abstr*, 31(1), 9–22.

Schmidtke, R.H. & Lajtai, E.Z. (1985) The long term strength of Lac du Bonnet granite. *Int J Rock Mech Min Sci & Geomech Abstr*, 22, 461–465.

Shao, J.F. & Henry, J.P. (1991) Development of an elastoplastic model for porous rock. *Int J Plasticity*, 7(1), 1–13.

Shao, J.F., Khazraei, R. & Henry, J.P. (1996) *Application of continuum damage theory to borehole failure modelling in brittle rock*. In: Aubertin, M., Hassani, F. & Mitri, H. (eds.) *Rock Mechanics: Tools and Techniques*. Rotterdam, Balkema. pp. 1721–1728.

Sheorey, P.R. (1997) *Empirical Rock Failure Criteria*. Rotterdam, Balkema.

Singh, J., Ramamurthy, T. & Rao, G.V. (1989) *Strength of rocks at great depth*. In: Maury, V. & Fourmaintraux, D. (eds.) *Rock at Great Depth*. Rotterdam, Balkema. pp. 37–44.

Skrzypek, J.J. & Hetnarski, R.P. (1993) *Plasticity and Creep—Theory, Examples, and Problems*. Boca Raton, CRC Press.

Theocaris, P.S. (1995) Failure criteria for isotropic bodies revisited. *Engineering Fracture Mechanics*, 51(2), 239–264.

Thorel, L. (1994) *Plasticité et Endommagement des Roches Ductiles - Application au Sel Gemme*. Ph.D. thesis, École des Ponts et Chaussés.

Wastiels, J. (1979) *Failure criteria for concrete under multiaxial stress states*. Proceedings of the Colloquium on Plasticity in Reinforced Concrete, International Association for Bridge and Structural Engineering, Lyngby. pp. 3–10.

Wong, R.H.C., Chau, K.T. & Wang, P. (1996) Microcracking and grain size effect in Yuen Long marbles. *Int J Rock Mech Min Sci & Geomech Abstr*, 33(5), 479–485.

Wong, T.F., Szeto, H. & Zhang, J. (1992) Effect of loading path and porosity on the failure mode of porous rocks. *Appl Mech Rev*, 45(8), 281–293.

You, M.Q. (2015) Strength criterion for rocks under compressive-tensile stresses and its application. *J Rock Mech Geotech Eng*, 7, 434–439.

Yu, M.H. (2002) Advances in strength theories for materials under complex stress state in the 20th Century. *Appl Mech Rev*, 55(3), 169–218.

Chapter 15

Unified Strength Theory (UST)

M.-H. Yu

School of Civil Engineering & Mechanics, Xi'an Jiaotong University, Xi'an, China

1 INTRODUCTION

Strength theory deals with the strength of material under the complex stress state (bi-axial stress or tri-axial stress). Sometimes, it is referred to as the yield criterion in metallic mechanics and computational mechanics, or as the failure criterion in rock-soil mechanics and concrete mechanics. Great effort has been devoted to the formulation of strength theories, failure criteria, yield criteria and many versions of these have been presented during the past 100 years (Zyczkowski, 1981; Yu, 2002). From the 1950s to the 1970s, the von Mises criterion was considered the best. The von Mises criterion, however, is only suitable for those materials that have identical strength both in tension and compression, and shear yield stresses equal to $\tau_y = 0.577\sigma_y$, where τ_y is shear yield strength and σ_y is uni-axial yield strength of materials.

Another important failure criterion is the Mohr–Coulomb failure criterion. The Mohr–Coulomb failure criterion, because of its simplicity, is widely used in practice. The mathematical expression of the Mohr–Coulomb failure criterion is

$$F = \tau_{13} + \beta\sigma_{13} = C \quad \sigma_1 - \alpha\sigma_3 = \sigma_t \tag{1}$$

where $\tau_{13} = (\sigma_1 - \sigma_3)/2$ is maximum principal shear stress, $\sigma_{13} = (\sigma_1 + \sigma_3)/2$ is normal stress acted on the same section on which the maximum shear stress τ_{13} exists. β and C are the material parameter, σ_t is tensile strength of material, α is ratio of tensile strength to compressive strength.

The main disadvantage of the Mohr–Coulomb criterion is that the intermediate principal stress σ_2 is not taken into account. Clearly, it is unsuitable for a three-dimensional problem. Most rock-soil material under a structure such as strip footing, circular footing, slope or excavation, are under a spatial stress. The three principal stresses σ_1, σ_2 and σ_3 exist in a stress element. Jaeger & Cook (1979) said that the effect of the intermediate principal stress should be solved, since it is a problem of great significance in theory and practical matters. They also pointed out that it could be surmised that the effect of increasing the intermediate principal stress is to increase the strength from that obtained in tri-axial stress conditions to a higher value. They felt that an analytical formulation of this transition was so complex that its meaning was not obvious.

The intermediate stress is taken into account in the Drucker–Prager criterion. However, the deviation of the Drucker–Prager criterion from the Mohr–Coulomb criterion is surprising, as indicated by Davis and Selvadurai (2002). Recently, Yu showed clearly that: 'The Drucker–Prager yield criterion has been used quite widely in geotechnical analysis. However, experimental research suggests that its circular shape on a deviatoric plane does not agree well with experimental data. For this reason care is needed when the Drucker–Prager plasticity model is used in geotechnical analysis' (Yu, 2010).

In 1991, ninety years after the establishment of the Mohr–Coulomb failure criterion, a new unified strength theory appeared. The mechanical model, experimental determination of the parameters, mathematical expression, and other formulation of the unified strength theory will be described in this chapter. The unified strength theory is a set of serial failure criteria. All the failure criteria of the unified strength theory conform to Drucker's postulation.

2 CONVEXITY OF THE LIMIT SURFACE

A postulate concerning the yield surfaces was proposed by Drucker, academician of the American Science Academy in 1951, with the convexity of yield surface determined. Since then the study of yield criterion has been developing on a more reliable theoretical basis. This postulate is considered as a fundamental theorem in plasticity and solid mechanics.

According to the convexity, all the failure criteria and yield criteria must be convex and located in the region between the two bounds, as shown in Figure 1.

3 MECHANICAL MODEL OF THE UNIFIED STRENGTH THEORY (UST)

The principal stress state (σ_1, σ_2, σ_3) can be converted into the solo-shear (or single-shear) stress state (τ_{13}, τ_{12} and τ_{23}), as shown in Figure 2.

There are only two independent principal shear stresses among the three principal shear stresses τ_{12}, τ_{23}, τ_{13}, because the maximum shear stress equals to the sum of the other two, i.e. $\tau_{13} = \tau_{12} + \tau_{23}$. Therefore, the twin-shear stress state is presented. The twin-shear stress

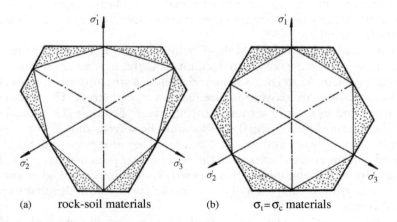

Figure 1 Two bounds and region of yield loci.

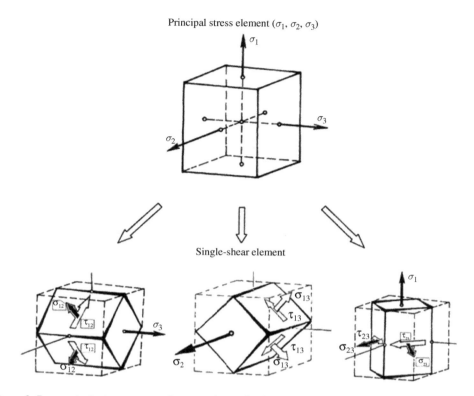

Figure 2 From principal stress state (σ_1, σ_2, σ_3) to solo-shear stress state (τ_{13}, τ_{12} and τ_{23}).

state can be converted from the solo-shear stress element, as shown in Figure 3. This element can be referred to as the twin-shear element. The two sets of shear stress and relevant normal stress (τ_{13}, τ_{12}; σ_{13}, σ_{12}) and (τ_{13}, τ_{23}; σ_{13}, σ_{23}) act on the element.

The twin-shear stress model is different from the regular octahedral model. The orthogonal octahedral model consists of two groups of four sections that are perpendicular to each other and are acted on by the maximum shear stress τ_{13} and the intermediate principal stress τ_{12} or τ_{23}. Based on the orthogonal octahedral element, a new strength theory can be developed.

Based on this twin-shear model, and taking into account the effects of all the stress components on the failure of materials, a new strength theory that has a unified mathematical expression was proposed by Mao-Hong Yu in 1990. It reflects the fundamental characteristics of materials in the complex stress state.

4 MATHEMATICAL MODELING AND EXPERIMENTAL DETERMINATION OF THE PARAMETERS FOR THE UNIFIED STRENGTH THEORY (UST)

Considering all the stress components acting on the twin-shear element and the different effects of various stresses on the failure of materials, the mathematical modeling of

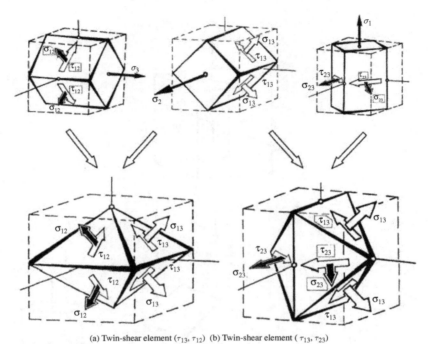

(a) Twin-shear element (τ_{13}, τ_{12}) (b) Twin-shear element (τ_{13}, τ_{23})

Figure 3 Twin-shear element.

the unified strength theory was established in 1990 as follows (Yu & He, 1991; Yu, 1992)

$$F = \tau_{13} + b\tau_{12} + \beta(\sigma_{13} + b\sigma_{12}) = C, \quad \text{when} \quad \tau_{12} + \beta\sigma_{12} \geq \tau_{23} + \beta\sigma_{23} \quad (2a)$$

$$F' = \tau_{13} + b\tau_{23} + \beta(\sigma_{13} + b\sigma_{23}) = C, \quad \text{when} \quad \tau_{12} + \beta\sigma_{12} \leq \tau_{23} + \beta\sigma_{23} \quad (2b)$$

where b is a parameter that reflects the influence of the intermediate principal shear stress τ_{12} or τ_{23} on the failure of the material; β is the coefficient that represents the effect of the normal stress on failure; C is a strength parameter of the material; τ_{13}, τ_{12} and τ_{23} are principal shear stresses; σ_{13}, σ_{12} and σ_{23} are the corresponding normal stresses acting on the sections where τ_{13}, τ_{12} and τ_{23} acted. They are defined as

$$\tau_{13} = \frac{1}{2}(\sigma_1 - \sigma_3), \quad \tau_{12} = \frac{1}{2}(\sigma_1 - \sigma_2), \quad \tau_{23} = \frac{1}{2}(\sigma_2 - \sigma_3) \quad (3)$$

$$\sigma_{13} = \frac{1}{2}(\sigma_1 + \sigma_3), \quad \sigma_{12} = \frac{1}{2}(\sigma_1 + \sigma_2), \quad \sigma_{23} = \frac{1}{2}(\sigma_2 + \sigma_3) \quad (4)$$

The magnitude of the parameters in the unified strength theory (UST) β and C can be determined by experimental results of uni-axial tension strength σ_t and uni-axial compression strength σ_c, the experimental conditions are:

$$\sigma_1 = \sigma_t, \quad \sigma_2 = \sigma_3 = 0 \quad (5a)$$

$$\sigma_1 = \sigma_2 = 0, \quad \sigma_3 = -\sigma_c \quad (5b)$$

Substituting Equation (5a) into Equation (2a) and Equation (5b) into Equation (2b), the material constants β and C can be determined as follows

$$\beta = \frac{\sigma_c - \sigma_t}{\sigma_c + \sigma_t} = \frac{1-\alpha}{1+\alpha}, \qquad C = \frac{(1+b)\sigma_c\sigma_t}{\sigma_c + \sigma_t} = \frac{(1+b)}{1+\alpha}\sigma_t \tag{6}$$

5 MATHEMATICAL EXPRESSION OF THE UNIFIED STRENGTH THEORY

Substituting Equation (6) into Equation (2a) and (2b), we obtain

$$F = \tau_{13} + b\tau_{12} + \frac{1-\alpha}{1+\alpha}(\sigma_{13} + b\sigma_{12}) = \frac{(1+b)\sigma}{1+\alpha}, \text{ when } \tau_{12} + \beta\sigma_{12} \ge \tau_{23} + \beta\sigma_{23} \tag{7a}$$

$$F = \tau_{13} + b\tau_{23} + \frac{1-\alpha}{1+\alpha}(\sigma_{13} + b\sigma_{23}) = \frac{(1+b)\sigma}{1+\alpha}, \text{ when } \tau_{12} + \beta\sigma_{12} \le \tau_{23} + \beta\sigma_{23} \tag{7b}$$

It is the mathematical expression of the unified strength theory (UST) in terms of the principal shear stress.

Substituting Equation. (3) and Equation (4) into Equation (7a) and Equation (7b), the UST in terms of the principal stress is now obtained. It can be expressed by the three principal stresses $(\sigma_1, \sigma_2, \sigma_3)$ as follows:

$$F = \sigma_1 - \frac{\alpha}{1+b}(b\sigma_2 + \sigma_3) = \sigma_t, \text{ when } \sigma_2 \le \frac{\sigma_1 + \alpha\sigma_3}{1+\alpha}, \tag{8a}$$

$$F' = \frac{1}{1+b}(\sigma_1 + b\sigma_2) - \alpha\sigma_3 = \sigma_t, \text{ when } \sigma_2 \ge \frac{\sigma_1 + \alpha\sigma_3}{1+\alpha}, \tag{8b}$$

This theory is a new system of strength theory and contains a series of new criteria. It takes into account the effect of all stress components on the failure of materials, gives a series of failure criteria, and establishes a relationship among various failure criteria. Though the mathematical expression of the UST is very simple and linear, it has rich and varied contents.

6 OTHER FORMULATIONS OF THE UNIFIED STRENGTH THEORY (UST)

The UST can be expressed in terms of the principal shear stresses and principal stresses have been described above. It can also be expressed in other terms, as described in the following subsections.

6.1 In terms of principal stress and cohesive parameter $F(\sigma_1, \sigma_2, \sigma_3, C_0, \varphi)$

In Equation (8a) and Equation (8b), we adopt the material constants σ_t and the tension-compression ratio α. In geomechanics and engineering, the cohesion C_0 and the coefficient φ reflect the material properties used. The relationships among the tensile strength σ_t, the tension-compression ratio α, the material parameter C_0 and φ can be obtained as follows:

$$\sigma_t = \frac{2C_0 \cdot \cos \varphi}{1 + \sin \varphi}, \qquad \alpha = \frac{1 - \sin \varphi}{1 + \sin \varphi} \tag{9}$$

By substituting Equation (11) into Equation (8a) and Equation (8b), the UST can be expressed in terms of C_0 and φ as

$$F = \sigma_1 - \frac{1 - \sin \varphi}{(1 + b)(1 + \sin \varphi)}(b\sigma_2 + \sigma_3) = \frac{2C_0 \cos \varphi}{1 + \sin \varphi},$$

$$\text{when} \quad \sigma_2 \leq \frac{1}{2}(\sigma_1 + \sigma_3) - \frac{\sin \varphi}{2}(\sigma_1 - \sigma_3) \tag{10a}$$

$$F' = \frac{1}{1 + b}(\sigma_1 + b\sigma_2) - \frac{1 - \sin \varphi}{1 + \sin \varphi}\sigma_3 = \frac{2C_0 \cos \varphi}{1 + \sin \varphi},$$

$$\text{when} \quad \sigma_2 \geq \frac{1}{2}(\sigma_1 + \sigma_3) - \frac{\sin \varphi}{2}(\sigma_1 - \sigma_3) \tag{10b}$$

6.2 In terms of stress invariant $F(I_1, J_2, \theta, \sigma_t, \alpha)$

$$F = (1 - \alpha)\frac{I_1}{3} + \frac{\alpha(1 - b)}{1 + b}\sqrt{J_2}\sin \theta + (2 + \alpha)\sqrt{\frac{J_2}{3}}\cos \theta = \sigma_t, \qquad 0° \leq \theta \leq \theta_b \tag{11a}$$

$$F' = (1 - \alpha)\frac{I_1}{3} + \left(\alpha + \frac{b}{1 + b}\right)\sqrt{J_2}\sin \theta + \left(\frac{2 - b}{1 + b} + \alpha\right)\sqrt{\frac{J_2}{3}}\cos \theta = \sigma_t, \quad \theta_b \leq \theta \leq 60° \tag{11b}$$

where I_1 is the first stress invariant (hydrostatic pressure), J_2 is the second deviatoric stress invariant and θ is the stress angle on the deviatoric plane. The stress angle at the corner θ_b can be determined by the condition $F = F'$.

$$\theta_b = \text{arctg}\frac{\sqrt{3}(1 + \beta)}{3 - \beta}, \quad \beta = \frac{1 - \alpha}{1 + \alpha} \tag{12}$$

6.3 In terms of stress invariant and cohesive parameter $F(I_1, J_2, \theta, C_0, \varphi)$

The UST can also be expressed by the stress invariant, stress angle and material parameters C_0 and φ.

$$F = \frac{2I_1}{3}\sin\varphi + \frac{2\sqrt{J_2}}{1+b}\left[\sin\left(\theta + \frac{\pi}{3}\right) - b\sin\left(\theta - \frac{\pi}{3}\right)\right] + \frac{2\sqrt{J_2}}{(1+b)\sqrt{3}}\cdot$$

$$\left[\sin\varphi\cos\left(\theta + \frac{\pi}{3}\right) + b\sin\varphi\cos\left(\theta - \frac{\pi}{3}\right)\right] = 2C_0\cos\varphi, \qquad 0° \le \theta \le \theta_b \qquad (13a)$$

$$F' = \frac{2I_1}{3}\sin\varphi + \frac{2\sqrt{J_2}}{1+b}\left[\sin\left(\theta + \frac{\pi}{3}\right) - b\sin\theta\right]$$

$$+ \frac{2\sqrt{J_2}}{(1+b)\sqrt{3}}\left[\sin\varphi\cos\left(\theta + \frac{\pi}{3}\right) + b\sin\varphi\cos\theta\right] = 2C_0\cos\varphi, \quad \theta_b \le \theta \le 60$$

$$(13b)$$

6.4 In terms of principal stresses and compressive strength parameter $F(\sigma_1, \sigma_2, \sigma_3, \alpha, \sigma_c)$

In soil and rock mechanics and engineering, the compressive strength σ_c is often adopted. Rewriting Equation (8a), Equation (8b) in terms of the principal stress and compressive strength σ_c, we have

$$F = \frac{1}{\alpha}\sigma_1 - \frac{1}{1+b}(b\sigma_2 + \sigma_3) = \sigma_c, \quad \text{when} \quad \sigma_2 \le \frac{\sigma_1 + \alpha\sigma_3}{1+\alpha} \qquad (14a)$$

$$F' = \frac{1}{\alpha(1+b)}(\sigma_1 + b\sigma_2) - \sigma_3 = \sigma_c, \quad \text{when} \quad \sigma_2 \ge \frac{\sigma_1 + \alpha\sigma_3}{1+\alpha} \qquad (14b)$$

6.5 In terms of stress invariant and compressive strength parameter $F(I_1, J_2, \theta, \alpha, \sigma_c)$

$$F = \frac{1-\alpha}{3\alpha}I_1 + \frac{1-b}{1+b}\sqrt{J_2}\sin\theta + \frac{2+\alpha}{\alpha\sqrt{3}}\sqrt{J_2}\cos\theta = \sigma_c \quad 0° \le \theta \le \theta_b \qquad (15a)$$

$$F' = \frac{1-\alpha}{3\alpha}I_1 + \frac{\alpha + \alpha b + b}{\alpha(1+b)}\sqrt{J_2}\sin\theta + \frac{2+\alpha+\alpha b - b}{\alpha\sqrt{3}(1+b)}\sqrt{J_2}\cos\theta = \sigma_c \quad \theta_b \le \theta \le 60° \qquad (15b)$$

The UST can also be expressed in other terms.

7 SPECIAL CASES OF THE UST FOR DIFFERENT PARAMETER B

The unified strength theory (UST) contains a family of the convex failure criteria. A series of convex failure criteria can be deduced from the UST by giving a certain value to parameter b. The series of convex yield criteria ($\alpha = 1$) is its special cases.

The parameter b reflects the influence of the intermediate principal shear stress τ_{12} or τ_{23} on the failure of a material. It also reflects the influence of the intermediate principal stress σ_2 on the failure of a material. As can be seen below, b is also the parameter that determines the formulation of a failure criterion. A series of convex failure criteria can

432 Yu

be obtained when the parameter varies in the range of $0 \leq b \leq 1$. The five types of failure criteria with the values of $b = 0$, $b = 1/4$, $b = 1/2$, $b = 3/4$ and $b = 1$ are introduced from the UST.

7.1 $b = 0$

The Mohr–Coulomb strength theory can be deduced from the UST with $b = 0$ as follows:

$$F = F' = \sigma_1 - \alpha\sigma_3 = \sigma_t \tag{16}$$

$$F = F' = \frac{1}{\alpha}\sigma_1 - \sigma_3 = \sigma_c \tag{17}$$

7.2 $b = 1/4$

A new failure criterion is deduced from the UST with $b = 1/4$ or $B = 1 + 4\alpha/5$ as follows:

$$F = \sigma_1 - \frac{\alpha}{5}(\sigma_2 + 4\sigma_3) = \sigma_t, \ \sigma_2 \leq \frac{\sigma_1 + \alpha\sigma_3}{1 + \alpha} \tag{18a}$$

$$F' = \frac{1}{5}(4\sigma_1 + \sigma_2) - \alpha\sigma_3 = \sigma_t, \ \sigma_2 \geq \frac{\sigma_1 + \alpha\sigma_3}{1 + \alpha} \tag{18b}$$

7.3 $b = 1/2$

A new failure criterion is deduced from the UST with $b = 1/2$ or $B = 1 + 2\alpha/3$ as follows:

$$F = \sigma_1 - \frac{\alpha}{3}(\sigma_2 + 2\sigma_3) = \sigma_t, \ \sigma_2 \leq \frac{\sigma_1 + \alpha\sigma_3}{1 + \alpha} \tag{19a}$$

$$F' = \frac{1}{3}(2\sigma_1 + \sigma_2) - \alpha\sigma_3 = \sigma_t, \ \sigma_2 \geq \frac{\sigma_1 + \alpha\sigma_3}{1 + \alpha} \tag{19b}$$

Since the Drucker–Prager criterion cannot match with the practice, this criterion is more reasonable and may be substituted for the Drucker–Prager criterion.

7.4 $b = 3/4$

A new failure criterion is deduced from the UST with $b = 3/4$ or $B = 1 + 4\alpha/7$ as follows:

$$F = \sigma_1 - \frac{\alpha}{7}(3\sigma_2 + 4\sigma_3) = \sigma_t, \ \sigma_2 \leq \frac{\sigma_1 + \alpha\sigma_3}{1 + \alpha} \tag{20a}$$

$$F = \frac{1}{7}(4\sigma_1 + 3\sigma_2) - \alpha\sigma_3 = \sigma_t, \ \sigma_2 \geq \frac{\sigma_1 + \alpha\sigma_3}{1 + \alpha} \tag{20b}$$

7.5 $b = 1$

A new failure criterion is deduced from the UST with $b = 1$ or $B = 1 + \alpha/2$ as follows:

$$F = \sigma_1 - \frac{\alpha}{2}(\sigma_2 + \sigma_3) = \sigma_t, \ \text{when} \ \sigma_2 \leq \frac{\sigma_1 + \alpha\sigma_3}{1 + \alpha} \tag{21a}$$

$$F' = \frac{1}{2}(\sigma_1 + \sigma_2) - \alpha\sigma_3 = \sigma_t, \text{ when } \sigma_2 \geq \frac{\sigma_1 + \alpha\sigma_3}{1 + \alpha} \tag{21b}$$

This is the generalized twin-shear strength model proposed by Mao-Hong Yu in 1983 (Yu, 1983; Yu et al., 1985).

8 LIMIT LOCI OF THE UST IN DEVIATORIC PLANE

The mathematical expression of the UST in terms of principal stresses is as follows:

$$F = \sigma_1 - \frac{\alpha}{1 + b}(b\sigma_2 + \sigma_3) = \sigma_t, \text{ when } \sigma_2 \leq \frac{\sigma_1 + \alpha\sigma_3}{1 + \alpha} \tag{22a}$$

$$F' = \frac{1}{1 + b}(\sigma_1 + b\sigma_2) - \alpha\sigma_3 = \sigma_t, \text{ when } \sigma_2 \leq \frac{\sigma_1 + \alpha\sigma_3}{1 + \alpha} \tag{22b}$$

The relationships between the coordinates of the deviatoric plane and hydrostatic stress axis z with the principal stresses are:

$$x = \frac{1}{\sqrt{2}}(\sigma_3 - \sigma_2), \quad y = \frac{1}{\sqrt{6}}(2\sigma_1 - \sigma_2 - \sigma_3), \quad z = \frac{1}{\sqrt{3}}(\sigma_1 + \sigma_2 + \sigma_3) \tag{23}$$

$$\sigma_1 = \frac{1}{3}(\sqrt{6}y + \sqrt{3}z);$$

$$\sigma_2 = \frac{1}{6}(2\sqrt{3}z - \sqrt{6}y - 3\sqrt{2}x); \tag{24}$$

$$\sigma_3 = \frac{1}{6}(3\sqrt{2}x - \sqrt{6}y + 2\sqrt{3}z)$$

By substituting Equation (23) and Equation (24) into Equation (8a) and Equation (8b), the equations of the UST in the deviatoric plane can be obtained:

$$F = -\frac{\sqrt{2}(1 - b)}{2(1 + b)}\alpha x + \frac{\sqrt{6}(2 + \alpha)}{6}y + \frac{\sqrt{3}(1 - \alpha)}{3}z = \sigma_t \tag{25a}$$

$$F' = -\left(\frac{b}{1 + b} + \alpha\right)\frac{\sqrt{2}}{2}x + \left(\frac{2 - b}{1 + b} + \alpha\right)\frac{\sqrt{6}}{6}y + \frac{\sqrt{3}(1 - \alpha)}{3}z = \sigma_t \tag{25b}$$

A great number of new failure criteria can be generated from the UST by changing α and b. The general shape of the limit loci of the UST on the deviatoric plane are shown in Figure 4.

Material parameters α and σ_t are the tension-compression strength ratio and the uni-axial tensile strength, respectively, and b is a material parameter that reflects the influence of intermediate principal shear stress. A series of limit surfaces can be obtained by varying b. Five special cases will be discussed with values of b from $b = 0$, $b = 1/4$, $b = 1/2$, $b = 3/4$ and $b = 1$.

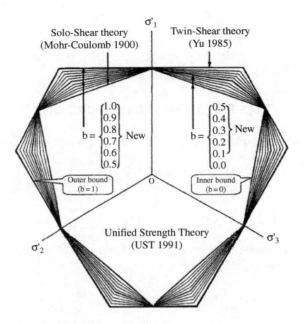

Figure 4 Limit loci of the UST on the deviatoric plane.

8.1 b = 0

Substituting $b = 0$ into Equation (25a) and Equation (25b) we have

$$F = F' = -\frac{\sqrt{2}}{2}ax + \frac{\sqrt{6}}{6}(2+\alpha)y + \frac{\sqrt{3}}{3}(1-\alpha)z = \sigma_t \tag{26}$$

This is the Mohr–Coulomb strength theory. The limit locus of the Mohr–Coulomb strength theory is the lower bound of the convex limit loci, as shown in Figure 4.

8.2 b = 1/4

Substituting $b = 1/4$ into Equation (25a) and Equation (25b) we have

$$F = -\frac{3\sqrt{2}}{10}ax + \frac{\sqrt{6}}{6}(2+\alpha)y + \frac{\sqrt{3}}{3}(1-\alpha)z = \sigma_t \tag{27a}$$

$$F' = -\left(\frac{1}{5}+\alpha\right)\frac{\sqrt{2}}{2}x + \left(\frac{7}{5}+\alpha\right)\frac{\sqrt{6}}{6}y + \frac{\sqrt{3}}{3}(1-\alpha)z = \sigma_t \tag{27b}$$

This is the limit surface of a new failure criterion.

8.3 b = 1/2

Substituting b = 1/2 into Equation (25a) and Equation (25b) we have

$$F = -\frac{\sqrt{2}}{6}ax + \frac{\sqrt{6}}{6}(2+a)y + \frac{\sqrt{3}}{3}(1-a)z = \sigma_t \tag{28a}$$

$$F' = -\left(\frac{1}{3}+a\right)\frac{\sqrt{2}}{2}x + (1+a)\frac{\sqrt{6}}{6}y + \frac{\sqrt{3}}{3}(1-a)z = \sigma_t \tag{28b}$$

This is a new failure criterion. It is intermediate between the Mohr–Coulomb strength theory and the twin-shear strength theory. The limit locus of the new criterion on the deviatoric plane is also shown in Figure 4.

8.4 b = 3/4

Substituting b = 3/4 into Equation (25a) and Equation (25b) we have

$$F = -\frac{\sqrt{2}}{14}ax + \frac{\sqrt{6}}{6}(2+a)y + \frac{\sqrt{3}}{3}(1-a)z = \sigma_t \tag{29a}$$

$$F' = -\left(\frac{3}{7}+a\right)\frac{\sqrt{2}}{2}x + \left(\frac{5}{7}+a\right)\frac{\sqrt{6}}{6}y + \frac{\sqrt{3}}{3}(1-a)z = \sigma_t \tag{29b}$$

This is the limit surface of a new failure criterion. The limit locus is close to the limit locus of the twin-shear strength theory.

8.5 b =1

Substituting b =1 into Equation (25a) and Equation (25b) we have

$$F = \frac{\sqrt{6}}{6}(2+a)y + \frac{\sqrt{3}}{3}(1-a)z = \sigma_t \tag{30a}$$

$$F' = -\left(\frac{1}{2}+a\right)\frac{\sqrt{2}}{2}x + \left(\frac{1}{2}+a\right)\frac{\sqrt{6}}{6}y + \frac{\sqrt{3}}{3}(1-a)z = \sigma_t \tag{30b}$$

This is the twin-shear strength theory proposed by Mao-Hong Yu in 1985. The limit locus of the twin-shear strength theory is the upper bound of the convex limit loci, as shown in Figure 4.

When the tensile strength and the compressive strength are identical, the tension–compressive strength ratio $a = \sigma_t/\sigma_c$ equals 1, or the friction angle $\varphi = 0$. The unified yield criterion (UYC) can be obtained as a special case of the UST. The mathematical expression of the unified yield criterion is expressed as follows. It also contains a series of yield criteria.

$$F = \sigma_1 - \frac{1}{1+b}(b\sigma_2 + \sigma_3) = \sigma_t, \quad \text{when} \quad \sigma_2 \leq \frac{\sigma_1+\sigma_3}{2}, \tag{31a}$$

$$F' = \frac{1}{1+b}(\sigma_1 + b\sigma_2) - \sigma_3 = \sigma_t, \quad \text{when} \quad \sigma_2 \geq \frac{\sigma_1+\sigma_3}{2}, \tag{31b}$$

The ratio between the tensile radius and the compressive radius on the deviatoric plane is given by

$$K = \frac{1+2\alpha}{2+\alpha} = \frac{3-\sin\phi}{3+\sin\phi} = 1 \qquad (32)$$

which means that the irregular dodecahedron is converted to a regular dodecahedron.

The equations of the unified yield criterion on the deviatoric plane can be obtained by Equation (25a) and Equation (25b)

$$F = -\frac{\sqrt{2}(1-b)}{2(1+b)}x + \frac{\sqrt{6}}{2}y = \sigma_t \qquad (33a)$$

$$F' = -\frac{\sqrt{2}(1+2b)}{2(1+b)}x + \frac{\sqrt{6}}{2(1+b)}y = \sigma_t \qquad (33b)$$

It is seen that the yield locus of the unified yield criterion has nothing to do with the value of z, and its shape is identical with different values of z. Therefore, the yield surfaces of the unified yield criterion are a series of infinite prisms. The equations of the yield loci of the unified yield criterion on the deviatoric plane with b = 0, 1/4, 1/2, 3/4 and b = 1 can be introduced. The relevant yield loci are illustrated in Figure 5.

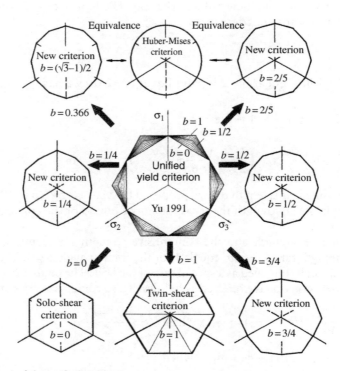

Figure 5 Yield loci of the unified yield criterion.

The unified yield criterion contains a series of new convex yield criteria, as shown in Figure 5. The solo-shear yield criterion and the twin-shear yield criterion can be given with $b=0$ and $b=1$, respectively.

9 CONCAVE STRENGTH THEORY ($b < 0$ OR $b > 1$)

The UST provides a completely new series of yield and failure criteria. It can also be extended to concave cases when $b < 0$ or $b > 1$, as shown in Figures 6 and 7. The concave limit loci and yield loci have seldom been studied before. The broken line in Figure 6 is the limit locus of the solo-shear strength theory (the Mohr–Coulomb strength theory). It is the lower bound of the convex limit loci. The limit loci of the UST when $b < 0$ are smaller than the lower bound of the convex failure criterion ($b < 0$, inside the lower bound). The broken line in Figure 7 shows the limit locus of the twin-shear strength theory (Yu et al., 1985). It is the upper bound of the convex limit loci. The limit loci of the UST when $b>1$ are larger than the upper bound of the convex failure criterion.

The meaning and application of the concave failure loci have not yet investigated. Until now, most experimental results for materials under complex stress states are chosen to show the convexity. So, the limit locus of a strength theory cannot be chosen arbitrarily. The property of convexity means that the limit loci of the failure criteria have to be situated between the two bounds, as shown in Figures 2 and 4. The solo-shear strength theory is the lower bound, and no admissible limit surface may exceed

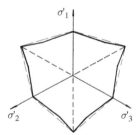

Figure 6 Concave yield criterion ($b = -1/3$).

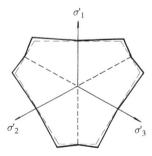

Figure 7 Concave yield criterion ($b = 5/4$).

that of the solo-shear strength theory (Mohr–Coulomb strength theory) from below. The twin-shear strength theory is the upper bound, and no admissible limit surface may exceed that of the twin-shear strength theory from above. The effect of failure criteria will be studied in the framework of convexity.

10 LIMIT SURFACES AND YIELD LOCI OF THE UST

10.1 Limit surfaces of the UST in principal stress space

The yield function can be interpreted for an isotropic material in terms of a geometrical representation of the stress state obtained by taking the principal stresses as coordinates. The advantage of such a space lies in its simplicity and visual presentation.

The yield surfaces in the stress space of the UST are usually a semi-infinite hexagonal cone with unequal sides and a dodecahedron cone with unequal sides, as shown in Figure. 8. The shape and size of the yield hexagonal cone depends on parameter b and on the tension-compression strength ratio α. The 3D computer images of yield surface for the UST in the stress space are given by Zhang (2005), as shown in Figure 8 and Figure 9.

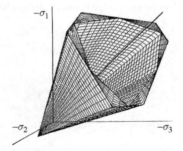

Figure 8 Yield surfaces of the UST in stress space.

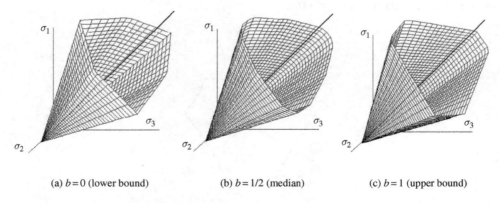

(a) $b = 0$ (lower bound)   (b) $b = 1/2$ (median)   (c) $b = 1$ (upper bound)

Figure 9 Three typical yield surfaces of the UST.

Unified Strength Theory (UST)   439

In engineering practice, the compressive strength of materials σ_c is often much greater than the tensile strength σ_t, since a region in tension becomes smaller, while it becomes larger in compression. Assuming the compressive strength is positive, the yield surfaces of the UST with different values of b are shown in Figure 8. The three typical yield surfaces of the UST with $b=0$, $b=1/2$, and $b=1$ are illustrated respectively in Figure 9.

10.2 Limit loci of the UST in the plane stress state

The limit loci of the UST in the plane stress state are the intersection line of the limit surface in the principal stress space and the $\sigma_1 - \sigma_2$ plane. Its shape and size depend on the values of b and a. It will be transformed into a hexagon when $b = 0$ or $b = 1$, and into a dodecagon when $0 < b < 1$.

The equations of the 12 limiting loci of the UST in the plane stress state can be given as follows:

$$
\begin{aligned}
& \sigma_1 - \frac{ab}{1+b}\sigma_2 = \sigma_t && \frac{\alpha}{1+b}(\sigma_1 + b\sigma_2) = \sigma_t \\[2mm]
& \sigma_2 - \frac{ab}{1+b}\sigma_1 = \sigma_t && \frac{\alpha}{1+b}(\sigma_2 + b\sigma_1) = \sigma_t \\[2mm]
& \sigma_1 - \frac{\alpha}{1+b}\sigma_2 = \sigma_t && \frac{1}{1+b}\sigma_1 - \alpha\sigma_2 = \sigma_t \\[2mm]
& \sigma_2 - \frac{\alpha}{1+b}\sigma_1 = \sigma_t && \frac{1}{1+b}\sigma_2 - \alpha\sigma_1 = \sigma_t \\[2mm]
& \frac{\alpha}{1+b}(b\sigma_1 + \sigma_2) = -\sigma_t && \frac{b}{1+b}\sigma_1 - \alpha\sigma_2 = \sigma_t \\[2mm]
& \frac{\alpha}{1+b}(b\sigma_2 + \sigma_1) = -\sigma_t && \frac{b}{1+b}\sigma_2 - \alpha\sigma_1 = \sigma_t
\end{aligned}
\tag{34}
$$

A series of new failure criteria and new limit loci in the plane stress state can be obtained from the UST.

10.2.1 Variation of the UST with b

The limit loci of the UST in the plane stress state with different values of b for $\alpha=1/2$ materials are shown in Figure 10.

Various limit loci of the UST in the plane stress state are shown in Figure 11. The unified yield criterion, the Mohr–Coulomb strength theory, the twin-shear strength theory and a series of new failure criteria can be obtained from the UST, as shown in Figure 11.

10.2.2 Limit locus of the UST by varying α

If the tensile strength is identical to the compressive strength, the UST will be transformed into the unified yield criterion (UYC). The yield loci of the unified yield criterion ($\alpha = 1$ materials) in the deviatoric plane has been shown in Figure 5.

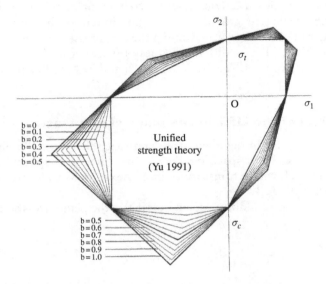

Figure 10 Variation of the limit loci of the UST in plane stress state ($\alpha = 1/2$).

The yield loci of the unified yield criterion ($\alpha = 1$ materials) in plane stress state are shown in Figure 11. The limit loci of the UST in the plane stress state with different values of α are shown in Figures 10 and 12. Figure 10 shows the limiting line of the UST in the $\sigma_1-\sigma_2$ plane with $\alpha = 1/2$. Figure 12 shows the limiting loci of the UST in the $\sigma_1-\sigma_2$ plane with $\alpha = 1/4$.

10.2.3 Limit loci of UST in meridian plane

Various expressions of the UST are given above. The UST can also be expressed in other terms, such as by the octahedral normal stress σ_8 and octahedral shear stress τ_8 in plasticity, or by the generalized normal stress σ_g and the generalized shear stress τ_g (or q) in geomechanics.

The relationships between the three principal stresses σ_1, σ_2, σ_3 and the cylindrical polar coordinates ξ, r, θ in the principal stress space are:

$$\begin{Bmatrix} \sigma_1 \\ \sigma_2 \\ \sigma_3 \end{Bmatrix} = \frac{1}{\sqrt{3}}\xi + \sqrt{\frac{2}{3}}r \begin{Bmatrix} \cos\theta \\ \cos(\theta - 2\pi/3) \\ \cos(\theta + 2\pi/3) \end{Bmatrix} \tag{35}$$

in which ξ is the major coordinate axis in the stress space, and r is the length of the stress vector in the π-plane. They are defined as follows:

$$\begin{aligned} \xi &= \frac{1}{\sqrt{3}}(\sigma_1 + \sigma_2 + \sigma_3) \\ r &= \frac{1}{\sqrt{3}}\sqrt{(\sigma_1 - \sigma_2)^2 + (\sigma_2 - \sigma_3)^2 + (\sigma_3 - \sigma_1)^2} \end{aligned} \tag{36}$$

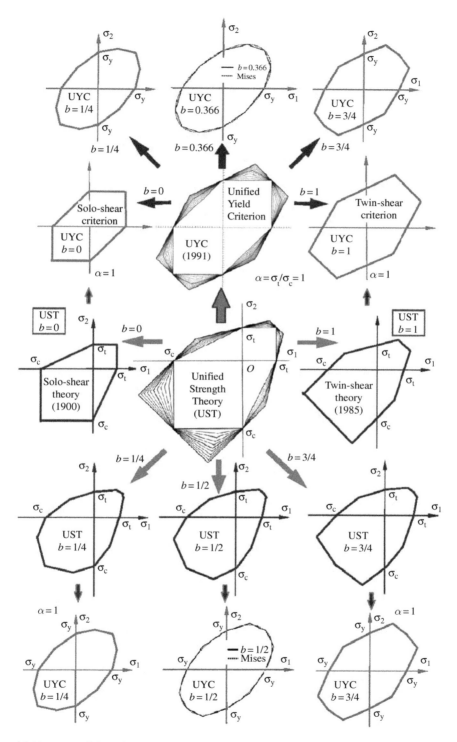

Figure 11 Variation of the UST in the plane stress state.

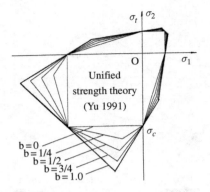

Figure 12 Limit loci of the UST in the plane stress state ($\alpha = 1/4$ material).

The relationship among the various variables is

$$\xi = \frac{1}{\sqrt{3}} I_1 = \sqrt{3}\sigma_8 = \sqrt{3}p = \sqrt{3}\sigma_m \tag{37}$$

$$r = \sqrt{2J_2} = \sqrt{3}\tau_8 = \sqrt{\frac{2}{3}}q = 2\tau_m$$

The principal stress can be expressed as

$$\begin{Bmatrix} \sigma_1 \\ \sigma_2 \\ \sigma_3 \end{Bmatrix} = \frac{1}{3} I_1 + \frac{2}{\sqrt{3}} \sqrt{J_2} \begin{Bmatrix} \cos \theta \\ \cos (\theta - 2\pi/3) \\ \cos (\theta + 2\pi/3) \end{Bmatrix} \tag{38}$$

$$\begin{Bmatrix} \sigma_1 \\ \sigma_2 \\ \sigma_3 \end{Bmatrix} = p + \frac{2}{3} q \begin{Bmatrix} \cos \theta \\ \cos (\theta - 2\pi/3) \\ \cos (\theta + 2\pi/3) \end{Bmatrix} \tag{39}$$

$$\begin{Bmatrix} \sigma_1 \\ \sigma_2 \\ \sigma_3 \end{Bmatrix} = \sigma_8 + \sqrt{2}\tau_8 \begin{Bmatrix} \cos \theta \\ \cos (\theta - 2\pi/3) \\ \cos (\theta + 2\pi/3) \end{Bmatrix} \tag{40}$$

$$\begin{Bmatrix} \sigma_1 \\ \sigma_2 \\ \sigma_3 \end{Bmatrix} = \sigma_m + \frac{2\sqrt{2}}{\sqrt{3}} \tau_m \begin{Bmatrix} \cos \theta \\ \cos (\theta - 2\pi/3) \\ \cos (\theta + 2\pi/3) \end{Bmatrix} \tag{41}$$

Substituting the above equations into Equation (8a) and Equation (8b), the UST can then be expressed in other terms.

In some books on geomechanics, $(\sigma_1-\sigma_3)$ is often used as a coordinate, then the variable $p \sim (\sigma_1-\sigma_3)$, $\sigma_8 \sim (\sigma_1-\sigma_3)$, or $\sigma_m \sim (\sigma_1-\sigma_3)$ can be drawn. In the case of tri-axial confined pressure experiments, the stress state is axial-symmetric, i.e., $\sigma_2 = \sigma_3$. The generalized shear stress q is

$$q = \sqrt{\frac{1}{2}[(\sigma_1 - \sigma_2)^2 + (\sigma_2 - \sigma_3)^2 + (\sigma_3 - \sigma_1)^2]} = \sigma_1 - \sigma_3$$

The $p{\sim}q$ coordinate and the $p{\sim}(\sigma_1{-}\sigma_3)$ coordinate are identical. It is worth noting, however, that they are not identical in other cases.

The strength behavior of material in the region of three-dimensional tensile stresses is more complex. The tri-axial tensile test is difficult. In this case a tension cut-off condition $F=\sigma_1=\sigma_t$ is needed. The mathematical expressions of the UST and tension cut-off are

$$F_1 = \sigma_1 - \frac{\alpha}{1+b}(b\sigma_2 + \sigma_3) = \sigma_t, \quad \text{when} \quad \sigma_2 \leq \frac{\sigma_1 + \alpha\sigma_3}{1+\alpha}, \tag{42a}$$

$$F_2 = \frac{1}{1+b}(\sigma_1 + b\sigma_2) - \alpha\sigma_3 = \sigma_t, \quad \text{when} \quad \sigma_2 \geq \frac{\sigma_1 + \alpha\sigma_3}{1+\alpha}, \tag{42b}$$

$$F_3 = \sigma_1 = \sigma_t, \quad \text{when} \quad \sigma_1 \geq \sigma_2 \geq \sigma_3 \geq 0 \tag{42c}$$

11 SIGNIFICANCE OF THE UST

The UST encompasses many well established criteria as its special cases or linear approximations. It also gives a series of new failure criteria. The relationship between the UST and the existing main strength theories can be illustrated in Figure 13.

The solo-shear strength theory, the twin-shear strength theory and a series of new failure criteria can be obtained from the UST in the range of $0 \leq b \leq 1, 0 \leq \alpha \leq 1$. The smooth-ridge models can also be approximated by the UST when $b = 1/2$ or $b = 3/4$. The convex failure criteria can be obtained by varying the value of α ($\alpha < 1$) and b ($0 \leq b \leq 1$). They can be used to suit various kinds of engineering materials.

A series of yield criteria and failure criteria can be introduced from the unified strength theory (UST).

The UST is a completely new theory system. The significance of the UST is summarized as follows:

1. It is suitable for various kinds of materials.
2. It contains various spread strength theories and forms a new system of yield criteria and failure criteria.
3. It agrees well with experimental results for various materials, such as metals, rock, soil, concrete and iron.
4. A series of new results can be obtained by using the UST.
5. The UST can be generalized to the unified elasto-plastic constitutive equations. It can be implemented in finite element code and forms a unified elasto-plastic program. It is convenient for elastic limit design, elasto-plastic analysis, and the plastic limit analysis of structures.

The UST is convenient for application to analytic solution of plasticity and engineering problems. Several unified solutions for plastic behavior of structures were introduced by using the UST. The research results show that the yield criterion has significant influence on the load-carrying capacities of structures. A series of analytical

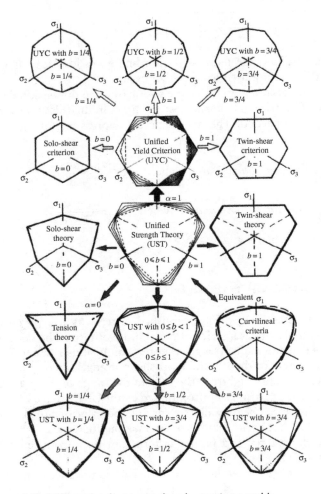

Figure 13 Variation of the UST on the deviatoric plane by varying α and b.

results are clearly illustrated to show the effects of yield criterion to elasto-plastic behavior, limit speed and dynamic behavior.

The concept of the UST can be generalized in many other branches. The UST can be generalized conveniently to multi-parameter strength theory, such as the three-parameter failure criterion and the five-parameter failure criterion. The recent result can be seen in a paper (Yu, 2002). A detailed description of the multi-parameter failure criterion can be found in the two books 'Twin-Shear Theory and its Applications' (in Chinese, Yu, 1998) and 'Unified Strength Theory and its Applications' (in English, Yu, 2004).

In general, the analytical results of structural strength and the computational results of numerical simulations of structural plasticity depend strongly on the choice of failure criterion.

A series of researches were carried out to show the effects of strength theory on the results of elasto-plastic analysis, load-carrying capacities of structures, deformation and discontinuous bifurcation, localization behavior and others.

The effects of failure criteria on the analytical results of slip field of plane strain problems, characteristics fields of plane stress problems and spatial axial-symmetric problems are summarized in 'Generalized Plasticity' (Yu, 2006). The choosing of strength theory has significant influence on these results.

12 APPLICATIONS OF THE UST

The unified strength theory has been widely used in several fields. It can be seen in literature. Three monographs were presented for the application of UST in three aspects (Yu, 2006; Yu et al., 2009; Yu & Li, 2012). Two examples are described briefly here.

Example 1: Trapezoid structure

The trapezoid specimen with a top angle 2ξ is considered. The uniform distributed load is applied on the top of the specimen, as shown in Figure 14. The physical properties of the material are: Young's modules: 368 kN/cm^2, specific gravity: 1.35-1.45g/cm^3, Poisson's ratio: 0.27, tensile strength: 5.89 kN/cm^2, compressive strength: 7.58 kN/cm^2. Determine the limit load.

Solution

The friction angle and cohesion of this material can be determined by

$$\varphi_0 = \sin^{-1}\frac{\sigma_c - \sigma_t}{\sigma_c + \sigma_t} = 7.208^0$$
$$C_0 = \frac{\sigma_t(1 + \sin \varphi_0)}{2\cos \varphi_0} = 3.339 \quad (43)$$

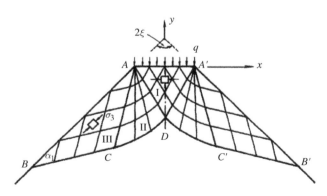

Figure 14 Slip lines field of a trapezoid structure.

It is assumed that the surfaces AA' are smooth and there is no friction. It is also assumed that there is a constant pressure on the top. The rest of the boundary is stress-free. The slip line field is shown in Figure 14.

The limit loading of a trapezoid specimen is obtained by using the UST and unified slip field theory as follows (Yu, 2006):

$$q = C_{UST} \cdot \cot\varphi_{UST} \left[\frac{1 + \sin \varphi_{UST}}{1 - \sin \varphi_{UST}} \exp(2\xi \cdot \tan \varphi_{UST}) - 1 \right] \tag{44}$$

With different choices of unified yield criterion parameter b, a series of limit loading are obtained as shown in Figure 15. The serial results for the top angles $2\xi = 120°$ are shown. Figure 16 is the variation of the slip angles with the variation of unified yield criterion parameter b.

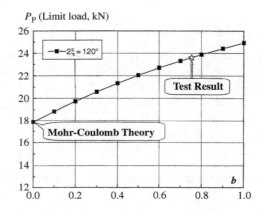

Figure 15 Unified solutions of limit loading.

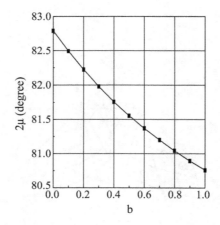

Figure 16 Variation of slip angle 2μ with the unified yield criterion parameter b.

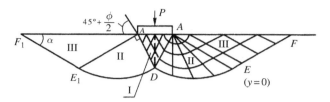

Figure 17 Slip lines field of a strip footing.

It is worth mentioning that not only the limit loading q but also the slip angle 2μ are different. They are different for different materials. The variation of slip angle 2μ with the unified yield criterion parameter b is shown in Figure 16. The result of the unified slip line field theory ($2\mu = 81.04°$ for $b = 0.8$) is much closer to the experimental result than that of the Mohr–Coulomb ($2\mu = 82.8°$).

It is a strip footing when the top angles of the trapezoid structure equals $2\xi = 180°$. The limit load can be deduced from Equation (44) when $2\xi = \pi$, a serial results can be given. The slip field is shown in Figure 17.

$$q = C_{UST} \cdot \cot\varphi_{UST} \left[\frac{1 + \sin\varphi_{UST}}{1 - \sin\varphi_{UST}} \exp(\pi \cdot \tan\varphi_{UST}) - 1 \right] \quad (45)$$

Example 2

The application of the unified strength theory by using the numerical analysis method can be illustrated by an example as follows. A series of shear strain cloud chart using the UST with several b for a slope under the same condition are given in Figure 18.

Figure 18 shows different shear strain cloud charts using the UST with several b for a slope under the same condition. The shear strain cloud chart using the UST with $b = 0$ is maximal, this result is the same as the result by using the solo-shear theory (Mohr–Coulomb theory). The result by using the twin-shear theory (UST with $b = 1$) is minimal.

Recently, a comprehensive and useful monograph entitled 'Solutions Manual to Design Analysis in Rock Mechanics' (Pariseau, 2008) was published giving the design analysis of slope stability, shafts, tunnels, entries in stratified ground, pillars in stratified ground, three-dimensional excavations, and subsidence. The Mohr–Coulomb condition is used in most cases. More results can be obtained if we use the unified strength theory instead of the Mohr–Coulomb condition.

13 SUMMARY

Based on the twin-shear mechanical model and a new type of mathematical modeling, a new UST was established by Yu in 1991 (Yu & He, 1991; Yu, 1992). This UST is not a single yield criterion suitable only for one kind of material, it is a completely new system. It embraces many well established criteria as its special or approximate cases, such as the Tresca yield criterion, the Huber-von Mises yield criterion, and the Mohr–Coulomb strength theory, as well as the twin-shear yield criterion, the generalized

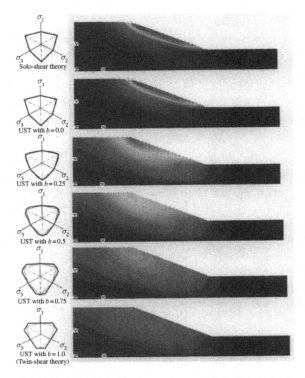

Figure 18 Shear strain cloud charts with several *b* for a slope under the same condition.

twin-shear strength theory (for SD materials, Yu *et al.*, 1985), and the unified yield criterion (for non-SD materials). The UST forms an entire spectrum of convex criteria, which can be used to describe many kinds of engineering materials. The UST has a unified mechanical model and a simple and unified mathematical expression, which can be adapted to the various experimental data. It is easy to use in both research and engineering practice.

The SD effect, hydrostatic stress effect, normal stress effect, effect of the intermediate principal stress, and the effect of intermediate principal shear stress are all taken into account in the UST. The UST establishes a very clear and simple relation among the various yield criteria. It also provides a method to choose the appropriate yield criterion.

The mathematical expression of the UST can be expressed in various forms. More than five kinds of expressions are discussed in this chapter.

The parameters of UST are the same as the parameters used in the Mohr–Coulomb strength theory (1900), Drucker–Prager criterion (1952), and other two-parameters criteria. Tensile strength σ_t, compressive strength σ_c (or σ_t, α) or friction angle φ and cohesion C_0 are the most used material parameters in engineering.

The yield surfaces and yield loci of the unified yield criterion, the twin-shear strength criterion, the twin-shear yield criterion, the solo-shear strength criterion (Mohr–Coulomb theory), the solo-shear yield criterion (Tresca yield criterion) and many empirical failure criteria are special cases or linear approximations of the yield surface of the UST. A series of new yield surfaces and yield loci can also be drawn based on the UST.

Postscript: A paper entitled 'Remarks on Model of Mao-Hong Yu' has been written by Altenbach and Kolupaev (2008). Two reviews of 'UST and its Applications' were published by Shen (2004) and Teodorescu (2006).

REFERENCES

Altenbach, H. and Kolupaev, V.A. (2008). *'Remarks on the model of Mao-Hong Yu'*. In Tong Yi, Zhang, Biao, Wang and Xi-Qiao, Feng eds. *The Eighth Int. Conference on Fundamentals of Fracture (ICFF VIII)*, 270–271.

Chen, W.-F. and Saleeb, A. F. (1994). *Constitutive Equations for Engineering Materials. Vol. 1: Elasticity and Modeling*, Revised edn. Elsevier, Amsterdam, 259–304, 462–489.

Chen, W.-F., A.F. Saleeb, W.O. McCarron and E. Yamaguchi (eds) (1994). *Constitutive Equations for Engineering Materials. Vol. 2: Plasticity and Modeling.* Elsevier, Amsterdam.

Davis, R.O. and Selvadurai, A.P.S. (2002). *Plasticity and Geomechanics.* Cambridge University Press, Cambridge, 74–75.

Drucker, D.C.(1951). 'A more foundational approach to stress-strain relations'. In: *Proc. of First U.S. National Cong. Appl. Mech.* ASME, 487–491.

Drucker, D.C.and Prager, W. (1952). 'Soil mechanics and plastic analysis for limit design'. *Q. Appl. Math.*, 10, 157–165.

Jaeger, J.C. and Cook, N.G.W. (1979). *Fundamentals of Rocks Mechanics*, 3rd edn. Chapman and Hall, London.

Kolupaev, V.A.and Altenbach, H. (2010). 'Einige Überlegungen zur Unified Strength Theory von Mao-Hong Yu' (Considerations on the Unified Strength Theory due to Mao-Hong Yu). *Forschung im Ingenieurwesen*, 74, 135–166 (in German).

Mohr, O. (1900). 'Welche Umstände bedingen die Elastizitatsgrenze und den Bruch eines Materials'. *Zeitschrift des Vereins deutscher Ingenieure*, 44, 1524–1530; 1572–1577.

Mohr, O. (1905). *Abhandlungen aus dem Gebiete der Technischen Mechanik.* Verlag von Wilhelm Ernst and Sohn, 1905, 1913, Third edn. 1928.

Pariseau, W.G. (2008). *Solutions Manual to Design Analysis in Rock Mechanics.* Taylor & Francis, London.

Paul, B. (1961). 'A modification of the Coulomb–Mohr theory of fracture'. *J. Appl. Mech.*, 28, 259–268.

Pisarenko, G.S. and Lebedev, A.A. (1976). *Deformation and Strength of Material under Complex Stressed State.* Naukova Dumka, Kiev (in Russian).

Teodorescu, P.P. (2006). 'Review of 'Unified strength theory and its applications. Springer, Berlin, 2004''. *Zentralblatt MATH 2006*, Cited in Zbl. Reviews, 1059.74002 (02115115).

Water Conservation Encyclopedia China (2006, Second edition). Beijing: Chinese Water Conservation and Hydraulic Electricity Press, Vol. 2, 940, 1077.

Yu, H.-S. (2010). *Plasticity and Geotechnics.* Springer, New York. p. 80.

Yu, M.-H., He, L.-N.and Song, L.-Y. (1985). 'Twin shear stress theory and its generalization'. *Scientia Sinica (Sciences in China)*, English edn. Series A, Vol. 28, No. 11, 1174–1183.

Yu, M.-H. and He, L.N. (1991). *'A new model and theory on yield and failure of materials under complex stress state'*.In: *Mechanical Behaviour of Materials-6*, Vol. 3. Pergamon Press, Oxford, 841–846.

Yu, M.-H. (1992). *A New System of Strength Theory.* Xian Jiaotong University Press, Xi'an (in Chinese).

Yu, M.-H. (2002) 'Advances in strength theories for materials under complex stress state in the 20th century'. *Appl. Mech. Rev. ASME*, 55, 3, 169–218.

Yu, M.-H., Zan, Y.-W., Zhao, J.and Yoshimine, M. (2002). 'A unified strength criterion for rock'. Int. *J. Rock Mech. Min. Sci.*, 39, 975–989.

Yu, M.-H. (2004). *Unified Strength Theory and Its Applications*. Springer, Berlin.

Yu, M.-H. *et al.* (2009). 'Basic characteristics and development of yield criteria for geomaterials'. *J. Rock Mech. Geotech. Eng.*, 1, 1, 71–88.

Zhang, C.-Q., Zhou, H. and Feng, X.-T. (2008). 'Numerical format of elastoplastic constitutive model based on the unified strength theory in FLAC~(3D)'. *Rock Soil Mech.*, 29, 3, 596–602 (in Chinese, English Abstract).

Zhang, L.-Y. (2005). 'The 3D images of geotechnical constitutive models in the stress space'. *Chinese J. Geotech. Eng.*, 27, 1, 64–68 (in Chinese, English Abstract).

Zhao, G.-H. (ed.) (2006). *Engineering Mechanics, Soil and Rock Mechanics, Engineering Structure and Material Fascicle*. Chinese Water Conservation and Hydraulic Electricity Press, Beijing, pp. 16, 20–21, 389 (in Chinese).

Zienkiewicz, O.C. and Pande, G.N. (1977). 'Some useful forms of isotropic yield surfaces for soil and rock mechanics'. In Gudehus G (ed.) *Finite Elements in Geomechanics*. Wiley, London, 179–190.

Zyczkowski, M. (1981). *Combined Loadings in the Theory of Plasticity*. Polish Scientific Publishers, PWN and Nijhoff.

Chapter 16

Failure criteria for transversely isotropic rock

Y.M. Tien, M.C. Kuo & Y.C. Lu
Department of Civil Engineering, National Central University, Taoyuan, Taiwan

Abstract: In this chapter, a new failure criterion for the transversely isotropic rocks is presented. The new criterion is based on two distinct failure modes; one is the sliding mode where the failure is caused by sliding along the discontinuity, and the other is the non-sliding mode where the failure is controlled by the rock material and is not dependent on discontinuity. This failure criterion is defined with seven material parameters. The physical meanings of, and the procedures for determining, these parameters are described. Both the original Jaeger criterion and the extended Jaeger criterion are shown to be special cases of the proposed criterion. The accuracy and applicability of the proposed failure criterion are examined using the published experimental data. The data used cover various types of transversely isotropic rocks, different orientation angels and confining pressures. The predicted strength behaviors of the transversely isotropic rocks agree well with the experimental data from various investigators. The accuracy and applicability of the proposed empirical failure criterion are demonstrated in this paper.

1 INTRODUCTION

The constitutive laws and failure criteria of rock materials and rock masses are required in most rock engineering analyses that are based on solid mechanics. Due to the preferred fabric orientation or the existence of non-random discontinuity, anisotropic behaviors of rock masses should be fully accounted for in the analysis. Several types of rocks such as sedimentary rocks and metamorphic rocks may be transversely isotropic. Most foliated metamorphic rocks, such as schist, slates, gneisses, and phyllites contain fabric with preferentially parallel arrangements of flat or long minerals. Metamorphism changes the initial fabric of rocks with the directional structure. Foliation induced by the non-random orientation of macroscopic mineral, parallel fracture or microscopic mineral plates, such as fracture cleavage, slaty cleavage, bedding cleavage, lepidoblastic schistosity, nematoblastic schistosity or lineation leads to rock properties that are highly direction-dependent (Goodman, 1993). Stratified sedimentary rocks like sandstone, shale or sandstone–shale alteration often display anisotropic behaviors. The anisotropy may also be found in the isotropic rocks, such as granite and basalt, if cut by regular discontinuities (Amadei, 1983; Wittke, 1990).

Over the past several decades, many authors have devoted considerable efforts to the study of rock anisotropy, from both the theoretical and the experimental points of view. Many scholars have investigated mechanical properties of both nature and synthetic transversely isotropic rocks under varied confining pressures

452 Tien et al.

(Donath, 1964; Hoek, 1964; Chenevert & Gatlin, 1965; McLamore & Gray, 1967; Horino & Ellickson, 1970; Attewell & Sandford, 1974; Brown et al., 1977; Niandou et al., 1997; Lai et al., 1997; Lai et al., 1999; Tien & Tsao, 2000) The shape of the curve of compression strength and the orientation angle (the angle between the discontinuity and the direction of major principal stress) are the most common representation of the nature of strength anisotropy. Most transversely isotropic rocks are found to have their maximal compression strength at an orientation angle $\beta = 0°$ or $90°$, and their minimal compression strength at an orientation angle in the range of 30–45 °. As the confining pressure is increased, the anisotropic rocks become more ductile, and the effect of the strength anisotropy is usually reduced.

In general, the characteristic of strength anisotropy of rocks can be presented by:

The relations the maximum and minimum uniaxial compressive strength at specific β. Ramamurthy (1993) defined the anisotropy ratio as,

$$R_C = \frac{\sigma_{c\ max}}{\sigma_{c\ min}} \quad (1)$$

where $\sigma_{c\ max}$ is the maximum minimum uniaxial compressive strength; $\sigma_{c\ min}$ is the minimum uniaxial compressive strength.

The relation between β and uniaxial compressive strength.

Table 1 lists anisotropy ratios for various kinds of rocks; it is obviously observed that the anisotropy ratios in metamorphic rocks (slate, phyllite, and coal) are larger than sedimentary rocks (shale and sandstone).

Based upon the analysis of the shape of the anisotropy curve, Ramamurthy (1993) classified the anisotropy of rocks into three groups, namely, U type, undulatory type, and shoulder type anisotropy (see Figure 1).

Table 1 Anisotropy ratios for various kinds of rocks (Ramamurthy, 1993).

Rock type	Value of β for $\sigma_{c\ max}$	Anisotropic ratio, R_c*	Source
Martinsburg slate	90°	13.5	Donath (1964)
Fractured sandstone	90°	6.37	Horino & Ellickson (1970)
Barnsley Hard coal	90°	5.18	Pomeroy et al. (1971)
Penrhyn slate	90°	4.85	Attewell & Sandford (1974)
South African slate	0°	3.68	Hoek (1964)
Texas slate	90°	3.00	McLamore & Gray (1967)
Permian shale	90°	2.33	Chenevert & Gatlin (1965)
Green River shale I	0°	1.62	McLamore & Gray (1967)
Green River shale II	0°	1.41	McLamore & Gray (1967)
Green River shale	0°, 90°	1.37	Chenevert & Gatlin (1965)
Kota sandstone	0°	1.12	Rao (1984)
Arkansas sandstone	0°	1.10	Chenevert & Gatlin (1965)
Chamera phyllites			
Quartizitic	90°	2.19	Singh (1988)
Carbonaceous	90°	2.19	Singh (1988)
Micaceous	90°	6.00	Singh (1988)

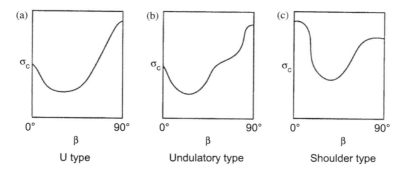

Figure 1 Classification of anisotropy for transversely isotropic rocks (after Ramamurthy, 1993).

2 FAILURE MODES

In the development of a failure criterion, it is important to observe the failure modes of rock specimens with different orientation angles and under different confining pressures. An ideal failure criterion should be able to predict not only the state of stress at failure but also the failure mode. The failure mode of anisotropic rocks under triaxial compression is influenced by the orientation of the stresses, as well as the confining pressure. Hence, it is far more complicated than that of isotropic rocks. Many scholars (Donath, 1964; McLamore & Gray, 1967; Niandou et al., 1997) have described in detail the failure modes of the transversely isotropic rocks under various confining pressures. Jaeger (1960) simplified the failure of transversely isotropic rocks into two modes: (1) sliding along the discontinuity, and (2) fracture through the rock materials. Jaeger's criterion is mainly based on the simplified assumption of failure modes described above. Recently, some scholars (Tien & Tsao, 2000; Tien et al., 1995a; Tien et al., 1995b; Tien et al., 1996; Tien et al., 1997) have developed a sample preparation technique for synthetic layered rocks. The overall mechanical properties of synthetic layered rocks are found to be very similar to those of transversely isotropic natural rocks (Tien et al., 2006).

An ideal failure criterion should predict not only the state of stress at failure, but also the failure mode. The observation of failure processes and failure modes may provide the feedback necessary to verify a new failure criterion. Therefore, it is an important task to observe the failure processes and failure modes of rocks at different stress levels. The failure modes of anisotropic rocks depend on not only the confining pressure, but also the orientation of the specimen.

In order to examine the failure behavior and failure mechanism of transversely isotropic rock, it is essential to observe and record the failure process at different stress–strain levels under compression. The rotary scanner (Tien & Chu, 2004) designed previously, which cannot scan the unrolled image during loading, is applied. A "rotatable CCD sensor" has been designed, which circumnavigates the rock specimen during the uniaxial compressive test (see Figure 2).

After the specimen had been placed on the platen of the MTS, a reference image was scanned before any loads were applied. The platen was then raised to the

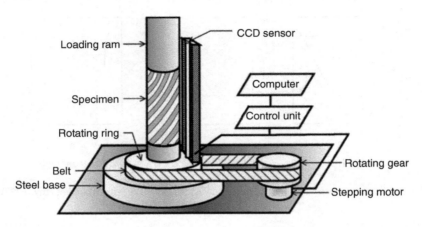

Figure 2 Schematic diagram of the modified rotary scanner for scanning of a cylindrical specimen during the loading process (Tien et al., 2006).

predetermined displacement (strain level) and fixed at that level, scanning then followed. This procedure was repeated until the failure of the specimen occurred. For distinction between bedding plane and crack, the digital unrolled images were processed with software to enhance the cracks and the edges of layers. The observations during uniaxial compressive tests (Tien et al., 2006) are shown in Figure 3. Based on observations of crack initiation and propagation, the failure processes and stress conditions of the transversely isotropic rock under uniaxial compression are summarized in Table 2.

Not only β can affect failure modes but also confining pressure can affect it. Tien and Kuo (2001) prepared series of syntactic transversely isotropic rocks to investigate their failure modes under triaxial compressive tests. Figure 4 shows a series of photos that depict the failure modes of the samples in triaxial compression tests. For samples with $\beta = 0°$ or $90°$ and loaded without confining pressure, fracture through both white and brown layers was observed. When they were loaded under confining pressures, behavior of ductile deformation (*i.e.*, axial strain accumulation) was observed. For samples prepared at $\beta = 45°$ and loaded without confining pressure, the failure mode was that of sliding along the discontinuity. When they were loaded under a confining pressure up to 0.8MPa, the artificial layered rocks behave more like ductile materials, and the sliding mode was suppressed.

It is obvious from the above observations, the failure of transversely isotropic rocks may be divided into two failure modes: (1) sliding mode in which the plane discontinuity predominated, (2) non-sliding mode in which the material strength dominated. Jaeger treated the former failure mode by the theory of plane of weakness, which yielded fairly accurate and reasonable prediction of the strength. However, the plateau of constant strength at low values of β, or high values of β predicted by Jaeger's criterion is not always present in the experimental strength data when the sliding mode is prevented. This suggests that the assumption of rock

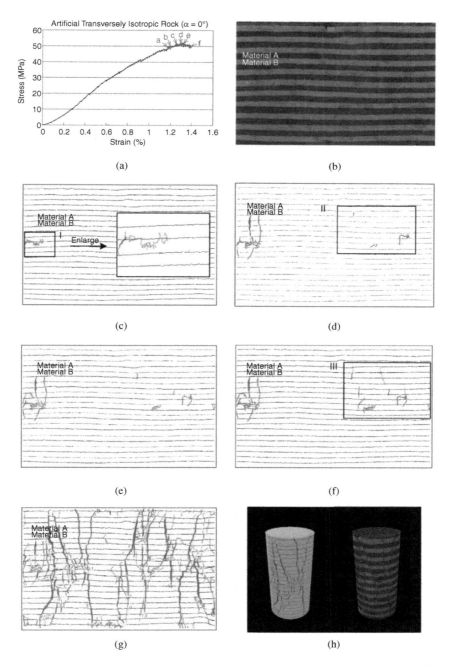

Figure 3 (a) Stress–strain curve during the uniaxial compressive test of a low dip specimen ($\beta = 90°$). (b) Unrolled image of a low-dip specimen—corresponding to point a marked. (c-g) Enhanced image of a low-dip specimen—corresponding to point b-f marked. (h) Reconstructed 3-D image of a specimen—corresponding to point f marked (Tien et al., 2006).

Table 2 Failure processes of simulated transversely isotropic rock under uniaxial compressive tests (Tien *et al.*, 2006).

Mode type	*Failure process (stress-strain condition)*				
Tensile fracture across discontinuous (TM mode) ($\alpha = 0°, 15°, 30°$)	No obvious crack ($< 95\%\ \sigma_p$, $< 91\%\ \varepsilon_p$)	Microcracks initiate within softer layer ($96\%\ \sigma_p$, $91\%\ \varepsilon_p$)	Microcracks propagate into stiffer layer and new cracks initiate within softer layer ($99\%\ \sigma_p$, $96\%\ \varepsilon_p$)	Cracks cross the layers and new cracks initiate within stiffer layer ($100\%\ \sigma_p$, $100\%\ \varepsilon_p$)	Post-peak failure occurs with macrocracks parallel to the direction of axial force ($96\%\ \sigma_p$, $105\%\ \varepsilon_p$)
Sliding failure along discontinuities (SD mode) ($\alpha = 45°, 60°, 75°$)		No obvious crack ($< 100\%\ \sigma_p$, $< 100\%\ \varepsilon_p$)		No obvious crack ($100\%\ \sigma_p$, $100\%\ \varepsilon_p$)	Post-peak failure occurs suddenly along the discontinuity ($98\%\ \sigma_p$, $100\%\ \varepsilon_p$)
Tensile-split along discontinuities (TD mode) ($\alpha = 90°$)		No obvious crack ($< 100\%\ \sigma_p$, $< 100\%\ \varepsilon_p$)		No obvious crack ($100\%\ \sigma_p$, $100\%\ \varepsilon_p$)	Post-peak failure occurs with macrocracks parallel to the direction of axial force ($99\%\ \sigma_p$, $101\%\ \varepsilon_p$)

|—— Pre-peak ——|—— Peak ——|—— Post-peak ——|

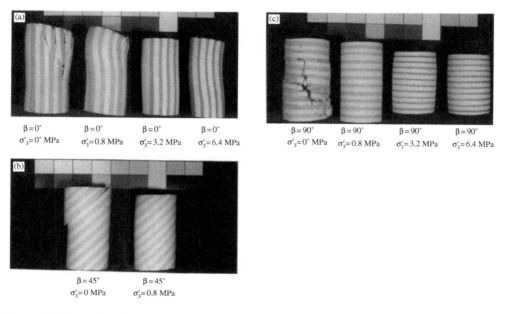

Figure 4 Deformation characteristic of saturated synthetic layered rocks after triaxial test (a) $\beta = 0°$ (b) 45° and (c) 90° (Tien & Kuo, 2001).

as an isotropic material in Jaeger's criterion results in an oversimplified representation of strength, when the failure is controlled by the rock materials. According to experimental observations of syntactic rock mass (Tien et al., 2006), the failure modes can be categorized into (a) tensile fracture across discontinuities mode (TM mode); (b) tensile-split along discontinuities mode (TD mode); (c) sliding failure along discontinuities mode (SD mode); (d) sliding failure across discontinuities mode (SM mode) (Figure 5).

The failure modes of experiments conducted by previous investigations (McLamore & Gray, 1967; Donath, 1964; Niandou et al., 1997) are presented in Table 3. The SD mode is synonymous with that described by Donath (1964) as the slip-on-the-discontinuity mode, shear along-the-discontinuity mode defined by McLamore & Gray (1967), and shearing-on-the-discontinuity mode defined by Niandou (1997). The TD mode is synonymous with Niandou's (1997) extension mode. The SM mode is synonymous to the shear-across-the-discontinuity mode (McLamore & Gray, 1967; Donath, 1964), and the shearing-in-shale-matrix mode (Niandou et al., 1997).

Discrepancies between results of this study and those of previous researchers (McLamore & Gray, 1967; Donath, 1964; Niandou et al., 1997) are found in the TM, PD (Plastic flow along discontinuity mode) and KK (Kinking) modes. The uniaxial compressive test was not conducted in previous studies (McLamore & Gray, 1967; Donath, 1964; Niandou et al., 1997), thus, the existence of a TM mode was not present

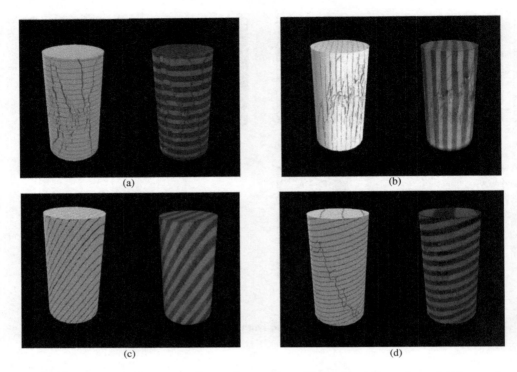

Figure 5 Reconstructed 3-D image of specimens, (a) tensile fracture across discontinuities mode, TM mode ($\beta = 90°$, $\sigma_3 = 0$ MPa); (b) tensile-split along discontinuities mode, TD mode ($\beta = 0°$, $\sigma_3 = 0$ MPa); (c) sliding failure along discontinuities mode, SD mode ($\beta = 30°$, $\sigma_3 = 35$ MPa); (d) sliding failure across discontinuities mode, SM mode ($\beta = 75°$, $\sigma_3 = 6$ MPa).

in those findings. For high-confining pressures, the SD and SM modes may be transformed to either the PD or the KK mode. Failure criterion as developed by Tien & Kuo (2001) has been used to predict the failure modes of the transversely isotropic rock mass under various confining pressures.

3 STRESS-STRAIN RELATION OF LINEAR ELASITICITY

To understand transversely isotropic rocks mechanical behaviors, establish stress-strain relation is a fundamental basic. According to *generalized Hooke's law*, the stress-strain relation of linear elasticity can be presented as,

$$\sigma_{ij} = C_{ijkl}\varepsilon_{kl} \qquad (2)$$

where σ_{ij} is stress; ε_{kl} is strain; C_{ijkl} is elastic constant matrix, includes 81 constants.

For *generally anisotropy* media, the constitutive law can be expressed as follows (Lekhnitskii, 1963),

Table 3 The relationship between failure mode, orientation of the discontinuity, and confining pressure (Tien *et al.*, 2006).

Failure mode		Conditions of failure modes	Rock type	References
		Modes: specimen orientation α/ confining		
Sliding failure along discontinuities (SD mode)	Discontinuity / Fracture	SD mode: $45° \leq \alpha \leq 75°$/ $3.5\ \text{MPa} \leq \sigma_3 \leq 100\ \text{MPa}$	Martinsburg slate	Donath (1964)
Sliding failure across discontinuities (SM mode)	Discontinuity / Fracture	SM mode: $0 \leq \alpha \leq 30$ and $\alpha = 90°$/ $3.5\ \text{MPa} \leq \sigma_3 \leq 200\ \text{MPa}$		
Plastic flow along discontinuities (PD mode)	Discontinuity	PD mode: $45° \leq \alpha \leq 60°$/ $100\ \text{MPa} \leq \sigma_3 \leq 200\ \text{MPa}$		
Kinking (KK mode)	Discontinuity	KK mode: $\alpha = 75°$/ $100\ \text{MPa} \leq \sigma_3 \leq 200\ \text{MPa}$		
Sliding failure along discontinuities (SD mode)	Discontinuity / Fracture	SD mode: $60° \leq \alpha \leq 80°$/ $34.5\ \text{MPa} \leq \sigma_3 \leq 103.4\ \text{MPa}$	Austin slate	Mclamore and Gray (1967)
		SM mode: $0° \leq \alpha \leq 50$ and $\alpha = 90°$/ $34.5\ \text{MPa} \leq \sigma_3 \leq 103.4\ \text{MPa}$ PD mode: $0° \leq \alpha \leq 50°$/ $103.4\ \text{MPa} \leq \sigma_3 \leq 275.8\ \text{MPa}$ KK mode: $60° \leq \alpha \leq 90°$/ $103.4\ \text{MPa} \leq \sigma_3 \leq 275.8\ \text{MPa}$		
Sliding failure across discontinuities (SM mode)	Discontinuity / Fracture	SD mode: $60° \leq \alpha \leq 70°$/ $6.9\ \text{MPa} \leq \sigma_3 \leq 172.4\ \text{MPa}$	Green river shale-1	
Plastic flow along discontinuities (PD mode)	Discontinuity	SM mode: $0° \leq \alpha \leq 50°$ and $80° \leq \alpha \leq 90°$/ $6.9\ \text{MPa} \leq \sigma_3 \leq 172.4\ \text{MPa}$		
Kinking (KK mode)	Discontinuity	SD mode: $50° \leq \alpha \leq 70°$/ $6.9\ \text{MPa} \leq \sigma_3 \leq 69\ \text{MPa}$ and $50° \leq \alpha \leq 60°$/ $69\ \text{MPa} \leq \sigma_3 \leq 172.4\ \text{MPa}$ SM mode: $0° \leq \alpha \leq 30°$ and $80° \leq \alpha \leq 90°$/ $6.9\ \text{MPa} \leq \sigma_3 \leq 69\ \text{MPa}$ and $\alpha = 0°$/ $69\ \text{MPa} \leq \sigma_3 \leq 172.4\ \text{MPa}$ PD mode: $\alpha = 30°$/$69\ \text{MPa} \leq \sigma_3 \leq 172.4\ \text{MPa}$ KK mode: $70 \leq \alpha \leq 90°$/ $69\ \text{MPa} \leq \sigma_3 \leq 172.4\ \text{MPa}$	Green river shale-2	
Tensile-split along discontinuities (TD mode)	Fracture / Discontinuity	TD mode: $75° \leq \alpha \leq 90°$/ low confining pressure	Tournemire shade	Niandou *et al.* (1997)
Sliding failure along discontinuities (SD mode)	Discontinuity / Fracture	SD mode: $30° \leq \alpha \leq 75°$/low confining pressure: $30° \leq \alpha \leq 75°$/high confining pressure (fracture may across the discontinuity)		
Sliding failure across discontinuities (SM mode)	Discontinuity / Fracture	SD mode: $30° \leq \alpha \leq 90°$/low confining pressure: $0° \leq \alpha \leq 35°$ and $75° \leq \alpha \leq 90°$/high confining pressure		

$$
\begin{Bmatrix} \varepsilon_x \\ \varepsilon_y \\ \varepsilon_z \\ \gamma_{yz} \\ \gamma_{xz} \\ \gamma_{xy} \end{Bmatrix} =
\begin{bmatrix}
\dfrac{1}{E_x} & \dfrac{-\nu_{yx}}{E_y} & \dfrac{-\nu_{zx}}{E_z} & \dfrac{\eta_{x,yz}}{G_{yz}} & \dfrac{\eta_{x,xz}}{G_{xz}} & \dfrac{\eta_{x,xy}}{G_{xy}} \\[2mm]
\dfrac{-\nu_{xy}}{E_x} & \dfrac{1}{E_y} & \dfrac{-\nu_{zy}}{E_z} & \dfrac{\eta_{y,yz}}{G_{yz}} & \dfrac{\eta_{y,xz}}{G_{xz}} & \dfrac{\eta_{y,xy}}{G_{xy}} \\[2mm]
\dfrac{-\nu_{xz}}{E_x} & \dfrac{-\nu_{yz}}{E_y} & \dfrac{1}{E_z} & \dfrac{\eta_{z,yz}}{G_{yz}} & \dfrac{\eta_{z,xz}}{G_{xz}} & \dfrac{\eta_{z,xy}}{G_{xy}} \\[2mm]
\dfrac{\eta_{yz,x}}{E_x} & \dfrac{\eta_{yz,y}}{E_y} & \dfrac{\eta_{yz,z}}{E_z} & \dfrac{1}{G_{yz}} & \dfrac{\mu_{yz,xz}}{G_{xz}} & \dfrac{\mu_{yz,xy}}{G_{xy}} \\[2mm]
\dfrac{\eta_{xz,x}}{E_x} & \dfrac{\eta_{xz,y}}{E_y} & \dfrac{\eta_{xz,z}}{E_z} & \dfrac{\mu_{xz,yz}}{G_{yz}} & \dfrac{1}{G_{xz}} & \dfrac{\mu_{xz,xy}}{G_{xy}} \\[2mm]
\dfrac{\eta_{xy,x}}{E_x} & \dfrac{\eta_{xy,y}}{E_y} & \dfrac{\eta_{xy,z}}{E_z} & \dfrac{\mu_{xy,yz}}{E_{yz}} & \dfrac{\mu_{xy,xz}}{E_{xz}} & \dfrac{1}{G_{xy}}
\end{bmatrix}
\begin{Bmatrix} \sigma_x \\ \sigma_y \\ \sigma_z \\ \tau_{yz} \\ \tau_{xz} \\ \tau_{xy} \end{Bmatrix}
\tag{3}
$$

where γ is the engineering strain which is equal to 2ε; E_x, E_y, E_z are the *Young's modulus* on x, y, z directions, respectively; G_{yz}, G_{xz}, G_{xy} are *shear modulus* on yz, xz, xy planes, respectively; v_{ij} is the *Poisson's ratio* determining the ratio of strain in the j direction to the strain in i direction due to stress acting in the i direction; $\mu_{ij,kl}$ is the shear in the plane parallel to the one defined by indices ij that induces the tangential stress in the plane parallel to the one defined by indices kl; $\eta_{k,ij}$ characterizes the stretching in the direction parallel to k induced by the shear stress acting within a plane parallel to the one defined by indices ij; $\eta_{ij,k}$ characterizes a shear in the plane defined by indices ij under the influence of a normal stress acting in the k direction (Amadei, 1983).

Equation 3 also called the *compliance matrix*. From elasticity, the compliance matrix and the stiffness matrix exists an inversible relation. In addition, the compliance matrix and the stiffness matrix are symmetric, where,

$$\frac{v_{ij}}{E_i} = \frac{v_{ji}}{E_j}, \quad \frac{\eta_{ij,k}}{E_k} = \frac{\eta_{k,ij}}{G_{ij}}, \quad \frac{\mu_{ij,kl}}{G_{kl}} = \frac{\mu_{kl,ij}}{G_{ij}} \tag{4}$$

If composition of rock mass is symmetry, the number of independent constants will be less than 21. Two of anisotropy elastic symmetry are introduced as follows.

1. One plane of elastic symmetry

   The illustration of one plane of elastic symmetry is shown in Figure 6. Equation 4 can be reduced to the following one,

$$
\begin{Bmatrix} \varepsilon_x \\ \varepsilon_y \\ \varepsilon_z \\ \gamma_{yz} \\ \gamma_{xz} \\ \gamma_{xy} \end{Bmatrix} =
\begin{bmatrix}
\dfrac{1}{E_x} & \dfrac{-v_{yx}}{E_y} & \dfrac{-v_{zx}}{E_z} & 0 & 0 & \dfrac{\eta_{x,xy}}{G_{xy}} \\
 & \dfrac{1}{E_y} & \dfrac{-v_{zy}}{E_z} & 0 & 0 & \dfrac{\eta_{y,xy}}{G_{xy}} \\
 & & \dfrac{1}{E_z} & 0 & 0 & \dfrac{\eta_{z,xy}}{G_{xy}} \\
 & & & \dfrac{1}{G_{yz}} & \dfrac{\eta_{yz,xz}}{G_{xz}} & 0 \\
 & & & & \dfrac{1}{G_{xz}} & 0 \\
 & & & & & \dfrac{1}{G_{xy}}
\end{bmatrix}
\begin{Bmatrix} \sigma_x \\ \sigma_y \\ \sigma_z \\ \tau_{yz} \\ \tau_{xz} \\ \tau_{xy} \end{Bmatrix} \tag{5}
$$

The number of independent constants is reduced to 13.

Amadei (1983) transformed the 13 elastic constants by using coordinate transformation (Figure 6), the Equation 5 can be rewritten as,

$$
\begin{Bmatrix} \varepsilon_x \\ \varepsilon_y \\ \varepsilon_z \\ \gamma_{yz} \\ \gamma_{xz} \\ \gamma_{xy} \end{Bmatrix} =
\begin{bmatrix}
K_{11} & K_{12} & K_{13} & 0 & 0 & K_{16} \\
 & K_{22} & K_{23} & 0 & 0 & K_{26} \\
 & & K_{33} & 0 & 0 & K_{36} \\
 & & & K_{44} & K_{45} & 0 \\
 & & & & K_{55} & 0 \\
 & & & & & K_{66}
\end{bmatrix}
\begin{Bmatrix} \sigma_x \\ \sigma_y \\ \sigma_z \\ \tau_{yz} \\ \tau_{xz} \\ \tau_{xy} \end{Bmatrix} \tag{6}
$$

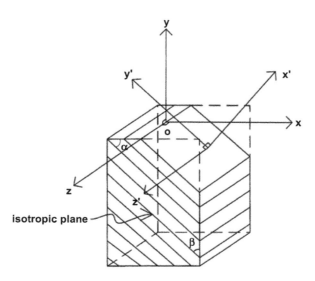

Figure 6 Coordinate systems used for transformation law for material constants of a transversely isotropic medium (Tien & Kuo, 2001).

where

$$K_{11} = \frac{1}{E_x} = \frac{\sin^4\alpha}{E_{x'}} + \frac{\cos^4\alpha}{E_{y'}} + \frac{\sin^2 2\alpha}{4}\left(\frac{1}{G_{x'y'}} - \frac{2\nu_{y'xx}}{E_{y'}}\right)$$

$$K_{12} = -\frac{\nu_{yx}}{E_y} = -\frac{\nu_{y'x'}\sin^4\alpha}{E_{y'}} - \frac{\nu_{y'x'}\cos^4\alpha}{E_{y'}} + \frac{\sin^2 2\alpha}{4}\left(\frac{1}{E_{y'}} + \frac{1}{E_{x'}} - \frac{1}{G_{x'y'}}\right)$$

$$K_{13} = -\frac{\nu_{zx}}{E_z} = -\frac{\nu_{z'x'}\sin^2\alpha}{E_{y'}} - \frac{\nu_{z'x'}\cos^2\alpha}{E_{z'}}$$

$$K_{22} = \frac{1}{E_y} = \frac{\cos^4\alpha}{E_{x'}} + \frac{\sin^4\alpha}{E_{y'}} + \frac{\sin^2 2\alpha}{4}\left(\frac{1}{G_{x'y'}} - \frac{2\nu_{x'y'}}{E_{y'}}\right)$$

$$K_{23} = -\frac{\nu_{zy}}{E_z} = -\frac{\nu_{z'x'}\cos^2\alpha}{E_{z'}} - \frac{\nu_{z'y'}\sin^2\alpha}{E_{z'}}$$

$$K_{33} = \frac{1}{E_z} = \frac{1}{E_{z'}} \quad K_{44} = \frac{1}{G_{yz}} = \frac{\sin^2\alpha}{G_{y'z'}} + \frac{\cos^2\alpha}{G_{x'z'}} \quad K_{55} = \frac{1}{G_{xz}} = \frac{\cos^2\alpha}{G_{y'z'}} + \frac{\sin^2\alpha}{G_{x'z'}}$$

$$K_{66} = \frac{1}{G_{xy}} = \frac{\cos^2 2\alpha}{G_{x'y'}} + \sin^2 2\alpha\left(\frac{1}{E_{x'}} + \frac{1}{E_{y'}} + \frac{2\nu_{y'x'}}{E_{y'}}\right)$$

$$K_{45} = \frac{\mu_{yz,xz}}{G_{xz}} = \frac{\sin 2\alpha}{2}\left(\frac{1}{G_{x'z'}} - \frac{1}{G_{y'z'}}\right) \quad K_{36} = \frac{\eta_{z,xy}}{G_{xy}} = \sin 2\alpha\left(\frac{\nu_{z'y'}}{E_{z'}} + \frac{\nu_{z'x'}}{E_{z'}}\right)$$

$$K_{16} = \frac{\eta_{x,xy}}{G_{xy}} = \sin{}^2\alpha \sin 2\alpha\left(\frac{1}{E_{x'}} + \frac{\nu_{x'y'}}{E_{y'}}\right) - \cos{}^2\alpha \sin 2\alpha\left(\frac{1}{E_{y'}} + \frac{\nu_{x'y'}}{E_{y'}}\right)$$

$$+ \frac{\sin 2\alpha \cos 2\alpha}{2G_{x'y'}}$$

$$K_{26} = \frac{\eta_{y,xy}}{G_{xy}} = \sin 2\alpha \cos{}^2\alpha\left(\frac{1}{E_{x'}} + \frac{\nu_{x'y'}}{E_{y'}}\right) - \sin 2\alpha \sin{}^2\alpha\left(\frac{1}{E_{y'}} + \frac{\nu_{x'y'}}{E_{y'}}\right)$$

$$+ \frac{\sin 2\alpha \cos 2\alpha}{2G_{x'y'}} \tag{7}$$

2. One axis of elastic symmetry of rotation

If rock mass has the same behaviors on direction of $y'z'$ plane (Figure 6). The x' axis can be defined as *axis of radial elastic symmetry of axis of elastic symmetry of rotation* (Amadei, 1983). This type of elastic symmetry is called *transverse isotropy,*

$$
\begin{Bmatrix} \varepsilon_x \\ \varepsilon_y \\ \varepsilon_z \\ \gamma_{yz} \\ \gamma_{xz} \\ \gamma_{xy} \end{Bmatrix} =
\begin{bmatrix}
\frac{1}{E} & \frac{-\nu'}{E} & \frac{-\nu'}{E'} & 0 & 0 & 0 \\
\frac{-\nu'}{E} & \frac{1}{E} & \frac{-\nu'}{E'} & 0 & 0 & 0 \\
\frac{-\nu'}{E'} & \frac{-\nu'}{E'} & \frac{1}{E'} & 0 & 0 & 0 \\
0 & 0 & 0 & \frac{1}{G'} & 0 & 0 \\
0 & 0 & 0 & 0 & \frac{1}{G'} & 0 \\
0 & 0 & 0 & 0 & 0 & \frac{1}{G}
\end{bmatrix}
\begin{Bmatrix} \sigma_x \\ \sigma_y \\ \sigma_z \\ \tau_{yz} \\ \tau_{xz} \\ \tau_{xy} \end{Bmatrix} \tag{8}
$$

where

$$
\begin{aligned}
E &= E_x = E_y, & E' &= E_z, \\
\nu &= \nu_{xy}, & \nu' &= \nu_{zx} = \nu_{zy}, \\
G &= G_{xy} = \frac{E}{2(1+\nu)}, & G' &= G_{yz} = G_{xz},
\end{aligned} \tag{9}
$$

The number of independent constants is reduced to 5.

In natural, the inherent structure of rocks sometimes present stratified. The macro mechanical behaviors of these kinds of rocks can be regarded as transversely isotropy. Salamon (1968) showed that 5 elastic constants can be expressed in terms of the elastic properties and thickness of layers (Equation 10) by employed *equivalent homogeneous continuum* and *volumetric average* approaches.

Failure criteria for transversely isotropic rock **463**

However, Equation 10 is only valid under certain conditions (Amadei, 1983):

- *all layers are bounded by parallel planes and no relative displacement takes place on these planes,*
- *the thickness and elastic properties of the layers vary randomly with respect to dimension perpendicular to the bounding planes. The randomness is necessary to ensure overall homogeneity of the equivalent material,*
- *a representative sample of the stratified rock mass on the basis of which the equivalent homogeneous properties are calculated must contain a large number of layers.*

$$E = \left(1 - \nu^2\right) \sum \frac{\lambda_i E_i}{1 - \nu_i^2}$$

$$E' = \frac{1}{\sum \frac{\lambda_i}{E_i} \left(\frac{E_i}{E'_i} - \frac{2\nu_i'^2}{1 - \nu_i}\right) + \frac{2\nu'^2}{E(1 - \nu)}} \tag{10}$$

$$\nu = \frac{\sum \frac{\lambda_i \nu_i E_i}{1 - \nu_i^2}}{\sum \frac{\lambda_i E_i}{1 - \nu_i^2}} \quad \nu' = (1 - \nu) \sum \frac{\lambda_i \nu_i'}{1 - \nu_i} \quad G' = \frac{1}{\sum \frac{\lambda_i}{G_i}}$$

where λ_i is thickness ratio of ith layer.

4 LITERATURE REVIEW OF FAILURE CRITERIA

The purpose of failure criteria establishment is to obtain quantified rock strength at limited stress state under known conditions. For isotropic and homogeneous media, the failure criteria usually take the form of major principal stress σ_1, confining stress σ_3, or triaxial stress states as mathematical formulas. However, because of some preferred orientation of fabric (microstructure, interlayer, cleavages, or joints), the mechanical behavior of rock may belongs anisotropy. Because of computational complexity and determining elastic modulus difficulty, the simplest anisotropic form, transverse isotropy, is widely used to establish anisotropic failure criteria. Section 2 notes that β plays an important role in failure mechanism. Hence, the transversely isotropic failure criteria not only take σ_1 and σ_3 as a consideration but also take orientation of plane of isotropy β (foliation or weakness planes) as a parameter. The form of anisotropic failure criteria is shown in Equation 11.

$$\sigma_1 = f(\sigma_3, \beta) \tag{11}$$

Recently, numerous scholars have developed and proposed failure criteria for anisotropic media, which are listed in Table 3. The categories of failure criteria for anisotropic media can be classified according to their theoretical assumptions. Three main categories herein are: (1) continuous criteria based on mathematical approach; (2) continuous criteria based on empirical approach; (3) discontinuous criteria based on plane of weakness theory. In this Chapter, common failure criteria of anisotropic rock are introduced the following section.

464   Tien et al.

Table 4 Anisotropic failure criteria classification (Duveau & Shao, 1998).

Continuous criteria		Discontinuous criteria
Mathematical approach	*Empirical approach*	
• Von Mises (1928) • Hill (1950) • Olszak & Urbanowicz (1956) • Goldenblat (1962) • Goldenblat & Kopnov (1966) • Boehler & Sawczuk (1970, 1977) • Tsai & Wu (1971) • Pariseau (1972) • Boehler (1975) • Dafalias (1979, 1987) • Allirot & Boehler (1979) • Nova (1980, 1986) • Nova & Sacchi (1982) • Boehler & Raclin (1982) • Raclin (1984) • Kaar *et al.* (1989) • Cazacu *et al.* (1995)	• Casagrande & Carillo (1944) • Jaeger (variable cohesive strength theory) (1960) • McLamore & Gray (1967) • Hoek & Brown (1980) • Ramamurthy (1993)	• Jaeger (single plane of weakness theory) (1960) • Walsh & Brace (1964) • Murrell (1965) • Hoek (1964, 1968, 1983) • Barron (1971) • Ladanyi & Archambault (1972) • Bieniawski (1974) • Duveau & Shao (1998)

4.1 Analytical approach

4.1.1 Jaeger (1960) failure criterion

(1)   Sliding along weakness plane:

In this failure mode, the sliding failure will occur at weakness plane. Jaeger (1960) assumed that each weakness plane has a limited shear strength which is dominated by Coulomb's criterion. By this definition, the relation of σ_1, σ_3, and β can be expressed as follows,

$$\sigma_1 = \sigma_3 + \frac{2(c_w + \sigma_3 \tan\phi_w)}{(1 - \tan\phi_w \cot\beta)\sin 2\beta} \tag{12}$$

where, c_w is the cohesion of weakness plane; ϕ_w is the friction angle of weakness plane.
(2)   Failure of intact rock:

The strength of intact rock can be obtained from Mohr-Coulomb failure criterion,

$$\sigma_1 = \sigma_3 \tan^2\left(45° + \frac{\phi}{2}\right) + 2c\tan\left(45° + \frac{\phi}{2}\right) \tag{13}$$

where, c is the cohesion of intact rock; ϕ is the friction angle of intact rock. Under this assumption, Jaeger's criterion yields the same compression strength at $\beta = 0°$ and $90°$.

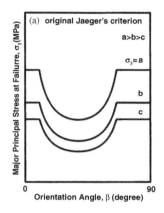 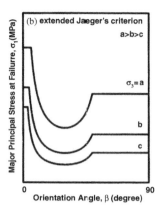

Figure 7 Schematic view of strength variation versus b (a) original Jaeger's criterion (b) extended Jaeger's criterion (Tien & Kuo, 2001).

In this failure criterion, four physical material parameters, c_j, ϕ_j, c, and ϕ can be easily obtained from compressional tests, and provided a fundamentally theoretical research for anisotropic rock. However, the published experimental data show that in some rocks, the maximum strength occurs at $\beta = 0°$, while in other rocks, it occurs at $\beta = 90°$ (Borecki & Kwasniewski, 1981; Kwasniewski, 1993; Sheorey, 1997). In order to account for this discrepancy (possible strength discrepancy at $\beta = 0°$ and $90°$), other researchers have modified Jaeger's criterion by adding two more parameters. Such modification is referred to herein as the extended Jaeger's criterion (Figure 7b). Furthermore, Duveau & Shao (1998) provided yet another modification by replacing the Mohr–Coulomb criterion with a nonlinear model to express the strength along discontinuity. Their criterion used seven parameters to describe the failure strength for transversely isotropic rocks.

4.2 Empirical approach

4.2.1 McLamore & Gray (1967) failure criterion

McLamore & Gray (1967) proposed a continuous failure criterion (Equations 14–15) for transversely isotropic rocks based on empirical approach, which modified the discontinuous phase of Jaeger's criterion into continuous phase.

$$c = A - B[\cos 2(\beta - \beta_{\min,C})]^m \tag{14}$$

$$\tan\phi = C - D[\cos 2(\beta - \beta_{\min,\phi})]^n \tag{15}$$

where, A, B, C, D, m, and n are constants obtained from experiments; $\beta_{\min,c}$ is the inclination angle at which c has the minimum value; $\beta_{\min,\phi}$ is the inclination angle at which ϕ has the minimum value.

4.2.2 Hoek & Brown (1980) failure criteria

Hoek & Brown (1980) employed "rock classification" and statistics to process hundreds of data from triaxial tests, and proposed empirical strength formula for isotropic (fragment) rock mass,

$$\sigma_1 = \sigma_3 + \sqrt{m\sigma_3\sigma_c + s\sigma_c^2} \tag{16}$$

where, σ_c is uniaxial compressive strength of intact rock; m is a parameter related to particles composition in rock mass; s is a parameter related to rock fragmentation.

Hoek & Brown (1980) proposed an empirical conjunction between m, s, and β for which rocks have significant weakness planes. The modified equations of m and s are as follows,

$$m = m_i \left(1 - Ae^{-\theta^4} \right) \tag{17}$$

$$s = 1 - Pe^{-\zeta^4} \tag{18}$$

where, A and P are constants obtained from experiments; m_i is m value of intact rock;

$$\theta = \frac{\beta - \xi_m}{A_2 + A_3\beta} \tag{19}$$

$$\zeta = \frac{\beta - \xi_s}{P_2 + P_3\beta} \tag{20}$$

A_2, A_3, P_2, and P_3 are constants obtained from experiments; ξ_m is the inclination angle at which m has the minimum value; ξ_s is the inclination angle at which s has the minimum value.

Although this failure criterion can predict strength anisotropy precisely, there are 10 parameters should be determined, and only one parameter σ_c has its physical meaning.

Saroglou & Tsiambaos (2008) also proposed modified Hoek & Brown failure criterion for which rocks exhibit "inherent" anisotropy, shown in Equation 21. Comparing with the original Hoek & Brown failure criterion, there is a new parameter k_β which can consider the effect of strength anisotropy.

$$\sigma_1 = \sigma_3 + \sigma_{c\beta} \sqrt{k_\beta m \frac{\sigma_3}{\sigma_{c\beta}} + 1} \tag{21}$$

where, k_β is the parameter describing the anisotropy effect; $\sigma_{c\beta}$ can be obtained by Donath (1961),

$$\sigma_{c\beta} = A - D[\cos 2(\beta_m - \beta)] \tag{22}$$

where A and D are constant; β_m is the angle at which the uniaxial compressive strength is minimum.

4.2.3 Ramamurthy (1993) failure criterion

Ramamurthy (1985) proposed an empirical nonlinear strength criterion for intact rock by modifying the Mohr-Coulomb failure criterion. Ramamurthy (1993) modified his failure criterion to predict strength anisotropy for jointed rock or anisotropic rock mass,

$$\frac{\sigma_1 - \sigma_3}{\sigma_3} = B_j \left(\frac{\sigma_{cj}}{\sigma_3}\right)^{n_j} \tag{23}$$

$$\frac{n_j}{n_{90}} = \left(\frac{\sigma_{cj}}{\sigma_{c90}}\right)^{1-n_{90}} \tag{24}$$

$$\frac{B_j}{B_{90}} = \left(\frac{n_{90}}{n_j}\right)^{0.5} \tag{25}$$

where σ_{cj} is the uniaxial compressive strength of jointed rock; σ_{c90} is the uniaxial compressive strength at $\beta = 90°$; n_{90} and B_{90} is the material parameters at $\beta = 90°$.

4.2.4 Saeidi et al. (2014) failure criterion

Saeidi *et al.* (2014) modified Rafiai (2011) proposed empirical failure criterion (Equation 26 which is for isotropic rocks) to predict strength anisotropy for transversely isotropic rocks.

$$\frac{\sigma_1}{\sigma_{ci}} = \frac{\sigma_3}{\sigma_{ci}} + \left[\frac{1 + A(\sigma_3/\sigma_{ci})}{1 + B(\sigma_3/\sigma_{ci})}\right] - r \tag{26}$$

where σ_{ci} is the uniaxial compressive strength of intact rock; A and B are constants depending on rock properties; r is the strength reduction factor to which rock has been fractured (zero is for intact rocks, one is for heavily fractured rocks).

Saeidi *et al.* (2014) employed Equation 26 for transversely isotropic rock fitting work by numerous experimental data, and considered rocks as intact (r is zero). The modified failure criterion is shown below,

$$\sigma_1 = \sigma_3 + \sigma_{c\beta} \left[\frac{1 + A(\sigma_3/\sigma_{c\beta})}{\alpha + B(\sigma_3/\sigma_{c\beta})}\right] \tag{27}$$

where $\sigma_{c\beta}$ is the uniaxial compressive strength of intact rock at inclination angle β; α is the strength reduction parameter related to rock anisotropy. Furthermore, in Saeidi *et al.* (2014) determined the goodness of fittings by correlation coefficient and root mean square errors (RMSE), and A and B can be determined by

$$A = \frac{\sum XY \sum YZ - \sum Y^2 \sum XZ}{\left(\sum XY\right)^2 - \sum X^2 \sum Y^2} \tag{28}$$

$$B = \frac{\sum X^2 \sum YZ - \sum XY \sum XZ}{\left(\sum XY\right)^2 - \sum X^2 \sum Y^2} \tag{29}$$

where

$$X = \frac{\sigma_3}{\sigma_{c\beta}} \qquad Y = \frac{\sigma_3}{\sigma_{c\beta}}\left(\frac{\sigma_1}{\sigma_{c\beta}} - \frac{\sigma_3}{\sigma_{c\beta}}\right) \qquad Z = \alpha\left(\frac{\sigma_1}{\sigma_{c\beta}} - \frac{\sigma_3}{\sigma_{c\beta}}\right) - 1 \tag{30}$$

X, Y, Z are satisfied the form $Z = AX - BY$, which is obtained from rewritten form of Equation 27.

Although the prediction of strength anisotropy in Saeidi *et al.* (2014) model is more accurate than Hoek-Brown (1980) and Ramamurthy (1993) failure criteria, this criterion is limited for intact anisotropic rocks.

The strength criteria for the transversely isotropic rocks developed by McLamore & Gray (1967), Hoek & Brown (1980), Ramamurthy (1993), and Saeidi *et al.* (2014) generally provide fairly accurate simulation of the experimental data. However, these approaches all require a wide range of tests and/or a considerable amount of curve fitting work.

5 PROPOSED FAILURE CRITERION

Tien & Kuo (2001) failure criterion is developed for transversely isotropic rocks based on the Jaeger's criterion (1960) and the maximum axial strain theory. The axial strain is calculated from the constitutive law of the transversely isotropic rocks.

5.1 Methodology

(1)   Sliding along weakness plane:
Jaeger (1960) derived the shear strength induced by sliding along the discontinuity. The major principal stress for sliding along the discontinuity (Equation 12).

(2)   Failure of intact rock:
As the values of β approaching $0°$ or in the range of $(90°-\phi_w)\sim90°$, the sliding failure along the discontinuity will not occur. In such case, the strength of rocks is dominated by the rock materials and is independent of the discontinuity. The major principal stress at failure under a given confining pressure must be controlled by the strength of the isotropic material and will not vary with the orientation angle β. However, the constant strength at low values of β, or high values of β, predicted by the Jaeger's criterion is not supported by experimental data. Borecki & Kwasniewski (1981) have collected dozens of values of $\sigma_{c(0°)} = \sigma_{c(90°)}$, which shows the ratio to be in range of 0.6–1.33. Thus, Jaeger's criterion is indeed an oversimplified representation of the strength of rock specimens whose failure is controlled by rock material.

To correctly reflect the difference between the strength of rocks at $\beta = 0°$ and $90°$, the proposed criterion treats the rock material as transversely isotropic material. Thus, rocks at $\beta = 0°$ and $90°$ have different strength, and both are assumed to follow the failure criterion by Hoek & Brown (1980):

$$S_{1(0°)} = \sigma_{1(0°)} - \sigma_3 = \sqrt{m_{(0°)}\sigma_3\sigma_{c(0°)} + \sigma^2_{c(90°)}} \tag{31}$$

$$S_{1(90°)} = \sigma_{1(90°)} - \sigma_3 = \sqrt{m_{(90°)}\sigma_3\sigma_{c(90°)} + \sigma^2_{c(90°)}} \tag{32}$$

where $\sigma_{c(0°)}$ and $\sigma_{c(90°)}$ are the uniaxial compression strength of rock samples at $\beta = 0°$ and $90°$, respectively; $m_{(0°)}$ and $m_{(90°)}$ are the m values in the Hoek & Brown criterion for the rock samples at $\beta = 0°$ and $90°$, respectively.

Treating rocks at $\beta = 0°$ and $90°$ as two different materials is in some way similar to the extended Jaeger's criterion (Figure. 6b). However, in the latter Mohr–Coulomb criterion was used, while in the proposed criterion, the Hoek–Brown criterion is adopted. The rationale for adopting the Hoek–Brown criterion in the present study lies in the fact that the σ_1–σ_3 relationship is generally nonlinear, particularly when the range of σ_3 under consideration is large. The Hoek–Brown criterion can fit the experimental data in both brittle and ductile regions better than the Mohr–Coulomb criterion does. Since both criteria require two parameters, using the Hoek–Brown criterion in the present study does not increase the number of parameters required in the proposed model.

From section 3, ductile deformation due to axial strain accumulation is another important failure mode of transversely isotropic rocks, in addition to the failure mode of sliding along the discontinuity. The axial strain may be calculated using the theory of elasticity of an anisotropic medium. The constitutive laws of linearly elastic, transversely isotropic medium in the local coordinate system (x', y', z') is shown in Equation 8.

The constitutive equations of the transversely isotropic medium in the global coordinate system (x, y, z), defined in Figure 6, can be obtained by tensor transformation in Equations 6–7.

The state of stresses at failure, when subjected to the triaxial loading, can be decomposed into the hydrostatic and deviatoric stress components,

$$\begin{bmatrix} \sigma_x \\ \sigma_y \\ \sigma_z \\ \tau_{yz} \\ \tau_{xz} \\ \tau_{xy} \end{bmatrix} = \begin{bmatrix} \sigma_3 \\ \sigma_1 \\ \sigma_3 \\ 0 \\ 0 \\ 0 \end{bmatrix} = \begin{bmatrix} \sigma_3 \\ \sigma_3 \\ \sigma_3 \\ 0 \\ 0 \\ 0 \end{bmatrix} + \begin{bmatrix} 0 \\ \sigma_1 - \sigma_3 \\ 0 \\ 0 \\ 0 \\ 0 \end{bmatrix} = \begin{bmatrix} \sigma_3 \\ \sigma_3 \\ \sigma_3 \\ 0 \\ 0 \\ 0 \end{bmatrix} + \begin{bmatrix} 0 \\ S_1 \\ 0 \\ 0 \\ 0 \\ 0 \end{bmatrix} \tag{33}$$

The strain tensor during the application of the deviatoric stress can be obtained by

$$\begin{bmatrix} \varepsilon_{xx} \\ \varepsilon_{yy} \\ \varepsilon_{zz} \\ \gamma_{yz} \\ \gamma_{xz} \\ \gamma_{xy} \end{bmatrix} = \begin{bmatrix} K_{11} & K_{12} & K_{13} & 0 & 0 & K_{16} \\ K_{21} & K_{22} & K_{23} & 0 & 0 & K_{26} \\ K_{31} & K_{32} & K_{33} & 0 & 0 & K_{36} \\ 0 & 0 & 0 & K_{44} & K_{45} & 0 \\ 0 & 0 & 0 & K_{54} & K_{55} & 0 \\ K_{61} & K_{62} & K_{63} & 0 & 0 & K_{66} \end{bmatrix} \begin{bmatrix} 0 \\ S_1 \\ 0 \\ 0 \\ 0 \\ 0 \end{bmatrix} \tag{34}$$

470 Tien et al.

Thus, the axial strain is

$$\varepsilon_{yy} = K_{22}S_1 \tag{35}$$

$$\text{where} \quad K_{22} = \frac{1}{E_y} = \frac{\cos^4\beta}{E} + \frac{\sin^4\beta}{E'} + \cos^2\beta\sin^2\beta\left(\frac{1}{G'} - \frac{2\nu'}{E'}\right) \tag{36}$$

The failure criteria for anisotropic materials may be categorized into three basic types: (1) stress dominated; (2) strain dominated (3) interactive (Halpin & Brown, 1984). The ductile deformation due to the axial strain accumulation is referred to herein as the strain-dominated criterion. To account for also the non-sliding mode in the proposed criterion, it is assumed that the failure occurs when the axial strain exceeds its maximum limiting value, ε_{yf} under a specific confining pressure. This failure criterion is referred to herein as the maximum axial strain criterion.

With the maximum axial strain criterion,

$$S_{1(\beta)} = E_y\varepsilon_{yf} \tag{37}$$

$$S_{1(90°)} = E_{(90°)}\varepsilon_{yf} \tag{38}$$

The value of axial strain at failure ε_{yf} is varied with different confining pressures and independent of orientation angle. This paper adopted Hooke's law to calculate the axial strains and the strength ratio of specimens with various orientation angles under a specified confining pressure. According to the coordinate system of Figure 6,

$$E = E_{(0°)} \tag{39}$$

$$E' = E_{(90°)} \tag{40}$$

From Equations 36–38,

$$\frac{S_{1(\beta)}}{S_{1(90°)}} = \frac{1}{\left[\dfrac{\cos^4\beta}{E_{(0°)}} + \dfrac{\sin^4\beta}{E_{(90°)}} + \cos^2\beta\sin^{2\beta}\left(\dfrac{1}{G'} - \dfrac{2\nu'}{E_{(90°)}}\right)\right]E_{(90°)}} \tag{41}$$

By introducing the strength ratio, k (Borecki & Kwasniewski, 1981) and the transversal anisotropy parameter, n:

$$k = E_{(0°)}/E_{(90°)} = S_{1(0°)}/S_{1(90°)} \tag{42}$$

$$n = \left(E_{(90°)}/2G'\right) - \nu' \tag{43}$$

Equation 41 becomes

$$\frac{S_{1(\beta)}}{S_{1(90°)}} = \frac{\sigma_{1(\beta)} - \sigma_3}{\sigma_{1(90°)} - \sigma_3} = \frac{k}{\cos^4\beta + k\sin^4\beta + 2n\sin^2\beta\cos^2\beta} \tag{44}$$

Equation 44 represents the failure condition of the transversely isotropic rocks for the non-sliding mode.

5.2 Determination of the material parameters

The Tien & Kuo (2001) failure criterion is based on two distinct failure modes, and thus, the model parameters of the proposed criterion can be categorized into two groups:

- Strength parameters of discontinuity (c_w and ϕ_w), related to the sliding failure mode,
- Strength parameters of rock material ($m_{(0°)}$; $\sigma_{c(0°)}$; $m_{(90°)}$; $\sigma_{c(90°)}$; n), related to the non-sliding failure mode.

The Tien & Kuo (2001) failure criterion is a seven-parameter model. Therefore, seven experimental data points are required in order to determine these parameters. The material parameters of proposed model can be obtained by conducting triaxial tests for at least four orientation angles, say $\beta = 0°, 30°, 60°,$ and $90°$.

The parameters, c_w and ϕ_w, are the cohesion and friction of the discontinuity, respectively. By setting the derivative of $\sigma_{1(\beta)}$ with respect to β equal to 0, the orientation angle at which the minimum strength occurs is obtained

$$\tan\phi_w = \cot 2\beta_{\min} \tag{45}$$

$$\beta_{\min} = \frac{\pi}{4} - \frac{\phi_w}{2} \tag{46}$$

In principle, ϕ_w can be obtained from Equation 46 if the orientation angle that corresponds to a minimum strength is determined. One possible approach to determining the shear strength parameters of discontinuity (c_w and ϕ_w) is conducting triaxial tests on specimens with $\beta = 20°$–$50°$, in which the sliding mode is expected. By transforming the stress state into the normal stress (σ_n) and the shear stress (τ) on the discontinuity, and then plotting of σ_n versus τ at failures, the parameters c_w (intercept) and ϕ_w (slope) can be determined.

The four Hoek–Brown parameters of rock material ($m_{(0°)}$; $\sigma_{c(0°)}$; $m_{(90°)}$; $\sigma_{c(90°)}$) should be determined from triaxial tests on the rock specimens that are prepared with $\beta = 0°$ and $90°$, respectively. Hoek–Brown criterion is a two-parameter model. Thus, each set of triaxial tests should be conducted under at least two confining pressures. Hoek & Brown (1997) provided guidance for selecting confining pressures and procedures for determining these material parameters. The test should be carried out over a range of confining pressures from zero to $0.25\sigma_{c(0°)}$ (or $0.25\sigma_{c(90°)}$), with eight equally spaced value of confining pressure (Hoek & Brown, 1997).

The strength ratio, k under a specified confining pressure, can be expressed in terms of Hoek–Brown parameters ($m_{(0°)}$; $\sigma_{c(0°)}$; $m_{(90°)}$; $\sigma_{c(90°)}$) according to Equations 31, 32, and 42. It should be noted that the value of strength ratio, k may vary slightly with confining pressure, thus k is not a basic material parameter in the proposed criterion.

The transversal anisotropy parameter n is the unique new parameter introduced in the Tien & Kuo (2001) failure criterion. It plays a critical role to describe the strength variation when the sliding failure cannot occur (usually in the range of $\beta = 0°$–$10°$ and $\beta = 60°$–$90°$). The transversal anisotropy parameter n can be determined by performing triaxial tests at $\beta = 60°$ (or $\beta = 75°$ alternatively) and $90°$. In Equation 44, The term $\cos^4\beta$ becomes negligible if β is in the range of $60°$–$90°$ (for example, $\cos^4 60° = 0.0625$, $\cos^4 90° = 0.0045$). Thus, Equation 44 becomes approximately

$$\frac{S_{1(\beta)}}{S_{1(90°)}} = \frac{k}{\sin^2\beta\left(\sin^2\beta + 2(n/k)\cos^2\beta\right)} \tag{47}$$

In Equation 47, the strength ratio $S_{1(\beta)}/S_{1(90°)}$ is a function of n/k. It is, however, independent of k. Rewrite Equation 44 in terms of n/k and let $k = 1$, the strength ratio becomes

$$\frac{S_{1(\beta)}}{S_{1(90°)}} = \frac{1}{1 + 2[(n/k) - 1]\sin^2\beta\cos^2\beta} \tag{48}$$

For $\beta = 60°$, Equation 48 becomes

$$\frac{S_{1(60°)}}{S_{1(90°)}} = \frac{1}{1 + 0.375[(n/k) - 1]} \tag{49}$$

For $\beta = 75°$, Equation 48 becomes

$$\frac{S_{1(75°)}}{S_{1(90°)}} = \frac{1}{1 + 0.125[(n/k) - 1]} \tag{50}$$

Figure 8 shows a plot of $S_{1(60°)}/S_{1(90°)}$ and $S_{1(75°)}/S_{1(90°)}$ versus the ratio n/k. This plot is based on Equation 44 with $k = 1$, which is the same as Equations 49, 50. While not shown in this figure, the difference in the obtained curves using different k values is negligible, as implied by Equation 47. Thus, the relationship between the strength ratio $S_{1(60°)}/S_{1(90°)}$ and $S_{1(75°)}/S_{1(90°)}$ and the parameter ratio n/k shown in Figure 8 is valid for different k values.

Given the strength ratio $S_{1(60°)}/S_{1(90°)}$ (or $S_{1(75°)}/S_{1(90°)}$ alternatively), which may be obtained from triaxial tests at orientation angle $\beta = 60°$ (or 75° alternatively) and 90°, the transversal anisotropy parameter n can be determined from Figure 8 or Equations 49, 50.

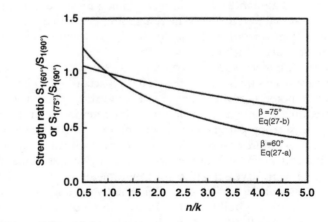

Figure 8 Chart for determination of parameter n (Tien & Kuo, 2001).

5.3 Evaluation of failure criterion

To evaluate the capabilities of the Tien & Kuo criterion, comparisons are made between experimental data taken from the literatures and the predictions obtained from the proposed criterion based on Equations 44 and 29. The Tien & Kuo failure criterion is examined by comparing model predictions with experimental data from the literature. Figure 9 (a)–(i) show these comparisons. The material parameters for each transversely isotropic rock, natural or artificial, are listed in Table 5. In each figure, the solid lines correspond to the predictions obtained from the proposed failure criterion, while data points represent experimental results. These rocks include slates, shales, limestone, and artificial layered rocks and exhibited three types of anisotropies, namely U type, shoulder type and undulatory type as defined by Ramamurthy (1993). The proposed criterion is shown to be applicable to all types of anisotropies. Duveau & Shao (1998) adopted a nonlinear model (proposed by Barton for rock joint) to modify the Mohr–Coulomb criterion for discontinuity. Generally, a nonlinear model provides flexibility for describing the shear strength along discontinuity. However, it requires more parameters than does the linear Mohr–Coulomb criterion. In the present study, the linear Mohr–Coulomb criterion is adopted for its good balance of model simplicity with the accuracy.

The transversal anisotropy parameter n reflects the strength variation for the region where non-sliding failure occurs. The value of n varies from 1.0 to 4.0 for most of the transversely isotropic rocks. When the value of n is in the range of 1.0–2.0 (for example, Figures 9 (c)–(e), (h) and (i)), the strength is roughly constant in the neighborhood of $\beta = 0°$ or 90°. Those rocks may be classified into shoulder type anisotropy. As value of n increases, the region of "shoulder" disappear gradually, the strength variation around the neighborhood of $\beta = 0°$ or 90° is more significant as shown in Figure 9 (a) and (f). The type of anisotropy for such rocks may be referred to as U type or undulatory type.

The failure criteria for anisotropic rocks can be categorized into two groups: (1) discontinuous models (*e.g.* Jaeger's criterion and extended Jaeger's criterion) and (2) continuous models (*e.g.* Pariseau's criterion, Cazacu *et al.* criterion), depending upon the continuous and discontinuous characteristics of the corresponding anisotropy (Amadei, 1983). Compared to the experimental observations, the discontinuous models predict relatively well the strength behavior of a rock cut by joint. However, the continuous models are more suitable for the continuous rocks. The discontinuous models divide failure modes into the sliding and non-sliding modes. The sharp corner exists in the plot of failure stress as a function of orientation angle implies the transition point of two distinct failure modes. On the other hand, the continuous models treat the transversely isotropic rock as a continuous medium, ignoring the existence of the sharp corner and evading the failure mode problem.

Because the relationship between the failure stresses and the orientation angle of the transversely isotropic rocks obtained from the experiment is discrete, it is generally difficult, by the experimental approach alone, to identify whether the sharp corner exists or not. It is more meaningful to discuss this issue from both experimental and theoretical approaches, and by considering both strength variation and failure mode simultaneously. Whether a sharp corner exists depends on the rock type and the confining pressure. For an anisotropic rock that can be treated as a continuous medium at the

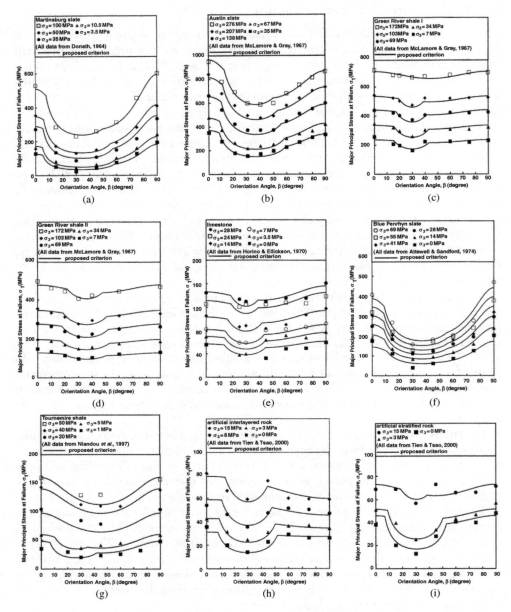

Figure 9 Comparison of Tien & Kuo failure criterion and experimental data (a) Martinsburg slate (Donath, 1964); (b) Austin slate (McLamore & Gray, 1967); (c) Green River shale I (McLamore & Gray, 1967); (d) Green River shale II (McLamore & Gray, 1967); (e) limestone (Horino & Ellickson, 1970); (f) Blue Penrhyn slate (Attewell & Sandford, 1974); (g) Tournemire shale (Niandou et al., 1977); (h) artificial interlayered rock (Tien & Tsao, 2000); (i) artificial stratified rock (Tien & Tsao, 2000). (Tien & Kuo, 2001).

Failure criteria for transversely isotropic rock **475**

Table 5 Material parameters of the proposed criterion for various rocks.

Rock	c_w (MPa)	ϕ_w (°)	$m_{(0°)}$	$\sigma_{c(0°)}$ (MPa)	$m_{(90°)}$	$\sigma_{c(90°)}$ (MPa)	n	Data source
Martinsburg slate	9	21	16.4	97	14.2	155	3.9	Donath (1964)
Austin slate	31	17	6.0	249	4.6	234	2.7	McLamore & Gray (1967)
Green River shale I	49	28	6.7	208	6.7	208	1.2	McLamore & Gray (1967)
Green River shale II	29	18	4.4	106	4.4	106	1.3	McLamore & Gray (1967)
Limestone	11	30	5.9	58	7.1	63	1.3	Horino & Ellickson (1970)
Blue Penrhyn slate	22	16	7.9	148	8.9	177	3.7	Attewell & Sandford (1974)
Tournemire shale	4	36	4.4	45	4.4	45	2.5	Niandou et al. (1977)
Artificial interlayered rock	4	29	6.5	31	3.1	27	1.1	Tien & Tsao (2000)
Artificial stratified rock	5	29	1.8	46	1.8	46	1.4	Tien & Tsao (2000)

sample scale, a continuous variation of strength with the orientation angle is expected. For the continuous rock or the discontinuous rock under high confining pressure, the effect of discontinuity is fully suppressed; the sharp corner is not significant. The phenomenon of suppression of discontinuity effect (or anisotropy) as the confining pressure increases has been identified by the experimental evidence (Ramamurthy, 1993; Kwasniewski, 1993). This criterion is a discontinuous model at lower confining pressure, and as confining pressure increases, it is gradually transformed into a continuous model. For example, as shown in Figure 9 (g), when $\sigma_3 > 20$ MPa, the Tien & Kuo criterion (Equation 44) becomes a continuous model.

From the results shown in Figure 9, the Tien & Kuo failure criterion is shown to be able to accurately predict the compression strength of transversely isotropic rocks of various types, prepared at different orientation angels and under various confining pressures. As a final note, the proposed criterion is a hybrid of the two well-known criteria in the field of rock mechanics, the Hoek–Brown and the Mohr–Coulomb criteria. Both the Hoek–Brown and the Mohr–Coulomb formulations are expressed in terms of major and minor principal stresses, neglecting the effect of the intermediate principal stress. Thus, this criterion inherits this limitation. Further research to improve the proposed criterion considering three-dimensional stress conditions is worth undertaking.

6 SUMMARY

The Tien & Kuo (2001) failure criterion is based on two distinct failure modes; one is the sliding mode where the failure is caused by sliding along the discontinuity, and the other is the nonsliding mode where the failure is controlled by the rock material and is not dependent on discontinuity.

The proposed failure criterion consists of seven material parameters. They are the cohesion and the friction angle of the discontinuity (c_w, ϕ_w), Hoek–Brown's parameters $(m_{(0°)}; \sigma_{c(0°)}; m_{(90°)}; \sigma_{c(90°)})$ and the transversal anisotropy parameter (n). The physical meanings of, and the procedures for determining, these parameters are described.

When $n = 1$, the proposed failure criterion is very similar to the extended Jaeger's criterion. With additional condition that $k = 1$, which implies that $m_{(0°)} = m_{(90°)}$, $\sigma_{c(0°)} = \sigma_{c(90°)}$, the proposed criterion becomes the original Jaeger's criterion.

The predictions of the strength behaviors of various types of the transversely isotropic rocks with different orientation angels and under various confining pressures agree well with experimental data from various investigators. The accuracy and the versatility of the proposed failure criterion are demonstrated.

REFERENCES

Amadei, B. (1983) *Rock Anisotropy and the Theory of Stress Measurements*. Heidelberg: Springer-Verlag.

Attewell, B. & Sandford, M.R. (1974) Intrinsic shear strength of a brittle anisotropic rock. I. Experimental and mechanical interpretation. II. Textural data acquisition and processing. III. Textural interpretation of failure. *Int J Rock Mech Min Sci*, 11, 423–430, 431–438, 439–451.

Borecki, M. & Kwasniewski, M.A. (1981) Experimental and analytical studies on compressive strength of anisotropic rocks. *Proceedings of Seventh Plenary Scientific Session of the International Bureau of Rock Mechanics, Katowice*. pp. 23–49.

Brown, E.T., Richards, L.R. & Barr, M.V. (1977) Shear strength characteristics of Delabole slate. *Proceedings of the Conference on Rock Engineering*. Newcastle-upon-Tyne. pp. 31–51.

Chenevert, M.E. & Gatlin, C. (1965) Mechanical anisotropies of laminated sedimentary rocks. *Soc Petrol Eng J*, 5, 67–77.

Donath, F.A. (1964) A strength variation and deformational behavior of anisotropic rocks. In: *State of Stress in the Earth's Crust*. New York: Elsevier. pp. 281–297.

Duveau, G. & Shao, J.F. (1998) A modified single discontinuity theory for the failure of highly stratified rocks. *Int J Rock Mech Min Sci*, 35(6), 807–813.

Goodman, R.E. (1993) *Engineering Geology-rock in Engineering Construction*. New York: John Wiley and Sons, Inc. pp. 293–332.

Halpin, J.C. (1984) *Primer on Composite Materials: Analysis*. Lancaster: Technomic Pub. Co., Inc. pp. 67–98.

Hoek, E. (1964) Fracture of anisotropic rock. *J S Afr Inst Min Metall*, 64(10), 510–518.

Hoek, E. & Brown, E.T. (1980) *Underground Excavation in Rock*. London: Institution of Mining and Metallurgy. pp. 157–162.

Horino, F.G. & Ellickson, M.L. (1970) *A Method of Estimating the Strength of Rock Containing Planes of Weakness*. US Bureau of Mines. Report Investigation: 7449.

Jaeger, J.C. (1960) Shear failure of anisotropic rocks. *Geol Mag*, 97, 65–72.

Kwasniewski, M.A. (1993) Mechanical behavior of anisotropic rocks. In: Hudson, J.A. (eds) *Comprehensive Rock Engineering, vol. 1. Fundamentals*. Oxford: Pergamon Press. pp. 285–312.

Lai, Y.S., Wang, C.Y. & Tien, Y.M. (1997) Micromechanical analysis of imperfectly bonded layered media. *J Eng Mech ASCE*, 123(10), 986–995.

Lai, Y.S., Wang, C.Y. & Tien, Y.M. (1999) Modified Mohr–Coulomb-type micromechanical failure criteria for layered rocks. *Int J Numer Anal Meth Geomech*, 23, 451–460.

Lekhnitskii, S.G. (1963) *Theory of Elasticity of an Anisotropy Elastic Body*. In: Fern P, translator. San Francisco: Holden-Day Inc.

McLamore, R. & Gray, K.E. (1967) The mechanical behavior of anisotropic sedimentary rocks. *J Eng Ind Trans ASME*, 89, 62–73.

Niandou, H., Shao, J.F., Henry, J.P. & Fourmaintraux, D. (1997) Laboratory investigation of the mechanical behavior of Tournemire shale. *Int J Rock Mech Min Sci*, 34, 3–16.

Rafiai, H. (2011) New empirical polyaxial criterion for rock strength. *Int J Rock Mech Min Sci*, 48, 922–931.

Ramamurthy, T. (1993) Strength and modulus responses of anisotropic rocks. In: Hudson, J.A. (eds.) *Comprehensive Rock Engineering, vol. 1. Fundamentals*. Oxford: Pergamon Press. pp. 313–329.

Saeidi, O., Rasouli, V., Vaneghi, R.G., Gholami, R. & Torabi, S.R. (2014) A modified failure criterion for transversely isotropic rocks. *Geosci. Front.*, 5, 215–225.

Salamon, M.D.G. (1968) Elastic moduli of a stratified rock mass. *Int J Rock Mech Min Sci*, 5(6), 519–527.

Saroglou, H. & Tsiambaos, G. (2008) A modified Hoek-Brown failure criterion for transversely isotropic intact rock. *Int J Rock Mech Min Sci*, 45, 223–234.

Sheorey, P.R. (1997) *Empirical Rock Failure Criteria*. Rotterdam: A.A. Balkema.

Tien, Y.M., Wang, C.Y., Huang, T.Y. & Lai, Y.S. (1995a) *Constitutive Laws and Failure Criterion for Interstratified Rock Mass*. National Science Council of ROC, Taiwan. Report No. NSC 84-2611-E008-004.

Tien, Y.M., Wang, C.Y., Wang, R.Z. & Lai, Y.S. (1995b) *Preparation and Mechanical Behavior of Artificial Anisotropic Rock Mass (I)*. National Science Council of ROC, Taiwan. Report No. NSC 84-2611-E008-004.

Tien, Y.M., Wang, C.Y., Huang, T.Y. & Lai, Y.S. (1996) *Preparation and Mechanical Behavior of Artificial Anisotropic Rock Mass (II)*. National Science Council of ROC, Taiwan. Report No. NSC 85-2211-E008-036.

Tien, Y.M., Tsao, P.F. & Young, S.H. (1997) *Preparation and Mechanical Behavior of Artificial Anisotropic Rock Mass (III)*. National Science Council of ROC, Taiwan. Report No. NSC 86-2621-E008-009.

Tien, Y.M. & Tsao, P.F. (2000) Preparation and mechanical properties of artificial transversely isotropic rock. *Int J Rock Mech Min Sci*, 37, 1001–1012.

Tien, Y.M. & Kuo, M.C. (2001) A failure criterion for transversely isotropic rocks. *Int J Rock Mech Min Sci*, 38(3), 399–412.

Tien, Y.M. & Chu, C.A. (2004) Rotary scanner for cylindrical specimens. *Proceedings of the Fifth Asian Young Geotechnical Engineers Conference, Taipei*. pp. 181–186.

Tien, Y.M., Kuo, M.C. & Juang, C.H. (2006) An experimental investigation of the failure mechanism of simulated transversely isotropic rocks. *Int J Rock Mech Min Sci*, 43, 1163–1181.

Wittke, W. (1990) *Rock Mechanics-theory and Applications with Case History*. Heidelberg: Springer-Verlag. pp. 5–170.

Chapter 17

Use of critical state concept in determination of triaxial and polyaxial strength of intact, jointed and anisotropic rocks

Mahendra Singh
Department of Civil Engineering, IIT Roorkee, Roorkee, Uttarakhand, India

Abstract: Engineering structures in rocks are always subjected to triaxial or polyaxial stress fields. Analysis and design of these structures involves assessment of rock strength subject to the prevailing stress state. Mohr-Coulomb criterion is the most widely used strength criterion in rock engineering problems. However, in its conventional form, the criterion treats the strength behavior as linear function of confining pressure. Also, the effect of intermediate principal stress σ_2, which is quite substantial in rock engineering problems, is ignored. The present article suggests modified forms of the Mohr-Coulomb criterion to incorporate non-linearity in triaxial strength behavior of intact isotropic, jointed anisotropic and intact anisotropic (transversely isotropic) rocks. Barton's critical state concept for rocks has been employed to correctly define the shape of the strength criterion. The triaxial strength criterion is further extended to polyaxial stress conditions for intact isotropic and jointed anisotropic rocks. The applicability of the proposed MMC criteria has also been verified by applying them to database of experimental test results compiled from worldwide literature.

1 INTRODUCTION

Design of engineering structures in rocks like underground excavations, foundations and slopes essentially involves assessment of the strength of the rock or rock mass subject to prevailing stress state. Strength criteria are used to define the effect of confinement on the strength of rocks for a given stress state. An ideal strength criterion should have good predictive capability even with least triaxial test data for obtaining the criterion parameters. Several criteria have been proposed during past and many investigators are still working on this topic (Chang & Haimson, 2012; Shen *et al.*, 2014; Singh *et al.*, 2015; Langford & Diederichs, 2015). Despite the large number of criteria available in the literature, the Mohr-Coulomb (MC) criterion is the criterion most favored by geotechnical engineers in the field. The primary reason behind this popularity is its simplicity; especially in assessing the criterion parameters. The parameters, namely cohesion c and friction angle ϕ, do carry a physical feel and an experienced designer can easily assign values to these parameters once the rock or rock mass is characterized through simple tests and/or classification techniques. The MC criterion, however, suffers from two major limitations (Singh *et al.*, 2011):

- It expresses the strength of the rock as a linear function of confining pressure or normal stress. If the parameters are obtained from triaxial tests at low confining pressure, the predicted strength at high confining pressure will deviate substantially from the actual value.
- In its conventional form, the criterion ignores the effect of intermediate principal stress σ_2. There is ample evidence available that, the intermediate principal stress does have substantial influence on the strength of rocks, barring a few cases of non-dilatant rocks.

The present article attempts to overcome the limitations of the conventional MC criterion and suggests a Modified-Mohr-Coulomb (MMC) criterion for defining non-linear triaxial strength for intact isotropic, jointed anisotropic and intact anisotropic rocks. The suggested MMC is also extended to polyaxial stress conditions for intact isotropic and jointed anisotropic rocks. Mohr-Coulomb shear strength parameters, as used in the conventional form, are retained and the same values are used in MMC criteria. The MMC criteria have been deduced through Barton's critical state concept for rocks.

2 CRITICAL STATE CONCEPT FOR ROCKS

When a rock is tested under low confining pressure, it fails in a dilatant and brittle manner due to opening up of the pre-existing micro cracks; as a consequence a high value of instantaneous friction angle ϕ is obtained. If the tests are conducted at higher confining pressure, the tendency of dilation is suppressed; the failure mechanism shifts from brittle to ductile and a relatively lower value of ϕ is obtained. With further increase in confining pressure, the rock becomes completely ductile and at sufficiently high confining pressure the rock enters the critical state. The failure envelope of rock plotted in τ-σ space is non-linear and concave upward (Fig. 1a). The tangential gradient of the envelope is steep where crossing the shear stress axis and tends to become asymptotic to a horizontal line at critical state. Barton (1976) states *"critical state for an initially intact rock is defined as the stress condition under which Mohr envelope of peak shear strength of the rock reaches a point of zero gradient. This condition represents the maximum possible shear strength of the rock. For each rock, there will be a critical effective confining pressure above which the shear strength cannot be made to increase"*. Figure 1a shows how an intact rock passes through brittle, brittle-ductile transition, ductile and critical stress states. Figure 1b drawn for Indiana Limestone (data from Hoek, 1983) also confirms to the concept that at sufficiently high confining pressure, ϕ approaches almost a zero value. This characteristic of Mohr failure envelope approaching horizontal has been used to define the correct shape of the failure criterion (Singh & Singh, 2005, 2012; Singh *et al.*, 2011, 2015).

3 MMC STRENGTH CRITERION FOR INTACT ROCKS

3.1 Triaxial conditions

Consider an intact rock which is tested under low confining pressure ($\sigma_3 \rightarrow 0$) and the corresponding resulting MC parameters are c_{i0} and ϕ_{i0} respectively. Figure 2 shows a

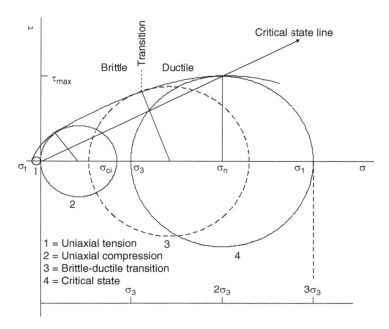

Figure 1a Barton's critical state concept for rocks (Singh et al., 2011).

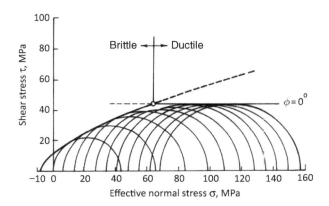

Figure 1b Critical state for Indiana limestone (Schwartz, 1964; redrawn from Hoek, 1983).

plot of MC criterion in $(\sigma_1-\sigma_3)$ vs. (σ_3) space. The MC criterion may be expressed in terms of σ_3 and σ_1 as follows.

$$(\sigma_1 - \sigma_3) = \frac{2c_{i0} \cos \phi_{i0}}{1 - \sin \phi_{i0}} + \frac{2 \sin \phi_{i0}}{1 - \sin \phi_{i0}} \sigma_3 \qquad (1)$$

Where, the term $(\sigma_1-\sigma_3)$ is the deviatoric stress at failure; σ_3 and σ_1 are the minor and major effective principal stresses at failure.

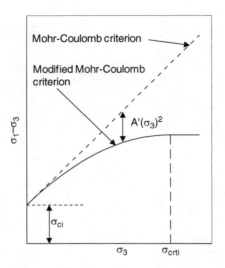

Figure 2 Modified Mohr-Coulomb criterion (Intact rocks).

Due to non-linear response the actual plot will deviate from the linear behavior. Considering the deviation to be a second degree term, a correction equal to $A'\sigma_3^2$ is applied to the MC criterion, where A' is an empirical constant for the rock type under consideration. The non-linear Mohr-Coulomb criterion is now written as:

$$(\sigma_1 - \sigma_3) = \frac{2c_{i0} \cos \phi_{i0}}{1 - \sin \phi_{i0}} + \frac{2 \sin \phi_{i0}}{1 - \sin \phi_{i0}} \sigma_3 - A' \sigma_3^2 \tag{2}$$

To obtain the empirical coefficient A', critical state concept (Barton, 1976, 2013) is employed. The gradient of the curve (Fig. 2) should approach 0 when σ_3 approaches critical confining pressure. If the critical confining pressure for the rock type is σ_{crti}, then differentiating Equation 2 and putting the condition $\partial(\sigma_1 - \sigma_3)/\partial \sigma_3 \to 0$ for $\sigma_3 \to \sigma_{crti}$, the value of A' is obtained and the Modified Mohr-Coulomb (MMC) triaxial strength criterion is written as:

$$(\sigma_1 - \sigma_3) = \sigma_{ci} + \frac{2 \sin \phi_{i0}}{1 - \sin \phi_{i0}} \sigma_3 - \frac{1}{\sigma_{crti}} \frac{\sin \phi_{i0}}{(1 - \sin \phi_{i0})} \sigma_3^2 \quad \text{for } 0 \le \sigma_3 \le \sigma_{crti} \tag{3}$$

Where σ_{ci} = UCS of intact rock = $(2c_{i0} \cos \phi_{i0})/(1 - \sin \phi_{i0})$ \hfill (4)

The above MMC (Equation 3) will be applicable only up to the critical state ($\sigma_3 \le \sigma_{crti}$). At $\sigma_3 = \sigma_{crti}$ the shear strength of rock will reach its maximum, the deviatoric stress at failure ($\sigma_1 - \sigma_3$) will be a constant for $\sigma_3 \ge \sigma_{crti}$.

$$(\sigma_1 - \sigma_3) = \sigma_{ci} + \{\sin \phi_{i0}/(1 - \sin \phi_{i0})\} \times \sigma_{crti} \quad \text{for } \sigma_3 \ge \sigma_{crti} \tag{5}$$

The MMC criterion was initially reported by Singh & Singh (2005) as a simple parabolic equation in the following form.

$$(\sigma_1 - \sigma_3) = A (\sigma_3)^2 + B(\sigma_3) + \sigma_{ci}; \ 0 \leq \sigma_3 \leq \sigma_{ci} \qquad (6)$$

$$\text{Where,} \ A = -\sin \phi_{i0}/\{\sigma_{crti} \times (1 - \sin\phi_{i0})\}; \quad B = 2 \sin\phi_{i0}/(1 - \sin\phi_{i0}) \qquad (7)$$

To use MMC (Equation 3), two parameters namely σ_{crti} and ϕ_{i0} will be required. This form of the criterion is termed as "two parameter criterion". Theoretically a minimum of two triaxial tests will be required to determine the criterion parameters. Singh *et al.* (2011) collected a database of triaxial tests conducted worldwide on more than 150 rock types. This database comprises of more than about 1100 triaxial test data. The authors carried out statistical analysis and back analyzed critical confining pressure. It was inferred that for the applicability of the proposed strength criterion, a value of the average critical confining pressure could be taken nearly equal to σ_{ci}. The actual critical confining pressure for a given rock type may depend on lithology and may vary. However, from statistical point of view, using $\sigma_{crti} \approx \sigma_{ci}$ is not likely to introduce error of engineering significance in rock strength prediction. Taking critical confining pressure equal to σ_{ci}, the MMC criterion (Equation 3) is written as:

$$(\sigma_1 - \sigma_3) = \sigma_{ci} + \frac{2\sin\phi_{i0}}{1 - \sin\phi_{i0}}\sigma_3 - \frac{1}{\sigma_{ci}}\frac{\sin \phi_{i0}}{(1 - \sin\phi_{i0})}\sigma_3^2 \quad \text{for} \ 0 \leq \sigma_3 \leq \sigma_{ci} \qquad (8)$$

For confining pressure range $\sigma_3 \geq \sigma_{ci}$ the deviatoric stress at failure $(\sigma_1 - \sigma_3)$ will remain constant as given below:

$$(\sigma_1/\sigma_3) = \sigma_{ci}/(1 - \sin\phi_{i0}) \quad \text{for} \ \sigma_3 \geq \sigma_{ci} \qquad (9)$$

The ratio of major to minor principal stress at the onset of critical state, as per the present criterion, will be

$$(\sigma_1 - \sigma_3) = (2 - \sin \phi_{i0})/(1 - \sin\phi_{i0}) \qquad (10)$$

The application of MMC (Equation 8) will require assessment of only one criterion parameter (ϕ_{i0}). The criterion is, therefore, termed as "single parameter criterion". Theoretically, only one triaxial test should be sufficient to determine the criterion parameter. If more than one triaxial test data are available, the optimized value of ϕ_{i0} may be obtained through the following expressions:

$$A_i = \frac{\sum(\sigma_1 - \sigma_3 - \sigma_{ci})}{\sum(\sigma_3^2 - 2\sigma_{ci}\sigma_3)} \quad \text{for} \ 0 < \sigma_3 \leq \sigma_{ci}; \ B_i = -2A_i\sigma_{ci}; \ \sin \phi_{i0} = B_i/(2 + B_i)$$

$$(11)$$

Alternatively, if triaxial tests are conducted at very low confining pressure ($\sigma_3 \rightarrow 0$), optimized c_{i0} and ϕ_{i0} may be obtained from linear MC criterion and Equation 4 may be used to obtain σ_{ci}.

3.2 Performance of MMC criterion for intact rocks

Performance of a failure criterion may be judged based on its goodness of fit, robustness and predictive capability (Shen *et al.*, 2014). The performance of the proposed strength

484   Singh

Table I Triaxial test data for Indiana limestone (Schwartz, 1964; source: Hoek, 1983).

σ_3, MPa	0	6.5	13.7	20.3	27.9	34.4	41.2	48.4	55.4	62.3	68.4
σ_1, MPa	44	66	85	99	109	119	128.2	135.1	141.9	149.1	156.5

criterion was compared with two other most widely used strength criteria namely, conventional Mohr-Coulomb (MC) criterion, and Hoek & Brown (1980) criterion (Singh *et al.*, 2011). The Hoek-Brown (HB) criterion for intact rocks is expressed as:

$$\sigma_1 = \sigma_3 + \sqrt{m_i\ \sigma_{ci}\sigma_3 + \sigma_{ci}^2} \tag{12}$$

where m_i and σ_{ci} are criterion parameters and are determined by statistical analysis of triaxial test data considering the UCS test also as a triaxial test.

A very popular triaxial test data set in rock engineering literature, comprising of triaxial strength of Indiana limestone for a wide range of confining pressure (Schwartz, 1964) was used to compare the performance of the proposed criterion. The data is presented in Table 1 and there are eleven data points including the UCS test.

3.2.1 Robustness

Shen *et al.* (2014) have considered robustness as one of the most important attributes for comparing the performance of different strength criteria. Robustness indicates that the parameters of a failure criterion, which are derived from available data, should be insensitive to the range of confining stress used for fitting the criterion. The best fitting parameters of the three different criteria were obtained a) by considering only first three data points (including UCS); b) by considering first four data points, and so on *i.e.* by considering triaxial test data at increasing confining pressures. Equations 11 were used to get the parameter ϕ_{i0}, and σ_{ci} was considered to be known. Criterion parameters of HB and conventional MC criteria were obtained through least square method. Table 2 presents the criterion parameters obtained along with the respective average and standard deviation. Table 2 indicates very wide variation in MC and HB parameters when different numbers of experimental data points are used to evaluate the criterion parameters. Against a standard deviation of 3.62 and 6.01 in conventional MC parameters, and 1.41 and 5.85 in HB parameters, the standard deviation of MMC parameter ϕ_{i0} is only 0.98. This indicates that, MMC is very robust and there is very small influence of range of confining pressure used in triaxial tests for assessing parameters.

Shen *et al.* (2014) collected an extensive database comprising 1579 triaxial test data by extending the triaxial database originally compiled by Singh *et al.* (2011). They compared four failure criteria namely Hoek–Brown (1980), parabolic criterion (Singh & Singh 2005), Modified Mohr-Coulomb (MMC) criterion (Singh *et al.*, 2011) and the one criterion which was proposed by themselves (Shen *et al.*, 2014). It was concluded that the parameter m of the Hoek-Brown criterion is the most sensitive parameter to the range of stress employed for fitting, whereas ϕ_{i0} parameter of the MMC criterion is the least sensitive to such stress range.

Use of critical state concept in determination of triaxial and polyaxial strength   485

Table 2 Best fitting parameters for different criteria for Indiana limestone (Singh *et al.*, 2011).

Number of triaxial test data points used	Mohr-Coulomb		Hoek-Brown (1980)		MMC
	c, MPa	$\phi°$	m_i	σ_{ci}, *MPa*	$\phi°_{io}$
3	12.94	29.97	5.16	44.44	33.28
4	13.93	27.53	4.63	45.39	32.57
5	15.56	24.10	3.53	48.28	31.52
6	16.84	21.70	2.97	50.24	30.95
7	18.06	19.7	2.54	52.14	30.67
8	19.46	17.62	2.07	54.60	30.48
9	20.87	15.69	1.68	57.03	30.50
10	22.18	14.0	1.40	59.13	30.77
11	23.25	12.7	1.21	60.66	31.32
Average	18.12	20.33	2.80	52.43	31.34
Standard deviation	3.62	6.01	1.41	5.85	0.98

Barton (2013) while discussing MMC comments *"the curvature of peak shear strength envelopes is now more correctly described, so that few triaxial tests are required and need only be performed at low confining stress, in order to delineate the whole strength envelope. This simplicity does not of course apply to M-C, nor does it apply to non-linear criteria including H-B, where triaxial tests are required over a wide range of confining stress, in order to correct the envelope, usually to adjust to greater local curvature"*.

3.2.2 Predictive capability

Shen *et al.* (2014) state that a criterion should have good predictive capabilities even with least triaxial data available for obtaining criterion parameters. To compare predictive capabilities, the criterion parameters obtained from only first three triaxial test data (Table 2) were used and σ_1 values were predicted for rest of the confining pressure values. The expressions for the three criteria are given below:

$$\text{MC criterion: } c_i = 12.94 \text{ MPa}, \phi_i = 29.97°; \ (\sigma_1 - \sigma_3) = 44.79 + 1.99\sigma_3 \tag{13}$$

$$\text{HB criterion: } m_i = 5.16; \sigma_{ci} = 44.44 \text{ MPa.}; \ \sigma_1 = \sigma_3 + \sqrt{229.31 \ \sigma_3 + 44.44^2} \tag{14}$$

MMC criterion: c_{i0} = 12.94 MPa, ϕ_{i0} = 29.97°

$$(\sigma_1 - \sigma_3) = 44.79 + 1.99\sigma_3 - 0.0223\sigma_3^2 \text{ for } 0 \le \sigma_3 \le 44.79 \text{ MPa} \tag{15}$$

$$(\sigma_1 - \sigma_3) = 89.18 \text{ for } \sigma_3 \ge 44.79 \text{MPa} \tag{16}$$

Figure 3 presents a comparison of the predicted strength values from different criteria. The conventional MC criterion is observed to predict values with maximum

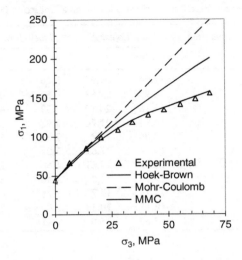

Figure 3 Comparison of strength predicted through different criteria (Rock: Indiana Limestone, data from Schwartz, 1964).

deviation from experimental results with an over prediction of about 59.3% for confining pressure of 68.4 MPa. The HB criterion also predicts very high values and exhibits a deviation of about 28.6% at confining pressure of 68.4 MPa. The MMC criterion gives the best predictive capability as the predicted values closely match with the experimental values. The maximum deviation is of the order of 4% and deviation at confining pressure of 68.4 MPa was about 0.9%.

The predictive capabilities of the different strength criteria, when triaxial data from which their parameters can be fitted, were also compared by Shen *et al.* (2014). The authors used two cases of confining pressure range *i.e.* when tests in 'low' stress range ($\sigma_3 \leq \sigma_{ci}$), when tests in high stress range ($\sigma_3 > \sigma_{ci}$), are employed. It was shown that for the range $0 \leq \sigma_3 \leq \sigma_{ci}$, the MMC (Singh *et al.*, 2011) gave best results; though for the range $\sigma_3 > \sigma_{ci}$ the criterion proposed by Shen *et al.* (2014) gave better results. It may be noted that for majority of rock engineering problems in civil, mining and tunnel engineering, the confining pressure σ_3 is likely to be less than σ_{ci}, for which MMC criterion was found to be the most suitable criterion for intact rocks.

3.3 Polyaxial stress condition

In real life situations, the rocks are subjected to polyaxial stress conditions and the intermediate principal stress σ_2 substantially affects the strength σ_1 (Fig. 4) except for some non-dilatant rocks (Chang & Haimson, 2005). Effect of intermediate principal stress on strength of rocks has been discussed by many researchers (Murrell, 1963; Wiebols & Cook, 1968; Mogi, 1971; Chang and Haimson, 2000, 2005; Haimson & Chang, 2000; Colmenares & Zoback, 2002; Al-Ajmi & Zimmerman, 2005; Haimson 2009). It is shown by some of the studies that the strength increases with increase in σ_2, reaches a maximum at some intermediate value of σ_2 and then decreases when σ_2 approaches a value equal to σ_1. Some investigators, however, did not observe any clear

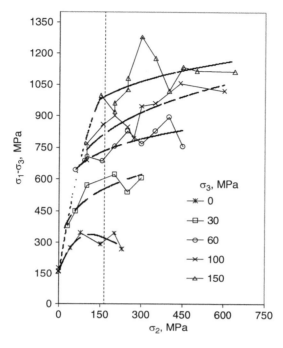

Figure 4 Effect of σ_2 on deviatoric stress at failure (data from Chang & Haimson, 2000).

cut trend of decreasing strength at higher σ_2 and rather observed a plateau which indicates a phenomenon similar to critical state concept. Figure 5 shows average increase in deviatoric stress at failure due to σ_2 over the conventional triaxial strength for KTB amphibolite (data from Chang & Haimson, 2000). The plot indicates that effect of σ_2 is large at low σ_3 and decreases with increase in σ_3. Beyond a value of σ_3 nearly equal to the UCS, σ_2 has no further influence on the strength. A condition of critical state is thus reached when the level of confining stress σ_3 is nearly equal to the UCS of the intact rock (Singh et al., 2011). Singh et al. (2011) extended the single parameter MMC triaxial strength criterion (Equation 8) to polyaxial stress conditions as follows:

$$(\sigma_1 - \sigma_3) = \sigma_{ci} + \frac{2\sin\phi_{i0}}{1-\sin\phi_{i0}}\left(\frac{\sigma_2+\sigma_3}{2}\right) - \frac{1}{\sigma_{ci}}\frac{\sin\phi_{i0}}{1-\sin\phi_{i0}}\left(\frac{\sigma_2^2+\sigma_3^2}{2}\right)$$

$$0 \leq \sigma_3 \leq \sigma_2 \leq \sigma_{ci} \qquad (17)$$

Some of the important features of the criterion are: a) For higher stress range, if σ_2 or σ_3 exceed σ_{ci}, its value in RHS should be replaced with σ_{ci}. b) The proposed polyaxial strength criterion assigns higher weightage to σ_3 as compared to σ_2 as it is an observed fact that an increase in σ_3 has greater influence than that produced by the same increase in σ_2. c) An increase in σ_2, while keeping σ_3 constant will increase deviatoric stress at failure. As a result, the ratio σ_1/σ_3 will increase with increase in σ_2. This will make

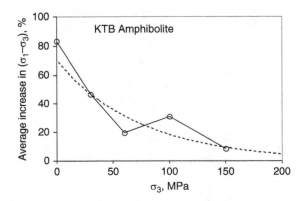

Figure 5 Diminishing effect of σ_2 at higher σ_3 (KTB amphibolites; data from Chang & Haimson, 2000).

failure more brittle and will also result in increase in extent of linear elastic deformation as observed by Haimson & Chang (2000) and Chang & Haimson (2005). d) Both σ_2 and σ_3 follow the concept that principal stress difference $(\sigma_1-\sigma_3)$ reaches its maximum when the corresponding stress becomes equal to σ_{ci} *i.e.*

$$\partial(\sigma_1 - \sigma_3)/\partial\sigma_3 = 0, \text{ for } \sigma_3 = \sigma_{ci}; \; \partial(\sigma_1 - \sigma_3)/\partial\sigma_2 = 0, \text{ for } \sigma_2 = \sigma_{ci} \qquad (18)$$

3.3.1 Application to polyaxial test data

Singh *et al.* (2011) evaluated the applicability of the suggested criterion by applying it to the available polyaxial strength database. The database comprised of polyaxial test data on ten rock types as summarized in Table 3. The complete data is available in Singh *et al.* (2011). The data include tests performed under the condition $\sigma_2=\sigma_3$. First three such data points were utilized to obtain c_{i0} and ϕ_{i0} using least square method. UCS σ_{ci} was obtained from Equation 4 and the criterion was applied to obtain σ_1 values for given σ_2 and σ_3 values. To quantify the accuracy in prediction, the percent error in prediction was obtained as

$$\text{pe} = \{(\sigma_{1\text{cal}} - \sigma_{1\text{exp}})/\sigma_{1\text{exp}}\} \times 100 \;\; \text{percent} \qquad (19)$$

Where pe is the percent error in prediction, $\sigma_{1\text{exp}}$ and $\sigma_{1\text{cal}}$ are the experimental and the predicted values of the polyaxial strength of the rock. The average percent error (avpe) for a given data set for a given rock type was computed as:

$$\text{avpe} = \sqrt{\frac{1}{\text{npt}}\sum_{i=1}^{\text{npt}}(\text{pe}^2)} \qquad (20)$$

where npt is the number of data points in the set.

To compare the proposed criterion with those already existing, five popular polyaxial strength criteria were selected: a) Modified Lade criterion (Colmenares &

Use of critical state concept in determination of triaxial and polyaxial strength 489

Table 3 Polyaxial test database (Singh *et al.*, 2011).

Sl	Details of rock	Source	σ_3 (MPa), number in parentheses indicates the number of polyaxial tests conducted at this σ_3
1	Dunham dolomite, Mogi (1967)	Yu *et al.* (2002)	0(1), 25(7), 45(8), 65(6), 85(5), 105(6), 125(6), 145(4)
2	Solenhofen limestone Mogi (1971)	Colmenares & Zoback (2002)	20(6), 40(7), 60(7), 80(9)
3	Mizuho trachyte Mogi (1967)	Yu *et al.* (2002)	0(1), 15(1), 30(1), 45(8), 60(7), 75(7), 100(6)
4	Dense marble Michelis (1985)	Yu *et al.* (2002)	0(4), 3.45(7), 6.89(6), 13.79(7), 20.68(5), 27.58 (5), 55.16(1)
5	Yubari shale Takahashi & Koide (1989)	Colmenares & Zoback (2002)	25(15), 50(11)
6	Shirahama sandstone, Takahashi & Koide (1989)	Colmenares & Zoback (2002)	5(5), 8(5), 15(5), 20(8), 30(7), 40(8)
7	KTB Amphibolite Chang & Haimson (2000)	Chang & Haimson (2000)	0(8), 30(6), 60(10), 100(10), 150(11)
8	Westerly granite Haimson & Chang (2000)	Haimson & Chang (2000)	0(5), 2(6), 20(10), 38(8), 60(5), 77(5), 100(6)

Zoback, 2002); b) Modified Weibols and Cook criterion (Colmenares & Zoback, 2002); c) Inscribed Druker-Prager criterion (Colmenares & Zoback, 2002); d) Circumscribed Drucker-Prager criterion (Colmenares & Zoback, 2002); e) Mogi-Coulomb criterion (Al-Ajmi & Zimmerman, 2005). Only the first three triaxial tests data points were considered for obtaining the criterion parameters and polyaxial strength values were predicted. The average percent error for each rock type was computed. The overall response of various criteria was obtained by an overall average of the percent error. A comparison of the overall average percent error is shown in Fig. 6. Out of all the criteria considered for comparison, the MMC polyaxial criterion is found to predict results with minimum overall average percent error.

4 MMC CRITERION FOR JOINTED ROCKS

4.1 Triaxial conditions

Engineering behavior of jointed rocks is greatly influenced by the presence of discontinuities. Jointed rocks also exhibit non-linear response of strength with confining pressure σ_3 (Singh & Singh, 2012). Brown (1970) conducted a classical study on strength behavior of jointed rocks under triaxial condition. The variation of shear strength from this study is shown in Fig 7. An important outcome of the study is that at sufficiently high confining pressure the Mohr failure envelopes of jointed and intact rocks merge with each other. Singh & Singh (2012) argued that similar to intact rocks, the critical state concept is followed by jointed rock also. Figure 8 shows the proposed strength criterion for jointed rock along with the failure criterion of intact rock. The UCS of jointed rock σ_{cj} (*i.e.* σ_1 for $\sigma_3=0$) will always be less than the UCS of intact rock σ_{ci}. Considering the critical confining pressure for jointed rock

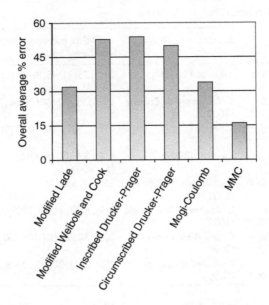

Figure 6 Comparison of overall average percent error in prediction of polyaxial strength by different criteria (Singh et al., 2015).

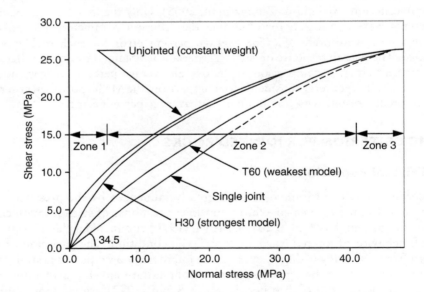

Figure 7 Mohr failure envelopes for intact and jointed specimens (source: Singh & Singh, 2012; redrawn from Brown, 1970).

Use of critical state concept in determination of triaxial and polyaxial strength 491

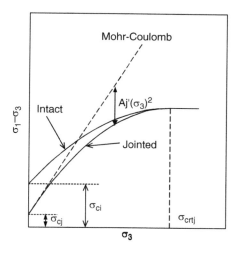

Figure 8 Modified Mohr Coulomb criteria for intact and jointed rocks.

equal to σ_{crtj}, Singh & Singh (2012) obtained the expression for MMC for jointed rock (Fig. 8) as:

$$(\sigma_1 - \sigma_3) = \sigma_{cj} + \frac{2\sin\phi_{j0}}{1-\sin\phi_{j0}} \sigma_3 - A'_j \sigma_3^2 \quad \text{for } 0 \leq \sigma_3 \leq \sigma_{crtj} \quad (21)$$

where σ_{cj} is the anisotropic UCS of the jointed rock in the direction of major principal stress and can be expressed in terms of c_{j0} and ϕ_{j0} similar to Equation 4; c_{j0} and ϕ_{j0} are the MC shear strength parameters of the anisotropic jointed rock at low confining pressure range ($\sigma_3 \to 0$); A'_j is an empirical constant which accounts for the deviation in strength of jointed rock from conventional linear MC criterion and can be obtained by differentiating Equation 21 with respect to σ_3 and equating gradient to zero at critical confining pressure, σ_{crtj}. The MMC criterion for jointed rocks is written as:

$$(\sigma_1 - \sigma_3) = \sigma_{cj} + \frac{2\sin\phi_{j0}}{1-\sin\phi_{j0}} \sigma_3 - \frac{1}{\sigma_{crtj}} \frac{\sin\phi_{j0}}{(1-\sin\phi_{j0})} \sigma_3^2 \quad \text{for } 0 \leq \sigma_3 \leq \sigma_{crtj} \quad (22)$$

Singh & Singh (2012) compiled a database comprising of more than 730 triaxial test data for variety of rocks (σ_{ci} = 9.5 to 123 MPa) from worldwide literature. The proposed criterion was fitted into the database and the critical confining pressure was back analyzed. It was concluded (Singh & Singh, 2012) that critical confining pressure of jointed rocks may be taken nearly equal to σ_{ci} for the application of the proposed criterion. Consequently the MMC criterion "single parameter form" for jointed rocks is expressed as:

$$(\sigma_1 - \sigma_3) = \sigma_{cj} + \frac{2\sin\phi_{j0}}{1-\sin\phi_{j0}} \sigma_3 - \frac{1}{\sigma_{ci}} \frac{\sin\phi_{j0}}{(1-\sin\phi_{j0})} \sigma_3^2 \quad \text{for } 0 \leq \sigma_3 \leq \sigma_{ci} \quad (23)$$

4.1.1 The parameter ϕ_{j0}

The failure envelopes of intact and jointed rocks will merge with each other beyond critical state. By commuting maximum deviatoric stresses at failure for jointed and intact rock the following relation may be obtained.

$$\sin \phi_{j0} = \{(1 - \text{SRF}) + \sin \phi_{i0}/(1 - \sin\phi_{i0})\}/\{(2 - \text{SRF}) + \sin \phi_{i0}/(1 - \sin\phi_{i0})\} \tag{24}$$

Where

$$\text{SRF} = \text{Strength reduction factor} = \sigma_{cj}/\sigma_{ci} \tag{25}$$

The friction angle ϕ_{j0} as obtained above will be slightly higher than ϕ_{i0}. The effect of interlocking between the blocks and dilation are indirectly considered by this expression.

4.1.2 Predictive capability

Singh & Singh (2012) used the triaxial test database compiled by them to assess the predictive capability of the proposed criterion. For each data set, the values of σ_{ci}, σ_{cj} and ϕ_{i0} were considered to be known and the triaxial strength values were predicted. Figure 9 shows cumulative distribution of the error in prediction for the compiled database. The plot indicates that there is a probability of 87.7% for the error to lie within ±20%. The proposed criterion may therefore be used with confidence if precise values of σ_{ci}, σ_{cj} and ϕ_{i0} are known.

4.1.3 How to obtain σ_{cj} in field?

Triaxial strength of jointed rock can be predicted using the proposed MMC criterion if σ_{ci}, σ_{cj} and ϕ_{i0} are known. The values of σ_{ci} and ϕ_{i0} will be available from laboratory tests on intact rocks; however determination of the UCS of jointed rock, σ_{cj} is a difficult task. Major factors governing σ_{cj} are strength of rock substance, kinematics and possible failure mode and characteristics of the discontinuities *e.g.* frequency, orientation, surface roughness, persistence and infilling. Various approaches are available in literature for assessing σ_{cj} and some of them are summarized are summarized in Table 4. Among rock mass classification systems, the Q system is most widely used for tunneling projects in India. Following expressions may be used to assess σ_{cj}:

$$\text{Singh } et\ al.(1997): \sigma_{cj} = 7\gamma Q^{1/3}\,\text{MPa}$$

$$(for\ Q < 10,\ 2 < \sigma_{ci} < 100\ MPa,\ SRF\ =\ 2.5) \tag{26}$$

$$\text{Barton (2002): } \sigma_{cj} = 5\gamma(Q\sigma_{ci}/100)^{1/3}\,\text{MPa} \tag{27}$$

where, γ is the unit weight of the rock mass in gm/cc, σ_{cj} and σ_{ci} are in MPa.

In the opinion of this writer, a reliable estimate of σ_{cj} can only be made through field testing. However, it is not feasible to stress a rock mass in the field to its ultimate strength. Alternatively Singh & Rao (2005) have suggested that the rock mass in the

Table 4 Approaches for assessing UCS of jointed rocks (Major source: Zhang, 2010).

Authors	Relation
Yudhbir & Prinzl (1983)	$\sigma_{cj}/\sigma_{ci} = \exp[\{7.65 \times (RMR - 100)\}/100]$
Ramamurthy et al. (1985) and Ramamurthy (1986)	$\sigma_{cj}/\sigma_{ci} = \exp\{(RMR - 100)/18.75\}$
Trueman (1988) and Asef et al. (2000)	$\sigma_{cj} = 0.5 \exp(0.06\ RMR)$ MPa
Kalamaras & Bieniawski (1993)	$\sigma_{cj}/\sigma_{ci} = \exp\{(RMR - 100)/24\}$
Sheorey (1997)	$\sigma_{cj}/\sigma_{ci} = \exp\{(RMR - 100)/20\}$
Aydan & Dalgic (1998)	$\sigma_{cj}/\sigma_{ci} = RMR/\{RMR + 6(100 - RMR)\}$
Zhang (2010)	$\sigma_{cj}/\sigma_{ci} = 10^{(0.013 RQD - 1.34)}$
Ramamurthy (1993); Ramamurthy & Arora (1994)	$\sigma_{cj}/\sigma_{ci} = \exp(-0.008 J_f)$
Singh (1997); Singh et al. (2002)	$\sigma_{cj}/\sigma_{ci} = \exp(-a\,J_f)$; a=0.0123 for splitting, 0.010 for shearing, 0.0250 for rotation and 0.0180 for sliding

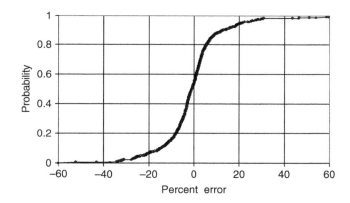

Figure 9 Probability of error in predicting triaxial strength of jointed rocks.

field may be tested up to a pre-determined stress level using test like uniaxial jacking test and reliable estimate of σ_{cj} can be made from the modulus of the mass. A correlation was suggested by Singh & Rao (2005) between the engineering properties of jointed and intact rocks as follows:

$$\sigma_{cj}/\sigma_{ci} = (E_j/E_i)^{0.63} \qquad (28)$$

Where E_j and E_i are the elastic moduli of the rock mass and the intact rock respectively in the principal stress direction.

4.2 Polyaxial stress conditions

Singh & Singh (2012) extended the MMC triaxial strength criterion to polyaxial stress conditions. The criterion for polyaxial strength is expressed as:

$$(\sigma_1 - \sigma_3) = \sigma_{cj} + \frac{2\sin\phi_{j0}}{1 - \sin\phi_{j0}}\left(\frac{\sigma_2 + \sigma_3}{2}\right) - \frac{1}{\sigma_{ci}}\frac{\sin\phi_{j0}}{(1 - \sin\phi_{j0})}\left(\frac{\sigma_2^2 + \sigma_3^2}{2}\right)$$

$$\text{for } 0 \leq \sigma_3 \leq \sigma_2 \leq \sigma_{ci} \qquad (29)$$

There is very limited data available on polyaxial strength of jointed rocks. Tiwari & Rao (2006, 2007) have reported results of fifty four tests conducted on a rock mass model under polyaxial stress condition. Singh & Singh (2012) evaluated the applicability of the polyaxial strength criterion to the results from Tiwari & Rao (2006, 2007). It was observed that with only input parameters σ_{ci}, ϕ_{i0} and σ_{cj}, the predictions are reasonably good for specimens with joint orientations θ = 0, 20, 40, 80 and 90° respectively, and have average percent error within about 15%. The orientation θ represents angle between joint plane and the major principal plane. For orientation θ = 60° the joints are critically inclined and the wedge formed by the intersecting joints is likely to slide under its own weight at low confining pressure. Rather than treating the rock mass as continuum, wedge analysis by incorporating joint shear strength models such as Barton & Choubey (1977) will be more appropriate for such cases. The proposed criterion is therefore not applicable where joints are critically oriented and single discontinuity governs the failure and the mass cannot be treated as continuum.

4.2.1 Rock burst conditions

Polyaxial strength criterion is very helpful in assessing the condition where rock burst conditions are likely to occur. As a thumb rule for Indian tunnels, squeezing is considered likely to occur if tunnel is under high overburden and the joint friction angle obtained from Q (= $\tan^{-1}(j_r/j_a)$) is less than 30°; whereas rock burst is considered likely to occur if the joint friction angle is more than 30° and the overburden is more than about 900 m. The failure under rock burst condition is brittle with instant release of high amount of energy. For an unsupported tunnel the minor principal stress σ_3 at the periphery will be nearly zero and the possibility of rock burst may be obtained by substituting σ_3=0 in the polyaxial criterion as follows:

$$\sigma_\theta \geq \sigma_{cj} + \frac{2\sin\phi_{j0}}{1 - \sin\phi_{j0}}\left(\frac{\sigma_2}{2}\right) - \frac{1}{\sigma_{ci}}\frac{\sin\phi_{j0}}{(1 - \sin\phi_{j0})}\left(\frac{\sigma_2^2}{2}\right) \quad \text{for } 0 \leq \sigma_2 \leq \sigma_{ci} \qquad (30)$$

Where σ_θ is the mobilized circumferential stress (after redistribution of stresses due to excavation) at the periphery of the tunnel. The expression indicates that if out of plane stress σ_2 is high, the strength of the rock mass will be high. If mobilized stress σ_θ exceeds this high strength it may lead to rock burst condition.

Singh & Singh (2012) have analyzed a case study of a power project (NJPC tunnel) in Indian Himalayas. During initial design, as per past practice, heavy rock burst conditions were feared for more than 900 m overburden. However, during excavation only slabbing and minor bursts were observed and minor supports were sufficient. Fifteen sections of the tunnel were analyzed using conventional Mohr-Coulomb and MMC polyaxial criteria and biaxial strength to stress ratio was computed (Singh & Singh, 2012). When conventional MC criterion (without considering the effect of σ_2) was used, the results indicated bursting conditions for all the fifteen section. However, when

Use of critical state concept in determination of triaxial and polyaxial strength 495

MMC polyaxial strength criterion was employed, the bursting criterion was satisfied only at few locations. The results of analysis by using MMC polyaxial criterion were more realistic.

5 TRANSVERSELY ISOTROPIC ROCKS

5.1 MMC criterion for transversely isotropic rocks

Rocks such as shale, slate, gneiss, schist and phyllite are inherently anisotropic and their strength is greatly influenced by the direction of testing. Modified Mohr Coulomb criterion for intact and jointed rocks has been derived by modifying the conventional Mohr-Coulomb criterion (Singh *et al.*, 2011; Singh & Singh, 2012). On similar lines, Singh *et al.* (2015) have suggested MMC criterion for intact anisotropic (transversely isotropic) rocks as given below:

$$(\sigma_1 - \sigma_3) = \sigma_{c\beta} + \frac{2 \sin \phi_{\beta 0}}{1 - \sin \phi_{\beta 0}} \sigma_3 - \frac{1}{\sigma_{crt}} \frac{\sin \phi_{\beta 0}}{(1 - \sin \phi_{\beta 0})} \sigma_3^2 \text{ for } 0 \le \sigma_3 \le \sigma_{crt}$$

(31)

Where $\sigma_{c\beta}$ is the UCS of anisotropic rock with planes of anisotropy oriented at an angle of β from major principal stress direction and can be represented in form of $c_{\beta 0}$, $\phi_{\beta 0}$ similar to Equation 4; $c_{\beta 0}$, $\phi_{\beta 0}$ are conventional MC parameters obtained from low triaxial tests at confining pressure ($\sigma_3 \to 0$); σ_{crt} is critical confining pressure for the anisotropic rock. For higher stress level the criterion is expressed as:

$$(\sigma_1 - \sigma_3) = \sigma_{c\beta} + \frac{\sin \phi_{\beta 0}}{1 - \sin \phi_{\beta 0}} \sigma_{crt} \text{ for } \sigma_3 > \sigma_{crt}$$

(32)

The UCS $\sigma_{c\beta}$, of the rock at a given orientation β, may be obtained by conducting UCS tests at orientations β = 0, 30 and 90° (Nasseri *et al.*, 2003). The above form of the criterion is termed as "two parameter criterion" (Singh *et al.*, 2015) as two parameters $\phi_{\beta 0}$ and σ_{crt} will be required to predict the strength of anisotropic rock. Singh *et al.* (2015) carried out statistical analysis and evaluated the proposed criterion by applying it to a compiled database comprising of 38 rock types with total number of 255 UCS and 1141 triaxial tests. Best fitting values of the parameters $\phi_{\beta 0}$ and σ_{crt} were obtained and these values were used to predict σ_1 values for given σ_3 values. It was shown by Singh *et al.* (2015) that the "two parameter criterion" exhibits excellent goodness of fit to the database. Probability distribution of percent error in prediction is shown in Fig. 10. The plot indicates that if triaxial test data are available for a rock type, the best fitting parameters of the proposed criterion will yield results such that there will be a probability of 0.9737 for error to be within ±20%. The very high probability of error to lie within small range of 20% indicates an excellent fitness of the proposed model to the database.

Back analysis of the database was also performed by Singh *et al.* (2015) to arrive at the most probable value of critical confining pressure for anisotropic rocks. It was shown statistically that an optimum value of the critical confining pressure equal to $1.25\sigma_{cmax}$ in the proposed criterion gives minimum error in prediction. Where σ_{cmax} is

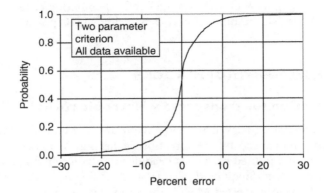

Figure 10 Probability distribution of error in prediction of triaxial strength of anisotropic rocks (Two parameter criterion; all data available for assessing parameters).

the maximum value of UCS when β is varied from 0 to 90°. Usually, the maximum UCS occurs for β=90°. It is recommended that if sufficient amount of triaxial test data is available, σ_{crt} may be obtained through best fitting and minimization of error. Where sufficient triaxial tests data is not available, an average value of the critical confining pressure may be taken equal to $1.25\sigma_{cmax}$. The MMC "single parameter criterion" for anisotropic rocks may be written as:

$$(\sigma_1 - \sigma_3) = \sigma_{c_\beta} + \frac{2\sin\phi_{\beta 0}}{1 - \sin\phi_{\beta 0}}\sigma_3 - \frac{1}{1.25\,\sigma_{cmax}}\frac{\sin\phi_{\beta 0}}{(1-\sin\phi_{\beta 0})}\sigma_3^2$$

$$\text{for } 0 \leq \sigma_3 \leq 1.25\,\sigma_{cmax} \qquad (33)$$

$$(\sigma_1 - \sigma_3)_{max} = \sigma_{c_\beta} + 1.25\frac{\sin\phi_{\beta 0}}{1-\sin\phi_{\beta 0}}\sigma_{cmax} \text{ for } \sigma_3 > 1.25\,\sigma_{cmax} \qquad (34)$$

It may be noted that for intact isotropic rocks and jointed anisotropic rocks, the critical confining pressure was statistically evaluated to be nearly equal to the UCS of the isotropic rock substance. However, in the present case of intact anisotropic rocks, the critical confining pressure works out to be about 1.25 times the maximum UCS exhibited by the anisotropic rock. Singh *et al.* (2015) attribute the possible reason for this difference to the chemical alteration of rock material during geological foliation process. To understand this phenomenon a jointed rock specimen may be considered which is tested for UCS by keeping loading direction normal to joint planes. The UCS of this specimen will be lower than the UCS of the rock substance, σ_{ci}. The strength reduction will depend on joint characteristics *e.g.* orientation, surface roughness and frequency. In a similar fashion, if an intact anisotropic rock is tested for UCS by keeping loading direction normal to the planes of anisotropy, the observed UCS should be lower than the real UCS of the rock substance. The analysis done by Singh *et al.* (2015) indirectly infers that the real UCS of the rock substance would be about 1.25 times the maximum UCS, σ_{cmax}. Critical state will reach when confining pressure approaches a value equal to $1.25\sigma_{cmax}$. The criterion parameter $\phi_{\beta 0}$ may be obtained by adopting the following steps.

$$A_\beta = \frac{\sum(\sigma_1 - \sigma_3 - \sigma_{c\beta})}{\sum(\sigma_3^2 - 2.5\,\sigma_{cmax}\,\sigma_3)}, \quad 0 \le \sigma_3 \le 1.25\,\sigma_{cmax};$$

$$B_\beta = -2.5 A_\beta \sigma_{cmax}\,;\, \sin \phi_{\beta 0} = B_\beta / (2 + B_\beta) \tag{35}$$

5.2 Predictive capability of single parameter criterion for anisotropic rocks

Singh *et al.* (2015) statistically evaluated the predictive capability of the single parameter criterion by applying it to the compiled database. The parameter $\phi_{\beta 0}$ was determined for the following three conditions and the σ_1 values were predicted for given confining pressure values. a) All triaxial test data are used to obtain $\phi_{\beta 0}$; b) Only UCS and two triaxial test data are used to obtain $\phi_{\beta 0}$; c) Only UCS vale is used to obtain $\phi_{\beta 0}$. The third case is an extreme case and not desirable, however may be unavoidable especially during preliminary feasibility studies. Singh *et al.* (2015) analyzed the triaxial test data available and found a correlation for term A_β (Equation 35) with a R^2 value of 0.893, as follows:

$$A_\beta = -4.75(\sigma_{crt})^{-1.22} \tag{36}$$

This expression may be used in Equation 35 to get a rough estimate of parameter $\phi_{\beta 0}$ in absence of triaxial tests data.

The criterion parameter was obtained for all the data sets for the three conditions stated above and strength values were predicted for given confining pressure values. The probability distributions of the percent error for the three conditions are presented in Fig. 11. The probability of error to be within ±20% has been found to be 94.66, 83.64 and 57.00% for cases a, b and c respectively. Excellent results are, therefore, obtained for condition (a) when all triaxial tests data are used for assessing the criterion

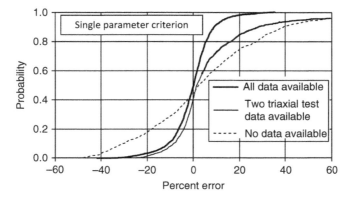

Figure 11 Probability distribution of error in prediction of triaxial strength of anisotropic rocks.

498 Singh

parameter. The results for condition (b) *i.e.* when only two triaxial test data are available are also reasonably good. The case (c) shows that even without any triaxial test data available for fitting, rough estimates of the strength are possible. The analysis indicates that depending on number of triaxial test data available, excellent accuracy in prediction can be achieved for field applications. However, during preliminary stages and feasibility studies, only UCS data may be available and a very rough estimate of the strength can also be made through the proposed criterion.

6 CONCLUDING REMARKS

6.1 Intact rocks

Modified Mohr Coulomb criterion has been proposed to predict triaxial strength of intact isotropic rocks as a non-linear function of confining pressure. The shear strength parameters c_{i0} and ϕ_{i0}, as used in conventional MC criterion, are retained and used in the MMC criterion. Barton's critical state concept for rocks was employed to define the shape of the criterion. The concept envisages that beyond critical confining pressure the Mohr failure envelope becomes asymptotic to a horizontal line. The "two parameter MMC criterion" for intact rocks considers critical confining pressure as one of the parameters. Back analysis of a database comprising of more than 150 triaxial test data sets from worldwide literature indicates that the critical confining pressure in the proposed MMC criterion may be taken nearly equal to the UCS of the intact rock σ_{ci}. The "single parameter MMC criterion" for intact rock has only single parameter ϕ_{i0} that can be obtained through few triaxial tests on intact rocks in low confining pressure range ($\sigma_3 \rightarrow 0$). The performance of the proposed MMC criterion has been compared with other popular criteria in vogue. It is recommended that the proposed MMC can be used with confidence over a wide range of confining pressure with an input parameter ϕ_{i0} obtained from low confining pressure range tests.

The proposed non-linear criterion, with parameter ϕ_{i0}, has further been extended to polyaxial stress conditions. The applicability of the MMC polyaxial criterion has been verified by applying it to polyaxial test data sets for ten rock types and comparing its performance with other popular polyaxial strength criteria in vogue. The MMC for triaxial and polyaxial stress conditions has been found to be the most promising criterion.

6.2 Jointed anisotropic rocks

Field engineers working in civil and mining applications invariably deal with rocks that are jointed in nature. Jointed rocks exhibit highly non-linear strength response which is not captured by the conventional form of the MC criterion. The conventional form has been modified to incorporate non-linearity in strength response by invoking critical state concept, and retaining the conventional shear strength parameters. The criterion was applied to triaxial test database comprising of more than 730 triaxial test results for variety of rocks (σ_{ci} = 9.5 to 123 MPa). It has been shown that Barton's critical state concept, which was originally suggested for intact rocks, is applicable to jointed rocks as well. The statistical analysis indicates that similar to intact rocks, the critical confining pressure for jointed rocks also can be taken nearly equal to the UCS of intact rocks. The parameter ϕ_{j0} will be required for assessing the triaxial strength. An expression has been derived to get ϕ_{j0} from intact rock friction angle ϕ_{i0}, and strength reduction factor σ_{cj}/σ_{ci}. It is recommended that this expression may be used to get the friction angle ϕ_{j0}. The UCS of jointed rock σ_{cj} may be obtained by using classification approaches and laboratory test results for intact rocks. It has, however, been recommended that more reliable estimates of σ_{cj} can be made if deformability test is conducted in the field and strength reduction factor is computed as a function of modulus reduction factor. Statistical analysis of the database indicates that with only input of σ_{cj}, σ_{ci} and ϕ_{i0} there is a probability of 87.7% that the predicted results would lie within $\pm20\%$ error.

The MMC triaxial strength criterion for jointed rocks has also been extended to polyaxial stress conditions. Very limited data is available in literature for polyaxial tests on jointed rocks. A set of 54 polyaxial test results were analyzed to verify the applicability of the criterion to polyaxial stress conditions in anisotropic jointed rocks. It is recommended that the criterion can be applied with confidence to those situations where assumption of equivalent continuum is valid. For situations where rigid wedge is likely to slide along joint planes, an appropriate criterion for rock joints (say Barton & Choubey, 1977) should be used. The polyaxial strength criterion has also been used to define a condition to assess the possibility of rock burst in the field.

6.3 Intact anisotropic rocks

Critical state concept based Modified Mohr Coulomb criterion has also been extended to inherently anisotropic (transversely isotropic) rocks. The "two parameter" form of the criterion uses UCS, $\sigma_{c\beta}$ for given orientation β, and two parameters *i.e.* $\phi_{\beta0}$ and σ_{crt} to estimate the triaxial strength. A statistical evaluation of the criterion was accomplished by applying it to a triaxial test database comprising of test results on 38 rock types with more than 250 UCS and 1140 triaxial tests. Excellent goodness of fit is observed.

Back analysis was performed to get an average value of critical confining pressure to be used in the criterion. The analysis revealed that the critical confining pressure for inherently anisotropic rocks, as applicable in the proposed MMC criterion, may be taken nearly equal to $1.25\sigma_{cmax}$. This appears to be in contradiction with intact isotropic and jointed anisotropic rock results where the critical confining pressure was evaluated to be nearly equal to σ_{ci}. The probable reason of difference in the critical confining pressure is attributed to the chemical alteration of rock material during geological foliation process. It is argued that the real UCS of the rock substance between foliation planes would be higher and on the order of about $1.25\,\sigma_{cmax}$.

The "single parameter" form of the MMC criterion uses single parameter $\phi_{\beta0}$ which can be obtained from few triaxial tests for given orientation β. The predictive capability of this form has been checked for three conditions for assessing fitting parameter $\phi_{\beta0}$ *i.e.* i) all triaxial test data are used, ii) only UCS and two triaxial test data are used, and iii) only UCS vale is used to obtain $\phi_{\beta0}$. The analysis reveals that excellent results are obtained when sufficient triaxial test results are available to assess the criterion parameter. Reasonably good predictions are possible with only two triaxial tests available. Also even without any triaxial tests data available for fitting, rough estimates for strength are possible.

ACKNOWLEDGMENT

The author is thankful to his PhD supervisors Prof. T. Ramamurthy and Prof. K.S. Rao of IIT Delhi for introducing him to the field of Rock Mechanics. Thanks are due to colleagues and researchers at Roorkee namely Prof. Bhawani Singh, Prof. M.N. Viladkar, Prof. N.K. Samadhiya, Dr. R.K. Goel and Dr. P. Maheshwari for their fruitful discussions related to rock mechanics and statistics. Many research scholars and students have directly or indirectly contributed toward the writer's understanding on the strength behavior of rocks. The author would like to acknowledge contributions from Dr. B.K. Agrawal, Dr. Ajit Kumar, Dr. Ajay Bindlish, Dr. Jaysing Choudhari, Lt. Col. Anil Raj, Dr. R.D. Dwivedi, Dr. Harsh Verma and Dr. L.P. Srivastava. Some of the research reported in this article was partly supported by Department of Science and Technology (DST), Government of India, New Delhi. The author acknowledges the support from DST and expresses thanks to Dr. Bhoop Singh for his cooperation. The author is also thankful to Prof. B. Haimson from University of Wisconsin, USA, for readily sharing polyaxial tests data for SAFOD granodiorite and TCDP siltstone.

REFERENCES

Al-Ajmi AMB, Zimmerman RW. (2005) Relation between the Mogi and the Coulomb failure criteria, Int. J. Rock Mech. Min. Sci., 42:431–439.

Asef MR, Reddish DJ, Lloyd PW (2000) Rock-support interaction analysis based on numerical modeling, Geotech. Geol. Eng., 18:23–37.

Aydan O, Dalgic S (1998) Prediction of deformation behaviour of 3-lanes Bolu tunnels through squeezing rocks of North Anatolian fault zone (NAFZ). In: Proceedings of regional symposium on sedimentary rock engineering, Taipei, 228–233.

Barton N (1976) Rock Mechanics Review: The shear strength of rock and rock joints, Int. J. Rock Mech. Min. Sci. & Geomech. Abstr., 13:255–279.

Barton N (2002) Some new Q–Value correlations to assist in site characteristics and tunnel Design, Int. J. Rock Mech. Min. Sci., 39:185–216.

Barton N (2013) Shear strength criteria for rock, rock joints, rockfill and rock masses: Problems and some solutions. J. Rock Mech. Geotech. Eng., 5(4):249–261.

Barton N, Choubey V (1977) The shear strength of rock joints in theory and practice. Rock Mech., 10:1–54.

Brown ET (1970) Strength of models of rock with intermittent joints, J. Soil Mech. & Found. Div. Proc. ASCE, 96(SM6):1935–1949.

Chang C, Haimson B (2000) True triaxial strength and deformability of the German Continental deep drilling program (KTB) deep holeamphibolite. J. Geophys. Res, 105:18999–19013.

Chang C, Haimson B (2005) Non-dilatant deformation and failure mechanism in two Long Valley Caldera rocks under true triaxial compression, Int. J. Rock Mech. Min. Sci., 42:402–414.

Chang C, Haimson B (2012) ISRM suggested method: A failure criterion for rocks based on true triaxial testing, Rock Mech. Rock Eng., 45:1007–1010.

Colmenares LB, Zoback MD (2002) A statistical evaluation of intact rock failure criteria constrained by polyaxial test data for five different rocks, Int. J. Rock Mech. Min. Sci., 39:695–729.

Haimson B, Chang C (2000) A new true triaxial cell for testing mechanical properties of rock, and its use to determine rock strength and deformability of Westerly granite, Int. J. Rock Mech. Min. Sci., 37:285–296.

Haimson B. (2009) A three-dimensional strength criterion based on true triaxial testing of rocks. In: Hudson *et al.*, editors. Proceedings of the ISRM-Sponsored International Symposium on Rock Mechanics: Rock Characterisation, Modelling and Engineering Design Methods, SINOROCK2009, pp 21–28.

Hoek E, Brown E. (1980) Empirical strength criterion for rock masses, J. Geotech. Eng. Div., 106 (GT9):1013–1035.

Hoek E. (1983) Strength of jointed rock masses, Géotechnique, 23(3):187–223.

Kalamaras GS, Bieniawski ZT (1993) A rock mass strength concept for coal seams. In: Proceedings of 12th conference ground control in mining, Morgantown, 274–283.

Langford JC, Diederichs MS (2015) Quantifying uncertainty in Hoek–Brown intact strength envelopes, Int. J. Rock Mech. Min. Sci., 74:91–102.

Michelis P (1985) Polyaxial yielding of granular rock, J. Eng. Mech., ASCE, 111(8):1049–1066.

Mogi K (1967) Effect of the intermediate principal stress on rock failure, J. Geophys. Res., 72:5117–5131.

Mogi K (1971) Fracture and flow of rocks under high triaxial compression, J. Geophys. Res., 76:1255–1269.

Murrell SAF (1963) A criterion for brittle fracture of rocks and concrete under triaxial stress, and the effect of pore pressure on the criterion. In: Fairhurst C, editor. Proceedings of the 5th Symposium on Rock Mechanics, University of Minnesota, Minneapolis, MN, 563–577.

Nasseri MHB, Rao KS, Ramamurthy T (2003) Anisotropic strength and deformation behaviour of Himalayan schists. Int. J. Rock Mech. Min. Sci. 40:3–23.

Ramamurthy T (1986) Stability of rock mass – Eighth Indian Geotechnical Society Annual Lecture, Indian. Geotech. J., 16:1–73.

Ramamurthy T, Rao GV, Rao KS (1985) A strength criterion for rocks. In: Proceedings of Indian geotechnical conference, Vol. 1, Roorkee, pp 59–64.

Ramamurthy T (1993) Strength and modulus response of anisotropic rocks. In: Comprehensive Rock Engineering, JA Husdon, editor. Oxford: Pergamon Press, Vol. 1, pp 313–329.

Ramamurthy T, Arora VK (1994) Strength prediction for jointed rocks in confined and unconfined states, Int. J. Rock Mech. Min. Sci. & Geomech. Abstr., 31(1):9–22.

Schwartz AE (1964) Failure of rock in the triaxial shear test, Proceedings of the 6th Symposium on Rock Mechanics, Rolla, Mo., 109–135.

Shen J, Jimenez R, Karakus M, Xu C (2014) A Simplified failure criterion for intact rocks based on rock type and uniaxial compressive strength. Rock Mech. Rock Eng., 47(2):357–369.

Sheorey PR (1997) Empirical rock failure criteria. Balkema, Rotterdam.

Singh B, Viladkar MN, Samadhiya NK, Mehrotra VK (1997) Rock mass strength parameters mobilised in tunnels, Tunnelling Underground Space Technol., 12(1):47–54.

Singh M (1997) Engineering behaviour of jointed model materials, Ph.D. Thesis, IIT, New Delhi, India.

Singh M, Raj A, Singh B (2011) Modified Mohr-Coulomb criterion for non-linear triaxial and polyaxial strength of intact rocks, Int. J. Rock Mech. Min. Sci., 48:546–555.

Singh M, Rao KS, Ramamurthy T (2002) Strength and deformational behaviour of jointed rock mass, Rock Mech. Rock Eng., 35(1):45–64.

Singh M, Rao KS (2005) Empirical methods to estimate the strength of jointed rock masses, Eng. Geol., 77:127–137.

Singh M, Samadhiya NK, Kumar A, Kumar V, Singh B (2015) A nonlinear criterion for triaxial strength of inherently anisotropic rocks, Rock Mech. Rock Eng., 48(4):1387–1405.

Singh M, Singh B (2005) A strength criterion based on critical state mechanics for intact rocks, Rock Mech. Rock Eng., 38(3):243–248.

Singh M, Singh B (2012) Modified Mohr–Coulomb criterion for non-linear triaxial and polyaxial strength of jointed rocks, Int. J. Rock Mech. Min. Sci., 51:43–52.

Takahashi M, Koide H (1989) Effect of the intermediate principal stress on strength and deformation behavior of sedimentary rocks at the depth shallower than 2000 m. In: Maury V, Fourmaintraux D, editors. Rock at great depth, Vol. 1, Rotterdam: Balkema, pp 19–26.

Tiwari RP, Rao KS (2006) Post failure behaviour of a rock mass under the influence of triaxial and true triaxial confinement, Eng. Geol., 84:112–129.

Tiwari RP, Rao KS (2007) Response of an anisotropic rock mass under polyaxial stress state. ASCE J. Mater Civ Eng., 19(5):393–403.

Trueman R (1988) An evaluation of strata support techniques in dual life gateroads. PhD thesis, University of Wales, Cardiff.

Wiebols G, Cook N (1968) An energy criterion for the strength of rock in polyaxial compression, Int. J. Rock Mech. Min. Sci., 5:529–549.

Yu Mao-Hong, Zana Yue-Wen, Jian Zhaob C, Yoshimined M (2002) A unified strength criterion for rock material, Int. J. Rock Mech. Min. Sci., 39:975–989.

Yudhbir WL, Prinzl F (1983) An empirical failure criterion for rock masses. In: Proceedings of 5th international congress on Rock Mechanics, Vol. 1, Melbourne, pp B1–B8.

Zhang L (2010) Estimating the strength of jointed rock masses. Rock Mech. Rock Eng., 44:391–402.

Chapter 18

Practical estimate of rock mass strength and deformation parameters for engineering design

M. Cai
Bharti School of Engineering, Laurentian University, Sudbury, Ontario, Canada

Abstract: Knowledge of the rock mass strength is required for the design of many engineering structures to be built in or on rocks. For this purpose, it is necessary to obtain design parameters such as deformation moduli, peak and residual strength parameters, and dilation angle for numerical modeling and design. The GSI system, proposed by Hoek *et al.* (1995), is now widely used for the estimation of the rock mass peak strength and the rock mass deformation parameters. A quantitative approach to assist in the use of the GSI system is presented. It employs the in-situ block volume and a joint condition factor as quantitative characterization factors to determine the peak GSI value. To use the GSI system to estimate the residual strength of jointed rock masses, the peak GSI can be adjusted to a residual GSIr value based on the two major controlling factors in the GSI system, *i.e.*, the residual block volume and the residual joint surface condition factor. Methods to estimate peak and residual block volumes and joint surface condition factors are presented. In addition, a detailed discussion on the determination of other design analysis input parameters, such as uniaxial compressive strength of intact rocks and Hoek-Brown constant m_i are given and a method to estimate dilation angles of rock masses is presented. The determined Hoek-Brow rock mass strength parameters, dilation angles, and deformation modulus can be used in numerical analyses for safe and cost-effective engineering design. Because of its quantitative nature, this approach allows the consideration of variability of rock mass strength and deformation parameters in design, using the Monte Carlo or point estimate method.

1 INTRODUCTION

Knowledge of the rock mass strength and deformation behaviors is required for the design of many engineering structures to be built in or on rocks, such as foundations, slopes, tunnels, underground caverns, drifts, and mining stopes. The strength and deformation modulus of a jointed rock mass depends on the strength of the intact rocks and the joint conditions. A better understanding of the mechanical properties of rock masses will facilitate cost-effective design of such structures.

The determination of the global mechanical properties of a jointed rock mass remains one of the most difficult tasks in rock mechanics. Anyone who had practiced in the geomechanics and geotechnical fields will not hesitate to admit that trying to estimate the deformation modulus and uniaxial compressive strength of a jointed rock mass is such a daunting difficult task. There are many reasons why it is so. The name of a rock mass is not CHILE (Continuous, Homogeneous, Isotropic, and Linearly-Elastic) but DIANE (Discontinuous, Inhomogeneous, Anisotropic, and Non-Elastic) (Hudson & Harrison, 1997). Complex spatial variation, scale effect, stress path dependency, and limited access to monitoring and measurement are other factors that render estimating the mechanical properties of a rock mass difficult.

Many researchers have developed constitutive models to describe the strength and deformation behaviors of jointed rock masses (*e.g.*, Oda, 1983; Amadei, 1988; Cai & Horii, 1992). Because there are so many parameters that affect the deformability and strength, it is generally impossible to develop a universal constitutive model that can be used to a priori predict the strength of a rock mass. In addition, model parameters need to be calibrated before the models can be used in design analysis.

Traditional methods to determine the mechanical property parameters include plate-loading tests for deformation moduli and in-situ block shear tests for strength parameters. These tests can only be performed when the exploration adits are excavated and the cost of conducting these tests is high. Although back-analyses, which are based on field measurement, are helpful in determining the strength and deformation parameters as a project proceeds (Cividini *et al.*, 1981; Sakurai & Takeuchi, 1983; Cai *et al.*, 2007), they do not provide design parameters at the pre-feasibility or feasibility study stages.

As computers become much more powerful and high performance computing is now easily accessible, there is a new trend to model the rock mass response using some basic measured mechanical and geometrical properties of the rock and joints as inputs. Jointing is considered using a stochastic discrete fracture network (DFN) simulation (Dershowitz & Einstein, 1988). Many equally significant realizations of the fracture network can be produced. One can pick a particular fracture network realization and then import it into some numerical packages (*e.g.*, ELFEN (Rockfield Software Ltd., 2003), PFC3D (Itasca, 2010)) to create a jointed rock mass model. The approach adopted in PFC3D is called the synthetic rock mass (SRM) approach (Pierce *et al.*, 2007; Mas Ivars *et al.*, 2011). Failure of the rock mass is simulated by considering intact rock fracturing, joint sliding, or the combination of the two. Using this approach, it not only helps us to understand better the failure mechanism of jointed rock masses, but also assists us to estimate rock mass strengths and deformabilities. While the approach is promising, it also bears some major deficiencies. The approach is not simple enough for site engineers to use, and it requires them to be able to run some of the most skill demanding software packages (ELFEN, PFC3D, FRACMAN (Dershowitz *et al.*, 1993)) in the geotechnical community. Computing time is long and there are many model parameters that cannot be directly measured and have to be calibrated using field monitoring data. In addition, the discrete fracture network generated is often a very rough representation of reality (*e.g.*, smooth joint in a large scale, circular or elliptical joint shape etc.), and the limitation of the approach created by the DFN model is often overlooked by some researchers and users.

Some attempts have been made to develop simple methods to characterize jointed rock masses to estimate the deformability and strength indirectly. The Geological Strength Index (GSI), developed by Hoek *et al.* (1995), is one of them. It uses properties of intact rock and conditions of jointing to determine/estimate the rock mass deformability and strength. GSI values can be estimated based on the geological description of the rock mass and this is well suited for rock mass characterization at the initial stage of a project. The GSI system concentrates on the description of two factors, rock structure and block surface conditions. Some efforts (Sonmez & Ulusay, 1999; Russo, 2009; Hoek *et al.*, 2013) were made to make the system more user friendly, but the approach presented in Cai *et al.* (2004), which employs the block volume (V_b) and a joint condition factor (J_c) as quantitative characterization factors, will be presented in this chapter. This approach adds quantitative means to facilitate use of the system, especially by inexperienced engineers. Because of its quantitative nature, it facilitates the use of probabilistic design approach to tunnel and cavern design using the GSI system (Cai & Kaiser, 2006a; Cai, 2011). Furthermore, the approach has been developed and tested for rock mass's residual strength estimation, by adjusting the peak *GSI* to the residual GSI_r value based on the two major controlling factors in the GSI system – the residual block volume V_b^r and the residual joint condition factor J_c^r (Cai *et al.*, 2007).

Although imperfect, the GSI system provides a simple and yet practical means to define a complete set of mechanical properties (peak and residual Hoek-Brown strength parameters m_b and s, or the equivalent Mohr-Coulomb strength parameters c and ϕ, as well as deformation modulus E) for design purpose.

Firstly some widely used rock mass classification systems and empirical relations to estimate rock mass strength and deformation modulus using these classification systems are reviewed. Next, a complete quantitative approach to determine peak and residual strength parameters of jointed rock masses using the Generalized Hoek-Brown failure criterion and the GSI system is presented. Other input parameters required for engineering design, such as deformation modulus and dilation angle, are also discussed. An example is given to illustrate the application of the proposed method.

2 ROCK MASS CHARACTERIZATION FOR ENGINEERING DESIGN

2.1 Brief summary of rock mass classification systems

Rock mass characterization is the process of collecting and analyzing qualitative and quantitative data that provide indices and descriptive terms of the geometrical and mechanical properties of a rock mass. It is a significant part in any field geological investigation involving rock engineering problems. This process requires the collection and recording as well as analyzing a sizable amount of geological and geotechnical data. Methods for rock mass characterization include core logging, borehole logging, scanline surveying, cell mapping, geologic structure mapping, and rock index testing. New technologies, such as digital image processing of fracture information and laser-based imaging of joint roughness, can be applied for rock mass characterization.

When all necessary data are collected, the rock masses are classified according to the emphasis of influence of certain index on the overall rock mass quality based on a classification system or scheme. Ideally rock mass classification should provide a quick means to estimate the support requirement and to estimate the strength and deformation properties of the rock mass. More specifically, a rock mass classification

506 Cai

scheme is intended to classify the rock masses, provide a basis for estimating deformation and strength properties, supply quantitative data for support estimation, and present a platform for communication between exploration, design and construction groups.

Many rock mass classification systems have been proposed and used in engineering practice, such as the Terzaghi's classification (Terzaghi, 1946), RQD (Deere, 1968), RSR (Wickham *et al.*, 1974), RMR (Bieniawski, 1976), Q (Barton *et al.*, 1974; Barton, 2002), GSI (Hoek *et al.*, 1995, 1998), and RMi system (Palmstrøm, 1996a,b). Some systems are based on the modification of the existing ones to suit a specific application. For examples, the RMR system was modified by Laubscher (1990) for mine design and by Kendorski *et al.* (1983) for drift support design in caving mines. The Q system was modified by Potvin (1988) for stope design.

2.2 Estimation of rock mass properties using rock mass classification systems

Rock mass classification systems have been used to estimate mechanical properties (*i.e.*, deformation modulus and uniaxial compressive strength) of jointed rock masses at the preliminary design stage of a project. Table 1 summarizes some of the widely used empirical equations for determining deformation moduli of rock masses from

Table 1 Empirical relations to estimate deformation modulus from a classification index.

No.	Deformation modulus	Reference	Note
1	$E = E_i\,(0.0231RQD-1.32)$	Coon & Merritt (1970)	RQD>57
2	$E = E_i\,(RQD-60)/40$	Deer et al. (1967)	RQD>60
3	$E = E_i 10^{0.0186RQD-1.91}$	Zhang & Einstein (2004)	
4	$E = 2RMR-100$	Bieniawski (1978)	Applicable for RMR > 50
5	$E = 0.1(RMR/10)^3$	Read et al. (1999)	
6	$E = 10^{\frac{RMR-10}{40}}$	Serafim & Pereira (1983)	Applicable for RMR < 50
7	$E = E_i\left(0.0028RMR^2 + 0.9e^{\frac{RMR}{22.82}}\right)$	Nicholson & Bieniawski (1990)	E_i is the elastic modulus of the intact rock
8	$E = 25\log Q$	Barton et al. (1980)	Applicable for Q > 1
9	$E = 10Q_c^{1/3}$, where $Q_c = Q\dfrac{\sigma_c}{100}$	Barton (2002)	σ_c is the strength of intact rock
10	$E = 5.6(RMi)^{0.375}$	Palmstrøm (1995)	For RMi > 0.1
11	$E = \sqrt{\dfrac{\sigma_c}{100}}10^{\frac{GSI-10}{40}}$	Hoek & Brown (1997)	Applicable for σ_c < 100
12	$E = E_i^{(sa)0.4}$, $E_i = 50$ GPa, $s = e^{\frac{GSI-100}{9}}$ $a = 0.5 + \dfrac{1}{6}\left(e^{-GSI/15} - e^{-20/3}\right)$	Sonmez et al. (2004)	GSI = RMR

Practical estimate of rock mass strength and deformation parameters for engineering design **507**

Table 2 Empirical relations to estimate the uniaxial compressive strength of jointed rock masses from a classification index.

No.	Rock mass to rock strength ratio	Reference	Note
1	$\dfrac{\sigma_{cm}}{\sigma_c} = 10^{0.013RQD-1.34}$	Zhang (2010)	For RQD < 60
2	$\dfrac{\sigma_{cm}}{\sigma_c} = e^{(-0.008J_f)}$ $J_f = \dfrac{J_n}{nr}$	Ramamurthy et al. (1985), Ramamurthy (1994)	J_n – number of joints per meter; n – inclination parameter (0.05 to 0.98) depending on the angel between the joint and $\sigma 1$. r – joint strength factor related to joint condition (= $\tan\phi$). For RQD < 60
3	$\dfrac{\sigma_{cm}}{\sigma_c} = 0.039 + 0.893e^{\left(\frac{-J_f}{160.99}\right)}$	Jade & Sitharam (2003)	For RQD < 60. The definition of J_f is the same as above.
4	$\dfrac{\sigma_{cm}}{\sigma_c} = e^{7.65\left(\frac{RMR-100}{100}\right)}$	Yudhbir et al. (1983)	
5	$\dfrac{\sigma_{cm}}{\sigma_c} = \dfrac{MRMR - \text{Rating for } \sigma_c}{106}$	Laubscher (1984)	
6	$\dfrac{\sigma_{cm}}{\sigma_c} = e^{\left(\frac{RMR-100}{18.75}\right)}$	Ramamurthy et al. (1985)	
7	$\dfrac{\sigma_{cm}}{\sigma_c} = e^{\left(\frac{RMR-100}{20}\right)}$	Sheorey et al. (1989)	RMR ≥ 18, 1976 version of RMR.
8	$\dfrac{\sigma_{cm}}{\sigma_c} = e^{\left(\frac{RMR-100}{24}\right)}$	Kalamaras & Bieniawski (1993)	
9	$\dfrac{\sigma_{cm}}{\sigma_c} = \dfrac{RMR}{RMR+6(100-RMR)}$	Aydan & Dalgiç (1998)	
10	$\sigma_{cm} = 7\gamma\, f_c Q^{1/3}$ MPa	Bhasin & Grimstad (1996)	$f_c = \sigma_c/100$ for Q > 10 and σ_c > 100 MPa, otherwise $f_c = 1$; and γ is the unit weight of the rock mass in g/cm^3.
11	$\sigma_{cm} = 5\gamma\left(Q\frac{\sigma_c}{100}\right)^{1/3}$ MPa	Barton (2002)	γ is the unit weight of the rock mass in g/cm^3.
12	$\dfrac{\sigma_{cm}}{\sigma_c} = JP = 0.2\sqrt{jC} \times V_b^{0.37jC^{-0.2}}$	Palmstrøm (1995)	V_b is the block volume and j_c is the joint condition factor

index values such as RQD, RMR, Q, and GSI. Table 2 summarizes some widely used empirical equations for the determination of uniaxial compressive strengths of jointed rock masses.

It is noted that these empirical equations try to link a rock mass classification index value to the deformation modulus and uniaxial compressive strength of a jointed rock mass. They do not provide a complete description of the failure envelopes which are needed in many design analyses. Although GSI appears a few times in the tables, it is fundamentally associated with the generalized Hoek-Brown failure criterion. This failure criterion, since its inception in 1980 (Hoek & Brown, 1980), has undergone a few revisions (Hoek, 1983; Hoek & Brown, 1988; Hoek et al., 1995, 1998, 2002; Hoek & Brown, 1997) and is widely accepted in the engineering community. In the following discussion, we present some extensions and refinements to the GSI system for the estimation of peak and residual strength parameters of jointed rock mass in the context of the generalized Hoek-Brown failure criterion.

3 ESTIMATION OF PEAK AND RESIDUAL STRENGTH PARAMETERS OF JOINTED ROCK MASS USING THE GSI SYSTEM

3.1 Peak strength parameters

3.1.1 Generalized Hoek-Brown failure criterion

The generalized Hoek-Brown failure criterion for jointed rock masses is (Hoek *et al.*, 2002)

$$\sigma_1 = \sigma_3 + \sigma_c \left(m_b \frac{\sigma_3}{\sigma_c} + s \right)^a \tag{1}$$

where m_b, s, and a are constants for the rock mass, and σ_c is the uniaxial compressive strength of the intact rock. To apply the Hoek-Brown failure criterion for estimating the strength of a jointed rock mass, three properties of the rock mass need to be obtained. The first one is the uniaxial compressive strength of the intact rock σ_c, the second is the value of the Hoek-Brown constant m_i for the intact rock, and the last one is the value of GSI for the rock mass. σ_c and m_i can be determined by statistical analysis of the results of a set of triaxial tests on carefully prepared core specimens. The GSI value can be obtained from a chart provided in Hoek *et al.* (1995) or other relevant references. Once the GSI value is known, other Hoek-Brown parameters m_b, s, a are given as (Hoek *et al.*, 2002)

$$m_b = m_i \exp\left(\frac{GSI - 100}{28 - 14D} \right) \tag{2}$$

$$s = \exp\left(\frac{GSI - 100}{9 - 3D} \right) \tag{3}$$

$$a = 0.5 + \frac{1}{6} \left(e^{-GSI/15} - e^{-20/3} \right) \tag{4}$$

where D is a factor between 0 and 1, which depends on the degree of disturbance to which the rock mass has been subjected by blast damage and stress relaxation. For a tunnel excavated by controlled blasting, manual excavation, or Tunnel Boring Machine (TBM) leading to excellent excavation quality, the disturbance to the confined rock mass surrounding the tunnel is minimal and $D = 0$ can be used. When very poor quality blasting in a hard rock tunnel results in severe local damage in the surrounding rock mass, $D = 0.8$. For very poor blasting in rock slopes leading to severe rock mass damage, $D = 1.0$. The D factor can affect the rock mass strength and deformability significantly and sufficient consideration must be given to the selection of this parameter.

3.1.2 Intact rock strength σ_c

Getting σ_c right is the first step toward getting the rock mass properties right. In addition to σ_c, intact rock properties such as tensile strength σ_t, elastic modulus E_i, and Poisson's ratio v are also needed in design. These parameters can be obtained from laboratory Brazilian test, uniaxial compression test, or triaxial test.

Practical estimate of rock mass strength and deformation parameters for engineering design 509

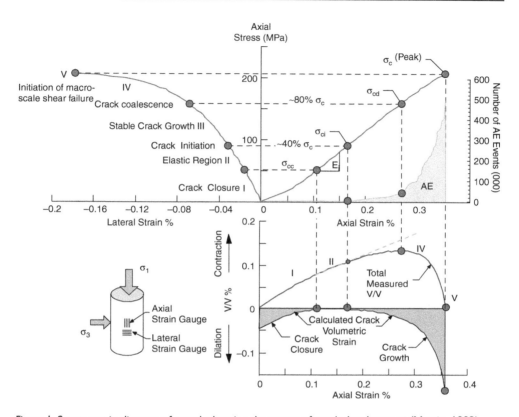

Figure 1 Stress-strain diagram of a rock showing the stages of crack development (Martin, 1993).

A typical stress-strain relations obtained from a uniaxial compression test is presented in Figure 1, where σ_{cc} is the crack closure stress, σ_{ci} is the crack initiation stress, σ_{cd} is the crack damage stress, and σ_c is the peak stress at failure. The three stress thresholds, i.e., σ_{ci}, σ_{cd}, and σ_c, represent important stages in the development of the macroscopic failure process of intact rocks.

In laboratory tests on intact rocks, the crack initiation stress is defined by the onset of stable crack growth or dilatancy, which can be identified from the stress – volumetric strain curve as the point of the departure of the volumetric strain observed at a given mean stress from that observed in hydrostatic loading to the corresponding pressure. Whenever possible, the value of σ_c should be determined by laboratory testing on cores of approximately 50 mm in diameter and 100 mm in length.

One important aspect in obtaining intact rock properties from laboratory tests is sample damage. This is particularly true for engineering design at depth because core samples taken from deep highly stressed ground are very much prone to sample damage, due to stress relaxation. Intact rock strength properties obtained using cores obtained from deep or high stress zones can underestimate the intact rock strength in-situ.

3.1.3 Practical estimate of m_i value

As seen in Equation 2, m_i is a Hoek-Brown model parameter for the intact rock, which can be obtained from triaxial test results (σ_c can also be obtained from triaxial test results). The triaxial test data can be processed using a program called RocData, available from Rocscience Inc.

When time or budget constraints do not allow a triaxial testing program to be carried out, m_i values can be estimated from some tables given in Hoek *et al.* (1995) and Hoek (2007). In the table given by Hoek (2007), possible m_i data ranges are shown by a variation range value immediately following the suggested m_i value. m_i values range from 4 to 33 for some commonly encountered rocks and an impression that m_i depends only on rock type can be seen from the table but this is not true. Rock type cannot be used directly to define the m_i value. The m_i value depends on many factors such as mineral content, foliation, and grain size (texture). There is a large variation range of m_i values and it presents a major challenge for engineers to choose a reasonably accurate m_i value for a particular rock.

Recently, a simple but yet practical method to estimate tensile strength and m_i values of brittle rocks was proposed by Cai (2010). According to the method, the tensile strength σ_t is equal to $\sigma_{ci}/8$; the m_i value is equal to $m_i = 8\,\sigma_c/\sigma_{ci}$ in low confinement zone, where σ_{ci} is the crack initiation stress which can be obtained from the uniaxial compression test, using either volumetric strain measurement or acoustic emission monitoring techniques (see Figure 1). It is found that $m_i = 12\,\sigma_c/\sigma_{ci}$ can be applied to strong, brittle rocks, applicable to high confinement zone. Because rock mass failure around excavation boundaries is governed by low confinement conditions, $m_i = 8\,\sigma_c/\sigma_{ci}$ is recommended for the estimation of m_i values.

m_i values inferred from the literature can only be used when there are no test data available at the initial design stage of a project. Whenever possible, simple laboratory uniaxial compression tests should be conducted to determine σ_{ci}, σ_c, and hence m_i values more accurately.

3.1.4 Quantitative determination of GSI value

Having σ_c and m_i values determined, we need to define the GSI value in order to use the generalized Hoek-Brown failure criterion for jointed rock masses. As discussed above, the GSI system has been developed and evolved over many years based on practical experience and field observations. The GSI value is estimated based on geological descriptions of the rock mass involving two factors, rock structure or block size and joint or block surface conditions (Hoek & Brown, 1997). Although careful consideration has been given to the precise wording for each category and to the relative weights assigned to each combination of structural and surface conditions, the use of the original GSI table/chart involves some subjectivity and long-term experiences and sound judgment are required to use the GSI system successfully.

A means to quantify the GSI chart by use of field measurement data, which employs the block volume V_b and a joint surface condition factor J_c as quantitative characterization factors, is presented by Cai *et al.* (2004). The resulting approach adds quantitative measures in an attempt to render the system more objective. By adding measurable quantitative input for quantitative output, the system becomes less dependent on experience while maintaining its overall simplicity. The block volume can be calculated from joint spacings of joint sets. The effect of joint persistence on the block volume can be considered using a joint persistence factor. The joint surface condition factor is obtained by rating joint roughness depending on the large-scale waviness, small-scale smoothness of joints, and joint alteration depending on the weathering and infillings in joints. The quantitative approach was validated using field test data and applied to the estimation of the rock mass properties at some project sites (Alejano *et al.*, 2009; Fischer *et al.*, 2010; Hashemi *et al.*, 2010; Gischig *et al.*, 2011; Ghafoori *et al.*, 2011; Soleiman Dehkordi *et al.*, 2013). This approach adds quantitative means to assist in the selection of modeling parameters and is of particular interest to site engineers.

3.1.4.1 Block volume

Block size, which is determined from the joint spacing, joint orientation, number of joint sets and joint persistence, is an extremely important indicator of rock mass quality. The block volume can be calculated from

$$V_b = \frac{s_1 s_2 s_3}{\sin \gamma_1 \sin \gamma_2 \sin \gamma_3 \sqrt[3]{p_1 p_2 p_3}} \tag{5}$$

where s_i, γ_i and p_i are the joint spacing, the angle between joint sets, and joint persistence factor, respectively. If the joints are not persistent, *i.e.*, with rock bridges, the rock mass strength is higher and the global rock stability is enhanced. This effect can be considered using the concept of equivalent block volume as suggested in Cai *et al.* (2004). The consideration of joint persistency has been verified using numerical simulation by UDEC and 3DEC (Kim *et al.*, 2007). For persistent joint sets, $p_i = 1$. Blocks defined by three joint sets are shown in Figure 2.

Random joints may affect the shape and size of the block. Statistically, joint spacing follows a negative exponential distribution. For a rhombohedral block, the block volume is usually larger than that of cubic blocks with the same joint spacings. However, compared with the variation in joint spacing, the effect of the intersection angle between joint sets is relatively small. Hence, for practical purpose, for a rock mass containing persistent joint sets, the block volume can be approximated as

$$V_b = s_1 s_2 s_3 \tag{6}$$

Traditional methods for obtaining discontinuity data (joint sets, orientation, spacing, length, etc.) in the field include core/borehole logging, scanline survey, and cell mapping. Core/borehole logging alone cannot provide joint length information so that face mapping is needed to compensate. Scanline surveys, which are time consuming, provide detailed information on the individual joint in each set that can be used in probabilistic design, whereas cell mapping, which are easier and more efficient, only provides average information about each joint set. Decisions have to be made to select the most appropriate method to obtain the required information for block volume and joint surface condition factor (see Section 3.1.4.2) determination.

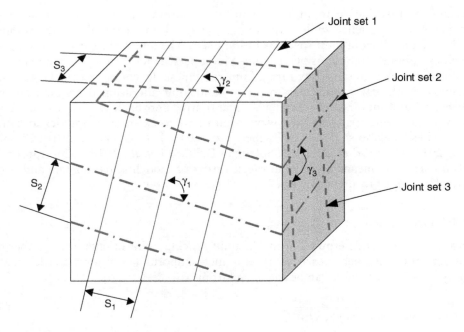

Figure 2 Block delimited by three joint sets.

3.1.4.2 Joint condition factor

In the GSI system, the joint surface condition is defined by the roughness, weathering, and infilling condition (Hoek et al., 1995; Cai et al., 2004). The combination of these factors defines the strength of a joint or block surface. The joint condition factor, J_c, used to quantify the joint surface condition, is defined as

$$J_C = \frac{J_W \cdot J_S}{J_A} \tag{7}$$

where J_W and J_S are the large-scale waviness (in meters from 1 to 10 m) and small-scale smoothness (in centimeters from 1 to 20 cm) and J_A is the joint alteration factor. The ratings for J_W, J_S, and J_A are listed in Table 3, Table 4, and Table 5, respectively.

3.1.4.3 Peak GSI value and strength parameters

The quantified GSI chart is presented in Figure 3. The descriptive block size is supplemented with the quantitative block volume (V_b) and the descriptive joint condition is supplemented with the quantitative joint condition factor (J_c). The influence of V_b and J_c on GSI was calibrated using published data (Cai et al., 2004).

Once V_b and J_c are determined, users can use Figure 3 or the following equation (Cai & Kaiser, 2006b) to determine the peak GSI value.

Table 3 Terms to describe large-scale waviness (Palmstrøm, 1995).

Waviness terms	Undulation	Rating for waviness J_W
Interlocking (large-scale)		3
Stepped		2.5
Large undulation	> 3 %	2
Small to moderate undulation	0.3 – 3 %	1.5
Planar	< 0.3 %	1

Undulation = a/D
D – length between maximum amplitudes

514   Cai

Table 4 Terms to describe small-scale smoothness (Palmstrøm, 1995).

Smoothness terms	Description	Rating for J_s
Very rough	Near vertical steps and ridges occur with interlocking effect on the joint surface	3
Rough	Some ridge and side-angle are evident; asperities are clearly visible; discontinuity surface feels very abrasive (rougher than sandpaper grade 30)	2
Slightly rough	Asperities on the discontinuity surfaces are distinguishable and can be felt (like sandpaper grade 30 − 300)	1.5
Smooth	Surface appear smooth and feels so to the touch (smoother than sandpaper grade 300)	1
Polished	Visual evidence of polishing exists. This is often seen in coating of chlorite and specially talc	0.75
Slickensided	Polished and striated surface that results from sliding along a fault surface or other movement surface	0.6 − 1.5

Table 5 Rating for the joint alteration factor J_A (Barton et al., 1974; Palmstrøm, 1995).

	Term	Description	JA
Rock wall contact	**Clear joints** Healed or "welded' joints (unweathered)	Softening, impermeable filling (quartz, epidote etc.)	0.75
	Fresh rock walls (unweathered)	No coating or filling on joint surface, except for staining	1
	Alteration of joint wall: slightly to moderately weathered	The joint surface exhibits one class higher alteration than the rock	2
	Alteration of joint wall: highly weathered	The joint surface exhibits two classes higher alteration than the rock	4
Filled joints with partial or no contact between the rock wall surfaces	**Coating or thin filling-** Sand, silt, calcite etc.	Coating of frictional material without clay	3
	− Clay, chlorite, talc etc.	Coating of softening and cohesive minerals	4
	− Sand, silt, calcite etc.	Filling of frictional material without clay	4
	− Compacted clay materials	"Hard" filling of softening and cohesive materials	6
	− Soft clay materials	Medium to low over-consolidated of filling	8
	− Swelling clay materials	Filling material exhibits swelling properties	8 − 12

$$GSI(V_b, J_c) = \frac{26.5 + 8.79 \ln J_c + 0.9 \ln V_b}{1 + 0.0151 \ln J_c - 0.0253 \ln V_b} \qquad (8)$$

where J_c is a dimensionless factor, and V_b is in cm^3. With GSI directly expressed as a function of V_b and J_c, the Hoek-Brown strength parameters (m_b, s, a) can also be directly expressed as a function of V_b and J_c. This convenience can facilitate the use of

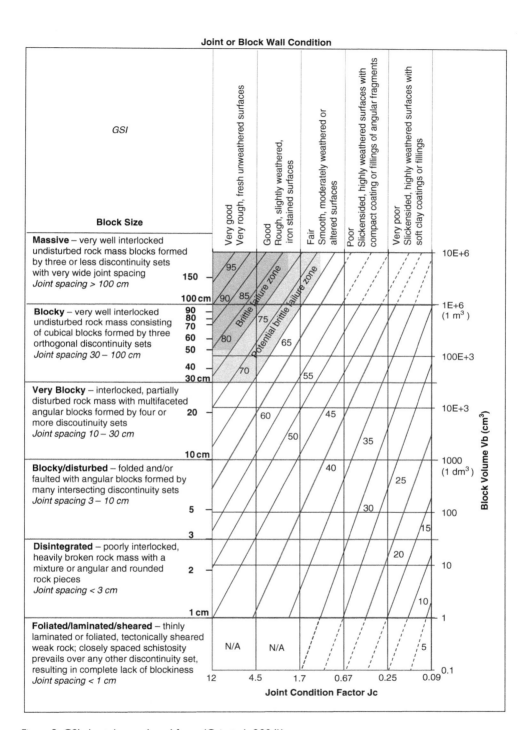

Figure 3 GSI chart (reproduced from (Cai et al., 2004)).

516 Cai

probabilistic design approach to tunnel and cavern design using the GSI system (Cai & Kaiser, 2006a; Cai, 2011).

3.2 Residual strength parameters

3.2.1 Generalized Hoek-Brown failure criterion for residual strength

It is observed that a rock mass in its residual state represents one particular kind of rock mass in the spectrum in the GSI chart. The rock mass spectrum is defined by the combination of the block volume spectrum and the joint surface condition factor spectrum. The generalized Hoek-Brown failure criterion for the residual strength of jointed rock masses can be written as

$$\sigma_1 = \sigma_3 + \sigma_c \left(m_r \frac{\sigma_3}{\sigma_c} + s_r \right)^{a_r} \tag{9}$$

where m_r, s_r, a_r are the residual Hoek-Brown constants for the rock mass. As for the intact rock properties, fracturing and shearing do not weaken the intact rocks so that the mechanical parameters (σ_c and m_i) should be unchanged. Cai et al. (2007) considered that constants m_r, s_r, and a_r can be determined from a residual GSI_r value using the same equations for peak strength parameters. Because the rock masses are in a damaged, residual state, $D = 0$ is used for the residual strength parameter calculation. According to the logic of the original GSI system, the strength of a rock mass is controlled by its block size and joint surface condition. The same concept is valid for fractured rock masses at their residual strength state. In other words, the residual GSI_r is a function of residual joint surface condition factor J_c^r and residual block volume V_b^r (Cai et al., 2007).

3.2.2 Residual GSI value and strength parameters

Once the residual block volume and joint surface condition factor are obtained, one can refer to the GSI chart or use the following equation to obtain the residual GSI value

$$GSI_r(V_b^r, J_c^r) = \frac{26.5 + 8.79 \ln J_c^r + 0.9 \ln V_b^r}{1 + 0.0151 \ln J_c^r - 0.0253 \ln V_b^r} \tag{10}$$

where J_c^r is a dimensionless factor, and V_b^r is in cm^3.

If a rock experiences post-peak deformation with sufficient straining, the rock in the broken zone is fractured and consequently turned into a "poor" and eventually "very poor" rock. For the residual block volume, it is observed that the post-peak block volumes are small because the rock mass has experienced tensile and shear fracturing with sufficient straining. After the peak load and with sufficient straining, the rock mass becomes less interlocked, and is heavily broken with a mixture of angular and partly-rounded rock pieces. Detailed examination of the rock mass damage state before and after the in-situ block shear tests at some underground cavern sites revealed that in areas that were not covered by concrete, the failed rock mass blocks were 1 to 5 cm in size. The rock mass was disintegrated along a shear zone in these tests. Hence, the residual block volumes can be considered independent of the original (peak) block volumes for most

Practical estimate of rock mass strength and deformation parameters for engineering design 517

strain-softening rock masses. The fractured residual rock mass will have more or less the same residual block volume in the shear band for intact rocks, moderately jointed and highly jointed rock masses. As an estimate, Cai *et al.* (2007) recommended that if the peak block volume V_b is greater than 10 cm^3, then, the residual block volume V_b^r in the disintegrated category can be taken to be 10 cm^3. If V_b is smaller than 10 cm^3, then, no reduction to the residual block volume is recommended, *i.e.*, $V_b^r = V_b$.

The major factor that alters the joint surface condition in the post-peak region is the reduction of joint surface roughness. Using the concept of ultimate mobilized joint roughness suggested by Barton *et al.* (1985), the large-scale waviness and the small-scale smoothness of joints can be calculated by reducing its peak value by half to calculate the residual *GSI* value. The residual joint surface condition factor J_c^r can be calculated from (Cai *et al.*, 2007)

$$J_c^r = \frac{J_W^r \cdot J_S^r}{J_A^r} \tag{11}$$

where J_W^r, J_S^r, and J_A^r are residual values for large-scale waviness, small-scale smoothness, and joint alteration factor, respectively. The residual values are obtained based on the corresponding peak values assessed from field mapping. The reduction of J_W^r and J_S^r, from their peak values J_W and J_S, are based on the concept of mobilized joint roughness, and the equations are given as

$$\text{If } \frac{J_W}{2} < 1, \ J_W^r = 1; \quad \text{Else } J_W^r = \frac{J_W}{2} \tag{12}$$

$$\text{If } \frac{J_S}{2} < 0.75, \ J_S^r = 0.75; \quad \text{Else } J_S^r = \frac{J_S}{2} \tag{13}$$

In a short period, joint alteration is unlikely to occur so that the joint alteration factor J_A can be considered as unchanged in most circumstances.

3.3 Discussion on the use of the generalized Hoek-Brown failure criterion

The generalized Hoek-Brown failure criterion is only applicable to intact rock or to heavily jointed rock masses that can be roughly considered as homogenous and isotropic. The criterion should not be applied to highly schistose rocks such as slates or to rock masses in which the properties are controlled by a single set of discontinuities such as bedding planes (Hoek *et al.*, 1995). The criterion works well for rock masses at low confinement conditions and it should not be used for defining rock mass strength at very high confinement conditions.

Because of the inherent uncertainty of the intact rock properties and jointing in the rock mass, any estimate of rock mass strength parameters from using the Hoek-Brown failure criterion should not be considered as final. The approach is particularly useful in the preliminary design stage where only limited ranges of site characterization and test data are available. As the project progresses, field monitoring should be conducted to verify previously obtained approximate estimates of rock mass strength parameters. This is a process important for proper use of not only the Hoek-Brown failure criterion but also other failure criteria.

Pelli *et al.* (1991) found that the parameters obtained from Equations 2 and 3 did not predict the observed failure locations and extent near a tunnel in a cemented sand or siltstone. They found that lower m_b and higher s values were required to match predictions with observations. Further analyses of underground excavations in brittle rocks eventually lead to the development of brittle Hoek-Brown parameters (Martin *et al.*, 1999; Kaiser *et al.*, 2000; Diederichs, 2007) for massive to moderately fractured rock masses with tight interlock that fail by spalling or slabbing rather than by shear failure. Accordingly, Equations 2 and 3 are clearly not applicable for GSI > 75 in massive to moderately or discontinuously jointed hard rocks. The zone of anticipated brittle failure conditions is highlighted in Figure 3 by the hatched near the upper left corner.

The Hoek–Brown failure criterion was initially derived based on triaxial test data of intact rocks. Many data in the high confinement range were included and hence the criterion assumed a shear failure mechanism by default. The generalization to jointed rock mass also inherited this assumption of shear failure mechanism. Hence, care must be given when using the criterion outside the range of applicability of the assumptions and data on which it was based, such as the modeling of brittle failure of hard rocks in low confinement conditions. Brittle failure of hard massive rocks is governed by a process of gradual cohesion loss and friction mobilization (Martin, 1997; Kaiser *et al.*, 2000; Hajiabdolmajid *et al.*, 2002). The fundamental mechanism of this is tensile crack (Griffith crack) initiation, propagation, and coalescence in low to zero confinement environments. Using conventional strength parameters derived from Hoek–Brown criterion to model brittle rock failure was found less useful and less successful because the failure zones around excavations could not be predicted satisfactorily. Specific brittle failure parameters, including both peak and residual ones, are required to model brittle failure of massive rocks adequately (Cai & Kaiser, 2014).

The generalized Hoek–Brown criterion may not be applied to weak rocks with, for σ_c < 15 MPa, because it has been found that, at these low strengths, the index, a, can be greater than the maximum value of about 0.65 given by Equation 4 and can approach one (Brown, 2008).

4 ESTIMATION DEFORMATION MODULUS

The deformation modulus of the jointed rock mass is required when carrying out numerical analysis in design. Traditional method to determine the deformation modulus is through in-situ plate loading tests or using back analysis based on measured displacements of excavations. As discussed in Section 2.2, there are many empirical equations to correlate rock mass deformation modulus with some rock mass classification indexes. One important thing to remember is the applicable boundary conditions for each individual equation (as show in Table 1 of the "Note" column). When the boundary is crossed, meaningless deformation modulus can be obtained.

Blasting tends to loosen rock mass and reduce its deformation modulus. Hence, including the factor D in the empirical equation allows us to consider the effects of blast damage and stress relaxation. The deformation modulus is related to the GSI value as (Hoek *et al.*, 2002)

Practical estimate of rock mass strength and deformation parameters for engineering design 519

$$E = \left(1 - \frac{D}{2}\right)\sqrt{\frac{\sigma_c}{100}}10^{\left(\frac{GSI-10}{40}\right)}, \ (GPa) \ \text{for} \ \sigma_c < 100\text{MPa} \tag{14}$$

The inclusion of σ_c in Equation 14 shows indirectly the influence of the modulus of the intact rock (E_i) on the deformation modulus of the rock mass, because there is a good correlation between E_i and σ_c (Deere, 1968). A more recent update on the deformation of modulus of jointed rock mass is (Hoek & Diederichs, 2006)

$$E = E_i\left(\frac{1 - D/2}{1 + e^{((75+25D-GSI)/11)}}\right) \tag{15}$$

Equation 15 considers the influence of intact rock modulus directly and avoids unrealistically high values of rock mass deformation moduli when the GSI value is high. In general, the rock mass deformation moduli can be highly anisotropic, and are also confining stress dependent (Barton, 2002; Cai & Kaiser, 2002). Those features need to be properly addressed in order to correctly predict deformation distribution in the rock mass around excavations. Unfortunately, there exist no simple equations that relate the deformation modulus to confining stress.

5 DILATION ANGLE

When a rock or a rock mass fails, its volume increases and this phenomenon is known as dilation. The excavation-induced rock failure and displacement near an underground opening boundary is closely associated with rock mass dilation. A better understanding of rock mass dilation around the excavation helps us to predict or anticipate displacements and failure extent and shape, and subsequently assist the design of proper ground support systems.

In addition to rock mass strength (both peak and residual) and deformation modulus, most numerical tools (*e.g.*, Phase2, FLAC) require another important input parameter – dilation angle. The dilation angle is not only a suitable parameter for the description of soil dilation, but also appears to be useful for rocks to describe rock dilation.

However, in rock engineering, when the dilation angle is taken into consideration, especially for numerical modeling studies, the approach employed by most researchers is often simplistic; it is generally assumed as either one of the two constants – zero in a non-associated flow rule and the same as the friction angle in an associated flow rule. In the most popular failure criteria, such as linear Mohr-Coulomb failure criterion and non-linear Hoek-Brown failure criterion, the rock dilation is assumed to remain as a constant when the rock mass is deformed.

Hoek and Brown (1997), based on wide engineering experience, suggest the use of constant dilation angle values that are dependent on rock mass quality. For very good rock, they recommended that the dilation angle is about 1/4 of the friction angle; for the average quality rock, the value suggested is 1/8, and poor rock seems to have a negligible dilation angle.

In reality, a constant dilation angle is an approximation that is clearly not physically sound. This constant dilation assumption is made largely because little is known about how the dilation of a rock changes past peak load. Some researchers (Detournay, 1986;

Alejano & Alonso, 2005) illustrate that it may be unrealistic and misleading to use a constant dilation angle. They also point out that dilation angle should be a function of plastic parameters and confining stress.

A few dilation models have been proposed for rocks, considering the influence of plastic strain (Detournay, 1986) and confining stress (Alejano & Alonso, 2005). A more recent empirical mobilized dilation angle model considers the influence of both confining stress and plastic shear strain (Cai & Zhao, 2010; Zhao & Cai, 2010a,b). The empirical dilation angle model was derived based on published data acquired from modified triaxial compression tests with volumetric strain measurement. Based on the model response and in combination with the grain size description and the uniaxial compressive strength of rocks, the model parameters for four rock types (coarse-grained hard rock, medium-grained hard rock, fine-medium-grained soft rock, and fine-grained soft rock) are suggested. New test data (Arzua & Alejano, 2013) support the validity of the mobilized dilation angle model.

For jointed rock masses, it is suggested to estimate the peak dilation angle from the peak friction angle of rock mass determined by the GSI system (Cai & Zhao, 2010; Zhao & Cai, 2010b). It is also assumed that the dilation behavior of jointed rock masses follow similar trend as observed for intact rocks so that the empirical relations established for intact rocks can be applied to jointed rock masses. In this fashion, plastic strain and confinement dependent dilation angles can be defined for jointed rock masses. One example is presented in Figure 4. The dilation angle is zero when there is no plastic deformation; it increases rapidly and reaches a peak value at a small plastic deformation when the confinement is low. When confining stress increases, a general trend is that the peak dilation angles decrease and the locations of peak dilation angle gradually shift toward right with more plastic shear straining. Confinement drastically reduces rock dilation. For example, a 5 MPa confinement can reduce the peak dilation angle of the intact rock from 53° at zero confinement to about 12°. As plastic deformation continues, the dilation angles decrease gradually until an asymptotic low value is reached. This makes sense as dilation rate will reduce as the rocks deform. At zero confinement, the peak dilation angle of the intact rock reduces from 53° to 38° for a jointed sandstone with GSI = 50.

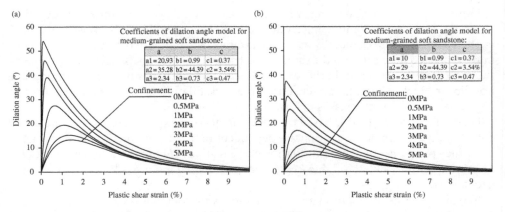

Figure 4 (a) Dilation angle variation for intact rock (medium-grained soft sandstone); (b) Dilation angle variation for a jointed sandstone with GSI = 50.

Practical estimate of rock mass strength and deformation parameters for engineering design 521

The mobilized dilation angle model can be easily implemented into some numerical tools such as FLAC and FLAC3D. Displacement distributions obtained from using the dilation angle model are more reasonable, when compared with the general trend measured underground. The generation of large deformations near the excavation boundary is attributed to the existence of low to zero confinements. The displacement decreases rapidly as confinement increases. Rock dilation behavior near the excavation boundary such as this can only be properly simulated when a dilation angle model, which considers the influence of both plastic shear strain and confinement, is used.

6 APPLICATION EXAMPLE

One of the long-standing challenges in analyzing rock mass strength and deformation data is attributed to the fact that these values are quite variable. The intact rock strength, joint spacing, and joint surface condition vary even within the same rock type zone. In rock engineering practice, geological and geotechnical data, because of the huge cost involved in their acquisition, are often incomplete and hence contain uncertainties. Uncertainties are inherent and unavoidable in the rock mass classification/characterization process. Common sources of uncertainties in rock engineering include the spatial and temporal variability of the rock mass properties; random and systematic errors in data mapping, logging, testing, and monitoring; analytical and numerical model simplification; human omissions and errors. In engineering design, the appropriate approach is to cope with the uncertainties, to assess and manage the risk associated with them, *i.e.*, to incorporate uncertainties into the design and decision-making process.

One advantage of the quantitative GSI system approach is that the variability of inherent parameters can be explicitly considered in the calculation process (Cai & Kaiser, 2006a). The closed-form solution to obtain GSI values from dependent variables such as joint spacing, orientation, persistence, surface condition factor, etc., makes it suitable for probabilistic analysis using the Monte Carlo method. The variability of strength and deformation parameters can be implemented in the design tools to calculate the variability of stress and deformations as well as anticipated loads in rockbolts and anchors.

To apply the GSI system for rock mass characterization, two groups of parameters need to be determined. One is the intact rock parameters, which includes σ_c and m_i. Another is the joint parameters, which is further divided into the joint geometry and strength subgroups. All these parameters can be considered as random variables. In general, a normal distribution with the mean and the coefficient of variation (COV) can be used to describe the probability distribution of σ_c and m_i.

Priest and Hudson (1976) stated that statistically, joint spacing follows a negative exponential distribution. However, some researchers consider that the joint spacing distribution is logarithmic. If the interaction of jointing corresponds to the multiplicatory process, lognormal distribution may result (Dershowitz & Einstein, 1988). The type of distribution seems to be affected by the minimum bin size used in the histogram analysis of joint spacing. In the example shown here, lognormal distribution is applied for joint spacing.

Joint orientation affects both the block shape and size, and it is usually defined by joint dip and dip direction. Joint orientations are stochastic but quite often cluster in preferred orientations to form joint sets. The joint orientation variability is thus governed by the degree of clustering within each set. The Fisher distribution is often used to describe the joint orientation distribution.

Joint sizes and trace lengths vary even a wide range and are difficult to be determined accurately. Several distributions, such as exponential, lognormal, hyperbolic, Gamma-1 distributions, have been proposed to describe the joint trace lengths. The variability in joint size includes the inherent natural randomness of this property and our limited ability to measure or model this property, rendering it one of the most difficult parameters in joint system modeling. Priest and Hudson (1981) used the negative exponential distribution to describe the joint lengths. Depending on the problem scale and bin size used, both negative exponential distribution and lognormal distribution are appropriate to describe the joint length distribution.

The joint surface condition factor is a measure of the joint strength against shearing. In practice, when the small-scale joint smoothness is, on average, "rough," there are possibilities that some portions of the joint are "very rough" while other portions are "slightly rough." This uncertainty can result from both spatial variability of the joint surface condition and human discrepancy in field mapping. A decision therefore has to be made about the rating range and the distribution type. For simplicity, the normal distribution can be used to represent the joint roughness and alteration variability. When the mean values are near the extreme values in the rating, a truncated normal distribution can be used.

In the following illustration example, we consider the rock mass classification for a rock mass at a large-scale hydropower cavern site in Japan and assume that the joint spacing follows a lognormal distribution. Details about the cavern construction project can be found in Koyama $et\ al.$ (1997). Three orthogonal joint sets exist and the average joint spacing for the three joint sets is 10, 25, and 50 cm and the standard deviations are 3, 7.5, and 15 cm, respectively. Using the Monte Carlo simulation technique available in @RISK, the probability distribution function (PDF) of the block volume for the rock mass is calculated using 5000 iterations and the result is presented in Figure 5(a). It is seen that the block volume follows a lognormal distribution.

It was determined from field mapping that the average values for large-scale joint waviness J_w, small-scale smoothness J_s, and joint alteration J_A are 2, 2, and 1, respectively, and the coefficients of variation for all three factors were assumed to be 8%. Truncated normal distributions are assumed for J_w, J_s, and J_A. The truncation is based on the minimum and maximum ratings for each parameter. For example, J_w should not be less than 1 and not greater than 3. J_w is thus described by a normal distribution with a mean of 2 and standard deviation of 0.16, truncated at 1 and 3. J_c thus calculated also follows a normal distribution as shown in Figure 5(b).

Equation 9 is used to calculate the GSI distribution based on V_b and J_c, again using the Monte Carlo method. Although V_b follows a lognormal distribution, the calculated GSI values follow a normal distribution, as shown in Figure 12(c). The average GSI is 59.9 with a standard deviation of 2.1. The probability density distributions for the Hoek-Brown strength parameters m_b and s are presented in Figure 12(d & e), and it is found that the m_b values follow a normal distribution. Although the s values are best

Practical estimate of rock mass strength and deformation parameters for engineering design 523

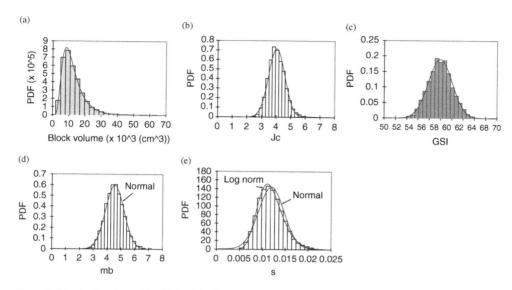

Figure 5 (a) Block volume Vb; (b) Jc; (c) GSI distributions; (d) m_b; (e) s simulated using @ RISK.

described by a lognormal distribution, it can also be approximated by a normal distribution as shown in the figure. It is seen from the results that the *GSI* value and hence the mechanical properties of the jointed rock masses exhibit variability. These properties are not just the average values, but have a distribution about the means, even under ideal conditions. The design will make more sense if these property distributions are properly considered.

Based on the variability information about the rock mass strength and deformability, stability analysis can be performed using the point estimate method (PEM) (Rosenblueth, 1981) in combination with a FEM analysis program, which considers the possible combinations of strength and deformation parameters as well as in-situ stress. PEM is an alternative to Monte Carlo simulation with models containing a limited number of uncertain inputs. In this method, the model is evaluated at a discrete set of points in the uncertain parameter space, with the mean and variance of model predictions computed using a weighted average of these functional evaluations.

As the output of the analysis, the probability distributions of yielding or loosening zones and total displacements in the roof and on the sidewalls of the cavern can be obtained. As an example, the probability density function of the yielding zone distribution around the cavern is presented in Figure 6, with the consideration of material variability alone. 15 m long pre-stressed (PS) anchors had been selected for cavern support. It is seen that the probability that the 15 m anchor may be shorter than the yielding zone depth on the right sidewall is 0.0077 %, and the probability that the anchor's 3 m anchorage length may fall in the yielding zone is 1.3 %. It is obvious that cost saving in terms of reducing the support quantities is achievable if a certain level of risk is acceptable. The simulation results allow us to better understand how uncertainty arises and how the rock support system design decision may be affected by it.

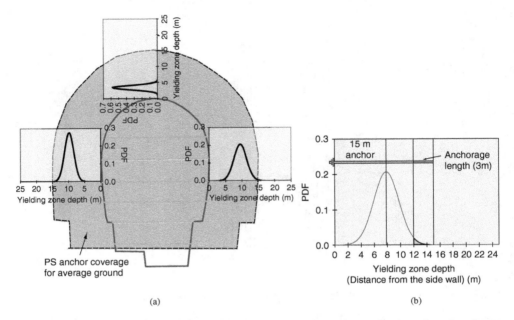

Figure 6 (a) Probability distributions of the yielding zone depths in the roof and on the sidewalls; (b) enlarged figure of probability distribution on the right sidewall. The cavern dimensions are: width 34 m, height 54 m, located 500 m underground.

7 CONCLUSION

Different from other rock mass classification systems, the GSI system is directly linked to engineering parameters such as Mohr-Coulomb or Hoek-Brown strength parameters or rock mass deformation modulus. The original GSI system, which is applied mainly for the estimation of the peak strength, is based on a descriptive approach, rendering the system somewhat subjective and difficult to use for inexperienced personnel. To assist the use of the GSI system, a supplementary quantified approach for the GSI system, which incorporates quantitative measures of block volume and joint surface condition factor, can be used. The block volume can be calculated, in most cases, from joint spacings of three dominant joint sets. The joint condition factor is obtained by rating joint roughness depending on the large-scale waviness, small-scale smoothness of joints, and joint alteration depending on the weathering and infillings in joints.

The concept of residual block volume V_b^r and residual joint surface condition factor J_c^r was used to extend the GSI system for the estimation of rock mass's residual strength. The residual strength parameters can be calculated using the same form of the generalized Hoek-Brown strength criterion by assuming that the intact rock properties such as σ_c and m_i remain unchanged as the rock mass changes from its peak state to its residual state.

The quantitative approach for peak and residual strength estimation extends the GSI system and adds quantitative means to determine the complete set of rock mass strength properties needed for design. In addition, the approach is built on the linkage between descriptive geological terms and measurable field parameters such as joint spacing and joint roughness, which are random variables. Because of its quantitative nature, it allows the evaluation of both the means and variances of strength and deformation parameters, using the Monte Carlo or point estimate method.

When using the generalized Hoek-Brown strength criterion in design, we should pay not only more attention to the determination of GSI values more objectively, but also sufficient attentions to the determination of other parameters such as σ_c, m_i, D, and dilation angle. It is suggested that at least simple uniaxial compression tests should be conducted to obtain σ_c and m_i values more accurately. Test specimens should be strain-gauged to define the crack initiation stress level σ_{ci} and then the m_i value can be estimated using $m_i = 8\,\sigma_c/\sigma_{ci}$.

Although this chapter provides a contemporary method for obtaining rock mass mechanical parameters needed in engineering design, its successful application relies heavily on the professional judgment, as is typically the case in rock mechanics and rock engineering.

ACKNOWLEDGMENTS

The author wishes to thank Tokyo Electric Power Services Co. Ltd (TEPSCO) and Tokyo Electric Power Company (TEPCO) for their financial support to this study, and the contributions and constructive comments provided by Dr. P. Kaiser of Laurentian University (Emeritus Professor), Dr. E. Hoek of Evert Hoek Consulting Engineer Inc., Dr. D. McCreath of Laurentian University (Emeritus Professor), Mr. Y. Tasaka and Dr. H. Uno of TEPSCO, Mr. M. Minami of TEPCO, and Dr. X. Zhao of Beijing Research Institute of Uranium Geology.

REFERENCES

Alejano, L. R. & Alonso, E. (2005) Considerations of the dilatancy angle in rocks and rock masses. Int. J. Rock Mech. Min. Sci. 42(4), 481–507.

Alejano, L. R., Rodriguez-Dono, A., Alonso, E., & Fdez.-Manin, G. (2009) Ground reaction curves for tunnels excavated in different quality rock masses showing several types of post-failure behaviour. Tunnelling Underground Space Technol. 24(6), 689–705.

Amadei, B. (1988) Strength of a regularly jointed rock mass under biaxial and axisymmetric loading. Int. J. Rock Mech. Min. Sci. Geomech. Abstr. 25(1), 3–13.

Arzua, J., & Alejano, L. R. (2013) Dilation in granite during servo-controlled triaxial strength tests. Int. J. Rock Mech. Min. Sci. 61, 43–56.

Aydan, O. & Dalgiç, S. (1998) Prediction of deformation behavior of 3-lanes Bolu tunnels through squeezing rocks of North Anatolian fault zone (NAFZ). In: Proceedings of Regional Symposium on Sedimentary Rock Engineering, Taipei, 228–233.

Barton, N. (2002) Some new Q-value correlations to assist in site characterisation and tunnel design. Int. J. Rock Mech. Min. Sci. 39(2), 185–216.

Barton, N. R., Bandis, S. C. & Bakhtar, K. (1985) Strength, deformation and conductivity coupling of joints. Int. J. Rock Mech. Min. Sci. Geomech. Abstr. 22(3), 121–140.

Barton, N. R., Lien, R. & Lunde, J. (1974) Engineering classification of rock masses for the design of tunnel support. Rock Mech. 6(4), 189–239.

Barton, N. R., Loset, F., Lien, R. & Lunde, J. (1980) Application of the Q-system in design decisions concerning dimensions and appropriate support for underground installations. In: Int. Conf. Subsurface Space, Rockstore, Stockholm, 553–561.

Bhasin, R. & Grimstad, E. (1996) The use of stress-strength relationships in the assessment of tunnel stability. Tunnelling Underground Space Technol. 11(1), 93–98.

Bieniawski, Z. T. (1976) Rock mass classification in rock engineering. In: Proc. Symp. on Exploration for Rock Engineering, Balkema, Cape Town, 97–106.

Bieniawski, Z. T. (1978) Determining rock mass deformability – experience from case histories. Int. J. Rock Mech. Min. Sci. Geomech. Abstr. 15(5), 237–247.

Brown, E. T. (2008) Estimating the mechanical properties of rock masses. In: SHIRMS 2008, 3–21.

Cai, M. (2010) Practical estimates of tensile strength and Hoek-Brown strength parameter mi of brittle rocks. Rock Mech. Rock Eng. 43(2), 167–184.

Cai, M. (2011) Rock mass characterization and rock property variability considerations for tunnel and cavern design. Rock Mech. Rock Eng. 44(4), 379–399.

Cai, M. & Horii, H. (1992) A constitutive model of highly jointed rock masses. Mech. Mater. 13, 217–246.

Cai, M. & Kaiser, P. K. (2002) Dependency of wave propagation velocity on rock mass quality and confinement. In: NARMS 2002, University of Toronto Press, 615–622.

Cai, M. & Kaiser, P. K. (2006a) Rock mass characterization and rock mass property variability considerations for tunnel and cavern design. In: Proc. 4th Asian Rock Mech. Symp. (ARMS 4), World Scientific, Singapore, Paper 144.

Cai, M. & Kaiser, P. K. (2006b) Visualization of rock mass classification systems. Geotech. Geol. Eng. 24(4), 1089–1102.

Cai, M. & Kaiser, P. K. (2014) In-situ rock spalling strength near excavation boundaries. Rock Mech. Rock Eng. 47(2), 659–675.

Cai, M., Kaiser, P. K., Tasaka, Y. & Minami, M. (2007) Determination of residual strength parameters of jointed rock masses using the GSI system. Int. J. Rock Mech. Min. Sci. 44(2), 247–265.

Cai, M., Kaiser, P. K., Uno, H., Tasaka, Y. & Minami, M. (2004) Estimation of rock mass strength and deformation modulus of jointed hard rock masses using the GSI system. Int. J. Rock Mech. Min. Sci. 41(1), 3–19.

Cai, M., Morioka, H., Kaiser, P. K., Tasaka, Y., Minami, M. & Maejima, T. (2007) Back analysis of rock mass strength parameters using AE monitoring data. Int. J. Rock Mech. Min. Sci. 44(4), 538–549.

Cai, M. & Zhao, X. G. (2010) A confinement and deformation dependent dilation angle model for rocks. In: Proc. 44th US Rock Mech. Symp. and 5th U.S.-Canada Rock Mech. Symp., Paper 459, Salt Lake City, UT.

Cividini, A., Gioda, G. & Jurina, L. (1981) Some aspects of 'characterization' problems in geomechanics. Int. J. Rock Mech. Min. Sci. Geomech. Abstr. 18, 487–503.

Coon, R. F. & Merritt, A. H. (1970) Predicting in situ modulus of deformation using rock quality indices. In: Determination of the In Situ Modulus of Deformation of Rock, ASTM STP 477, 154–173.

Deere, D. U. (1968). "Geological consideration." Rock Mechanics in Engineering Practice, K. G. Stagg and O. C. Zienkiewicz, eds., John Wiley & Sons, New York, 1–20.

Deere, D. U., Hendron, A. J., Patton, F. D. & Cording, E. J. (1967) Design of surface and near surface construction in rock. In: Failure and breakage in rock, 237–302.

Dershowitz, W. S. & Einstein, H. H. (1988) Characterizing rock joint geometry with joint system models. Rock Mech. Rock Eng. 21(1), 21–51.

Dershowitz, W. S., Lee, G., Geier, J., Hitchcock, S. & la Pointe, P. (1993). *Support of underground excavations in hard rock*. Golder Associates, Seattle.

Detournay, E. (1986) Elastoplastic model of a deep tunnel for a rock with variable dilatancy. Rock Mech. Rock Eng. 19(2), 99–108.

Diederichs, M. S. (2007) The 2003 Canadian Geotechnical Colloquium: Mechanistic interpretation and practical application of damage and spalling prediction criteria for deep tunnelling. Can. Geotech. J. 44(9), 1082–1116.

Fischer, L., Amann, F., Moore, J. R., & Huggel, C. (2010) Assessment of periglacial slope stability for the 1988 Tschierva rock avalanche (Piz Morteratsch, Switzerland). Eng. Geol. 116(1–2), 32–43.

Ghafoori, M., Lashkaripour, G. R. & Tarigh Azali, S. (2011) Investigation of the geological and geotechnical characteristics of Daroongar Dam, Northeast Iran. Geotech. Geol. Eng. 29(6), 961–975.

Gischig, V., Amann, F., Moore, J. R., Loew, S., Eisenbeiss, H., & Stempfhuber, W. (2011) Composite rock slope kinematics at the current Randa instability, Switzerland, based on remote sensing and numerical modeling. Eng. Geol. 118(1–2), 37–53.

Hajiabdolmajid, V., Kaiser, P. K. & Martin, C. D. (2002) Modelling brittle failure of rock. Int. J. Rock Mech. Min. Sci. 39(6), 731–741.

Hashemi, M., Moghaddas, S., & Ajalloeian, R. (2010) Application of rock mass characterization for determining the mechanical properties of rock mass: A comparative study. Rock Mech. Rock Eng. 43(3), 305–320.

Hoek, E. (1983) Strength of jointed rock masses. Geotechnique 33(3), 187–223.

Hoek, E. (2007) Practical Rock Engineering. Available online: www.rocscience.com, 342.

Hoek, E. and Brown, E. T. (1980) Underground excavations in rock. Institution of Mining and Metallurgy, London, 527.

Hoek, E. & Brown, E. T. (1988) The Hoek-Brown failure criterion – a 1988 update. In: Rock engineering for underground excavations, Proc. 15th Canadian Rock Mech. Symp., University of Toronto, Toronto, Toronto, Canada, 31–38.

Hoek, E. & Brown, E. T. (1997) Practical estimates of rock mass strength. Int. J. Rock Mech. Min. Sci. 34(8), 1165–1186.

Hoek, E., Carranza_Torres, C. & Corkum, B. (2002) Hoek-Brown failure criterion – 2002 edition. In: Proc. 5th North American Rock Mech. Symp., Toronto, Canada, 267–273.

Hoek, E., Carter, T. G. & Diederichs, M. S. (2013) Quantification of the geological strength index chart. In: 47th US Rock Mechanics/Geomechanics Symposium, San Francisco, CA, USA, ARMA 13–672.

Hoek, E. & Diederichs, M. S. (2006) Empirical estimation of rock mass modulus. Int. J. Rock Mech. Min. Sci. 43(2), 203–215.

Hoek, E., Kaiser, P. K. & Bawden, W. F. (1995) Support of underground excavations in hard rock. Taylor & Francis, New York, 215.

Hoek, E., Marinos, P. & Benissi, M. (1998) Applicability of the geological strength index (GSI) classification for very weak and sheared rock masses. The case of Athens Schist Formation. Bull. Eng. Geol. Env. 57, 151–160.

Hudson, J. A. & Harrison, J. P. (1997) Engineering rock mechanics – an introduction to the principles. Elsevier Science Ltd. UK, 444.

Itasca (2010) PFC3D-Particle Flow Code. Itasca Consulting Group Inc., 4.0.

Jade, S. & Sitharam, T. G. (2003) Characterization of strength and deformation of jointed rock mass based on statistical analysis. Int. J. Geomech. ASCE. 3, 43–54.

Kaiser, P. K., Diederichs, M. S., Martin, C. D., Sharp, J. & Steiner, W. (2000) Underground works in hard rock tunnelling and mining. In: Keynote lecture at GeoEng2000, Technomic Publishing Co., Melbourne, Australia, 841–926.

Kalamaras, G. S. & Bieniawski, Z. T. (1993) A rock mass strength concept for coal seams. In: Proceedings of 12th Conf. Ground Control in Mining, Morgantown, 274–283.

Kendorski, F. S., Cummings, R. A., Bieniawski, Z. T. & Skinner, E. H. (1983) Rock mass classification for block caving mine drift support. In: Proc. 5th ISRM, Melbourne, Australia, 51–63.

Kim, B. H., Cai, M., Kaiser, P. K. & Yang, H. S. (2007) Estimation of block sizes for rock masses with non-persistent joints. Rock Mech. Rock Eng. 40(2), 169–192.

Koyama, T., Nanbu, S. & Komatsuzaki, Y. (1997) Large-scale cavern at a depth of 500 m. Tunnel underground 28(1), 37–45. (in Japanese).

Laubscher, D. H. (1984) Design aspects and effectiveness of support system in different mining conditions. Trans. Inst. Min. Met. 93, A70–81.

Laubscher, D. H. (1990) A geomechanics classification system for the rating of rock mass in mine design. J. South Afr. Inst. Min. Metall. 90(10), 257–273.

Martin, C. D. (1993) The strength of massive Lac du Bonnet granite around underground opening. Ph.D. thesis, University of Manitoba, 278.

Martin, C. D. (1997) Seventeenth Canadian Geotechnical Colloquium: The effect of cohesion loss and stress path on brittle rock strength. Can. Geotech. J. 34(5), 698–725.

Martin, C. D., Kaiser, P. K. & McCreath, D. R. (1999) Hoek-Brown parameters for predicting the depth of brittle failure around tunnels. Can. Geotech. J. 36(1), 136–151.

Mas Ivars, D., Pierce, M. E., Darcel, C., Reyes-Montes, J., Potyondy, D., Young, R. P. & Cundall, P. (2011) The synthetic rock mass approach for jointed rock mass modelling. Int. J. Rock Mech. Min. Sci. 48(2), 219–244.

Nicholson, G. A. & Bieniawski, Z. T. (1990) A nonlinear deformation modulus based on rock mass classification. Int. J. Min. Geol. Eng. 8, 181–202.

Oda, M. (1983) A method for evaluating the effect of crack geometry on the mechanical behavior of cracked rock masses. Mech. Mater. 2, 163–171.

Palmstrøm, A. (1995) RMi – a rock mass characterization system for rock engineering purposes. Ph. D. thesis, University of Oslo, Norway, 400.

Palmstrøm, A. (1996a) Characterizing rock masses by the RMi for use in practical rock engineering, Part 1: The development of the rock mass index (RMi). Tunnelling Underground Space Technol. 11(2), 175–188.

Palmstrøm, A. (1996b) Characterizing rock masses by the RMi for use in practical rock engineering, Part 2: Some practical applications of the rock mass index (RMi). Tunnelling Underground Space Technol. 11(3), 287–303.

Pelli, F., Kaiser, P. K. & Morgenstern, N. R. (1991) An interpretation of ground movements recorded during construction of the Donkin-Morien tunnel. Can. Geotech. J. 28(2), 239–254.

Pierce, M., Cundall, P., Potyondy, D. & Mas Ivars, D. (2007) A synthetic rock mass model for jointed rock. In: Proc. 1st Canada-U.S. Rock Mechanics Symposium, Taylor & Francis, Vancouver, 341–349.

Potvin, Y. (1988) Empirical open stope design in Canada. Ph.D. thesis, Dept. Mining and Mineral Processing, University of British Columbia.

Priest, S. D. & Hudson, J. A. (1976) Discontinuity spacings in rock. Int. J. Rock Mech. Min. Sci. Geomech. Abstr. 13(5), 135–148.

Priest, S. D. & Hudson, J. A. (1981) Estimation of discontinuity spacing and trace length using scanline surveys. Int. J. Rock Mech. Min. Sci. Geomech. Abstr. 18(3), 183–197.

Ramamurthy, T. (1994) Strength predictions for jointed rocks in confined and unconfined states. Int. J. Rock Mech. Min. Sci. Geomech. Abstr. 31, 9–22.

Ramamurthy, T., Rao, G. V. & Rao, K. S. (1985) A strength criterion for rocks. In: Indian Geotech. Conf., Roorkee, 59–64.

Read, S., Richards, L. & Perrin, N. (1999) Applicability of the Hoek-Brown failure criterion to New Zealand greywacke rocks. In: 9th ISRM, 655–660.

Rockfield Software Ltd. (2003) ELFEN. 3.7.

Rosenblueth, E. (1981) Two-point estimates in probabilities. J. Appl. Math. Modell. 5(5), 329–335.

Russo, G. (2009) A new rational method for calculating the GSI. Tunnelling Underground Space Technol. 24(1), 103–111.

Sakurai, S. & Takeuchi, K. (1983) Back analysis of measured displacement of tunnels. Rock Mech. Rock Eng. 16, 173–180.

Serafim, J. L. & Pereira, J. P. (1983) Consideration of the geomechanical classification of Bieniawski. In: Proc. Int. Symp. on Engineering Geology and Underground Construction, Lisbon, 33–44.

Sheorey, P. R., Biswas, A. K. & Choubey, V. D. (1989) An empirical failure criterion for rocks and jointed rock masses. Eng. Geol. 26, 141–159.

Soleiman Dehkordi, M., Shahriar, K., Moarefvand, P., & Gharouninik, M. (2013) Application of the strain energy to estimate the rock load in squeezing ground condition of Eamzade Hashem tunnel in Iran. Arabian J. Geosci. 6(4), 1241–1248.

Sonmez, H., Gokceoglu, C. & Ulusay, R. (2004) Indirect determination of the modulus of deformation of rock masses based on the GSI system. Int. J. Rock Mech. Min. Sci. 41(5), 849–857.

Sonmez, H. & Ulusay, R. (1999) Modifications to the geological strength index (GSI) and their applicability to stability of slopes. Int. J. Rock Mech. Min. Sci. 36(6), 743–760.

Terzaghi, K. (1946) Rock defects and loads on tunnel supports. In: Rock Tunneling with Steel Supports, 17–99.

Wickham, G. E., Tiedemann, H. R. & Skinner, E. H. (1974) Ground support prediction model – RSR concept. In: Proc. 1st Rapid Excavation and Tunnelling Conference, Am. Inst. Min. Eng., 691–707.

Yudhbir, L. W., Lemanza, W. & Prinzl, F. (1983) An empirical failure criterion for rock masses. In: 5th ISRM, Balkema, B1–B8.

Zhang, L. (2010) Estimating the strength of jointed rock masses. Rock Mech. Rock Eng. 43(4), 391–402.

Zhang, L. & Einstein, H. H. (2004) Using RQD to estimate the deformation modulus of rock masses. Int. J. Rock Mech. Min. Sci. 41(4), 337–341.

Zhao, X. G. & Cai, M. (2010a) Influence of plastic strain and confinement dependent rock mass dilation on the failure and displacement near an excavation boundary. Int. J. Rock Mech. Min. Sci. 47(5), 723–738.

Zhao, X. G. & Cai, M. (2010b) A mobilized dilation angle model for rocks. Int. J. Rock Mech. Min. Sci. 47(3), 368–384.

Modeling Rock Deformation and Failure

Chapter 19

Constitutive modeling of geologic materials, interfaces and joints using the disturbed state concept

Chandrakant S. Desai

Regents' Professor (Emeritus), Department of Civil Engineering and Engineering Mechanics, University of Arizona, Tucson, AZ, USA

Abstract: The unified disturbed concept (DSC) for constitutive modeling of geologic materials and interfaces (joints) is presented with details including the relative intact (RI) state, fully adjusted (FA) state and disturbance. Attention is given to model parameters based on triaxial, multiaxial and interface tests, and validations at specimen level. The DSC is implemented in 2- and 3-D finite element procedures which are used to predict behavior of wide range problems including comparisons with field and laboratory simulated tests. The DSC is considered to be a unique and versatile procedure for modeling behavior of engineering materials and interfaces.

I INTRODUCTION

Behavior of materials such as soils, rocks, concrete, interfaces between structures and soils, and joints in rocks, plays a vital role for reliable solutions of geomechanical problems. Constitutive models based on appropriate bases of mechanics and testing define the mechanical response of solids, interfaces and joints. A great number of constitutive models, from simple to the advanced, have been proposed. Most of them account for specific characteristics of the material behavior. However, a deforming material may experience simultaneously many characteristics such as elastic, plastic and creep strains, loading (stress) paths, volume change, microcracking leading to failure, strain softening or degradation, liquefaction and healing or strengthening. Hence, there is a need for unified models that account for these characteristics. This Chapter presents a unique approach called the Disturbed State Concept (DSC) that includes a number of available constitutive models for solids and interfaces as special cases, and provides a unified model that allows the above factors simultaneously. The DSC includes models for the behavior of the continuum part of material; the hierarchical single surface (HISS) plasticity model has been often used for the continuum, hence, the model covered here is called DSC/HISS. Description of various constative models and the DSC/HISS are presented in various publications, *e.g.* Desai (2001).

Computer methods (*e.g.* Desai & Abel, 1972; Desai & Zaman, 2014) with appropriate constitutive models for behavior of geologic materials and interfaces have opened a new era for accurate and economic analysis and design for problems in geomechanics and geotechnical engineering. Such procedures account for many significant factors such as (1) Initial or in situ stress or strain; elastic, irreversible (plastic) and creep deformations; volume change under shear and its initiation during loading;

isotropic and anisotropic hardening; stress (load) path dependence; inherent and induced discontinuities; microstructural modifications leading to fracture and instabilities like failure and liquefaction; degradation or softening; static, repetitive and cyclic (dynamic) loading; forces: loads, temperature, moisture (fluid) and chemical effects; anisotropy, nonhomogeneities, and strengthening or healing (Desai *et al.*, 1998).

A number of general plasticity models have been proposed, *e.g.* Mroz, *et al.*, (1978), Pestana & Whittle (1999), Pastor *et al.*, (2000) and Elagamal *et al.*, (2002). These models can account for various factors beyond those in previously available (plasticity) models. They are based on continuum plasticity, and introduce modifications to simulate behavioral features such as degradation (softening). However, a deforming material can involve discontinuities resulting from microstructural modifications. The continuum plasticity models may not account for such discontinuous deformations. On the other hand, the DSC model accounts for the changing microstructure resulting into degradation or strengthening. It also allows intrinsically for the definition of instability or liquefaction based on the internal response (disturbance) of a deforming material. Furthermore, introduction of disturbance with continuum models (elasticity or plasticity) for RI behavior, may not involve any more effort than the modification of continuum plasticity models. Models such as MIT-S1 (Pestana & Whittle, 1999; Elgamal, 2002) based on the bounding surface plasticity may also involve a greater number of parameters compared to those in the DSC model for comparable capabilities (Desai, 2001).

Now, a description of the DSC is followed by typical geomechanical problems solved by using computer (finite element) method in which the DSC has been implemented.

2 THE DISTURBED STATE CONCEPT

Because of length limitations, comprehensive review of publications on constitutive models is not feasible. Hence, wherever appropriate, models such as based on elasticity, plasticity, viscoplasticity, damage, fracture and micromechanics have been referenced. Reviews of such constitutive models are given in (Desai, 2001, 2015a, 2015b; Desai *et al.*, 2011); the middle two references present details of DSC/HISS for a number of disciplines in engineering.

The idea of the DSC is considered to be relatively simple, and it can be easily implemented in computer procedures. It is believed that the DSC can provide a realistic and unified approach for constitutive modeling for a wide range of materials and interfaces/joints. It is a general approach that can accommodate most of the forgoing factors including discontinuities that influence the material behavior, and provide a hierarchical framework that can include many of the available models as special cases. One of the attributes of the DSC is that its mathematical framework can be specialized for interfaces and joints, thereby providing consistency in using the same model for both solids and interfaces (Desai, 2001).

In the DSC, a deforming material element is assumed to be composed of two or more components. Usually, for a dry solid, two components are assumed, Figure 1, a continuum part called the relative intact (RI) which is defined by using a theory from the continuum mechanics, and the disturbed part, called the fully adjusted (FA), which is defined based on the approximation of the ultimate asymptotic state of the material.

The DSC has been published in a number of papers and books, only a few are cited here (Desai, 2001, 2015a, 2015b); these works include application of the DSC by the

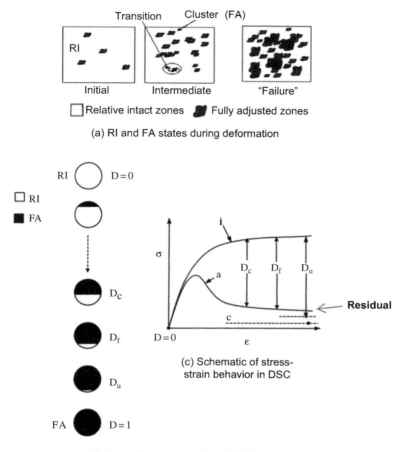

Figure 1 Schematics of DSC.

author and coworkers, and other researchers for materials such as soils, structured soil, masonry, concrete, asphalt concrete, fully and partially saturated materials, rock and rockfills, pavement materials, metals, alloys, ceramics, polymers and silicon, and interfaces and joints. It has been used for applications beyond material behavior, *e.g.*, developing expressions for earth pressures (Zhu et al., 2009), computation of pile capacity (in Desai, 2013), and free surface fluid flow [Desai, 1976; Desai & Li, 1983).

The origin of the DSC constitutive modeling can be traced to the papers by Desai (1974, 1976) on the subject of behavior of overconsolidated soils and free surface flow in porous materials, respectively. The DSC is based on rather a simple idea that the behavior of a deforming material can be expressed in terms of the behaviors of its components. Thus, the behavior of a dry material can be defined in terms of the continuum (called relative intact-RI, i) and microstructurally organized, *e.g.* microcracked part which approaches, in the limit, to the fully adjusted (FA, c) state; the latter can be essentially considered as collection of particles in failure or material in an invariant state like constant volume or density. The behavior of the FA part is unattainable (or unmanifested) in practice because it cannot be measured; therefore, a state, somewhere near the residual or ultimate, Figure 1, can be chosen as approximate FA state. The space between the RI and FA denoted by (i) and (c), respectively, can be called the domain of deformation, whose observed or average behavior (can be called manifested) occurs between the RI and FA states, Figures 1 and 2. The deviation of the observed state from the RI (or FA) states is called *disturbance*, and is denoted by D. It represents the difference between the RI and observed behavior or difference between the observed and FA behavior, which can be considered as a state parameter.

By defining the observed material behavior in terms of RI (continuum) and fully adjusted parts, the DSC provides for the coupling between two parts of the material behavior rather than on the behavior of particle(s) at the micro level. Thus, the emphasis is on the modeling of the *collective behavior of interacting mechanism in clusters of* RI and FA states, rather than on the particle level processes, thereby yielding a *holistic* model. These comments are similar to those in the self-organized criticality concept (Bak & Tang, 1989), which is used to simulate catastrophic events such as avalanches and earthquakes. In this context, the DSC assumes that as the loading (deformation) progresses, the material in the continuum state tends continuously into the FA state through transformations in the microstructure of the material. The definition of the DSC is *not* based on the behavior at the microlevel (say, as in micromechanics); rather it is based on the definition of the behavior of the material clusters in the RI and FA states defined from the measured behavior in those states, Figure 2.

Behaviors of the RI and FA can be defined from laboratory or field tests, and the observed behavior can be expressed in terms of the behaviors of the RI and FA parts. Assume that the material is continuous in the beginning and remains so during deformation; such a behavior is called that of the RI State, which contains no disturbance. As stated before, the fully adjusted behavior is related to the strength of the material in the FA state. Some of the ways to define RI and FA responses are given below. Figure 2(a) shows the continuum response as linear elastic, which can be considered as the RI state. However, the observed response can be nonlinear (elastic), due to factors such as existing cracks and cracking. The FA response can be assumed to have a small finite strength. The disturbance can be defined as the difference between linear elastic and nonlinear elastic responses. Figure 2b shows a strain softening response. Here the RI response can be assumed to be nonlinear elasticplastic and the FA response based on the critical state concept. Figure 2c shows cyclic response. Here the RI response can be adopted as the extended response of the first cycle. The FA response can be assumed to be asymptotic as the response become steady after a number of cycles.

2.1 Relative Intact (RI) state

Figures 3(a) to (d) show schematics of RI, observed and FA behaviors in terms of various measured quantities: stress vs. strain, volume or void ratio response, nondestructive behavior (velocity), and effective stress (or pore water pressure). Figure 4 shows schematic for disturbance vs. accumulated plastic strain ξ_D or number of cycles or time. In some cases, the RI behavior can be assumed to be linear elastic defined by the initial slope. However, if the material behavior is nonlinear and involves effect of other

Constitutive modeling of geologic materials 537

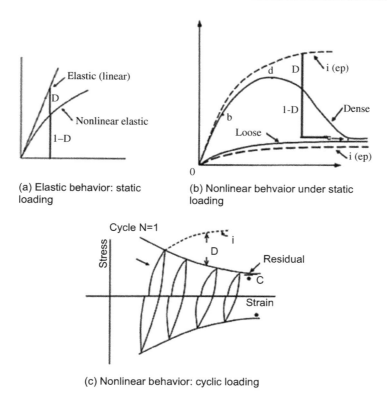

Figure 2 Schematics of RI (i), FA(c) behavior, and disturbance (D).

factors such as coupled volume change behavior, *e.g.* volumetric change under shear loading, such an assumption will not be realistic. Hence, very often, conventional or continuous yield or HISS plasticity is adopted as the RI response.

2.2 Fully Adjusted (FA) state

As a simple approach, it can be assumed that the material in the FA state has no strength, just like in the classical damage model (Kachanov, 1986); this assumption ignores interaction between RI and FA states, may lead to *local* models, and may cause computational difficulties. The second assumption is to consider that the material in the FA state can carry hydrostatic stress like a *constrained liquid*, in which case the bulk modulus (K) can be used to define the FA state. The FA material can be considered as of *liquid-solid* like in the critical state (Roscoe *et al.*, 1958; Desai, 2001), when after continuous yield, the material approaches a state at which there is no change in volume or density or void ratio under increasing shear stress. The equations for the strength of the material in the critical state (FA) are given below:

$$\sqrt{J_{2D}^c} = \overline{m} J_1^c \tag{1a}$$
$$e^c = e_o^e - \lambda \ln\{J_1^c/(3p_a)\} \tag{1b}$$

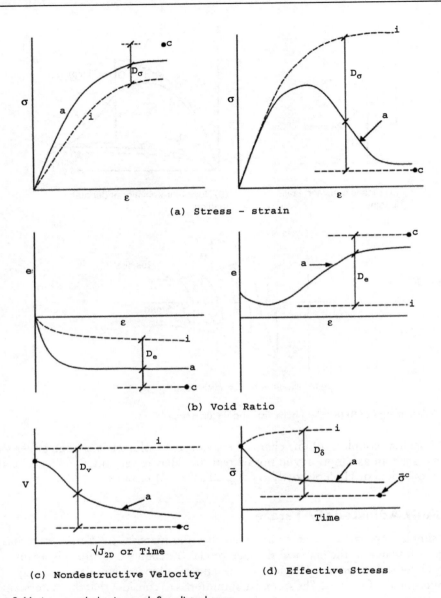

Figure 3 Various test behavior to define disturbance.

where c denotes at the critical state, J_{2D} is the second invariant of the deviatoric stress tensor, $\overline{m}$ is the slope of the critical state line, Figure 5, J_1 is the first invariant of the stress tensor, e is the void ratio, e_o is the initial void ratio, λ is the slope of the consolidation line, Figure 5, and p_a is the atmospheric pressure constant.

For fluid saturated materials with drainage with time, the RI behavior can be assumed to be that at time near zero, and the FA response can be assumed like that of a constrained liquid. A description of the DSC for partially saturated materials is given in (Desai, 2001).

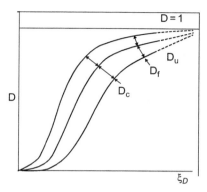

Figure 4 Disturbance vs. ξ_D (or number of cycles ot time).

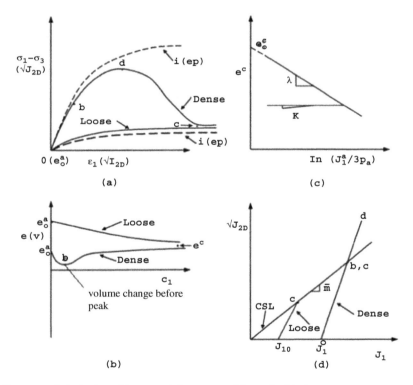

Figure 5 Stress-strain behavior of loose and dense materials, and critical state.

2.3 Disturbance

As stated before, disturbance defines the coupling between the RI and FA states, and is represented by the deviation (disturbance) of the observed behavior from RI or FA state. It can be determined based on the stress-strain behavior, Figure 3(a). It can be determined from other tests like void ratio vs. strain, Figure 3(b), nondestructive behavior such a for P- and S- waves velocities, Figure 3(c), fluid (pore) water pressure or effective stress ($\bar{\sigma}$) vs. time or number of cycles, Figure 3(d). Figure 4 shows the schematic of the disturbance (D) as function of ξ_D or number of cycles (N) or time (t); here D_c, D_f and D_u denote initiation of fracture, failure and ultimate disturbance, respectively.

The disturbance can defined in two ways, (1) from measurements, Figure 3, as stated before, and (2) by mathematical expression in terms of internal variables such as (ξ_D):

From measurements, for example:

$$\text{Stress-Strain behavior:} \quad D_\sigma = \frac{\sigma^i - \sigma^a}{\sigma^i - \sigma^c} \tag{2a}$$

$$\text{Nondestructive velocity:} \quad D_v = \frac{V^i - V^a}{V^i - V^c} \tag{2b}$$

where σ^a is the measured stress and V^a is the measured nondestructive velocity, and i and c represent RI and FA responses.

2.3.1 Mathematical expression for D

Disturbance, D, can be expressed using the (Weibull) function in terms of internal variable such as accumulated (deviatoric) plastic strains (ξ_D) or plastic work:

$$D = D_u \left[1 - \exp\left(-A\xi_D^Z\right)\right] \tag{3}$$

where A, Z and D_u are the parameters. The value of Du is obtained from the ultimate FA state, Figure 2. Equations 2 are used to find the disturbance, Figure 3, at various points on the response curves, which are substituted in Equation 3 to find the parameters. Note that the expression in Equation 3 is similar to that used in various areas such as biology to simulate birth to death, or growth and decay, and in engineering to define damage in classical damage mechanics, and disturbance in the DSC. However, the concept of disturbance is much different from damage; the former defines deviation of observed response from the RI (or FA) state, in the material treated as a mixture of interacting components, while the latter represents physical damage or cracks.

2.4 DSC equations

Once the RI and FA states and disturbance are defined, the incremental DSC equations based on equilibrium of a material element can be derived as (Desai, 2001):

$$d\sigma_{ij}^a = (1 - D)d\sigma_{ij}^i + Dd\sigma_{ij}^c + dD(\sigma_{ij}^c - \sigma_{ij}^i)$$
$$or \tag{4}$$
$$d\sigma_{ij}^a = (1 - D)C_{ijkl}^i d\varepsilon_{kl}^i + DC_{ijkl}^c d\varepsilon_{kl}^c + dD(\sigma_{ij}^c - \sigma_{ij}^i)$$

where σ_{ij} and ε_{ij} denote stress and strain tensors, respectively, C_{ijkl} is the constitutive tensor, and dD is the increment or rate of D.

Equation 4 represents DSC equation from which conventional continuum (elasticity, plasticity, creep, etc.) models can be derived as special cases by setting D=0, as

$$do_{ij}^a = C_{ijkl}^i \, d\varepsilon_{kl}^i \tag{5}$$

in which the observed and RI behavior are the same, and the constitutive tensor can be based on the appropriate continuum model. If $D \neq 0$, Equation 4 accounts for microstructural modifications in the material leading to fracture, and instabilities like failure and liquefaction (in saturated materials). The latter can be defined corresponding to the critical disturbance, D_c, Fig. 4, obtained from measured response of the material.

A major advantage of the DSC approach is that it is hierarchical and unified. Hence, one can extract available models as special cases from Equation 4. When the RI behavior is modeled by using the HISS plasticity, various conventional and continuous yield plasticity models can also be derived as specialization of the HISS model (Desai, 2001).

2.5 Hierarchical Single Surface (HISS) plasticity

The need for a unified and general plasticity model that can account for the factors mentioned before was the driving force for the development of the hierarchical single surface (HISS) plasticity model [Desai, 1980, 2001; Desai *et al.*, 1986a]; it is based on the continuum assumption, hence, it cannot account for discontinuities.

The yield surface, F, in HISS associative plasticity is expressed as (Figure 6a):

$$F = \overline{J_{2D}} - (-\alpha \overline{J_1^n} + \gamma \overline{J_1^2})(1 - \beta S_r)^{-0.5} = 0 \tag{6}$$

where $\overline{J_{2D}} = J_{2D}/p_a^2$ is the non-dimensional second invariant of the deviatoric stress tensor, $\overline{J_1} = (J_1 + 3R)/p_a)$ is the non-dimensional first invariant of the total stress tensor, R is the term related to the cohesive (or tensile) strength, $\overline{c}$, Figure 6a, $S_r = \frac{\sqrt{27}}{2} \frac{J_{3D}}{J_{2D}^{1.5}}$, n is the parameter related to the transition from compressive to dilative volume change, Figure 5, γ and β are the parameters associated with the ultimate surface, Figure 6(a), and α is the hardening or growth function; in a simple form, it is given by

$$\alpha = \frac{a_1}{\xi^{\eta_1}} \tag{7}$$

where a_1 and η_1 are the hardening parameters, and ξ is the accumulated or trajectory of plastic strains, given by

$$\xi = \xi_v + \xi_D \tag{8}$$

Here the accumulated volumetric plastic strain is given by

$$\xi_v = \frac{1}{\sqrt{3}} |\varepsilon_{ii}^p| \tag{9a}$$

and the accumulated deviatoric plastic strain is given by

$$\xi_D = \int \left(E_{ij}^p E_{ij}^p \right)^{1/2} \tag{9b}$$

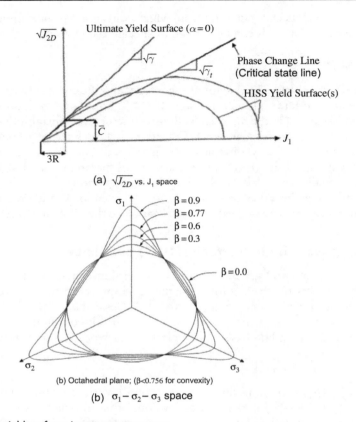

Figure 6 HISS yield surfaces in two stress spaces.

where ε_{ii}^p is the plastic volumetric strain, and E_{ij}^p is the plastic shear strain tensor. In the HISS model, the yield surface grows continuously and approaches the *ultimate* yield, Figure 6; it can include, as special cases, other conventional and continuous yield plasticity models (Desai, 2001).

For compression intensive materials (*e.g.* geologic, concrete, powders) the model and the yield surface, Figure 6, are relevant for compressive yield only in the positive $\sqrt{J_{2D}} - J_1$ space, in which $\bar{c}$ will be the compressive strength. Similarly, for tension intensive materials (*e.g.* metals and alloys), the model and yield surfaces are relevant for tensile yield only in the positive $\sqrt{J_{2D}} - J_1$ space, in which $\bar{c}$ would denote tensile strength. In both cases, the extension of yield surfaces in the negative J_1 - axis is not relevant; they are usually shown for convenience of plotting. Sometimes, the extended yield surfaces in the negative J_1 – axis have been used with an *ad hoc* model for materials experiencing tensile conditions, which may not be realistic. As discussed below (HISS-CT model), for example, when a material experiences tensile stress (during deformation), it would be realistic to use the model (*e.g.* HISS) defined on the basis of tensile tests, and *vice versa*.

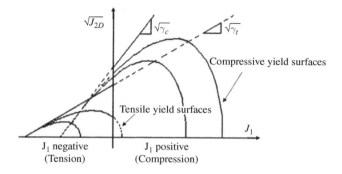

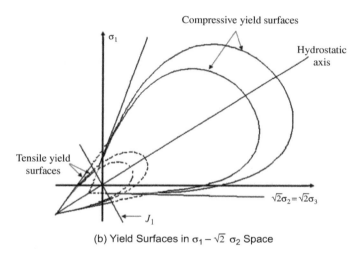

Figure 7 HISS yield surfaces for compressive and tensile yielding.

2.6 HISS for Compression and Tension (HISS-CT)

For some problems the material can be subjected to both compressive and tensile stress conditions. To develop and use the *same* model for both conditions is difficult, and perhaps not possible. However, the same model like HISS can be formulated for both conditions by obtaining parameters from separate compression and tension (extension) tests. Figure 7 shows the surfaces for compressive and tensile yields obtained from compression and tensile tests (Desai, 2007, 2009; Akhaveissy & Desai, 2013).

The HISS plasticity model allows for continuous yielding, volume change (dilation) before the peak, stress path dependent strength, effect of both volumetric and deviatoric strains on the yield behavior, and it does not contain any discontinuities in the yield surface. The HISS surface, Equation 6, represents a unified plastic yield surface, and most of the previous conventional and continuous yield surfaces can be derived from it as special cases (Desai, 2001). Also, the HISS model can be used for nonassociative and anisotropic hardening responses, etc. The idea of the single yield surface has been also used by Lade and coworkers (*e.g.* Lade & Kim, 1988), based on prior open yield surfaces (Matsuoka & Nakai, 1974).

2.7 Creep behavior

Many materials exhibit creep behavior, increasing deformations under constant stress or stress relaxation under constant strain (displacement). A number of models have been proposed for various types of creep behavior, viscoelastic (ve), viscoelasticplastic (vep) and viscoelasticviscoplastic (vevp); they are also based on the assumption of continuum material. A generalized creep model has been proposed under the disturbed state concept (DSC) (Desai, 2001). It is called Multicomponent DSC (MDSC) which includes ve, vep and vevp versions as special cases. Details of the creep models are given in (Desai & Zheng, 1987; Desai, 2001).

Models based on theories of elasticity, plasticity and creep assume that the material is initially continuous and remains continuous during deformation. However, it is realized that many materials contain discontinuities (microcracks, dislocations, etc.), initially and during loading. During deformations, they coalesce and grow, and separate, resulting in microcracks and fractures, with consequent failure. Since the stress at a point implies continuity of the material, theories of continuum mechanics may not be valid for such discontinuous materials.

2.8 Discontinuous materials

There a number of models available to consider discontinuities in a deforming material. Chief among those are them are considered to be fracture mechanics, damage mechanics, micromechanics, microcrack interaction, gradient and Cosserat theories (Mühlhaus, 1995). Most of them combine the effect of discontinuities and microcracks, with the continuum behavior. Descriptions of these models are presented in [Mühlhaus, 1995; Desai, 2001, 2015a, 2015b].

3 PARAMETRES

The parameters in the DSC model can be obtained from laboratory test such as triaxial, multiaxial, uniaxial and interface shear. The basic DSC model contains the following parameters:

Relative Intact (RI)

Elastic: Young's modulus, E, and Poisson's ratio, v, (or shear modulus, G and bulk modulus K), and

Plasticity: (a) von Mises: tensile yield /cohesion, c, or (b) Mohr Coulomb: cohesion, c and angle of internal friction, φ or (c) HISS plasticity: ultimate yield, γ and β; phase change (transition from compaction to dilation), n; continuous yielding, a_1 and η_1; and cohesive strength intercept, $\bar{c}$ (R).

Constitutive modeling of geologic materials 545

Fully Adjusted (FA)
For the critical state, the parameters are shown in Equation 1.

Disturbance
The parameter D_u can be obtained from Figure 1; often a value near unity can be used. Parameters A and Z are obtained by first determining various values of D from the test data by using Equations 2, and then plotting logarithmic form of Equation 3.

3.1 Comments

Most of the above parameters in the DSC have physical meanings, *i.e.* almost all are related to specific states in the material response, *e.g.* elastic modulus to the unloading slope of stress-strain behavior, and β to the ultimate state, and n to the transition from compactive to dilative volume change, Figure 5. Their number is equal or lower than that of previously available model of comparable capabilities. They can be determined from standard laboratory tests such as uniaxial, shear, triaxial and/or multiaxial. The procedures for the determination of the parameters are provided in various publications (*e.g.* Desai, 2001)

3.2 Softening (degradation) and stiffening (healing) response

Details of the softening and stiffening behavior are given various publications (Desai, 1974a, 2001; Desai *et al.*, 1998; Shao & Desai, 2000).

3.3 Mesh dependence and localization

A constitutive model including discontinuities should satisfy properties such as mesh dependence and localization. The DSC has been analyzed for localization and mesh dependence and details are presented in (Desai, 2001, 2015a, 2015b; Desai & Zheng, 1998).

4 INTERFACES AND JOINTS

Behavior at interfaces between two (different) materials and joints play a significant role in the overall response of an engineering system [Desai *et al.*, 1986b; Desai and Ma, 1992; Samtani *et al.*, 1996; Fakharian & Evgin, 2000). One of the main advantages of the DSC is that its mathematical framework for solids can be applied also for interfaces.

4.1 Relative Intact (RI) response

Schematics of two- and three-dimensional interfaces, disturbed states, and deformation modes are shown in Figure 8. Consider a two-dimensional interface, Figure 8(a). In the same way as was assumed in the solid material, an element for the interface is considered to be composed of RI and FA states, Figure 8(c). The RI behavior in the interface can be simulated by various models such as nonlinear elastic and plastic (conventional or

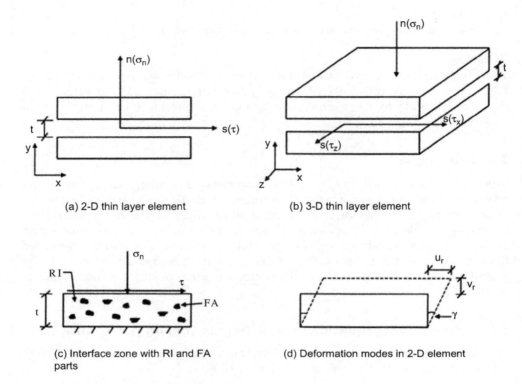

Figure 8 Schematics of 2- and 3-D interface elements and DSC.

continuous yield). Here, the HISS plasticity model is adopted for the RI part, as the specialized form of Equation 6 for solids. It can be calibrated from laboratory tests in terms of shear stress, τ vs. relative shear (horizontal) displacement, u_r, and relative normal (vertical) displacement v_r vs. u_r, Figures 9a and 9b, respectively.

The yield function specialized from Equation 6 for two-dimensional interface is given by (Figure 10):

$$F = \tau^2 + \alpha\sigma_n^n - \gamma\sigma_n^q = 0 \qquad (10)$$

where τ is the shear stress, σ_n is the normal stress, which can be modified as $\sigma_n + R$, R is the intercept along σ_n- axis which is proportional the adhesive strength, c_0, γ is the slope of ultimate response, n is the phase change parameter, which designates transition from compressive to dilative response, q governs the slope of the ultimate envelope (if the ultimate envelope is linear, q = 2), and α is the growth or yield function given by

$$\alpha = \frac{h_1}{\xi^{h_2}} \qquad (11)$$

where h_1 and h_2 are hardening parameters, and ξ is the trajectory of plastic relative horizontal (u_r) and vertical (normal) (v_r) displacements, given by

Constitutive modeling of geologic materials 547

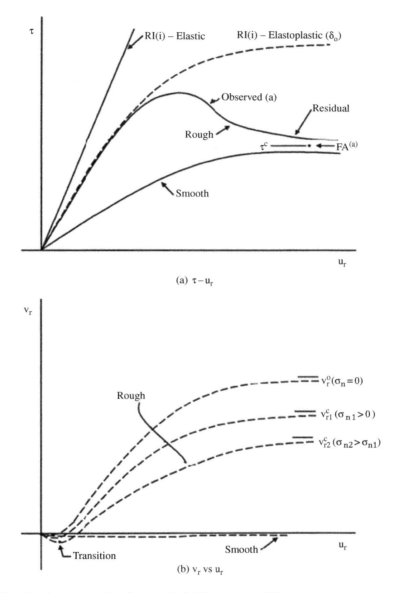

Figure 9 Test data for contact (interface or joint): (a) τ vs. u_r, and (b) v_r vs. u_r.

$$\xi = \int \left(du_r^p \cdot du_r^p + dv_r^p dv_r^p \right)^{1/2} = \xi_D + \xi_V \qquad (12)$$

ξ_D and ξ_V denote accumulated plastic shear and normal displacements, respectively, and superscript p denotes plastic.

The interface can reach the critical state irrespective of the initial roughness and normal stress (σ_n), as in the case of solids; when the relative normal displacement v_r

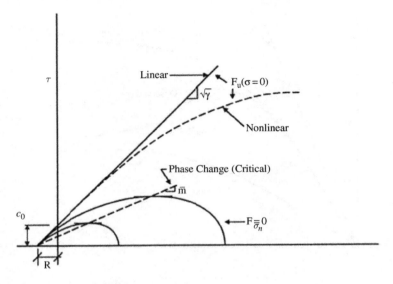

Figure 10 HISS yield surfaces for interfaces and joints.

tends to a steady state, Figure 9b. The equation for the material at the critical state, proposed by Archard (1959) is given by (Figure 11a)

$$\tau^c = c_o + c_1 \sigma_n^{(c)c_2} \tag{13a}$$

where c_0 is related to the adhesive strength and denotes the critical value of τ^c when σ_n = 0, σ_n^c is the normal stress at the critical state, and c_1 and c_2 are parameters related to the critical state.

The relation between the normal stress, σ_n, and the relative normal displacement at the critical state, v_r^c, was proposed by Schneider (1976), Figure 11b:

$$v_r^c = v_r^0 e^{-\lambda \sigma_n} \tag{13b}$$

where λ is a parameter and other quantities are shown in Figure 11.

Equations 11a and 11b for modeling interfaces/joints are similar to Equations 1a and 1b for solids.

Disturbance: Like in the case of solids, the disturbance for interfaces can be obtained from measured quantities as shown in Figure 12.

5 VALIDATIONS AND APPLICATIONS

In various publications, validations of the DSC model are presented in two parts: (a) Specimen Level 1 (individual) in which model predictions by integration of incremental constitutive equations, Equations 4, are compared with test data from which the parameters were determined, Specimen Level 2 with independent tests not used to find the parameters, and (b) Boundary Value Problem Level 3 in which finite element

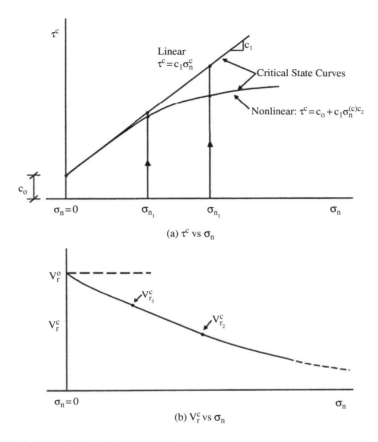

Figure 11 FA behavior of interface/joint at critical state.

(with the DSC model) predictions are compared with measurements in the field and from problems simulated in the laboratory.

The DSC and its special versions like HISS plasticity have been used by the author, coworkers, and other researchers, to model a wide range of materials such as geologic (sands, clays, rocks and concrete), asphalt concrete, metals, alloys (*e.g.* leaded and unleaded solders), polymers and silicon; and interfaces/joints; they are covered in various publications, *e.g.* Desai (2001). It has been implemented in computer (finite element) methods for nonlinear static and dynamic problems in structural- and geo-mechanics, coupled flow through porous media and composites in electronic packaging. Examples of only typical materials, particularly those containing complexities that are difficult to model by conventional models, are presented below.

Typical examples of validations at the specimen level are first presented using the DSC model. Then typical examples of validations are presented for practical problems using the finite element method in which the DSC has been implemented.

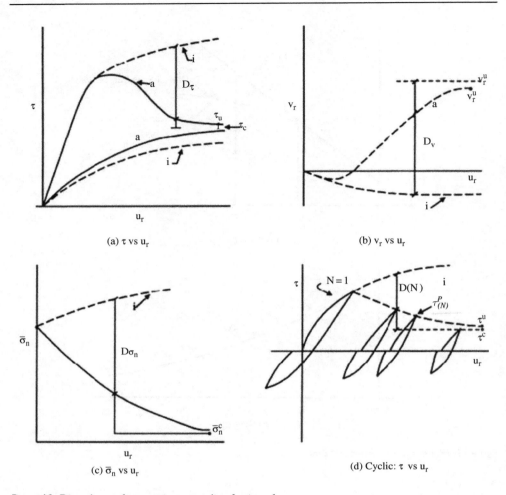

Figure 12 Disturbance from various test data for interface.

5.1 Example 1: Specimen level–cyclic behavior of sand

A series of laboratory three-dimensional (multiaxial) tests was performed on saturated Ottawa sand (specific gravity = 2.64; coefficient of curvature = 1.6; coefficient of uniformity = 2.0; maximum void ratio = 0.8; minimum void ratio = 0.46). The device allows variation (application) of three principal stresses, σ_1, σ_2, and σ_3. Two types of tests were performed. Specimen Level 1 involved sand with relative density of $D_r = 60\%$ under initial confining pressures, $\sigma'_o = 69$, 138, and 207 kPa (Gyi, 1996; Shao & Desai, 2000; Park & Desai, 2000). Typical test results for $\sigma'_0 = 138$ kPa, in terms of applied stress difference, $\sigma_d = \sigma_1 - \sigma_3$, measured axial strain, ϵ_1, measured axial strain, ϵ_1, vs. time, pore water pressure U_e vs. time, and σ_d vs. axial strain are shown in Figure 13. Specimen Level 2 validations included independent tests with $D_r = 40\%$

Constitutive modeling of geologic materials 551

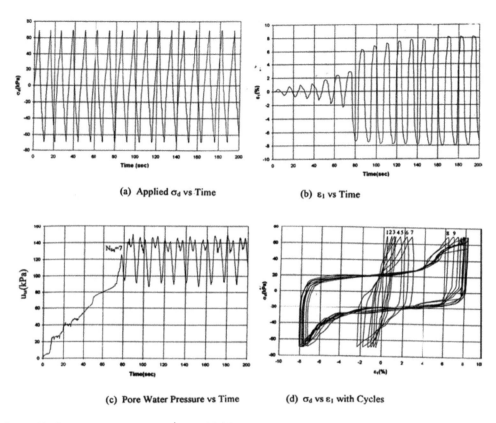

Figure 13 Cyclic multiaxial tests - $\sigma'_o = 138$ kPa: (a) Applied σ_d vs. time, (b) measured axial strain ε_1 vs. time, (c) measured pore water pressure U_w vs. time and (d) measured σ_d vs. axial strain ε_1.

for $\sigma'_o = 69$ kPa. The following material parameters were found by using the laboratory data for $D_r = 60\%$:

Elasticity: $E = 193,000$ kPa, $\nu = 0.38$;

Plasticity (HISS): $\gamma = 0.123$, $\beta = 0.000$, n = 2.45, $a_1 = 0.845$, $\eta_1 = 0.0215$;

Critical State: $\bar{m} = 0.15$, $\lambda = 0.02$, $e^c_o = 0.601$; **Disturbance:** $A = 4.22$, $Z = 0.43$, $D_u = 0.99$;

Unloading/reloading: $E^{EUL} = 177,6000$ kPa, $\varepsilon^p_1 = .0013$; the latter are used in simulating loading, unloading and reloading (Shao & Desai, 2000).

Finite element analysis was performed for the specimen (10×10×10 cm) to predict its laboratory behavior; the finite element mesh for a quarter of the specimen is shown in Figure 14. The simulated loading involved first application of the confining pressure, σ_3, and then cyclic axial (shear) loading, Figure 13a.

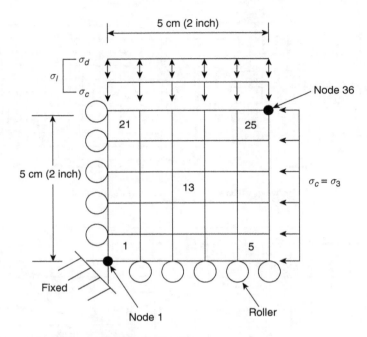

Figure 14 FE mesh, boundary conditions and loading.

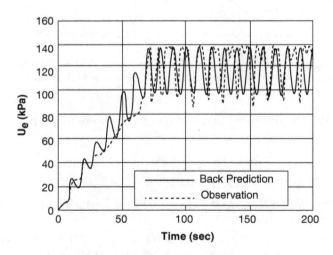

Figure 15 Comparisons between observed and predicted pore water pressure for $\sigma'_0 = 138$ kPa; $D_r = 60\%$.

Figure 15 show typical comparisons between predicted and measured pore water pressures vs. time for $\sigma'_0 = 138$ kPa, respectively and for Dr = 60%, which involved the tests used for finding the parameters. Figures 16 shows comparisons between predicted and measured effective stress, $\bar{\sigma}$ vs. time, and pore water pressure vs. time, respectively

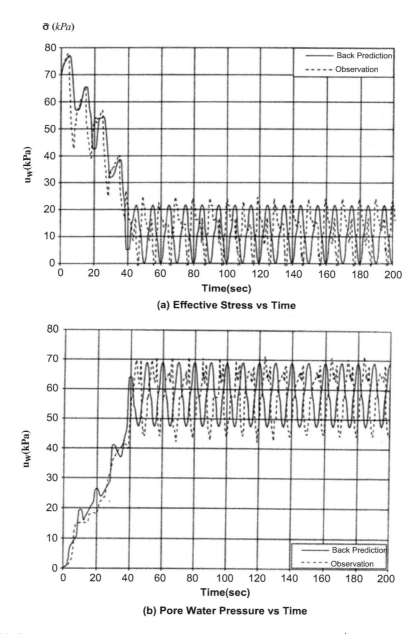

Figure 16 Comparisons between independent predictions and test data for $\sigma'_0 = 69\,\text{kPa}$ and $D_r = 40\%$: (a) effective stress vs. time; (b) pore water pressure vs. time.

554 Desai

for $D_r = 40\%$, which represents independent validation. It can be seen from Figures 15 and 16 that the correlations between predictions and measurements are highly satisfactory for behavior of fluid saturated materials, loaded cyclically.

5.2 Example 2: Specimen level–rockfill materials

Rockfill materials are composed of particles of different sizes, and represent a challenging problem for constitutive modeling. Here we consider the example of rockfill materials reported by Varadarajan *et al.*, (2003), which are often used for construction of dams. Rockfill material, obtained from Ranjit Sagar Hydropower in the state of Punjab, India consisted of alluvial with rounded/sub rounded particles up to 320 mm in size. The material from Purulia Pumped Storage Hydroelectric projects in West Bengal, India; it was obtained by blasting rock and consisted of angular to subangular particles up to 1,200 mm in size.

The rockfill materials were tested by using a triaxial device that can handle large sized specimens. Two specimen sizes were used, 381 mm diameter, and 813 mm long, and 500 mm diameter and 600 mm long; details are given in Varadarajan *et al.*, (2003). It was found that the Ranjit Sagar dam rockfill exhibited essentially compactive volume change, whereas the Purulia rockfill showed first compactive and then dilative volume change.

The DSC/HISS approach was used for modeling the rockfill. The yield surface, F, Equation 6, was used. The nonassociative behavior was simulated by expressing plastic potential surface, Q, as (Desai, 2001)

$$Q = \bar{J}_{2D} - \left(-\alpha_Q \bar{J}_1^n + \gamma \bar{J}_1^2\right)(1 - \beta S_r)^{-0.5} \tag{14}$$

where the function, α_Q, is given by

$$\alpha_Q = \alpha + \kappa(\alpha_o - \alpha)(1 - r_v) \tag{15}$$

α_o is the value of α at the beginning of shear loading, κ is a parameter and $r_v = \xi_v/\xi$, $\xi = \xi_v + \xi_D$; ξ_v and ξ_D are given in Equations 9a and b, respectively.

The disturbance function D was expressed as

$$D = \frac{\xi_D}{A + B\xi_D} \tag{16}$$

where A and B are the disturbance parameters and ξ_D is the accumulated plastic deviatoric strain. The elastic behavior was found to be dependent on the confining pressure, hence, the elastic modulus, E, was expressed as

$$E = kPa \left(\frac{\sigma_3}{p_a}\right)^{n'} \tag{17}$$

where k and n' are the parameters and p_a is the atmospheric pressure constant.

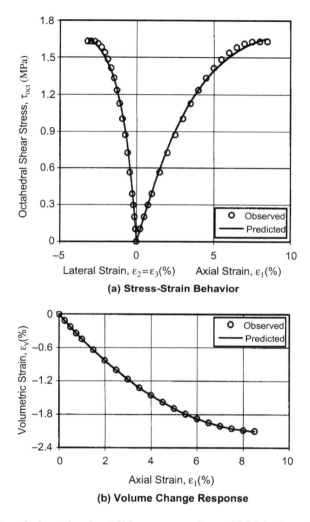

Figure 17 Comparison (independent Level 2) between test data and DSC for Ranjit Sagar rockfill: D_{max} = 25 mm and $\sigma_3 = 1.1$ MPa.

Details of constitutive parameters for the two rockfills are given by Varadarajan *et al.*, (2003).

5.2.1 Validations

The behavior of the rockfills was predicted by integrating Equation 4, and using the parameters given in Varadarajan *et al.*, (2003). Typical comparisons for only independent validations are shown Figure 17 for D_{max} = 25 mm and $\sigma_3 = 1.1$ MPa for Ranjit Sagar rockfill. In Figure 18 are shown similar comparisons for D_{max} = 50 mm and σ_3 = 0.60 MPa. It can be seen for the independent tests, Level 2 and also Level 1 (not presented here), the correlations are very good.

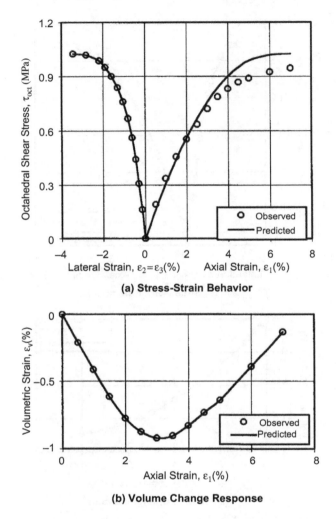

Figure 18 Comparison (independent Level 2) between test data and DSC predictions for Purulia rockfill, σ_3 = 0.60 MPa, D_{max} = 50mm.

5.3 Example 3: DSC molding of saturated sand–concrete interfaces

Saturated (Ottawa) sand-concrete interfaces were tested under various normal stresses and cyclic shear loading by using the Cyclic Multi-Multi Degree-of-Freedom (CYMDOF) device (Desai & Rigby, 1997). A section of the device is shown in Figure 19a. Simple shear deformations are allowed in the device by confining laterally the specimen by a (fixed) number of thin and smooth Teflon-coated rings, Figure 19b. Each ring has inner and outer diameters of 16.50 cm (6.5 in.) and 18.80 cm (7.4 in.), respectively, and thickness of 1.5 mm (0.059 in.).

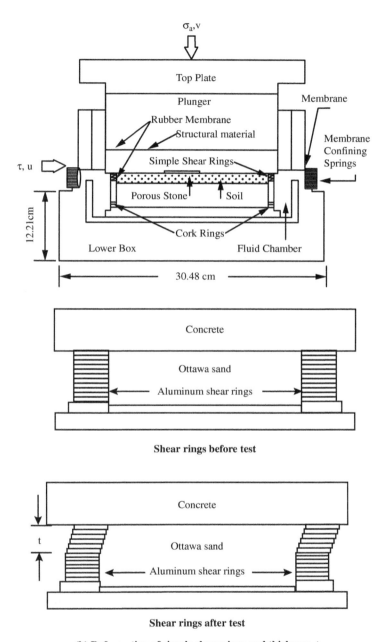

Figure 19 Cross- section of CYMDOF device and simple shear deformation.

558 Desai

Table 1(a) Parameters for the Ottawa sand-concrete interface.

Elasticity: E = 3183 kPa; ν = 0.42
Plasticity: γ = 0.109; n = 3.12; h_1 = 0.289; h_2 = 0.470
FA (critical state): $\overline{m}$ = 0.22; λ = 0.131; e_o^c = 0.598
Disturbance: D_u = 0.99; Z = 0.665; A = 0.595.

Table 1(b) Parameters for the Ottawa sand.

Elasticity: E = 193,000 kPa; ν = 0.38
Plasticity: γ = 0.123; n = 2.45; h_1 = 0.8450; h_2 = 0.0215
FA (critical state):= $\overline{m}$ 0.15; λ = 0.02; e_o^c = 0.601
Disturbance: D_u = 0.99; Z = 0.43; A = 4.22.

The specimen is installed in the lower box with relative density of 60%. The concrete specimen in the upper box has medium roughness, with the roughness R= 0.50 (Kulhawy & Peterson, 1979). Test were conducted under drained and undrained conditions with one-way (quasi-static) and two-way cyclic loading under constant initial normal stresses, σ_n = 35, 70, 207 kPa, and two relative densities. For the cyclic loading, the tests were performed with horizontal displacement amplitude of 5.00 mm with sinusoidal frequency of 0.38 Hz. Details of test results under various normal stresses are given in Desai *et al.*, (2005).

The parameters for the DSC model were found based on the laboratory tests using the procedure given in (Desai, 2001). The parameters (average) for the Ottawa sand-concrete interface and the sand are given in Table 1; details of the parameters, symbols, etc. are given in (Desai *et al.* 2005).

5.3.1 Validations

The predictions for the laboratory results were obtained by using the finite element procedure, DSC-DYN2D (Desai, 2000b). The FEM mesh is shown in Figure 20. Figure 21a shows comparisons between predictions and test data for shear stress vs. relative displacement for σ_n = 35 kPa and D_r = 60%. The comparisons between pore water pressures and time are shown in Figure 21b. Figure 21c shown comparisons between observed and predicted disturbance vs. accumulated plastic shear relative displacements, ξ_D. It can be seen the DSC model provides highly satisfactory correlation between predictions and tests data.

Typical examples of validations of practical boundary value problems using the DSC are presented below.

5.4 Example 4: BVP level–reinforced earth

Description of Wall: Forty-three geogrid-reinforced walls were constructed at Tanque Verde Road site for grade-separated interchanges on the Tanque Verde-Wrightstown-Pantano Road project in Tucson, Arizona, USA. This project represents the first use of geogrid reinforcement in mechanically stabilized earth (MSE) retaining walls in a major

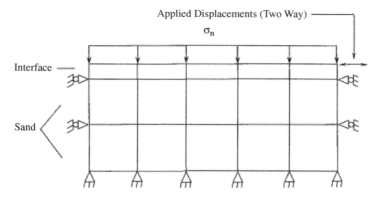

Figure 20 Finite element mesh for interface simulation.

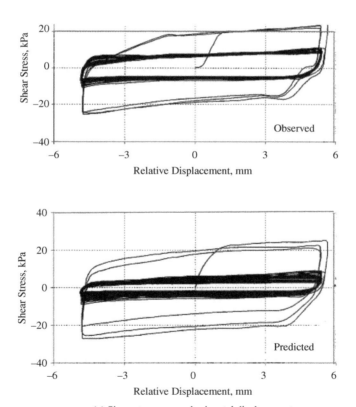

(a) Shear stress versus horizontal displacement

Figure 21 Comparisons between prediction and test data: sand-concrete interface.

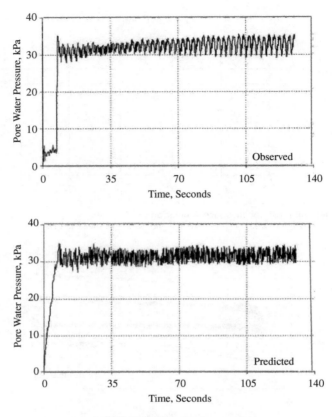

(b) Pore water pressure versus time

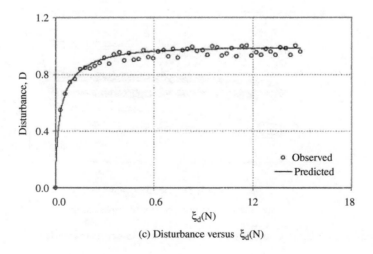

(c) Disturbance versus $\xi_d(N)$

Figure 21 (cont.)

Constitutive modeling of geologic materials 561

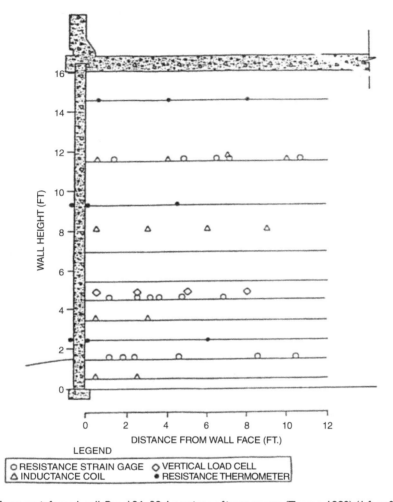

Figure 22 Tensar reinforced wall, Panel 26–32, Locations of instruments (Tensar, 1989) (1 ft. = 0. 305 m).

transportation-related application in North America (Tensar, 1989). Here, the behavior of the wall panel No. 26–32 is simulated using finite element method (FEM); this is one of the two-instrumented panels. The wall height is 4.88 m (16.0 ft). The reinforced soil mass was faced with 15.24 cm (6.0 in) thick and 3.05 m (10.0 ft.) wide precast reinforced concrete panels. Soil reinforced geogrid were mechanically connected to the concrete facing panels at elevations shown in Figure 22, and extended to a length of 3.66 m (12.0 ft). On the top of the wall fill, a pavement structure was constructed that consisted of 10.16 cm (4.0 in) base course covered by 24.13 cm (9.5 in) of Portland cement concrete. Details of the various geometries are reported by Berg et al., (1986), Fishman et al., (1991, 1993), and Fishman & Desai (1991); the latter presents a (linear) finite element analysis for the wall.

The soil reinforcement used was Tensar's SR2 structural geogrid; it is a uniaxial product that is manufactured from high-density polyethylene (HDPE) stabilized with

562 Desai

about 2.5% carbon black to provide resistance to attack by ultraviolet (UV) light (Tensar, 1989). It is reported to be resistant to chemical substances normally existing in soils (Fishman *et al.*, 1991). The geogrid have maximum tensile strength of 79 kN/m (5400 lb/ft) and a secant modulus in tension at 2% elongation of 1094 kN/m (75000 lb/ft). The allowable long-term tensile strength based on creep considerations is reported to be 29 kN/m (1986 lb/ft) at 10% strain after 120 years. This value was reduced by an overall factor of safety equal to 1.5 to compute a long-term tensile strength equal to 19 kN/m (1324 lb/ft).

5.4.1 Numerical modeling

The numerical analysis of the reinforced soil wall was performed using a finite element code called DSC-SST-2D, Desai (1998); the program allows for plane strain, plane stress and axisymmetric idealizations including simulation of construction sequences. Various constitutive models, elastic, elastic-plastic (Von Mises, Drucker-Prager, Mohr-Coulomb, Hoek-Brown, Critical State and Cap), hierarchical single surface (HISS), viscoelastic, plastic and disturbance (softening, DSC)) can be chosen for the analysis. The wall was modeled as a plane-strain, two-dimensional problem; since the Tensar reinforcement is continuous and normal to the cross section, Figure 22, the plane strain idealization is considered to be appropriate. The program was written to allow incremental fill placement to be simulated (*i.e.*, rows of elements added sequentially as the fill placement).

Two finite elements meshes, coarse and fine, were used. The coarse mesh contained 184 nodes and 167 elements including 10 wall facing, 18 interface elements between soil and reinforcement, and 9 bar (for reinforcement) elements; in the coarse mesh, only three layers of reinforcement were considered. The fine mesh contained 1188 nodes, and 1370 elements including 480 interface, 35 wall facing, and 250 bar elements; it contained eleven layers as in the field. The fine mesh was considered to contain a great number of nodes and elements; hence, intermediate and finer meshes were not analyzed. The dimensions for the fine mesh were the same as the coarse mesh; part of the fine mesh near the reinforcement is shown in Figure 23.

It was assumed that the relative motions between the backfill and reinforcement have significant effect on the behavior. Hence, interface elements were provided between backfill and reinforcement. It was also assumed that the relative motions between wall facing and backfill soil in this problem may not have significant influence; hence, interface elements were not provided. This is discussed later under Displacements. However, such relative motions can have influence, and in general, interface elements need to be provided.

The meshes involved four-node quadrilateral elements for soil, wall and interfaces, and one-dimensional elements for the reinforcement. Details of boundary constraints are given in Desai & El Hoseiny (2005). It was found that the fine mesh provided satisfactory and improved predictions compared to those from the coarse mesh. Hence, the results are presented here are for the fine mesh.

5.4.2 Construction simulation

The *in situ* stress was introduced in the foundation soil by adopting coefficient, $K_o = 0.4$. Then the backfill was constructed into eleven layers, Figure 23, as was done in the

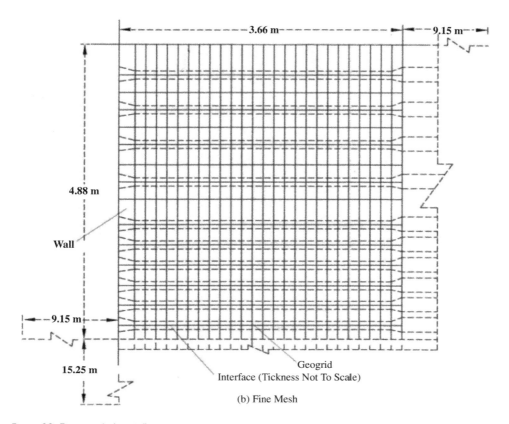

Figure 23 Fine mesh (partial).

field. The compacted soil was included in each layer, and the reinforcement was placed on a layer before the next layer was installed. The compacted soil in a given layer was assigned the material parameters according to the stress state induced after installing the layer. The completion of the sequences of construction is referred to as "end of construction." Then the surcharge load due to the traffic of 20 kPa was applied uniformly on the top of the mesh, Fig. 22; this stage is referred to as "after opening to traffic." The concrete pavement was not included in the mesh. However, since it can have an influence on the behavior of the wall, in general, it is desirable to include the pavement.

5.4.3 Testing and parameters

A comprehensive series of triaxial tests were performed on the backfill soils. The shear tests on reinforcement-soil interfaces were performed using the CYMDOF device. Details of the tests, typical results, parameters and validations for the DSC/HISS models for soils and interfaces are given in Desai & El-Hoseiny (2005).

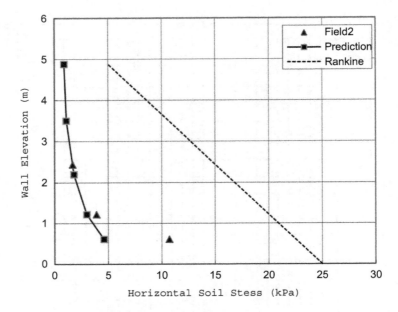

Figure 24 Lateral pressure on facing panel after opening to traffic.

5.4.4 Comparisons between predictions and field measurements

It was found that the results using the fine mesh provided improved correlation with the field test data. Hence, typical results presented here are for the fine mesh only.

Lateral Earth Pressure against Facing Panel: The distribution of lateral earth pressure on the wall facing was measured based on the four pressure cells located at or near the wall face, about 0.61, 1.22, 2.44 and 3.66 m. distance from the bottom of the wall. The earth pressure against the facing panel was obtained in the finite element analysis from the horizontal stresses in the soil elements near the facing. This pressure distribution is useful for evaluating the magnitude of the stresses exerted on the facing panels and the tension in the geogrid connection. Figure 24 shows the typical predicted and measured lateral soil pressure behind the facing panel after opening to traffic, along with the Rankine distribution. Predicted and measured horizontal soil stresses agree very well. The design procedure assumes that no significant lateral earth pressure should be transferred to the reinforcement. Except at the bottom of the wall, the low value of the horizontal soil stress on the wall panel approximately confirms this assumption.

Geogrid Strains: Measured and predicted reinforcement tensile strains at elevation of 4.42 m are shown in Figure 25. Agreement between the measured and predicted values is considered very well. The results demonstrate that tensile strains in the geogrid are less than 0.4% corresponding to 4.4-kN/m load in the geogrid. Comparison of this load to the maximum tensile strength of the geogrid, which is 79 kN/m, indicates that the grids are loaded to about 6.0% of the ultimate strength.

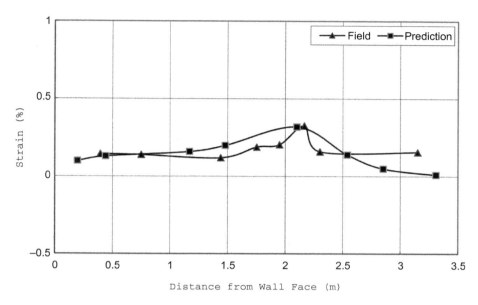

Figure 25 Comparison of Measured and predicted geogrid strains at Elev. 4.42 m.

Displacements: Figure 26 shows predicted and measured wall movements. The correlation is satisfactory near the lower heights of wall; however, it is not satisfactory elsewhere. For example, near the top of the wall the predicted value of about 42 mm is not in good agreement with the measured value of about 76 mm. The finite element analysis using linear elastic model reported the maximum displacement of about 30 mm (Fishman & Desai, 1991). With the present nonlinear soil and interface models, the maximum displacement increased to about 42 mm, Figure 26.

A main reason for the discrepancy is considered to be possible errors in the measurements. Since displacements of the wall are important in design, it is desirable to obtain more accurate measurements. It is believed that since other measurements compare well with the predictions, the displacements from the finite element prediction can be considered to be reasonable. The magnitude of the maximum wall displacement, δ_{max}, can be estimated from the following equation (Christopher et al., 1989):

$$\delta_{max} = \delta_r \times H/75 \tag{18}$$

where δ_r = relative displacement found from the chart based on L/H ratio, H = wall height and L = reinforcement length. According to Equation 18, the $\delta_{max} \approx 60$ mm, which also does not compare well with the measured value of about 78 mm?

From Fig. 26, it can be seen that the wall rotates about the toe of the wall. Also, the displacements of the wall and the soil strains were not high. The maximum displacement is about 1.5% with respect to the wall height. It appears from this behavior that there is no significant relative motion between the wall and soil for this problem. Hence, interface elements between the wall and backfill soil were not included.

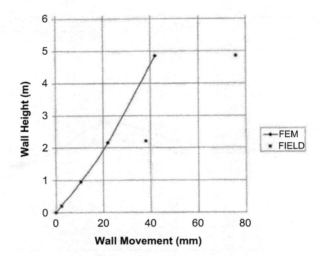

Figure 26 Comparison between predicted and measured wall face movements after opening to traffic.

5.5 Example 5: BVP level–powerhouse cavern in rock

This example contains testing and DSC modeling of rocks and joints in a Powerhouse cavern, river Sutlej, India under the Nathpa-Jhakri hydropower project, and computer (FE) analysis of the behavior of the cavern starting from the initial conditions to construction simulation, and comparisons of FE predictions with field measurements.

Figure 27 shows the powerhouse complex which consists of two major openings.*i.e.*, the machine hall 216 m×20 m×49 m (length×width×height) with an overburden of 262.5 m at crown and the transformer hall 198m×18m×29m which is located downstream of the machine hall. The longitudinal axis of the openings is in the N-S direction. The openings are located in the left bank about 500m away from the Sutlej River (NJPC, 1992; Bhasin *et al.*, 1995).

The *in situ* stress for the rock mass in the power house drift was determined using the hydraulic fracturing technique, by the Central Soil and Materials Research Station (CSMRS), New Delhi, India (Varadarajan *et al.*, 2001). The lateral pressure coefficient (K_o) was equal to about 0.8035 for the E-W cross section; it was used in the analysis.

5.5.1 Constitutive model

From the geology of the powerhouse complex, it was found that most of the rock mass consist of quartz mica schist and biotite schist. The quartz mica schist is weaker than biotite schist and it forms most of the rock mass around the cavern. Therefore, the properties of quartz mica schist are used in the analysis. The rock in the power house area is jointed and has discontinuities. The geologic study of the area shows that the average Rock Mass Rating, RMR, and Tunneling Quality Index, Q, are 50 and 2.7, respectively (Varadarajan *et al.*, 2001).

Constitutive modeling of geologic materials 567

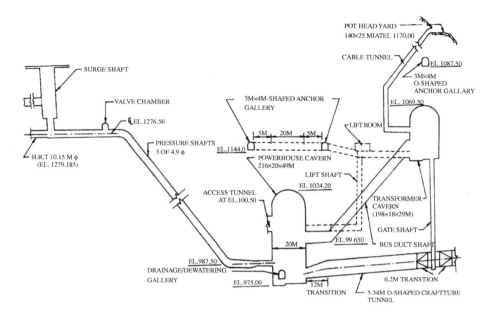

Figure 27 East-west section of surge shaft, pressure shaft and powerhouse caverns, Nathpa-Jhakri hydropower project (after NJPC).

The parameters for the DSC constitutive model for behavior of quartz mica schist rock samples were obtained from strain-controlled triaxial tests on the intact rock samples using a servo-controlled loading system. Material parameters have been obtained for the rock samples have been reported in Varadarajan *et al.*, (2001). The material parameters for the rock mass have to be determined to conduct the finite element analysis of the powerhouse cavern. At the moment, there is no known method which is available to determine the material parameters for the rock mass. In this study, it was proposed to use a procedure suggested by Ramamurthy (1993) to determine the strength and the Young's modulus of the jointed rock mass from the intact rock properties. The parameters for the intact rock and rock mass are given in Varadarajan *et al.*, (2001).

The other parameters viz., phase change, hardening and disturbance parameters for the rock mass have been assumed to be the same as for the intact rock. It is believed that this assumption may not significantly affect the results of the analysis for (i) the values of hardening parameters are small and (ii) the differences in the values of the parameters for disturbance for intact rock and rock mass may not be significant. It is observed that, (i) the ultimate parameters γ and β, for the rock mass show 33% and 17% reduction as compared to the intact rock, (ii) the Young's modulus and uniaxial compressive strength (UCS) for the rock mass are decreased by 22% and 16% as compared to the intact rock, and (iii) the value of bonding stress (3R) shows 11% reduction as compared to the intact rock.

The unit weight of the rock mass is taken to be the same as for the intact rock, *i.e.*, 27 kN/m^3.

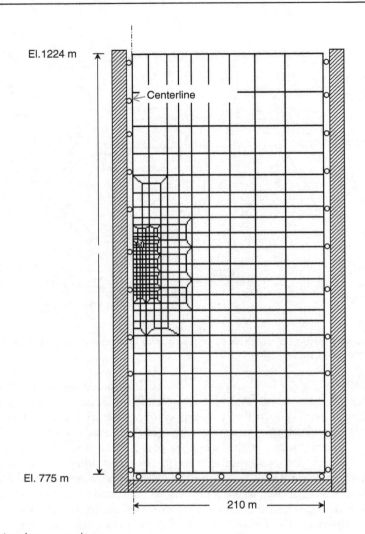

Figure 28 Finite element mesh.

5.5.2 Finite element analysis

The finite element analysis of powerhouse cavern was conducted for the loading due to excavation. The effect of the excavation of the transformer hall on the powerhouse (machine hall) has been found to be negligible (Varadarajan *et al.*, 2001), and therefore, only the excavation of the powerhouse cavern has been considered in the present study.

Since the loading and the geometry of the cavern are symmetric, only the half of the portion has been discretized. The cavern and the rock mass included for the discretization are shown in Fig. 28. The rock mass is discretized into 364 eight-noded elements and 1167 nodes keeping in view various stages of excavation of the cavern. The boundary conditions are also shown in the Fig. 28. The *in situ* vertical stresses have been calculated from the ground at 262.5 m from the crown and K_0 value of 0.8035 as found from the tests has been used.

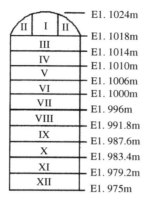

Figure 29 Excavation sequence for the powerhouse cavern.

The analysis has been carried out using the computer code DSC-SST-2D developed by Desai (1998). The analysis consists of the simulation of excavation of the powerhouse cavern. Twelve stages of excavation have been used in the study as shown in Figure 29.

In the computer code, at each stage of excavation the elements and nodes to be removed from the finite element mesh are deactivated, *i.e.*, their stiffness matrices and load vectors are not included in the global stiffness matrix and load vector. The complete details of the excavation procedure are presented elsewhere (Varadarajan *et al.*, 2001). The analysis is conducted using incremental-iterative procedure adopted by (Shao & Desai, 2000) in which the tangent stiffness matrix $[K_t]$ is updated at each load increment, and an iterative procedure is used during each load increment. The method proposed by Potts & Gens (1985) which considers hardening during the drift correction has been adopted.

The results of the analysis from the computer code DSC-SST-2D have been processed through a commercial package NISA (1993) and the contours of displacements, deformed shape and the variation of major and minor principal stresses around the cavern have been plotted for the full excavation of the cavern at 12th stage.

Instrumentation: The powerhouse cavern has been instrumented by National Institute of Rock Mechanics, NIRM, to study the movement in the rock mass during various stages of excavation. The instruments have been installed at various sections along the length of the cavern. The instrumentation scheme of a section in the middle of the cavern is available in Varadarajan *et al.*, (2001).

5.5.3 *Results and discussions*

The horizontal displacement contours are shown in Figure 30. Higher movements of the wall are noted around the mid-height of the wall. The maximum value of 42.6 mm movement is observed at the cavern face and this value decreases to 9.22 mm at a distance of 73 m from the face. The former value compares well with the predicted value of 45 mm.

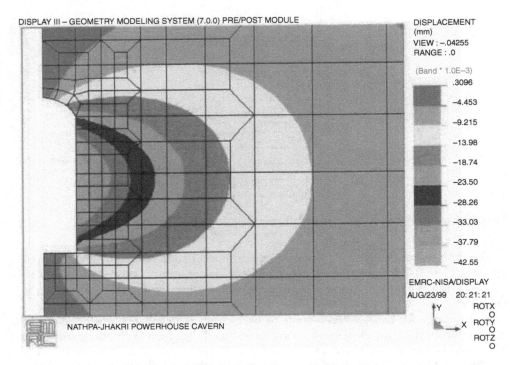

Figure 30 Contours of the horizontal displacement around cavern.

The variations of the major and minor principal stresses for typical section are shown in Fig. 31.

5.5.3.1 Comparison of observed and predicted displacements

The finite element analysis was conducted simulating the same excavation stages as shown in Fig. 29, and the displacements have been determined at various locations where the instruments have been installed. The predicted and observed values of the displacements are presented in Table 2. The predicted values lie within the range of the observed values of displacements at five out of six locations. It can be considered that, in general, the predictions are satisfactory.

5.6 Example 6: BVP level–pavement materials and application

Use of the DSC model for pavements materials and application for layered pavement system are presented in this example. Detailed analyses for two- and three-dimensional problems are presented in Desai (2007, 2009). Here typical example for three-dimensional analysis is presented. In addition to displacements, strains, stresses and other quantities like temperature and pore water pressure, the progressive failure and rutting are also important. Here, the DSC is used to identify microcracking and rutting. DSC can also allow for behavior both concrete and asphalt concrete.

Table 2 Comparison of Predicted (FEM) and Observed (instrumentation) Deformation at the Powerhouse Cavern Boundary.

Stage No.	Excavation Done		Instrumentation at El. (m)	Deformation (mm)	
	From El. (m)	To El. (m)		Predicted (FEM)	Observed (Instrumentation)
1	Widening of the Central drift		1024 (A)	10.4	13–18
2	Widening of the Central drift		1022 (B)	12	6–12
3	1018	1006	1022 (B)	0.6	−1.3 to +2.5
4	1006	1000	1018 (C)	3.5	1–4
5	1000	975	1006 (D)	23.7	10–45
6	983	975	996 (E)	9.4	1–3

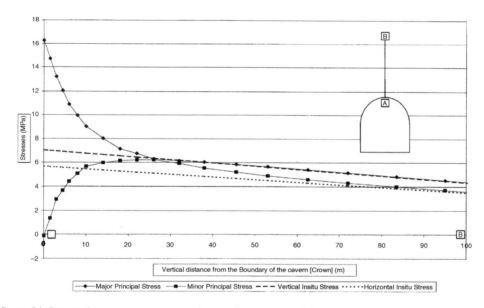

Figure 31 Principal stress variation in rock mass from crown of the powerhouse cavern.

5.6.1 Material parameters

Table 3 shows the material parameters obtained from the triaxial tests under various confining pressures and temperatures on asphalt concrete reported by (Monismith & Secor, 1962). These tests were comprehensive; however, they usually did not exhibit the softening behavior. Hence, the parameters were found for only the HISS model.

Scarpas et al. (1997) reported uniaxial tests for asphalt concrete in which the pre peak and post peak (softening) behavior was observed. These tests were used to evaluate the parameters for disturbance, D, in Table 3.

Table 3 Material Parameters for Pavement Materials for HISS Model and Disturbance (1 psi = 6.89 kPa).

Parameter	Asphalt concrete	Base	Subbase	Subgrade	Concrete
E	500000 psi	56532.85 psi	24798.49 psi	10013.17 psi	3×10^6 psi
ν	0.3	0.33	0.24	0.24	0.25
γ	0.1294	0.0633	0.0383	0.0296	0.0678
β	0.0	0.7	0.7	0.7	0.755
n	2.4	5.24	4.63	5.26	5.24
3R	121 psi	7.40 psi	21.05 psi	29.00 psi	8×10^3 psi
α_1	1.23E-6	2.0E-8	3.6E-6	1.2E-6	0.46×10^{-10}
η_1	1.944	1.231	0.532	0.778	0.826
D_u	1				0.875
A	5.176				668.0
Z	0.9397				1.502

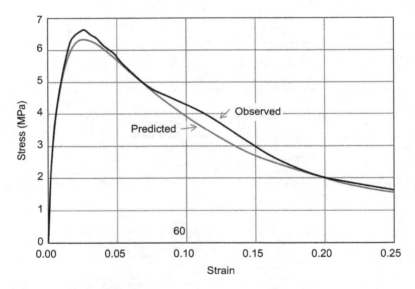

Figure 32 Comparison Between Predicted and Observed Test Data for $T = 25°C$, $\dot{\varepsilon} = 5$ mm/sec.

The validation of the stress-strain curves was obtained in two ways: by integrating the incremental constitutive Equations 4, and by using the finite element analysis. Comparisons between the observed and predicted stress-strain curves for strain rate = 1 in/min at three typical temperatures (T) and confining pressures (σ_3): (a) T = 40°F, σ_3 = 43.8 psi, (b) T = 77°F, σ_3 = 0.0 psi, (c) T = 140°F, σ_3 = 250 psi (Monismith & Secor, 1962) are given in Desai (2007,2009). Figure 32 shows typical comparison between predicted and observed typical stress-strain curve for T = 25°C and strain rate = 5 mm/sec (Scarpas et al., 1997). The above comparisons show that the DSC model can provide very good simulation of the behavior asphalt concrete.

5.6.2 Computer implementation

The DSC model has been implemented in two- and three-dimensional computer (finite element) procedures (Desai, 1998, 2000c, 2001). The computer codes allow for the nonlinear material behavior, *in situ* or initial stresses, static, repetitive and dynamic loading, thermal and fluid effects. They include computation of displacements, strain (elastic, plastic, and creep), stresses, pore water pressures and disturbance during the incremental and transient loading. Specifications of critical values of disturbance, D_c, permit identification of the initiation of microcracking leading to fracture and softening, and cycles to fatigue failure. Plots of the growth of disturbance, *i.e.*, microcracking to fracture, are obtained as a part of the computation. Accumulated plastic strains lead to the evaluation of the growth of permanent deformations and rutting.

Loading: The codes allow for quasistatic and dynamic loading for dry and saturated materials. The repetitive loading on pavement can involve a large number of cycles. An approximate procedure to handle a great number of cycles by performing FEM up to about ten reference cycles is described below.

5.6.3 Repetitive loading: Accelerated procedure

Computer analysis for 3- and 2-D idealizations can be time consuming and expensive, especially when significantly greater number of cycles of loading need to be considered. Therefore, approximate and accelerated analysis procedures have been developed from a wide range of problems in civil (pavements) (Huang, 1993; Lytton *et al.*, 1993), mechanical engineering and electronic packaging (Desai *et al.*, 1997, Desai & Whitenack, 2001). Here, the computer analysis is performed for only a selected initial cycles (say, 10, 20), and then the growth of plastic strains is estimated on the basis of empirical relation between plastic strains and number of cycles obtained from laboratory test data. A general procedure with some new factors has been developed (Desai & Whitenack, 2001). This procedure is modified for pavement analysis and is described below.

From cyclic tests on an engineering material, the relation between plastic strain (in the case of DSC, the deviatoric plastic strain trajectory, ξ_D, Equation 9b, and the number of loading cycles can be expressed as

$$\xi_D(N) = \xi_D(N_r)\left(\frac{N}{N_r}\right)^b \tag{19}$$

where N_r = reference cycle, and b is a parameter, depicted in Figure 33. The disturbance Equation 3 can be written as

$$D = D_u[1 - \exp\left(-A\left\{\xi_D(N)^Z\right\}\right] \tag{20}$$

Substitution of $\xi_D(N)$ from Equation 19 into Equation 20 leads to

$$N = N_r\left[\frac{1}{\xi_D(N_r)}\left\{\frac{1}{A}\ell n\left(\frac{D_u}{D_u - D}\right)\right\}^{1/Z}\right]^{1/b} \tag{21}$$

Now, Equation 21 can be used to find the cycles to failure, N_f, for chosen critical value of disturbance = D_c (say, 0.50, 0.75, 0.80).

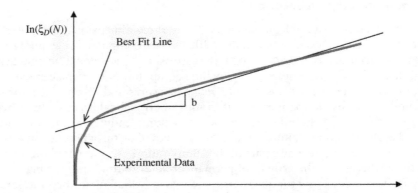

Figure 33 Accumulated Plastic Strain vs. Number of Cycles for Approximate Accelerated Analysis.

The accelerated approximate procedure for repetitive load is based on the assumption that during the repeated load applications, there is no inertia due to "dynamic" effects in loading. The inertia and time dependence can be analyzed by using the 3-D and 2-D procedures; however, for million cycles, it can be highly time consuming. Application of repeated load in the approximate procedure involves the following steps:

1. Perform full 2-D or 3-D FE analysis for cycles up to N_r, and evaluate the values of $\xi_D(N_r)$ in all elements (or at Gauss points).
2. Using Equation 19 compute $\xi_D(N)$ at selected cycles in all elements.
3. Compute disturbance in all elements using Equation 20.
4. Compute cycles to failure N_f by using Equation 21, for the chosen value of D_c.

The above value of disturbance allows plot of contours of D in the finite element mesh, and based on the adopted value of D_c, it is possible to evaluate extent of fracture and N_f.

Loading-Unloading-Reloading: Special procedures are integrated in the codes to allow for loading, unloading and reloading during the repetitive loads; details are given in Desai (2001).

5.6.4 Validations and applications

Generally, the pavement problem and wheel loading would require three-dimensional analysis, particularly to predict microcracking and fracture response. However, for an economic analysis, a two-dimensional procedure can provide satisfactory but approximate solutions for certain applications. A typical example for the three-dimensional case for four layered flexible pavement is presented here.

Figure 34 shows the 3-D problem and three dimensional mesh is shown in Fig. 35. The material properties used are shown in Table 3.

Analyses are performed by applying linearly monotonic loading and repetitive loading. The monotonic loading was applied up to 200 psi (1400 kPa) in 50 increments; details are given in Desai (2007,2009). For the repetitive loading, Figure 36, the load amplitude (P) was equal to 200 psi (1400 kPa). As discussed before, the cyclic load (loading, unloading, and reloading) was applied sequentially; however, time dependence was not included at this time.

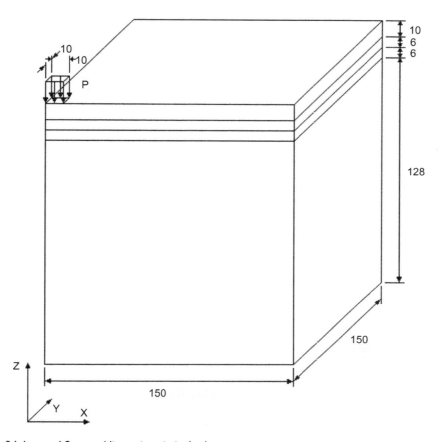

Figure 34 Layered System (dimensions in inches).

Full finite element analysis was performed for each load amplitude up to $N_r = 20$ cycles. Then the deviatoric plastic strain trajectory (ξ_D) at the given cycle (N) was computed for the subsequent cycles using Equation 19. The disturbance, D, was computed at the given cycles using Equation 20. This allowed analysis of disturbance with cycles and computation of the cycles to failure, N_f, depending upon the chosen criteria for critical disturbance, D_c, Equation 21.

Figures 37 (a) to (c) shows contours plots of disturbance after 10, 1000 and 20,000 cycles, respectively with load amplitude = 70 psi and b = 1.0. After about 20,000 cycles for $D_c = 0.8$, a portion of pavement has experienced fracture. Based on the contours of D and critical D_c, it is possible to trace the history of microcracking, cracking and fracture.

5.7 Example 7: BVP level–centrifuge testing of pile and liquefaction

This example presents use of the DSC model for cyclic behavior of saturated sand and FE predictions of pile tested using the centrifuge including computation for liquefaction

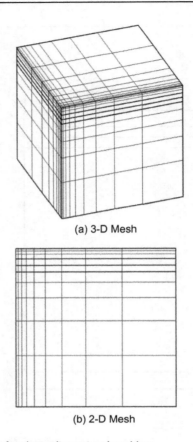

Figure 35 Finite element mesh for three-dimensional problem.

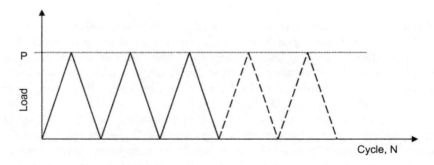

Figure 36 Schematic of repetitive loading.

by the DSC model (Desai, 2000a). The Nevada, California sand was used in the centrifuge. To determine the DSC parameters, soil test data for the Nevada sand was used obtained by The Earth Technology Corporation, Huntington Beach, California for VELACS (Arulmoli et al., 1992). For determining parameters, three undrained

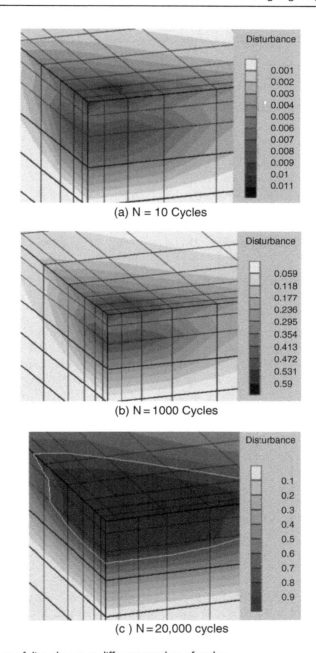

Figure 37 Contours of disturbance at different number of cycles.

monotonic triaxial tests, and three undrained cyclic triaxial test with confining pressures, σ_3, = 40, 80, and 160 kPa at 60 % relative density were used. Details of evaluating these parameters from the test data are given in (Pradhan & Desai, 2006). Table 4 lists parameters for the sand and interface.

578 Desai

Table 4 Parameters for Nevada Sand and Sand-Aluminum Interface.

Group	Subgroup	Parameters	Nevada Sand	Nevada Sand Aluminum Interface
Group 1	**Elastic** Parameters	E	40848.8 kPa	14.6 MPa
		ν	0.316	0.384
	Plasticity Parameters	γ	0.0675	0.246
		β	0.0	0.000
		3 R	0.0	0.0
		n	4.1	3.350
		h_1	0.1245	0.620
		h_2	0.0725	0.570
Group 2	*Critical* **State Parameters**	$\overline{m}$	0.22	0.304
		λ	0.02	0.0278
		e_c^0	0.712	0.791
Group 3	**Disturbance Parameters**	Du	0.99	0.990
		Z	0.411	1.195
		A	5.02	0.595

5.7.1 DSC model for interfaces

Interface test data for the aluminum-Nevada sand were not available for the prediction from the centrifuge tests, described subsequently. Hence, a neural network procedure was used (Pradhan & Desai, 2002). Here two sets of data were used for the training of the neural network. Set 1 consisted of the Ottawa sand parameters using the triaxial testing and the Ottawa sand-steel interface using the cyclic multi-degree of freedom (CYMDOF) device (Alanazy, 1996; Desai & Rigby, 1997). The second set consisted of the parameters for marine clay at Sabine, Texas, from triaxial and multiaxial testing (Katti & Desai, 1995) and marine clay-steel interface using the CYMDOF (Shao and Desai, 2000). It was assumed that the steel and aluminum behave approximately similar in interface with the soil. Also, the Nevada sand gradation curve falls between those of the Ottawa sand and the Sabine clay. The parameters for the Nevada sand found from the available triaxial data (Arulmoli, *et al.*, 1992) were used in the neural network, which provided estimates for Nevada sand-aluminum interface parameters. They are given in Table 4 and were used for the interfaces in the pile test using the centrifuge device, described later.

The model parameters were used to back predict specimen level laboratory tests (D_r = 60%) that were employed to find parameters and also an independent test (D_r = 40%) whose results were not used to find parameters (Pradhan & Desai, 2006). The predictions were obtained using DSC-DYN2D program developed by Desai (2000b). The predictions compared very well with the test data.

5.7.2 Centrifuge test

Prediction of the behavior of pile foundations in liquefiable sands under earthquake loading can be a challenging problem. The best way to understand such a problem and develop a design and analysis procedure is comparison with full-scale field data, which can be expensive. Hence, centrifuge testing represents a useful development in studying fundamental mechanisms of soil-pile structure interaction. The centrifuge test data used herein were performed at the National Geotechnical Centrifuge at University of California, Davis (Kutter et al., 1991, 1994; Wilson et al., 1997a, b, c). It has a radius of 9.0 m and is equipped with large shaking table. It has a maximum model mass of about 2500 kg, and available bucket area of 4.0 m^2 and a maximum centrifugal acceleration of 50 g. Details of the centrifuge can be found in (Kutter et al., 1991, 1994; Wilson et al. 1997a, b, c).

Five containers of soil- structure systems were tested at a centrifugal acceleration of 30 g. Full details for each test can be found in (Wilson et al., 1997a, b). In this study, earthquake event J in the model, referred to as CSP3, was simulated. The soil profile in this container consisted of two horizontal layers. The upper 9.3 m layer was medium – dense Nevada sand (D_r = 55 %) and lower 11.4 m thick layer was dense Nevada sand (D_r = 80 %), Figure 38. Details of the structure and pile systems can be found in Wilson et al., (1997a, b, c). Foundation models included single pile foundations, four pile group and a nine pile group, with superstructure mass typically being 500 kN per each supporting pile.

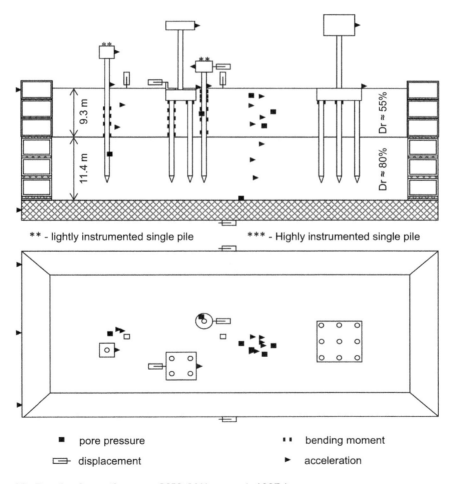

Figure 38. Details of centrifuge test CSP3 (Wilson et al., 1997c).

In this analysis, a single prototype pile near the middle, Figure 38, made of aluminum pipe with 0.67 m diameter, 72 mm wall thickness, and 20.6 m and 16.8 m embedment depth was used. Linear elastic properties were used for the pile: E = 70.0 GPa and ν = 0.33. The DSC model was used for the sand and interface; the parameters are given in Table 4.

It would be appropriate to use a three-dimensional finite element (FE) for the problem; however, such procedures were not readily available. Hence, as an approximation, the single pile was modeled by using two-dimensional FE procedure (Desai, 2000b), with plane strain idealization. The plane strain idealization for such a pile is considered to be more appropriate than the axisymmetric idealization, partly because the latter allows mainly the axial behavior. Similar plane strain idealization for piles has been used previously (Desai, 1974b; Anandarajah, 1992; Ellis, 1997; Fujii, et al., 1998). The finite element mesh is shown in Fig. 39 and the applied loading in Fig. 40.

Initial *in situ* stresses and pore pressures at centers of all elements were introduced by using the following expressions:

$$\sigma'_v = \gamma_s h \tag{22a}$$

$$\sigma'_h = K_j \sigma'_v \tag{22b}$$

$$K_o = \nu/(1-\nu) \tag{22c}$$

$$p = \gamma_w h \tag{22d}$$

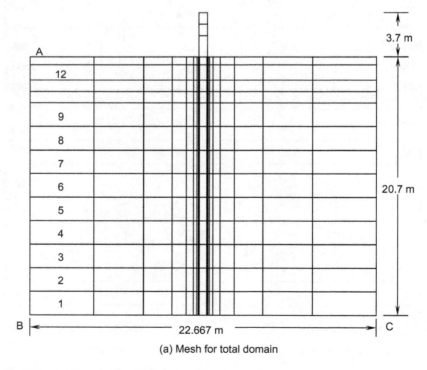

(a) Mesh for total domain

Figure 39 Finite element mesh for single pile: Full and partial.

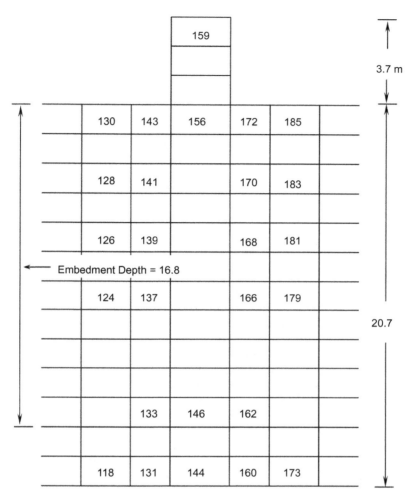

Figure 39 (cont.)

where σ'_v and σ'_h = effective vertical and horizontal stresses at depth, h, γ_s = submerged unit weight of soil, γ_w = unit weight of water, p = initial pore water pressure and K_o = co-efficient of earth pressure at rest.

Comparison of Results: Two finite element analyses were performed, *with and without interfaces*. The predictions and observed pore water pressure were compared for a number of elements: 1, 5, 8, 9, 12 and 139, Figure 39. Typical results for element 139 near the pile, element 9 away from the pile but at the same level as element 139, and

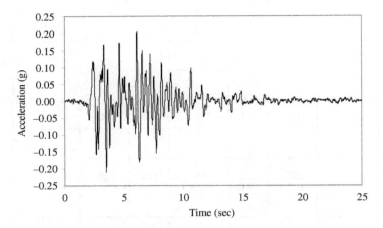

Figure 40 Base motion acceleration (Wilson et al., 1997c).

element 5 away from the pile but at a lower level are presented in Figures 41–43, respectively. It can be seen from these figures that the model with the interface provides improved correlation with the test results. Pore water pressure, with interface, in element 139 reaches the initial vertical effective stress, σ'_v, after about 9 seconds indicating liquefaction, while that without interface does not indicate liquefaction. Furthermore, the predictions for both with and without interface, show similar values and trends for elements away from the pile. This can be mainly because the influence of the relative motions at interface diminishes away from the pile.

Table 5 compares the time when the disturbance reaches its critical value, $D_c = 0.86$, with the conventional procedure in which liquefaction occurs when the excess pore water pressure, U_w, reaches the initial effective pressure (σ'_v), for typical elements. The times to liquefaction correlate well. However, those from the DSC model are lower than those from the conventional procedure, implying that the microstructural instability can occur earlier than the time to liquefaction obtained from the conventional procedure. The conventional procedure implies that the soil does not possess any strength when liquefaction occurs. In fact, it has been reported, *e.g.*, by Desai (2000a), that the soil retained a (small) strength after the Port Island, Kobe, Japan earthquake. Thus, in the DSC, the time to liquefaction ($D_c \approx 0.86$) is lower than that at complete loss of strength. Also, it is believed that the microstructural instability or liquefaction can occur before the complete collapse of the microstructure. Further investigation regarding this aspect is desirable.

Figure 44 shows typical variation of disturbance with time in the interface element No. 139 and in the adjacent element No. 126. These figures indicate that the liquefaction for this problem can occur in the interface element earlier than in the surrounding elements.

The top bucket ring at which the displacements were measured was not included in the pile mesh. Hence, comparisons for displacements were not available. However, Fig. 45 shows comparisons between predicted accelerations near the top of the pile from analysis with interfaces and the measured values. The peak values differ by about 12.5%; overall, the correlation is considered to be satisfactory.

Constitutive modeling of geologic materials 583

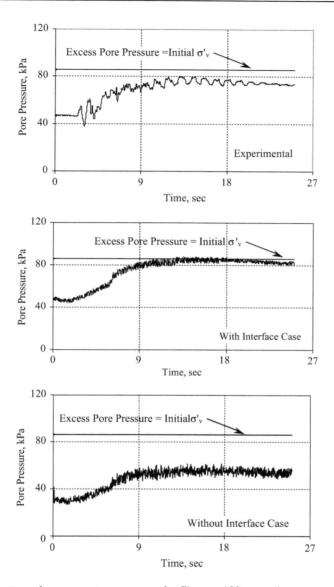

Figure 41 Comparisons for pore water pressures for Element 139 near pile.

Practical Use: For practical application for analysis and design, the following ratios can be defined:

$$R_1 = \frac{V^\ell}{V^s} \tag{23a}$$

$$R_2 = \frac{V^\ell}{V^p} \tag{23b}$$

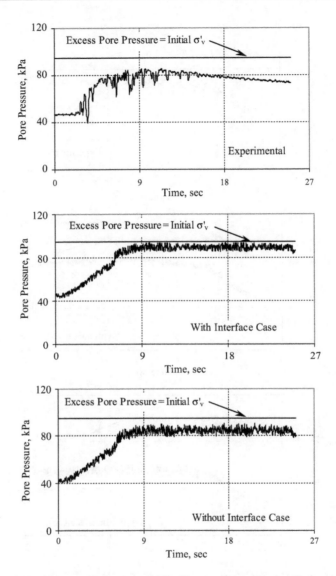

Figure 42 Comparisons of pore water pressures for Element 9, away from pile on same level as I39.

where V^ℓ = liquefied volume, V^s = total soil volume, and V^p = pile volume. For this problem, R_1 and R_2 are defined based on the fixed geometry and dimensions. However, for a field problem involving semi-infinite extent, it will be difficult to define R_1. Hence, only R_2 involving the volume of the pile (or the structure) can be used. Figure 46 shows the variation with time for R_1 and R_2. Considering that liquefaction occurs after about 9 secs, the critical values are: R_1 = 0.24 and R_2 = 8.0. Thus, the finite element predictions can be used to define critical ratios such as R_1 or R_2 for analysis and design for liquefaction.

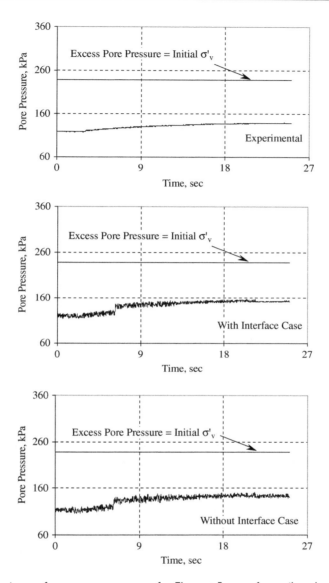

Figure 43 Comparisons of pore water pressures for Element 5, away from pile at lower level.

5.8 Example 8: DSC for active and passive earth pressures

The DSC is a general and unified approach which can used to model many physical phenomena. In an interesting approach, Zhu et al. (2009) has used the DSC to derive general expressions for active and passive earth pressures.

The RI is assumed to be the state at which (rigid) retaining wall moves toward or away from the backfill with an associated parameter S_0 that can assume value of

Table 5 Times to Liquefaction from Disturbance and Conventional Methods.

Element	$U_w = \sigma'_v$	$D = D_c$
143	1.74 Seconds	1.23 Seconds
130	1.83 Seconds	1.29 Seconds
104	2.55 Seconds	1.92 Seconds
78	3.36 Seconds	2.67 Seconds
52	3.81 Seconds	2.94 Seconds

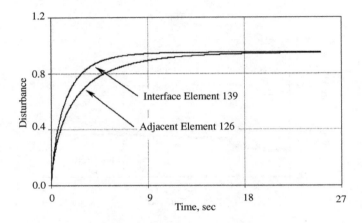

Figure 44 Disturbance versus time in elements 126 and 139.

positive, negative or zero; in most cases, the at-rest condition is assumed to be the RI state, hence, S_0 is zero. The FA state can be considered the asymptotic state at which the system reaches a state at which the (soil) properties are different from those in the RI state.

The disturbance function D as the function of the parameter S, related to the wall movement, is expressed as

$$D = D(S) = \frac{2 \arctan\left[\frac{b(s-s_o)}{(s-s_a)(s-s_p)}\right]}{\pi} \qquad (24)$$

where S are defined in Figure 47, subscripts 0, a and p denote at rest condition (for rigid wall), active Rankine state and passive Rankine state, respectively, and b denotes the material parameter related to the relative density of the soil back fill.

The expression for lateral pressure, p_D, against the vertical rigid wall is derived as (Zhu et al., 2009):

$$p_D = \gamma h K_D \qquad (25)$$

where γ is the unit weight of the backfill, h is the height of the calculated point, and h ranges from 0 to H, H being the total height of the retaining wall.

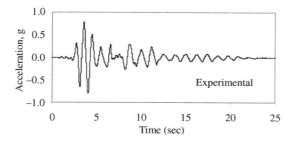

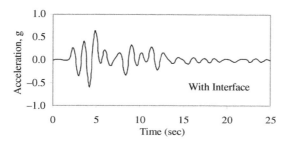

Figure 45 Comparisons between predicted and experimental acceleration near the pile top.

$$K_D = \tan^2(45° - \frac{\varphi_D}{2}) \tag{26a}$$

where φ_D, related to the disturbance in the backfill is given by

$$\varphi_D = 90° - 2\arctan\sqrt{A} \tag{26b}$$

$$A = \tan^2[45° - \frac{D\varphi}{2}] \times \{1 - \sin[(1-D)(1+D)\varphi]\} \tag{26c}$$

Zhu et al., (2009) conducted laboratory tests to measure earth pressures, in which details of the set up and properties of soil backfill have been presented. Comparisons between prediction by using Equation 25 and the measurements are also included in Zhu et al., (2009).

6 SUMMARY AND CONCLUSIONS

The disturbed state concept is used to characterize the behavior of geologic materials and interfaces (joints) between structural materials (piles) and joints. The parameters for the DSC model are determined based on the available triaxial and multiaxial tests. Those for interface are obtained usually based on tests for interfaces by using devices such as CYMDOF. Finite element procedures with the DSC model based on dry materials and saturated materials using the generalized Biot's theory are used to validate the DSC model, in which FE predictions and measurements are compared for displacements, stresses and disturbance; the latter can be used to identify liquefaction.

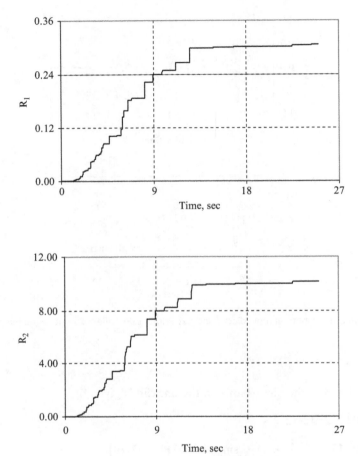

Figure 46 Variation of disturbance ratios R_1 and R_2.

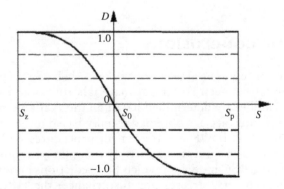

Figure 47 Disturbance, D, and wall movement (S).

Based on the results reported herein, the DSC model can provide satisfactory and improved predictions for the behavior of a wide range of problems in geotechnical engineering and Geomechanics. It is believed the DSC can provide a unique and unified constitutive model for a wide range of engineering materials, interfaces and joints.

REFERENCES

Akhaveissy, A. &. Desai, C.S. (2013) FE analysis of RC structures using DSC model with yield surfaces for tension and compression. *Comput. Concr.*, 11(2), 123–148.

Alanazy, A.S. (1996). Testing and modeling of sand-steel interfaces under static and cyclic loading. *PhD Dissertation*, Department of Civil Engineering and Engineering Mechanics, University of Arizona, Tucson, AZ.

Anandarajah, A. (1992) Fully coupled analysis of a single pile founded in liquefiable sands. *Report*, Department of Civil Engineering, Johns Hopkins University, Baltimore, MD.

Archard, J.L. (1959) Elastic deformation and the laws of friction. *Proc. R. Soc. London, Ser. A*, 243, 190–205.

Arulmoli, K., Muraleetharan, K.K., Hossain, M.M. & Fruth, L.S. (1992) VELACS: verification of liquefaction analysis by centrifuge studies. *Soil Data Report*, Earth Technology Corporation, Irvine, CA.

Bak, P. & Tang, C. (1989) Earthquakes as a self-organized critical phenomenon. *J. Geophy. Res.*, 94, B1.

Berg, R.R., Bonaparte, R., Anderson, R.P. & Chouery, V.E. (1986) Design, construction, and performance of two reinforced soil retaining walls. *Proc. 3rd Int. Conf. on Geotextiles*, Vienna, Austria, 2, 401–406.

Bhasin, R., Barton, N., Grimstad, E. & Chryssanthakis, P. (1995) Engineering characterization of low strength anisotropic rocks in the Himalayan region for assessment of tunnel support. *Eng. Geol.*, 40, 169–193.

Christopher, B.R. *et al.* (1989) Reinforced soil structure. Report No. FHWA-RD-89-043, Volume II, Federal Highway Administration, McClean, VA, USA, 158.

Desai, C.S. (1974) A consistent finite element technique for work-softening behavior. In: Oden, J.T. *et al.* (Editors), *Proc., Int. Conf. on Comp. Meth. in Nonlinear Mechanics*, University of Texas, Austin, TX.

Desai, C.S. (1976) Finite element residual schemes for unconfined flow. *Tech. Note. Int. J. Numer. Methods Eng.*, 10(6), 1415–1418.

Desai, C.S. (1980) A general basis for yield, failure and potential functions in plasticity. *Int. J. Numer. Anal. Method Geomech.*, 4, 361–375.

Desai, C.S. (1998) DSC-SST-2D: computer code for static, dynamic, creep and thermal analysis – solid, structure and soil-structure problems. *User's Manuals*, Tucson, AZ.

Desai, C.S. (2000a) Evaluation of liquefaction using disturbance and energy approaches. *J. Geotech. Geoenviron. Eng., ASCE*, 126(7), 618–631.

Desai, C.S. (2000b) DSC-DYN2D, Computer code for dynamic and static analysis: dry and saturated materials. *Reports Part I, II and III*, Tucson, AZ.

Desai, C.S. (2000c) DSC-SST3D code for three-dimensional coupled static, repetitive and dynamic analysis. *User's Manual I to III*, Tucson, AZ.

Desai, C.S. (2001) *Mechanics of materials and interfaces: the disturbed state concept*. CRC Press, Boca Raton, FL.

Desai, C.S. (2007) Unified DSC constitutive model for pavement materials with numerical implementation. *Int. J. Geomech., ASCE*, 7(2), 83–101.

Desai, C.S. (2009) Unified disturbed state constitutive modeling of asphalt concrete. Chapter 8 in *Modeling of Asphalt Concrete*, Kim, Y.R. (Editor), American Society of Civil Engineers, New York.

Desai, C.S. (2013) Disturbed state concept (DSC) for constitutive modeling of geologic materials and beyond. In: *Proc., Symposium on Constitutive Modeling by Geomaterials*, Tsinghua University, Beijing, China, Springer Series in Geomechanics and Engineering, 27–45.

Desai, C.S. (2015a) Constitutive modeling of materials and contacts using the disturbed state concept: Part 1 – Background and analysis. *Comput. Struct.*, 146, DOI: 10.1016/j.compstruc.2014.07.018.

Desai, C.S. (2015b) Constitutive modeling of materials and contacts using the disturbed state concept: Part 2 – Validations at specimen and boundary value problem levels. *Comput. Struct.*, 146, DOI: 10.1016/j.compstruc.2014.07.026.

Desai, C. S. & Abel, J.F. (1972) *Introduction to the finite element method*. Van Nostrand Reinhold Co., New York.

Desai, C.S., Chia, J., Kundu, T. & Prince, J. (1997) Thermomechanical response of materials and interfaces in electronic packaging: Parts I and II. *J. Elect. Packag.*, ASME, 119(4), 294–300, 301–309.

Desai, C.S., Dishongh, T. & Deneke, P. (1998) Disturbed state constitutive model for thermomechanical behavior of dislocated silicon with impurities. *J.Appl. Phys.*, 84(11), 5977–5984.

Desai, C.S. & El-Hoseiny, K.E. (2005) Prediction of field behavior of geosynthetic-reinforced soil wall. *J. Geotech. Geoenviron. Eng.*, ASCE, 131(6), 729–739.

Desai, C.S. & Li, G.C. (1983) A residual flow procedure and application for free surface flow in porous media. *Int. J. Adv. Water Res.*, 6(1), 27–35.

Desai, C.S. & Ma, Y. (1992) Modeling of joints and interfaces using the disturbed state concept. *Int. J. Numer. Anal. Methods Geomech.*, 16(9), 623–653.

Desai, C.S., Park, I.J. & Shao, C. (1998) Fundamental yet simplified model for liquefaction instability. *Int. J. Numer. Anal. Method Geomech.*, 22, 721–748.

Desai, C.S., Pradhan, S.K. & Cohen, D. (2005) Cyclic testing and constitutive modeling of saturated sand-concrete interfaces using the disturbed state concept. *Int. J. Geomech.*, ASCE, 5(4), 286–294.

Desai, C.S. & Rigby, D.B. (1997) Cyclic interface and joint shear device including pore pressure effects. *J. Geotech Geoenviron. Eng.*, ASCE, 123(6), 568–579.

Desai, C.S., Samtani, N.C. & Vulliet, L. (1995) Constitutive modeling and analysis of creeping slopes. *J. Geotech. Eng.*, ASCE, 121(1), 43–56.

Desai, C.S., Sane, S. & Jenson, J. (2011) Constitutive modeling including creep and rate dependent behavior and testing of glacial till for prediction of motion of glaciers. *Int. J. of Geomech.*, ASCE, 11(6), 465–476.

Desai, C.S. & Siriwardane, H.J. (1984) *Constitutive laws for engineering materials*. Prentice-Hall, Englewood Cliffs, NJ.

Desai, C.S., Somasundaram, S. & Frantziskonis, G. (1986a) A hierarchical approach for constitutive modeling of geologic materials. *Int. J. Numer. Anal. Methods Geomech*, 10(3), 225–257.

Desai, C.S. & Toth, J. (1996) Disturbed state constitutive modeling based on stress-strain and nondestructive behavior. *Int. J. Solids Struct.*, 33(1), 1619–1650.

Desai, C.S. & Whitenack, R. (2001) Review of models and the disturbed state concept for thermomechanical analysis in electronic packaging. *J. Elect. Packag.*, ASME, 123(1), 19–33.

Desai, C.S. & Zaman, M.M. (2014) *Advanced geotechnical engineering: soil-structure interaction using computer and material models*. CRC Press/Taylor and Francis, Boca Raton, FL.

Desai, C.S., Zaman, M.M., Lightner, J.G. & Siriwardane, H.J. (1986b) Thin-layer element for interfaces and joints. *Int. J. Numer. Anal. Methods Geomech*, 14(1), 1–18.

Desai, C.S. & Zhang, W. (1998) Computational aspects of disturbed state constitutive models. *Int. J. Comput. Methods Appl. Mech. Eng.*, 151, 361–376.

Desai, C.S. & Zheng, D. (1987) Visco-plastic model with generalized yield function. *Int. J. Num. Anal. Methods Geomech.*, 11, 603–620.

Elgamal, A., Yang, Z. & Parra, E. (2002). Computational modeling of cyclic mobility and post-liquefaction site response. *Soil Dyn. Earthquake Eng.*, 22, 259–271.

Ellis, E.A. (1997). Soil-structure interaction for full-height piled bridge abatements constructed on soft clay. *Ph.D. Thesis*, Cambridge University, UK.

Fakharian, K. & Evgin, E. (2000) Elasto-plastic modeling of stress-path development behavior of interfaces. *Int. J. Numer. Anal. Methods Geomech.*, 24(2), 183–199.

Fujii, S., Cubrinovski, M., Tokimatsu, K. & Hayashi, T. (1998) Analyses of damaged and undamaged pile foundations in liquefied soils. *Geotech. Sp. Publ. 050.75, Proc. Geotech. Earthquake Engineering and Soil Dynamics III, ASCE*, Seattle, WA.

Fishman, K.L. & Desai, C.S. (1991) Response of geogrid earth reinforced retaining wall with full height precast concrete facing. *Proc. Geosynthics-91*, Atlanta, GA, 2.

Fishman, K.L., Desai, C.S. & Berg, R.R. (1991). Geosynthetic-reinforced soil wall: 4-year history. In: *Behavior of Jointed Rock Masses and Reinforced Soil Structures, Transportation Research Record 1330*, TRB, Washington, DC, 30–39.

Fishman, K.L., Desai, C.S. & Sogge, R.L. (1993). Yield behavior of instrumented reinforced wall. *J. Geotech. Eng., ASCE*, 119(8), 1293–1307.

Gyi, M.M. (1996) Multiaxial testing of saturated Ottawa sand. *M.S. Thesis*, Department of Civil Engineering and Engineering Mechanics, University of Arizona, Tucson, AZ.

Huang, Y.H. (1993) *Pavement analysis and design*. Prentice Hall, Englewood Cliffs, New Jersey, USA.

Kachanov, L.M. (1986) *Introduction to continuum damage mechanics*. Martinus Nijhoft Publishers, Dordrecht, The Netherlands.

Katti, D.R. & Desai, C.S. (1995) Modeling and testing of cohesive soil using disturbed state concept. *J. Eng. Mech., ASCE*, 121(5), 648–658.

Kulhawy, F.H. & Pewterson, M.S. (1979) Behavior of sand-concrete interfaces. *Proc. 6th Pan American Conf. on Soil Mech. Found. Eng.*, Brazil 2, 225–230.

Kutter, B.L., Idriss, I.M., Kohnke, T., Lakeland, J., Li, X.S., Sluis, W., Zeng, X., Tauscher, R., Goto, Y. & Kubodera, I. (1994) Design of large earthquake simulator at UC Davis. *Proc., Centrifuge 94*, Leung, Lee and Tan (Editors), Balkema, Rotterdam, 169–175.

Kutter, B.L., Li, X.S., Sluis, W. & Cheney, J.A. (1991). Performance and instrumentation of the large centrifuge at Davis. *Proc., Centrifuge 91*, Ko and Mclean (Editors), Balkema, Rotterdam, 19–26.

Lade, P.V. & Kim, M.K. (1988) Single hardening constitutive model for frictional materials. *Comput. Geotech.*, 6, 31–47.

Lytton, R.L. *et al.* (1993) Asphalt concrete pavement distress prediction: laboratory testing, analysis, calibration and validation. *Report No. A357, Project SHRP RF. 7157-2*, Texas A&M University, College Station, TX.

Mroz, Z., Norris, V.A. & Zienkiewicz, O.C. (1978) An anisotropic hardening model for soils and its application to cyclic loading. *Int. J. Numer. Anal. Methods Geomech.*, 2, 208–221.

Matsuoka, H. & Nakai, T. (1974) Stress-deformation characteristics of soil under three different principal stresses. *Proc., Jpn. Soc. Civil Eng.*, 232, 59–70.

Monismith, C.L. & Secor, K.E. (1962) Viscoelastic behavior of asphalt concrete pavements. *Proc. Conf. Association of Asphalt Paving Technologists*.

Mühlhaus, H.B. (Editor) (1995). *Continuum models for materials with microstructure*. John Wiley, Chichester, UK.

NISA (1993) *User's manual for NISA*. Engineering Mechanics Research Corporation, Detroit, Michigan (USA), December, 1993.

NJPC (1992) *Nathpa-Jhakri Hydroelectric Project, Executive Summary*. Nathpa Jhakri Power Corporation (NJPC), New Delhi, India.

Park, I.J. & Desai, C.S. (2000) Cyclic behavior and liquefaction of sand using disturbed state concept. *J. Geotech. Geoenviron. Eng., ASCE*, 126(9), 834–846.

Pastor, M., Zienkiewicz, O.C. & Chan, A.H.C. (1999) Generalized plasticity and the modeling of soil behavior. *Int. J. Numer. Anal. Methods Geomech.*, 14(3), 151–190.

Pestana, J.M. & Whittle, A.J. (1999) Formulation for unified constitutive model for clays and sands. *Int. J. Numer. Anal. Methods Geomech.*, 23(12), 1215–1243.

Potts, D.M. & Gens, A. (1985) A critical assessment of methods for correcting for drift from yield surface in elastoplastic finite element analysis. *Int. J. Numer. Anal. Methods Geomech.*, 9(2), 149–159.

Pradhan, S.K. & Desai, C.S. (2006) DSC model for soil and interface including liquefaction and prediction of centrifuge test. *J. Geotech Geoenviron. Eng., ASCXE*, 132(2), 214–222.

Ramamurthy, T. (1993) Strength and modulus responses of anisotropic rocks. Chapter 13 in *Comprehensive Rock Engineering*, Vol. I. Pergamon Press, Oxford, UK.

Roscoe, K.H., Schofield, A. & Wroth, C.P. (1958) On yielding of soils. *Geotechnique*, 8, 22–53.

Samtani, N.C., Desai, C.S. & Vulliet, L. (1996) An interface model to determine viscoplastic behavior. *Int. J. Numer. Anal. Methods Geomech.*, 20(4), 231–252.

Scarpas, A., Al-Khoury, R., Van Gurp, C.A.P.M. & Erkens, S.M.J.G. (1997) Finite element simulation of damage development in asphalt concrete pavements. *Proc. 8th Int. Conf. on Asphalt Pavements*, University of Washington, Seattle, Washington, DC, pp. 673–692.

Schneider, H.J. (1976) The friction and deformation behavior of rock joint. *Rock Mech.*, 8, 169–184.

Shao, C. & Desai, C.S. (2000) Implementation of DSC model and application for analysis of field pile tests under cyclic loading. *Int. J. Numer. Anal. Methods Geomech.*, 24(6), 601–624.

Tensar Geogrid Reinforced Soil Wall (1989). *Experimental Project 1, Ground Modification Techniques*, FHWA-EP-90-001-005, U.S. Department of Transportation, Washington, DC.

Varadarajan, A., Sharma, K.G., Desai, C.S. & Hashemi, M. (2001) Analysis of a powerhouse cavern in the Himalaya: Parts 1 and 2. *Int. J. Geomech.*, 1(1), 109–127.

Varadarajan, A., Sharma, K.G., Venkatachamy, K. & Gupta, A.K. (2003) Testing and modeling of two rockfills. *J. Geotech. Geoenviron. Eng.*, 129(3), 206–218.

Wilson, D.W., Boulanger, R.W. & Kutter, B.L. (1997a) Soil-pile-superstructure interaction at soft or liquefiable soil sites – centrifuge data report for CSP1. *Report No. UCD/CGMDR – 97/02*, Center for Geotechnical Modeling, Department of Civil and Environmental Engineering, University of California, Davis, CA.

Wilson, D.W., Boulanger, R.W. & Kutter, B.L. (1997b). Soil-pile-superstructure interaction at soft or liquefiable soil sites – centrifuge data report for CSP5. *Report No. UCD/CGMDR – 97/06*, Center for Geotechnical Modeling, Department of Civil and Environmental Engineering, University of California, Davis, CA.

Wilson, D.W., Boulanger, R.W. & Kutter, B.L. (1997c) Soil-pile-superstructure interaction at soft or liquefiable soil sites – centrifuge data report for CSP3. *Report No. UCD/CGMDR – 97/04*, Center for Geotechnical Modeling, Department of Civil and Environmental Engineering, University of California, Davis, CA.

Zhu, J.F., Xu, R.Q., Zhu, K. & Li, X.R. (2009) Application of disturbed state concept in calculation of earth pressure. *Electron. J. Geotech. Eng.*, 14(L), 1–15.

Chapter 20

Modeling brittle failure of rock

Vahid Hajiabdolmajid
BC Hydro, Burnaby, BC, Canada

Abstract: Observations of brittle failure at the laboratory scale indicate that the brittle failure process involves the initiation, growth, and accumulation of micro-cracks. Around underground openings, observations have revealed that brittle failure is mainly a process of progressive slabbing resulting in a revised stable geometry that in many cases take the form of V-shaped notches. Continuum models with traditional failure criteria (*e.g.* Hoek-Brown or Mohr-Coulomb) based on the simultaneous mobilization of cohesive and frictional strength components have not been successful in predicting the extent and depth of brittle failure. This chapter presents a continuum modeling approach that captures an essential component of brittle rock mass failure, that is, cohesion weakening and frictional strengthening (CWFS) as functions of plastic strain. A Strain-dependent rock brittleness index is defined which can represent the mobilized strength during brittle failure of rock.

I INTRODUCTION

Brittle failure is the product of the creation, growth and accumulation of micro- and macro-cracks. Many researchers have reported slabbing and spalling initiated on the boundary of the excavation as a dominant failure mode around underground excavations in massive to moderately jointed rock masses subjected to high in situ stresses (Ortlepp, 1997; Martin, 1997, 2014). Unlike openings at shallow depth, or at low in situ stress, in which failure is controlled by discontinuities, at greater depth, the extent and depth of failure is predominantly a function of the in situ stress magnitudes relative to the rock mass strength. Laboratory-scale observations have indicated that tensile cracking is present in inducing damage during the brittle failure of hard rocks (*e.g.* Brace & Bombolakis, 1963; Hallbauer *et al.*, 1973; Fredrich & Wong, 1986; Myer *et al.*, 1992; Martin, 1997). Excavation-scale observations also suggest that shallow spalling are generally associated with tensile failure which causes the thin slabs to initiate and peel off the Excavation surfaces. However, many field observations have indicated that the stress required to initiate spalling occurs at magnitudes that are considerably less than the peak strength of intact samples obtained in the laboratory (*i.e.* Uniaxial Compressive Strength, UCS) (Martin, 1997, 2014).

Traditional approaches of modeling rock mass failure are often based on a linear Mohr-Coulomb failure criterion or on a non-linear criterion such as the Hoek-Brown failure criterion. In both criteria, it is implicitly assumed that the cohesive and the

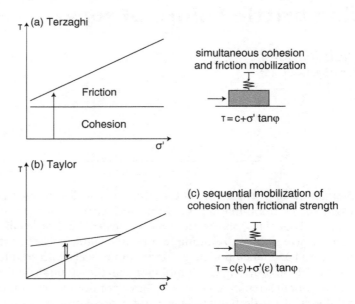

Figure 1 (a) Terzaghi's model and (b) Taylor's model for soil (after Schofield, 1998); and (c) analog for rock.

normal stress-dependent frictional strength components are mobilized simultaneously, *i.e.*, they are assumed to be additive as illustrated by the Terzaghi model in Figure 1a.

Even when strain-softening models with residual strength parameters are chosen, the two strength components are assumed to be simultaneously mobilized and then lost in the post-peak range. These approaches with typical strength parameters have not been successful in predicting the depth and extent of failed rock in deep underground openings in hard rocks (Wagner, 1987; Pelli *et al.*, 1991; Martin, 1997; Hajiabdolmajid *et al.*, 2000; Martin, 2014).

1.1 Strain-softening soil

As Schofield (1998) pointed out "there is no true cohesion on the dry side of critical state". In dense soil pastes, the peak strength is due to interlock and friction among particles and not due to the chemistry of bonds. While friction is immediately mobilized and the frictional strength component is proportional to the normal or confining stress, at low normal stress, the interlock resistance can be mobilized and then lost, leading to the typically observed strain-softening behavior of dense soils (the Taylor model; Figure 1b). In other words, the Coulomb criterion should be written in a form whereby both strength terms are a function of plastic strain ε (Equation 1):

$$\tau = c(\varepsilon) + \sigma'(\varepsilon) \tan \phi \tag{1}$$

The pioneering work of Schmertmann & Osterberg (1960) showed that in some soils, these two strength components (cohesive and frictional) are not necessarily mobilized simultaneously. They showed that the maximum of the cohesive component of strength

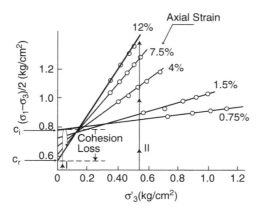

Figure 2 Bilinear envelope for stiff clay (developed as a function of axial strain) illustrating transition from mostly cohesive (I) to an almost exclusively frictional yield mode (II) (modified after Schmertmann & Osterberg, 1960). C_i and C_r, represent the initial and residual cohesion, respectively.

$c(\varepsilon)$ was mobilized early in the test, while the frictional component $\sigma'(\varepsilon) \tan \phi$ required 10 to 20 times more straining to reach full mobilization as shown in Figure 2.

1.2 Strain-softening rock

Unlike in ductile materials where shear slip surfaces form in such a manner that continuity of material is maintained, brittle failure is a process whereby continuity is disrupted to create kinematically feasible failure. From a mechanistic point of view, what happens during the brittle failure process of rock is the destruction of the strength derived from bonds between grains (cohesive strength). The frictional strength component gradually mobilizes as the disintegrated blocks readjust and deform by shearing at newly created surfaces. Therefore in brittle failure of rocks at relatively low confinement, the cohesive strength component is gradually lost when the rock is strained beyond its peak strength. This is illustrated by the gap model of Figure 1c, representing an analog for brittle rock. This analog illustrates that the cohesion at the bottom of the sliding wedge must be overcome before the frictional strength can be mobilized when the gap between the two wedges is closed. Only after this gap-closure deformation has taken place, will the normal stress, symbolized by the spring, be activated and a strain-dependent effective stress builds up to create a frictional resistance. In the low confinement range, the stress path will retract after reaching the cohesive strength surface as illustrated in Figure 1b. In fact it is the effective normal stress $\sigma'_n(\varepsilon)$ causing a gradual development of the frictional strength component (Figure 3). Martin & Chandler (1994) demonstrated that the frictional strength component of granite is only mobilized after a significant amount of the rock's cohesive strength is lost. Consequently, for brittle failure of rocks at low confinement the Mohr-Coulomb and Hoek-Brown failure criteria must be written in the form of Equation 2 and Equation 3, respectively. In Equation 2 and Equation 3, the strength component (1) is the plastic strain dependent cohesive strength, and the strength component (2) is the plastic strain dependent frictional strength.

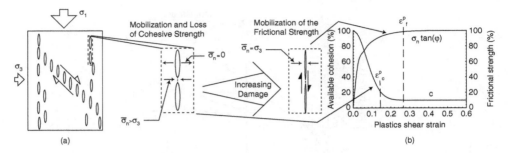

Figure 3 Strain-dependent cohesive and frictional strength mobilization during the brittle failure process of rock.

$$\tau = \underbrace{c(\varepsilon)}_{1} + \underbrace{\sigma_n(\varepsilon) \tan \phi}_{2} \qquad (2)$$

$$\frac{\sigma_1}{\sigma_c} = \frac{\sigma_3(\varepsilon)}{\sigma_c} + \left(\underbrace{s}_{1} + \underbrace{m\sigma_3(\varepsilon)/\sigma_c}_{2} \right)^{0.5} \qquad (3)$$

2 BRITTLE FAILURE OF ROCK

The laboratory testing of rocks is traditionally carried out to determine the laboratory samples peak strengths *i.e.* UCS; however, it is well known that in low porosity crystalline rocks, three distinct loading stages can be identified (Martin & Chandler, 1994): (1) crack initiation, (2) unstable crack growth, *i.e.*, crack coalescence, and (3) peak strength (Figure 3 and Figure 4).

Laboratory-scale observations have indicated that tensile cracking is dominant mechanisms during the brittle failure of hard rocks. A most important element of the tensile induced cracking process is that the (normal) stresses at points of friction mobilization are not constant. In other words, the effective normal stress, $\bar{\sigma}_n$ at the contact points changes gradually as the disintegrating rock mass is deformed (Figure 3).

Hajiabdolmajid (2001) demonstrated that the brittle failure process of rock masses in a low confinement condition can be modeled using a continuum modeling approach called cohesion weakening-frictional strengthening (CWFS) model. In this model, the mobilized strength components (cohesive and frictional) are plastic strain (damage) dependent (Figure 3, and Figure 4). In Figure 3, ε_c^p and ε_f^p are the plastic strain (damage) levels necessary for cohesion loss and frictional strengthening, respectively.

Failure of underground openings in hard rocks is a function of the in situ stress magnitudes, the mining-induced stress, and the degree of natural fracturing (jointing) of the rock mass. At low in situ stress magnitudes, the continuity and distribution of the natural fractures in the rock mass control the failure process. However, at elevated stress levels, the failure process is affected and eventually dominated by new stress-induced fractures growing preferentially parallel to the excavation boundary. This fracturing is

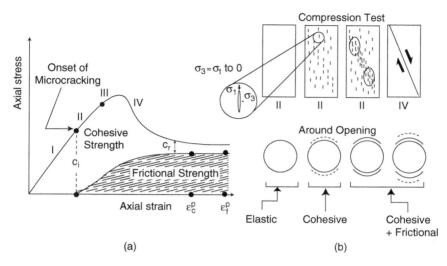

Figure 4 Strain-dependent mobilization of the strength components in laboratory and in situ. Note that the frictional strength may need higher levels of damage in order to reach its full mobilization, c_i and c_r represent the initial and residual cohesion, respectively.

generally referred to as brittle rock failure by slabbing or spalling. In-situ, the act of excavation removes some or all of the confining pressure and gives rise to circumstances in which not all strength components are always equally mobilized (*i.e.* they are not at their maximum effectiveness) in all stages of the failure process (Figure 4).

Figure 4a illustrates the concept of non-simultaneous mobilization of the strength components as functions of induced damage. Figure 4b compares the mobilization of the strength components in a compression test with the one around a circular tunnel in hard rocks at different stages of the loading process. The pre-peak stress-strain curve in uniaxial loading of the hard rocks in Figure 4a shows three distinct stages (I) crack initiation, (II) unstable crack growth, *i.e.*, crack coalescence, and (III) peak strength. The crack initiation stress level corresponds to the start of cohesion loss process by micro-cracking inside the sample. In Figure 4a, ci and cr represent the initial and residual cohesion, respectively.

3 COHESION WEAKENING-FRICTIONAL STRENGTHENING (CWFS) MODEL

Due to the limitation of the continuum modeling, the plastic strain dependency of effective confinement cannot be directly considered. However, the effect of this dependency can be represented by plastic strain dependency of the frictional strength component ($\bar{\sigma}_n \tan\phi$) in the CWFS model by making friction angle (ϕ) a plastic strain-dependent property (Hajiabdolmajid, 2001). The CWFS model is characterized by its yield function, strengthening/weakening functions, and flow rule. An effective plastic strain parameter, $\bar{\varepsilon}^p$ defined by Equation 4 was used to represent the accumulated plastic strains and the weakening/strengthening parameters (*i.e.* ε_c^p and ε_f^p in Figure 3).

The parameter $\bar{\varepsilon}^p$ is calculated from the principal plastic strain increments and is essentially a measure of the plastic shear strain (Hill, 1950; Vermeer & de Borst, 1984). Equations 5 and 6 express the adapted Mohr-Coulomb failure criterion used in the *CWFS* model.

$$\bar{\varepsilon}^p = \int \sqrt{\frac{2}{3}(d\varepsilon_1^p d\varepsilon_1^p + d\varepsilon_2^p d\varepsilon_2^p + d\varepsilon_3^p d\varepsilon_3^p)}\, dt \qquad (4)$$

where $d\varepsilon_1^p$ $d\varepsilon_2^p$ and $d\varepsilon_3^p$ are the increments of principal plastic strains.

$$\tau = c(\varepsilon_c^p)\sigma_n \tan\phi\left(\varepsilon_f^p\right) \qquad (5)$$

$$f(\sigma) = f(c, \bar{\varepsilon}^p) + f\left(tan(\phi), \bar{\varepsilon}^p\right)\sigma_n \qquad (6)$$

The general shapes, representing cohesion weakening and frictional strengthening as a function of plastic strain can be expressed by the following equations:

$$f(c, \bar{\varepsilon}^p) = (c_i - c_r)\left(exp\left[-\left(\frac{\bar{\varepsilon}^p}{\varepsilon_c^p}\right)^2\right]\right) + c_r \qquad (7)$$

$$f\left(tan(\phi), \bar{\varepsilon}^p\right) = \left\{ \begin{array}{ll} \left(2\frac{\sqrt{\bar{\varepsilon}^p \varepsilon_f^p}}{\bar{\varepsilon}^p + \varepsilon_f^p}\sigma_n\right)tan\phi & if\ \bar{\varepsilon}^p \leq \varepsilon_f^p \\ tan\phi & if\ \bar{\varepsilon}^p = \varepsilon_f^p \end{array} \right\} \qquad (8)$$

4 AECL MINE-BY EXPERIMENT

Between 1990 and 1995 Atomic Energy Canada Limited (AECL) carried out a Mine-by Experiment in Manitoba, Canada. This well documented experiment involved the excavation of a 3.5-m-diameter circular test tunnel in massive granite (Read, 1994). The primary objective of the experiment was to investigate brittle failure processes. To achieve this objective the tunnel was excavated by 0.5 to 1 meter rounds using a line-drilling technique and displacements, strains, stress changes and micro-seismic emissions were monitored with state-of-the art instruments. Martin *et al.* (1997) reported observations of the brittle failure process resulting in classic V-shaped notches, in the region of maximum compressive stress (Figure 5).

Read (1996) showed in an extensive characterization report of the damage zone around the tunnel that the extent of the compressive stress-induced damage was confined to the notch regions. Outside the notch little damage could be visually observed. Based on low P-wave velocities and acoustic emission measurements, Read (1996) noted that the tensile zones in the side walls of the tunnel were damaged, even though no cracking or fracturing could be visually observed.

One of the objectives of the Mine-by Experiment was to assess the predictive capability of numerical models in capturing the extent and shape of the failed zone. For this purpose, the in situ stresses near the tunnel were determined accurately: $\sigma_1 = 60 \pm 3MPa$,

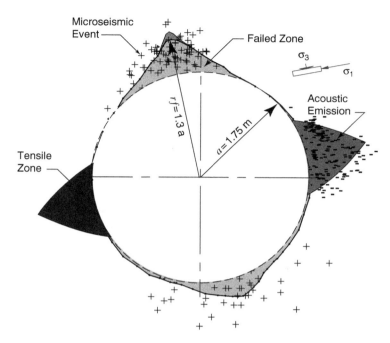

Figure 5 Shape of the failed zone observed around the circular test tunnel. Also shown are microseismic events locations in the notch area (+), and acoustic emission locations (−) in the tensile failure zone (after Read, 1996).

$\sigma_2 = 45 \pm 4 MPa$, $\sigma_3 = 11 \pm 2 MPa$ (Martin & Read, 1996). Extensive laboratory testing was carried out and the Hoek-Brown failure parameters were defined: ($\sigma_c = 224 MPa$, m = 28.11, s = 1).

This failure criterion is shown in Figure 6 together with the stress required to initiate damage (acoustic emissions) in the laboratory ($\sigma_{ci} = 71 \pm 1.5\ \sigma_3$) and in situ ($\sigma_1 - \sigma_3 = 70 MPa$). Given the well-defined stress state, the simple circular geometry of the excavation, the essentially intact, massive granite, predicting the extent of failure should be a trivial task. However, as many attempts (Martin et al., 1999) have shown and will be documented below this is not the case.

5 MODELING BRITTLE FAILURE-AECL MINE-BY TUNNEL

The common approach to simulate brittle rock failure is to adopt an elastic-brittle-plastic or strain softening model (Figure 7).

Table 1 indicates Lac du Bonnet granite parameters derived for the Mine-by Experiment in Lac du Bonnet granite by Martin (1997). The rock mass strength parameters provided above are used to explore the post-peak response for modeling the brittle failure observed in the Mine-by tunnel. This approach is then compared to a strain-dependent strength mobilization model (CWFS) utilizing the numerical code FLAC 2D.

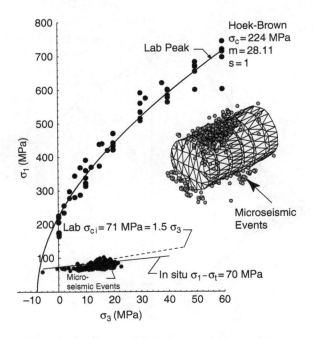

Figure 6 Hoek-Brown failure parameters for Lac du Bonnet granite, and the stress required initiating damage (after Martin, 1997).

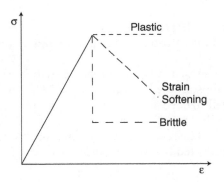

Figure 7 Various post-peak responses used in continuum models.

5.1 Elastic models

One simple way of estimating the depth and extent of the failed zone is by an elastic stress analysis determining the induced stresses and comparing them with the rock mass strength based on *GSI*. Figure 8 shows the elastic major principal stress distribution around the test tunnel, which reaches a maximum value of approximately 150 MPa at the roof. Considering the rock mass uniaxial compressive

Table 1 Lac du Bonnet granite parameters derived for the Mine-by Experiment in Lac du Bonnet granite by Martin (1997).

Rock Type	Lac du Bonnet granite
Intact compressive strength	σ_{ci} = 224 MPa
Intact tensile strength	σ_{ti} = 10 MPa
Hoek-Brown Constant	m_i = 28.11
Geological Strength Index	GSI = 90
Friction angle	ϕ = 48°
Cohesive strength	c = 25 MPa
Hoek-Brown Constant	m_b = 19.67
Hoek-Brown Constant	s = 0.329
Compressive strength of rock mass	σ_{cm} = 128 MPa
Tensile strength of rock mass	σ_{tm} = −3.7 MPa
Rock mass modulus	E = 60 GPa
Poisson's ratio	ν = 0.2

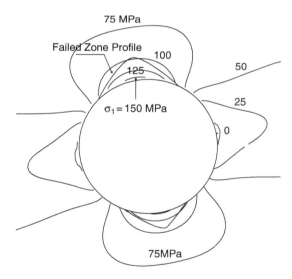

Figure 8 Major principal stress distribution around the circular test tunnel.

strength $\sigma_{cm} = 128 MPa$, this stress is sufficient to create a very thin failed zone in the region of the notch, which significantly underestimates the depth, and extent of the actual breakout or failed zone (Figure 5 and Figure 8). One method that is often used to overcome the limitation of elastic analyses is to simulate the progressive nature of slabbing and spalling by successive removals of failed elements (Zheng et al., 1989). This approach was used by Read (1994) and Martin (1997) with different criteria for element removal but in both cases the depth of breakout zone was overestimated by a factor of 2 to 3.

5.2 Elastic-perfectly-plastic model

An elastic-perfectly-plastic constitutive law includes the effects of plastic straining and related stress redistribution on the depth of failure, but is hardly appropriate for a brittle rock because the obvious material weakening is ignored. It therefore provides an upper bound (minimum depth of failure with stress redistribution). Figure 9 shows the predicted failed zone for rock without dilation ($\psi = 0°$), which indicates that this approach still does not predict the failed zone.

5.3 Elastic-brittle model

In this model the slabbing process is simulated by decreasing the Hoek-Brown parameter m and s to very small values in the post peak range to represent (a) a rapid loss in cohesion of the rock mass to about 20% of peak, and (b) a reduction in friction to the rock basic friction angle.

Figure 10 shows results from FLAC 2D with $m_r = 1$ and $s_r = 0.01$. As with the elastic-perfectly-plastic model, this approach underestimates the depth of failure. However, it overestimates the lateral extent of failure in the roof. Read & Martin (1996) combined this approach with an element removal scheme, which over predicted the depth and extent of failure by a factor of 2.

In summary, the conventional approaches commonly adopted for rock failure modeling all failed to predict both the shape and extent of the failed zone that developed around a circular tunnel in massive, brittle granite. None of these traditional approaches could be used to predict the failed zone *a priori* with any degree of confidence.

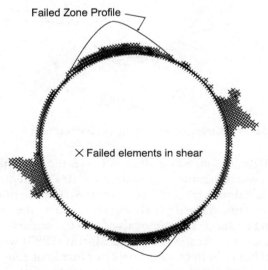

Figure 9 Extent of damage with an elastic-perfectly-plastic constitutive model; * indicate elements presently in the yielding state in shear, + indicate elements that previously yielded.

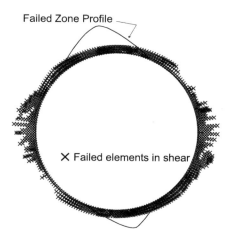

Figure 10 Extent of damage with an elastic-brittle constitutive model with $m_r=1$ and $s_r=0.01$.

6 COHESION WEAKENING-FRICTIONAL STRENGTHENING MODEL (CWFS MODEL)

The plasticity model in FLAC 2D was used to study the effect of non-simultaneous mobilization of the frictional and cohesive strength components on the mobilized strength of rock under various loading conditions (Hajiabdolmajid, 2001). In CWFS model the plastic strain limits at which the cohesive component of strength reaches a residual value, and the frictional strength component mobilizes are two material properties that in reality depend primarily on heterogeneity and grain characteristics. However, they should be calibrated on laboratory and *in situ* failure cases. For the Mine-by tunnel model, the laboratory damage-controlled tests on Lac du Bonnet granite reported by Martin & Chandler (1994) were used to establish the plastic strain limit for cohesion loss. From back analyses of the failure zone by slabbing around the Mine-by tunnel, it was found that the plastic strain (or damage, ε_c^p) necessary for the destruction of the cohesive strength is in general lower than the plastic strain required (*i.e.* ε_f^p) for the full mobilization of the frictional strength. For Lac du Bonnet granite the strain-dependent cohesive and frictional strength mobilization were linearized as illustrated in Figure 11 and Figure 12 and introduced into the continuum modeling code FLAC 2D.

Hajiabdolmajid (2001) argued that while the plastic strain limit for cohesion loss can be considered a true material property, the circumstances (strain limit) under which the frictional strength reaches its full mobilization depends to some extent on the loading system characteristics (geometry and loading rate). He attributed the very low strength observed around the Mine-by tunnel to a delayed mobilization of the frictional strength (when $\varepsilon_f^p \rangle \varepsilon_c^p$), compared to the high strength obtained in the laboratory compression tests in which the frictional strength reaches its full mobilized capacity with less damage or cohesion loss.

The plastic strain limits and the strength parameters listed in Table 2 were used to simulate the brittle failure of Lac du Bonnet granite, near the Mine-by tunnel. The

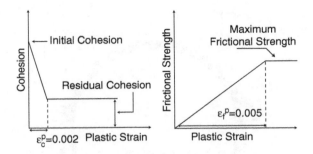

Figure 11 Illustration of the cohesion-loss and frictional strength mobilization as a function of plastic strain.

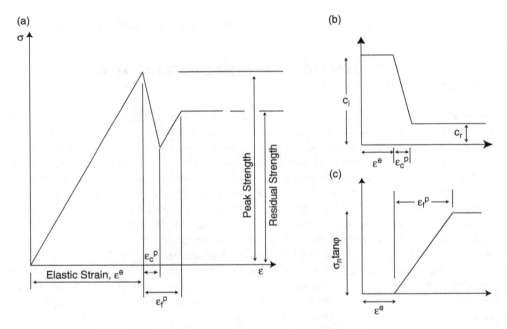

Figure 12 Stress-strain curve in the CWFS model when the frictional strengthening is delayed ($\varepsilon_f^p > \varepsilon_c^p$).

cohesion was reduced from its initial peak value of 50 MPa to its residual value 15 MPa when $\bar{\varepsilon}^p = \varepsilon_c^p = 0.002$. The initial cohesion was taken as 22% of UCS or the anticipated long-term cohesion without friction mobilization. Dilation was held constant at 30°, and the tension cut off was 10 MPa. Figure 13 shows the simulated failed zone, using the CWFS model with the parameters listed in Table 2.

Figure 13 shows FLAC 2D result using the material properties in Table 2 and the CWFS modeling approach indicating the spalled areas around the Mine-by tunnel.

Table 2 Parameters of cohesion weakening-frictional strengthening model.

	Cohesion (MPa)	Friction angle	Dilation angle
Initial	50	0°	30°
Residual	15	48°	30°
Plastic strain	$\varepsilon_c^p = 0.002$	$\varepsilon_f^p = 0.005$	zero

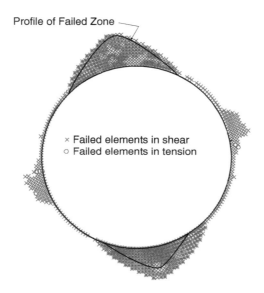

Figure 13 Prediction of the failed zone around AECL's Mine-by test tunnel using FLAC 2D and cohesion weakening-frictional strengthening model, with parameters listed in Table 2; o indicates elements failed in tension. Contours of plastic strain (i.e. cohesion loss or damage) inside the notch are shown in Figure 14.

Despite several simplifications, such as ignoring stiffness softening, the predicted depth and extent of the failed zone is in excellent agreement with the measured failed zone shown in Figure 5 and Figure 13. The non-symmetry in observed notch shapes in the roof and floor is attributed to excavation effects (muck in the floor, Martin, 1997). As expected the induced damage decreases when moving from the tunnel boundary toward the notch tip (Figure 14).

This is in general agreement with the characterization results reported by Read (1996) which demonstrated that outside the notch the rock mass was essentially undamaged. Most importantly, this approach properly predicts the arrest of the failure process that is difficult if not impossible to simulate with traditional models. Various observations suggest that once the maximum depth of the spall has been achieved, they remain stable due to more complicated involvement of both tensile and probably shear failure at the notch tip (Martin, 2014) leading to the arrest of progressive brittle fracturing of the rock. Using the CWFS concept, the arrest of the observed spalling

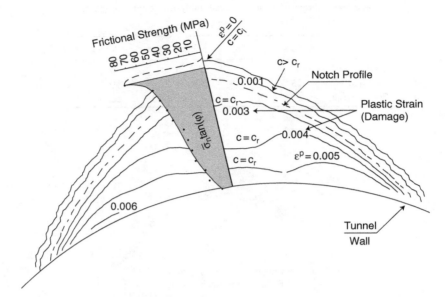

Figure 14 Contours of plastic strain (full cohesion loss at $\varepsilon_c^p = 0.002$ to c_r). The progressive frictional strengthening inside the notch leading to failure arrest at the notch tip is illustrated by the profile of $((\bar{\sigma}n\,tan\,\phi))$ superimposed on the contour plot.

process after a new, more stable, geometry is reached, can be explained by an increase in confinement (progressive frictional strengthening), coupled with a decrease in the induced damage (plastic strain) and thus a decrease in cohesion loss, *i.e.* arresting of progressive spalling. In Figure 14 the mobilized frictional strength $\bar{\sigma}n\,tan(\phi)$ is calculated using principal stresses from the FLAC 2D model and the strain dependent friction angle. Beyond the damaged zone (beyond the notch in the intact rock) where there is no plastic straining there is no mobilized frictional strength and the cohesive strength is not affected.

7 BRITTLENESS OF ROCK

In back analyzing the brittle failure of granite around the Mine-by tunnel and the breakout zone (Figure 14) the initial cohesion (c_i = 50 MPa) was reduced to its residual value (c_r = 15 MPa) when the accumulated plastic strain $\bar{\varepsilon}^p = \varepsilon_c^p = 0.002$. The frictional strength needed more than two times plastic straining (damage) in order to reach its full capacity ($\bar{\varepsilon}^p = \varepsilon_f^p = 0.005$), *i.e.* the cohesive strength reduces to 30% of its original value by the time the frictional strength reaches only 40% of its full capacity. Figure 15 demonstrates a simple application of the *CWFS* model in compression tests. Figure 15a represents a simple linear cohesion weakening process ($\bar{\varepsilon}^p = \varepsilon_c^p = 0.002$) with a fully and instantaneous frictional strength mobilization ($\varepsilon_f^p = 0$), using the material properties of the Lac-du Bonnet granite (Table 2).

A peak strength UCS=260 MPa was obtained which is about 15% higher than the uniaxial compressive strength obtained in the laboratory compression tests

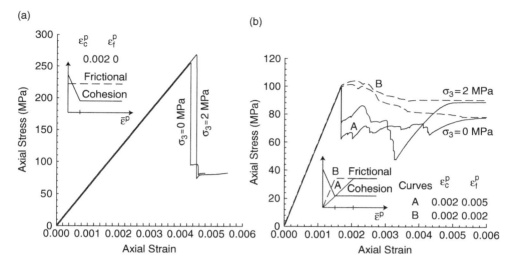

Figure 15 Effect of frictional strengthening rate on the mobilized strength, a) with an instantaneous frictional strengthening ($\varepsilon_f^p = 0$), and b) with a delayed frictional strengthening.

(UCS=224 MPa in undamaged samples of rock). Figure 15b demonstrates the same tests as in Figure 15a using a non-simultaneous mobilization of the cohesive and frictional strength components. In both scenarios (A and B in Figure 15b), the material properties are the same (Table 2) except for the plastic strain limit at which the frictional strength is fully mobilized (ε_f^p) differs. In test series B, the strain limits are equal ($\varepsilon_c^p = \varepsilon_f^p = 0.002$). In test series A, the plastic strain limits are chosen in accordance with the results of back analysis of the breakout zone around the Mine-by tunnel ($\varepsilon_c^p = 0.002$ and $\varepsilon_f^p = 0.005$).

A much lower strength (see Figure 15b) is obtained in the case of non-simultaneous frictional strength mobilization which is in the order obtained for the crack initiation stress level, (*i.e.* about 0.45 of the UCS; Martin, 1997, 2014). Of particular interest in Figure 15b (Tests A) is the stress-strain curve when two different strain limits for cohesion weakening and frictional strengthening are chosen. A more brittle post-peak behavior is obtained when $\varepsilon_f^p \rangle \varepsilon_c^p$, *i.e.* a non-instantaneous/non-simultaneous mobilization of the frictional strength (CWFS concept) can be used to define the entire failure process (crack initiation, unstable crack growth, peak and post-peak strengths). This was accomplished by defining a strain-dependent brittleness index ($I_{B\varepsilon}$, Equation 9 and insert in Figure 16) using the plastic strain limits ε_c^p and ε_f^p.

$$I_{B\varepsilon} = \frac{\varepsilon_f^p - \varepsilon_c^p}{\varepsilon_c^p} \qquad (9)$$

The strain-dependent brittleness index *i.e.* Equation 9 explicitly considers the contribution of the cohesive and frictional strength components during the failure process. This definition of brittleness implicitly considers the ease of micro-cracking process during the failure process by considering the rate at which the strength components are

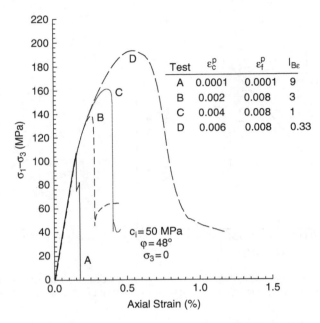

Figure 16 Sensitivity of the mobilized strength and stress-strain curve to the plastic strain limits and to the strain-dependent (non-linear) functions for cohesion loss and friction mobilization.

mobilized as functions of damage (plastic strain). From the physical point of view the brittleness indexes as defined by Equation 9 reflects the presence of both tensile and shear mechanisms during the brittle failure of rocks. The rock mass properties and loading system characteristics are expected to influence the brittleness (*i.e.* cohesion loss and frictional strengthening rates *i.e.* $\varepsilon_f^p - \varepsilon_c^p$). For instance, two loading systems, one in laboratory compression test and another around a large underground opening may rise to different conditions for cohesion loss and frictional strengthening rates. Fast mobilization of the frictional strength component (or slow cohesion loss rate) is represented by smaller brittleness numbers. However, the effects of material properties such as, lithology, fabric, mineralogy, and foliation should also be considered (see Section 9). This can be better understood by considering the various micro-mechanism processes involved in initiation, propagation and coalescence of micro-cracks, hard rocks. The micro-cracks in hard rocks may initiate from: pores, point loading, and local stiffness mismatch, these mechanisms promote a more brittle crack initiation and propagation, accompanied with less plastic straining (fast cohesion loss and/or slow friction mobilization rate, *i.e.* large $\varepsilon_f^p - \varepsilon_c^p$), leading to a minimum contribution of (micro) friction (frictional strengthening). The processes, which involve the initiation and propagation of cracks at the grain boundaries (frictional cracks), cleavage, foliation, and soft inclusion, most likely need more plastic straining, and higher degree of (micro) frictional strengthening (*i.e.* small $\varepsilon_f^p - \varepsilon_c^p$, less brittle failure and higher mobilized frictional strength).

The strain sensitivity of the mobilized strength and the entire failure process (pre- to post-peak stages) to the non-simultaneous mobilization of the strength components were further investigated using the non-linear functions for cohesion loss and friction mobilization. Figure 16 illustrates the results of the simulation of the uniaxial compression tests using various rates for cohesion loss and frictional strengthening (ε_c^p and ε_f^p in insert in Figure 16). In all these models, the material properties remain the same, however, both the mobilized strength (peak strength) and the entire stress-strain curve (pre- to post-peak) have been affected by changing the strain limits (ε_c^p and ε_f^p). The most brittle behavior is associated with the lowest cohesion loss strain limit (small ε_c^p, i.e. high cohesion loss rate) and highest frictional strength strain limit (large ε_f^p, i.e. low strengthening rate).

8 FAILURE ENVELOPE

The strength envelopes that are commonly used to assess stability of underground openings are either the linear Mohr-Coulomb envelope or the non-linear Hoek-Brown envelope with a downward curvature. However, the CWFS model results in an upward-bent bilinear envelope of the form shown in Figure 17 and Figure 18. Figure 17 was obtained by tracking the history of several grid points located in the V-shaped notch region of the test tunnel in Figure 13. The peak strength of this region is only a function of the rock mass cohesion while the post-peak strength degrades toward the residual frictional strength. Figure 17 shows that some points (open squares) in the notch region have reached the residual level with others still possess higher cohesion strength component.

The notion that the yield envelope for cohesive materials is bilinear is not new. Schofield & Worth (1966) demonstrated that this type of yield envelope was appropriate for stiff over-consolidated clays and used this notion to lay the foundations for critical state soil mechanics. Taylor (1948) also suggested this type of yield envelope for interlocked sands (Figure 19).

Figure 20 compares the strength envelopes associated with various brittleness indices (illustrated in insert in Figure 19). The in situ strength envelope is illustrated by the bilinear failure envelope which corresponds to $I_{B\varepsilon} = 1.5$ (characterizing the failure

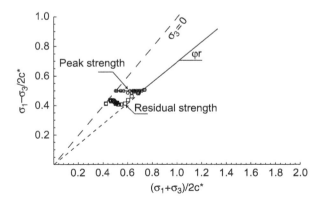

Figure 17 Bilinear yield envelope developed resulting from the cohesion weakening frictional strengthening model. C*=2c.

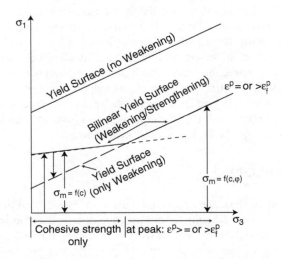

Figure 18 Failure surfaces: with no weakening, with only cohesion weakening, and with cohesion weakening-frictional strengthening behaviors (bilinear), σ_m represents the peak mobilized strength.

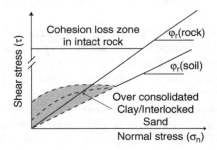

Figure 19 Schematic yield envelope for geomaterials.

process of the Mine-by tunnel). The significant effect of the simultaneous-instantaneous frictional strengthening ($I_{B\varepsilon} = -1$) can be noticed by realizing how much (about 160 MPa) the peak strength has increased at $\sigma_3 = 0$ when compared with the mobilized strength when the frictional strengthening is delayed (*i.e.* $\varepsilon_c^p = 0.002$ $\varepsilon_f^p = 0.005$, $I_{B\varepsilon}=1.5$, in situ bilinear) in Figure 20.

9 DEPTH OF SPALLING

It was found that the brittleness has a direct relation with the depth of spalling or failure (*i.e.* d_f in Figure 21), which is expressed by Equation 10. A similar relationship was found for the lateral extent (*i.e.* α in Figure 21 and Figure 22) of breakout zone as expressed by Equation 11.

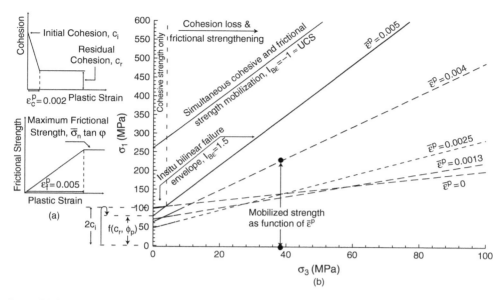

Figure 20 The CWFS model used in back analyzing the Mine-by tunnel spalling (a) and bilinear failure envelope found by implication of the concept of brittleness index in the CWFS constitutive model (b).

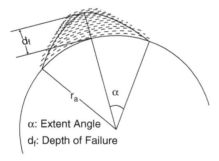

Figure 21 Geometric characteristics of the failed zone used in Equations 10 and 11.

$$d_f = 25(1 + I_{B\varepsilon})^{0.5} \qquad (10)$$

$$\alpha = 25(1 + I_{B\varepsilon})^{0.4} \qquad (11)$$

Figure 22 illustrates the associated brittleness indices with the measured depths of failure in the roof and floor around the Mine-by tunnel and for the roof when tunnel pass through granodiorite. It is expected that different rocks possess different strain limits for cohesion loss and frictional strengthening (different brittleness $I_{B\varepsilon}$, i.e. $\varepsilon_f^p - \varepsilon_c^p$). Thus, creating the same opening in the same far field stress environment inside different rocks with different brittleness will lead to different failed zones. The Mine-by tunnel nicely illustrates this where the tunnel passes from granite to granodiorite of the Lac du Bonnet formation. Figure 23a depicts the distribution of gray granite and granodiorite

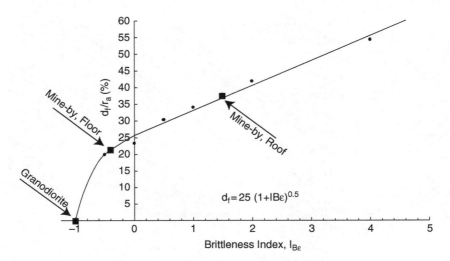

Figure 22 Depths of failures in the roof and floor of the Mine-by tunnel, and in granodiorite sections of the Mine-by tunnel and their associated brittleness indices in the CWFS model.

on an unfolded perimeter map of the Mine-by Tunnel. Figure 23b illustrates the depth of the breakout zone in contour form (in the roof and floor) on an unfolded perimeter map. Spalling (depth of notch) is almost eliminated where the tunnel passes through the granodiorite. Figure 23c illustrates the brittleness indices associated with the observed depths of failure. The brittleness indices associated with the observed extent angle (α), are also shown (by squares).

Using the concept of brittleness index ($I_{B\varepsilon}$) one can explain the differences observed, between the granodiorite and granite of Lac du Bonnet around the Mine-by tunnel, while the in-situ, stresses and environmental conditions can be considered constant along the length of tunnel (Figure 23b). The stability in the granodiorite or the slabbing in the granite can be related to differences in their brittleness ($I_{B\varepsilon}$). The two materials (granite and granodiorite) are reported to have very similar strength and deformational (laboratory) properties but different grains size distributions (granite is coarser than granodiorite). In the granodiorite the size of all grains is roughly equal (1 mm) and somewhat more uniformly distributed. It is likely that the presence of larger grains in granite contributes to a faster cohesion loss rate (with straining) and/or slower frictional strengthening (*i.e.* higher brittleness) in-situ. Therefore, the mobilized strength of the granodiorite in situ is higher than the mobilized strength in-situ in the granite (Figure 20, and Figure 22). This cannot be simulated and considered in the models in which the failure initiation and arrest are simulated using purely stress-based criteria. It follows that the brittleness ($I_{B\varepsilon}$) is a dominant factor, often more than stress, in controlling the breakout zone shape. This explains the failure of methods adopted by many researchers in establishing stress related breakout prediction models.

In mapping the failed zones around the Mine-by tunnel, it was noticed that the breakout zone in the floor was smaller than the one in the roof (Figure 5 and Figure 13). Figure 22 illustrates the associated brittleness indices with the measured depths of

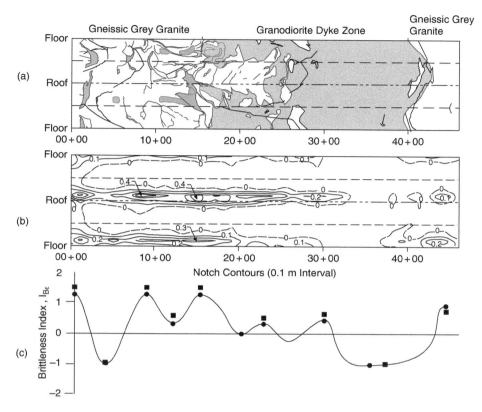

Figure 23 Perimeter maps of geology (a), the depth of the breakout (notch in meters), granite is shown as the light regions and granodiorite as the dark regions on the geology map (b) (after Martin et al., 1997), (c) associated brittleness indices.

failure in the roof and floor around the Mine-by tunnel. The difference was explained by the effect of muck pressure and/or the effect of having two different stress paths followed by the regions in the floor and in the roof (Martin, 1997). In order to clarify this uncertainty, a three-dimensional implication of the CWFS model was used (Figure 24) to predict the formation of the failed zones around the Mine-by tunnel in which the non-alignment of the tunnel axis with the axial stress (which is the cause for the stress path difference between the roof and floor) is considered. Figure 24 illustrates the formation of the failed zones in the roof and floor in a FLAC 3D model, which are similar. Hajiabdolmajid & Kaiser (2003) considered the effect of muck pressure in limiting the progression of the spalling zone in the invert using the CWFS concept. They argued that support pressure around the openings in hard brittle rocks aids the rock to mobilize more of its strength capacity during the failure process, and consequently limit the extent of progressive spalling.

In summary the CWFS modeling approach implicitly captures the complex phenomena involved in the brittle failure of rock *i.e.* transition from continuum to discontinuum due to the formation of tensile cracks which precedes a kinematically feasible failure in

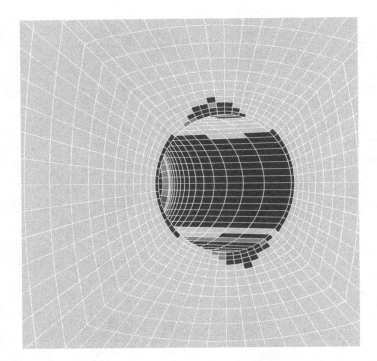

Figure 24 Simulation of the failed zones around the Mine-by tunnel in a FLAC 3D model.

low porosity hard rocks, it also captures the arrest of progressive brittle fracturing process. Diederichs *et al.* (2010) adopted the concept of CWFS to model the rock spalling using Hoek-Brown strength parameters. Barton (2014) and Barton & Pandey (2011) compared the implication of the CWFS modeling concept and Q-system based parameters with other conventional strain softening approaches based on Mohr-Coulomb and Hoek-Brown parameters in FLAC 3D to analyze the underground mining stopes. They demonstrated that the adoption of CWFS model together with the Q-system based parameters provide the most realistic match to in situ observations.

10 GROUND REACTION CURVES

Ground reaction curves (GRC) are often used in stability analyses of tunnels and for evaluation of the ground-support interaction (Brown *et al.*, 1983). Most of the previously reported works in this field use closed form solutions or numerical methods with simple and often unrealistic material behaviors to calculate the ground reaction curves for tunnels in hard brittle rocks.

The CWFS model can be used to calculate the ground reaction curves. Figure 25 illustrates the GRC with the listed parameters in Table 3 which demonstrates the effect of cohesion loss rate on GRC. Figure 26 illustrates the GRC with the listed parameters in Table 4, which demonstrates the effect of frictional strengthening rate on GRC.

As can be seen using the same material properties but different brittleness results in a completely different GRC. This demonstrates the significance of rock brittleness on the

Table 3 The CWFS model parameters used in cases presented in Figure 25.

Test	ε_c^p	ε_f^p	Dilation angle
A	0.002	0.002	0°
B	0.002	0.002	30°
C	0.0006	0.002	0°
D	0.0006	0.002	30°

Table 4 The CWFS model parameters in Figure 26.

Test	ε_c^p	ε_f^p	Dilation angle
A	0.0001	0.0	0°, 30°
B	0.0001	0.002	0°
C	0.0001	0.002	30°

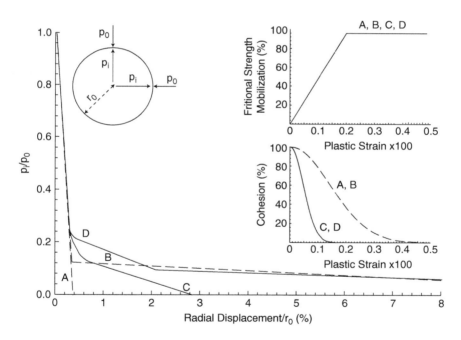

Figure 25 Ground reaction curves and the effect of cohesion loss rate in the CWFS model.

interpretation and prediction of the observed displacements around openings and determination of the support system necessary for the excavation stability.

In calculating the ground reaction curves in Figures 25 and 26 a constant dilation angle (30°) was used. It should be mentioned that using a strain-dependent function for dilation angle (*i.e.* $\psi(\varepsilon)$) as it was used for the cohesive and frictional strength

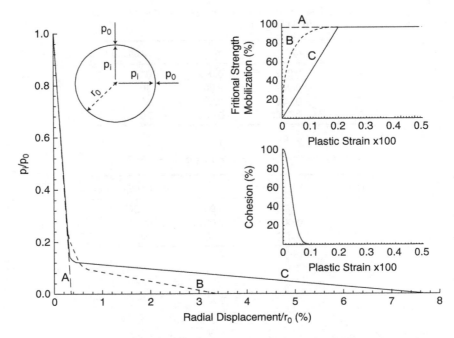

Figure 26 Ground reaction curves and the effect of frictional strengthening rate on the observed displacements in the CWFS model.

components, will affect the induced inelastic strains and displacements in the failing rocks around the tunnel wall and different ground reaction curves will be produced.

11 MODELING BRITTLE FAILURE IN JOINTED ROCK

From a practical point of view, the failure of jointed rock masses can be categorized under two distinct categories. The first category deals with the failure mechanisms, which are mainly structurally controlled. Failures that involve wedges/blocks along the intersection of at least two major discontinuities, or the failures that consist of plane shears and tension cracks in rock slopes belong to these structurally controlled failures. The most commonly used design approaches, which deal with the design of these failure mechanisms, adopt stability analyses by various limit equilibrium methods. The second category of failure includes non-structurally controlled failure surfaces in which some or the entire failure surface do not pass through discontinuities on the scale of considered structure. However, often the failure of rock masses involves both failure along the pre-existing discontinuities and failure of the intact rock (*i.e.* failure of smaller intact rock bridges/blocks precedes a kinematically feasible failure of larger jointed rock mass). For instance, the failure of rock slopes involving the failure of both intact rock and pre-existing discontinuities are shown in Figure 27.

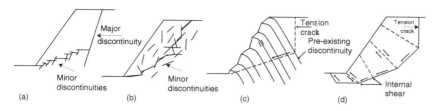

Figure 27 Rock slopes failure mechanisms involving both intact rock and pre-existing discontinuities.

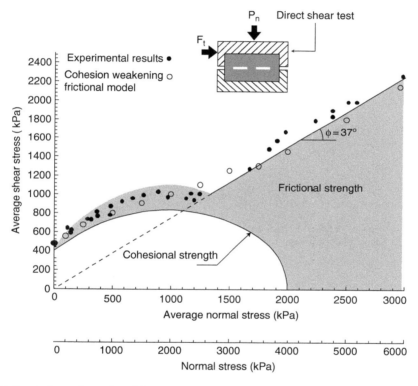

Figure 28 Results from direct shear laboratory tests using solid blocks containing voids, laboratory data from Lajtai (1969); model result from Hajiabdolmajid (2000).

Lajtai (1969) carried out direct shear tests on solid plaster blocks containing two voids to represent cracks or fracture. The rock bridges between the voids make up 50% of the plane through the voids (Figure 28). The plaster material had a tensile strength of approximately 1.1 MPa, a uniaxial compressive strength of 4.1 MPa and a basic friction angle 37°. The test results of Lajtai (1969); shown in Figure 28 illustrate the transition from pre-dominantly cohesive to frictional strength controlled failure. Lajtai's results showed that at low normal stresses, rock subjected to direct shear loading, fails by tension-induced damage or cohesion loss and at high

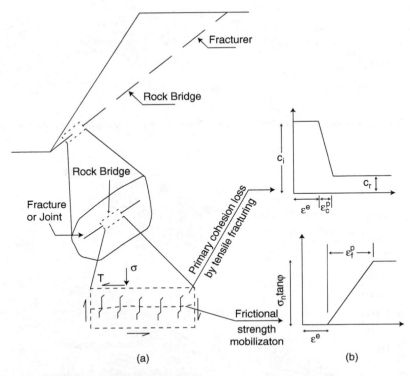

Figure 29 Mechanisms of failure in rock bridges along the failure planes of rock slopes (a) and modeling the failure of rock bridges using the CWFS model.

normal load in shear with full frictional strength mobilization. This data clearly support the notion of a bilinear failure envelope for brittle and jointed rock masses. The direct shear test as shown in Figure 28 was simulated, using the continuum-modeling code FLAC 2D using the CWFS model. Instead of a bilinear model for the frictional strengths and cohesion loss, a non-linear model for full cohesion loss and relatively rapid frictional strength mobilization characteristics was adopted (Hajiabdolmajid, 2001).

Following the implication of the CWFS concept in modeling brittle failure of intact rocks, similarly, it is argued that in jointed rock masses the intact rocks along the failure plane would fail at much smaller strain, followed by mobilization of friction along fracture surfaces, joints and discontinuities at increasingly larger strain (Figure 29). Therefore, when failure of jointed rock masses involves failure of intact rock bridges in low confinement environments, the simultaneous and full mobilization of the cohesive and frictional strength components cannot be supported. The failure of rock bridges between discontinuities is known to occur primarily by tensile fracturing. In these circumstances, the tensile fracturing first destroy the cohesive strength before that the frictional component of strength can contribute to the mobilized strength (Figure 29). Thus, conventional modeling approaches either underestimate ($c = 0$ approach) or overestimate (simultaneous mobilization of cohesion and friction).

Hajiabdolmajid & Kaiser (2002) used the CWFS modeling approach to analyze the stability of rock slopes; they demonstrated that the ultimate stable geometry and the location of the failure plane within a rock slope are functions of the strain sensitivity of the rock mass.

Barton (2014) and Barton & Pandey (2011) employed the CWFS modeling concept and Q-system based parameters for modeling the jointed rock masses in two Indian underground mines. They concluded that the failure of jointed rock masses occur as a strain-dependent process, starting with the failure of intact rock bridges, followed by sliding on the newly formed fracture surfaces (*i.e.* high roughness and dilation) and then progresses to surrounding joint sets (usually lower roughness components) and then to nearby clay-filled discontinuities.

The understanding and implication of the strain dependent mobilization of the strength components should be considered in the design and stability analyses of surface and underground excavations.

12 CONCLUSIONS

Brittle failure results from the growth and accumulation of tensile cracks. Around underground openings this progressive failure process manifests itself in the form of spalling or slabbing. This transition from continuum to discontinuum behavior is extremely difficult to capture in numerical models despite advances in discontinuum modeling. Traditional continuum modeling approaches to this class of problems assume that the mobilization of the cohesive and frictional strength components is simultaneous and instantaneous. This approach overlooks a fundamental aspect of brittle failure, that the formation of tensile cracks precedes failure in shear. The constitutive model (CWFS model) introduced in this chapter with strain-dependent cohesion weakening and frictional strengthening implicitly captures these phenomena. The CWFS modeling approach captured both the notch formation and its stabilization around AECL's Mine-by tunnel in Lac du Bonnet granite.

Different rocks are expected to have different strain limits for cohesion weakening and frictional strengthening. These strain limits for the cohesion weakening-frictional strengthening model can be established by laboratory tests and from back-analyses of excavations that have experienced brittle failure. A strain-dependent rock brittleness was introduced which can be used to describe the pre- to post-failure stages of a failing rock.

Adopting a continuum modeling approach to model a discontinuum process will certainly not capture all the subtleties of brittle failure process. However, understanding and implication of the strain dependent mobilization of the strength components should be considered in the design and stability analyses of excavations in hard rocks as from the practical point of view what is of paramount importance to the designer is the maximum extent of brittle fracturing as it is directly related to the support requirements. The proposed modeling approach is capable of filling this need reliably.

Where the stability of jointed rock masses is not completely controlled by the continuous discontinuities (*i.e.* when intact rock and asperities are involved), the stain-dependent mobilization of cohesive, frictional and roughness components of strength provides more realistic analysis approach.

REFERENCES

Barton, N. (2014) Shear strength of rock, rock joints, and rock masses: problems and some solutions. R. Alejano, Áurea Perucho, Claudio Olalla, & Rafael Jiménez (Eds.), Rock Engineering and Rock Mechanics: Structures in and on Rock Masses, CRC Press.

Barton, N. & Pandey, S.K. (2011) Numerical modelling of two stoping methods in two Indian mines using degradation of c and mobilization f φ based on Q-parameters. International Journal of Rock Mechanics and Mining Sciences, 48(7), 1095–1112.

Brace, W.F. & Bombolakis, E.G. (1963) A note on brittle crack growth in compression. Journal of Geophysical Research, 68(12), 3709–3713.

Brown, E.T., Bray, J.W., Ladanyi, B. & Hoek, E. (1983) Characteristics line calculations for rock tunnels. Journal of Geotechnical Engineering Division, ASCE, 109: 15–39.

Diederichs, M. S., Carter, T. G. & Martin, C. D. (2010) Practical rock spall prediction in tunnels. In E. Eberhardt (Ed.), Proceedings ITA World Tunnel Congress, Vancouver, Volume CD-ROM. International Tunnelling Organization.

Fredrich, J.T. & Wong, T. (1986) Micromechanics of thermally induced cracking in three crustal rocks. Journal of Geophysical Research, 91, 12743–12764.

Hallbauer, D.K., Wagner, H. & Cook, N.G.W. (1973). Some observations concerning the microscopic and mechanical behavior of quartzite specimens in stiff, triaxial compression tests. International Journal of Rock Mechanics and Mining Sciences, Pergamon Press, 10, 713–726.

Hajiabdolmajid, V.R. (2001) Mobilization of strength in brittle failure of rock, Ph.D. Thesis, Department of Mining Engineering, Queen's University, Kingston, Canada, 268p.

Hajiabdolmajid, V. & Kaiser, P. (2002) Modelling slopes in brittle rock. Proceedings of the Narms-TAC 2002, Mining and Tunnelling Innovation and Opportunity, Toronto, Vol. 1, 331–338.

Hajiabdolmajid, V. & Kaiser, P.K. (2003) Brittleness of rock and stability assessment in hard rock tunneling. Tunnelling and Underground Space Technology, 18, 35–48.

Hajiabdolmajid, V., Martin, C.D. & Kaiser, P.K. (2000) Modelling brittle failure of rock. Proceedings of the 4th North American Rock Mechanics Symposium, Seattle, Washington. A. A. Balkema, Rotterdam, 991–998.

Hill, R. (1950) The Mathematical Theory of Plasticity. Oxford, Clarendon Press.

Lajtai, E.Z. (1969) Mechanics of second order faults and tension gashes. Geological Society of America Bulletin, 80, 2253–2272.

Martin, C.D. (1997) Seventeenth Canadian Geotechnical Colloquium: The effect of cohesion loss and stress path on brittle rock strength. Canadian Geotechnical Journal, 34(5), 698–725.

Martin, C.D. (2014) The impact of brittle behaviour of rocks on excavation design. In R. Alejano, Áurea Perucho, Claudio Olalla, & Rafael Jiménez (Eds.), Rock Engineering and Rock Mechanics: Structures in and on Rock Masses, CRC Press, 51–62.

Martin, C.D. & Chandler, N. (1994) The progressive fracture of Lac du Bonnet granite. International Journal of Rock Mechanics and Mining Sciences, Pergamon Press, 31(6), 643–659.

Martin, C.D., Kaiser, P.K. & McCreath, D.R. (1999) Hoek-Brown parameters for predicting the depth of brittle failure around tunnels. Canadian Geotechnical Journal, 36(1), 136–151.

Martin, C.D. & Read, R.S. (1996) AECL's Mine-by Experiment: A test tunnel in brittle rock. In M. Aubertin, F. Hassani, & H. Mitri (Eds.), Proceedings of the 2nd North American Rock Mechanics Symposium, Montreal, Rotterdam, A.A. Balkema, Vol. 1, 13–24.

Martin, C.D., Read, R.S. & Martino, J.B. (1997) Observations of brittle failure around a circular test tunnel. International Journal of Rock Mechanics and Mining Sciences, Pergamon Press, 34(7), 1065–1073.

Myer, L.R., Kemeny, J. M., Zheng, Z., Suarez, R., Ewy, R.T. & Cook, N.G.W. (1992) Extensile cracking in porous rock under differential compressive stress. Micromechanical modelling of quasi-brittle materials behaviour. Applied Mechanics Reviews, 45(8), 263–280.

Ortlepp, W.D. (1997) Rock Fracture and Rockbursts-An Illustrative Study. Monograph Series M9. The South African Institute of Mining and Metallurgy, Johannesburg, 98p.

Pelli, F., Kaiser, P.K. & Morgenstern, N.R. (1991) An interpretation of ground movements recorded during construction of Donkin-Morien tunnel. Canadian Geotechnical Journal, 28(2), 239–254.

Read, R.S. (1994) Interpreting excavation-induced displacements around a tunnel in highly stressed granite. Ph.D. Thesis, Department of Civil and Geological Engineering, University of Manitoba, Manitoba, Canada, 328p.

Read, R.S. (1996) Characterizing excavation damage in highly stressed granite at AECL's Underground Research Laboratory. In J.B. Martino and C.D. Martin (Eds.), Proceedings of the International Conference on Deep Geological Disposal of Radioactive Waste, Winnipeg, Toronto, Canadian Nuclear Society, 35–46.

Read, R.S. & Martin, C.D. (1996) Technical Summary of AECL's Mine-by Experiment Phase 1: Excavation Response. AECL Report AECL-11311, Atomic Energy of Canada Limited.

Schmertmann, J.H. & Osterberg, J.H. (1960) An experimental study of the development of cohesion and friction with axial strain in saturated cohesive soils. Research Conference on Shear Strength of Cohesive Soils, Boulder, Colorado, New York, American Society of Civil Engineers, 643–694.

Schofield, A.N. (1998) Don't use the C word – Mohr Coulomb error correction. Ground Engineering, 29–32.

Schofield, A.N. & Worth, C.P. (1966) Critical State Soil Mechanics. Maidenhead, McGraw Hill.

Taylor, D.W. (1948) Fundamentals of Soil Mechanics. New York, John Wiley Sons, Inc., 700p.

Vermeer, P.A. & de Borst, R. (1984). Non-associated plasticity for soils, concrete and rock. Heron, 29(3), 65p.

Wagner, H. (1987) Design and support of underground excavations in highly stressed rock. In G. Herget & S. Vongpaisal (Eds.), Proceedings of the 6th ISRM Congress on Rock Mechanics, Montreal, A.A. Balkema, Rotterdam, Vol. 3, 1443–1457.

Zheng, Z., Kemeny, J. & Cook, N.G.W. (1989) Analysis of borehole breakouts. Journal of Geophysical Researches, 94, 7171–7182.

Chapter 21

Pre-peak brittle fracture damage

E. Eberhardt[1], M.S. Diederichs[2] & M. Rahjoo[1]

[1]Geological Engineering/EOAS, University of British Columbia, Vancouver, Canada
[2]Geological Sciences & Geological Engineering, Queen's University, Kingston, Canada

Abstract: This chapter reviews several key milestones in the study of pre-peak brittle fracture damage in rock, including both experimental and in situ observations and the conceptual advances that have arisen from these. What is recognized is that the strength of rock relative to brittle damage and failure mechanisms like spalling is considerably lower than the peak strength values measured through laboratory testing. Instead, pre-peak brittle fracture damage parameters such as crack initiation and crack damage serve as better indicators of the lower and upper *in situ* strength limits for brittle rocks. This understanding has led to the development of improved predictive tools for assessing the potential depth of brittle failure around deep underground excavations, as described in this chapter, together with the important role confinement plays in limiting and managing the brittle damage and failure process.

1 INTRODUCTION

The past 50 years have seen considerable advances in both our fundamental understanding of brittle rock fracture and its application in assessing and safely managing highly stressed rock. This has been essential for enabling underground mines to reach unprecedented depths in response to global demands for mineral resources as near surface resources are exhausted. The Mponeng, TauTona and Savuka Gold Mines southwest of Johannesburg in South Africa are approaching or have surpassed 4000 m depth. The Kidd Creek and Creighton Mines in northern Ontario, Canada are approaching 3000 m depth – with *in situ* stresses where the horizontal stresses are more than twice the vertical overburden stress. Similarly, transportation and hydroelectric tunnels are reaching unprecedented depths as infrastructure development increasingly expands into mountainous regions. The Olmos Trans-Andean Tunnel in Peru experienced overburdens of up to 1930 m. The Gotthard Base Tunnel in Switzerland was constructed with overburdens of up to 2300 m. The tunnels of Jinping II reached maximum overburdens of 2525 m.

Rising to meet these challenges has required a transition in the state-of-practice toward a more detailed accounting of the complexity of pre-peak brittle fracture damage and failure under higher stresses. During excavation, openings in highly stressed rock masses experience significant strength degradation and bulking due to stress-induced brittle fracturing, slip along discontinuities and surface buckling leading to extensive deformation and loading of the support system. This has significant implications for support design.

624   Eberhardt et al.

It was the development of the deep level gold mines in South Africa in the 1960's that bridged the study of brittle fracture in rock to earlier studies involving metals and glass, for example those by Griffith (1920, 1924). These led to studies of brittle fracture propagation under compression and stable and unstable crack propagation as a proxy for spalling and rockbursting (*e.g.*, Cook, 1965). The second step-change came in the 1990's and 2000's with research conducted for nuclear waste repositories examining excavation damaged zones (EDZ) and spalling failures in test tunnels (*e.g.*, Martin, 1997). This work has continued to gain momentum as the lessons learned from underground research laboratories have come full circle with new understanding and practical applications for deep mining (Kaiser *et al.*, 2000)and deep tunneling (Diederichs, 2007) with respect to designing support systems to manage brittle failure and safely construct the next generation of deep underground excavations.

This chapter will review several key milestones in the study of pre-peak brittle fracture damage in rock, including both experimental and *in situ* observations and the conceptual advances that arose from these.

2 MECHANISTIC UNDERSTANDING

A common starting point for discussing pre-peak brittle fracture damage in rock begins with the work of Griffith (1920, 1924). This built on findings by Inglis (1913) and Hopkinson (1921) who showed that the presence of an elliptical crack in a stressed plate results in localized stress concentrations at the tips of the crack. Hopkinson concluded that these stress concentrations would cause a material to fail at a lower applied stress than the same material without a crack. Griffith (1920, 1924) drew on this work to show why materials fail at stress levels lower than their theoretical strength, setting the basis for mechanistic theories that explain brittle rock failure through the initiation of damage and propagation of fractures owing to material heterogeneity.

2.1 Damage initiation

At the atomic scale, damage initiation is considered to occur through the tensile breakage of bonds resulting in the formation of new crack surfaces; in compression, an atomic structure can theoretically generate infinitely large resistance as there is no limit to the magnitude of the repulsive forces between adjacent atoms (Figure 1). The two newly developed tensile surfaces require a certain amount of energy as surface energy, which is satisfied from the release of strain energy. On this basis, the theoretical tensile strength of materials can be calculated as:

$$T = \sqrt{\frac{\gamma E}{d_o}} \tag{1}$$

where T is the theoretical tensile strength, γ is the surface energy, and d_o is the atomic spacing. Assuming typical values for the surface energy and atomic spacing, this relationship simplifies to:

$$T = \frac{E}{10} \tag{2}$$

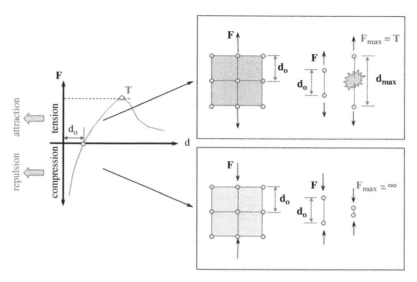

Figure 1 Development of interatomic forces due to loading. In extension, failure occurs when the attractive force F is exhausted at the theoretical tensile strength T. In contrast, compressional displacement is countered by an inexhaustible repulsive force.

Assuming a Young's modulus of 50 GPa, for example for granite, this would suggest a theoretical tensile strength of 5 GPa, which is at least three orders of magnitude higher than experimental values.

Recognizing this discrepancy in the failure of glass, Griffith (1920) postulated that the weakness of brittle materials relative to their theoretical strengths is due to the presence of small discontinuities or flaws that act as stress concentrators (these are commonly referred to as Griffith cracks). He then proposed an energy balance between the energy input absorbed to create the new crack surfaces and the energy liberated by the release of elastic strain energy as the regions adjacent to the crack become unloaded. This can be written in terms of the tensile stress for fracture, T_o, as:

$$T_o = \sqrt{\frac{2\gamma E}{\pi a}} \qquad (3)$$

where γ is the surface energy (the factor of 2 reflects that two new free surfaces have been formed), E is the Young's modulus, and a is the crack half length. Reformulating Equation 3 for a biaxial compressive stress field, Griffith (1924) surmised that the applied compressive stress, C_o, required for fracture is eight times that required for tension, or:

$$C_o = 8 * \sqrt{\frac{2\gamma E}{\pi a}} \qquad (4)$$

This relationship was later modified by McClintock & Walsh (1962) to allow for normal stress and friction acting across the surface of the closing crack, and by Hoek & Bieniawski (1965) who defined a modified Griffith fracture locus in terms of a parabolic Mohr envelope.

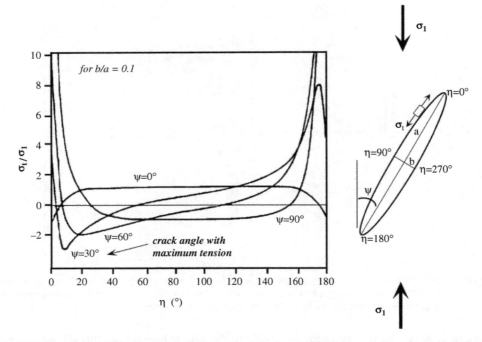

Figure 2 Stress concentrations along the periphery of an elliptical crack for varying crack angles, ψ, with respect to the direction of the major principal stress. Modified from Lajtai (1971).

In deriving Equation 4, Griffith (1924) noted that the presence of inclined cracks can give rise to localized tensile stress gradients even when the applied stress is compressive. This was further illustrated by Lajtai (1971) who calculated the stresses around an inclined elliptical crack in a compressive stress field and showed that: i) tensile stress concentrations develop locally close to the crack tip, and ii) the crack orientation at which the highest tensile stress concentrations develop is at 30° to the maximum principal stress (Figure 2). Although the latter may be the first to initiate, it may be assumed that the crack population is randomly distributed and orientated so that with incremental increases in the applied load, other crack angles will become critical.

Griffith (1920) also noted that Equation 3 indicates that material strength is inversely proportional to the square root of the crack length, and therefore, the smaller the crack length the stronger the material should be. Numerous studies have confirmed that the peak strength of rock decreases inversely with the square root of the grain size (Brace, 1961; Olsson, 1974; Fredrich et al., 1990; Wong et al., 1996; Hatzor & Palchik, 1997; Eberhardt et al., 1999c; Nicksiar & Martin, 2013). In these cases, the Griffith crack length can be taken as being roughly proportional to the grain size, whereby the grain boundary is assumed to act as the stress concentrating crack. Simmons & Richter (1976) and Kranz (1983) divide the petrographic characteristics of Griffith cracks into four types: i) grain boundary cracks, ii) intragranular cracks which lie within the mineral grain, iii) intergranular cracks which extend from a grain boundary crossing into another grain, and iv) multigranular cracks, which cross several grains and grain boundaries. Direct observations of microfractures using either optical microscopes or scanning electron microscopes (SEM) suggest that damage initiation usually occurs along grain boundaries with secondary initiation occurring within weaker grains along cleavage planes and at points where harder minerals induce a point load on neighboring softer minerals (Brace, 1961; Wawersik & Brace, 1971; Brace *et al.*, 1972; Bombolakis, 1973; Sprunt & Brace, 1974; Mosher *et al.*, 1975; Tapponnier & Brace, 1976; Kranz, 1979).

2.2 Crack propagation

Once damage initiates, the crack will then propagate in either a stable or unstable fashion depending on how much energy is available to drive the crack extension onwards. Examination of Griffith's criterion (Equation 4) reveals that a number of factors may influence the strength threshold of a Griffith crack, most notably crack size and crack orientation. This was confirmed by Mosher *et al.* (1975) who found that grain size (*i.e.* crack length) and crack orientation determines which cracks propagate and which do not. However, Brace & Bombolakis (1963) note that although Griffith theory specifies the stress at which cracks will initiate, it provides no subsequent information on the rate or direction of crack propagation. They then go on to explain that in tension, the most critically stressed cracks are those aligned perpendicular to the applied stress and that crack propagation will likewise occur perpendicular to the applied stress. In compression, the most critically stressed cracks are those inclined at 30° to the direction of compression, but propagation occurs parallel to the maximum applied stress. In their experiments, this took the form of "wing cracks", where crack growth would initiate at the tips of the inclined crack and then rotate to propagate in the direction of the maximum applied stress (Figure 3).

In effect, crack propagation occurs in the direction of the major principal stress (σ_1); *i.e.*, crack face opening occurs in the direction of the minor principal stress (σ_1). This has been observed experimentally in a variety of brittle materials including glass (Brace & Bombolakis, 1963; Hoek & Bieniawski, 1965), hard acrylics (Nemat-Nasser & Horii,

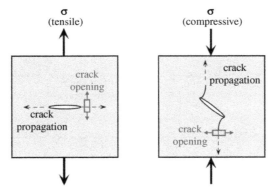

Figure 3 Direction of crack propagation (normal to crack opening) for critically oriented cracks under tensile and compressive applied stresses.

1982; Cannon et al., 1990), plaster (Lajtai, 1971), ice (Schulson et al., 1991), and rock (Wawersik & Fairhurst, 1970; Peng & Johnson, 1972; Huang et al., 1993). Pollard & Aydin (1988) explain this directionality as cracks propagating to align themselves with the direction that produces the maximum propagation energy (otherwise known as the energy release rate, G, which is the energy dissipated per unit of newly created crack surface area).

Bieniawski (1967a) further discussed the energy release rate, G, in terms of stable and unstable crack propagation. In formulating the critical condition for fracture, Griffith made assumptions which effectively ignored the behavior of the moving crack. Griffith's energy balance accounted for the stored elastic strain energy and the crack surface energy only. Bieniawski (1967a), however, noted that several other forms of energy loss occur through which part of the elastic strain energy is transformed:

- kinetic energy;
- plastic energy (including visco-plastic losses);
- energy dissipated on the breakdown of atomic bonds at the tips of extending cracks;
- energy changes due to excavation (*e.g.*, heat removal due to ventilation, etc.).

Among these, kinetic energy associated with the movement of the crack faces is considered the more significant (outside of what can be controlled through the tunneling or mining operations). Craggs (1960) showed that the stress required to maintain crack propagation decreases as the crack velocity increases. Bieniawski (1967a) emphasized that this wasn't accounted for in Griffith's energy balance and turned to crack velocity as a means to distinguish between stable and unstable crack propagation (Figure 4). Stable crack propagation is the process by which crack extension occurs in step with a load increase; *i.e.*, the energy released from the stored elastic strain energy

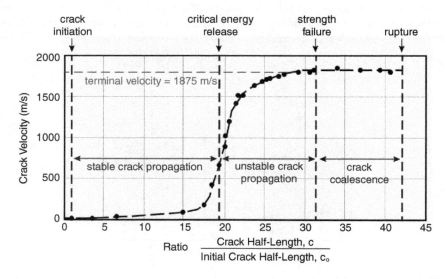

Figure 4 Crack velocity versus crack length ratio determined experimentally for a norite. Modified from Bieniawski (1967b).

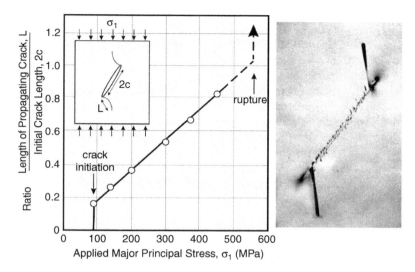

Figure 5 Relationship between stable crack length and applied uniaxial compressive stress for testing of inclined crack in glass plate. Modified from Hoek & Bieniawski (1965).

occurs at the same rate as the energy used to form additional crack surface area. Hoek & Bieniawski (1965) showed this through experiments on inclined cracks in glass plates where the lengths of the stable cracks were found to increase linearly with the applied uniaxial compressive stress (Figure 5). Crack propagation becomes unstable when the relationship between crack extension and loading ceases, and other quantities like crack growth velocity start to play an important role (Bieniawski, 1967a). In such cases, the crack will propagate uncontrollably even if no further increase in load is applied (*i.e.*, under constant load), and failure will quickly follow.

2.3 Influence of confinement

McClintock & Walsh (1962) considered the case of a confining stress being applied to an inclined Griffith crack loaded in compression. For this, they assumed the crack would partially close resulting in frictional forces that in turn would reduce the stress concentrations at the ends of the crack (see also Lajtai, 1971). In Griffith theory, the crack is assumed to be open and no forces are carried across the faces of the crack. After modifying the Griffith criterion to account for friction and normal and shear stresses, McClintock & Walsh (1962) demonstrated that the effect of confining stress on crack initiation and propagation is considerable. Wawersik & Fairhurst (1970) noted that the addition of moderate confining pressures could eliminate the development of slabbing in the compression testing they carried out. As noted by Bieniawski *et al.* (1969), this has practical implications as rock under higher confining pressures should be less prone to spalling and violent rupture than those under uniaxial compression.

The importance of confinement in limiting crack propagation has been demonstrated experimentally and numerically by various authors, as summarized by Hoek & Martin

(2014). Figure 6a presents a comparison they carried out specific to the initiation and propagation of wing cracks from a Griffith crack subjected to both confinement and compressive loading (compiled from Hoek, 1965b; Ashby & Hallam, 1986; Kemeny & Cook, 1987; Germanovich & Dyskin, 1988; Martin, 1997; Cai *et al.*, 1998). These results show that a confining stress of just 10-20% of the major principal stress can reduce the length cracks propagate by more than 80%. Similarly, Eberhardt *et al.* (1998a) showed in a numerical study of stress shadows arising in a multiple crack array, that although interactions between adjacent cracks may help promote crack initiation, the presence of confining stress subsequently suppresses the extent these cracks can propagate (Figure 6b). The results from their study suggest that rock under high confinement will contain a large number of small cracks, contributing to volumetric dilation, friction mobilization and plastic shear behavior with failure; whereas rock under low confinement will exhibit fewer but longer cracks, contributing to geometric dilation, spalling and slabbing.

3 EXPERIMENTAL UNDERSTANDING

Early studies investigating brittle fracture damage utilized controlled experiments involving individual inclined cracks embedded in thin plates of glass, acrylic and plaster to directly observe crack initiation and propagation responses to applied compressive loads. Focus was primarily placed on linking these to Griffith's theory. These were soon followed by studies involving uniaxial compression testing of brittle rock to investigate the evolution of brittle fracture damage in the form of stress-induced microfracturing (*e.g.*, Brace, 1964; Bieniawski, 1967b; Wawersik & Fairhurst, 1970; Lajtai & Lajtai, 1974; Martin & Chandler, 1994; Eberhardt *et al.*, 1998b; Diederichs, 2003; Diederichs, 2007). The general consensus of these studies has been that several stages of brittle fracture damage behavior can be identified, which lead to macroscopic failure of rock under low confinement.

3.1 Key thresholds for pre-peak brittle fracture damage

Brace (1964) and Bieniawski (1967a) postulated that the brittle failure of rock involves several stages, which can be identified in the measured stress-strain response of rock undergoing compressive loading. These are (Figure 7):

1) crack closure;
2) linear elastic deformation;
3) crack initiation and stable crack propagation;
4) crack damage and unstable crack propagation;
5) failure (peak strength); and
6) post-peak softening and rupture.

Crack closure occurs during the initial stages of loading when pre-existing cracks orientated roughly normal to the applied load close. During crack closure, the stress-strain response is non-linear. The extent of this non-linear region is dependent on the initial crack density and geometrical characteristics of the crack population. Once the majority of pre-existing cracks have closed, linear elastic deformation takes place.

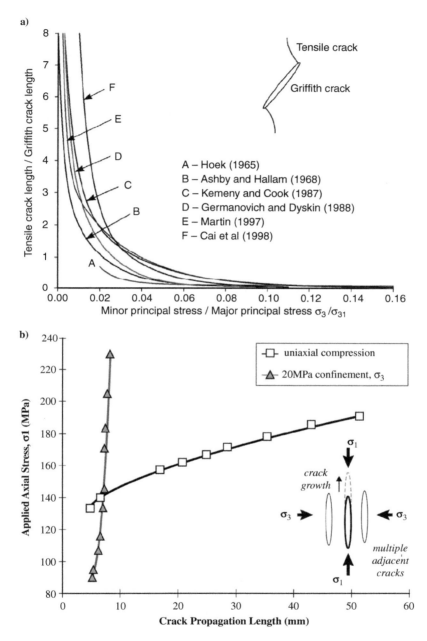

Figure 6 Influence of confining stress on limiting stable crack propagation, as shown by: a) Hoek & Martin (2014) for data compiled from experimental and numerical investigations of an inclined Griffith crack, and b) Eberhardt *et al.* (1998a) for numerical modelling of a multiple array of elliptical axial cracks.

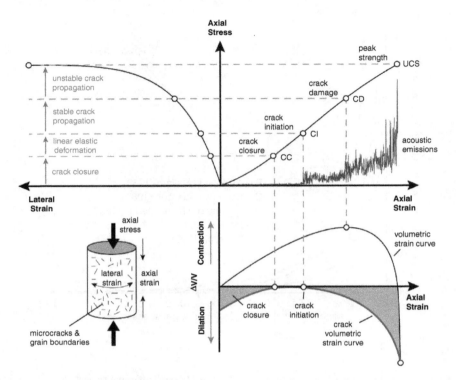

Figure 7 Stress-strain diagram showing the stages and thresholds of brittle fracture development. Note that the volumetric strain is calculated from the axial and lateral strain, and the crack volumetric strain is calculated from the volumetric strain, Young's modulus and Poisson's ratio. Modified from Martin (1993) and Martin & Christiansson (2009).

The Young's modulus and Poisson's ratio are calculated from the linear portion of the stress-strain curve that follows.

Crack initiation signifies the stress level where grain-scale microfracturing begins. Crack propagation at this point is stable with cracks propagating parallel to the applied axial load (Bieniawski, 1967b). Eberhardt et al. (1998b) and Diederichs et al. (2004) note that crack initiation is influenced by heterogeneity, and therefore the first cracks to initiate, not associated with sample end effects, will be those most critically aligned or involving statistical outliers of localized weak links within the sample. Hallbauer et al. (1973), for example, found that point loading of grains by other grains was a common source of crack initiation, and it follows that weaker minerals adjacent to stronger minerals would be the source of the first intergranular cracks. Accordingly, Diederichs et al. (2004) and Diederichs (2007) suggest that a more representative measure of crack initiation as a damage parameter is the detection of the onset of continuous, systematic cracking (i.e., damage).

As loading continues, propagating cracks will accumulate to the point that they will begin to interact. This interaction then becomes extremely complex as stress shadows overlap (Eberhardt et al., 1998a; Diederichs, 2003). Eberhardt et al. (1998b) postulated that as the population of propagating cracks increase, both in number and size,

Pre-peak brittle fracture damage **633**

they will begin to step out and coalesce incorporating an element of shearing (*e.g.*, en echelon cracks; see Lajtai *et al.*, 1994). They termed this threshold crack coalescence, which Diederichs (2007) noted correlates with the definition of yield used in other branches of material science (see ASTM, 2015).

Under low confinement, stable crack propagation will eventually reach a critical level and transition to unstable crack propagation. As previously mentioned, Bieniawski (1967a) defines unstable crack propagation as the condition that occurs when the relationship between the applied stress and the crack length ceases to exist and other parameters, such as the crack growth velocity take control of the propagation process. Thin section analysis of rock samples loaded up to and beyond this point revealed that this threshold coincides with a sudden increase in volumetric strain due to considerable damage taking place in the form of grain shattering (Bieniawski, 1967b). Martin (1993) renamed this threshold the "crack damage stress" as loads above this stress level result in damage to the rock that cannot be further tolerated. Craggs (1960) and Bieniawski (1967b) suggest that this involves the propagating cracks reaching a terminal velocity (Figure 4), which then leads to the cracks bifurcating and branching in order to dissipate the additional energy. Failure quickly follows. The peak strength marks the ultimate capacity of the sample and almost universally is used to establish the failure strength envelope of rock (Martin & Chandler, 1994).

Numerous studies, including those by Hudson *et al.* (1972), Peng & Johnson (1972) and Martin (1993) have shown that the peak strength of rock is significantly influenced by the specimen shape, loading rate, and boundary conditions of the test. However, Hudson *et al.* (1972) and later Martin (1993) concluded from this that peak strength is not an inherent material property but represents an arbitrary stage in the rock fracture and progressive failure process. Instead Martin (1993) found that crack initiation (CI) and crack damage (CD) were essentially independent of the loading conditions, and therefore serve as better material properties for brittle rock strength. Subsequent work by Martin (1997) and Diederichs (2007) have equated CI with the lower bound and CD with the upper bound *in situ* rock strength. Accordingly, considerable effort has gone into establishing means to measure these brittle fracture damage parameters based on laboratory testing.

3.2 Determination of crack initiation

A number of techniques have been developed to detect brittle fracture processes through laboratory testing. The most common of these involves the use of strain measurements (strain gauges, displacement transducers, etc.) to detect slight changes in sample deformation that can be correlated to the closing, opening and coalescence of cracks (Brace *et al.*, 1966; Bieniawski, 1967b; Martin & Chandler, 1994; Eberhardt *et al.*, 1998b, 1999b). The opening of crack faces parallel to the applied load and the closure of crack faces perpendicular to the load causes certain changes in the relative lateral and axial deformations, respectively. These changes appear as inflections in the stress-strain curves which, in turn, can be used to identify the different stages of brittle fracture damage. For example, because crack closure involves the preferential closure of crack faces parallel to the direction of applied load, its effect on the measured axial strain is significantly more pronounced than its effect on the lateral strain. Accordingly, the crack closure stress threshold, CC, is determined as the point where the axial stress-axial strain curve transitions from non-linear to a linear response (Figure 7).

The crack initiation stress threshold, CI, represents the onset of new damage. It was initially defined as the point where the lateral strain curve departs from linearity in response to new cracks opening normal to their propagation direction parallel to the applied axial stress (Brace *et al.*, 1966; Bieniawski, 1967b; Lajtai & Lajtai, 1974). However, detecting this change through strain measurements can be very subjective as the change is very subtle and often of the same magnitude as the measurement resolution. Noting this difficulty, Martin (1993) suggested using the calculated crack volumetric strain to identify CI (Figure 7). For a cylindrical sample tested in uniaxial compression, crack volume is determined by subtracting the elastic component of the volumetric strain from the total:

$$\varepsilon_{V\ crack} = \varepsilon_V - \varepsilon_{V\ elastic} \tag{5}$$

$$\varepsilon_{V\ elastic} = \frac{1 - 2v}{E}\sigma_{axial} \tag{6}$$

where E and v are the elastic constants and σ_{axial} is the applied axial stress. Volumetric strain is typically calculated from the measured axial (ε_{axial}) and lateral ($\varepsilon_{lateral}$) strains, given by:

$$\varepsilon_V = \varepsilon_{axial} + 2\varepsilon_{lateral} \tag{7}$$

Eberhardt *et al.* (1998b) found, however, that this method introduces considerable subjectivity and variability into the assessment of CI because of its dependence on the use of the elastic constants E and v. Although the Young's modulus, E, can be determined with a reasonably high degree of consistency, the nonlinearity of the lateral strain response complicates the measure of Poisson's ratio, v (Eberhardt, 1998). Testing of Westerly granite by Walsh (1965) showed that Poisson's ratio is not constant but continuously increases throughout loading, varying from 0.2 to 0.3 between 30 and 60% peak strength where elastic behavior is typically assumed. Figure 8a shows that this uncertainty can result in CI values that differ by up to ±40% for a change of ±0.05 in the Poisson's ratio assumed.

Instead, Eberhardt *et al.* (1998b) proposed that CI could be more accurately determined using an approach that combined the use of acoustic emissions (AE) with a moving point regression analysis performed on the stress-strain curves (to determine inflections). AE originate through the sudden release of stored elastic strain energy accompanying the initiation and propagation of microfractures. These can then be detected by AE sensors placed at the boundary of the sample (Figure 9), which can measure the number, amplitude and energy of the AE events. Numerous researchers have demonstrated that AE response provides a unique and direct method for studying brittle fracture processes. Scholz (1968) found that characteristic AE patterns in rock correlate closely with stress-strain and brittle fracture behavior identified by Brace *et al.* (1966). However, most of the success in correlating AE activity to microfracturing has involved the latter stages of crack development close to peak stress (Scholz, 1968; Sondergeld *et al.*, 1984; Rao, 1988; Lockner *et al.*, 1991; Shah & Labuz, 1995; Thompson *et al.*, 2006). This is due to the fact that the majority of AE events occur just prior to failure. Conversely, Eberhardt (1998) showed that if the AE monitoring frequency and detection sensitivity (*e.g.*, amplifier gain) are calibrated toward crack initiation, at the expense of recording extra noise and encountering data censoring due

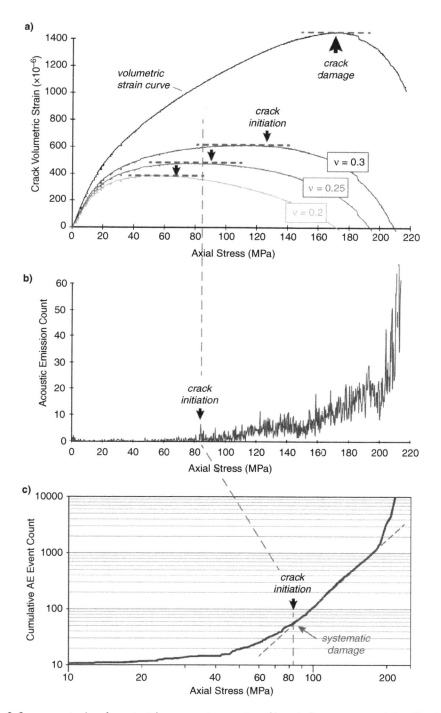

Figure 8 Stress-strain data for uniaxial compression testing of Lac du Bonnet granite (after Eberhardt, 1998). a) Sensitivity of crack volumetric strain curve to Poisson's ratio and resulting variability in crack initiation stress. Volumetric strain and crack damage under uniaxial conditions are shown for comparison. b) Acoustic emission event count for the same test showing detected crack initiation stress. c) Application of Diederich's (1999) method for determining the crack initiation upper limit representing the onset of systematic damage.

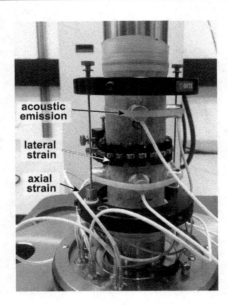

Figure 9 Instrumented sample showing use of LVDT's to measure axial strain, circumferential roller LVDT assembly to measure lateral strain, and multiple AE sensors for measuring acoustic emissions.

to overlapping event signals near the peak stress, then CI could be accurately and reliably determined (Figure 8b). The use of AE combined with strain proved even more robust and repeatable (±5% over 20 samples of Lac du Bonnet granite) when other AE parameters such as ringdown count, event duration, and event energy rate were used (Eberhardt *et al.*, 1998b). As discussed by Diederichs *et al.* (2004), detection of isolated AE events may be seen earlier representing noise and low amplitude random cracking (*e.g.*, note that a number of small events can be seen before the increase in AE labeled as crack initiation in Figure 8b).

Diederichs & Martin (2010) thus define CI as being the first point at which a systematic increase in AE is observed (Figure 8c). Given differences in detection limits between strain and AE based measurements, CI can be viewed as ranging between upper and lower bound values (see Ghazvinian *et al.*, 2015). The lower bound can be used for sensitive or long-term projects, whereas the upper bound would be suitable for normal projects. Typical values for CI have been reported as ranging between 0.4 and 0.5 of the uniaxial compressive strength (UCS); Figure 10. Diederichs & Martin (2010) subsequently emphasize the importance of CI as a design parameter as it marks the stress at which new cracks initiate and begin to propagate. Near the excavation boundary where confining stresses are low, crack propagation can easily transition from stable to unstable crack propagation and spalling. CI is therefore seen as representing the lower bound long-term *in situ* strength of rock; *i.e.*, only at low confinements, such as at an excavation boundary, would spalling failure be observed (Martin, 1997; Kaiser *et al.*, 2000; Hajiabdolmajid *et al.*, 2002; Diederichs, 2003; Diederichs *et al.*, 2004; Diederichs, 2007; Carter *et al.*, 2008).

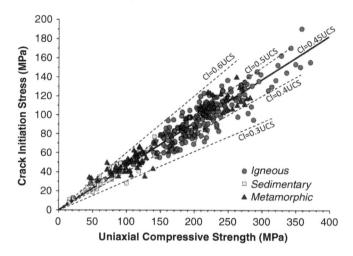

Figure 10 Crack initiation stresses determined by Nicksiar & Martin (2013) for a range of different rock types. Modified from Hoek & Martin (2014).

3.3 Determination of crack damage

Bieniawski (1967b) correlated the threshold marking the transition from stable to unstable crack propagation with the point of reversal in the volumetric strain curve (Figure 7, 8a). Diederichs (2003) showed both numerically and statistically that this threshold also marked the onset of crack interaction (with or without crack propagation) and the transition between compression parallel crack accumulation before CD and inclined crack growth after (Figure 11). Hallbauer et al. (1973) noted that thin section analysis of rocks loaded beyond the volumetric strain reversal, but stopped prior to peak strength, showed pronounced structural changes that involved stepwise coalescence of propagating microcracks. Martin & Chandler (1994) suggest that the dominant mechanism at this stage is sliding along inclined cracks, which explains the strong departure from linear behavior observed in the axial strain. This also explains the sharp increase seen at the same time in the lateral strain rate, which surpasses the axial strain rate to become the dominant component in the volumetric strain calculation, thus resulting in the reversal of the volumetric strain curve. Martin & Chandler (1994) note that this stress level has particular significance in the concrete industry as it is used to establish the long-term strength of concrete. Results from Schmidtke & Lajtai (1985) for Lac du Bonnet granite showed a similar threshold for long-term strength. Based on this, Martin & Chandler (1994) used the point of volumetric strain reversal to define the crack damage threshold.

However, Diederichs et al. (2004) and Diederichs (2007) suggest that because volumetric strain reversal is delayed until lateral strain begins to increase substantially, it over predicts the long-term *in situ* strength and therefore the crack damage threshold. They further explain that in laboratory testing, sample end effects and circumferential stresses induce internal confinement that constrains the ability of propagating cracks to dilate, thus suppressing unstable crack propagation. These effects are not present *in situ* around

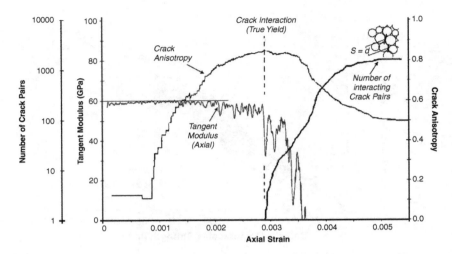

Figure 11 PFC modelling of crack interaction showing relationship between tangent modulus (deviation from axial stress-strain linearity) and number of interacting crack pairs (cracks within one mean particle diameter of each other, i.e., S ≤ d). Also shown is crack anisotropy (normalized second invariant of crack tensor; 0 for isotropy, 1 for parallel cracks). Modified from Diederichs (2003).

underground excavations, where under low confinement unstable crack propagation can more easily develop. Thus, Diederichs (2007) recommends the use of the onset of yield, defined as the first significant departure from linearity of the axial strain curve, as being a better indicator of CD. Eberhardt et al. (1998b), referring to this as crack coalescence, showed it can easily be determined by plotting the moving point regression of the slope of the axial strain curve (Figure 12). Diederichs (2007) subsequently notes that although volumetric strain reversal and yield are coincident for uniaxial loading conditions, they diverge for higher confining stresses (Figure 13). Accordingly, axial strain non-linearity and volumetric strain reversal serve as limiting bounds for CD (Diederichs et al., 2004), with CD representing the upper bound long-term in situ strength of rock.

Values of CD have been reported to vary between 0.7 and 0.9 of the UCS. In contrast to CI, CD has been shown to be sensitive to differences in grain size (i.e., initial crack length) and sampling damage (i.e., initial crack intensities). Increasing grain size was found to reduce CD owing to the longer grain boundaries facilitating the propagation of longer cracks and a more rapid acceleration to failure once these longer cracks began to interact and coalesce (Eberhardt et al., 1999c). Diederichs (2003) used a statistical simulation to likewise show that longer initial cracks lead to premature crack interactions, crack damage and failure (Figure 14a). With respect to crack intensities, Martin & Chandler (1994), Eberhardt et al. (1999a,b), and Diederichs (2003) showed that the presence of increased crack intensities (via sampling damage) reduced the crack damage stress, presumably by providing an increased number of planes of weakness for propagating cracks to coalesce with, resulting in their coalescence, and ultimately yield and failure at lower stresses. Diederichs (2003) demonstrated this through PFC modeling, finding that critical interactions between propagating cracks develop at lower stresses with increasing initial crack intensities, where those intensities were achieved through

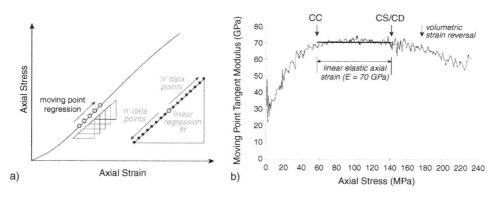

Figure 12 a) Moving point regression technique applied to an axial strain curve for Lac du Bonnet granite, and b) the corresponding moving point tangent modulus showing crack closure (CC) and crack coalescence (CS), i.e. crack damage (CD), based on the deflection of the axial strain curve from linear to non-linear. Modified after Eberhardt et al. (1998b).

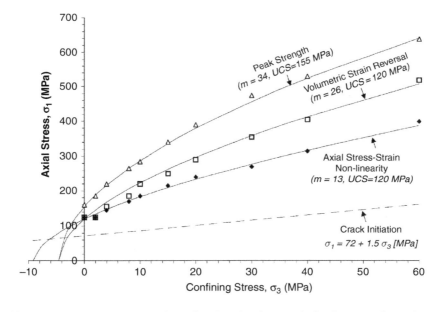

Figure 13 Brittle fracture damage envelopes fitted to data from cyclic load tests performed on Lac du Bonnet granite. Modified from Diederichs (1999).

either increased numbers of cracks or by increased extension of cracks (Figure 14b). He concluded that while CI was related to a stress-based threshold, CD is related to a critical probability of crack interaction which in turn is associated with a critical amount of accumulated lateral extension strain (normal to the maximum compression). Diederichs (2003) also demonstrated through fracture mechanics analysis that cracks propagated more readily near a flat free surface, in part due to low confinement but also due to

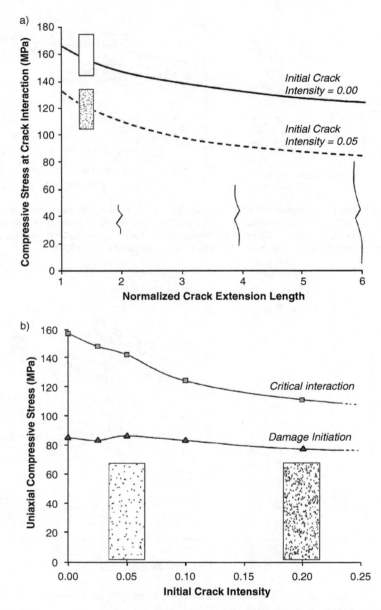

Figure 14 a) Statistical modelling of crack interaction stress as a function of crack length and initial crack intensity, and b) PFC simulation of damage initiation and crack interaction thresholds as a function of initial crack intensity. Modified after Diederichs (2007).

geometrical feedback with the surface. He also discusses the mechanisms by which crack propagation is suppressed within a standard cylindrical laboratory compression sample facilitating a CD threshold in the lab that does not correspond with that in the field.

3.4 Stress heterogeneity and confining stress

A final key piece in extending laboratory test data targeting brittle fracture damage to *in situ* observations of stress-induced brittle failure (*i.e.*, spalling) involves stress heterogeneity and its sensitivity to confinement. Diederichs (1999, 2003, 2007) and Diederichs *et al.* (2004) propose that CI represents a lower bound for *in situ* rock strength in massive rock and CD interaction represents an upper bound. Together these two thresholds delineate the region of stress-induced spalling that can be expected in high stress environments. This is discussed in more detail in Section 4. Stress heterogeneity plays a central role in that it generates grain interactions which promote early damage initiation, as well as stress shadows that lead to localized confinement that can arrest crack propagation (Fredrich *et al.*, 1990; Eberhardt *et al.*, 1998a; Cho *et al.*, 2002). As demonstrated by Hoek (1968), propagating cracks are hypersensitive to even very low confinement. For true crack propagation to occur the confining stress (normal to the extending cracks) must be near to or less than zero.

Figure 15 shows PFC modeling results from Diederichs (1999), in which the particle diameters and bond stiffnesses were randomly varied to create a heterogeneous assembly. Different levels of confining stress were then added. The results show that under low confinement, localized zones within the sample are in tension (Figure 15a). These would represent zones of enhanced crack propagation. Diederichs (2003) notes that, as would be expected, the spatial coverage of the tensile zones was seen to increase with increasing deviatoric stress and decrease with increasing confinement. Figure 15b plots the results of the spatial extent of the local tension zones for different confining stresses along with the known model thresholds for crack initiation, interaction and failure. Diederichs *et al.*

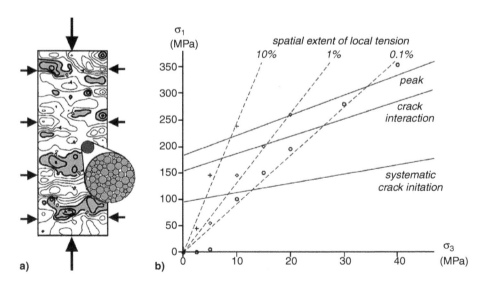

Figure 15 a) PFC modelling of influence of material heterogeneity on generating localized stress heterogeneity. The shaded zones inside the heavy black contours are those in tension. b) Plot of PFC results for different confining stresses, showing spalling limits (dashed lines) corresponding to percentages of local tensile stress occurrence within nominally confined samples at elevated deviatoric stress. Superimposed are the established brittle fracture thresholds (solid lines). Modified after Diederichs (2007).

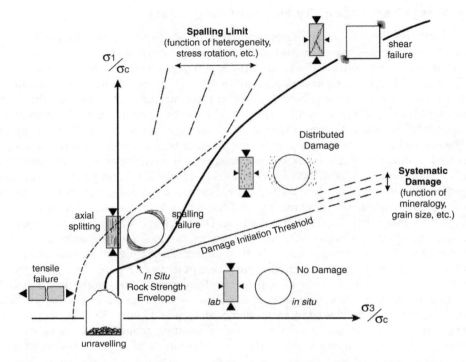

Figure 16 Composite in situ strength envelope for brittle rock (solid curve), composed of segments corresponding to upper bound strength (high confinements), lower bound strength or damage initiation (low confinements), and a transition zone related to the spalling limit. Modified after Diederichs (2002).

(2004) note that the stress ratio ranges indicated for the three coverage limits shown are comparable to those first proposed by Hoek (1968). They go on to define these limits as the spalling limit. Based on this plot (Figure 15b), a stress state (σ_1, σ_3) above the CI threshold and to the right of this spalling limit has the potential for premature yield due to strength reduction caused by unstable crack propagation. In practice, the slope of the critical spalling limit will vary according to the degree of heterogeneity within the rock, together with a number of other external factors including damage and stress rotation.

Thus as shown in Figure 16, heterogeneity controls the degree to which the rock strength degrades from its upper bound to its lower bound. Diederichs (2007) proposes that the *in situ* strength for stress-induced brittle failure can be estimated by adopting the CI stress at low confinements (lower bound brittle rock strength), and the CD threshold at higher confinements (upper bound brittle rock strength). The transition between these is defined by the spalling limit as previously discussed.

4 CONCEPTUAL ADVANCES AND PRACTICAL APPLICATION

The early work on brittle fracture damage by Cook (1965), Hoek & Bieniawski (1965) and Fairhurst & Cook (1966), was largely directed at better understanding rockburst hazards and observations of spalling and slabbing in deep South African mines. Similar to observations from laboratory compression testing, stress-induced brittle failure underground was also observed to involve the initiation and propagation of brittle fractures parallel to the direction of maximum compression. Hoek (1965a) observed that initiation began with the formation of a notch in the zone of maximum compression. Fairhurst & Cook (1966) observed that because confining stresses increase into the rock mass, these are very effective in inhibiting the propagation of cracks. Similar observations were subsequently made involving spalling failure around circular openings and deep underground excavations, including Gay (1973), Stacey & de Jongh (1977), Ortlepp & Gay (1984), Ewy & Cook (1990) and Carter *et al.* (1991).

These studies were followed by an extensive body of research conducted by the Atomic Energy of Canada Limited (AECL) at the Underground Research Laboratory (URL) in Pinawa, Manitoba, that led to a step-change in efforts to relate brittle fracture damage to the development of stress-induced spalling. This began with observations of spalling in the massive granite host rock in response to high *in situ* stress anomalies, during extension of the access shaft from the 240 m Level to a new 420 m Level (Martin, 1988). Given that one of the justifications for considering massive crystalline rock for a nuclear waste repository is its extremely low permeability, the development of stress-induced damage and its effects on permeability required a high level of scientific study.

4.1 The AECL-URL mine-by experiment

Especially noteworthy among the numerous *in situ* experiments at the URL, was the focused effort of the Mine-By Experiment (Figure 17a). This involved the installation of a dense array of extensometers, convergence arrays, triaxial strain cells, and accelerometers for microseismic (MS) detection prior to excavation of a test tunnel on the 420 m Level so that the complete excavation response of the rock mass could be monitored (Martin & Read, 1996). Read & Martin (1992) describe the planning of the experiment in which the tunnel axis was selected to align with the intermediate principal stress to maximize the stress ratio in the plane of the tunnel profile to promote the development of the deepest excavation-damaged zone. Excavation was carried out by first line-drilling the complete outer perimeter of the test tunnel, and then progressively breaking out the interior with hydraulic rock splitters. This eliminated any blast-induced damage.

The results of this experiment led to several key observations, the first of which was the classic V-shaped notches that formed in the tunnel roof and floor, perpendicular to the direction of the major principal stress (Figure 17b). Considerable MS events were recorded ahead of the tunnel face, and as the tunnel face was advanced, MS events were seen to locate in the roof and floor concentrating where the V-shaped notch was developing (Young & Martin, 1993). The progressive nature of the damage process continued even after completion of the excavation advance, and only ended once the notch that was forming stopped in response to the increasing confining stress experienced at the tip of the notch.

In terms of stress path, it was observed that damage initiation (observed as MS events ahead of the tunnel face), occurred at 33-50% of the UCS. Martin (1997) noted that this was very similar to the stress magnitudes associated with crack initiation (CI) in laboratory testing. Analysis of the MS source mechanisms indicated

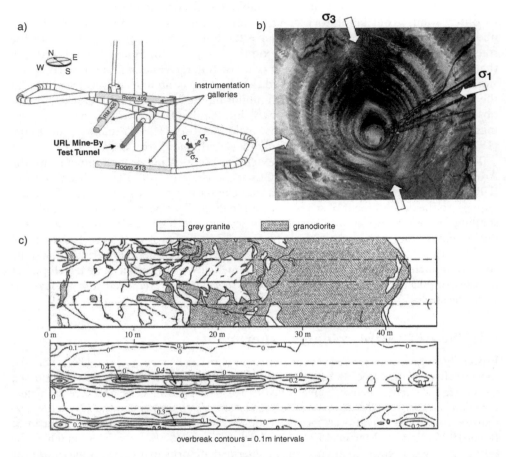

Figure 17 a) AECL-URL Mine-By Experiment, highlighting the 420 m level instrumentation galleries and test tunnel. b) Notch development in the test tunnel relative to the *in situ* stress field. c) Perimeter maps showing correlation between geology (i.e. grain size) and the depth of spalling in the test tunnel. Modified after Martin et al. (1997).

that the events were dominated by tensile modes of failure (Feignier & Young, 1992). Spalling associated with the notch formation was observed by Read *et al.* (1998) as developing just behind the tunnel face at stress levels of about 60% of UCS. This corresponds to the crack damage (CD) stress threshold in laboratory testing. Martin *et al.* (1997) further note that the spalling observed was more pronounced when in granite than when in the finer-grained, and therefore stronger, granodiorite (Figure 17c). Collins & Young (2000) made similar observations with respect to the detection of MS being predominantly in the granite. Comparison of the extent and depth of spalling in the granite and granodiorite emphasized the influence of grain size on the CI and CD stress magnitudes. This was also seen in laboratory testing of the same rock types (Eberhardt *et al.*, 1999c).

4.2 Cohesion loss and friction mobilization

Brace (1960) postulated that for brittle rock failure, it is unlikely that friction and cohesion act simultaneously and instead, cohesion likely drops to zero after the onset of motion, at which point friction comes into play. Martin & Chandler (1994) carried out a series of damage-controlled tests to investigate cohesion loss and friction mobilization. They used a series of load-unload cycles to first incrementally reach 75% of the expected peak strength, which was in excess of the CI stress and approximated the CD stress. The load-unload cycles were then performed in small increments of circumferential deformation (0.063 mm relative to a 63 mm diameter sample) using axial-strain control. What was observed was that with each load-unload cycle, damage accumulated but this had no effect on CI. However, it did act to degrade CD. It was proposed that up to the CD stress, the strength of the granite sample was derived only from cohesion, and therefore, with each load cycle that exceeded the cohesive strength (*i.e.*, CD stress level), fracture damage was introduced resulting in cohesion loss and friction mobilization (Figure 18). Based on this work, Martin & Chandler (1994) suggested that the peak friction angle of brittle rock is only reached when most of the cohesion is lost.

The concept of cohesion loss with frictional strengthening was subsequently used by several researchers to better model the depth of brittle failure using continuum-based numerical methods. Conventional strain-weakening models (SW) assume the simultaneous degradation of cohesion and frictional strength, as shown in Figure 19a. This was suggested by Hoek *et al.* (1995) as a means to model spalling by assigning pre-peak Hoek-Brown strength values based on CD (equated to the long-term strength), together with very low residual strength values (m = 1, s = 0.01) to simulate an elastic-brittle-plastic failure response. This concept was later used by Cai *et al.* (2007) to provide an estimation for the residual strength of rock masses. Martin (1997), however, showed that a strain-weakening constitutive model was unable to correctly estimate the extent and depth of brittle failure observed for the URL Mine-By Experiment. Instead, he proposed a Cohesive-Brittle-Frictional (CBF) model in which the extensional brittle failure

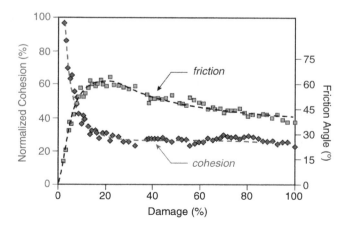

Figure 18 Mobilization of friction with cohesion loss as a function of percent damage. Modified after Martin *et al.* (1997).

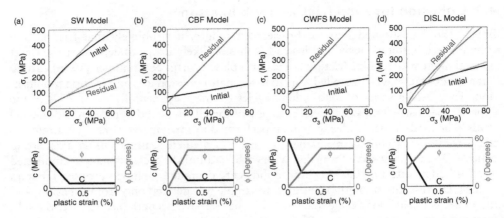

Figure 19 Yielding models used for modelling depth of brittle failure (*i.e.*, spalling) observed in the AECL-URL Mine-By Experiment test tunnel: a) conventional Strain-Weakening model used by Hoek *et al.* (1995) and Martin (1997), b) Cohesive-Brittle-Frictional model introduced by Martin (1997), c) Cohesion-Weakening and Friction-Strengthening model introduced by Hajiabdolmajid *et al.* (2002), d) Damage-Initiation and Spalling-Limit model introduced by Diederichs (2007). The cohesion and friction mobilization graphs show the evolution of cohesive and frictional resistance with respect to cumulative plastic strain (*i.e.*, damage). The SW and DISL models that were originally defined using the Hoek-Brown criterion, are converted here to Mohr-Coulomb for comparison purposes.

mechanism begins with cohesion loss without friction mobilization (Figure 19b). With continued loading, frictional resistance gradually mobilizes as the extensionally-formed spalling slabs make contact and shear relative to each other. In implementing the CBF model, the initial yielding parameters were set based on CI and the residual yielding parameters were set to simulate cohesion drop and friction mobilization. The transition between the initial and residual state is simultaneous.

Hajiabdolmajid *et al.* (2003) introduced the Cohesion-Weakening and Friction-Strengthening (CWFS) model, as shown in Figure 19c. The major difference between the CBF and CWFS models is that the latter assumes that frictional strength doesn't fully mobilize until significant damage, in excess of that required for cohesion to drop to its residual value, has accumulated. Numerical implementation of the CWFS model is done by defining cohesion and friction angel values as functions of plastic strain. Like the CBF model, the initial and residual yielding states of CWFS model have an intersection in $\sigma_1 - \sigma_3$ space (Figure 19c) which implies a softening behavior in low confinement and a hardening behavior in higher confinements. In comparison to the SW and CBF models, the CWFS model has been shown to be more successful in capturing the extent and depth of brittle failure around openings (Hajiabdolmajid *et al.*, 2002; Diederichs, 2007; Xia-Ting *et al.*, 2007; Lee *et al.*, 2012; Walton & Diederichs, 2015). Walton & Diederichs (2015) show how the impact of dilation can be included in this brittle analysis technique.

Diederichs (2007) introduced a similar model called the Damage Initiation and Spalling Limit (DISL) model, as shown in Figure 19d. The DISL model, designed for use with analysis software with fixed parametric input, was defined using the generalized Hoek-Brown criterion (Hoek *et al.*, 2002), and includes an initial yielding state based on the CI and CD thresholds, transitioning to a residual state defined by the spalling limit (Figure 16). The DISL and CBF models are similar with respect to assuming cohesion loss and friction mobilization occur simultaneously (Figure 19b, d). Although this differs from the CWFS model (Figure 19c), the DISL model produces similar results if the input parameters are assigned correctly (Diederichs, 2007; Cai, 2013; Perras *et al.*, 2014; Perras & Diederichs, 2016).

4.3 Depth of damage and spalling failure

Determining the depth of stress-induced damage in brittle rock is a key requirement for the design of nuclear waste repositories in terms of assessing permeability change in the near-field host rock, as well as for the design of deep tunnels and mine excavations in terms of assessing the potential for rock mass spalling and bulking and the rock support necessary to manage these. The terminology related to damage zones around underground excavations has varied based on different experiences and improving knowledge of the responses observed, resulting in several commonly used acronyms. Siren *et al.* (2015) and Perras & Diederichs (2016) provide recent updates and descriptions of the terminology used, with emphasis placed on the use of EDZ to refer to the Excavation Damage Zone as being specific to stress-induced brittle fracture damage. This is differentiated from the Construction Damage Zone (CDZ), which relates to the excavation method, for example the extent of blast-induced damage. Martino & Chandler (2004) refer to the depth of the EDZ as being defined by the extent of measurable and permanent changes to the mechanical and hydraulic-transport properties of the rock surrounding the excavation. Beyond this, the Excavation Influence Zone (EIZ) involves only elastic changes (Diederichs & Martin, 2010; Perras & Diederichs, 2016). EIZ has sometimes been referred to as the "disturbed zone" although this is inconsistent with conventional geotechnical definitions of disturbed as the changes in this zone are fully recoverable.

Several early studies investigated the relationship between the *in situ* stress state and the depth of brittle failure around highly stressed underground excavations (*e.g.*, Fairhurst & Cook, 1966; Hoek & Brown, 1980; Detournay & St. John, 1988). Martin *et al.* (1999) found that studies using traditional failure criteria based on frictional strength did not meet with much success in either predicting the initiation or maximum depth of brittle failure (*e.g.*, Wagner, 1987; Pelli *et al.*, 1991; Castro *et al.*, 1996; Martin, 1997). In comparison, criterion that considered cohesive strength degradation were successful (Martin, 1997). Martin *et al.* (1999) suggested that the depth of brittle failure can be found using an elastic stress analysis superimposed with a cohesion-based Hoek-Brown criterion ($m = 0$, $s = 0.11$); Diederichs (2007) warns that this criterion should not be used for inelastic analysis as it significantly overpredicts the extent of the yield zone in this application.

Using several case histories involving underground excavations that reported brittle failure, Martin *et al.* (1999) proposed a linear empirical relationship for depth of spalling around a tunnel (Figure 20a):

$$\frac{r_f}{a} = 0.49 + 1.25 \frac{\sigma_{max}}{UCS} \tag{8}$$

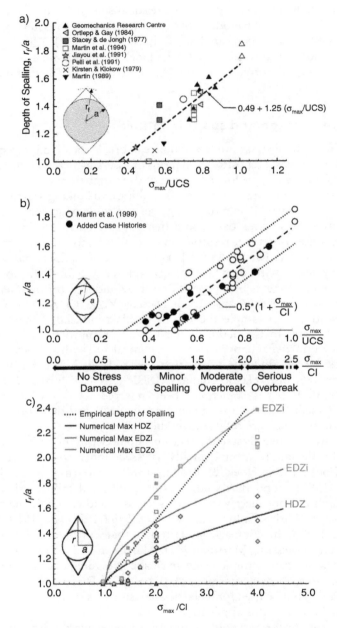

Figure 20 Relationship between depth of spalling and the maximum tangential stress at the boundary of an underground excavation, as reported by: a) Martin et al. (1999), b) Diederichs et al. (2010), and d) Perras & Diederichs (2016).

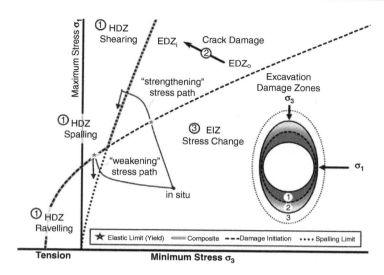

Figure 21 HDZ, EDZ and EIZ damage zones mapped to the DISL approach of Diederichs (2003, 2007). From Perras & Diederichs (2016).

where r_f is the depth of failure, which is normalized to the tunnel radius a. σ_{max} is the maximum tangential compressive stress at the excavation boundary and UCS is the laboratory measured uniaxial compressive strength. Diederichs et al. (2010) updated this analysis by adding several more case histories (Figure 20b), and presented an alternative formulation based on CI from direct test measurements:

$$\frac{r_f}{a} = 0.5\left(\frac{\sigma_{max}}{CI} + 1\right) \text{ for } \sigma_{max} > CI \quad (9)$$

Perras & Diederichs (2016) compared these empirical trends to a series of 2-D finite-element models that used the DISL procedure (Diederichs, 2007) to examine a broad spectrum of input parameters and stress scenarios. The analysis also used the DISL model to differentiate between the development of a highly damaged zone (HDZ) at the excavation boundary, involving interconnected macro-fractures, and the transition to an inner EDZ (EDZ_i), where the damage is interconnected, and then to an outer EDZ (EDZ_o), where the damage is only partially connected or isolated (see Bossart et al., 2002); Figure 21. Based on these results, Perras & Diederichs (2016) defined a relationship for depth of failure adaptable for each zone:

$$\frac{r_f}{a} = 1 + B\left(\frac{\sigma_{max}}{CI} - 1\right)^D \quad (10)$$

where B and D were determined through best fit nonlinear regression analyses for the model results specific to the different damage zones (B = 0.2, 0.4, 0.6 respectively for HDZ, EDZ_i, and EDZ_o; D = 0.7, 0.5, 0.6, respectively). The findings of Perras & Diederichs (2016) suggest that the empirical depth of failure limit of Martin et al. (1999) and Diederichs et al. (2010) should be used with caution at higher maximum tangential stress to CI strength ratios, as the numerical results indicate a nonlinear relationship with the depth of failure (Figure 20c).

5 CONCLUSIONS

Kaiser & Kim (2008) caution that the two most important lessons to be learned from deep mining and tunneling experiences are that: (1) rock at depth is much less forgiving, and (2) costly mistakes can be made when the failure mechanism is not properly understood. Kaiser (2016) continues by emphasizing that it is essential to anticipate the rock mass behavior correctly, specifically that brittle fracture damage requires underground excavation designs to consider and account for stress-induced rock mass degradation and deformation (*i.e.*, bulking) responses. The significance of stress-induced brittle fracture damage and spalling is often not recognized on deep mining and tunneling projects, despite the accumulation of experiences where brittle failure has been encountered imposing difficult and hazardous conditions.

First, it must be recognized that the brittle failure process begins at stresses that are approximately half of the UCS. As discussed in this chapter, the work of Martin (1997), Eberhardt *et al.* (1999b), Kaiser *et al.* (2000), Diederichs (2003, 2007) and Diederichs *et al.* (2004) have contributed to the characterization and use of crack initiation (CI) and crack damage (CD) as design parameters that, respectively, describe the lower-bound *in situ* rock strength where spalling damage initiates and the upper-bound *in situ* strength where spalling under higher confinements deeper into the rock mass can develop. By applying the CI and CD strength envelopes, the depth of spalling can be assessed empirically (Martin *et al.*, 1999; Diederichs *et al.*, 2010) or numerically (Diederichs, 2003, 2007). These predictive tools represent the evolvement of the mechanistic understanding of pre-peak brittle fracture damage, from the theoretical studies of Griffith cracks to more than 50 years' worth of laboratory and *in situ* observations of brittle fracture damage and spalling.

Second, confining stress plays an important and effective role in limiting crack propagation. Thus, in terms of support design in highly stressed brittle rock, adding just a small amount of support pressure can have a significant impact on excavation performance. As discussed in this chapter, the depth of brittle fracture damage and spalling is sensitive to the confinement feedback generated by dilational yielding and bulking. This feedback is only active if the yielding rock mass is held in place by an effective support system. Although a detailed discussion of rock support in stress-induced fractured rock is outside the scope of this chapter, the work of Kaiser *et al.* (2000), Diederichs (2007) and Kaiser (2016) explain the importance of understanding the brittle failure process in order to effectively control it through support design. The brittle fracture damage process and the rock mass degradation it causes (via spalling) effectively reduces the stand-up time to near zero, requiring near face support to properly reinforce the rock mass (to control bulking) and retaining elements (*e.g.*, mesh) to prevent unraveling and to hold the rock in place to provide confinement to the rock mass behind (Kaiser *et al.*, 2000; Kaiser, 2016).

REFERENCES

Ashby, M.F. & Hallam, D. (1986) The failure of brittle solids containing small cracks under compressive stress. *Acta Metallurgica*, 34(3), 497–510.

ASTM (2015) E6-15. *Standard terminology relating to methods of mechanical testing*. West Conshohocken, PA, ASTM International.

Bieniawski, Z.T. (1967a) Mechanism of brittle rock fracture: Part I – Theory of the fracture process. *International Journal of Rock Mechanics and Mining Sciences*, 4(4), 395–406.

Bieniawski, Z.T. (1967b) Mechanism of brittle rock fracture: Part II – Experimental studies. *International Journal of Rock Mechanics and Mining Sciences*, 4(4), 407–423.

Bieniawski, Z.T., Denkhaus, H.G. & Vogler, U.W. (1969) Failure of fractured rock. *International Journal of Rock Mechanics and Mining Sciences*, 6, 323–341.

Bombolakis, E.G. (1973) Study of the brittle fracture process under uniaxial compression. *Tectonophysics*, 18, 231–248.

Bossart, P., Meier, P.M., Moeri, A., Trick, T. & Mayor, J.C. (2002) Geological and hydraulic characterisation of the excavation disturbed zone in the Opalinus Clay of the Mont Terri Rock Laboratory. *Engineering Geology*, 66(1–2), 19–38.

Brace, W.F. (1960) An extension of the Griffith theory of fracture to rocks. *Journal of Geophysical Research*, 65(10), 3477–3480.

Brace, W.F. (1961) Dependence of fracture strength of rocks on grain size. *Bulletin of the Mineral Industries Experiment Station, Mining Engineering Series, Rock Mechanics*, 76, 99–103.

Brace, W.F. (1964). Brittle fracture of rocks. In: Judd, W.R. (ed.) *State of Stress in the Earth's Crust: Proceedings of the International Conference, Santa Monica*. American Elsevier Publishing Co., New York. pp. 110–178.

Brace, W.F. & Bombolakis, E.G. (1963) A note on brittle crack growth in compression. *Journal of Geophysical Research*, 68(12), 3709–3713.

Brace, W.F., Paulding, B.W., Jr. & Scholz, C. (1966) Dilatancy in the fracture of crystalline rocks. *Journal of Geophysical Research*, 71(16), 3939–3953.

Brace, W.F., Silver, E., Hadley, K. & Goetze, C. (1972) Cracks and pores: A closer look. *Science*, 178(4057), 162–164.

Cai, M. (2013) Realistic simulation of progressive brittle rock failure near excavation boundary. In: Yang, Q., Zhang, J.-M., Zheng, H. and Yao, Y. (ed.) *Constitutive Modeling of Geomaterials*. Berlin, Springer. pp. 303–312.

Cai, M., Kaiser, P.K. & Martin, C.D. (1998) A tensile model for the interpretation of microseismic events near underground openings. *Pure and Applied Geophysics*, 153, 67–92.

Cai, M., Kaiser, P.K., Tasaka, Y. & Minami, M. (2007) Determination of residual strength parameters of jointed rock masses using the GSI system. *International Journal of Rock Mechanics and Mining Sciences*, 44(2), 247–265.

Cannon, N.P., Schulson, E.M., Smith, T.R. & Frost, H.J. (1990) Wing cracks and brittle compressive fracture. *Acta Metallurgica et Materialia*, 38(10), 1955–1962.

Carter, B.J., Lajtai, E.Z. & Petukhov, A. (1991) Primary and remote fracture around underground cavities. *International Journal for Numerical and Analytical Methods in Geomechanics*, 15, 21–40.

Carter, T.G., Diederichs, M.S. & Carvalho, J.L. (2008) Application of modified transition relationships for assessing strength and post yield behaviour at both ends of the rock competency scale. *Journal of South African Institute of Mining and Metallurgy*, 108, 325–338.

Castro, L.A.M., McCreath, D.R. & Oliver, P. (1996) Rockmass damage initiation around the Sudbury Neutrino Observatory cavern. In: Aubertin, M., Hassani, F. and Mitri, H. (eds.) *Rock Mechanics and Tools: Proceedings of the 2nd North American Rock Mechanics Symposium*, Montreal. A.A. Balkema, Rotterdam. pp. 1589–1595.

Cho, N., Martin, C.D. & Christiansson, R. (2002) Suppressing fracture growth around underground openings. In: Hammah, R., Bawden, W.F., Curran, J. and Telesnicki, M. (eds.) *NARMS-TAC 2002, Mining and Tunnelling Innovation and Opportunity; Proceedings of the 5th North American Rock Mechanics Symposium and the 17th Tunnelling Association of Canada Conference Toronto*. University of Toronto Press. pp. 1151–1158.

Collins, D.S. & Young, R.P. (2000) Lithological controls on seismicity in granitic rocks. *Bulletin of the Seismological Society of America*, 90(3), 709–723.

Cook, N.G.W. (1965) The failure of rock. *International Journal of Rock Mechanics and Mining Sciences & Geomechanics Abstracts*, 2(4), 389–403.

Craggs, J.W. (1960) On the propagation of a crack in an elastic-brittle material. *Journal of the Mechanics and Physics of Solids*, 8(1), 66–75.

Detournay, E. & St. John, C.M. (1988) Design charts for a deep circular tunnel under non-uniform loading. *Rock Mechanics and Rock Engineering*, 21(2), 119–137.

Diederichs, M.S. (1999) *Instability of Hard Rockmasses: The Role of Tensile Damage and Relaxation*. Ph.D. Thesis, University of Waterloo, Waterloo, Canada, 566p.

Diederichs, M.S. (2002) Stress induced damage accumulation and implications for hard rock engineering. In: Hammah, R., Bawden, W.F., Curran, J. and Telesnicki, M. (eds.) *NARMS-TAC 2002: Mining and Tunnelling Innovation and Opportunity, Proceedings of the 5th North American Rock Mechanics Symposium and the 17th Tunnelling Association of Canada Conference, Toronto*. University of Toronto Press. pp. 3–12.

Diederichs, M.S. (2003) Rock fracture and collapse under low confinement conditions. *Rock Mechanics and Rock Engineering*, 36(5), 339–381.

Diederichs, M.S. (2007) Mechanistic interpretation and practical application of damage and spalling prediction criteria for deep tunnelling. *Canadian Geotechnical Journal*, 44(9), 1082–1116.

Diederichs, M.S., Carter, T. & Martin, D. (2010) Practical rock spall predictions in tunnels. In: *Tunnel Vision Towards 2020: Proceedings, World Tunnel Congress 2010, Vancouver*. Tunnelling Association of Canada. pp. 1–8.

Diederichs, M.S., Kaiser, P.K. & Eberhardt, E. (2004) Damage initiation and propagation in hard rock tunnelling and the influence of near-face stress rotation. *International Journal of Rock Mechanics and Mining Sciences*, 41(5), 785–812.

Diederichs, M.S. & Martin, C.D. (2010) Measurement of spalling parameters from laboratory testing. In: Zhao, J., Labiouse, V., Dudt, J.P. and Mathier, J.F. (eds.) *Rock Mechanics in Civil and Environmental Engineering, Proceedings of the European Rock Mechanics Symposium (EUROCK 2010), Lausanne*. CRC Press, London. pp. 323–326.

Eberhardt, E. (1998) *Brittle Rock Fracture and Progressive Damage in Uniaxial Compression*. Ph.D. Thesis, University of Saskatchewan, Saskatoon, Canada, 334p.

Eberhardt, E., Stead, D. & Stimpson, B. (1999a) Effects of sampling disturbance on the stress-induced microfracturing characteristics of brittle rock. *Canadian Geotechnical Journal*, 36(2), 239–250.

Eberhardt, E., Stead, D. & Stimpson, B. (1999b) Quantifying progressive pre-peak brittle fracture damage in rock during uniaxial compression. *International Journal of Rock Mechanics and Mining Sciences*, 36(3), 361–380.

Eberhardt, E., Stead, D., Stimpson, B. & Lajtai, E.Z. (1998a) The effect of neighbouring cracks on elliptical crack initiation and propagation in uniaxial and triaxial stress fields. *Engineering Fracture Mechanics*, 59(2), 103–115.

Eberhardt, E., Stead, D., Stimpson, B. & Read, R.S. (1998b) Identifying crack initiation and propagation thresholds in brittle rock. *Canadian Geotechnical Journal*, 35(2), 222–233.

Eberhardt, E., Stimpson, B. & Stead, D. (1999c) Effects of grain size on the initiation and propagation thresholds of stress-induced brittle fractures. *Rock Mechanics and Rock Engineering*, 32(2), 81–99.

Ewy, R.T. & Cook, N.G.W. (1990) Deformation and fracture around cylindrical openings in rock – II. Initiation, growth and interaction of fractures. *International Journal of Rock Mechanics and Mining Sciences & Geomechanical Abstracts*, 27(5), 409–427.

Fairhurst, C. & Cook, N.G.W. (1966). The phenomenon of rock splitting parallel to the direction of maximum compression in the neighbourhood of a surface. In: *Proceedings of the First Congress of the International Society of Rock Mechanics, Lisbon*. Laboratório Nacional de Engenharia Civil, Lisbon. pp. 687–692.

Feignier, B. & Young, R.P. (1992) Moment tensor inversion of induced microseismic events: Evidence of non-shear failures in the -4 < M < -2 moment magnitude range. *Geophysical Research Letters*, 19(14), 1503–1506.

Fredrich, J.T., Evans, B. & Wong, T.-F. (1990) Effect of grain size on brittle and semibrittle strength: Implications for micromechanical modelling of failure in compression. *Journal of Geophysical Research*, 95(B7), 10907–10920.

Gay, N.C. (1973) Fracture growth around openings in thick-walled cylinders of rock subjected to hydrostatic compression. *International Journal of Rock Mechanics and Mining Sciences & Geomechanical Abstracts*, 10(3), 209–233.

Germanovich, L.N. & Dyskin, A.V. (1988) A model of brittle failure for materials with cracks in uniaxial loading. *Mechanics of Solids*, 23(2), 111–123.

Ghazvinian, E., Diederichs, M.S., Labrie, D. & Martin, C.D. (2015) An investigation on the fabric type dependency of the crack damage thresholds in brittle rocks. *Geotechnical and Geological Engineering*, 33, 1409–1429.

Griffith, A.A. (1920) The phenomena of rupture and flow in solids. *Philosophical Transactions of the Royal Society of London, Series A, Mathematical and Physical Sciences*, 221(587), 163–198.

Griffith, A.A. (1924). The theory of rupture. In: Biezeno, C.B. and Burgers, J.M. (ed.) *Proceedings of the First International Congress for Applied Mechanics, Delft*. J. Waltman Jr., Delft. pp. 55–63.

Hajiabdolmajid, V., Kaiser, P.K. & Martin, C.D. (2002) Modelling brittle failure of rock. *International Journal of Rock Mechanics and Mining Sciences*, 39(6), 731–741.

Hajiabdolmajid, V., Kaiser, P.K. & Martin, C.D. (2003) Mobilised strength components in brittle failure of rock. *Geotechnique*, 53(3), 327–336.

Hallbauer, D.K., Wagner, H. & Cook, N.G.W. (1973) Some observations concerning the microscopic and mechanical behaviour of quartzite specimens in stiff, triaxial compression tests. *International Journal of Rock Mechanics and Mining Sciences & Geomechanics Abstracts*, 10(6), 713–726.

Hatzor, Y.H. & Palchik, V. (1997) The influence of grain size and porosity on crack initiation stress and critical flaw length in dolomites. *International Journal of Rock Mechanics and Mining Sciences*, 34(5), 805–816.

Hoek, E. (1965a) Rock fracture around mining excavations. In: *Fourth International Conference on Strata Control and Rock Mechanics, Proceedings, New York*. Columbia University Press. pp. 334–348.

Hoek, E. (1965b) *Rock Fracture Under Static Stress Conditions*. Pretoria, South Africa, National Mechanical Engineering Research Institute, Council for Scientific and Industrial Research.

Hoek, E. (1968) Brittle failure of rock. In: Stagg, K.G. and Zienkiewicz, O.C. (ed.) *Rock Mechanics in Engineering Practice*. New York, John Wiley & Sons. pp. 99–124.

Hoek, E. & Bieniawski, Z.T. (1965) Brittle fracture propagation in rock under compression. *International Journal of Fracture Mechanics*, 1(3), 137–155.

Hoek, E. & Brown, E.T. (1980) *Underground Excavations in Rock*. London, The Institution of Mining and Metallurgy.

Hoek, E., Carranza-Torres, C.T. & Corkum, B. (2002). Hoek-Brown failure criterion – 2002 edition. In: Hammah, R., Bawden, W., Curran, J. and Telesnicki, M. (ed.) *Proceedings of the*

Fifth North American Rock Mechanics Symposium (NARMS-TAC), Toronto. University of Toronto Press. pp. 267–273.

Hoek, E., Kaiser, P.K. & Bawden, W.F. (1995) *Support of Underground Excavations in Hard Rock.* Rotterdam, A.A. Balkema.

Hoek, E. & Martin, C.D. (2014) Fracture initiation and propagation in intact rock – A review. *Journal of Rock Mechanics and Geotechnical Engineering,* 6(4), 287–300.

Hopkinson, B. (1921) Brittleness and ductility; Lecture delivered January 24th, 1910 to the Sheffield Society of Engineers and Metallurgists. In: Ewing, J.A. and Larmor, J. (ed.) *The Scientific Papers of Bertram Hopkinson.* Cambridge University Press, pp. 64–78.

Huang, J., Wang, Z. & Zhao, Y. (1993) The development of rock fracture from microfracturing to main fracture formation. *International Journal of Rock Mechanics and Mining Sciences & Geomechanics Abstracts,* 30(7), 925–928.

Hudson, J.A., Brown, E.T. & Fairhurst, C. (1972). Shape of the complete stress-strain curve for rock. In: Cording, E.J. (ed.) *Stability of Rock Slopes, Proceedings of the 13th U.S. Symposium on Rock Mechanics, Urbana, Illinois.* American Society of Civil Engineers, New York. pp. 773–795.

Inglis, C.E. (1913) Stresses in a plate due to the presence of cracks and sharp corners. *Transactions of the Institution of Naval Architects,* 55, 219–230.

Kaiser, P.K. (2016) *Ground support for constructability of deep underground excavations: Challenge of managing highly stressed ground in civil and mining projects.* Muirwood Lecture, Longrine, France, International Tunnelling Association. ISBN: 978-2-9701013-8-3: 1–31.

Kaiser, P.K., Diederichs, M.S., Martin, D., Sharpe, J. & Steiner, W. (2000). Underground works in hard rock tunnelling and mining. In: *GeoEng2000, Proceedings of the International Conference on Geotechnical & Geological Engineering, Melbourne.* Technomic Publishing Company, Lancaster. pp. 841–926.

Kaiser, P.K. & Kim, B.-H. (2008) Rock mechanics challenges in underground construction and mining. In: Potvin, Y., Carter, J., Dyskin, A. and Jeffrey, R. (eds.) *Proceedings of the First Southern Hemisphere International Rock Mechanics Symposium, Perth.* Australian Centre for Geomechanics, Perth. pp. 23–38.

Kemeny, J.M. & Cook, N.G.W. (1987) Crack models for the failure of rock under compression. In: Desai, C.S., Krempl, E., Kiousis, P.D. and Kundu, T. (eds.) *Proceedings of the 2nd International Conference on Constitutive Laws for Engineering Materials, Theory and Applications, Tuscon.* Elsevier, New York. pp. 879–887.

Kranz, R.L. (1979) Crack-crack and crack-pore interactions in stressed granite. *International Journal of Rock Mechanics and Mining Sciences & Geomechanics Abstracts,* 16(1), 37–47.

Kranz, R.L. (1983) Microcracks in rocks: A review. *Tectonophysics,* 100(1–3), 449–480.

Lajtai, E.Z. (1971) A theoretical and experimental evaluation of the Griffith theory of brittle fracture. *Tectonophysics,* 11, 129–156.

Lajtai, E.Z., Carter, B.J. & Duncan, E.J.S. (1994) En echelon crack-arrays in potash salt rock. *Rock Mechanics and Rock Engineering,* 27(2), 89–111.

Lajtai, E.Z. & Lajtai, V.N. (1974) The evolution of brittle fracture in rocks. *Journal of the Geological Society of London,* 130(1), 1–18.

Lee, K.-H., Lee, I.-M. & Shin, Y.-J. (2012) Brittle rock property and damage index assessment for predicting brittle failure in excavations. *Rock Mechanics and Rock Engineering,* 45(2), 251–257.

Lockner, D.A., Byerlee, J.D., Kuksenko, V., Ponomarev, A. & Sidorin, A. (1991) Quasi-static fault growth and shear fracture energy in granite. *Nature,* 350(6313), 39–42.

Martin, C.D. (1988) Shaft excavation response in a highly stressed rock mass. In: Dormuth, K. W. (eds.) *Proceedings of the NEA Workshop on Excavation Response in Geological Repositories for Radioactive Waste, Winnipeg.* Organization for Economic Co-operation and Development. pp. 331–340.

Martin, C.D. (1993) *Strength of Massive Lac du Bonnet Granite Around Underground Openings*. Ph.D. Thesis, University of Manitoba, Winnipeg, Canada, 278p.

Martin, C.D. (1997) Seventeenth Canadian Geotechnical Colloquium: The effect of cohesion loss and stress path on brittle rock strength. *Canadian Geotechnical Journal*, 34(5), 698–725.

Martin, C.D. & Chandler, N.A. (1994) The progressive fracture of Lac du Bonnet granite. *International Journal of Rock Mechanics and Mining Sciences & Geomechanics Abstracts*, 31(6), 643–659.

Martin, C.D. & Christiansson, R. (2009) Estimating the potential for spalling around a deep nuclear waste repository in crystalline rock. *International Journal of Rock Mechanics and Mining Sciences*, 46(2), 219–228.

Martin, C.D., Kaiser, P.K. & McCreath, D.R. (1999) Hoek-Brown parameters for predicting the depth of brittle failure around tunnels. *Canadian Geotechnical Journal*, 36(1), 136–151.

Martin, C.D. & Read, R.S. (1996). AECL's Mine-by Experiment: A test tunnel in brittle rock. In: Aubertin, M., Hassani, F. and Mitri, H. (ed.) *Proceedings of the 2nd North American Rock Mechanics Symposium: Rock Mechanics Tools and Techniques, Montreal*. A.A. Balkema, Rotterdam. pp. 13–24.

Martin, C.D., Read, R.S. & Martino, J.B. (1997) Observations of brittle failure around a circular test tunnel. *International Journal of Rock Mechanics and Mining Sciences*, 34(7), 1065–1073.

Martino, J.B. & Chandler, N.A. (2004) Excavation-induced damage studies at the Underground Research Laboratory. *International Journal of Rock Mechanics and Mining Sciences*, 41(8), 1413–1426.

McClintock, F.A. & Walsh, J.B. (1962). Friction on Griffith cracks in rocks under pressure. In: Rosenberg, R.M. (ed.) *Proceedings of the Fourth U.S. National Congress of Applied Mechanics, Berkeley*. The American Society of Mechanical Engineers, New York. pp. 1015–1021.

Mosher, S., Berger, R.L. & Anderson, D.E. (1975) Fracturing characteristics of two granites. *Rock Mechanics*, 7(3), 167–176.

Nemat-Nasser, S. & Horii, H. (1982) Compression-induced nonplanar crack extension with application to splitting, exfoliation and rockburst. *Journal of Geophysical Research*, 87(B8), 6805–6821.

Nicksiar, M. & Martin, C.D. (2013) Crack initiation stress in low porosity crystalline and sedimentary rocks. *Engineering Geology*, 154, 64–76.

Olsson, W.A. (1974) Grain size dependence of yield stress in marble. *Journal of Geophysical Research*, 79(32), 4859–4862.

Ortlepp, W.D. & Gay, N.C. (1984) Performance of an experimental tunnel subjected to stresses ranging from 50 MPa to 230 MPa. In: Brown, E.T. and Hudson, J.A. (eds.) *ISRM Symposium on the Design and Performance of Underground Excavations, Cambridge, UK*. British Geological Society, London. pp. 337–346.

Pelli, F., Kaiser, P.K. & Morgenstern, N.R. (1991) An interpretation of ground movements recorded during construction of the Donkin-Morien tunnel. *Canadian Geotechnical Journal*, 28(2), 239–254.

Peng, S. & Johnson, A.M. (1972) Crack growth and faulting in cylindrical specimens of Chelmsford granite. *International Journal of Rock Mechanics and Mining Sciences & Geomechanics Abstracts*, 9(1), 37–86.

Perras, M.A. & Diederichs, M.S. (2016) Predicting excavation damage zone depths in brittle rocks. *Journal of Rock Mechanics and Geotechnical Engineering*, 8(1), 60–74.

Perras, M.A., Wannenmacher, H. & Diederichs, M.S. (2014) Underground excavation behaviour of the Queenston Formation: Tunnel back analysis for application to shaft damage dimension prediction. *Rock Mechanics and Rock Engineering*, 48(4), 1647–1671.

Pollard, D.D. & Aydin, A. (1988) Progress in understanding jointing over the past century. *Geological Society of America Bulletin*, 100, 1181–1204.

Rao, M.V.M.S. (1988) A study of acoustic emission activity in granites during stress cycling experiments. *Journal of Acoustic Emission*, 7(3), S29–S34.

Read, R.S., Chandler, N.A. & Dzik, E.J. (1998) *In situ* strength criteria for tunnel design in highly-stressed rock masses. *International Journal of Rock Mechanics and Mining Sciences*, 35(3), 261–278.

Read, R.S. & Martin, C.D. (1992) Monitoring the excavation-induced response of granite. In: Wawersik, W.R. and Tillerson, J.R. (eds.) *Rock Mechanics: Proceedings of the 33rd U.S. Symposium, Santa Fe, New Mexico*. A.A. Balkema, Brookfield. pp. 201–210.

Schmidtke, R.H. & Lajtai, E.Z. (1985) The long-term strength of Lac du Bonnet granite. *International Journal of Rock Mechanics and Mining Sciences & Geomechanics Abstracts*, 22(6), 461–465.

Scholz, C.H. (1968) Microfracturing and the inelastic deformation of rock in compression. *Journal of Geophysical Research*, 73(4), 1417–1432.

Schulson, E.M., Kuehn, G.A., Jones, D.A. & Fifolt, D.A. (1991) The growth of wing cracks and the brittle compressive failure of ice. *Acta Metallurgica et Materialia*, 39(11), 2651–2655.

Shah, K.R. & Labuz, J.F. (1995) Damage mechanisms in stressed rock from acoustic emission. *Journal of Geophysical Research*, 100(B8), 15527–15539.

Simmons, G. & Richter, D. (1976) Microcracks in rocks. In: Strens, R.G.J. (ed.) *The Physics and Chemistry of Minerals and Rocks*. Toronto, John Wiley & Sons. pp. 105–137.

Siren, T., Kantia, P. & Rinne, M. (2015) Considerations and observations of stress-induced and construction-induced excavation damage zone in crystalline rock. *International Journal of Rock Mechanics and Mining Sciences*, 73, 165–174.

Sondergeld, C.H., Granryd, L.A. & Estey, L.H. (1984). Acoustic emissions during compression testing of rock. In: Hardy, H.R. and Leighton, F.W. (ed.) *Proceedings of the Third Conference on Acoustic Emission/Microseismic Activity in Geologic Structures and Materials, University Park, Pennsylvania*. Trans Tech Publications, Germany. pp. 131–145.

Sprunt, E.S. & Brace, W.F. (1974) Direct observation of microcavities in crystalline rock. *International Journal of Rock Mechanics and Mining Sciences & Geomechanics Abstracts*, 11(4), 139–150.

Stacey, T.R. & de Jongh, C.L. (1977) Stress fracturing around a deep-level bored tunnel. *Journal of the South African Institute of Mining and Metallurgy*, 78(5), 124–133.

Tapponnier, P. & Brace, W.F. (1976) Development of stress-induced microcracks in Westerly granite. *International Journal of Rock Mechanics and Mining Sciences & Geomechanics Abstracts*, 13(4), 103–112.

Thompson, B.D., Young, R.P. & Lockner, D.A. (2006) Fracture in Westerly granite under AE feedback and constant strain rate loading: Nucleation, quasi-static propagation, and the transition to unstable fracture propagation. *Pure and Applied Geophysics*, 163(5), 995–1019.

Wagner, H. (1987). Design and support of underground excavations in highly stressed rock. In: Herget, G. and Vongpaisal, S. (ed.) *Proceedings of the 6th International Congress on Rock Mechanics, Montreal*. A.A. Balkema, Rotterdam. pp. 1443–1457.

Walsh, J.B. (1965) The effect of cracks in rocks on Poisson's ratio. *Journal of Geophysical Research*, 70(20), 5249–5257.

Walton, G. & Diederichs, M.S. (2015) A mine shaft case study on the accurate prediction of yield and displacements in stressed ground using lab-derived material properties. *Tunnelling and Underground Space Technology*, 49, 98–113.

Wawersik, W.R. & Brace, W.F. (1971) Post-failure behavior of a granite and diabase. *Rock Mechanics*, 3(2), 5–85.

Wawersik, W.R. & Fairhurst, C. (1970) A study of brittle rock fracture in laboratory compression experiments. *International Journal of Rock Mechanics and Mining Sciences*, 7(5), 561–575.

Wong, R.H.C., Chau, K.T. & Wang, P. (1996) Microcracking and grain size effect in Yuen Long marbles. *International Journal of Rock Mechanics and Mining Sciences & Geomechanics Abstracts*, 33(5), 479–485.

Xia-Ting, F., Guoshao, S. & Quan, J. (2007) Intelligent stability analysis and design optimum of a large hydraulic cavern group under high geo-stress condition. In: Ribeiro e Sousa, L., Olalla, C. and Grossmann, N. (eds.) *Proceedings of the 11th Congress of the International Society for Rock Mechanics, Lisbon.* Taylor & Francis, London. pp. 931–934.

Young, R.P. & Martin, C.D. (1993) Potential role of acoustic emission/microseismicity investigations in the site characterization and performance monitoring of nuclear waste repositories. *International Journal of Rock Mechanics and Mining Sciences & Geomechanical Abstracts*, 30(7), 797–803.

Chapter 22

Numerical rock fracture mechanics

M. Fatehi Marji & A. Abdollahipour
Mining and Metallurgical Engineering Faculty, Yazd University, Yazd, Yazd, Iran

Abstract: The analysis of fracture of rock or other materials has been developed since the mid 1940s. Although dealing with and exploiting rock fracturing has been a part of mining engineering for hundreds of years, rock fracture mechanics has been developed into an engineering discipline since the mid 1960s. Rock fracture mechanics mainly deals with a thorough understanding of what will happen to the rock formations in the subsurface when subjected to in situ stresses. In the fracturing process of rock, a number of important parameters are to be considered *i.e.* fracture toughness, in situ stress, Poisson's ratio, Young's modulus, etc. It should be noted that rock formations cannot often be treated as isotropic and homogeneous bodies. For example, in the case of hydrocarbon reservoirs, the porous and fluid filled nature of rock requires poroelastic theory for some problems.

Although some of rock fracture mechanics problems can be solved analytically, due to the complexity of these problems, some well sophisticated numerical methods have been developed to cope with the difficulties. The mesh-based numerical methods may include; Finite Element Method (FEM), eXtended Finite Element Method (XFEM), Boundary Element Method (BEM), Displacement Discontinuity Method (DDM), Discrete Element Method (DEM), the combined methods (such as the Combined Finite Discrete Element Method (CFDEM)), etc. In addition to the mesh-reduction methods such as BEM and combined methods; mesh-less interpolation methods are developed to overcome the drawbacks of mesh-based methods. A brief explanation of these numerical methods as applied to the rock fracture mechanics is given.

Keywords: Rock fracture mechanics; Numerical methods; FEM; BEM; DDM; DEM; Mesh-less methods

I INTRODUCTION

Rocks are brittle or quasi-brittle materials. They have a discontinuous combination of solid matter, pores, cracks and fractures. A crack in rock fracture mechanics may be defined as a rock separation by opening or sliding. Cracks vary in size from faults (very large scale cracks) with length of hundreds or thousands of meters to intra-granular cracks (infinitesimal cracks) as little as thousandth of centimeters (Jaeger *et al.*, 2009).

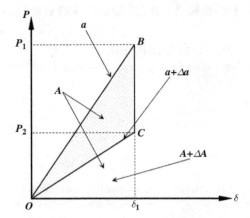

Figure 1 Load-displacement curve for a crack propagated from length a to a+Δa (assuming constant displacement condition).

Brittle rocks and glasses have a very low tensile strength which is due to stress concentration at the ends of infinitesimal internal or surface flaws (Griffith, 1921).

Irwin (1948) developed the concept of stress intensity factor for brittle fracture based on linear elastic fracture mechanics (LEFM). He extended the concept of energy release during progressive fracture and provided a means to measure material resistance to fracture using compliance approach (Sanford, 2003).

Figure 1 illustrates the load-deflection diagram for a "system isolated" fracture event in a linearly elastic body assuming constant displacement. The load deformation for the internal crack of area A is shown by line OB and that of the propagated crack (with change in crack area of ΔA) is shown by OC. The energy lost during load-unload process under the "system isolated" condition is represented by the triangle OBC. The energy must be related to the crack extension area ΔA. It can be related to the spatial rate of stored strain energy G as:

$$G = -\frac{\partial U}{\partial A}\bigg|_{\delta} \qquad (1)$$

where U is the strain energy of the system and the negative sign is introduced to make the G (material property) a positive quantity. For a linear elastic body

$$U = \frac{1}{2}\Delta p \delta \text{ and } \delta = C\Delta p \qquad (2)$$

here C is compliance which is reciprocal of the load-deflection slope. Therefore,

$$G = \frac{P^2}{2}\frac{\partial C}{\partial A} \qquad (3)$$

1.1 Linear elastic fracture mechanics

Materials with relatively low fracture resistance fail below their collapse strength and can be analyzed on the basis of elastic concepts through the use of linear elastic fracture mechanics (LEFM). Practically all high strength materials used in aerospace industry, high strength low-alloy steels, cold worked stainless steels, most rocks, and other construction materials like ceramics and concrete, etc. can be treated by LEFM. Many other materials can be analyzed by means of Elastic-Plastic Fracture Mechanics (EPFM), in which the understanding of its concept needs familiarity with LEFM concepts. Analysis of crack by LEFM approach is based on the stresses at the crack tip and the concept of stress intensity factor K.

Assuming Hooke's law to be valid without limitations to stresses and strains, enables the dissipative region to be shrunk to the crack edge.

LEFM uses stresses rather than loads in engineering analyses. The residual strength shown in Figure 2 is normally based on σ_{res}, the stress (instead of load) that a structure can sustain before fracture occurs.

The residual strength is like the tensile strength or the yield strength of the material and not a stress. If the stress equals σ_{res} fracture occurs. Figure 2 shows the relation between σ_{res} and crack size a. For a maximum permissible cracks length a_p there is a maximum permissible stress σ_p that can be applied to the structure which is much lower than the yield strength σ_{yield}. The residual strength of the structure decreases progressively with increasing crack size (see Figure 2).

Rocks are usually fragmented into two or more parts under the action of stresses which is called fragmentation. The separation distance is significantly smaller than the separation extent – the crack length. The existence of a crack in a homogeneous rock reduces the strength of the structure considerably. Any load acting on the body is magnified several times at the tip of such a discontinuity and when the stress concentration at the tip of the crack reaches a critical level, it propagates.

The problem of crack initiation in rock structures such as tunnels and boreholes has been approached in practice with elementary methods for a long time. Usually the crack

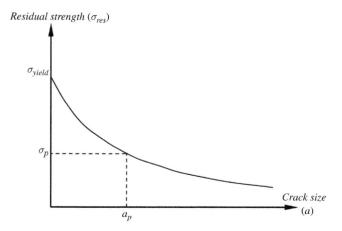

Figure 2 Residual strength curve.

has been expected to propagate instantly based on elastic-brittle behavior if the computed tensile stress is larger than the tensile strength of the rock. In this manner no information regarding the length of propagated crack could be obtained. It then became clear that the behavior of rock is dependent on the scale of the problem. The aforementioned method neglects the considerable additional energy absorption that can take place after reaching peak tensile strength of the rock. Application of fracture mechanics concepts in rock materials compensates for this reserve capacity to be accounted for in rock mechanics analyses.

1.2 Stress intensity factor and fracture toughness

Irwin introduced the concept of stress intensity factor K (SIF), as a measure of the strength of singularity at the crack tip (Irwin, 1957). He showed that $K \propto \sigma\sqrt{\pi r}$ controls the local stress quantity, where r is the radius of the circular stress field σ at a very close distance to the crack tip (Mohammadi, 2008).

Basically three modes of loading are considered for the analysis of cracks as shown in Figure 3, and are described as follows:

1. Mode I loading or opening mode in which the crack tip is subjected to a normal stress σ, and the crack faces are perpendicular to the crack plane.
2. Mode II or in-plane shearing mode, where the crack tip is subjected to an in-plane shear stress τ_i and the crack surfaces are sliding relative to each other, so that their displacements occur in the crack plane and normal to the crack front.
3. Mode III, or tearing mode, where, the crack tip is subjected to an out of plane (anti-plane) shear stress τ_o and the crack faces are moving relative to each other so that the displacement of the crack surfaces are in the crack plane but parallel to the crack front.

K_I, K_{II} and K_{III} are described as follows for $\theta = 0$,

$$K_I = \lim_{r \to 0} \sigma_{yy}\sqrt{2\pi r} \tag{4}$$

$$K_{II} = \lim_{r \to 0} \sigma_{xy}\sqrt{2\pi r} \tag{5}$$

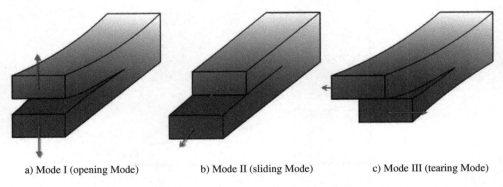

a) Mode I (opening Mode) b) Mode II (sliding Mode) c) Mode III (tearing Mode)

Figure 3 The three basic modes of loading for a crack and corresponding crack surface displacements (Fatehi Marji, 2015).

$$K_{III} = \lim_{r \to 0} \sigma_{yz}\sqrt{2\pi r} \tag{6}$$

r, θ are defined in Figure 4.

Consider a body of arbitrary shape with a crack of arbitrary size, subjected to arbitrary tensile and shear loading, bending and shear loading, or both as long as the loading is in the mixed form of loading *i.e.* in mode I (opening mode) and mode II (in-plane shearing mode) as shown in Figure 4. The details of the derivation of stresses and displacements near the crack tip are found in extensive text books (such as Broek, 1985; Whittaker *et al.*, 1992). The stresses in Cartesian coordinates are formulated as:

$$\sigma_{xx} = \frac{K_I}{\sqrt{2\pi r}}\cos\frac{\theta}{2}\left(1 - \sin\frac{\theta}{2}\sin\frac{3\theta}{2}\right) - \frac{K_{II}}{\sqrt{2\pi r}}\sin\frac{\theta}{2}\left(2 + \cos\frac{\theta}{2}\cos\frac{3\theta}{2}\right)$$

$$\sigma_{yy} = \frac{K_I}{\sqrt{2\pi r}}\cos\frac{\theta}{2}\left(1 + \sin\frac{\theta}{2}\sin\frac{3\theta}{2}\right) + \frac{K_{II}}{\sqrt{2\pi r}}\sin\frac{\theta}{2}\cos\frac{\theta}{2}\cos\frac{3\theta}{2}$$

$$\sigma_{xy} = \frac{K_I}{\sqrt{2\pi r}}\cos\frac{\theta}{2}\sin\frac{\theta}{2}\cos\frac{3\theta}{2} + \frac{K_{II}}{\sqrt{2\pi r}}\cos\frac{\theta}{2}\left(1 - \sin\frac{\theta}{2}\sin\frac{3\theta}{2}\right) \tag{7}$$

in which, σ_{ij} ($i,j = x,y$) are the near crack tip stresses at distance r from the crack tip.

The general quantity K (*i.e.* K_I and K_{II} for the mixed mode I and II) is essentially a measure of stress intensity at the tip of a crack, or a measure of the crack tip elastic stress field. Based on LEFM an unstable fracture occurs when one of the K_i (i = I, II, III) or a mixed mode of them reaches a critical value, K_{ic} known as fracture toughness. Table 1 shows the fracture toughness of some rocks. It presents the resistance ability of a material to crack propagation in a given stress field. The propagation angle of an unstable crack may be computed using one of the fracture criteria (*e.g.* maximum tangential stress (Fatehi Marji, 2014)).

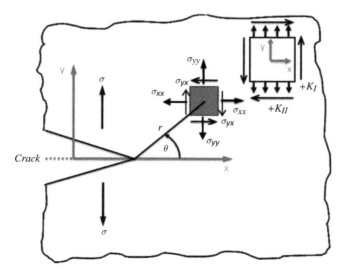

Figure 4 A crack in an infinite plane under mixed mode loading I and II (Fatehi Marji, 2015).

664 Fatehi Marji & Abdollahipour

Table 1 Fracture toughness of different rocks (Whittaker *et al.*, 1992).

Rock (Location/comment)	Fracture toughness (MPa√m)
Andesite (Tampomas)	1.26–1.68
Basalt	1.73
Basalt	3.01
Dolerite (Whin sill)	3.26
Gabro (Kallax/series 1)	2.58
Gabro (Kallax/series 2)	3.23
Granite (Bohus)	1.46
Granite (Bohus)	2.4
Limestone (Bedford)	1.1
Limestone (/white)	2.21
Marble (/Fine grain)	0.96
Marble (Ekeberg)	2.62
Sandstone	0.67
Sandstone (Pennant)	2.56
Shale (Anvil Points/oil)	0.63–1.04
Shale (Rulison Field)	0.17–2.61
Siltstone (Rulison Field)	0.17–2.61

Generally, numerical methods are used to evaluate and predict stress intensity factors in engineering structures with complex geometry which is evolving as cracks propagate in the structure. The finite element method (FEM) and the boundary element method (BEM) are two well-established numerical methods in fracture mechanics.

1.3 Fracture criteria

There are mainly three classic fracture criteria as used in rock fracture mechanics literature *i.e.* 1) maximum tangential stress criterion (or σ-criterion) (Erdogan & Sih, 1963) 2) strain energy release rate criterion (or G-criterion) (Hussain *et al.*, 1974) and 3) minimum strain energy criterion (or S-criterion) (Sih, 1974). Several modified version of these criteria have also been used in the literature *e.g.* F-criterion which is a modified version of G-criterion (Stephansson, 2002).

1.3.1 σ-criterion (Maximum tangential stress criterion)

This fracture criterion postulates that crack growth takes place in a direction perpendicular to the maximum principal stress. Hence, the fracture criterion requires that the maximum principal stress be a tensile stress for opening the crack along its plane.

The mixed mode criterion for a crack angle θ can then be defined as

$$K_{Ic} = K_I \cos^3 \frac{\theta}{2} - \frac{3}{2} K_{II} \cos \frac{\theta}{2} \sin \theta \qquad (8)$$

where θ and θ_0 are shown in Figure 5.

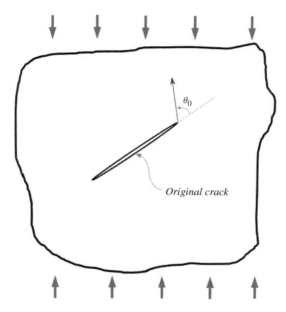

Figure 5 The propagation angle θ_0.

1.3.2 G-criterion (Maximum energy release rate criterion)

The original Griffith fracture criterion was extended to the general angled crack problems in tension (Palaniswamy, 1972). This extension is known as G-criterion. It states that when the energy release rate in the direction of the maximum G-value reaches the critical value G_C, the fracture tip will propagate in that direction.

Assume that the basic three mode interact on an elastic component, thus

$$G_I = \frac{K_I^2}{E'} + \frac{K_{II}^2}{E'} + \frac{(1+\nu)K_{III}^2}{E} \quad (9)$$

where $E' = E$ for plane stress and $E' = E/(1-\nu^2)$ for plane strain for pure mode I loading at fracture, the fracture toughness expression is

$$G_{IC} = G_I = \frac{K_{IC}^2}{E'} \quad (10)$$

Substituting Equations 10 into 9 gives the fracture criterion equation which is named as the G-Criterion

$$K_{IC}^2 = K_I^2 + K_{II}^2 + \frac{E'(1+\nu)K_{III}^2}{E} \quad (11)$$

The above Equation suggests that any combination of the stress intensity factors may give the value for fracture toughness K_{IC}.

1.3.3 S-criterion (Minimum strain energy density criterion)

This criterion is proposed for two-dimensional stress field and states that the initial crack growth takes place in the direction along which the strain-energy-density reaches a minimum stationary value. The final form of the criterion can be defined as:

$$\frac{8\mu}{(\kappa-1)}\left[a_{11}\left(\frac{K_I}{K_{Ic}}\right)^2 + 2a_{12}\left(\frac{K_I K_{II}}{K_{Ic}^2}\right) + a_{22}\left(\frac{K_{II}}{K_{Ic}}\right)^2\right] = 1 \tag{12}$$

where

$$a_{11} = \frac{1}{16\mu}[(1+\cos\theta)(\kappa-\cos\theta)]$$

$$a_{12} = \frac{1}{16\mu}\sin\theta[2\cos\theta-(\kappa-1)] \tag{13}$$

$$a_{22} = \frac{1}{16\mu}[(\kappa+1)(1-\cos\theta)+(1+\cos\theta)(3\cos\theta-1)]$$

v is Poisson's ratio and $\kappa=(1-3v)$ for plane strain and $\kappa=(3-v)/(1+v)$ for plane stress problems and $\mu=E/(2(1+v))$.

1.3.4 F-criterion (The modified energy release rate criterion)

The G-criterion does not distinguish between Mode I and Mode II fracture toughness of energy (G_{IC} and G_{IIC}). Both tensile (Mode I) and shear (Mode II) failure mechanisms are common in rock masses. Shen & Stephansson extended and improved the G-criterion to include both Mode I and Mode II fracture propagation (Shen & Stephansson, 1993). They suggested a fracture propagation criterion, which states that in an arbitrary direction (θ) at a fracture tip the F-value is calculated as:

$$F(\theta) = \frac{G_I(\theta)}{G_{IC}} + \frac{G_{II}(\theta)}{G_{IIC}} \tag{14}$$

where G_{IC} and G_{IIC} are the critical strain energy release rates for Mode I and Mode II fracture propagation, respectively. $G_I(\theta)$ and $G_{II}(\theta)$ are the respective strain energy release rates due to the potential Mode I and Mode II fracture growth of a unit length. If the maximum F-value reaches 1.0, fracture propagation will occur.

The criterion is known as suggested by its parameter F as F-criterion. The direction of fracture propagation will correspond to the direction where F reaches its maximum value. Summation of normalized G-values in the F-criterion is used to determine the failure load and its direction. G_I and G_{II} can be expressed as shown in Figure 6. If a fracture grows a unit length in an arbitrary direction and the new fracture opens without any surface shear displacement, the released strain energy in the surrounding body due to the fracture growth would be G_I. Similarly, if the new fracture has only a surface shear displacement and no opening in the normal direction, the released strain energy would be G_{II} (Shen & Stephansson, 1993).

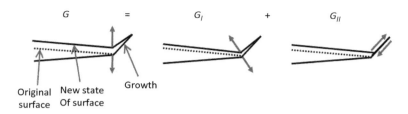

Figure 6 Definition of G$_I$ and G$_{II}$ for fracture growth. (a) G: the growth has both open and shear displacement; (b) G$_I$: the growth has only normal displacement showing increase in aperture; (c) G$_{II}$: the growth has only shear displacement (Shen & Stephansson, 1993).

2 COMPUTATIONAL FRACTURE MECHANICS

The main goal of fracture mechanics *i.e.* SIF computation may be accomplished using analytical, semi-analytical and numerical methods.

2.1 Analytical methods

The modern view of analytical fracture mechanics uses singular problems to represent elastic behavior of bodies containing one or more cracks (Sanford, 2003). Williams developed the general solution to a particular singular problem (Williams, 1957, 1952). The Williams approach represents a complete solution to problems of a single edge crack with stress-free crack faces subjected to smoothly varying boundary conditions over finite boundaries. The solution to other classes of crack problems such as multiple cracked bodies and central crack problems may be obtained using complex variables (Muskhelishvili, 1953).

The complex variable method introduced by Muskhelishvili (1953) offers solutions to a vast range of singular problems. However, his approach requires a broad knowledge of mathematics. Fortunately Westergaard (1939) proposed another complex variable approach that offers advantages of a complex formulation and in a meanwhile avoids Muskhelishvili's approach disadvantages. The generalized Westergaard approach using complex variables provides an infinite number of potential Airy stress functions that may be utilized in fracture mechanics (Sanford, 2003). This method applies the semi-inverse method to the Airy stress function expressed in the complex domain. The original method can only be used for infinite problems with uniform remote boundary conditions (Sanford, 2003). After some modifications and introduction of a second analytical function (Etfis & Liebowitz, 1972; Sanford, 1979; Sih, 1966) the generalized Westergaard method can now be applied to both finite and infinite body problems with arbitrary boundary conditions as long as the crack(s) is (are) constrained to lie along the y=0 plane.

Alternating method, compounding method and superposition are almost similar methods in SIF computation. These methods compute the SIF of a complex geometry as the sum of simpler problems with known SIFs. Figure 7 illustrates the superposition method.

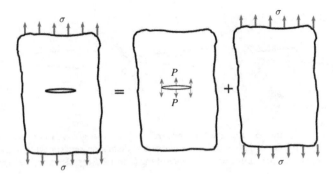

Figure 7 Superposition for a central crack in an infinite body under uniaxial far-field tension.

2.2 Numerical methods

It has been established that SIF plays an important role in LEFM and study of cracked bodies. Where a solution cannot be obtained directly from analytical methods, numerical methods should be applied to most practical crack problems.

3 FINITE ELEMENT METHOD

In mathematics, the FEM is a numerical technique for finding approximate solutions to boundary value problems for partial differential equations. It uses subdivision of a whole problem domain into simpler parts, called finite elements, and variational methods from the calculus of variations to solve the problem by minimizing an associated error function.

Considering a plane stress or a plane strain problem the following steps are used in the finite element modeling of elastostatic problems.

1. Discretization of the continuous domain into sub-regions called *finite elements* of arbitrary size, shape and orientation.
2. Elements are connected to the other neighboring elements through a finite number of discrete points called *nodes*.
3. The displacements at the nodes are assumed as the basic unknowns of the problem. The total number of these nodal displacement components is called the number of degrees of freedom of the finite element model. The larger this number, the more accurate is the solution, although computationally more expensive (which means, more equations to store and solve, and hence needing more computer resources).
4. It is assumed that by some means, displacements at all the nodes are obtained. Using interpolation and element nodal displacement field, the interior displacement field within each element may be found:

$$\{u\} = [N]\{u^e\} \qquad (15)$$

Numerical rock fracture mechanics 669

where $\{u\}$ is the displacement vector within an element, $[N]$ is called interpolation matrix or the shape function matrix and $\{u^e\}$ element nodal displacement vector.

5. Once the displacement field within the element is known, the strain field can be obtained by making use of the strain-displacement relations:

$$\{\varepsilon\} = [L]\{u\} = [L][N]\{u^e\} = [B]\{u^e\} \tag{16}$$

where $\{\varepsilon\}$ is the strain vector, the matrix $[B]$ is the strain-displacement matrix and $[B]=[L][N]$ in which the operator matrix $[L]$ is given by:

$$[L] = \begin{bmatrix} \partial/\partial x & 0 \\ 0 & \partial/\partial y \\ \partial/\partial y & \partial/\partial x \end{bmatrix} \tag{17}$$

6. Using stress-strain relationships, the stress field $\{\sigma\}$ in the element is obtained as

$$\{\sigma\} = [D]\{\varepsilon\} = [D][B]\{u^e\} \tag{18}$$

where $[D]$ is the constitutive matrix. In case of plane strain problems:

$$[D] = \frac{E}{1 - \nu^2} \begin{bmatrix} 1 & \nu & 0 \\ \nu & 1 & 0 \\ 0 & 0 & \dfrac{1 - \nu}{2} \end{bmatrix} \tag{19}$$

7. After determining the stress field $\{\sigma\}$, the principle of virtual work might be used to obtain the basic finite element equations (Ameen, 2005). Eventually the following equation is achieved:

$$[k^e]\{u^e\} = \{r^e\} \tag{20}$$

where

$$[k^e] = \int_{V_e} B^T D B \, dV_e \tag{21}$$

$[k^e]$ is the element stiffness matrix, V_e the element volume, and $\{r^e\}$ is element load vector

$$\{r^e\} = \int_{V_e} N^T \{b\} dV_e + \int_{S_e} N^T \{p\} dS_e \tag{22}$$

$\{b\}$ and $\{p\}$ are the body force and surface force vectors, respectively.

8. The summation of all element stiffness matrices and element load vectors results in a linear system of algebraic equations as

$$[K]\{U\} = \{R\} \tag{23}$$

where $[K]$ is the global stiffness matrix, $\{U\}$ is the global displacement vector, and $\{R\}$ is the global load vector.

Crack growth modeling using traditional finite element framework is cumbersome due to the need for the mesh to be updated to match geometry of the discontinuity. This becomes a major difficulty when treating problems with evolving discontinuities where the mesh must be regenerated at each step. Moreover, the crack tip singularity needs to be accurately represented by approximate functions.

3.1 SIF determination

For many practical problems, lack of analytical solution results in numerical determination of the stress intensity factor. The most popular *FEM* techniques for SIF determination includes (Mohammadi, 2008):

1. Classical methods using the finite element method solely as a continuum based analytical tool.
2. Techniques in which the SIFs are directly evaluated as part of the global stiffness matrix.
3. Techniques through which the SIF can be computed from a standard finite element analysis via a special purpose post-processing technique.
4. Methods in which the singularity of the stress field at the crack tip is modeled.

These methods are explained in details in (Mohammadi, 2008).

For the FEM to approximate SIF a very fine mesh at the crack tip is needed. SIFs can be determined at different radial distances from the crack tip by equating the numerically obtained displacements with their analytical expression in terms of the SIF. In most cases, FEM programs do not include SIF in their formulation. Conceptually we could obtain K_I from the σ_y stress ahead of the crack using the defined relation

$$K_I = \lim_{x^+ \to 0} \sigma_y \Big|_{\theta=0} \cdot \sqrt{2\pi x} \tag{24}$$

where x is the crack tip coordinate along the crack plane and the limit is taken from the material side. Alternatively we could use the crack opening displacement behind the crack tip u, to compute K from the relation

$$K_I = \lim_{x^- \to 0} u \Big|_{\theta=\pi} \cdot \frac{1}{\sqrt{2\pi x}} \frac{\pi}{2} E' \tag{25}$$

where $E'=E$ for plane stress and $E'=E/(1-v^2)$ for plane strain conditions.

For plane stress problems in the xy plane (Mohammadi, 2008).

$$K_I = \mu \sqrt{\frac{2\pi}{r}} \frac{u_y^b - u_y^a}{2(1-\nu)} \tag{26}$$

$$K_{II} = \mu \sqrt{\frac{2\pi}{r}} \frac{u_x^b - u_x^a}{2(1-\nu)} \tag{27}$$

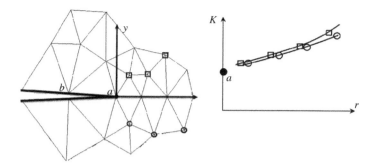

Figure 8 Stress intensity factor calculation in FEM using extrapolation.

$$K_{III} = \mu\sqrt{\frac{2\pi}{r}}\left(u_z^b - u_z^a\right) \quad (28)$$

A simple extrapolation technique, as depicted in Figure 8, can then be used to approximately evaluate the value of SIF at the crack tip. The same procedure for stresses can also be used, although it is likely to yield less accurate predictions (Mohammadi, 2008).

4 EXTENDED FINITE ELEMENT METHOD (XFEM)

The XFEM was originally proposed by Belytscho & Black (1999). They presented a method for enriching finite element approximations with minimal remeshing requirement in problems with evolving geometries. Dolbow and others (Dolbow, 1999; Dolbow *et al.*, 2000) and Moës *et al.* (1999) introduced a much more elegant technique by adapting an enrichment that includes the asymptotic near tip field and a Heaviside function H(x). Since then it has been used to model the propagation of various discontinuities: such as cracks, material interfaces etc. The idea behind XFEM is to retain most advantages of mesh-free methods without dealing with disadvantages.

The XFEM, is based on a standard Galerkin procedure, and the partition of unity method (PUM) (Melenk & Babuska, 1996) to accommodate the internal boundaries in the discrete model. It extends the classical FEM approach by enriching the solution space for solutions to differential equations with discontinuous functions.

4.1 Crack-tip enrichment

Theoretically, enrichment can be regarded as the principal of increasing the order of completeness. Computationally, it seeks higher accuracy of the approximation using the analytical solution. The choice of the enriched functions depends on the *a priori* solution of the problem. For instance, in a crack analysis this is equivalent to an increase in accuracy of the approximation if analytical near crack tip solutions are somehow included in the enrichment terms (Mohammadi, 2008).

The essential idea in the extended finite element method, which is closely related to the GFEM (Duarte *et al.*, 2000; Strouboulis *et al.*, 2000), is to add discontinuous enrichment functions to the finite element approximation using the partition of unity:

$$u(x) = \sum_{i=1}^{n} N_i(x) \left(u_i + \sum_{j=1}^{ne(i)} a_{ij} F_j(r, \theta) \right) \tag{29}$$

where $N_i(x)$ are the standard finite element shape functions and (r, θ) is a polar coordinate system with origin at the crack tip. The enrichment coefficient a_{ji} is associated with nodes and $ne(i)$ is the number of coefficients for node i e.g. it is chosen to be four for all nodes around the crack tip and zero at all other nodes.

The crack-tip enrichment functions in isotropic elasticity $F_i(r, \theta)$ are obtained from the asymptotic displacement fields:

$$[F_j[r, \theta]]_{j=1}^{4} = \left[\sqrt{r} \sin\left(\frac{\theta}{2}\right); \sqrt{r} \cos\left(\frac{\theta}{2}\right); \sqrt{r} \sin\left(\frac{\theta}{2}\right) \sin\theta; \sqrt{r} \cos\left(\frac{\theta}{2}\right) \sin\theta \right] \tag{30}$$

Note that the first function in the above equation is discontinuous across the crack. It represents the discontinuity near the tip, while the other three functions are added to get accurate result with relatively coarse meshes.

4.2 Heaviside function

The Heaviside function is a discontinuous function across the crack surface and is constant on each side of the crack (*i.e.* +1 on one side and –1 on the other). After intruding this jump function, the approximation will be changed to the following formula (Yazid *et al.*, 2009):

$$U = \sum_{i \in I} u_i N_i + \sum_{j \in J} o_j N_j H(x) + \sum_{k \in K1} N_k \left(\sum_{l=1}^{4} c_k^{l1} F_l^1(x) \right) + \sum_{k \in K2} N_k \left(\sum_{l=1}^{4} c_k^{l2} F_l^2(x) \right) \tag{31}$$

where N_i is the shape function associated to node i, I is the set of all nodes of the domain, J is the set of nodes whose shape function support is cut by a crack, K is the set of nodes whose shape function support contains the crack front, u_i are the classical degrees of freedom (*i.e.* displacement) for node i, o_j account for the jump in the displacement field across the crack at node j. If the crack is aligned with the mesh, o_j represent the opening of the crack, $H(x)$ is the Heaviside function, c_k^l are the additional degrees of freedom associated with the crack-tip enrichment functions F_l and F_l is an enrichment which corresponds to the four asymptotic functions in the development expansion of the crack-tip displacement field in a linear elastic solid.

4.3 Numerical integration

There are two difficulties for the integration of XFEM functions: the discontinuity along the crack, and the singularity at the crack tip. The numerical integration of cut elements is generally performed by partitioning them into standard sub-elements. To get accurate results, in the sub-elements, high order Gauss quadrature rule must be used. In earlier investigations, each side of a cut element was triangulated to form a set of sub-triangles. Some authors adopt a slightly different approach, choosing instead to partition cut elements into sub-quadrilaterals.

5 BOUNDARY ELEMENT METHOD

The boundary element method may be regarded as mesh-reducing method. It is divided into two main groups of direct and indirect methods. Direct boundary integral method use direct approach and fictitious stress and displacement discontinuity methods use an indirect approach in the boundary element methods.

In direct method a fundamental solution is used in the reciprocal theorem (Sokolnikoff, 1956) which results in Somigliana's identity (Cruse, 1989). The boundary integral equation is obtained by moving the source point of the fundamental solution to the boundary of the interested problem.

The key to direct boundary element method in linear elasticity is reciprocal theorem. This theorem links the solutions to two different boundary value problems for the same region.

$$\int_C \left(\sigma_s u'_s + \sigma_n u'_n \right) ds = \int_C \left(\sigma'_s u_s + \sigma'_n u_n \right) ds \tag{32}$$

where C is the boundary of the problem. The primed set $(\sigma'_s, \sigma'_n, u'_s, u'_n)$ is a known state of the problem and the unprimed set $(\sigma_s, \sigma_n, u_n, u_s)$ is an unknown set meant to be determined. Equation 32 is an integral equation that relates unspecified boundary parameters of the problem to the specified boundary parameters and the solution to another problem for the same region. For any boundary value problem half of the problem parameters are specified as the boundary conditions. Equation 32 can be used to write down a system of $2N$ algebraic equations. For a stress boundary value problem $\overset{i}{\sigma}_s = (\overset{i}{\sigma}_s)_0$ and $\overset{i}{\sigma}_n = (\overset{i}{\sigma}_n)_0$ are known for all boundary elements. The LHS of the equations

$$\sum_{j=1}^{N} \overset{ij}{B}_{ss} (\overset{j}{\sigma}_s)_0 + \sum_{j=1}^{N} \overset{ij}{B}_{sn} (\overset{j}{\sigma}_n)_0 = \sum_{j=1}^{N} \overset{ij}{A}_{ss} \overset{j}{u}_s + \sum_{j=1}^{N} \overset{ij}{A}_{sn} \overset{j}{u}_n$$

$$\sum_{j=1}^{N} \overset{ij}{B}_{ns} (\overset{j}{\sigma}_s)_0 + \sum_{j=1}^{N} \overset{ij}{B}_{nn} (\overset{j}{\sigma}_n)_0 = \sum_{j=1}^{N} \overset{ij}{A}_{ns} \overset{j}{u}_s + \sum_{j=1}^{N} \overset{ij}{A}_{nn} \overset{j}{u}_n \tag{33}$$

will then be known, the displacements $\overset{i}{u}_n$ and $\overset{i}{u}_n$, $i=1$ to N can be found by solving these equations. A similar set of equations can be formed for the case that the displacements $\overset{i}{u}_n = (\overset{i}{u}_n)_0$ and $\overset{i}{u}_s = (\overset{i}{u}_s)_0$ are known for all N boundary elements; the stresses $\overset{i}{\sigma}_s$ and $\overset{i}{\sigma}_n$, $i=1$ to N, are then the unknowns in the system.

The governing system of algebraic equations for any type of boundary value problem may be represented as

$$\left. \begin{array}{l} \overset{i}{Y}_s = \sum_{j=1}^{N} \overset{ij}{C}_{ss} \overset{j}{X}_s + \sum_{j=1}^{N} \overset{ij}{C}_{sn} \overset{j}{X}_n \\ \overset{i}{Y}_n = \sum_{j=1}^{N} \overset{ij}{C}_{ns} \overset{j}{X}_s + \sum_{j=1}^{N} \overset{ij}{C}_{nn} \overset{j}{X}_n \end{array} \right\}, \quad i = 1 \ to \ N \tag{34}$$

where $\overset{i}{Y_s}$ and $\overset{i}{Y_n}$, i =1 to N, are certain linear combinations of the known boundary parameters, and C_{ss}^{ij}, etc., are the appropriate influence coefficients associated with the unknown boundary parameters $\overset{j}{X_s}$ and $\overset{j}{X_n}$.

The close relation between the boundary integral equation (BIE) and full equilibrium differential equations results in a major limitation for BIE analysis of fracture mechanics problems. For non-symmetric crack problems, the solution of general crack problems cannot be achieved with the direct application of the BEM (for symmetric crack problems one can model just one side of the crack), because the coincidence of the crack surfaces gives rise to a singular system of algebraic equations (Cruse, 1989). An identical set of equations will be achieved for each certain point on both surfaces of the crack with the same co-ordinates. Some special techniques have been devised to overcome this difficulty including: the crack Green's function (Snyder & Cruse, 1975), the sub-regions method (Blandford *et al.*, 1981), the dual boundary element method (Hong & Chen, 1988) and the displacement discontinuity method (Crouch, 1976a).

5.1 Crack Green's function

The crack Green's function method eliminates the need for discretization of the crack by considering a traction-free crack. Fundamental solution to such a problem is known as crack Green's function and since it contains the correct behavior of stresses and displacements accurate values of SIF can be expected from numerical solutions. However, it is limited to problems with a single straight traction-free crack. In order to evaluate SIF using a crack Green's function the following boundary element equation is solved using the crack fundamental solutions and the tractions and displacements are evaluated along all boundaries except the crack (Aliabadi & Rooke, 1991)

$$
\begin{aligned}
C_{ij}(x')u_j(x') = & \int_\Gamma U_{ij}(x',x)t_j(x)d\Gamma(x) - \int_\Gamma T_{ij}(x',x)u_j(x)d\Gamma(x) \\
& + \int_\Omega U_{ij}(x',X)b_j(X)d\Omega(X)
\end{aligned}
\tag{35}
$$

where

$$
C_{ij}(x') = \delta_{ij} + \lim_{\varepsilon \to 0} \left\{ \int_{\Gamma'_\varepsilon} T_{ij}(x',x)d\Gamma(x) \right\}
\tag{36}
$$

where the fundamental solutions U_{ij} and T_{ij} are the Kelvin's displacement and traction fundamental solutions in the j direction at point X due to a unit point force acting in i direction at X', b_j is body force component and Γ'_ε represents the boundary a circular inclusion of radius ε (see Figure 9).

Once $t_j(x)$ and $u_j(x)$ are obtained, the SIF can be computed from the interior stresses (Blanford *et al.*, 1981) or interior displacements (Mews, 1987), in the limit as the internal source point approaches the crack tip.

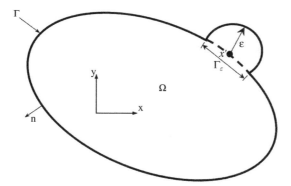

Figure 9 Source point located on the boundary surrounded by a circular arc boundary.

While theoretically the Green's function method can be applied to wide range of problems, the method is limited to two dimensional problems. Essentially the Green's function solutions are superior to any numerical approach, because the crack field is exactly contained in the internal Somigliana identities.

5.2 Sub-region method

To deal with crack modeling problem in the BEM, the sub-region method divides the domain into subdomains, so that the crack surfaces are not in the same domain. For multiple-crack problems sub-regions must be introduced to separate each individual crack and in the case of crack propagation re-meshing will be needed, because the domain boundaries are defined by the crack surfaces. For a body with multiple cracks this can be time consuming. The main drawback of this method is that the introduction of artificial boundaries is not unique, and thus cannot be easily implemented into an automatic procedure (Portela *et al.*, 1992). In addition, the method generates a larger system of algebraic equations than is strictly required.

5.3 Dual boundary element method

The basic task in any boundary element analysis of fracture mechanics is to solve the problem that arises because two nodes on opposite sides of a crack have equal coordinates. The dual BEM incorporates two independent boundary integral equations. It uses the displacement equation to model one of the crack boundaries and the traction equation to model the other. Consequently, mixed-mode crack problems can be solved with a single region formulation (Aliabadi & Rooke, 1991; Chen & Hong, 1999; Hu & Chen, 2013; Leme & Aliabadi, 2012; Yun & Ang, 2010). A major advantage of the dual BEM is its computational efficiency in modeling crack growth where new elements need to be created as the crack propagates. Figure 10 shows dual BEM and DDM modeling of an edge crack problem. In dual BEM the two sides of the crack can be separately discretized (*i.e.* the direct BEM) or they may be discrete together (as a single crack line) in the indirect BEM. The indirect BEM for solving elastic problems is divided to fictitious stress method (FSM) and displacement discontinuity method (DDM)

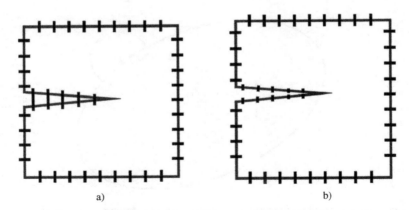

Figure 10 (a) Displacement discontinuity method, b) Dual boundary element method.

(Crouch & Starfield, 1983). Figure 10 shows the difference in discretization of an edge crack problem in dual BEM and DDM.

Following integral Equations are the basis of the dual boundary element method.

$$C_{ij}(x')u_j(x') + \int_\Gamma T_{ij}(x',x)u_j(x)d\Gamma(x) = \int_\Gamma U_{ij}(x',x)t_j(x)d\Gamma(x), \tag{37}$$

$$\frac{1}{2}t_j(x') + n_j(x')\int_\Gamma S_{ijk}(x',x)u_k(x)d\Gamma(x) = n_j(x')\int_\Gamma D_{ijk}(x',x)t_k(x)d\Gamma(x) \tag{38}$$

The general modeling strategy in the dual BEM can be summarized as follows

- The displacement Equation 37 is applied for collocation on one of the crack boundaries.
- The traction Equation 38 is applied for collocation on the opposite crack boundary and remaining boundaries.
- The crack boundaries are discretized.
- The remaining boundaries of the problem domain are discretized except the intersection between a crack and an edge where discontinuous or semi-discontinuous elements are required in order to avoid nodes at the intersection.

A wide range of literature of dual BEM exists which the interested readers may refer to (e.g. (Aliabadi & Rooke, 1991; Chen & Hong, 1999; Cisilino & Aliabadi, 2004; Davies & Crann, 2006; Dirgantara & Aliabadi, 2001; Fedelinski et al., 1993; Portela et al., 1992; Simpson & Trevelyan, 2011).

Recently displacement discontinuity method (DDM) has been used in rock fracture mechanics approaches. Therefore the DDM is explained in the following section.

5.4 Displacement discontinuity method

In displacement discontinuity method (DDM), the crack is considered as single surface across which the displacements are discontinuous. The method is based on the

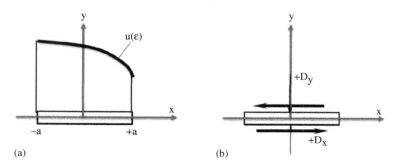

Figure 11 a) Displacement discontinuity element and the distribution of u(ε). b) Constant element displacement discontinuity (position sign convention) (Fatehi Marji, 2015).

analytical solution to the problem of a discontinuity in displacement over a finite line segment in the x, y plane of an infinite elastic solid.

Consider a displacement discontinuity element of length 2a along the x-axis, which is characterized by a general displacement discontinuity distribution $u(\varepsilon)$. By taking the u_x and u_y components of the general displacement discontinuity $u(\varepsilon)$ to be constant and equal to D_x and D_y respectively, in the interval $(-a, +a)$ as shown in Figure 11. Two displacement discontinuity element surfaces can be distinguished, one on the positive side of y ($y=0_+$) and another one on the negative side ($y=0_-$). The displacement undergoes a constant change in value when passing from one side of the displacement discontinuity element to the other side are given as follows:

$$D_x = u_x(x, 0_-) - u_x(x, 0_+)$$
$$D_y = u_y(x, 0_-) - u_y(x, 0_+)$$
(39)

The positive sign convention of D_x and D_y is shown in Figure 11 (b). It demonstrates that when the two surfaces of the displacement discontinuity overlap, D_y is positive, which leads to a physically impossible situation. This conceptual difficulty is overcome by considering that the element has a finite thickness in its un-deformed state, which is small compared to its length, but bigger than D_y.

The solution to the problem is given by Crouch (Crouch, 1976a,b). The displacements and stresses can be expressed as

$$u_x = D_x[2(1-\nu)f_{,y} - yf_{,xx}] + D_y[-(1-2\nu)g_{,x} - yg_{,xy}]$$
$$u_y = D_x[(1-2\nu)f_{,x} - yf_{,xy}] + D_y[2(1-\nu)g_{,y} - yg_{,yy}]$$
(40)

and

$$\sigma_{xx} = 2\mu D_x[2f_{,xy} + yf_{,xyy}] + 2\mu D_y[g_{,yy} + yg_{,yyy}]$$
$$\sigma_{yy} = 2\mu D_x[-yf_{,xyy}] + 2\mu D_y[g_{,yy} - yg_{,yyy}]$$
$$\tau_{xy} = 2\mu D_x[2f_{,yy} + yf_{,yyy}] + 2\mu D_y[-yg_{,xyy}]$$
(41)

where $f_{,x}, g_{,x}, g_{,y}$, etc. are the partial derivatives of the single harmonic functions f(x,y) and g(x,y) with respect to x and y, which are given as:

678 Fatehi Marji & Abdollahipour

$$f(x,y) = \frac{-1}{4\pi(1-\nu)} \int_{-a}^{a} u_x(\varepsilon) \ln\left[(x-\varepsilon)^2 + y^2\right]^{\frac{1}{2}} d\varepsilon$$

$$g(x,y) = \frac{-1}{4\pi(1-\nu)} \int_{-a}^{a} u_y(\varepsilon) \ln\left[(x-\varepsilon)^2 + y^2\right]^{\frac{1}{2}} d\varepsilon \tag{42}$$

The functions $f(x,y)$ and $g(x,y)$ for a constant element displacement discontinuity can be written in term of a single integral function, $I_0(x,y)$ as follows (Fatehi Marji, 2015).

$$f(x,y) = I_0(x,y)D_x$$

$$g(x,y) = I_0(x,y)D_y \tag{43}$$

where the integral function $I_0(x,y)$ is

$$I_0(x,y) = \int \ln[(x-\varepsilon)^2 + y^2]^{\frac{1}{2}} d\varepsilon \tag{44}$$

$$= y(\theta_1 - \theta_2) - (x-a)\ln(r_1) + (x+a)\ln(r_2) - 2a$$

The terms θ_1, θ_2, r_1, r_2 are defined as

$$\theta_1 = \arctan\left(\frac{y}{x-a}\right), \quad \theta_2 = \arctan\left(\frac{y}{x+a}\right),$$

$$r_1 = [(x-a)^2 + y^2]^{\frac{1}{2}}, \text{ and } r_2 = [(x+a)^2 + y^2]^{\frac{1}{2}} \tag{45}$$

The displacement discontinuity functions $u_x(\varepsilon)$ and $u_y(\varepsilon)$ can be used either in a constant element form or in a higher element form as follows, to solve the displacements and stresses of Equations 40 and 41. A system of algebraic equations is formed by considering boundary conditions for each element. For a boundary (*e.g.* the ith element) with prescribed stress conditions

$$\left(\overset{i}{\sigma}_s\right)_0 = \sum_{j=1}^{N} \overset{ij}{A}_{ss} \overset{j}{D}_s + \sum_{j=1}^{N} \overset{ij}{A}_{sn} \overset{j}{D}_n$$

$$\left(\overset{i}{\sigma}_n\right)_0 = \sum_{j=1}^{N} \overset{ij}{A}_{ns} \overset{j}{D}_s + \sum_{j=1}^{N} \overset{ij}{A}_{nn} \overset{j}{D}_n \tag{46}$$

where $\overset{ij}{A}_{ss}$, $\overset{ij}{A}_{sn}$, ... are boundary influence coefficients for the stresses. For instance $\overset{ij}{A}_{sn}$ gives the shear stress (σ_s) at the midpoint of the ith element due to a constant unit normal displacement discontinuity over the jth element (D_n=1). Similarly if displacements u_s and u_n are prescribed then ith element equations are

$$\left(\overset{i}{u}_s\right)_0 = \sum_{j=1}^{N} \overset{ij}{B}_{ss} \overset{j}{D}_s + \sum_{j=1}^{N} \overset{ij}{B}_{sn} \overset{j}{D}_n$$

$$\left(\overset{i}{u}_n\right)_0 = \sum_{j=1}^{N} \overset{ij}{B}_{ns} \overset{j}{D}_s + \sum_{j=1}^{N} \overset{ij}{B}_{nn} \overset{j}{D}_n \tag{47}$$

Mixed boundary conditions in which either u_n and σ_s or u_s and σ_n are prescribed are managed by selecting the appropriate ones of Equations 46 and 47. Following the same procedure for i=1 to N, a system of algebraic equations in $2N$ unknown displacement discontinuity components.

$$\left.\begin{array}{l} \left(\overset{i}{b_s}\right)_0 = \sum_{j=1}^{N} \overset{ij}{C_{ss}} \overset{j}{D_s} + \sum_{j=1}^{N} \overset{ij}{C_{sn}} \overset{j}{D_n} \\ \left(\overset{i}{b_n}\right)_0 = \sum_{j=1}^{N} \overset{ij}{C_{ns}} \overset{j}{D_s} + \sum_{j=1}^{N} \overset{ij}{C_{nn}} \overset{j}{D_n} \end{array}\right\} \quad i = 1\, to\, N \tag{48}$$

where $\overset{i}{b_s}$ and $\overset{i}{b_n}$ stand for the known boundary values of stress or displacement, and $\overset{ij}{C_{ss}}$, etc., are the corresponding influence coefficients from Equations 46 and 47. Equations 45 give $2N$ equations with $2N$ unknowns (*i.e.* displacement discontinuities) which can be solved using common methods.

Constant displacement discontinuity elements are simple to use and are widely utilized in analyzing engineering problems (Behnia *et al.*, 2012; Fatehi Marji *et al.*, 2011; Hadi Haeri *et al.*, 2013a,b; Kim & Pereira, 1997; Shou & Napier, 1999). However, these models, fail to accurately predict the stresses and displacements for field points near boundaries. Also due to the singularity variations $1/\sqrt{r}$, and $\sqrt{r}$ for the stresses and displacements near the crack ends, the accuracy of the displacement discontinuity method at the vicinity of the crack tip decreases (Courtesen, 1979; Kutter & Fairhurst, 1971). These shortcomings have been resolved by using higher order displacement discontinuity (HODD) elements (Crawford & Curran, 1982). Since then HODD have been developed and used to achieve higher accuracy in LEFM problems (Ash, 1985; Crouch & Starfield, 1983; Fatehi Marji & Hosseini-nasab, 2005; Fatehi Marji, 2014, 1997; Hosseini Nasab & Fatehi Marji, 2007; Itou, 2013; Shou & Crouch, 1995; Shou *et al.*, 1997). Higher order elements increase the accuracy of the numerical results, but the problem of crack tip singularities cannot be solved efficiently. The use of crack tip elements can substantially improve the accuracy of the method for crack analyses (Courtesen, 1979; Crouch, 1976a).

Constant, linear and quadratic element formulations of the displacement discontinuity are already defined and used in the literature (Crouch & Starfield, 1983; Fatehi Marji *et al.*, 2006; Shou *et al.*, 1997). The cubic element formulation of the displacement discontinuity is based on analytical integration of cubic collocation shape functions over collinear, straight-line displacement discontinuity elements (Fatehi Marji *et al.*, 2007). Figure 10 shows the cubic displacement discontinuity distribution, which can be written in a general form as:

$$D_i(\varepsilon) = N_1(\varepsilon)D_i^1 + N_2(\varepsilon)D_i^2 + N_3(\varepsilon)D_i^3 + N_4(\varepsilon)D_i^4, \, i = x, y \tag{49}$$

where, D_i^1, D_i^2, D_i^3 and D_i^4 are the cubic nodal displacement discontinuities, and,

$$\begin{aligned} N_1(\varepsilon) &= -(3a_1^3 - a_1^2\varepsilon - 3a_1\varepsilon^2 + \varepsilon^3)/(48a_1{}^3), \\ N_2(\varepsilon) &= (9a_1^3 - 9a_1^2\varepsilon - a_1\varepsilon^2 + \varepsilon^3)/(16a_1{}^3), \\ N_3(\varepsilon) &= (9a_1^3 + 9a_1^2\varepsilon - a_1\varepsilon^2 - \varepsilon^3)/(16a_1{}^3), \\ N_4(\varepsilon) &= -(3a_1^3 + a_1^2\varepsilon - 3a_1\varepsilon^2 - \varepsilon^3)/(48a_1{}^3) \end{aligned} \tag{50}$$

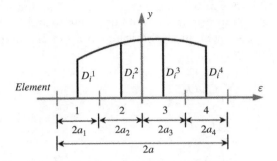

Figure 12 Cubic collocation for the cubic element displacement discontinuity (Fatehi Marji et al., 2007).

are the cubic collocation shape functions using $a_1 = a_2 = a_3 = a_4$. A cubic element has 4 nodes, which are at the centers of its four sub-elements as shown in Figure 12.

The potential functions for the cubic element case can be found from:

$$f(x,y) = \frac{-1}{4\pi(1-\nu)} \sum_{j=1}^{4} D_x^j F_j(I_0, I_1, I_2, I_3),$$
$$g(x,y) = \frac{-1}{4\pi(1-\nu)} \sum_{j=1}^{4} D_y^j F_j(I_0, I_1, I_2, I_3) \tag{51}$$

in which, the common function F_j, can be defined as:

$$F_j(I_0, I_1, I_2, I_3) = \int N_j(\varepsilon) \ln[(x-\varepsilon) + y^2]^{\frac{1}{2}} d\varepsilon, \quad j = 1 \text{ to } 4 \tag{52}$$

The integrals I_0, I_1 and I_2 are the same as those already given in literature for quadratic element case (Fatehi Marji et al., 2006), but I_3 can be expressed as:

$$I_3(x,y) = \int_{-a}^{a} \varepsilon^3 \ln[(x-\varepsilon)^2 + y^2]^{\frac{1}{2}} d\varepsilon = -xy(x^2 - y^2)(\theta_1 - \theta_2) + \\ 0.25(3x^4 - 6x^2y^2 + 8a^2x^2 + a^4 - y^4)[\ln(r_1) - \ln(r_2)] - \\ 2ax(x^2 + a^2)[\ln(r_1) + \ln(r_2)] + 1.5ax^3 - 3axy^2 + 7a^3x/6 \tag{53}$$

5.4.1 SIF computation

Considering a body of arbitrary shape with a crack of arbitrary size, subjected to arbitrary tensile and shear loadings (i.e. the mixed mode loading I and II), the stresses and displacements near the crack tip are given in the related text books (Aliabadi & Rooke, 1991; Broek, 1989; Sanford, 2003; Whittaker et al., 1992). The following formulations are suitable:

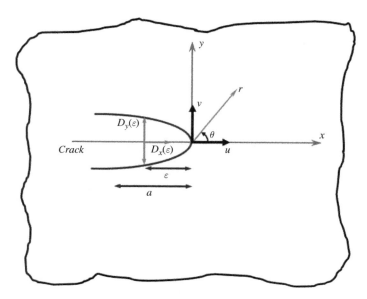

Figure 13 Displacement correlation technique for the special crack tip element (Fatehi Marji, 1990).

$$u_x = \frac{K_I}{4G}\sqrt{\frac{r}{2\pi}}\left[(2\kappa-1)\cos\frac{\theta}{2}-\cos\frac{3\theta}{2}\right] + \frac{K_{II}}{4G}\sqrt{\frac{r}{2\pi}}\left[(2\kappa+3)\sin\frac{\theta}{2}+\sin\frac{3\theta}{2}\right]$$

$$u_y = \frac{K_I}{4G}\sqrt{\frac{r}{2\pi}}\left[(2\kappa-1)\sin\frac{\theta}{2}-\sin\frac{3\theta}{2}\right] - \frac{K_{II}}{4G}\sqrt{\frac{r}{2\pi}}\left[(2\kappa-3)\cos\frac{\theta}{2}+\cos\frac{3\theta}{2}\right]$$

(54)

where, K_I and K_{II} are the stress intensity factors in Mode I and Mode II respectively; and r and θ are defined in Figure 13.

Based on LEFM theory, the Mode I and Mode II stress intensity factors K_I and K_{II} can be written in terms of the normal and shear displacement discontinuities as (Shou & Crouch, 1995):

$$K_I = \frac{\mu}{4(1-\nu)}\left(\frac{2\pi}{a}\right)^{\frac{1}{2}}D_y(a), \quad \text{and} \quad K_{II} = \frac{\mu}{4(1-\nu)}\left(\frac{2\pi}{a}\right)^{\frac{1}{2}}D_x(a) \qquad (55)$$

Once the SIFs are computed the propagation angle may be obtained using following relation based on the maximum tangential stress criterion (Erdogan & Sih, 1963).

$$\theta_0 = 2\tan^{-1}\left[\frac{1}{4}\left(\frac{K_I}{K_{II}}\right) - \frac{1}{4}\sqrt{\left(\frac{K_I}{K_{II}}\right)^2 + 8}\right] \qquad (56)$$

The angle θ_0 is shown in Figure 5.

5.4.2 Crack tip element formulation

The displacement discontinuity method permits the crack surfaces to be discretized and computes the crack opening displacement (normal displacement discontinuity), and crack sliding displacement (shear displacement discontinuity) directly as a part of the solution for each element (Fatehi Marji, 2014; Fatehi Marji *et al.*, 2006; Guo *et al.*, 1992, 1990; Scavia, 1995, 1990). Due to the singularity of the stresses and displacements near the crack ends, the accuracy of the displacement discontinuity method at the vicinity of the crack tip decreases, and usually a special treatment of the crack at the tip is necessary to increase the accuracy and make the method more efficient. The hybrid elements can be implemented in a general higher order displacement discontinuity method using cubic displacement discontinuity elements with three special crack tip elements at the end of each crack. To use three special crack tip elements, the displacement discontinuity variation along the element can be written in the following form (Fatehi Marji *et al.*, 2006) as:

$$D_i(\varepsilon) = C_1 \varepsilon^{\frac{1}{2}} + C_2 \varepsilon^{\frac{3}{2}} + C_3 \varepsilon^{\frac{5}{2}} \tag{57}$$

Equation 57 can be rearranged in the following form:

$$D_i(\varepsilon) = [N_{C1}(\varepsilon)]D_i^1(a) + [N_{C2}(\varepsilon)]D_i^2(a) + [N_{C3}(\varepsilon)]D_i^3(a) \tag{58}$$

The crack tip element has a length $a = a_3 + a_2 + a_1$. Considering $a_1 = a_2 = a_3$, the complicated shape functions $N_{C1}(\varepsilon)$, $N_{C2}(\varepsilon)$ and $N_{C3}(\varepsilon)$ can be written as:

$$N_{C1}(\varepsilon) = \frac{15\varepsilon^{\frac{1}{2}}}{8a_1^{\frac{1}{2}}} - \frac{\varepsilon^{\frac{3}{2}}}{a_1^{\frac{3}{2}}} + \frac{\varepsilon^{\frac{5}{2}}}{8a_1^{\frac{5}{2}}},$$

$$N_{C2}(\varepsilon) = \frac{-5\varepsilon^{\frac{1}{2}}}{4\sqrt{3}a_1^{\frac{1}{2}}} + \frac{3\varepsilon^{\frac{3}{2}}}{2\sqrt{3}a_1^{\frac{3}{2}}} - \frac{\varepsilon^{\frac{5}{2}}}{4\sqrt{3}a_1^{\frac{5}{2}}}, \tag{59}$$

$$N_{C3}(\varepsilon) = \frac{3\varepsilon^{\frac{1}{2}}}{8\sqrt{5}a_1^{\frac{1}{2}}} - \frac{\varepsilon^{\frac{3}{2}}}{2\sqrt{5}a_1^{\frac{3}{2}}} + \frac{\varepsilon^{\frac{5}{2}}}{8\sqrt{5}a_1^{\frac{5}{2}}}$$

The potential functions $f_C(x, y)$ and $g_C(x, y)$ for the crack tip can be expressed as:

$$f_C(x, y) = \frac{-1}{4\pi(1 - \nu)} \int_{-a}^{a} D_x(\varepsilon) \ln[(x - \varepsilon)^2 + y^2]^{\frac{1}{2}} d\varepsilon$$

$$g_C(x, y) = \frac{-1}{4\pi(1 - \nu)} \int_{-a}^{a} D_y(\varepsilon) \varepsilon^{\frac{5}{2}} \ln[(x - \varepsilon)^2 + y^2]^{\frac{1}{2}} d\varepsilon \tag{60}$$

Inserting Equations 59 in Equations 58 and the first part of Equations 60, gives:

$$f_C(x,y) = \frac{-1}{4\pi(1-v)} \left\{ \left[\int_{-a}^{a} N_1(\varepsilon) \ln\left[(x-\varepsilon)^2 + y^2\right]^{\frac{1}{2}} d\varepsilon \right] D_x^1 \right.$$

$$\left. + \left[\int_{-a}^{a} N_2(\varepsilon) \ln\left[(x-\varepsilon)^2 + y^2\right]^{\frac{1}{2}} d\varepsilon \right] D_x^2 + \left[\int_{-a}^{a} N_3(\varepsilon) \ln\left[(x-\varepsilon)^2 + y^2\right]^{\frac{1}{2}} d\varepsilon \right] D_x^3 \right\}$$

(61a)

$$f_C(x,y) = \left(-\frac{1}{4\pi(1-v)}\right)(I_{C1}(x,y)D_x^1 + I_{C2}(x,y)D_x^2 + I_{C3}(x,y)D_x^3)$$

(61b)

$$f_C(x,y) = -\frac{1}{4\pi(1-v)} \sum_{j=1}^{2} D_x^j F_{Cj}(I_{C1}, I_{C2}, I_{C3})$$

(61c)

Similarly,

$$g_C(x,y) = -\frac{1}{4\pi(1-v)} \sum_{j=1}^{2} D_y^j F_{Cj}(I_{C1}, I_{C2}, I_{C3})$$

(62)

from Equations 61c and Equation 62, the common function $F_{Cj}(I_{C1}, I_{C2}, I_{C3})$ can be obtained as:

$$F_{Cj}(I_{Cj}) = \int_{-a}^{a} N_{Cj}(\varepsilon) \ln[(x-\varepsilon)^2 + y^2]^{\frac{1}{2}} d\varepsilon, \quad j = 1, 2 \text{ and } 3$$

(63)

From which the following integrals are deduced:

$$I_{C1}(x,y) = \int_{-a}^{a} \varepsilon^{\frac{1}{2}} \ln[(x-\varepsilon)^2 + y^2]^{\frac{1}{2}} d\varepsilon,$$

$$I_{C2}(x,y) = \int_{-a}^{a} \varepsilon^{\frac{3}{2}} \ln[(x-\varepsilon)^2 + y^2]^{\frac{1}{2}} d\varepsilon$$

(64)

$$I_{C3}(x,y) = \int_{-a}^{a} \varepsilon^{\frac{5}{2}} \ln[(x-\varepsilon)^2 + y^2]^{\frac{1}{2}} d\varepsilon$$

The integrals I_{C1}, I_{C2} and I_{C3} with their corresponding derivatives are given in a recent work by the authors (Fatehi Marji *et al.*, 2007, 2005).

6 DISCRETE ELEMENT METHOD

Geo-materials like rocks, exhibit discontinuous and inhomogeneous nature leads to complex mechanical behaviors which can be difficult to tackle with classical numerical

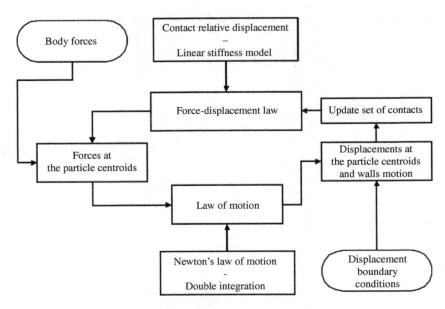

Figure 14 General DEM calculation cycle (Monteiro Azevedo & Lemos, 2006).

methods. Among these complex features which need to be reproduced as cracks, their nucleation, propagation and interaction.

Most of the numerical methods used in geo-mechanics have an implicit representation of the discontinuities, where their deformability or strength, are considered as equivalent continua through constitutive laws of the discontinuities. Since the introduction of joint elements in the FEM (Goodman, 1976), continuum mechanics based methods are still in progressing in the fracture mechanics literature in form of the eXtended Finite Element Method (XFEM), (Belytschko & Black, 1999; Waisman & Belytschko, 2008).

However, Figure 14 demonstrates the general calculation cycle in DEM where the explicit force-displacement formulation is being used.

Discrete-based methods can be used as an alternative to continuum based methods which represent the material as an assemblage of independent elements interacting with one another. The discrete element methods (DEM) are often applied to investigate the mechanical behavior of geo-materials, by approximating the geometry of materials as discrete elements bonded together by different models of cohesive forces or cementation effects.

In geotechnical field, different DEMs are used, the two major discrete based methods are: the Universal Distinct Element Code (UDEC) and Discontinuous Deformation Analysis (DDA).

UDEC as pioneered by Cundall & Strack (1979), is basically an algorithm involving Force-Displacement method and the Molecular Dynamics (MD) formalism, which are considered as smooth contact method. There are other DEMs which are referred to as "non-smooth contact methods" such as i) the event-driven integrators or the

Event-Driven Method (EDM) (Luding *et al.*, 1996) and ii) the time-stepping integrators or the Contact Dynamics Method (CD method) (Jean, 1995; Moreau, 1994).

While the DEM, EDM or CD often considered as the discrete elements for non-deformable bodies in time-explicit numerical schemes, their performances is somewhat limited when the discrete elements are themselves deformable. Then, the Discontinuous Deformation Analysis (DDA) (Shi & Goodman, 1988) can be used to overcome these deficiencies. DDA uses an FEM based solver for stress and deformation filed inside the Discrete Elements, and simultaneously accounts for the interaction of independent elements along discontinuities. In this section the discontinuous based methods are being discussed therefore, DDA which is considered as a continuous approach will not be considered here. For a detailed and complete description of DDA the interested reader is referred to Jing & Stephansson's book (2008).

7 MESH-LESS METHODS

Mesh-less methods are viewed as the next generation of the computational techniques. The conventional grid based methods such as FEM, BEM and DEM have some evident limitation in dealing with problems of fracture mechanics and large deformation processes (Daxini & Prajapati, 2014).

Grid based methods are widely used for analyzing various engineering problems. Eulerian and Lagrangian grid are two fundamental approaches in grid based methods. New combined approaches are developed to strengthen the advantages of each approach and avoid their limitation (Liu & Liu, 2003).

7.1 General features

Problems with moving material discontinuity, large deformation with excessive mesh distortion (*i.e.* the fracture mechanics problems) are not well suited to be treated by grid based methods.

The mesh-less methods were developed by modifying the internal structure of grid based methods. The mesh-less methods are more adaptive, versatile and robust. Therefore they can be used for problems when conventional methods are not suitable.

The basic concept of the mesh-less methods is to provide accurate and stable numerical solution for integral equations or partial differential equations (PDEs) with any possible boundary and initial conditions using a set of arbitrarily distributed nodes without defining mesh which connect these nodes (Liu & Liu, 2003). Element free galerkin (EFG) and mesh-less local Petrov-Galerkin (MLPG) methods are two well-known mesh-less methods. Mesh-less methods are categorized based on use of global or local weak form to derive system matrices. EFG method is based on global weak form, while MLPG method is based local symmetric weak form (LSWF) (Belytschko *et al.*, 1994b). These methods use the moving least square (MLS) approximants.

Some of the important features of mesh-less methods may be categorized as

1. No mesh alignment sensitivity (background mesh is only for integration purpose).
2. Node connectivity is not predefined by mesh.

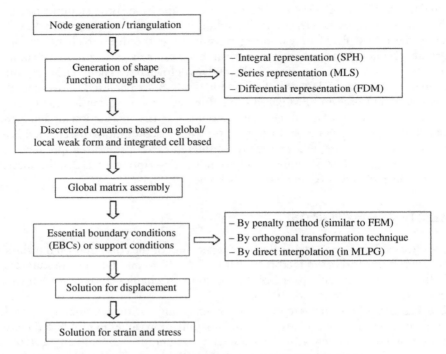

Figure 15 Procedural steps in mesh-less methods (Daxini & Prajapati, 2014).

3. No re-meshing is needed (especially for large deformation and moving discontinuity Problems).
4. Order of continuity in the required shape function can be constructed.
5. The smooth derivatives of unknowns require no post-processing.

Figure 15 shows the general calculation procedure in mesh-less methods.

7.2 Fracture mechanics applications

EFG and MLPG methods have been used in many static and dynamic fracture mechanics problems (Belytschko & Fleming, 1999; Brighenti, 2005; Gato, 2010; Kaiyuan *et al.*, 2006; Krysl & Belytschko, 1999; Parvanova, 2012; Rao & Rahman, 2000). EFG is mostly applied to LEFM problems. Several criteria has been proposed for handling discontinuities, among them are the "Visibility criterion", diffraction method, transparency method, and "see through" method or continuous line criterion (Belytschko & Fleming, 1999; Belytschko *et al.*, 1994a; Sukumar *et al.*, 2000). For dynamic fracture problems (with constant velocity of crack propagation and constant value of dynamic fracture toughness) also some EFG approaches were developed (Belytschko & Tabbara, 1996; Krysl & Belytschko, 1999). One of the most efficient MLPG variant, MLPG5, was used for analysis of elasto-dynamic deformations near crack tip (Kaiyuan *et al.*, 2006).

The enriching EFG approximations are mainly used for LEFM problems. Two of these methods are extrinsic and intrinsic enrichment approximations (Belytschko &

Fleming, 1999; Belytschko *et al.*, 1994a). The moving least square (MLS) shape function and truncated Gaussian weight functions can be employed (Krysl & Belytschko, 1999). The mixed mode dynamic crack growth in brittle solids using EFG methods in fracture process zone (FPZ) technique uses dynamic fracture mechanics formulation instead of those of LEFM. Based on this approach the numerical values of mode I and mixed mode SIFs in rock like materials can be estimated. (Belytschko *et al.*, 2000). For example, a MATLAB code is prepared along with intrinsic basis enrichment to precisely model the singular stress field around the crack tip for two-dimensional elasticity problems (Parvanova, 2012).

8 MIXED METHODS

The combined methods are some advanced methods using a hybridized form of the conventional continuum based method such as FEM and BEM (in the case of deformability) with discontinum based methods such as DEM and DDA. However any combination of the grid based methods with the mesh-less methods can also be regarded as a combined method. In this section a brief review of some of the most versatile combined methods cited in the literature are given. The hybridized FE-BEM, the combined finite-discrete element method (CFDEM) and the FE-mesh-less method are briefly discussed.

8.1 Coupling finite element and boundary element methods

The finite element and the boundary element methods are among the most popular methods in fracture mechanics. The finite element method is best suited for finite regions with inhomogeneities and nonlinearities. On the other hand, the boundary element method is the better alternative for solving problems dealing with infinite or semi-infinite bodies and problems with steep stress gradients and singularities such as fracture mechanics problems. The mathematic of FEM-BEM coupling is well reviewed in (Stephan, 2004).

Fracture mechanics problems especially in rock mechanics often occur in which there exists certain form of coupling between various parts. Soil-rock and water-rock or oil-rock interaction problems are typical examples of this category. Fluid related problems may need to be handled in poro-elastic framework. In such instances it may be advantageous to couple the finite element and the boundary element methods to use both methods benefits.

The coupling of finite element and boundary element methods is usually accomplished at the level of the discretized equations. Various methods are used and developed to combine these methods which may be distinguished as

1. Transformation of the boundary element equations to a suitable form to be embedded in a finite element formulation.
2. Transformation of the finite element equations to a suitable form to be embedded in a boundary element formulation.
3. Combining the methods at the basic weighted residual statement.

The first method is widely used but it may be erroneous sometimes and energy arguments are put forth to symmetrize the boundary element equations (Ameen, 2005). The

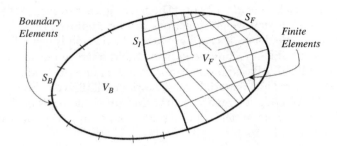

Figure 16 Domain divided into boundary element and finite element sub-domains (after Ameen, 2005).

second method is recommended when the finite element domain is much smaller in comparison with the boundary element part.

The finite element system leads to a set of algebraic equations (Equation 23). In Equation 23 the matrix *[K]* is the global stiffness matrix and *[R]* is the global load vector. The stiffness matrix *[K]* is, in general, symmetric and banded. Rewriting Equations 23 in the following general form for BEM formulation

$$[F]\{u\} = [G]\{p\} \tag{65}$$

The matrices [F] and [G] are influence coefficients and are, in general, fully populated and un-symmetric.

Consider a problem domain consisting of two distinct domains as in Figure 16. Where V_F and V_B represent the finite element and boundary element domains. The domain V_F which its external boundary defined as S_F is modeled using FEM. The same procedure goes on for V_B and S_F. Along the interfacial boundary between V_F and V_B (*i.e.* S_I) the following compatibility and equilibrium conditions must be satisfied respectively.

$$u_{IF} = u_{IB} = u_I \quad \text{on } S_I \tag{66}$$

$$f_{IF} + f_{IB} = \{f\}_I \quad \text{on } S_I \tag{67}$$

where u_I is displacements and f_I is the tractions along the S_I.

Multiplying both side of Equation 65 (which represents V_B in Figure 16) by $[N][G]^{-1}$ ([N] is the matrix of interpolation polynomials) results in

$$[N][G]^{-1}[F]\{u\} = [N]\{p\} \tag{68}$$

which has transformed Equation 65 into an equivalent finite element system of equations. Equation 68 can be rewritten as

$$K^B u = f^B \tag{69}$$

where K^B is the equivalent stiffness matrix and f^B the equivalent nodal force vector. Equation 69 can now be directly assembled into the global matrices of FEM. It should be noted that K^B is un-symmetric due to its BEM nature. There are several methods to symmetrize the matrix K^B (see for instance (Ameen, 2005)).

8.2 The combined Finite Element-Discrete Element Method (FE-DEM)

Fracture mechanics of heterogeneous media may be studied by means of a combination of the FE-DEM. The classical DEM has some limitation to be used for rock fracture analysis due to the high number of particles that are necessary in the discretization, which limits the use of particle systems in larger rock structures. The hybridized methods such as the combined FE-DEM may be modified to take the advantages of both FEM and DEM simultaneously. DEM algorithm discretizes the fracture zones and FEM discretization scheme is used for the surrounding areas. The incorporation of the FEM technology in DEM increases the deformation capabilities of the rigid block mesh (Ghaboussi, 1988; Munjiza *et al.*, 1995; Petrinic, 1996; Potapov & Campbell, 1996).

In this section, a coupling algorithm is introduced applying a DEM circular particle discretization only on the fracture zone and a FEM based discretization on the other zones assuming a linear elastic behavior (Monteiro Azevedo & Lemos, 2006).

The coupling of the FEM and the DEM can be accomplished by adopting special interface elements which enable for example the interaction of the circular particles with the edge of the contacting quadrilateral finite elements.

The rigid wall element is used to set boundary conditions where the wall is given by a line segment and a particle/line segment contact. The same concept can be used for the finite elements edges after the element deformation. The edge motion is defined in order to set the relative displacement at the contact interfaces and the distribution of forces from these interfaces to the adjacent edge nodes.

The centered difference algorithm may be used for explicitly integrating both the FEM and the DEM equilibrium equations. The smallest time increment of the two discretization schemes is used and the influence of the interface element is also accounted in order to determine the time increment for numerical stability (Monteiro Azevedo, 2003).

In the global problem (FEM+DEM), the equation for the translational and rotational motion with no damping, is given by

$$\mathbf{F}_i^{(t)} = m\left(\ddot{\mathbf{x}}_i - \mathbf{g}_i\right), \tag{70}$$

$$M_3^{(t)} = I\dot{\omega}_3 = \left(\beta\, mR^2\right)\dot{\omega}_3 \tag{71}$$

where $\mathbf{F}_i^{(t)}$ is the total applied force at time t, m is the total mass of the particle or the node, $\ddot{\mathbf{x}}_i$ is the particle translational acceleration, $\mathbf{g}_i$ is the body force acceleration, $M_3^{(t)}$ is the total applied moment at time t, I is the moment of inertia, $\dot{\omega}_3$ is the particle angular acceleration, and β is a coefficient reflecting the shape of the particle which is 2/5 for a spherical shape and 1/2 for a disk shaped particle.

The internal forces resulting from the finite elements deformation are due to the total applied force at a given instant t for a given node involves the contact forces of the particles interacting through common edges and the exterior forces applied at the nodes.

Wall/edge elements are used to represent the boundary edges of the finite elements that are allowed to interact with the particle assembly. Figure 17 shows the linear associated edge shape functions which are equivalent to a truss finite element model with two degree of freedom per node with a given length L (Monteiro Azevedo & Lemos, 2006).

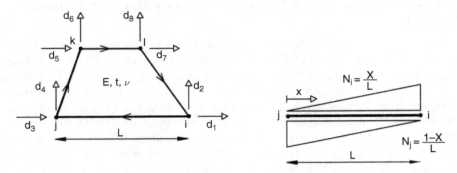

Figure 17 FEM geometry and associated edge shape functions (Monteiro Azevedo & Lemos, 2006).

The interaction of a particle with a given edge of a plane finite element obeying the force displacement law is being used. The interaction forces at the contact interface are developed which are related to the incremental contact displacement at the interface. The mechanism through which these forces are transferred to the edge nodal points and to the center of gravity of the rigid circular particles can also be modeled.

The wall/edge length, L and the wall/edge axial direction, a at a given instant are given in function of the nodal positions by

$$L = \sqrt{\left(X_i^{node,j} - X_i^{node,i}\right)\left(X_i^{node,j} - X_i^{node,i}\right)}, \qquad (72)$$

$$a_i = \frac{X_i^{node,j} - X_i^{node,i}}{L}, \qquad (73)$$

The wall/edge transverse/normal direction, n_j, is defined based on the wall/edge axial direction through n = $(-a_2, a_1)$.

The definition of the wall/edge velocity at the contact point $\dot{x}_j^{[C]}$ and the relation of the contact forces to the nodal forces are the main issues.

8.3 The combined finite element-mesh-less method

Although mesh-less methods seem attractive for crack propagation problems, their computational cost which often exceeds the cost of a regular FEM sets a major limitation to the applicability of these methods. In addition to that, considering the comprehensive capabilities of FEM, it is often advantageous to use mesh-less methods only in the sub-domains. It is more effective to apply mesh-less methods at the fracture zones and FEM in the remainder of the domain. Therefore, a combination of the mesh-less and finite element methods may utilize the benefits of both methods.

The element-free Galerkin method (EFGM) which is a mesh-less method may be integrated with the traditional finite element method (FEM) for LEFM problems. The EFGM may be used to model material behavior close to cracks (fracture zone) and the FEM for the surrounding areas (the same as coupled FE-DEM). The shape functions interpolating the interface region are a combination of both EFGM and FEM shape functions, ensuring convergence of the method. This coupled FE-Mesh-less method significantly saves computational effort compared with the existing mesh-less methods. Due to the partly mesh-less nature of the method, only a distributed set of nodal points is required in the fracture zone. Crack propagation can be modeled by extending the free surfaces, which correspond to the crack.

For a detailed formulation of FE-Mesh-less methods readers are encouraged to consult several proposed combinations of these methods (Hegen, 1996; Huerta & Fernandez Mendez, 2000; Krongauz & Belytschko, 1996; Liu et al., 1997).

8.3.1 Calculation of stress-intensity factors

The SIFs in a combined FE-Mesh-less method can be evaluated by converting the interaction integral (Yau et al., 1980) into a domain form (Moran & Shih., 1987). For example

$$K_I = \frac{E'}{2} M^{(1,2)} \tag{74}$$

is the effective elastic modulus, and the interaction integral, $M^{(1,2)}$ is

$$M^{(1,2)} = \int_A \left[\sigma_{ij}^{(1)} \frac{\partial u_i^{(2)}}{\partial x_1} + \sigma_{ij}^{(2)} \frac{\partial u_i^{(1)}}{\partial x_1} - W^{(1,2)} \delta_{1j} \right] \frac{\partial q}{\partial x_j} dA \tag{75}$$

which contains the mixed mode state for the given boundary conditions (superscript 1) and the super-imposed near-tip mode I auxiliary state (superscript 2), σ_{ij} and u_i are the components of the stress tensor and displacement vector, respectively, $W^{(1,2)}$ is the mutual strain energy from the two states and q is a weight function (equal to 1 at the crack tip, 0 along the boundary of the domain and arbitrary elsewhere). A similar procedure may be applied for the calculation of K_{II}, the only difference is that the near-tip mode II state is chosen as the auxiliary state while computing $M^{(1,2)}$.

The crack propagation using FE-Mesh-less method may be accomplished through the following steps:

1. Numerical prediction of stress and strain fields
2. SIFs calculation using the results of step 1.
3. Calculation of crack initiation angle (θ_0).
4. Updating the new crack-tip location (using an arbitrary element length for crack-tip).
5. Splitting the mesh-less node into two nodes locating on the opposite sides of the new crack-tip.
6. Adding new mesh-less nodes to improve discretization of the domain (optional).

9 SUMMARY

The topic of numerical rock fracture mechanics can be summarized as the mesh-based methods such as FEM and DEM, mesh-reducing methods such as BEM and mesh-less methods.

The FEM is a numerical technique for finding approximate solutions to rock fracture mechanics problems. It uses subdivision of a whole problem domain into simpler parts, called finite elements. The variational methods from the calculus of variations may be used to solve the problem by minimizing an associated error function. The XFEM, is an especially modified FEM to solve fracture mechanics problems. It is based on a standard Galerkin procedure, and the partition of unity method (PUM) to accommodate the internal boundaries in the discrete model.

Rock fracture mechanics problems, exhibit discontinuous and inhomogeneous nature which leads to complex mechanical behaviors can be difficult to tackle with the classical continuum-based numerical methods. DEM is an explicit discrete-based method is used as an alternative to continuum based methods which can be applied to investigate the mechanical behavior of geo-materials as discrete elements bonded together by different models of cohesive forces or cementation effects.

The mesh-reducing boundary element methods are divided into two main groups of direct and indirect methods. Direct boundary integral method uses direct approach and fictitious stress and displacement discontinuity methods use an indirect approach in the boundary element methods.

The dual BEM is a direct BEM which incorporates two independent boundary integral equations; it uses the displacement equation to model one of the crack boundaries and the traction equation to model the other boundary.

The indirect BEM incorporating the displacement discontinuities, along the cracks is known as displacement discontinuity method. It is based on the analytical solution to the problem of a discontinuity in displacement over a finite line segment in the x, y plane of an infinite elastic solid. Higher order elements using linear, quadratic and cubic displacement discontinuity elements increases the accuracy of SIF calculation near the crack-tips. The use of crack tip elements can also substantially improve the accuracy of the DDM for crack analyses.

Mesh-less methods as the next generation of the computational techniques were developed by modifying the internal structure of grid based methods. The mesh-less methods are more adaptive, versatile and robust. The mesh-less methods provide accurate and stable numerical solution for integral equations or partial differential equations (PDEs) with any possible boundary and initial conditions using a set of arbitrarily distributed nodes without defining mesh which connect these nodes. Element free Galerkin (EFG) and mesh-less local Petrov-Galerkin (MLPG) methods are two well-known mesh-less methods.

A combination of the mentioned numerical methods may be used in order to solve the more complicated fracture mechanics problems. The combined methods may be modified to take the advantages of both methods simultaneously. The FE-DEM, FE-BEM and FE-Mesh-less methods are the combined methods can be used in numerical rock fracture mechanics.

REFERENCES

Aliabadi, M.H., Rooke, D.P. (1991) Numerical Fracture Mechanics. Computational Mechanics Publications, Southampton, U.K.

Ameen, M. (2005) Computation of Elasticity. Alpha Science International Ltd., Calicut.

Ash, R.L. (1985) Flexural Rapture as a Rock Breakage Mechanism in Blasting. In: Fragmentation by Blasting. Society of Experimental Mechanics, Bethel, pp. 371–378.

Behnia, M., Goshtasbi, K., Marji, M.F., Golshani, A. (2012) The effect of layers elastic parameters on hydraulic fracturing propagation utilizing displacement discontinuity method. Anal. Numer. Methods Min. Eng. 3, 1–14.

Belytschko, T., Black, T. (1999) Elastic crack growth in finite elements with minimal remeshing. Int. J. Numer. Methods Eng. 45, 601–620.

Belytschko, T., Fleming, M. (1999) Smoothing, enrichment and contact in the element-free Galerkin method. Comput. Struct. 71, 173–195.

Belytschko, T., Gu, L., Lu, Y.Y. (1994a) Fracture and crack growth by element free Galerkin methods. Model. Simul. Mater. Sci. Eng. 2, 519–534.

Belytschko, T., Lu, Y.Y., Gu, L. (1994b) Element-free Galerkin methods. Int. J. Numer. Methods Eng. 37, 229–256.

Belytschko, T., Organ, D., Gerlach, C. (2000) Element-free Galerkin methods for dynamic fracture in concrete. Comput. Methods Appl. Mech. Eng. 187, 385–399.

Belytschko, T., Tabbara, M. (1996) Dynamic fracture using element-free Galerkin methods. Int. J. Numer. Methods Eng. 39, 923–938.

Blandford, G.E., Ineraffea, A.R., Lizeett, J.A. (1981) Two-dimensional stress intensity factor computations using the boundary element method. Int. J. Numer. Methods Eng. 17, 387–404.

Brighenti, R. (2005) Application of the element-free Galerkin meshless method to 3-D fracture mechanics problems. Eng. Fract. Mech. 72, 2808–2820.

Broek, D. (1985) Elementary Engineering Fracture Mechanics, 4th ed. Nijhoff.

Broek, D. (1989) The Practical Use of Fracture Mechanics. Kluwer Academic Publishers, Netherland.

Chen, J.T., Hong, H.K. (1999) Review of dual boundary element methods with emphasis on hyper singular integrals and divergent series. Appl. Mech. Rev. ASME 52, 17–33.

Cisilino, A.P., Aliabadi, M.H. (2004) Dual boundary element assessment of three-dimensional fatigue crack growth. Eng. Anal. Bound. Elem. 28, 1157–1173.

Courtesen, D.L. (1979) Cavities and Gas Penetration from Blasts in Stressed Rock with Flooded Joints. Acta Astron. 341–363.

Crawford, A.M., Curran, J.H. (1982) Higher order functional variation displacement discontinuity elements. Int. J. Rock Mech. Min. Sci. Geomech. Abstr. 19, 143–148.

Crouch, S.L. (1976a) Solution of plane elasticity problems by the displacement discontinuity method. I. Infinite body solution. Int. J. Numer. Methods Eng.

Crouch, S.L. (1976). Analysis of stresses and displacements around underground excavations: An application of the displacement discontinuity method. Department of Civil and Mineral Engineering, University of Minnesota.

Crouch, S.L., Starfield, A.M. (1983) Boundary Element Methods in Solid Mechanics. George Allen & Unwin, London.

Cruse, T.A. (1989) Boundary Element Analysis in Computational Fracture Mechanics. Kluwer Academic Publishers, Dordrecht.

Cundall, P.A., Strack, O.D.L. (1979) A discrete numerical model for granular assemblies. Geotechnique 29.

Davies, A.J., Crann, D. (2006) A Laplace transform boundary element solution for the biharmonic diffusion equation. Bound. Elem. Other Mesh Reduct. Methods XXVIII, WIT Transactions on Modelling and Simulation, 42(1), 243–252.

Daxini, S.D., Prajapati, J.M. (2014) A Review on Recent Contribution of Meshfree Methods to Structure and Fracture Mechanics Applications. Sci. World J., 2014, Article ID 247172, 13. doi:10.1155/2014/247172.

Dirgantara, T., Aliabadi, M.H. (2001) Dual boundary element formulation for fracture mechanics analysis of shear deformable shells. Int. J. Solids Struct. 38, 7769–7800.

Dolbow, J. (1999) An Extended Finite Element Method with Discontinuous Enrichment for Applied Mechanics. Northwestern University.

Dolbow, J., Moes, N., Belytschko, T. (2000) Discontinuous enrichment infinite elements with a partition of unity method. Finite Elem. Anal. 235–260.

Duarte, C., Babuska, I., Oden, J. (2000) Generalized finite element methods for three dimensional structural mechanics problems. Comput. Struct. 77, 215–232.

Erdogan, F., Sih, G. (1963) On the crack extension in plates under loading and transverse shear. J. Basic Eng. 85, 519–527.

Etfis, J., Liebowitz, H. (1972) On the modified Westergaard equations for certain plane crack problems. Int. J. Fract. Mech. 8, 383–392.

Fatehi Marji, M. (1990) Crack propagation modeling in rocks and its application to indentation problems. Middle East Technical University, Ankara, Turkey.

Fatehi Marji, M. (1997) Modelling of cracks in rock fragmentation with a higher order displacement discontinuity method. Middle East Technical University, Ankara, Turkey.

Fatehi Marji, M. (2014) Rock Fracture Mechanics with Displacement Discontinuity Method. LAP Lambert Academic Publishing, Germany.

Fatehi Marji, M. (2015) Higher order displacement discontinuity method in rock fracture mechanics. Yazd University.

Fatehi Marji, M., Gholamnejad, J., Eghbal, M. (2011) On the Crack Propagation Mechanism of Brittle Substances under Various Loading Conditions. In: 11th International Multidisciplinary Scientific Geo-Conference, Albena, Bulgaria.

Fatehi Marji, M., Hosseini-nasab, H. (2005) Application of Higher Order Displacement Discontinuity Method Using Special Crack Tip Elements in Rock Fracture Mechanics. In: 20th World Mining Congress & Expo. Tehran, Iran, pp. 699–704.

Fatehi Marji, M., Hosseini Nasab, H., Kohsary, A.H. (2006) On the uses of special crack tip elements in numerical rock fracture mechanics. Int. J. Solids Struct. 43, 1669–1692.

Fatehi Marji, M., Hosseini Nasab, H., Kohsari, A.H. (2007) A new cubic element formulation of the displacement discontinuity method using three special crack tip elements for crack analysis. J. Solids Struct. 1, 61–91.

Fatehi Marji, M., Hosseini-nasab, H., Kohsary, A.H. (2005) Two Dimensional Displacement Discontinuity Method Using Cubic Elements for Analysis of Crack Problems. In: 20th World Mining Congress & Expo. Tehran, Iran, pp. 653–658.

Fedelinski, P., Aliabadi, M.H., Rooke, D.P. (1993) The dual boundary element method in dynamic fracture mechanics. Eng. Anal. Bound. Elem. 12, 203–210.

Gato, C. (2010) Mesh-less analysis of dynamic fracture in thin-walled structures. Thin-Walled Struct. 48, 215–222.

Ghaboussi, J. (1988) Fully deformable discrete element analysis using a finite element approach. Comput. Geotech. 5, 175–195.

Goodman, R.E. (1976) Methods of geological engineering in discontinuous rocks. West Publishing Company, San Francisco.

Griffith, A.A. (1921) The phenomena of rupture and flow in solids. Phil. Trans. R. Soc. London 221, 163–197.

Guo, H., Aziz, N.I., Schmitt, L.C. (1990) Linear elastic crack tip modeling by displacement discontinuity method. Engin. Fract. Mech. 36, 933–943.

Guo, H., Aziz, N.I., Schmitt, L.C. (1992) Rock cutting study using linear elastic fracture mechanics. Engin. Fract. Mech. 41, 771–778.

Haeri, H., Shahriar, K., Fatehi Marji, M., Moaref Vand, P. (2013a) Simulating the Bluntness of TBM Disc Cutters in Rocks using Displacement Discontinuity Method. In: 13th International Conference on Fracture. China.

Haeri, H., Shahriar, K., Fatehi Marji, M., Moarefvand, P. (2013b) An experimental and numerical study of crack propagation and cracks coalescence in the pre-cracked rock-like disc specimens under compression. Int. J. Rock Mech. Min. Sci. Geomech. Abstr. 67, 20–28.

Hegen, D. (1996) Element-free Galerkin methods in combination with finite element approaches. Comput. Methods Appl. Mech. Eng. 135, 143–166.

Hong, H., Chen, J. (1988) Derivations of integral equations of elasticity. J. Eng. Mech. ASCE 114, 1028–1044.

Hosseini Nasab, H., Fatehi Marji, M. (2007) A Semi-infinite Higher Order Displacement Discontinuity Method and Its Application to the Quasi-static Analysis of Radial Cracks Produced by Blasting. J. Mech. Mater. Struct. 2(3), 439–458.

Hu, K., Chen, Z. (2013) Pre-kinking of a moving crack in a magnetoelectroelastic material under in-plane loading. Int. J. Solids Struct. 50, 2667–2677.

Huerta, A., Fernandez Mendez, S. (2000) Enrichment and coupling of the finite element and meshless methods. Int. J. Numer. Meth. Eng. 48, 1615–1636.

Hussain, M.A., Pu, S.L., Underwood, J. (1974) Strain Energy Release Rate for a Crack under Combined Mode I and Mode II. In: Fracture Analysis ASTM-STP. pp. 560–562.

Irwin, G.R. (1948) Fracture dynamics. Fract. Met. Am. Soc. Met. 147, 166, 56–63.

Irwin, G.R., (1957) Analysis of stresses and strains near the end of a crack traversing a plate. J. Appl. Mech. Trans. ASME 24, 361–364.

Itou, S. (2013) Effect of couple-stresses on the Mode I dynamic stress intensity factors for two equal collinear cracks in an infinite elastic medium during passage of time-harmonic stress waves. Int. J. Solids Struct. 50, 1597–1604.

Jaeger, J., Cook, N., Zimmerman, R. (2009) Fundamentals of rock mechanics. John Wiley & Sons.

Jean, M. (1995) Frictional contact in collections of rigid or deformable bodies: numerical simulation of geomaterial motions. Studies in Applied Mechanics, 42, 463–486.

Jing, L., Stephansson, O. (2008) Fundamentals of discrete element methods for rock engineering: theory and applications (developments in geotechnical engineering). Elsevier, Amesterdam.

Kaiyuan, L., Shuyao, L., Guangyao, L. (2006) A simple and less-costly meshless local Petrov-Galerkin (MLPG) method for the dynamic fracture problem. Eng. Anal. Bound. Elem. 30, 72–76.

Kim, K., Pereira, J.P. (1997) Rolling friction and shear behaviour of rock discontinuities filled with sand. Int. J. Rock Mech. Min. Sci. 34, 244.e1–244.e17.

Krongauz, Y., Belytschko, T. (1996) Enforcement of essential boundary conditions in meshless approximations using finite elements. Comput. Methods Appl. Mech. Eng. 131, 133–145.

Krysl, P., Belytschko, T. (1999) The element free Galerkin method for dynamic propagation of arbitrary 3-D cracks. J. Numer. Methods Eng. 44, 767–800.

Kutter, H.K., Fairhurst, C. (1971) On the Fracture Process in Blasting. Int. J. Fract. 8, (3), 189–202.

Leme, S.P.L., Aliabadi, M.H. (2012) Dual boundary element method for dynamic analysis of stiffened plates. Theor. Appl. Fract. Mech. 57, 55–58.

Liu, W.K., Li, S., Belytschko, T. (1997) Moving least square kernel Galerkin method -Part I: methodology and convergence. Comput. Methods Appl. Mech. Eng. 143, 422–33.

Liu, G.R., Liu, M.B. (2003) Smoothed particle hydrodynamics: a meshfree particle method. World Scientific.

Luding, S., Clement, E., Rajchenbach, J., Duran, J. (1996) Simulations of pattern formation in vibrated granular media. Europhys. Lett. 36, 247–252.

Melenk, J.M., Babuska, I. (1996) The partition of unity finite element method: basic theory and applications. Comput. Methods Appl. Mech. Eng. 139, 289–314.

Mews, H. (1987) Calculation of stress intensity factors for various crack problems with the boundary element method. In: Brebbia, C.A., Wendland, W.G., Kuhn, G. (Eds.), 9th International Conference on BEM. Southampton, U.K., pp. 259–278.

Moës, N., Dolbow, J., Belytschko, T. (1999) A finite element method for crack growth without remeshing. Int. J. Numer. Methods Eng. 46, 131–150.

Mohammadi, S. (2008) Extended finite element method for Fracture Analysis of Structures. John Wiley & Sons.

Monteiro Azevedo, N. (2003) A rigid particle discrete element model for the fracture analysis of plain and reinforced concrete. Doctoral dissertation, Heriot-Watt University.

Monteiro Azevedo, N., Lemos, J.V. (2006) Hybrid discrete element/finite element method for fracture analysis. Comput. Methods Appl. Mech. Eng. 195, 4579–4593.

Moran, B., Shih., C.F. (1987) Crack tip and associated domain integrals from momentum and energy balance. Eng. Fract. Mech. 27, 615–642.

Moreau, J.-J. (1994) Some numerical methods in multibody dynamics: application to granular materials. Eur. J. Mech. A/Solids 13, 93–114.

Munjiza, A., Owen, D., Bicanic, N. (1995) A combined finite-discrete element method in transient dynamics of fracturing solids. Eng. Comput. 12, 145–174.

Muskhelishvili, N.I., 1953. Some basic problems of the mathematical theory of elasticity, 3rd ed. Groningen, Holland.

Palaniswamy, K. (1972) Crack propagation Under General, In-plane Loading. California University of Technology.

Parvanova, S. (2012) Calculation of stress intensity factors based on force-displacement curve using element free Galerkin method. J. Theor. Appl. Mech. 42, 23–40.

Petrinic, N. (1996) Aspects of discrete element modelling involving facet-to-facet contact detection and interaction. University of Wales.

Portela, A., Aliabadi, M.H., Rooke, D.P. (1992) The dual boundary element method, effective implementation for crack problems. Int. J. Numer. Meth. Eng. 22, 1269–1287.

Potapov, A., Campbell, C. (1996) A hybrid finite-element simulation of solid fracture. Int. J. Mod. Phys. 7, 155–180.

Rao, B.N., Rahman, S. (2000) Efficient meshless method for fracture analysis of cracks. Comput. Mech. 26, 398–408.

Sanford, R.J. (1979) A critical re-examination of Westergaard method for solving opening mode crack problems. Mech. Res. Commun. 6, 289–294.

Sanford, R.J. (2003) Principles of fracture mechanics. Prentice Hall Upper Saddle River, NJ, 416 pages.

Scavia, C. (1990) Fracture mechanics approach to stability analysis of crack slopes. Eng. Fract. Mech. 35, 889–910.

Scavia, C. (1995) A method for the study of crack propagation in rock structures. Geotechnique 45, 447–463.

Shen, B., Stephansson, O. (1993) Modification of the G-criterion of crack propagation in compression. Int. J. Eng. Fract. Mech. 47, 177–189.

Shi, G., Goodman, R. (1988) Discontinuous deformation analysis – a new method for computing stress, strain and sliding of block systems. In: Balkema, A.A. (Ed.), 29th US Symposium on Rock Mechanics. Rotterdam.

Shou, K.J., Crouch, S.L. (1995) A higher order displacement discontinuity method for analysis of crack problems. Int. J. Rock Mech. Min. Sci. Geomech. Abstr. 32, 49–55.

Shou, K.-J., Napier, J.A.L. (1999) A two-dimensional linear variation displacement discontinuity method for three-layered elastic media. Int. J. Rock Mech. Min. Sci. 36, 719–729.

Shou, K.-J., Siebrits, E., Crouch, S.L. (1997) A higher order displacement discontinuity method for three-dimensional elastostatic problems. Int. J, Rock Mech. Min. Sci. 34, (2), 317–322.

Sih, G.C. (1966) On the Westergaard method on crack analysis. Int. J. Fract. Mech. 2, 628–631.

Sih, G.C. (1974) Strain energy density factor applied to mixed mode crack problems. Int. J. Fract. Mech. 10, 305–321.

Simpson, R., Trevelyan, J. (2011) A partition of unity enriched dual boundary element method for accurate computations in fracture mechanics. Comput. Methods Appl. Mech. Eng. 200, 1–10.

Snyder, M.D., Cruse, T.A. (1975) Boundary integral equation analysis of cracked anisotropic plates. Int. J. Fract. 11, 315–328.

Sokolnikoff, I. (1956) Mathematical Theory of Elasticity. McGraw-Hill Book Company, New York.

Stephan, E.P. (2004) Coupling of boundary element methods and finite element methods. Encyclopedia of Computational Mechanics. John Wiley & Sons, pp. 375–412.

Stephansson, O. (2002) Recent rock fracture mechanics developments. In: 1st Iranian Rock Mechanics Conference. Tehran, Iran.

Strouboulis, T., Babuska, I., Copps, K. (2000) The design and analysis of the generalized finite element method. Comput. Methods Appl. Mech. Eng. 181, 43–69.

Sukumar, N., Moes, N., Moran, B., Belytschko, T. (2000) Extended finite element method for three-dimensional crack modeling. Int. J. Numer. Methods Eng. 48, 1549–1570.

Waisman, H., Belytschko, T. (2008) Parametric enrichment adaptivity by the extended finite element method. Int. J. Numer. Methods Eng. 72, 1671–1692.

Westergaard, H.M. (1939) Bearing pressures and cracks. J. Appl. Mech. 6, A49–A53.

Whittaker, B.N., Singh, R.N., Sun, G. (1992) Rock Fracture Mechanics, Principles Design and Applications. Elsevier Science Publishers B.V., Netherland.

Williams, M.L. (1952) Stress singularities resulting from various boundary conditions in angular corners of plates in extension. J. Appl. Mech. 19, 526–528.

Williams, M.L. (1957) On the stress distribution at the base of a stationary crack. J. Appl. Mech. 24, 109–114.

Yau, J.F., Wang, S.S., Corten, H.T. (1980) A mixed-mode crack analysis of isotropic solids using conservation laws of elasticity. J. Appl. Mech. 47, 335–341.

Yazid, A., Abdelkader, N., Abdelmadjid, H. (2009) A state-of-the-art review of the X-FEM for computational fracture mechanics. Appl. Math. Modell. 33, 4269–4282.

Yun, B.I., Ang, W.T. (2010) A dual-reciprocity boundary element approach for axisymmetric nonlinear time-dependent heat conduction in a nonhomogeneous solid. Eng. Anal. Boundary Elem. 34, 697–706.

Chapter 23

Linear elasticity with microstructure and size effects

G. Exadaktylos
School of Mineral Resources Engineering, Technical University of Crete, Chania, Greece

Abstract: The revisiting of some fundamental problems of rock mechanics, such as cracks in stressed rocks, propagation of surface waves, beam bending and axial splitting among others, viewed in the light of a strain gradient elasticity theory, reveals the necessity of enriching elasticity of rocks with length parameters to model surface energy of free surfaces and predict non-classical dispersion phenomena and size effects. After a brief review of the formalism and applications of a linear elasticity theory with microstructure for the study of static and dynamic problems, two problems are further presented here, namely the bending of beams and the axial splitting of deep geological layers. It is demonstrated in both studied problems, that the consideration of internal length scales are responsible for the manifestation of size effects.

1 INTRODUCTION

1.1 Brief notes on the size effects of strength of materials

The size effect exhibited by the strength of solids for otherwise geometrically similar specimens, is not new in the context of the strength of brittle materials. Long before Griffith (1921) presented his theory, Karmarsch[1] in 1858 has proposed an empirical size effect law based on a best-fitting procedure of experimental data of tension tests on cylindrical metal wires with different diameters. This size effect is mentioned in the celebrated Griffith's paper and was applied successfully by him to fit experimental data referring to tension tests of glass fibers presented in Table V of his paper, namely

$$\sigma_t = 154.44 + \frac{17.27}{d} \tag{1}$$

in which the diameter of the rods d is expressed in 0.001 mm and the tensile strength σ_t in MPa. The best-fitted size-effect law given by Equation 1 on the experimental data is shown in Figure 1.

In most technical brittle materials, such as rocks, concretes and ceramics, the domains in the vicinity of the highly stress point participate in the force transmission more intensively than according to the local linear theory of elasticity (the term 'local' is explained below); this self-support-effect of a stress raiser is taken into account by the

---

1 "Mittheilungen des gew. Ver. Für Hannover", 1858, pp. 138–155.

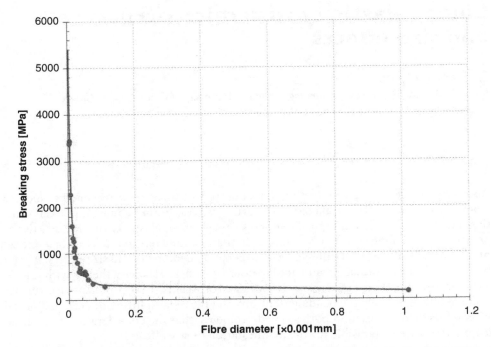

Figure 1 Size effect exhibited by the tensile strength of glass fibers (circles) tested by Griffith (1921) and best-fitted inverse diameter relationship (continuous line).

stress-mean-value theory and the more elegant gradient elasticity theory with surface energy that is presented later.

First, let us make a remark on the averaging procedure that is inherent in all local continuum mechanics theories. A simple example is to consider an one-dimensional case of a field, y=f(x), whose mean value is computed over a small but finite averaging length L – corresponding to the representative elementary volume – around a point x, that is

$$\langle y \rangle = \frac{1}{L} \int_{-L/2}^{L/2} f(x + \xi) d\xi \qquad (2)$$

If the field f(x) varies linearly in the considered region around x, then it is approximated locally by a linear function, using an 1-term Taylor series expansion of the function f around point x, *i.e.*

$$f(x + \xi) \approx f(x) + \xi f'(x) \qquad (3)$$

In the trivial case of a constant field, then the first and all higher derivatives vanish and indeed the local value coincides with the average value. Also it is true in case when the field varies locally linearly. Indeed we may then identify the field with its mean value over the considered averaging length, because by following the 'trapezoidal' integration rule, the value of a linearly varying field in the midpoint of the sampling interval is equal to its mean value in that interval

$$y = \langle y \rangle \tag{4}$$

that is to say, in this case the 'local' value y and the 'non-local' value $\langle y \rangle$ coincide. In the classical theories of elasticity, plasticity and damage mechanics, the failure criterion is expressed in terms of stresses and strains, and no characteristic length scale L is present. Hence, they are all "local theories". However, for quadratically varying fields, we have to approximate the stress function at least by a two-term Taylor series expansion around point x, *i.e.*

$$f(x+\xi) \approx f(x) + f'(x)\xi + \frac{1}{2}f''(x)\xi^2 + \frac{1}{6}f'''(x)\xi^3 + o(\xi^4) \tag{5}$$

We notice that in the midpoint integration rule the effect of the first derivative is null. Thus for 'quadratically' varying fields, computational rule described by Equation 4 must be enhanced, so as to incorporate the effect of the curvature

$$y = \langle y \rangle - \frac{L^2}{24} \frac{d^2 y}{dx^2}\bigg|_x + O(L^4) \tag{6}$$

Field theories which are based on averaging rules that include the effect of higher gradients are called higher gradient or 'nonlocal' theories. In particular, the above rule of Equation 6 represents a 2nd gradient or grade-2 rule, and can be readily generalized in two and three dimensions. One of the first researchers who proposed a gradient theory based on the mean value of the nominal stress along the potential fracture path was Neuber (1936). More specifically Neuber proposed a stress-mean-value taken over a finite length L normal to the surface within the range of high stress concentration. This so-called 'fictive' length of the elastic material represents an additional material constant apart say, from the two elasticity constants for a linear elastic and isotropic material. According to this argument the nominal stress σ_n can be found from the formula

$$\sigma_n = \frac{1}{L} \int_{r=R}^{R+L} \sigma dr \tag{7}$$

in which r denotes the radial distance from the notch tip, r=R is the notch boundary, and σ is the so-called 'comparison' stress that enters a suitable strength hypothesis.

In the case of mode-III (anti-plane shear) crack the nominal stress is derived by the following formulae according to definition of Equation 7 and the valid asymptotic expression for the comparison stress σ_{yz}

$$\sigma_n = \frac{1}{L} \int_{r=0}^{L} \sigma_{yz}(r,0)dr; \quad \sigma_{yz} = \frac{K_{III}}{(2\pi r)^{1/2}} \cos\frac{\theta}{2} \tag{8}$$

where K_{III} denotes the mode-III Stress Intensity Factor (SIF) and $Or\theta$ the polar coordinate system with origin at the crack tip. The direct evaluation of the above integral for $\theta = 0^o$ gives the result

$$\sigma = \sqrt{\frac{2}{\pi}} \frac{K_{III}}{\sqrt{L}} + o\left(\sqrt{L}\right) \tag{9}$$

On the other hand, the exact expression for the comparison stress by employing the Westergaard stress function reads as follows

$$\sigma_{yz} = \text{Re}(Z_{III}), \quad Z_{III} = \frac{\tau_\infty(z + \alpha)}{[(\alpha + z)^2 - \alpha^2]^{1/2}}; \quad z = re^{i\theta} \tag{10}$$

and consequently the exact stress along the Ox-axis is given by the expression

$$\sigma_{yz}\Big|_{\theta=0} = \frac{\tau_\infty(r + \alpha)}{(r^2 + 2\alpha r)^{1/2}} \tag{11}$$

where τ_∞ represents the far-field shear stress. In turn, the nominal stress in this case may be found to be

$$\sigma_n = \frac{1}{L} \int_{r=0}^{L} \sigma_{yz}\Big|_{\theta=0} dr = \tau_\infty \sqrt{1 + 2\frac{\alpha}{L}} \tag{12}$$

By requiring that both approaches should lead to the same result, *i.e.*

$$\underset{r \to 0}{Lim}\left[\sqrt{\frac{2}{\pi}} \frac{K_{III}}{\sqrt{L}} = \tau_\infty \sqrt{1 + 2\frac{\alpha}{L}}\right] \tag{13}$$

there results

$$\hat{K}_{IIIC} = \sqrt{1 + \frac{1}{2(\alpha/L)}}; \quad K_{IIIC} = \tau_\infty \sqrt{\pi \alpha} \tag{14}$$

where $\hat{K}_{IIIC}$ represents the normalized fracture toughness that is derived by dividing the expression for the critical stress intensity factor with the fracture toughness K_{IIIC} predicted by the classical theory. The variation of the normalized fracture toughness $\hat{K}_{IIIC}$ with the ratio α/L – *i.e.* the size effect exhibited by fracture toughness - is illustrated in Figure 2. This means that for long cracks relative to the scale length L one gets the result of Linear Elastic Fracture Mechanics (LEFM); on the other hand, for relatively short cracks the fracture toughness is larger than that predicted by classical LEFM.

1.2 Brief historical remarks on non-local elasticity theories

The classical theory of elasticity requires that the forces between the atoms to fulfill a very strong condition, namely that the range of these forces must be small enough so that the stress (strain) measured at a point depends in the desired approximation only on the stress (strain) in the volume element around this point; hence the term 'local' theory. Obviously, if interatomic forces did not reach farther than one atomic distance, a reaction against micro-deformation gradient would not exist and the theory does not

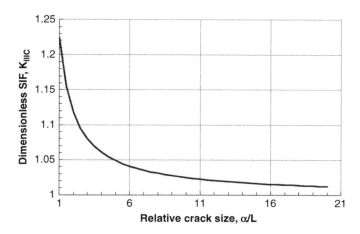

Figure 2 Size effect of the normalized fracture toughness $\hat{K}_{IIIC}$ predicted by the stress-mean-value theory.

have an intrinsic length scale; this in turn leads to the undesirable result that a 10 cm slab behaves the same as a 10 m geological bed, and there is no difference between a microcrack and a geological fault. However, since interatomic forces do, in principle, reach farther than one atomic distance, a resistance against micro-deformation gradient will be present, and therefore it is of no question whether gradient-dependent elasticity exists or not. The question is rather *how large* this effect might be.

The fundamental idea of considering not only the first, but also the higher gradients of the displacement field in the expression for the strain energy function of an elastic solid, can be traced back to J. Bernoulli (1654–1705) and L. Euler (1707–1783) in connection with their work on beam theory. In elementary beam theory there are associated two sets of kinematical quantities (a deformation vector and a rotation vector) and two sets of surface loads (tractions and bending couples) with a section of the bar. In plate theory the situation is similar. With the noticeable monograph of Cosserat brothers, Eugéne and François (1909), this concept was extended to a 3D continuum, where each point of the continuum is supplied with a set of mutually perpendicular rigid vectors (triad). Generalization of elasticity theory by incorporating the effect of higher gradients of the displacement field into the strain energy density function was systematically studied by them. The novel feature of their theory was the appearance of couple stresses in the equations of motion. An oriented continuum of this type was noted earlier by Voigt (1887) in connection with polar molecules in crystallography. Higher-order gradient and oriented media theories were rediscovered fifty years later in various special forms and degree of complexity. Fifty years after the first publication of the original work of the Cosserat brothers, the basic kinematic and static concepts of the 'Cosserat' continuum were reworked in a milestone paper by Guenther (1958). Guenther's paper marks the rebirth of continuum micro-mechanics in the late 50's and early 60's. Following this publication, several hundred papers were published all over the world on that subject. A variety of names have been invented and given to theories of various degrees of rigor and complexity: Cosserat continua or micro-polar

704   Exadaktylos

media, oriented media, continuum theories with directors, multi-polar continua, micro-structured or micro-morphic or non-local continua and others. A systematic treatment of elasticity with gradients was given in milestone papers by Mindlin & Tiersten (1962), Mindlin (1964) and Mindlin & Eshel (1968). The common feature of all these studies is that they relate the higher gradients of the displacement field to higher order stresses. Mindlin's work is noteworthy in that his aim was specifically targeted at understanding phenomenologically the effect of microstructure on the deformation of solids. Mindlin's cohesive elasticity theory accounts in a phenomenological manner for molecular forces of cohesion acting upon a body - which are not considered by the classical linear elasticity theory - by including in the potential energy density of an elastic solid the 'modulus of cohesion', which is essentially an initial, homogeneous, self-equilibrating triple stress. However, Mindlin's isotropic grade-3, linear elasticity theory with surface energy, which was further explored, as far as its mathematical potential is concerned, in a comprehensive paper by Wu (1992), includes sixteen material constants plus the classical Lamè's constants. The state-of-the-art at this time was reflected in the collection of papers presented at the historical IUTAM Symposium on the *"Mechanics of Generalized Continua"*, in Freudenstadt and Stuttgart in 1967. At the same time practically of publication of the pioneering papers by Mindlin, Professor Germain has encouraged the communication to the French Academy of Sciences of the ideas of Casal (1961) which in turn seem to have inspired Germain's (1973a,b) fundamental papers on the continuum mechanics structure of the grade-2 or higher grade theories. In our paper we want to give full credit to Casal's original idea, who was first to see the connection between surface tension effects and the anisotropic gradient elasticity theory. For this reason we provide here the simplest possible generalization of Casal's constitutive theory that accounts for only two additional material constants having the dimension of length: One, say ℓ, responsible for volumetric energy strain-gradient terms, and another, ℓ', responsible for surface energy strain-gradient terms. Casal considered the effect of the granular, polycrystalline and atomic nature of materials on their macroscopic response through the concept of internal and superficial capillarity expressed by the material lengths ℓ, ℓ', respectively, rather than through intractable statistical mechanics concepts. The concept that the surfaces of liquids are in a state of tension is a familiar one, and it is widely utilized. Actually it is known that no skin or thin foreign surface really is in existence at the surface, and that the interaction of surface molecules causes a condition analogous to a surface subjected to tension. The surface tension concept is therefore an analogy, but it explains the surface behavior in such satisfactory manner that the actual molecular phenomena need not be invoked. Of course such ideas are amenable to generalizations of various degrees of complexity. However, one should keep in mind that already the determination of the two material lengths ℓ and ℓ' constitutes a formidable experimental challenge.

The Casal-Mindlin grade-2 theory has been applied for the revisit of several static and dynamic boundary-value problems in Rock Mechanics (Vardoulakis & Sulem, 1995; Vardoulakis *et al.*, 1996; Exadaktylos *et al.*, 1996; Exadaktylos & Vardoulakis, 1998; Exadaktylos, 1998; Exadaktylos & Vardoulakis, 2001a; Aravas, 2011 among others). The consideration of the surface energy in this theory, leads to a constitutive character of the boundary conditions. This strengthens Aifantis' (1992) conjecture of the constitutive character of boundary constraints in materials with microstructure. Hence, the problem of constitutive boundary conditions deserves further attention

Linear elasticity with microstructure and size effects 705

from the theoretical, as well as the experimental point of view. Exadaktylos & Vardoulakis (2001a) have shown that the proposed theory is capable: (a) to capture scale effects in indentation and uniaxial tension testing of rocks, and (b) to predict cusping of cracks without recourse to extra assumptions. The present anisotropic gradient elasticity theory although it is basically a grade-2 theory gives rise to surface tension phenomena similar to those captured by Mindlin's (1964; 1965) grade-3 theory. This is demonstrated in Paragraph 1.5.

1.3 Formalism of the Casal-Mindlin microelasticity theory

In this sequel the basic formalism of the grade-2 theory of elasticity are outlined. With respect to a fixed Cartesian coordinate system $Ox_1x_2x_3$, the following *ansatz* for the elastic strain energy density with respect to three kinematic quantities is assumed in an ad hoc manner

$$v = v(\varepsilon_{qr}, \gamma_{qr}, \kappa_{qrs}) \tag{15}$$

where $\varepsilon_{qr} \equiv (1/2)(\partial_r u_q + \partial_q u_r)$ is the usual symmetric infinitesimal macro-strain tensor defined in terms of the displacement vector u_q, $\partial_s \equiv \partial/\partial x_s$, the indices (q,r,s) span the range (1,2,3), $\gamma_{qr} \equiv \partial_q u_r - \psi_{qr}$ is the relative deformation with ψ_{qr} denoting the micro-deformation of a particle in the form of a grain or crystal for a granular or crystalline rock, respectively, (Figures 3a, b), and $\kappa_{qrs} \equiv \partial_q \psi_{rs}$ is the micro-deformation gradient. Then, appropriate definitions for the stresses follow from the variation of v, *i.e.*

$$\tau_{qr} \equiv \frac{\partial v}{\partial \varepsilon_{qr}}, \quad \alpha_{qr} \equiv \frac{\partial v}{\partial \gamma_{qr}}, \quad \mu_{qrs} \equiv \frac{\partial v}{\partial \kappa_{qrs}}, \tag{16}$$

in which $\tau_{qr}, \alpha_{qr}, \mu_{qrs}$ denote the Cauchy stress (symmetric), relative stress (asymmetric), and double stress tensors, respectively. The twenty-seven components μ_{kij} have the character of double forces per unit area. The first subscript of a double stress μ_{kij} designates the normal to the surface across which the component acts; the second and third subscripts have the same significance as the two subscripts of σ_{ij}. The eight components of the deviator of the couple-stress or couples per unit area formed by the combinations $(1/2)(\mu_{pqr} - \mu_{prq})$ are all equal to zero in the present gradient dependent elasticity theory, whereas all the remaining ten independent combinations $(1/2)(\mu_{pqr} + \mu_{prq})$ are self-equilibrating (Mindlin, 1964). Double force systems without moments are stress systems equivalent to two oppositely directed forces at the same point; such systems have direction but not net force and no resulting moment.

In particular, the theory utilized here can be considered as one of the simplest versions of Casal-Mindlin theory corresponding to the following elastic strain energy density function (Exadaktylos & Vardoulakis, 1998)

$$v = \frac{1}{2}\lambda\varepsilon_{ii}\varepsilon_{jj} + G\varepsilon_{ij}\varepsilon_{ji} + G\ell^2\partial_k\varepsilon_{ij}\partial_k\varepsilon_{ji} + G\ell_k'\partial_k(\varepsilon_{ij}\varepsilon_{ji}) \tag{17}$$

where $\lambda = E\nu/(1 - 2\nu)(1 + \nu)$ and $G = E/2(1 + \nu)$ are the standard constants of Lame, E, ν denote the Young's modulus and Poisson's ratio, and as was mentioned already, ℓ, ℓ' are additional characteristic lengths of the material, where

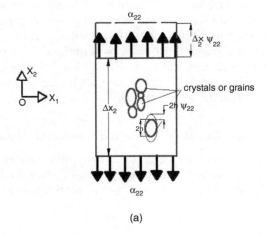

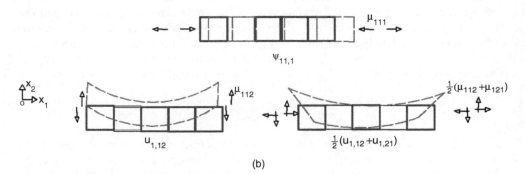

Figure 3 (a) Typical components of relative stress α_{ij} ($\alpha_{ij} \equiv \sigma_{ij} - \tau_{ij}$) displacement gradient $\partial_i u_j$, and micro-deformation ψ_{ij} for the simple case of uniaxial tension of a flat plate, and (b) various forms of micro-deformation gradients and associated double stresses.

$$\ell_k = \ell' \nu_k, \quad \nu_k \nu_k = 1 \tag{18}$$

is a director. The last term in Equation 17 has the meaning of surface energy, since by using the divergence theorem we get

$$\int_V \partial_r(\ell_r \varepsilon_{pq} \varepsilon_{qp}) dV = \ell' \int_{\partial V} (\varepsilon_{pq} \varepsilon_{qp})(\nu_r n_r) dS \tag{19}$$

wherein n_k is the outward unit normal on the boundary ∂V.

1.4 Stress equations of equilibrium

Germain (1973a, b) suggested a general framework for the foundation of consistent higher grade continuum theories on the basis of the virtual work principle. This approach starts from the definition of the variation of the total potential energy in a volume V of the body with arbitrary variation of the macro-strain ε_{ij}. A restricted Mindlin continuum, is a micro-homogeneous material for which the macroscopic strain coincides with the micro-deformation, $\gamma_{qr} = 0$ which in turn leads to the following relations

$$\psi_{qr} \equiv \partial_q u_r, \quad \hat{\kappa}_{qrs} \equiv \partial_q \varepsilon_{rs} = (1/2)(\partial_q \partial_r u_s + \partial_q \partial_k u_r) = \hat{\kappa}_{qsr} \tag{20}$$

and

$$\mu_{qrs} \equiv \partial v / \partial \hat{\kappa}_{qrs} = \mu_{qsr} \tag{21}$$

In the particular case the variation of the strain energy potential is modified from that in Equation 15 as follows (Mindlin, 1964)

$$\delta \int_V v dV = \int_V (\tau_{ij} \delta \varepsilon_{ij} + \mu_{ijk} \partial_i \delta \varepsilon_{jk}) dV \tag{22}$$

where

$$\tau_{ij} = \frac{\partial v}{\partial \varepsilon_{ij}}, \mu_{ijk} = \frac{\partial v}{\partial(\partial_i \varepsilon_{jk})} \tag{23}$$

The second order stress tensor τ_{ij}, is dual in energy to the macroscopic strain and is symmetric (*i.e.* $\tau_{ij} = \tau_{ji}$), whereas the third order stress tensor μ_{ijk}, is dual in energy to the strain-gradient. To prepare for the formulation of a variational principle, we apply to Equation 22 the chain rule of differentiation and the divergence theorem; furthermore, we resolve $\partial_i u_j$ on the boundary ∂V of V into a in plane – gradient and a normal-gradient as follows

$$\partial_i \delta u_j \equiv D_i \delta u_j + n_i D \delta u_j, \quad D_i \equiv (\delta_{ik} - n_i n_k) \partial_k, \quad D \equiv n_k \partial_k. \tag{24}$$

where δ_{ij} is the Kronecker delta. The final expression for the variation in potential energy of a smooth boundary ∂V reads

$$\begin{aligned}
\delta U = \int_V \delta v dV &= -\int_V \partial_j (\tau_{jk} - \partial_i \mu_{ijk}) \delta u_k dV + \int_{\partial V} n_j (\tau_{jk} - \partial_i \mu_{ijk}) \delta u_k dS \\
&+ \int_{\partial V} \left(\left[\frac{1}{R_1} + \frac{1}{R_2} \right] n_j - D_j \right) n_i \mu_{ijk} \delta u_k dS + \int_{\partial V} n_i n_j \mu_{ijk} D \delta u_k dS
\end{aligned} \tag{25}$$

where $(1/R_1 + 1/R_2)$ is the mean curvature of the bounding surface. Looking at the structure of Equation 25 we now postulate the following form for the variation of work U_e done by external forces (*i.e.* body forces, tractions and double-tractions, respectively)

$$\delta U_e = \int_V f_k \delta u_k dV + \int_{\partial V} (\tilde{P}_k \delta u_k + \tilde{R}_k D \delta u_k) dS \tag{26}$$

where f_k is the body force per unit volume, $\widetilde{P}_k$, $\widetilde{R}_k$ are the specified tractions and double tractions, respectively, on the smooth surface ∂V. Then, from the variational principle, the stress-equilibrium equations in the volume V is found in the following manner

$$\partial_i(\tau_{ij} - \partial_k\mu_{ijk}) + f_j = 0 \tag{27}$$

The workless second order relative stress tensor α_{ij} in a restricted Mindlin continuum is in equilibrium with the double stress (Mindlin, 1964)

$$\alpha_{jk} + \partial_i\mu_{ijk} = 0 \tag{28}$$

Next, by defining the 'total stress tensor' σ_{ij}

$$\sigma_{ij} = \tau_{ij} + \alpha_{ij} = \tau_{ij} - \partial_k\mu_{ijk} \tag{29}$$

the stress-equilibrium Equation 27 takes the following final form in the volume V

$$\partial_j\sigma_{ij} + f_i = 0 \tag{30}$$

One may notice that according to Equation 30 the *total stress tensor* is identified with the common (macroscopic) equilibrium stress tensor. Although the above results are obtained for static cases, there is no essential difficulty to derive their dynamic counterpart.

1.5 Boundary conditions

The surface ∂V of the considered volume V is divided into two complementary parts ∂V_u and ∂V_σ such that on ∂V_u kinematic data whereas on ∂V_σ static data are prescribed. In classical continua these are constraints on displacements and tractions, respectively. For the stresses the following set of boundary conditions on a smooth surface ∂V_σ is also derived from the virtual work principle (Wu, 1992; Exadaktylos & Vardoulakis, 2001a)

$$n_j\tau_{jk} - n_j\partial_i\mu_{ijk} + \left(\left[\frac{1}{R_1} + \frac{1}{R_2}\right]n_j - D_j\right)n_i\mu_{ijk} = \widetilde{P}_k \tag{31}$$

$$n_in_j\mu_{ijk} = \widetilde{R}_k \tag{32}$$

Since second-grade or grade-2 models introduce second strain gradients into the constitutive description, additional kinematic data must be prescribed on ∂V_u. With the displacement already given in ∂V_u, only its normal derivative with respect to that boundary is unrestricted. This means that on ∂V_u the normal derivative of the displacement should also be given, *i.e.*

$$u_i = w_i \, on \quad \partial V_{u1} \quad and \quad Du_i = r_i \quad on \quad \partial V_{u2} \tag{33}$$

1.6 Constitutive relations

From Equations 17 and 23 follow the constitutive relations for the total stress, Cauchy stress and double stress tensors, respectively

Linear elasticity with microstructure and size effects 709

$$\left.\begin{array}{l}\sigma_{ij} = \lambda\delta_{ij}\varepsilon_{kk} + 2G(\varepsilon_{ij} - \ell^2\nabla^2\varepsilon_{ij}) \\[2mm] \tau_{ij} = \lambda\delta_{ij}\varepsilon_{kk} + 2G\varepsilon_{ij} + 2G\ell_k\partial_k\varepsilon_{ij} \\[2mm] \mu_{kij} = 2G\ell_k\varepsilon_{ij} + 2G\ell^2\partial_k\varepsilon_{ij}\end{array}\right\} \qquad (34)$$

From the last of the above relations we may note that the double stress is symmetric in the last two indices as is also depicted by Equation 21.

In closing this exposition of basic notions and relations, we may prove that positive definiteness of the strain-energy density is valid provided the following restrictions of the material constants hold true

$$(3\lambda + 2G) > 0, \quad G > 0, \quad \ell^2 > 0, \quad -1 < \frac{\ell'}{\ell} < 1 \qquad (35)$$

The third inequality simply means that ℓ should be a real and not imaginary number.

1.7 Skin effect and surface free energy

Our purpose here is to show that a basic feature of the present strain gradient elasticity theory with surface energy is the appearance of a skin effect associated with the volume energy parameter ℓ. Furthermore, it will be shown that the effect of the relative surface energy parameter ℓ'/ℓ is equivalent to the effect of initial stresses in presence of an infinite, plane boundary.

The deformation of an isotropic semi-infinite body $x_1 \geq 0$ due to a large uniform tensile stress $\sigma_{22} = \sigma, (\sigma > 0)$, parallel with the surface with outward unit normal vector $(n_1\, n_2\, n_3)=(-1\,0\,0)$ with the Cartesian coordinates be x_1, x_2, and x_3, is considered as was done in (Exadaktylos & Vardoulakis, 1998). Starting from a stress-free configuration, C_0, the body is stressed uniaxially under plane strain conditions, and C is the resultant configuration. Then, the pre-stressed body is incrementally deformed and let its current configuration state to be that of C'. The problem under consideration is formulated in terms of the first Piola-Kirchhoff stress π_{ij} with respect to current configuration C', with $\Delta\pi_{ij}$ being its increment referred to the deformed initially stressed state C. Assuming infinitesimal strain elasticity, the Jaumann stress increments $\Delta°\sigma_{ij}$ of the total stress are related directly to the strain increments through constitutive Equations 34. For the traction-free surface of the half-space the following inceremetal boundary conditions are valid

$$\Delta\pi_{11} = \Delta\pi_{21} = \mu_{111} = \mu_{112} = 0 \quad \text{on } x_1 = 0 \qquad (36)$$

It is possible to assume, without loss of generality (it can be shown that, in this problem, the quantities u_1, u_3 do not couple with u_2; these quantities satisfy homogeneous equations with homogeneous boundary conditions and therefore vanish identically) the following one-dimensional displacement field

$$u_2 = u_2(x_1), u_1 = u_3 = 0 \qquad (37)$$

and the only non-zero initial stress σ_{22} to act along x_2-axis. Upon substituting the strain-displacement relation into the stress-strain relations and the resulting expressions for the stresses into the stress-equation of equilibrium $\partial_j\Delta\pi_{ij} = 0$, we find only the following surviving displacement equation of equilibrium

$$\left(1 - \frac{\ell^2}{(1+\xi)\,dx_1^2}\frac{d^2}{dx_1^2}\right)\frac{d^2}{dx_1^2}u_2 = 0 \tag{38}$$

where we have set $\xi = -\sigma_{22}/2G$. The solution of Equation 38, vanishing at infinity, is

$$u_2(x_1) = c\exp(-\frac{\sqrt{1+\xi}}{\ell}x_1) \tag{39}$$

where c denotes an integration constant. The first three boundary conditions described by Equations 36 are satisfied identically, whereas the only remaining boundary condition along $x_1 = 0$ takes the form

$$\mu_{112} = 2G\left\{\ell\frac{d}{dx_1} + \ell^2\frac{d^2}{dx_1^2}\right\}u_2 = 0 \qquad \text{on } x_1 = 0 \tag{40}$$

which holds true for $\nu_r \equiv -n_r$ and gives the following equation

$$c\left[-\frac{\ell'}{\ell} + \sqrt{1+\xi}\right] = 0 \tag{41}$$

From Equation 41 one may deduce that the only case which gives non-zero and exponentially decaying displacement with distance from the surface of the solid, that is $c \neq 0$, is the following

$$\frac{\ell'}{\ell} = \sqrt{1+\xi} \Longleftrightarrow \xi = -\left[1 - \left(\frac{\ell'}{\ell}\right)^2\right] \tag{42}$$

The above relation elucidates the importance of the surface strain gradient term ℓ' in determining surface effects. Equation 42 depicts that the effect of the surface energy parameter is equivalent to the effect of an *initial stress*. The dependence of initial stress ξ on the relative surface energy parameter ℓ'/ℓ is shown in Figure 4. From this figure it may be seen that if $\ell'/\ell = 0$ the half-space is under surface tension, with this surface tension to be maximum. As ℓ'/ℓ increases from the value of zero the initial tension or in other words the *surface tension* of the medium decreases reaching the value of zero for $\ell'/\ell = 1$. At $\ell'/\ell = 1$ the initial stress changes sign and for $\ell'/\ell > 1$ becomes compressive in nature. That is, for values of the relative surface energy parameter higher than the value of one, the medium is under surface compression and it is no longer in a state of elastic equilibrium, or in other words as it is also shown by the inequality of Equation 35 its strain energy density function is negative definite.

The elastic strain energy density of the considered 1D configuration is given by

$$v = G\left\{\varepsilon^2 + \ell^2\nabla\varepsilon\nabla\varepsilon + 2\ell'\varepsilon\nabla\varepsilon\right\}, \quad \nabla \equiv d/dx_1 \tag{43}$$

Substituting in Equation 43 the values for the strain and the strain-gradient, we find

$$\hat{v} = \left\{1 - \left(\frac{\ell'}{\ell}\right)^2\right\}\left(\frac{\ell'}{\ell}\right)^2\left(\frac{c}{\ell}\right)^2\exp\left(-2\frac{\sqrt{1+\xi}}{\ell}x_1\right), \quad \hat{v} = v/G \tag{44}$$

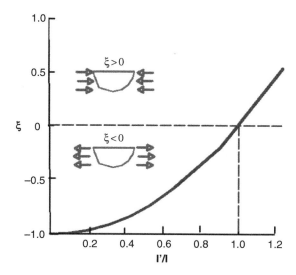

Figure 4 Graphical representation of the relation of the dimensionless pre-stress ξ with the relative surface energy parameter ℓ'/ℓ (Exadaktylos & Vardoulakis, 1998).

By adopting the following definition for the average surface stress (or surface free energy)

$$\gamma_{se} = \int_V v dV/A \tag{45}$$

where A is the area of the free surface, we may find after some manipulations

$$\gamma_{se} = \frac{G}{2}\left\{1 - \left(\frac{\ell'}{\ell}\right)^2\right\}\frac{\ell' c^2}{\ell^2} \tag{46}$$

This is also, for each surface, the energy per unit area required to separate the body along a plane and $\gamma_{se} > 0$ if inequality described by the last of Equations 35 holds true.

1.8 Anti-plane shear (SH) surface waves

There are a number of cases in Rock Mechanics where stresses and strains are of dynamic nature – as in the case of earthquakes, rock blasting and rock bursting - and the propagation of these stresses and strains through the rock mass should be studied (Jaeger et al., 2007). In this context the propagation and interaction of elastic waves with interfaces in the rock mass (like joints, interfaces of geological layers etc) are important. When an incident wave is a shear wave whose displacement vector is parallel to the interface then there are produced anti-plane shear or SH waves since for the case of an interface that is horizontal these waves are polarized in the horizontal plane (Jaeger et al., 2007).

In the next we consider *SH* motions in a gradient elastic half-space with surface energy. With respect to a fixed Cartesian coordinate system $Oxyz$, the half-space occupies the region $(-\infty < x < \infty, y \geq 0)$ and is thick enough in the z-direction to allow an anti-plane shear state when the loading acts in the same direction. In this case and assuming additionally a time-harmonic *steady state*, any problem can be described by the displacement field $u_x = u_y = 0$, $u_z \equiv w(x, y, t) = w(x, y) \cdot \exp(-i\omega t) \neq 0$, with $i \equiv (-1)^{1/2}$ is the unit imaginary number and ω being the frequency. In the case of SH waves, the only surviving equations of motion are one written for the total stresses $(\sigma_{xz}, \sigma_{yz})$ that are given by the constitutive Equations 34a and two written for the double stresses $(\mu_{xxz}, \mu_{xyz}, \mu_{yxz}, \mu_{yyz})$ that are given by constitutive Equations 34c. Vardoulakis & Georgiadis (1997) have shown that the field equation for such a state in terms of displacements is

$$\ell^2 \nabla^4 w - g \nabla^2 w - k^2 w = 0, \tag{47}$$

where ∇^2 and ∇^4 are the Laplace and biharmonic operators, $g = 1 - (\omega^2 \hat{I}/G)$, $k = \omega/V$, $V = (G/\rho)^{1/2}$ is the shear wave velocity in the absence of gradient effects, $\hat{I} = (1/3)\rho h^2$ is the micro-inertia coefficient, ρ is the mass density, and h is the half-length of the crystal (*e.g.* Figure 3a). Further, operating with the two-sided Laplace transform on Equation 47 yields an o.d.e. for the transformed displacement $w^*(p, y)$. The general solution of the latter equation that is bounded at infinity is

$$w^*(p, y) = B(p) \cdot \exp(-\beta y) + C(p) \cdot \exp(-\gamma y);$$

$$\beta(p) \equiv \beta = i(p^2 + \sigma^2)^{1/2}, \quad \gamma(p) \equiv \gamma = (\tau^2 - p^2)^{1/2},$$

$$\sigma = \sqrt{\frac{\sqrt{g^2 + 4\ell^2 k^2} - g}{2\ell^2}}, \quad \tau = \sqrt{\frac{\sqrt{g^2 + 4\ell^2 k^2} + g}{2\ell^2}}, \tag{48}$$

where p is the Laplace-transform variable and B, C are obtainable through enforcement of the boundary conditions.

As is well known, the criterion for *surface waves* is that the displacement decays exponentially with distance from the free surface (Achenbach, 1973). Thus, if we consider *plane-wave* solutions of the form $\exp[i(qx - \omega t)]$ with a dispersion relation $\omega = \omega(q)$, a distinct harmonic component of propagation of the SH surface wave satisfying the equations for a grade-2 continuum in the half-space $y \geq 0$ will be expressed as

$$\overline{w}_s(x, y, t) = [B(q) \cdot \exp(-|\beta|y) + C(q) \cdot \exp(-|\gamma|y)] \cdot \exp[iq(x - C_{ph}t)]$$

$$C_{ph} = \frac{\omega}{q}, \tag{49}$$

where C_{ph} is the phase velocity, $q \equiv (p/i)$ is the wave number which should be a real quantity such that $-\infty \langle q \langle -\sigma$ or $\sigma \langle q \langle \infty$ in order for surface waves to exist, $|\beta| = (q^2 - \sigma^2)^{1/2}$ and $|\gamma| = (q^2 + \tau^2)^{1/2}$. Next, the appropriate *dispersion* (or *frequency*) *equation* can be obtained by enforcing the pertinent boundary conditions at the half-space surface. These are zero traction conditions which in the transform domain provide a linear homogeneous system. This has a nontrivial solution if and only if

$$-(\sigma_d^2(q_d^2 - \sigma_d^2)^{1/2} + \tau_d^2(q_d^2 + \tau_d^2)^{1/2}) + m\alpha_d^2 = 0, \sigma_d < |q_d| < \infty;$$

$$q_d = \ell q, \sigma_d = \ell\sigma, \tau_d = \ell\tau, m = \ell'/\ell, \alpha_d^2 = \ell^2\alpha^2 = (g^2 + 4\ell^2 k^2)^{1/2}. \tag{50}$$

Equation 50 constitutes the dispersion relation for surface waves. Since this is an irrational algebraic equation, a single mode of SH waves may exist that is directly related to the parameter m. Another immediate observation is that SH surface waves do exist only when $\ell \neq 0$ and $m > 0$. This finding means that the inclusion of the surface energy strain gradient term ℓ', that expresses an *anisotropy in the microscale*, is necessary for predicting surface SH waves. In order to obtain numerical results for the relation between the phase velocity C_{ph} and the wavenumber q (or, equivalently, the wavelength $\lambda = 2\pi/q$), one has generally to *numerically* solve Equation 50. Here, however, we chose to work in a different manner and obtain some representative *exact* results, which can be obtained for the particular case $\ell = (1/\sqrt{3})\,h$. The latter is equivalent to the relation $\omega_d^2 = \ell^2 k^2$. Then, Equation 50 takes the form

$$-\omega_d^2\sqrt{q_d^2 - \omega_d^2} - \sqrt{q_d^2 + 1} + m(1 + \omega_d^2) = 0, \tag{51}$$

where $\omega_d^2 = 1 - g$. Further, the above irrational equation possesses four roots, three of which are extraneous and, therefore, possess no physical meaning. Also, the appearance of complex roots marks cut-off frequencies. It can be shown that the following root is the only one satisfying the original Equation 51,

$$q_d = \begin{cases} \dfrac{\sqrt{\omega_d^6 - 2\omega_d^4 + 2\omega_d^2 - 1 + m^2(1 + \omega_d^4) - 2m\omega_d^2\sqrt{\omega_d^2 + (m^2 - 1)}}}{|1 - \omega_d^2|}, & \omega_d \neq 1 \\[4mm] \dfrac{1}{2}\dfrac{\sqrt{1 + 4m^4}}{m}, & \omega_d = 1 \end{cases} \tag{52}$$

For high frequencies the first of the Equation 52 assumes the asymptotic expansion $q_d = \omega_d + (1/2)m^2(\omega_d^{-1}) + O(\omega_d^{-2}); \omega_d \to \infty$, whereas the following relations are generally valid for the particular case $\ell = (1/\sqrt{3})h$

$$\frac{C_{ph}}{V} = \frac{\omega_d}{q_d}, \quad \frac{\lambda}{h} = \frac{2\pi}{\sqrt{3}}\frac{1}{q_d} \tag{53}$$

and facilitate the creation of the graphs illustrated in Figure 5. From these curves it can be seen that there is a minimum velocity. We also note that the graphical form of the dispersion relation reminds the one found by Coulson (1958) for surface waves in liquids that possess *surface tension*.

1.9 Rayleigh waves in grade-2 elastic solids

The possibility of a wave traveling along the free surface of an elastic half-space, under conditions of plane stress or plane strain, such that the disturbance is largely confined to the neighborhood of the boundary was first considered by Lord Rayleigh (1887). The classical theory of linear elasticity does not predict any dispersion for these motions;

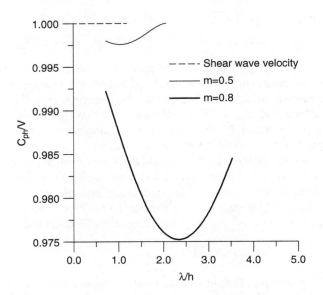

Figure 5 Dispersion curves for the propagation of SH surface waves showing the variation of the normalized phase velocity C_{ph}/V with the normalized wavelength λ/h.

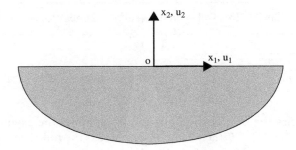

Figure 6 Half-space and coordinates.

only by including *viscoelastic* (Currie et al., 1977) or *thermoelastic* (Georgiadis et al., 1998) effects in the constitutive behavior leads to *dispersive* Rayleigh waves. In order to explain the occurrence of dispersion of Rayleigh waves, Vardoulakis (1981) has considered a graded half-space, that is a material with stiffness increasing with depth. Here, we take another point of view and consider the propagation of Rayleigh waves in a *gradient-elastic, macrohomogeneous and isotropic* half-space $x_2 \geq 0$ (Figure 6) having as an objective examining the possibility of dispersive behavior.

In particular, the theory utilized here can be considered one of the simplest versions of Mindlin's theory containing only the volumetric length scale corresponding to the following strain-energy density function

$$v = (1/2)\lambda\varepsilon_{qq}\varepsilon_{rr} + G\varepsilon_{qr}\varepsilon_{rq} + G\ell^2(\partial_s\varepsilon_{qr})(\partial_s\varepsilon_{rq}) \tag{54}$$

The displacement equation of motion in the absence of body force may be derived in the following manner

$$G\overline{D}^2\nabla^2\mathbf{u} + (\lambda + G\overline{D}^2)\nabla\nabla\cdot\mathbf{u} = \rho\ddot{\mathbf{u}} - \frac{1}{3}\rho h^2\nabla^2\ddot{\mathbf{u}} \quad (55)$$

where we have used the operator $\overline{D}^2 \equiv 1 - \ell^2\nabla^2$ (also known as Schroedinger's operator in quantum mechanics). The boundary conditions for the problem at hand, for $h/\ell \to 0$ and for the two cases of boundary conditions (Case I refers to the approximate and II to the exact boundary conditions, respectively) take the form

$$\sigma_{22} = \sigma_{21} = 0 \ (Case\ I)$$

$$\sigma_{22} - \frac{\partial\mu_{221}}{\partial x_1} = 0, \quad \sigma_{21} - \frac{\partial\mu_{211}}{\partial x_1} = 0 \quad (Case\ II),$$

$$\mu_{222} = \mu_{221} = 0 \ (Case\ I,\ II), \quad -\infty < x_1 < \infty, \quad x_2 = 0 \quad (56)$$

Then we fix the wave numbers, as well as Poisson's ratio, and we construct the equation for the characteristic determinant of the problem at hand as an equation for the dimensionless frequency $\Omega = \omega\ell/c_T$ (Stavropoulou et al., 2003). It is not difficult to verify that for $\ell = 0$, the determinant equation reduces to the classical Rayleigh function whose roots are given by Eringen & Suhubi (1975) for various Poisson's ratios ν. It is also clear that the root of the determinant equation is a function of ℓ, consequently in contrast to the classical theory the Rayleigh wave velocity predicted by the proposed gradient elasticity theory is dispersive. Figure 7 illustrates the relation exhibited between the normalized phase velocity of Rayleigh wave with the normalized frequency in the framework of the present theory for the case of rock with small grain size, *i.e.* h/ℓ<<1.

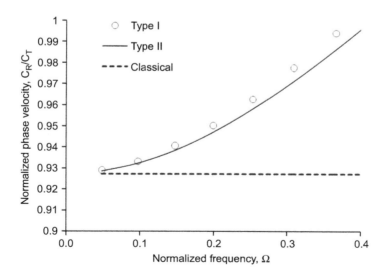

Figure 7 Dispersion curve for the propagation of Rayleigh surface waves showing the variation of the normalized phase velocity with the normalized frequency for h/ℓ tending to zero and $\nu = 0.3$ (Stavropoulou et al., 2003).

716 Exadaktylos

The following conclusions can be drawn from the analysis given above:

- If the volumetric gradient length scale ℓ is small compared to the characteristic wavelength of the Rayleigh wave then the results obtained from gradient and classical elasticity theories coincide.
- For increasing relative frequencies the gradient theory predicts larger Rayleigh wave velocities than the classical theory in a monotonic manner. This property - which is due to the fact that $h/\ell << 1$ – may be used to establish the gradient parameter ℓ through carefully performed Rayleigh wave propagation experiments. This has been demonstrated with the analysis of Rayleigh wave experiments in Pentelikon marble used for the construction of Parthenon monument in Athens (Stavropoulou *et al.*, 2003).
- A new material parameter may be defined as the product $G\ell$ with dimensions of $[FL^{-1}]$, where F denotes force and L denotes length. This new parameter is called 'crack stiffness' and influences the magnitude of mode-I, -II and –III crack deformation under given stress in rocks. It was demonstrated (Stavropoulou *et al.*, 2003) that this parameter may be experimentally determined through carefully performed in situ Rayleigh wave measurements.
- The results obtained by applying the two types of boundary conditions do not differ appreciably in the whole range of normalized frequencies of Rayleigh waves.

1.10 Size effect of the fracture toughness of the pressurized crack

The possible size effect exhibited by hydraulic fractures – *i.e.* the dependence of the resistance of fracture to propagation with increasing crack length – is a very important problem in hydraulic fracturing of rocks. Exadaktylos (1998) has postulated the following criterion for mode-I fracture propagation subjected to constant internal pressure (assuming zero diffusivity of the rock)

$$\Psi\left(\alpha - \eta; \sigma_0, \ell, \ell'\right) \geq \beta, \quad \Psi\left(\alpha - \eta; \ell, \ell'\right) = \frac{\pi\alpha}{8}\frac{\sigma_0^2}{G}\hat{\psi}(\alpha - \eta) = \frac{K_I^2}{8G}\hat{\psi}(\alpha - \eta) \tag{57}$$

where the function Ψ depends on the applied pressure on the crack lips σ_0, and the two strain gradient length scales. The quantity β has the dimensions of specific volume energy or stress $[FL^{-2}]$, that was called 'modulus of cohesion' and is assumed to be a constant material parameter. The symbol η is a small length with respect to the semi-crack length α, in order to remove the weak logarithmic singularity of the function $\hat{\psi}(t)$ at $t = \alpha$; the latter function is given in closed form. This criterion was applied after the solution of the relevant boundary value problem in the frame of grade-2 Casal-Mindlin theory. This solution revealed that the crack shape is no longer elliptical as is predicted by the classical theory but the crack lips take a cusp shape such as shown in Figure 8a.

By setting $K_I = K_{IC}$ in the above criterion of Equation 57 we obtain the following expression for the fracture toughness

$$K_{IC} = \sqrt{\frac{8\beta G}{\hat{\psi}(\alpha - \eta)}} \tag{58}$$

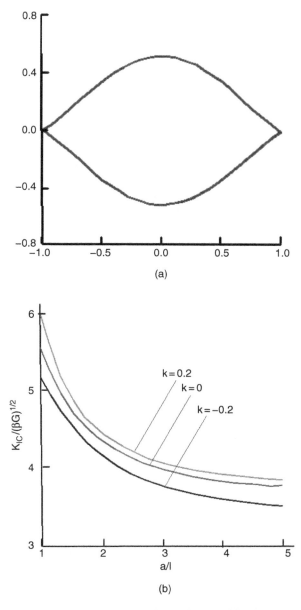

Figure 8 (a) Deformed mode-I crack with tips in the form of cusps of first kind, and (b) size effect of the normalized mode-I fracture toughness K_{IC} for three values of the material length ratio $k = \ell'/\ell$ and for Poisson's ratio of the material $\nu = 1/4$ (Exadaktylos, 1998).

The size-effect of the fracture toughness is demonstrated in Figure 8b for the various values of the relative surface energy parameter $k = \ell'/\ell$. It may be observed that: (a) as the surface energy length scale increases the fracture toughness increases due to the surface tension effect mentioned in Paragraph 1.5 above, and (b) the size effect resembles that of the simple stress-mean-value theory presented in Paragraph 1.1 (*e.g.* Figure 2).

2 A GRADE-2 ENGINEERING BEAM THEORY WITH SURFACE ENERGY

2.1 Introductory remarks

The experimental analysis for mechanical parameters identification like modulus of elasticity and tensile strength of rocks and other brittle structural materials, as well as theoretical models of the deformability and strength of beams, beams-columns and plates, are of great practical interest in many applications in rock and structural engineering. They depict the serviceability and strength of such types of engineering structures. Beam elements occupy a wide range of technological applications and length scales. For example, in Civil Engineering applications beams from timber, steel, concrete, aluminum etc are used as structural elements in buildings and bridges, at the scale of several meters to several tenths of meters. In Monumental Constructions and restoration works one may mention m*arble or limestone beams* in temples resting on marble columns with spans of the order of several meters. In Mining and Tunneling one may encounter artificial span support beams (from timber, cast iron, concrete etc) or beams and plates overhanging above underground openings in mines, tunnels and caverns. For example a beam may be formed by a rock layer at tunnel's roof with one end free (entrance) and the other hinged (tunnel's face). Beams are also encountered in biomechanical applications: *e.g.* micro-cantilever sensors at the scale of $1 \div 10$ μm, and in nanomechanical applications in thin films technology, biosensors and atomic force microscopes at the scale of $10 \div 100$ nm. For this purpose, there is a growing interest of proper theories incorporating additional to the characteristic macroscale also smaller length scales (these are called micromechanical theories and they include discrete and distinct element theories among others).

Here, Timoshenko's engineering beam bending theory of linear elastic materials is extended by considering surface energy effects that have been discussed in Section 1. A beam bending micromechanical theory with surface energy is formulated that is based on a modified strain energy function of a material with microstructure that includes the classical Bernoulli-Euler term, the shape correction length scale ℓ_v introduced by Timoshenko to account for the effect of shear forces, and another extra new length scale ℓ_s introduced here, that is associated with surface energy effects.

2.2 Fundamentals of the technical beam theory

The longitudinal section of the beam is referred to a Cartesian coordinate system O(x,y, z) positioned on the neutral axis – which is the locus of centroids of cross-sections - with its origin at mid-span and with the Ox-axis directed along the neutral axis of the beam while Oz-axis extending vertically downwards. Deformation quantities are assumed as infinitesimal, and the corresponding displacements of points in a cross-section along Ox and Oz directions are denoted by the symbols u, w respectively. Let the infinitesimal normal strains ε_{xx}, ε_{zz} and the engineering shear strain γ_{xz} in the plane xOz to be defined as follows

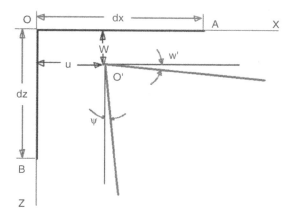

Figure 9 Deformations of vertical and horizontal beam sections.

$$\varepsilon_{xx} = \varepsilon = u_{,x}, \quad \varepsilon_{zz} = w_{,z}, \quad \gamma_{xz} = \gamma = u_{,z} + w_{,x} \tag{59}$$

where ψ denotes the rotation (considered to be a small quantity) of the cross-section A of the beam at position x (Figure 9) and the comma denotes differentiation w.r.t. the variable after the comma. It may be easily shown that the representation of the strain energy density (potential) of the beam in the context of Timoshenko's beam bending theory is given by the following *ansatz*

$$v_T = \frac{1}{2} EI \left(\kappa^2 + \frac{\gamma^2}{\ell_v^2} \right) \tag{60}$$

where the term EI denotes the flexural rigidity or stiffness of the beam, I denotes the moment of inertia of the cross-section A of the beam, ℓ_v stands for a microstructural length scale of the beam material that considers the effect of the transverse shear stress contributing to the deflection $w = w(x)$ of the beam, the symbol κ we denote the gradient of the rotation angle (bending curvature) of the cross-section, that is

$$\kappa \equiv \psi_{,x} \tag{61}$$

In the frame of this technical theory the horizontal strain is simplified as follows

$$\varepsilon = \kappa z \tag{62}$$

Also, the bending curvature $\kappa = 1/R$ is found as $\partial \varepsilon / \partial z = \psi_{,x}$.

The following constitutive relationships for the bending moment and transverse shear force may be deduced from Equation 60

$$M = \frac{\partial v_T}{\partial \kappa} = EI\kappa, \quad Q = \frac{\partial v_T}{\partial \gamma} = EI \frac{\gamma}{\ell_v^2} \tag{63}$$

where the shear forces and bending moments, are denoted as Q, M, respectively, The first of Equations 63 forms the Bernoulli-Euler theorem depicting the analogy of the bending moment with the bending curvature of the beam, while the second is due to

720 Exadaktylos

Timoshenko that considers the effect of the transverse shear forces on the beam deflection. The characteristic length scale ℓ_v is related to the dimensionless quantity ℓ_T in the following manner

$$\ell_v^2 = (\ell_T L)^2 \tag{64}$$

For example for a rectangular cross-section with height H, we get that ℓ_T essentially compares with the inverse of the aperture ratio of the beam, that is to say for a rectangular cross-section of the beam, Timoshenko (1921) found that ℓ_T compares with the inverse of the length to height ratio

$$\ell_T^2 = \frac{1}{5}(1+\nu)\left(\frac{H}{L}\right)^2 << 1 \quad for \quad H < L \tag{65}$$

Accordingly for long prismatic beams ($\ell_T << 1$) or ($H/L << 1$) Bernoulli-Euler elementary beam theory is recovered.

2.3 Formulation of the kappa-gamma beam technical theory

Herein an engineering beam bending theory that has been previously presented by Vardoulakis *et al.* (1998) containing two material length scales and aiming at capturing the size effect exhibited by beams in bending, is reformulated. In fact we change the strain energy density (or elastic potential energy density) *ansatz* for an elastic material with microstructure initially proposed in our previous work, with the following straightforward expression

$$v_E = \frac{1}{2}EI\left(\kappa^2 + \frac{1}{\ell_v^2}\gamma^2 + \frac{2}{\ell_s}\kappa\gamma\right) \tag{66}$$

So, Bernoulli-Euler theory which leads to the proportionality of the bending moment with curvature kappa (κ), is expressed only by the first term, whereas Timoshenko's beam bending theory that explains the effect of shear forces (gamma) on beam deflection and bending curvature of the beam is expressed by the first two terms. The third term in the above strain energy density function has not been obtained arbitrarily, but rather on the simple and straightforward argument, namely that since the curvature and shear strain are already included by Bernoulli-Euler and Timoshenko, respectively, then their product should be also included for completeness of the representation. This argument introduces an additional material length scale ℓ_s. It may be easily shown that the positive-definiteness of the strain energy density is guaranteed if the following inequalities are valid

$$-1 < \frac{\ell_v}{\ell_s} < 1 \tag{67}$$

wherein from Equations 64 & 65

$$\ell_v^2 = (\ell_T L)^2 = \frac{1}{5}(1+\nu)H^2 \tag{68}$$

that is, for positive strain energy density Timoshenko's shape factor ℓ_v must not vanish if surface or scale effects are going to be taken into account. The above *ansatz* described by Equation 66 contains the last term that considers surface energy effects through the microstructural length scale ℓ_S, and also contains as a special case Timoshenko's beam bending theory through the length scale ℓ_v as may be observed from Equation 60. In fact, by applying Gauss' divergence theorem the total elastic strain energy of the beam takes the form

$$
\begin{aligned}
U_E &= \frac{1}{2}EI\int_0^L\left(\kappa^2 + \frac{1}{\ell_v^2}\gamma^2 + \frac{2}{\ell_s}\kappa\gamma\right)dx \\
&= \frac{1}{2}EI\int_0^L\left(\kappa^2 + \frac{1}{\ell_v^2}\gamma^2 + \frac{2}{\ell_s}\kappa w'\right)dx + \frac{1}{2\ell_s}EI\int_0^L \nabla\psi^2 dx \\
&= \frac{1}{2}EI\int_0^L\left(\kappa^2 + \frac{1}{\ell_v^2}\gamma^2 + \frac{2}{\ell_s}\kappa w'\right)dx + \frac{1}{2\ell_s}EI[\psi^2]_0^L
\end{aligned}
\tag{69}
$$

where we have set $(\cdot)' \equiv \nabla(\cdot) \equiv d/dx$. The constitutive equations for the shear force Q and bending moment M, are also easily derived from the potential of Equation 66 as follows

$$
Q \equiv \frac{\partial v_E}{\partial \gamma} = EI\left(\frac{1}{\ell_v^2}\gamma + \frac{1}{\ell_s}\kappa\right) = EI\left(\frac{1}{\ell_v^2}\left[w' + \psi\right] + \frac{1}{\ell_s}\psi'\right)
\tag{70}
$$

$$
M \equiv \frac{\partial v_E}{\partial \kappa} = EI\left(\kappa + \frac{1}{\ell_s}\gamma\right) = EI\left(\psi' + \frac{1}{\ell_s}\left[w' + \psi\right]\right)
\tag{71}
$$

Vardoulakis *et al.* (1998) who studied the size effect exhibited by the flexural strength of marble beams in laboratory tests employed the following ansatz

$$
v_{VE} = \frac{1}{2}EI[\kappa^2 + \ell_v^2(\nabla\kappa)^2 + \ell_s\nabla(\kappa^2)]
\tag{72}
$$

This is a gradient *almost* B-E theory enhanced with two length-scales where the second term accounts for the shear strain effect

$$
\gamma \approx \ell_v^2 \nabla\kappa
\tag{73}
$$

Papargyri-Beskou *et al.* (2003) assumed only one surviving surface length scale $\ell_x = \ell'v_x$ along the axial direction of long beams (very small height to span ratio) according to Casal's theory, and a volumetric length scale denoted in their paper by the symbol g, and made the reasonable assumption of null transverse normal strain $\varepsilon_{zz} = 0$ of the technical beam theory, according to the notation used in the present paper

$$
v_{PB} = \frac{1}{2}EI\left[\left(w''\right)^2 + g^2\left(w'''\right)^2 + 2\ell_x w'' w'''\right], \quad \ell_x << 1
\tag{74}
$$

Our kappa-gamma model for long beams (*i.e.* B-E theory) such that the following approximations to be valid

$$\kappa = \frac{d\psi}{dx} \approx -\frac{d^2w}{dx^2}, \quad \gamma = \frac{dw}{dx} + \psi \approx -\ell_T^2 \frac{d^3w}{dx^3} \tag{75}$$

gives

$$\upsilon_{PB} \approx \frac{1}{2}EI\left[\left(w''\right)^2 + \left(\frac{\ell_T^2}{\ell_\upsilon}\right)^2\left(w'''\right)^2 + \frac{2\ell_T^2}{\ell_s}w''w'''\right] = \upsilon_b, \quad \ell_T^2 \ll 1 \tag{76}$$

that is exactly the same to the elastic potential proposed by Papargyri-Beskou *et al.* (2003).

Later on, Vardoulakis & Giannakopoulos (2006) have proposed the following potential or strain energy density (energy per unit beam length) for the beam

$$\upsilon_{VG} = \frac{1}{2}EI\left(\kappa^2 + \frac{1}{\ell_\upsilon^2}\gamma^2 + \frac{2}{\ell_s}\left(\kappa'\right)^2\right) \tag{77}$$

It may be noted that the first two terms of the kappa-gamma potential have the same form with those appearing in the ansatz given by Equation 77, although they differ in the last term.

2.4 Generalization of the gradient beam theory with surface energy

A general expression of the elastic strain energy density of a gradient elastic solid with surface energy with two additional length scales has been given by Equation 17. Applying the following simplifications

$$\varepsilon_{xx} = \kappa z, \quad \varepsilon_{xz} = \frac{1}{2}\gamma \tag{78}$$

and elaborating on the expressions, the final expression of the strain energy density for the beam is composed from three distinct parts. Firstly, the classical part of the elastic potential energy may be found in the following manner,

$$\upsilon^{clas} = \iint_A \left[\left(\frac{1}{2}[\lambda + 2G]z^2\right)\kappa^2 + \frac{1}{2}G\gamma^2\right]dA = \frac{1}{2}[\lambda + 2G]I\kappa^2 + \frac{1}{2}G\gamma^2 A,$$

$$I = \iint_A z^2 dA \tag{79}$$

This expression is composed from two terms appearing also in the elastic potential energy Equations 66, 72, 76 and 77. Secondly, one may find a volumetric part that is associated with the volumetric length scale which does not give a scale effect, namely

$$\upsilon^{vol-grad} = G\ell^2 \iint_A \left[z^2\left(\frac{\partial\kappa}{\partial x}\right)^2 + \frac{1}{2}\left(w_{,xx} + \kappa\right)^2 + \kappa^2\right]dA \Leftrightarrow$$

$$\upsilon^{vol-grad} = G\ell^2\left[I\left(\frac{\partial\kappa}{\partial x}\right)^2 + \frac{1}{2}A\nabla\gamma\nabla\gamma + A\kappa^2\right] \tag{80}$$

Linear elasticity with microstructure and size effects 723

One may note that the first term of the last expression is the same with the thrid term considered by Vardoulakis & Giannakopoulos (2006) in Equation 77, and that is why their model does not predict a scale effect. Finally, it may be found a surface energy part associated with the only surviving surface energy length scale $\ell_x = \ell' \nu_x$ that gives a size effect, *i.e.*

$$
\begin{aligned}
v^{surf-grad} &= 2G\ell' \iint_A \left[z^2 \kappa \frac{\partial \kappa}{\partial x} + \frac{1}{2}(\kappa + w_{,xx})\gamma \right] dA \Leftrightarrow \\
v^{surf-grad} &= 2G\ell' \left[I\kappa \frac{\partial \kappa}{\partial x} + \frac{A}{2}\kappa\gamma + \frac{A}{2}w_{,xx}\gamma \right]
\end{aligned}
\tag{81}
$$

The first term of the expression above $\kappa\nabla\kappa$ has been adopted by Vardoulakis *et al.* (1998) *i.e.* the third term in Equation 72, while the second term of the expression above is the third term of the kappa-gamma beam theory. In Vardoulakis *et al.* (1998) and in the present publication it is demonstrated that both technical theories are capable to predict scale effects of beams.

The transverse shear force and bending moment expressions resulting from the 3D gradient theory may be formally derived in the following manner and are equivalent with Equations 70 and 71, respectively,

$$
Q = \frac{\partial v^{2ndgr}}{\partial \gamma} = G\left(\gamma + \ell'\kappa + \ell' w_{,xx}\right) \iint_A dA,
\tag{82}
$$

$$
\begin{aligned}
M = \frac{\partial v^{2ndgr}}{\partial \kappa} &= 2[\lambda + 2G]\kappa \iint_A z^2 dA + G\ell'\gamma \iint_A dA + \\
&+ G\ell^2 \left(\frac{\partial \gamma}{\partial x} + 2\kappa\right) \iint_A dA + +2G\ell' \frac{\partial \kappa}{\partial x} \iint_A z^2 dA
\end{aligned}
\tag{83}
$$

By comparing the above two sets of relationships 82 & 70 it may be observed that the kappa-gamma theory does not contain the kinematical term w_{xx} in the expression for the transverse shear force, and the terms $\partial\gamma/\partial x, \partial\kappa/\partial x$ in the expression for the bending moment (*e.g.* compare Equations 83 & 71).

2.5 Closed-form solution of 3PB simply supported beam

Next we proceed with the solution of the 3PB configuration employing the simpler engineering beam theory. It may be shown that the equilibrium equations for the beam

$$
\frac{dQ}{dx} = 0, \quad -Q + \frac{dM}{dx} = 0
\tag{84}
$$

are automatically satisfied, with the following expressions for the bending moment and transverse shear force along the beam subjected to concentrated loading P at its midspan (natural boundary condition

$$
M = \frac{PL}{4}\left(1 - \frac{2x}{L}\right), \quad Q = -\frac{P}{2} \quad 0 \le x \le \frac{L}{2}
\tag{85}
$$

724   Exadaktylos

Substituting the values of Q, M given by the above Equations 85 into Equations 84 the following system of linear odes is obtained

$$\left.\begin{array}{l} \ell_r(w' + \psi) + \psi' = -\lambda, \\ \ell_s\psi' + w' + \psi = \dfrac{\lambda L}{4}\left(1 - 2\dfrac{x}{L}\right), \end{array}\right\} \qquad 0 \le x \le \dfrac{L}{2} \tag{86}$$

wherein we have set the following normalized variables with units $[L^{-1}]$

$$\lambda = \frac{P\ell_s}{EI}, \quad \ell_r = \frac{\ell_s}{\ell_v^2} \tag{87}$$

The closed form solution of the above system of ode's has as follows

$$\psi = \frac{1}{4(1 - \ell_r\ell_s)}[\lambda(L + 4\ell_s) + 4C_2(\ell_r\ell_s - 1) - \lambda(4 + \ell_r L)x + \lambda\ell_r x^2],$$

$$w = \frac{1}{24(\ell_r\ell_s - 1)}[24C_1(\ell_r\ell_s - 1) + 24C_2(\ell_r\ell_s - 1)x - 3\lambda(2 + \ell_r L)x^2 + 2\lambda\ell_r x^3]$$

$$0 \le x \le \frac{L}{2} \tag{88}$$

in which C_1, C_2 are integration constants to be found from appropriate boundary conditions. The *essential boundary conditions* refer to the vertical displacement at the supported end of the beam, as well as the rotation at the mid-span; both of them should vanish, *i.e.*

$$w = 0, \quad x = \frac{L}{2}$$
$$\psi = 0, \quad x = 0 \tag{89}$$

Substituting the deflection from Equation 88_2 and the rotation from Equation 88_1 into the two kinematical conditions described by Equations 89 we may easily obtain the expressions for the two constants in the following manner

$$C_1 = \frac{\lambda L^2}{48}\frac{\left(L\ell_r + 12\dfrac{\ell_s}{L} + 6\right)}{\eta^2 - 1}, \quad C_2 = 0 \tag{90}$$

where we have set the dimensionless parameter

$$\eta^2 = \ell_r\ell_s = \left(\frac{\ell_s}{\ell_v}\right)^2, \quad \eta^2 > 1 \tag{91}$$

Subsequently, the expression for the deflection may be found from Equation 88_2 and the above expressions for the constants, that is to say

$$w = \frac{w_c}{\eta^2 - 1}\left\{6\hat{\ell}_s\left(1 + 2\hat{\ell}_s\right) + \eta^2 - 12\hat{\ell}_s\left(1 + 2\hat{\ell}_s\right)\xi - 6\eta^2\xi^2 + 4\eta^2\xi^3\right\},$$

$$0 \le \xi \le \frac{1}{2} \tag{92}$$

Linear elasticity with microstructure and size effects 725

where we have used the following dimensionless quantity

$$w_c = \frac{PL^3}{48EI} \tag{93}$$

w_c represents the maximum (*i.e.* mid-span) deflection derived from *Bernoulli-Euler* beam theory. Also, the rotation of the initially vertical cross-section of the beam could be found from Equations 85_1 and 90, *i.e.*

$$\psi = \frac{\bar{\lambda}}{\eta^2 - 1} \xi \left\{ 2\hat{\ell}_s + \eta^2 - \eta^2 \xi \right\}, \quad 0 \le \xi \le \frac{1}{2} \tag{94}$$

wherein

$$\bar{\lambda} = \frac{PL^2}{4EI}, \quad \hat{\ell}_s = \frac{\ell_s}{L} \tag{95}$$

The engineering shearing strain could be also found in the following manner

$$\gamma = w' + \psi = \frac{\bar{\lambda}}{\eta^2 - 1} \left\{ -\hat{\ell}_s [1 + 2\hat{\ell}_s] + 2\hat{\ell}_s \xi \right\}, \qquad 0 \le \xi \le \frac{1}{2} \tag{96}$$

In addition, the bending curvature of the beam may be found by formal differentiation of Equation 94 as follows

$$\kappa = \frac{\bar{\lambda}}{L} \frac{1}{\eta^2 - 1} \left\{ 2\hat{\ell}_s + \eta^2 - 2\eta^2 \xi \right\}, \quad 0 \le \xi \le \frac{1}{2} \tag{97}$$

As, it may be seen from Equation 97, in contrast to classical theory, the present gradient theory with surface energy predicts always for any value of η^2 a finite and larger value of the beam curvature at its supporting ends (*i.e.* for $\xi = 1/2$). This is due to the presence of the surface energy term $2\hat{\ell}_s$ in the expression for the curvature that also is responsible for the inequality $\kappa \ne - \partial^2 w / \partial x^2$.

2.6 Numerical results

Various beam deflection curves obtained from the theory are shown in Figures 10 a÷c. For this purpose use was made of Equation 92 and of the following expression

$$\eta^2 = \left(\frac{\ell_s}{\ell_v} \right)^2 = \frac{5\hat{\ell}_s^2}{(1 + \nu)\left(\frac{H}{L} \right)^2} \quad , \quad \eta^2 > 1 \tag{98}$$

As was expected the gradient theory, *i.e.* for the relative length scale η comparable to unity, predicts always larger beam deflections compared to the classical B-E theory for $\eta \gg 1$. This is clearly illustrated in Figure 10a. According to Equation 98 the latter case is approached for beams of very large span (L) compared to their height (H) and vanishing surface energy length scale. The effect of Poisson's ratio on beam deflection for vanishing surface energy term and H/L=1/10 is displayed in Figure 10b. Finally, the effect of the surface energy length on the beam's deflection for constant H/L=1/10 and Poisson's ratio of 0.3 is shown in Figure 10c. It is clear from Figure 10c that as the relative surface energy length increases, the beam deflection decreases w.r.t. that predicted by Timoshenko's beam theory, which is an indication of a "beam rigidity effect". This effect is attributed to the surface energy term that as in the case of the half-space problem treated in Paragraph 1.5 gives rise to a pre-tensioning of the beam.

2.7 Size effect of beam strength

Assuming that the *Poncelet - Saint Venant (PSV) failure hypothesis* is valid for granular brittle materials, then the fracture of the beam will occur when the horizontal extension strain at the mid-span of the bottom face of the beam denoted here as ε_{xx}^{max}, reaches the limit strain ε_f

$$\varepsilon_{xx}^{max} \geq \varepsilon_f \quad at \quad \xi = 0, \quad z = \frac{H}{2} \tag{99}$$

where H is the height of the beam. Considering that $\varepsilon_{xx} = \kappa z$, then substituting the value of the bending curvature at mid-span found by Equation 97 and multiplying with the modulus of elasticity E, the failure stress at the lower fiber of the beam is found as follows

$$\sigma_{bu} = E\varepsilon_f = E\frac{H}{2}\kappa(0) = \sigma_{bu}^{B-E} \frac{\eta^2}{\eta^2 - 1}\left\{1 + \frac{2\ell_s}{\eta^2}\frac{1}{L}\right\} \tag{100}$$

wherein σ_{bu}^{B-E} denotes the well-known quantity of Modulus of Rupture of the beam that is given by the formula of the classical Bernoulli-Euler beam bending theory by assuming again the validity of the PSV *failure hypothesis*

$$\sigma_{bu}^{B-E} = E\varepsilon_{xx}^{max} = \frac{P_f LH}{8I} \tag{101}$$

In the formula above, P_f denotes the value of the concentrated load at failure. For constant beam aperture ratio L/H, the following three observations could be made from Equation 100, *i.e.*: (i) Timoshenko's theory does not predict a size effect and simply modifies the modulus of rupture, (ii) the extended beam bending theory accounting for surface effects, predicts a (−1)- power of the beam length dependence of the flexural strength of the beam, and (iii) this size effect law resembles Karmarsch's empirical law also used later by Griffith (*i.e.* Equation 1).

The above size effect law was investigated with a series of 3PB experiments with prismatic marble beams of a square cross-section (*i.e.* B=H) of Dionysos marble of the same aperture $L/H \cong 4$ but with various spans L ranging from 7.4 cm up to 1 m. This range is considered to be significant for standard rock mechanics tests. Strains at various locations on the beams including their lower surface at the mid-span, were recorded by virtue of electrical strain-gages. More data referring to 3PB experiments on Dionysos marble are provided in Exadaktylos *et al.* (2001b). As is illustrated in Figure 11, the rupture strength calculated according to Equation 101, apart from some dispersion of results for a given aperture ratio that is expected for crystalline brittle

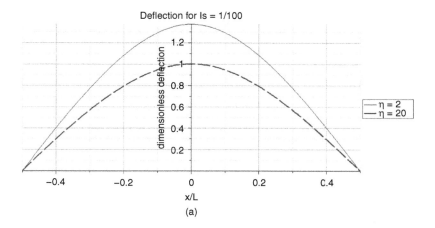

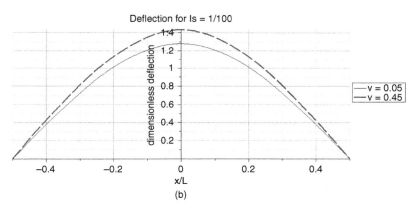

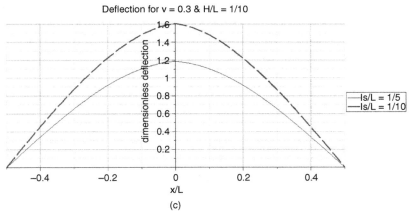

Figure 10 Distribution of beam deflection in 3PB; (a) effect of (H/L) on beam deflection curve for constant Poisson's ratio and surface energy term (i.e. H/L=1/10 for the continuous line and H/L=1/1000 for the dashed line), (b) effect of Poisson's ratio on beam deflection curve for constant (H/L=1/10) and surface energy term, and (c) effect of surface energy term on beam deflection curve for constant (H/L) and Poisson's ratio.

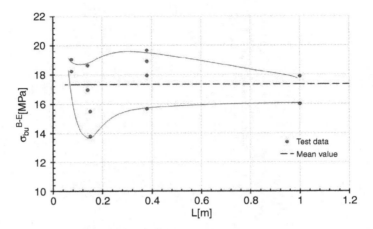

Figure 11 Variation of the modulus of rupture of Dionysos marble for various beam lengths for constant beam aspect ratio $L/H \cong 4$.

materials, was found to be independent of the length of the beam with an average value of $\sigma_{bu}^{B-E} = 17.4\ MPa$.

Then using Equation 97 and with an elastic modulus of Dionysos marble $E = 85\ GPa$ (Exadaktylos *et al.*, 2001b), the best-fitted curve on the experimental data assuming the validity of the inverse length of the beam size effect law, was found to have the following form

$$\sigma_{bu} \cong 17.4 + \frac{0.84}{L/m}, \quad \sum_{1}^{11}(\sigma_a - \sigma_m)^2 = 0.21041 \tag{102}$$

in which the length of the beam is expressed in m and σ_{bu} in MPa, and the sum in the right indicates the sum of squared differences between the "actual" data (subscript "a") and the "model" predictions (subscript "m").

From Figure 12 it may be seen that the above size effect law fits well the experimental results apart from some overestimation of the relative strength in particular of one of the two tests at L=1 m that gives $\sigma_{bu}/\sigma_{bu}^{B-E} \cong 0.7$. This may be attributed to the lower value of $\sigma_{bu}^{B-E} = 16\ MPa$ found in this test compared to the mean value of $\sigma_{bu} = 17.4\ MPa$ assumed for the whole size range. However, even this correction gives $\sigma_{bu}/\sigma_{bu}^{B-E} \cong 0.8$, which is much lower than the predicted value of $\sigma_{bu}/\sigma_{bu}^{B-E} \cong 1.0$.

3 FORMATION OF AXIAL SPLITTING CRACKS IN A DEEP ROCK LAYER

3.1 Introduction

Axial splitting phenomena in rocks, *i.e.* tensile fractures, also called joints, which run parallel to the major compression axis, are important in mining and petroleum

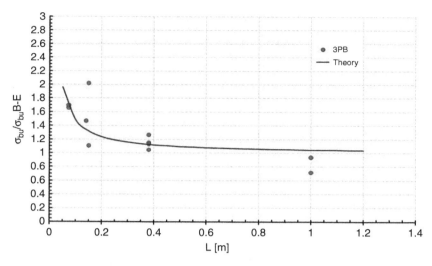

Figure 12 Size effect exhibited by the modulus of rupture of Dionysos marble for constant beam aspect ratio $L/H = 4$.

engineering practices. For example one may mention that deep underground mining results occasionally to explosive "rock bursts" at stope faces in the form of longwall or room and pillar etc. In geological setting, on the other hand, limestone deposits embedded between thin shale layers are characterized by periodic axial splitting, the spatial frequency or spacing of which is very important for permeability estimates. From joint mapping in the field (Bock, 1971, 1980) there are evidences that joints in a geological layer display some kind of periodicity. These layers are transected usually by two main (*haupt*) joint sets that are mutually orthogonal to each other and with spacings exhibiting periodicity. Depending on the case the ratio of the spacing of joints, S, to the thickness of the layer, T, is constant, which means that these two geometrical quantities obey a certain relationship. This ratio S/T varies in most of the cases around the value of two (Bock, 1971, 1980).

In this chapter we consider this problem using two approaches. One refers to the LEFM, and the other refers to the application of bifurcation theory to internal buckling of geological layers under initial stress (Biot, 1965; Vardoulakis & Sulem, 1995). The latter approach is based on the assumption that the critical buckling stress of a continuous medium is that which causes a radical change of the deformational field without a change of the boundary conditions. It is assumed that brittle fracture is affected by strain gradients. The corresponding bifurcation problem is formulated and solved numerically for a rock layer with anisotropic macrostructure and microstructure.

3.2 LEFM model of axial splitting joints in an isolated rock layer

It is assumed that a deep rock layer is uniaxially compressed under the action of in situ vertical stress σ_V as is illustrated in Figure 13. If the layer behaves in a linear elastic fashion and is situated far from free surfaces (like mountain slopes, workings, caverns,

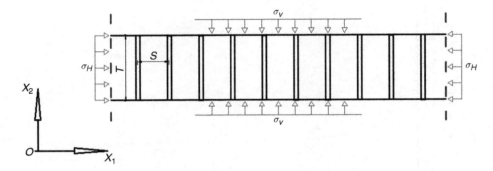

Figure 13 Sketch of a system of parallel periodic axial joints in a horizontal rock layer and Cartesian system of coordinates.

holes etc) deformations in the horizontal directions cannot be realized; therefore the deformation is a constrained uniaxial compression with zero lateral strain *i.e.*

$$\varepsilon_H \approx 0; \quad \sigma_H = K\sigma_V, \quad K = \frac{\nu}{1-\nu} \tag{103}$$

where K denotes the lateral stress ratio.

Based on micromechanical experimental evidences it may be said that in polycrystalline or granular materials like rocks the nonhydrostatic compressive loads generate locally tensile stresses. These local tensile stresses arise from material property mismatches and grain boundary irregularities (Tapponnier & Brace, 1976). In turn, these tensile stresses cause the initiation and propagation of mode-I cracks that are aligned with the major compressive principal stress *i.e.* along Ox_2 axis as is shown in Figure 13 (compressive stresses are considered positive quantities unless stated otherwise). In the configuration of the rock bed subjected to geostatic stresses the mean stress may be found as follows

$$p = \frac{1+\nu}{1-\nu}\frac{\sigma_V}{3} \tag{104}$$

Then it may be shown that the principal deviatoric stresses along the horizontal Ox_1 and the vertical Ox_2 axes are given by the following formulae, respectively

$$s_1 = -\frac{1}{3}\frac{1-2\nu}{1-\nu}\sigma_V, \quad s_2 = \frac{2}{3}\frac{1-2\nu}{1-\nu}\sigma_V \tag{105}$$

It may be observed that the horizontal deviatoric stress is tensile, while the vertical deviatoric stress is compressive, which could explain the alignment of the axial splitting cracks along Ox_2-axis based on the above consideration of local stress concentrations at grain scale.

As is shown in Figure 14a, for the periodic parallel crack problem the mode-I SIF K_I is assumed to be the superposition of the tensile deviatoric stress s_1 properly amplified and of the all-around uniform compression p in the following fashion (from now on we consider tensile stresses and extensional strains as positive quantities)

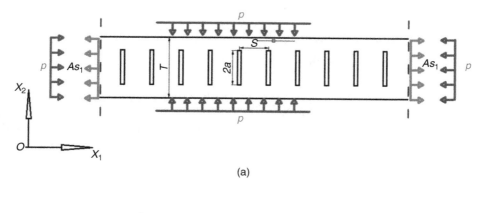

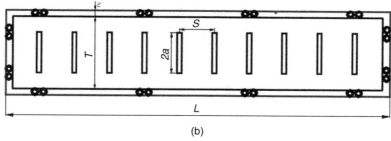

Figure 14 Plane strain model of the infinite layer weakened by parallel periodic axial splitting cracks; (a) Array of periodic and parallel cracks in an elastic layer, and (b) array of parallel joints in a layer of length L with fixed boundaries.

$$K_I = Y\left(\frac{S}{T}, \frac{2a}{T}\right)[A \cdot s_1 + p]\sqrt{\pi a} \tag{106}$$

where Y is a configuration correction factor that is a function of the spacing-to-thickness and crack length-to-thickness ratios, and A is an amplification factor between the local tensile stress and the applied deviatoric stress s_1 (Costin, 1983) that is assumed to be a constant in this model. The SIF due to the deviatoric stress s_2 that acts in a direction parallel with the cracks is obviously null. Chen (2004) has solved the stress boundary value problem of an infinite strip weakened by an array of periodic parallel cracks and has presented numerical values of the mode-I SIF acting on the crack tips of the axial cracks for various values of the ratios S/T and $2a/T$. The regression analysis of his results performed herein has been accomplished by using the following interpolation function for the configuration correction factor Y

$$Y\left(\frac{S}{T}, \frac{2a}{T}\right) = x_1\left(\frac{2a}{T}\right)^2 + x_2\left(\frac{2a}{T}\right) + x_3\left(\frac{2a}{T}\right)\left(\frac{S}{T}\right) + x_4\left(\frac{S}{T}\right) + x_5\left(\frac{S}{T}\right)^2 + x_6 \tag{107}$$

The regression analysis showed that the following values of the constant coefficients result in a mean error of 3% in the range of values of the two ratios also used to find the numerical values by Chen, *i.e.* $S/T \in [0.4, 2]$ and $2a/T \in [0.1, 0.8]$,

$$
\begin{aligned}
x_1 &= 2.4369, \quad x_2 = -1.4272, \quad x_3 = 0.1497, \\
x_4 &= 0.5390, \quad x_5 = -0.2058, \quad x_6 = 0.8533
\end{aligned}
\tag{108}
$$

The above solution is not representative for the bed that lies at great depth below the free surface. Instead the most appropriate boundary conditions for the bed are fixed displacements as is shown in Figure 14b. The compliance of the cracked layer with clamped boundaries is depicted partly from the cracks and partly from the intact rock,

$$
\varepsilon = C\sigma + \frac{1}{E}\sigma, \quad \sigma = As_1 + p
\tag{109}
$$

where C is the compliance of the cracked bed. The additional strain energy due to N cracks is given by the formula below

$$
\Delta U_a = \frac{TL}{2}\sigma\varepsilon = \frac{TL}{2}C\sigma^2
\tag{110}
$$

Irwin (1957) has proved the following relationship that is valid under fixed grips, plane strain conditions and mode-I cracks

$$
\frac{\partial \Delta U_a}{\partial a} = \frac{(1-v^2)}{E}K_I^2
\tag{111}
$$

By virtue of Equations 110 and 111 and integrating we get the expression for the compliance of the elastic layer due to $N = \frac{L}{s}$ cracks

$$
C = \frac{2\pi(1-v^2)N}{TLE}\left(-A\frac{1}{3}\frac{1-2v}{1-v}+\frac{1}{3}\frac{1+v}{1-v}\right)^2 \int_0^a aY^2 da
\tag{112}
$$

Subsequently from Equation 109 the stress-strain relationship may be derived as follows

$$
\sigma = \frac{E\varepsilon}{1+\dfrac{2\pi(1-v^2)N}{TL}\left(-A\dfrac{1}{3}\dfrac{1-2v}{1-v}+\dfrac{1}{3}\dfrac{1+v}{1-v}\right)^2 \displaystyle\int_0^a aY^2 da}
\tag{113}
$$

The definite integral appearing in the above expression may be easily computed in closed form from the polynomial Equation 109 of the configuration correction factor. Finally, the SIF may be derived from Equations 106 and 113 in the following manner

$$
\begin{aligned}
K_I = Y\sqrt{\pi a} \times \\
\times \frac{E\varepsilon}{1+\dfrac{2\pi(1-v^2)N}{TL}\left(-A\dfrac{1}{3}\dfrac{1-2v}{1-v}+\dfrac{1}{3}\dfrac{1+v}{1-v}\right)^2 \displaystyle\int_0^a aY^2 da}
\end{aligned}
\tag{114}
$$

A dimensionless SIF may be defined by setting $\sigma_0 = E\varepsilon$ in the following fashion $K_I/\sigma_0\sqrt{\pi T}$. Using Equation 110 the elastic strain energy of the cracked geological layer with N joints of length $2a$ under certain macro-strain ε becomes

$$U_a = \frac{1}{2}\frac{ELT\varepsilon^2}{1 + \frac{2\pi(1-\nu^2)N}{TL}\left(-A\frac{1}{3}\frac{1-2\nu}{1-\nu} + \frac{1}{3}\frac{1+\nu}{1-\nu}\right)^2\int_0^a \alpha Y^2 da} \tag{115}$$

The determination of the equilibrium crack length in the elastic bed with fixed displacement is then based on the following three criteria proposed by Kemeny & Cook (1985),

$$K_I = K_{IC},$$

$$\frac{\partial K_I}{\partial a} < 0, \tag{116}$$

$$\min U_a$$

The third ad hoc criterion postulated by Kemeny & Cook (1985) and is shown in Equation 116 means that for a given strain ε applied to the rock layer, rock parameters ν, A, K_{IC} and bed thickness T, the optimum configuration will be such that minimizes the stored elastic strain energy U_a. Figure 15 illustrates the variation of the dimensionless SIF with the crack semi-length to bed thickness ratio for various values of crack spacing to bed thickness ratios and for a constant Poisson's ratio $\nu = 1/3$ and amplification factor $A = 20$. As was expected the SIF under fixed-grips conditions is initially increasing with crack length, reaches a peak and then decreases monotonically since the stress is released due to increasing crack length. As it may be seen from Figure 15 below as the number N of cracks increases – for fixed bed length L this means a decreasing S/T ratio – the equilibrium crack length decreases. According to the second crack propagation criterion the equilibrium crack length is found by the intersection of the respective curve with the fracture toughness line that is parallel to the horizontal axis. Employing the third criterion of the minimization of the stored strain energy then as may be seen in Table 1. it turns out that the optimum configuration is established for $S/T \cong 2.2$ as is also observed in reality and crack length comparable to bed thickness. It is remarked here that this model does not predict a size effect, that is to say dependence of the critical strain or stress on the bed thickness, for fixed S/T ratio.

3.3 Internal buckling of a single layer of rock under initial stress

The same problem is considered here as a non-homogeneous, plane-strain deformation of a horizontal layer of thickness T and very large horizontal extent, due to constant vertical compression σ_v as shown in Figure 13. The theory used in this alternative analysis is based on incremental plane strain deformations superimposed on the large strain of a uniform compression. For the considered non-homogeneous deformation mode, we seek such a displacement field that displays certain periodicity along Ox_1 and Ox_2 axes. An appropriate periodical deformation field would be such that: a) along the vertical axis Ox_2 the joints open, b) the deformations attenuate until the middle of the

Table 1 Dependence of the dimensionless elastic strain energy for the various values of crack spacing to bed thickness and crack length to bed thickness ratios.

S/T	$2a/T$	$U_a/E\varepsilon^2 T^2$
0.4	0.15	0.918
1	0.36	1.326
2	0.82	0.972
2.2	0.92	0.876

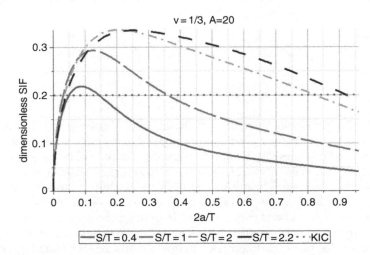

Figure 15 Dependence of the dimensionless SIF on the crack length to bed thickness ratio for various number of cracks or crack spacing to bed thickness ratio for $\nu = 0.3$, A=20 and $K_{IC}/\sigma_0\sqrt{\pi T} = 0.2$.

distance to the neighboring cracks, c) it corresponds to the locations of the joints, and d) it is given in terms of two unknown amplitude functions of the dimensionless coordinate x_2.

The sine and cosine functions are the most appropriate to describe the deformational field prescribed above (Biot, 1965), that is essentially the deformational pattern of a "standing wave". Hence the following expressions for the displacement components are employed

$$V_1 = \Delta u_1 = A\sin(a_n x_1)\cos(b_m x_2) \\ V_2 = \Delta u_1 = B\cos(a_n x_1)\sin(b_m x_2)$$ (117)

where we have set

$$a_n = \pi\frac{n}{S}, \quad b_m = \pi\frac{m}{T}$$ (118)

with *n, m being even natural numbers*, and the dimensionless coefficients A, B denoting the displacement amplitudes. Figures 16a illustrate the deformation modes of the geological layer weakened by periodic parallel axial splitting joints for the cases *n=1*,

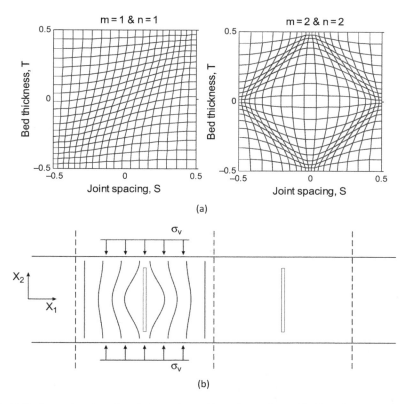

Figure 16 (a) Deformational modes of the geological layer for n=1, m=1 (left) and n=2, m=2 (right), and (b) lines of equal horizontal strain around the joint.

$m=1$ and για $n=2$, $m=2$. Figure 16b shows the deformation field around each vertical axial joint.

It is assumed that the elasticity of the geological layer displays a cubic symmetry that is described with three elasticity constants instead of the usual two constants of isotropic elasticity (Landau & Lifshitz, 1975). Cubic materials posses a shear modulus denoted here with the symbol G that is not related to the Young's modulus and Poisson's ratio with the usual relation of isotropic elasticity. The ratio G/G^* is used here as a measure of anisotropy of the geological material, i.e.

$$\frac{2(1+v)}{E}G = \frac{G}{G^*} = \xi_2 \qquad (119)$$

This anisotropy of the macrostructure is essential for the modeling of the axial splitting fracture of the layer, as is the deviatoric stresses with large enough amplification factor A assumed in the frame of the LEFM approach presented previously. Assuming infinitesimal strain elasticity, the Jaumann stress increments $\Delta^\circ \sigma_{ij}$ of the total stress are related directly to the strain increments $\Delta\varepsilon_{ij}$ through the constitutive relations of linear elastic materials, perturbated properly in order to account for higher order strain gradients and anisotropy in the microstructure (Exadaktylos & Vardoulakis, 1998)

736 Exadaktylos

$$\overset{\circ}{\Delta}\sigma_{11} = \frac{2G^*}{1-2v}\{(1-v)\Delta\varepsilon_{11} + v\Delta\varepsilon_{22}\} - 2G\ell^2\nabla^2\Delta\varepsilon_{11}$$

$$\overset{\circ}{\Delta}\sigma_{22} = \frac{2G^*}{1-2v}\{v\Delta\varepsilon_{11} + (1-v)\Delta\varepsilon_{22}\} - 2G\ell^2\nabla^2\Delta\varepsilon_{22}$$

$$\overset{\circ}{\Delta}\sigma_{12} = \overset{\circ}{\Delta}\sigma_{21} = 2G\left(\Delta\varepsilon_{12} - \ell^2\nabla^2\Delta\varepsilon_{12}\right) \tag{120}$$

In this first attempt we simplify considerably the problem at hand by assuming that the strain gradients affect only the horizontal stress increment in the following manner,

$$\overset{\circ}{\Delta}\sigma_{11} = \frac{2G^*}{1-2v}\{(1-v)\Delta\varepsilon_{11} + v\Delta\varepsilon_{22}\} - G\ell^2\Delta\varepsilon_{11,11}$$

$$\overset{\circ}{\Delta}\sigma_{22} = \frac{2G^*}{1-2v}\{v\Delta\varepsilon_{11} + (1-v)\Delta\varepsilon_{22}\}$$

$$\overset{\circ}{\Delta}\sigma_{12} = \overset{\circ}{\Delta}\sigma_{21} = 2G\Delta\varepsilon_{12} \tag{121}$$

where ℓ is an internal length scale that is used for the consideration of the strain gradient only in the horizontal component of stress, and $\Delta\varepsilon_{ij}$ designates the incremental infinitesimal strain tensor

$$\Delta\varepsilon_{ij} = \frac{1}{2}\left(\Delta u_{i,j} + \Delta u_{j,i}\right) \tag{122}$$

Considering that the layer of infinite lateral extent has fixed upper and lower boundaries (internal buckling problem) while the horizontal displacements along the cracks cancel out, plus the symmetry conditions, then the boundary conditions of the internal buckling problem are imposed in the following fashion,

$$V_1 = 0, \quad \Delta\varepsilon_{12} = 0 \quad \forall x_2 \text{ and } x_1 = \pm l/2$$

$$V_2 = 0, \quad \Delta\varepsilon_{12} = 0 \quad \forall x_1 \text{ and } x_2 = \pm h/2 \tag{123}$$

The model given be Equation 117 satisfies the boundary conditions for n=2 and m=2,

$$V_1(\pm l/2, x_2) = A\sin(\pm\pi)\cos(b_m x_2) = 0$$

$$\Delta\varepsilon_{12}(\pm l/2, x_2) = -\frac{1}{2}(Ba_n + Ab_m)\sin(\pm\pi)\sin(b_m x_2) = 0 \tag{124}$$

and

$$u_2(x_1, \pm h/2) = B\cos(a_n x_1)\sin(\pm\pi) = 0$$

$$\varepsilon_{12}(x_1, \pm h/2) = -\frac{1}{2}(Ba_n + Ab_m)\sin(a_n x_1)\sin(\pm\pi) = 0 \tag{125}$$

For continuing linear equilibrium in plane strain conditions (*i.e.* $\partial_3 = 0$) and in the coordinate system of principal axes of initial stress σ_{ij} in the plane of the deformation, the stress equilibrium equations take the following form (Biot, 1963)

$$\overset{\circ}{\Delta}\sigma_{11,1} + \overset{\circ}{\Delta}\sigma_{12,2} + \sigma_v\Delta\omega_{21,2} = 0$$

$$\overset{\circ}{\Delta}\sigma_{21,1} + \overset{\circ}{\Delta}\sigma_{22,2} + \sigma_v\Delta\omega_{21,1} = 0 \tag{126}$$

Linear elasticity with microstructure and size effects   737

where $\Delta\dot{\omega}$ being the incremental rotation (spin) tensor

$$\Delta\omega_{ij} = \frac{1}{2}\left(\Delta u_{i,j} - \Delta u_{j,i}\right) \tag{127}$$

Substituting Equations 121 & 117 in the equilibrium Equations 126 the following equations are obtained

$$\begin{aligned}
&C_{11}\Delta\varepsilon_{11,1} + C_{12}\Delta\varepsilon_{22,1} - G\ell^2\Delta\varepsilon_{11,111} + 2G\Delta\varepsilon_{12,2} + \sigma_v\Delta\omega_{,2} = 0 \\
&2G\Delta\varepsilon_{12,1} + C_{21}\Delta\varepsilon_{11,2} + C_{22}\Delta\varepsilon_{22,2} + \sigma_v\Delta\omega_{,1} = 0
\end{aligned} \tag{128}$$

where we have set

$$C_{11} = C_{22} = \frac{2G^*(1-v)}{1-2v}, \quad C_{12} = C_{21} = \frac{2G^*v}{1-2v} \tag{129}$$

Finally, by employing the strain-displacement and the rotation-displacement Equations 122 and 127, respectively, we obtain the following system of algebraic equations

$$\begin{cases}
-\left(A\left(G\ell^2 a_n^4 + C_{11}a_n^2 - \left(\frac{\sigma_v}{2} - G\right)b_m^2\right) + B\left(C_{12} + G + \frac{\sigma_v}{2}\right)a_n b_m\right) = 0 \\
-\left(A\left(C_{21} + G - \frac{\sigma_v}{2}\right)a_n b_m + B\left(C_{22}b_m^2 + \left(G + \frac{\sigma_v}{2}\right)a_n^2\right)\right) = 0
\end{cases} \tag{130}$$

The above system of Equations 130 is further simplified by dividing both equations with the term a_n^2 and with the shear modulus G^*, that is

$$\begin{cases}
\left(A\left(\ell_a^2 + c_{11} - (\xi_1 - \xi_2)r^2\right) + B\left(c_{12} + (\xi_1 + \xi_2)\right)r\right) = 0 \\
\left(A\left(c_{12} - (\xi_1 - \xi_2)\right)r + B\left(c_{11}r^2 + (\xi_1 + \xi_2)\right)\right) = 0
\end{cases} \tag{131}$$

where we have set

$$\begin{aligned}
&c_{11} = C_{11}/G^* = C_{22}/G^*, \quad c_{12} = C_{12}/G^* = C_{21}/G^*, \\
&\xi_1 = \frac{\sigma_v}{2G^*}, \quad \xi_2 = \frac{G}{G^*} \\
&\ell_a^2 = \ell^2 \xi_2 a_n^2, \quad r = \frac{b_m}{a_n} = \frac{S\,m}{n\,T} = \frac{S}{T}
\end{aligned} \tag{132}$$

For non-trivial solution in terms of A and B, the determinant of the system of Equations 131 must vanish. This leads to the following biquadratic equation for the aspect ratio of axial joints r, $i.e.$

$$\begin{aligned}
&r^4 + 2mr^2 + k^2 = 0, \\
&2m = \frac{p_2}{p_4} = \frac{c_{11}\ell_a^2 + c_{11}^2 - c_{12}^2 - 2c_{12}\xi_2}{c_{11}(\xi_2 - \xi_1)}, \\
&k^2 = \frac{p_0}{p_4} = \frac{(\ell_a^2 + c_{11})(\xi_2 + \xi_1)}{c_{11}(\xi_2 - \xi_1)}
\end{aligned} \tag{133}$$

The roots of Equation 133 are

$$r_1^2 = -m + \sqrt{m^2 - k^2},$$
$$r_2^2 = -m - \sqrt{m^2 - k^2} \qquad (134)$$

A solution is possible if there exists a real root r_i; that is, if either r_1^2 or r_2^2 or both are positive. This occurs in the following cases: (Case 1) $m > 0$, $k^2 < 0$ in which the root r_1^2 is positive and r_1 is real; (Case 2) $m < 0$, $m^2 > k^2 > 0$ in which both r_1^2 and r_2^2 are positive and so r_1 and r_2 are real.

The critical internal buckling stress is then found as the minimum load ξ_1 for which Equation 133 has real roots. This is illustrated in Figure 17 that presents the dependence of the buckling load on the aspect ratio of the jointed layer for four cases of anisotropy, namely $\xi_2 = G/G^*$ equal to 2.5, 3, 5, and 7, a constant Poisson's ratio $\nu = 0.49$ – since rock masses at great depths behave in an almost incompressible manner – and a dimensionless microstructural length scale $\ell_a = 0.1$. It may be observed that as the macroscopic anisotropy of the layer becomes more pronounced, then both the crack spacing to bed thickness ratio and the buckling load decrease. In general, both the aspect ratio of axial splitting cracks and of the buckling load decrease with the increase of the macrostructural anisotropy of the bed or equivalently with the amplification factor A of the tensile stresses acting on the crack tips.

Plots of several spectra of the buckling stress with aspect ratio for various values of the dimensionless microstructural length scale ℓ_a are illustrated in Figure 18a. From these

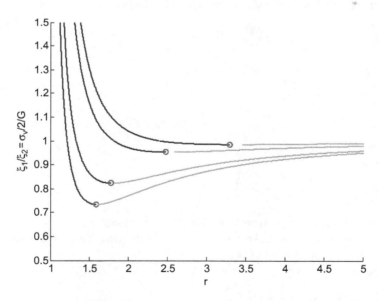

Figure 17 Dependence of the dimensionless buckling load on the aspect ratio of the crack spacing to bed thickness for $\nu=0.49$ and for four shear moduli $\xi_2 = G/G^*$ equal to 2.5, 3, 5, and 7, respectively and constant internal length scale. The global minima represented by the circles are moving toward the left of the diagram as ξ_2 increases.

Linear elasticity with microstructure and size effects 739

plots it may be drawn the interesting result referring to the dependence of the buckling load on the crack spacing to bed thickness ratio for any specified degree of anisotropy of the macrostructure, shown in Figure 18b, that is a manifestation of a size effect.

Indeed, a size effect law of the following form was best-fitted on the numerical data presented in Figure 18b,

$$\xi_1 = C_1 \left(\frac{r}{\ell_a} \right)^{-\beta} + C_0 \tag{135}$$

where C_1, C_0, β are constant factors. In fact C_0 is the buckling load for $\ell_a \to 0$. Then by combining Equation 135 and the third of Equation 132, it may be seen that the size effect of the buckling load on the spacing of cracks for constant aspect ratio, anisotropy ratio and internal length scale takes the form

$$\xi_1 = \left[C_1 \left(\frac{S}{T} \right)^{-\beta} \left(2\pi \sqrt{\xi_2} \ell \right)^{\beta} \right] S^{-\beta} + C_0 \tag{136}$$

The regression analysis of the numerical data by using the power-law given by Equation 135 gave the following values of the constant coefficients

$$C_1 = 1.4476, \ \beta = 2.1748, \ C_0 = 6.048 \tag{137}$$

It may be seen that the exponent is relative large and explains the fact that in real situations it is very rare that the ratio of axial splitting joints spacing relatively to the bed thickness is less than unity.

4 SUMMARY

After a brief overview of some applications of the Casal-Mindlin microelasticity or grade-2 or second gradient of strain theory with surface energy, for the study of fundamental static and dynamic problems, two problems are thoroughly presented here, namely the bending of beams and the axial splitting of deep geological layers. In all cases that were reviewed and examined, it is demonstrated that the consideration of internal length scales are responsible for the manifestation of size effects in static problems and non-classical dispersion phenomena in dynamic problems.

More specifically, it was illustrated that the surface energy term of the technical beam theory is responsible for a size effect exhibited by the flexural strength of beams in three-point bending, namely the dependence of the flexural strength on the inverse length of the beam for the same aspect ratio. Based on the assumption that the failure extensional strain in bending is equal to the failure extensional strain in direct or indirect tension, and the assumption of a linear elastic behavior of the brittle material up to failure, then also a L^{-1} size effect of the tensile strength of quasi-brittle solids in direct as well as in indirect tensile tests, has been derived. The size effect predicted by the proposed theory is validated against experimental results of beam bending of Dionysos marble.

It has been also found the interesting result referring to the dependence of the buckling load of a rock bed transected by periodic system of axial splitting cracks on

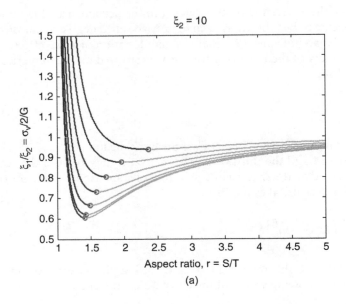

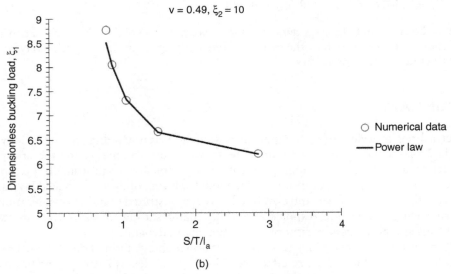

Figure 18 (a) Dependence of the dimensionless buckling load represented by the circles on the aspect ratio of joints for various microstructural length scales (as the microstructural scale parameter increases the critical buckling load decreases) for $\xi_2 = 10$ and Poisson's ratio $\nu = 0.49$, and (b) size effect exhibited by the buckling load for $\nu = 0.49$ and $\xi_2 = 10$.

the bed thickness, for fixed crack spacing to bed thickness ratio and a specified degree of anisotropy of the macrostructure.

Hence, there is ample space for further applications of this theory both for the development of computational codes and new experimental techniques that could take into account the effect of rock microstructure on rock behavior. For example in a series of papers Exadaktylos & Xiroudakis (2009, 2010a,b) have developed a special grade 2 constant displacement discontinuity method – *i.e.* with one collocation point per element – for the accurate solution of plane crack problems. Furthermore, it is important to further develop this theory in the context of nonlinear elasticity and plasticity theories and the development of new failure theories of rocks. Other technologically important problems that may be considered in the frame of the present theory, are the elastic wave propagation in earthquakes and seismic wave characterization of rock masses, and modeling of the mechanical behavior of rock joints and size effects among many others.

REFERENCES

Achenbach J.D. (1973). *Wave Propagation in Elastic Solids*. Amsterdam, North-Holland.
Aifantis E.C. (1992). On the role of gradients in the localization of deformation and fracture. *Int. J. Eng. Sci.*, 30, 1279–1299.
Aravas N. (2011). Plane-strain problems for a class of gradient elasticity models—A stress function approach. *J. Elast.*, 104, 45–70.
Biot M.A. (1963). Internal buckling under initial stress in finite elasticity. *Proc. R. Soc. London, Ser. A*, A273, 306–328.
Biot M.A. (1965). *Mechanics of Incremental Deformations*. John Wiley & Sons, New York (1965).
Bock H. (1980). Das Fundamentale Kluftsystem. *7dt. Geol. Ges.*, 131, 627–650, Hannover.
Bock H. (1971). Gelandebeobachtungen ueber Klueftung in tectonisch wenig beansrpuchten Sedimentgesteinen, Int. Symposium Soc., Mecanique des Roches, Nancy, pp. 1–11.
Casal P. (1961). La capilaritè, interne. *Cahier du Groupe Français d'Etudes de Rhèologie* C.N.R. SVI no. 3, 31–37.
Chen Y-Z (2004). Stress analysis for an infinite strip weakened by periodic cracks. *Appl. Math. Mech.*, 25(11), 1298–1303.
Cosserat E. & Cosserat F. (1909). *Théorie des corps déformables*. A. Herman et Fils, Paris.
Costin L.S. (1983). A microcrack model for the deformation and failure of brittle rock. *J. Geophys. Res.*, 88(B11), 9485–9492.
Coulson C.A. (1958) *Waves*. Oliver and Boyd, Edinburgh and London, Reprinted 1958.
Currie D.K., Hayes M.A. & O'Leary P.M. (1977). Viscoelastic rayleigh waves. *Q. Appl. Math.*, 35, 35–53.
Eringen A.C. & Suhubi E.S. (1975). *Elastodynamics*, Volume II, Academic Press, New York and London.
Exadaktylos G., Vardoulakis I. & Aifantis E. (1996). Cracks in gradient elastic bodies with surface energy. *Int. J. Fract.*, 79, 107–119.
Exadaktylos G. & Vardoulakis I. (1998). Surface instability in gradient elasticity. *Int. J. Solids Struct.*, 35, 2251–2281.
Exadaktylos G. (1998). Gradient elasticity with surface energy: Mode-I crack problem. *Int. J. Solids Struct.*, 35, 421–456.
Exadaktylos G. & Vardoulakis I. (2001a). Microstructure in linear elasticity and scale effects: A reconsideration of basic rock mechanics and rock fracture mechanics. *Tectonophysics*, 335, 81–109.
Exadaktylos G.E., Vardoulakis I. & Kourkoulis S.K. (2001b). Influence of nonlinearity and double elasticity on flexure of rock beams-II. Characterization of Dionysos marble. *Int. J. Solids Struct.*, 38, 4119–4145.

Exadaktylos G. & Xiroudakis G. (2009). A G2 constant displacement discontinuity element for analysis of crack problems. *Comput. Mech.*, 45, 245–261.

Exadaktylos G. & Xiroudakis G. (2010a). The G2 constant displacement discontinuity method – Part I: Solution of plane crack problems. *Int. J. Solids Struct.*, 47, 2568–2577.

Exadaktylos G. & Xiroudakis G. (2010b). The G2 constant displacement discontinuity method – Part II: Solution of half-plane crack problems. *Int. J. Solids Struct.*, 47, 2578–2590.

Georgiadis H.G., Brock L.M. & Rigatos A.P. (1998). Transient concentrated thermal/mechanical loading on the faces of a crack in a coupled-thermoelastic solid. *Int. J. Solids Struct.*, 35, 1075–1097.

Germain P. (1973a). La mèthode des puissances virtuelles en mècanique des milieux continus. Part I. *Journal de Mècanique*, 12, 235–274.

Germain P. (1973b). The Method of virtual power in continuum mechanics. Part 2: Microstructure. *SIAM, J. Appl. Math.*, 25, 556–575.

Griffith A.A. (1921) The phenomena of rupture and flow in solids. *Philos. Trans. R. Soc. London, Ser. A*, 221, 163–198.

Guenther W. (1958) Zur Statik und Kinematik des Cosseratschen Kontinuums. In: Abhandlungen der *Braunschweigischen* Wissenschaftlichen Gesellschaft, Göttingen, Verlag E. Goltze, 101, 95–213.

Irwin G.R. (1957). Analysis of stresses and strains near the end of a crack traversing a plate. *J. Appl. Mech.*, 24, 361–364.

Jaeger J.C., Cook N.G.W. & Zimmerman R.W. (2007) *Fundamentals of Rock Mechanics*. 4th Edition. Blackwell Publishing.

Kemeny J. & Cook N.G.W. (1985). Formation and stability of steeply dipping joint sets. *Proceedings of 26th U.S. Symposium on Rock Mechanics*. South Dakota Schools of Mines and Technology, Rapid city, Dakota, 26–28, June, 1985, pp. 471–478.

Landau L. M. & Lifshitz E.M. (1975). *Theory of Elasticity*. 2nd Edition, Pergamon Press.

Mindlin R.D., Tiersten H.F. (1962) Effects of couple-stresses in linear elasticity. *Arch. Ration. Mech. Anal.*, 11, 415–448.

Mindlin, R.D. (1964). Microstructure in linear elasticity. *Arch. Ration. Mech. Anal.*, 10, 51–77.

Mindlin, R.D. & Eshel, N.N. (1968). On first strain-gradient theories in linear Elasticity. *Int. J. Solids Struct.*, 4,109–124.

Papargyri-Beskou S., Tsepoura K.G., Polyzos D. & Beskos D.E. (2003) Bending and stability analysis of gradient elastic beams. *Int. J. Solids Struct.*, 40, 385–400.

Rayleigh L. (1887). On waves propagated along the plane surface of an elastic solid. *Proc. London Math. Soc.*, 17, 4–11.

Stavropoulou M., Exadaktylos G.E., Papamichos E., Larsen I. & Ringstad, C. (2003). Rayleigh wave propagation in intact and damaged geomaterials. *Int. J. Rock Mech. Min. Sci.*, 40, 377–387.

Tapponnier P. & Brace W.F. (1976). Development of stress-induced microcracks in westerly granite. *Int. J. Rock Mech. Min. Sci. & Geomech. Abstr.* 13, 103–112.

Timoshenko, S.P. (1921). On the correction for shear of the differential equation for transverse vibrations of prismatic beams. *Philos. Mag.*, sec. 6., 41, 744–746.

Vardoulakis I. (1981). Surface waves in a half-space of submerged sand, Earthquake Eng. Struct. Dyn., 9, 329–342.

Vardoulakis, I. & Sulem, J. (1995). *Bifurcation Analysis in Geomechanics*. Blackie Academic and Professional.

Vardoulakis I., Exadaktylos G. & Aifantis, E., (1996). Gradient elasticity with surface energy: Mode III crack problem. *Int. J. Solids Struct.*, 33(30), 4531–4559.

Vardoulakis I. & Georgiadis H.G. (1997). SH surface waves in a homogeneous gradient-elastic half-space with surface energy. *J. of Elasticity*, 47, 147–165.

Vardoulakis I., Exadaktylos G.E. & Kourkoulis S.K. (1998). Bending of marble with intrinsic length scales: A gradient theory with surface energy and size effects. *J. Phys. IV France*, 8, 399–406.

Vardoulakis I. & Giannakopoulos A.E. (2006). An example of double forces taken from structural analysis. *Int. J. Solids Struct.*, 43, 4047–4062.

Voigt W. (1887). Theoretische Studien ueber die Elasticitaetsverhaeltnisse der Krystalle. *Abh. Ges. Wiss. Goettingen.*, 34, 100 pages monograph.

Wu C.H. (1992). Cohesive elasticity and surface phenomena. *Q. Appl. Math.*, L(1), 73–103.

Chapter 24

Rock creep mechanics

F.L. Pellet
MINES ParisTech, Geosciences and Geoengineering Department, Fontainebleau, France

Abstract: Creep is a time-dependent process which leads to deformations that are strongly related to a rock's petrophysical properties. This chapter explains the causes of creep at the microscopic scale. Micro-mechanisms of deformation are described according to the atomic structure of crystals. Relevant laboratory tests, namely creep tests, relaxation tests and strain rate-controlled loading tests are outlined and their performance is discussed. Thereafter, the main rate-dependent constitutive models are listed with comments on their adequacy for different rock types. A summary of the theory of viscoplasticity is provided, along with a procedure for identifying constitutive parameters from test results. In the last section, time-dependent numerical modeling of underground openings is presented with the help of case studies. Particular emphasis is placed on the analysis of time-dependent ground-support interaction.

1 INTRODUCTION

In mechanics, creep relates to the slow time-dependent deformations of a given body subjected to a constant state of stress over time. This deformations process can lead to large delayed strains and, eventually to the rupture of the studied body.

In rock engineering, creep is a major concern as it can cause facilities to be out of service for a few weeks, or a few years after their construction. A perfect example to illustrate this observable fact is the extreme strains developed in squeezing rocks, which can produce a large convergence in underground openings. Creep can also trigger rock slope instability or detrimental settlement of constructions such as buildings, bridge foundations and dam abutments. Creep sometimes occurs with, or is superimposed on other time-dependent phenomena, such as aging and weathering processes. More specifically, time-dependent deformation can be attributed to different physical or chemical transformations; in fact, these physical changes develop at different time scales, which may range from hours to several years, decades or centuries. Therefore, before analyzing a creep problem it is of the utmost importance to assess the characteristic time involved in each process in order to determine which need to be accounted for. Multiphysics and Chemo-Thermo-Hydro-Mechanics coupling analyses will most probably have to be performed.

Creep is also very sensitive to temperature and rock water content. For instance, soft rock or soils consolidation resulting from water drainage, and the associated pore pressure dissipation, produces significantly delayed deformation of the ground.

Another example is rock swelling, which can occur due to chemical transformation (*e.g.* anhydrite versus gypsum) or to water sorption in argillaceous rock (especially those which contain smectites). All these transformations result in changes to the thermal, chemical and hydraulic regime. The study of these phenomena requires a comprehensive understanding and knowledge of the petrophysics of the rock being considered.

In this chapter we will exclusively focus on the time-dependent deformation resulting from mechanical action. Therefore, rate-dependent constitutive models will be presented, keeping in mind that experimental investigations are essential, both upstream to develop constitutive models with physical meaning, and downstream for model validation.

2 EXPERIMENTAL EVIDENCE

Creep is a time-dependent process; it is, in fact, a particular strain path that depends on the magnitude of the load. Any creep test requires the application of a load at a given rate, prior to performing the actual creep stage. It is therefore useful to classify strain rate regimes as proposed by Field & Walley (2013) as shown in Figure 1.

Following this classification, we will focus on strain rates smaller than 10^{-5} s^{-1} in the subsequent sections. It should be mentioned that very slow creep tests (down to 10^{-13} s^{-1}) performed on rock salt and on argillite have been reported by Bérest *et al.*, (2005).

2.1 Causes of creep at a microscopic scale

Before proceeding further, it might be helpful to examine what happens inside the rock material in the course of creep deformation. This will permit the identification of the deformation mechanisms at the micro-scale and, later, it will help in conceptualizing relevant constitutive models.

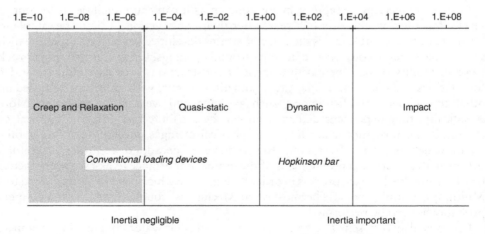

Figure 1 Schematic diagram of strain rate regimes (in reciprocal second) and the techniques that have been developed to obtain them (adapted from Field & Walley, 2013).

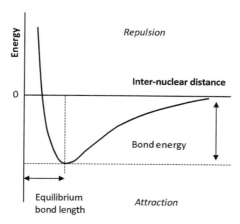

Figure 2 Bond potential energy as a function inter-nuclear distance.

2.1.1 Rock microstructures and potential energy

Rocks are composed of minerals or crystals formed by atoms arranged in lattices. As it is well known, in nature everything is "Matter or Energy". Therefore, the atoms (Matter) are positioned with respect to each other in order to minimize the inter-atomic forces which bind them (Energy). In this equilibrium configuration atoms are separated by a distance that is related to their potential energy. This configuration remains unchanged under steady conditions of temperature and pressure. If the rock body is then stretched or compressed, variations in the inter-atomic distances will develop to generate attraction or repelling forces, respectively (Figure 2). When the deformation (extension or contraction) is not too great, the deformation will be reversible. In other words, the rock body will, in time, recover its initial configuration when the stress responsible of the deformation is removed (this is called visco-elastic behavior). Of course, there is a limit beyond which the atomic bonds will progressively break.

The nature and the magnitude of the inter-atomic forces - or intermolecular forces - depends on the type of bond. Basically, there are 4 main types of bonds, some of which are illustrated in Figure 3:

- Ionic bond: atoms exchange pairs of electrons due to electrostatic charge. This very strong bond represents high potential energy. It is present in rock salt crystals made up of calcium chlorine (Halite, NaCl).
- Covalent bond: the atoms share electrons. This is also a strong bond that is present in diamond (Carbon, C).
- Metallic bond: some electrons are free to move. This moderate bond is not often encountered in rock materials.
- Hydrogen bonds are of two types - intermolecular and intra-molecular. Hydrogen bonds are weaker compared to primary bonds (ionic and covalent).

Beside these inter-atomic bonds, intermolecular attachment may develop. The most frequent in geomaterials are the van der Waals forces, which are due to the presence of

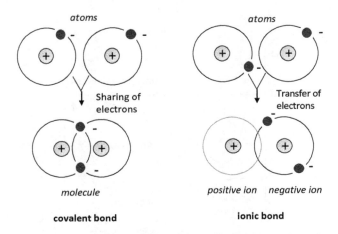

Figure 3 Schematic representation of covalent and ionic bonds.

water. Water molecules are linked by a hydrogen bond due to negative and positive electronic charges (dipole molecules). Several models have been proposed to evaluate these bonds, for example, the double layer theory.

The potential energy associated with each of these bonds is formally expressed in electron volts. The order of magnitude of covalent and ionic bonds is 10 ev per atom. The van der Waals bond is about 0.1 ev per mol. Since most rocks are polycrystalline, the average potential energy will depend on their mineralogical composition.

2.1.2 Micro-mechanisms of deformation

Crystal lattices formed by atoms always contain some defects. In the arrangement of the atoms, an atom is sometimes missing or, on the contrary, an extra atom might be present. Occasionally, some foreign substances or impurities (molecules or ions) may be present within the crystal lattice (Figure 4); any of these defects could be the perfect location to initiate bond breakage and the ensuing micro-crack nucleation and propagation.

Defects can also be linear or planar, as in dislocations (Davis & Reynolds, 1996). Under differential stress, atoms will jump from one site to another until the full array of atoms is shifted. Depending on the pressure and the temperature conditions, different creep mechanisms will occur (Figure 5). These range from dissolution creep, for low temperature and moderate differential stress, to diffusion creep for higher temperatures. When differential stress is elevated, dislocation creep will be the predominant mechanism leading to fracture and the production of cataclasite. This phenomenon could be advantageously described using dislocation mechanics (McClintock & Argon, 1966). Additionally, movement at the grain boundaries can also occur.

Deformation recovery will heal the material by rearranging the dislocation to the original configuration. More detailed information on the specifics of creep mechanisms at the micro-scale can be found in Davis & Reynolds (1996) and in Püsch (1993).

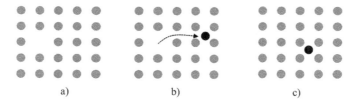

Figure 4 Representation of defects in a crystalline structure (the dots represent the atoms): a) one atom is missing, b) one atom migrates, c) presence of an impurity.

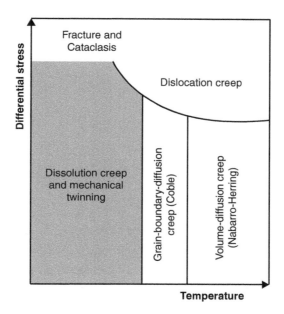

Figure 5 Simplified creep deformation mechanisms chart (adapted from Davies & Reynolds, 1996).

2.1.3 Evolution of electro-chemical bonds

In addition to stress and temperature conditions, the petro-physical nature of the rock plays an important role in creep susceptibility. For instance, moisture is an important parameter; as we have seen water molecules are dipole so the electrical resistivity may change with time. This is particularly true in argillaceous rocks such as shale, marl and mudstone. In this regard, the use of Archie's law permits to rock porosity to be related to the electrical resistivity.

A covalent bond can be weakened or broken due to oxidation-reduction reactions. This was observed in rock specimens that were subjected to high temperatures (Keshavarz et al., 2010). Rock salt (made up of halite crystals) and argillaceous rock are known to be prone to creep because the bonds evolved quickly. Weathering and aging are also responsible for changes in the nature of the bond due to changes in humidity, temperature and pressure conditions (Butenuth, 2001).

750 Pellet

Table 1 Activation energy for different single crystals (after Poirier, 1995; Palandri & Kharaka, 2004).

	Quartz	*Pyroxene*	*Plagioclase*	*Halite*	*Calcite*	*Kaolinite*
Q [kJ/mol]	90–117	54–94	18–70	7–10	24–35	13–18

2.1.4 Activation energy and enthalpy

From a micro-mechanical point of view, when a solid body is deformed the strain rate depends on the activation energy (see section 2.1.1). Based on thermodynamics considerations, the Arrhenius empirical equation can be used to compute the strain rate:

$$\dot{\varepsilon} = Ae^{-\frac{Q}{RT}}. \tag{1}$$

where: $\dot{\varepsilon}$, deformation rate
    A, a constant
    Q, activation energy
    R, universal ideal gas constant
    T, absolute temperature

Table 1 summarizes activation energy values published for different single crystals (Poirier, 1995; Palandri & Kharaka, 2004).

2 Mechanical testing

2.2.2 Strain rate controlled compression test

In general, rock materials prone to creep exhibit a viscous behavior; this means that their mechanical response is sensitive to the rate of loading. Figure 6a shows stress-strain curves for a specimen subjected to compression tests performed with different strain rates of loading, $\dot{\varepsilon}$. For a given strain rate loading ($\dot{\varepsilon} = \dot{\varepsilon}_0$), a typical stress-strain curve is observed. If the same specimen is then loaded at a higher rate ($\dot{\varepsilon} >> \dot{\varepsilon}_0$), the response is stiffer and the strength is increased. On the other hand, when the rate of loading is lower ($\dot{\varepsilon} << \dot{\varepsilon}_0$), the response is more compliant and the strength decreases. There is a limit curve under which no further changes are noticeable. This curve is obtained with very low rates of loading (typically less than 10^{-9} s^{-1}), while the deformation is developed in real time. It should also to be remembered that low-rate loading tests lead to ductile specimen failure whereas high-rate loading tests result in brittle failure (Peng, 1973). Indeed, in this case, there is not enough time to accommodate the deformation and dynamic effects will develop.

2.2.2 Creep test

Let us now imagine that at a given point on the stress-strain curve ($\dot{\varepsilon} = \dot{\varepsilon}_0$), the loading process is stopped when the stress $\sigma = \sigma_0$ (Figure 6a). The rock specimen is now subjected to a constant load and an increase in strain is measured with respect to time (Figure 6b). This is the strain creep path!

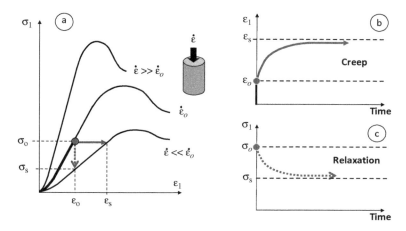

Figure 6 Test in compression: a) axial stress vs. axial strain (constant strain-rate test), b) axial strain vs. time (creep test), c) axial stress vs. time (relaxation test).

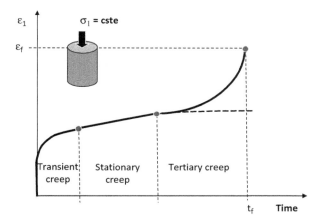

Figure 7 Creep test with the three successive stages: Primary (transient), Secondary (stationary) and tertiary (time-dependent damage and failure).

Looking in more detail at a typical strain-time curve resulting from a creep test, we observe that the curve can usually be clearly divided in three stages (Figure 7). During the first stage, called transient creep (or primary creep), the strain rate decreases with respect to time. This is due to a strain hardening process. During this stage, if the specimen is unloaded the strain will partially recover with respect to time. Another important feature is the existence of a creep threshold below which no delayed deformation is observed. The value of this threshold depends on the particular rock type. Of course, we also need to consider the instantaneous strain, ε_o, produced by the immediate load application. The following stage, known as stationary creep (or secondary creep), is characterized by a linear variation of strain versus time. The span of this stage also depends on the type of rock. Rock salt

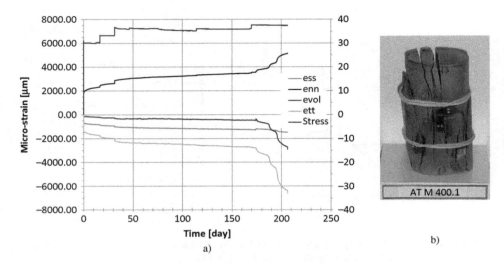

Figure 8 Multi stage creep test on a claystone specimen: a) Strains versus time, b) View of the specimen after testing (after Fabre & Pellet, 2006).

exhibits a long secondary creep stage but this is very short for argillaceous rocks (Pellet, 2015).

Finally, if the applied stress deviator is high enough, a third stage appears with an increase in strain-rate until creep failure occurs. Throughout this stage, micro-cracks grow until they coalesce, eventually forming macro-cracks and fractures. This tertiary creep stage induces time-dependent damage as reported by Fabre & Pellet (2006).

Figure 8 shows an example of a multistage creep test performed on a claystone specimen over more than 200 days (Fabre & Pellet, 2006). Because this rock was transversely isotropic, two pairs of 3 strain gauges were mounted on the specimen. Each pair is composed of an axial strain gauge, ε_{nn} and two transverse gauges, one perpendicular to the isotropic plane, ε_{tt}, and one parallel to this plane, ε_{ss}. The graph below clearly shows the tertiary stage, which materialized when there was a sudden increase in both axial strain and transversal strain across the isotropic plane. As a result, the volumetric strain shows dilation.

2.2.3 Relaxation test

Alternatively, we could, from the point $\sigma = \sigma_o$ (Figure 6a), envisage a different path. Instead of maintaining a constant load, the strain is kept constant. This can be achieved only with servo-controlled loading equipment. In this case, a decrease of stress with respect to time is observed (Figure 6c). This is the stress relaxation path!

Stress relaxation is due to the material accommodating the deformation. A good illustration of stress relaxation in a rock formation was given by Heim (1878) who stated that, at great depth, the state of stress is isotropic. In other words, in the long term rocks cannot bear differential stress (shear stress). Of course, this is not observed in the areas where tectonic thrusts are acting, for example, near major fault zones.

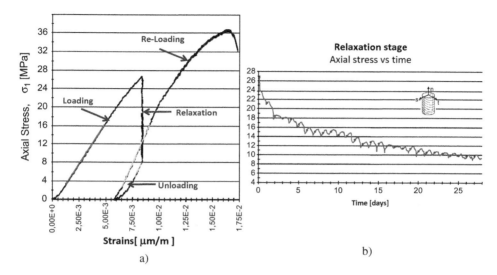

Figure 9 Relaxation creep test on a claystone specimen: a) stress versus strain, b) stress versus time during the relaxation stage (after Boidy & Pellet, 2000).

Figure 9 shows the results of a relaxation test carried out on a claystone. The specimen is first loaded at a strain rate of $10^{-6}\,s^{-1}$ (Figure 9a). Thereafter, the relaxation is performed over a period of 28 days. The stress decreases from 26 MPa to 9 MPa (Figure 9b). Stress oscillations, due to daily temperature variations, are clearly visible. At the end of the test, the specimen was completely unloaded and reloaded until failure occurred.

2.2.4 Technical specification for creep tests

In a rock engineering framework, creep tests are the most commonly performed test for characterizing time dependency. Besides the necessary care required to prepare and store specimens, creep tests need a perfectly temperature controlled room. This is to avoid temperature influences on the rock specimen itself and to eliminate any thermal effects on the loading device (see Figure 9b), which are generally made up of steel, a temperature-sensitive material.

In the case of multiple loading stage tests, each loading increment has to be carefully selected in order to avoid sudden specimen failure (Gasc-Barbier et al., 2004). To accelerate the creep process some authors (Fabre & Pellet, 2002) have suggested that oligo-cyclic tests be carried out. This consists of performing small amplitude cycles of loading-unloading to speed up the creep process. Eventually the rock specimen will fail due to fatigue.

Another valuable source of information for creep tests is the technical specifications for the hardware equipment and the rock specimen's instrumentation. Strain gages appropriately mounted on the specimen will provide linear strains (dilation or contraction) in the principal strain directions. Additionally, strain measurement will allow the computation of the volumetric strain, which is of primary importance for rock behavior characterization. Complementary to this, a record of the ultra-sonic velocity and acoustic emissions during the creep stage can give insights into the damage processes that develop within the specimen (Grgic & Amitrano 2009, Pellet & Fabre, 2007).

More information on test preparation and environmental conditions are described in a comprehensive method suggested and published by the International Society for Rock Mechanics (Aydan et al., 2014). This text is a useful guide for engineers who have to perform creep tests.

3 CONSTITUTIVE EQUATIONS

The first mathematical expressions established to calculate creep strain were proposed by Andrade (1910) for transient creep and by Norton (1929) for stationary creep; these were from results of tensile tests performed on steel. These mathematical expressions are creep laws (see section 3.3) rather than proper constitutive models since their parameters are directly inferred from test results. Later, numerous advanced viscoplastic constitutive models were proposed for different rock types, such as rock salt (Munson, 1997), sedimentary rocks (Cristescu & Hunsche, 1998) and tuff (Aydan et al., 2011).

3.1 Rheological analogical models

The first and the more intuitive approach to understanding creep behavior is based on rheological analogical models. Conventionally, the viscous behavior is schematically represented by a dashpot whereas plasticity corresponds to a slider and elasticity to a spring. In its simplest form, viscosity is considered as Newtonian. It means that the shear stress varies linearly with the shear strain-rate, as presented in Figure 10. Thus, the shear stress and the axial stress can be expressed as a function of the strain rate as follow:

$$\tau = \eta \cdot \dot{\gamma} \qquad (2)$$
$$\sigma = 3\eta \cdot \dot{\varepsilon} \qquad (3)$$

where: η is the kinematic viscosity [MPa.s]
$\dot{\gamma}$ is the shear strain-rate [s^{-1}]
$\dot{\varepsilon}$ is the axial strain-rate [s^{-1}]

Moreover, if it is assumed that no volumetric strain will be developed during the deformational process, the Poisson ratio, v, is equal to 0.5. Thus, the creep law (axial

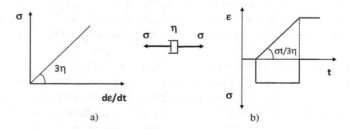

Figure 10 Newtonian viscosity: a) shear stress vs. shear strain rate, b) axial stress vs. axial strain rate.

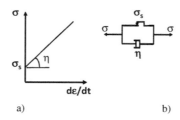

Figure 11 Bingham Model: a) rheological analog; b) stress versus strain-rate.

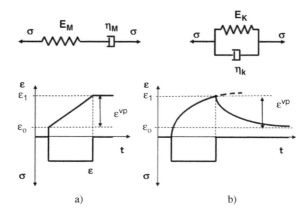

Figure 12 Strain and stress vs. time for Maxwell model a) and Kelvin model b).

strain vs. time, eq. 4) and the relaxation law (axial stress vs. time, eq. 5) will be established after time integration by the following equations:

$$\varepsilon = \frac{\sigma t}{2\eta(1+\vartheta)} = \frac{\sigma t}{3\eta} \quad (4)$$

$$\sigma = 3\eta\frac{\varepsilon}{t} \quad (5)$$

For most rock types creep strain develops only when the applied stress overcomes a threshold. Such a material obeys the Bingham model, which incorporates a threshold below which no creep strain occurs (Figure 11).

Using the elementary rheological components, different assemblies can be made up to reproduce the different stages of the creep curve. Primary creep (transient creep) can be modeled with a Kelvin unit that utilizes a dashpot and a spring in parallel (Figure 12b). Secondary creep (stationary creep) is well reproduced by the Maxwell model which is made up of a dashpot and a spring connected in series (Figure 12a). In this model the instant reversible strain can be accounted for by the spring.

It should also to be noted that, unlike the Kelvin model, the Maxwell model allows the stress relaxation to be modeled with respect to time. However, Kelvin's model is able to model time-dependent strain recovery following unloading (Figure 12b). Therefore, the

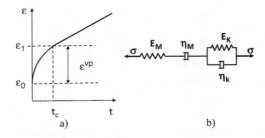

Figure 13 Strain vs. time for the Burger model; a) stress versus strain-rate; b) rheological analog.

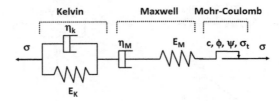

Figure 14 CVISC analog model (c, cohesion; ϕ, friction angle; ψ, dilatancy angle; σ_t, tensile strength).

Kelvin model is a visco-elastic model whereas the Maxwell model belongs to the viscoplastic family.

To model both primary and secondary creep with one equation, it is convenient to use the Burger model, which is composed by the association of Kelvin model and Maxwell model.

Figure 13 shows the analog and the strain versus time curve, which is expressed by the equation

$$\varepsilon_{(t)} = \frac{\Delta\sigma_1}{E_M} + \frac{\Delta\sigma_1 \cdot t}{2\eta_M(1+\nu)} + \frac{\Delta\sigma_1}{E_K}\left(1 - \exp\left(\frac{-E_K \cdot t}{2\eta_K(1+\nu)}\right)\right) \tag{6}$$

More recently, the CVISC model (Itasca, 2006) was proposed; this consists of the Burger model coupled to a plastic threshold. Therefore, the applied stress is limited by the introduction of a load surface, which in this case is the Mohr-Coulomb criterion. With the help of a flow rule, it is possible to compute instantaneous and delayed irreversible deformation. Figure 14 shows the basic analog of the model, which is a viscous elastoplastic one.

3.2 Viscoplasticity

Viscoplastic constitutive models are rate-dependent models developed within the framework of the theory of thermodynamics of irreversible processes. The classical assumptions of the mechanics of continuum media, such as the strain compatibility assumption and the objectivity principle must be fulfilled. Additionally, small strain transformations are frequently considered. Unlike the rheological analog models, viscoplastic constitutive models are based on the existence of one or several load or yield surfaces in the space of the three principal stresses.

The fundamental assumption is the partition of the total strain, $\dot{\varepsilon}$, into two parts.

$$\dot{\varepsilon} = \dot{\varepsilon}^e + \dot{\varepsilon}^{in} \tag{7}$$

The first term, $\dot{\varepsilon}^e$, represents the instantaneous reversible strain, which is computed using the generalized Hooke's law (elasticity). The second terms, $\dot{\varepsilon}^{in}$, corresponds to inelastic- or irreversible- strain. Irreversible strains encompass instantaneous plastic strain and delayed plastic strain (viscoplastic strain). We will see that plasticity is actually a particular case of viscoplasticity theory. Moreover, the occurrence of delayed reversible strain is treated through visco-elasticity, which will be disregarded in the following.

3.2.1 Viscoplastic potential and specific energy

Irreversible strains produce energy dissipation related to the micro-mechanisms of deformation, presented in section 2.2, which is basically, the free energy derived from entropy. Therefore, the Clausius-Duhem inequality must be satisfied:

$$\boldsymbol{\sigma} : \dot{\varepsilon} - \left(\Psi + s\dot{T}\right) - \vec{q}\,\frac{\overrightarrow{\operatorname{grad} T}}{T} \geq 0 \tag{8}$$

where: $\boldsymbol{\sigma}$ is the Cauchy stress tensor
ε is the strain rate tensor
ρ is the density
Ψ is the specific free energy
s is the specific entropy
T is the absolute temperature
q is the heat flux vector

This energy dissipation is related to a viscoplastic potential that is derived from a family of equipotential load surfaces represented in the space of the principal stresses in Figure 15. Viscoplastic deformation starts to develop when stresses hit the inner surface which is actually a viscoplastic threshold, sometimes called the initial yield surface with reference to plasticity theory. The outer surface corresponds to an infinite strain rate; in other words, it represents the dynamic material failure. The viscoplastic domain lies between these two surfaces.

The viscoplastic strain rate, $\dot{\varepsilon}^{vp}$ is expressed with the help of the viscoplastic potential, Ω, by equation 9. It should be pointed out that this equation implies the normality rule for the viscoplastic flow.

$$\dot{\varepsilon}^{vp} = \frac{\partial \Omega}{\partial \sigma} \tag{9}$$

The change of the surface in course of loading is due to the strain rate and the subsequent strain hardening effect (see section 2.2). The latter could be isotropic or kinematic. It has also to be noted that the consistency rule, which is mandatory in plasticity, is not required anymore.

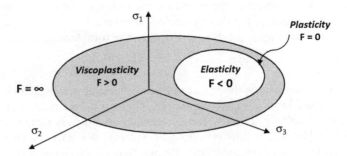

Figure 15 Load and yield surfaces in the deviatoric stress plane.

3.2.2 Load surface and hardening process

In this framework, Perzyna (1966) established the concept of overstress, relative to the limit curve (F < 0), which allows one to characterize the hardening state in the viscoplastic domain. The hardening function is related to the viscoplastic strain and therefore the strain rate is expressed as follow:

$$\dot{\varepsilon}_{ij}^{vp} = \frac{\partial \Omega(\sigma_{ij}, \varepsilon_{kl})}{\partial \sigma_{ij}} \tag{10}$$

Based on this concept, several expressions were proposed to define the viscoplastic dissipating potential Ω and its evolution. Using the von Misès yield surface, Lemaitre and Chaboche (1978) proposed the following:

$$\Omega = \frac{K}{N+1} \left\langle \frac{f(\sigma)}{K} \right\rangle^{N+1} p^{-N/M} \tag{11}$$

where: $f(\sigma)$ is the load surface
p is the hardening variable
K, M, N are the model parameters.

In this expression, the hardening variable is the cumulative plastic strain. Therefore unloading is not possible to model. The strain rate is expressed as

$$\dot{\varepsilon}_{ij}^{vp} = \frac{1}{K^N} \cdot \sigma^N (\varepsilon^{vp})^{-N/M} \tag{12}$$

The load surface – or yield surface – has to be appropriately chosen in accordance with the observed behavior of the material under study. Rock materials exhibit changes in volumetric strain which need to be described by the selected constitutive model; they, like any geomaterials, are sensitive to both the isotropic stress – first invariant of the stress tensor – and the differential stress – second invariant of the stress tensor. We have seen that, depending on their microstructure, the trend to dilation or contraction is more or less pronounced. Different rocks, for example rock salt, sandstone, and shale, will behave differently and will require different load surfaces. Moreover, the state of stress, which is mostly related to depth, also plays an important role in the changes in volumetric strain. One constitutive model proposed by Critescu & Hunsche (1998) is expressed as follows:

$$f(J_2, I_1) = \sigma_{eq} + \alpha\sigma_m \tag{13}$$

where: J_2 is the second invariant of the stress tensor load surface
I_1 is the first invariant of the stress tensor
α ισ τηε parameter ruling dilation-contraction trend.
Substituting the von Misès load surface in the Lemaitre law by the Drucker Prager one, Pellet *et al.*, (2005) proposed that the strain rate could be expressed as

$$\dot{\varepsilon}^{vp} = \left\langle \frac{\sigma_{eq} + \alpha\sigma_m}{Kp^{1/M}} \right\rangle^N \left(\frac{3}{2}\frac{S}{\sigma_{eq}} + \frac{\alpha I}{3} \right) \tag{14}$$

In this expression, **I** and **S** are respectively the identity tensor and the stress deviator and p is the cumulative viscoplastic strain that drives the hardening process, defined by:

$$\dot{p} = \sqrt{\frac{2}{3}\dot{\varepsilon}^{vp} : \dot{\varepsilon}^{vp}} \tag{15}$$

3.2.3 Choosing a suitable viscoplastic law for a rock type

In rock mechanics, it is essential to select a viscoplastic model in line with the type of rock being studied. Rock salt creep, which exhibits an extensive secondary creep phase, can be advantageously described by the specific laws proposed by Munson (1997). Using the same rock material, Hou & Lux (1999) proposed a viscoplastic constitutive model that accounted for creep diffusion and dislocation, strain hardening and recovery, and damage healing. Porous rocks, such as sandstone, require load surfaces with a cap to allow for compaction, as proposed by Cristescu & Hunshe (1998). Argillaceous rocks (shale, claystone) show an important primary creep phase and a short secondary stage, which is quickly replaced by tertiary creep and the associated micro-cracking (Fabre & Pellet, 2006). Moreover, anisotropic creep models should be considered as these rocks are very often transversely isotropic (Cristescu & Cazacu, 1995).

Finally creep in crystalline rocks is, under normal conditions of pressure and temperature, inextricably linked to the damage resulting from cracking. This phenomenon will be outlined below. Table 2 summarizes the characteristics of some viscoplastic constitutive equations.

3.3 Explicit laws for creep

Creep laws, which are directly derived from experimental observation, express changes in viscoplastic deformation with respect to time. Formally, creep laws are not constitutive models *stricto sensu*, as the time variable is explicitly accounted for. As a consequence, creep laws do not allow for the correct description of the development of the deformation in real loading paths, such as multi-axial loading or unloading; they are only valid for a specific strain path.

760 Pellet

Table 2 Characteristics of some viscoplastic constitutive equations.

	Norton (1929)	Lemaitre (1978)	Munson (1997)	Critescu (1998)	Hou-Lux (1999)	Pellet et al., (2005)
Load function	F = f	F = f	F = f	F = f/ κ −1	F = f	F = f
Load surface	Von Misès	Von Misès	Tresca	Tresca Mohr-Coulomb	Von Misès	Drucker Prager
Flow rule	Associated	Associated	Associated	Non associated	Non Associated	Non associated
Volumetric strain	No	No	No	Yes	Yes	Yes
Strain hardening	Viscoplastic	Isotropic	Isotropic or Kinematic	Isotropic or kinematic	Kinematic	Isotropic
Creep threshold	Possible	Possible	Possible	No	Possible	Possible

The creep power law function is particularly well suited to model the majority of creep mechanisms (with the exception of the diffusion creep process). This fact was recognized by Andrade (1910), who proposed that the primary creep of soft metals could be described by a relationship of the following form:

$$\varepsilon^{\text{vp}} = At^{1/3} \tag{16}$$

The exponential law, suggested by Norton (1929), was first expressed by equation 17. For crystalline materials, this has been extended to account for the influence of the temperature (eq 18).

$$\dot{\varepsilon}^{vp} = \left(\frac{\sigma}{K}\right)^N \tag{17}$$

$$\dot{\varepsilon}^{vp} = \left(\frac{A\sigma^N}{d^q T}\right) e^{\left(\frac{-Q}{RT}\right)} \tag{18}$$

where : σ is the loading stress, d is the crystal diameter, T the absolute temperature in [K], Q the activation energy of the thermal reaction in [Joule.mol^{-1}] and R is the constant of the ideal gas (see section 2.1.4). A, N and q are material parameters. Table 3 indicates the parameters of the Norton law suggested by Dusseault & Fordham, (1993) for different types of creep.

In rock mechanics, Lomnitz (1957) was one of the first to propose a logarithmic relationship based on creep measurements in igneous rocks. The creep law is expressed as follows:

$$\varepsilon^{vp} = C_2 \ln(1 + b_2 t) \tag{19}$$

where: C_2 and b_2 are parameters of the material.

The logarithmic law has also been successfully used by Kharchafi & Descoeudres (1995) for the modeling of tunnel convergences in marl. Bažant & Chern (1984) used logarithmic laws to describe creep of concrete. Table 4 summarizes different creep laws and their application domains.

Table 3 Suggested parameters of the Norton law for different creep mechanisms (after Dusseault & Fordham, 1993), see also Figure 5.

Creep Mechanisms	Exponent N	Exponent q
Creep diffusion (Nabarro-Herring)	1	2
Creep diffusion (Coble)	1	3
Dislocation creep	2–5	2–3
Dissolution creep-mechanical twinning	3–9	0

Table 4 Different creep laws and their domains of application.

Andrade (1910)	$\varepsilon_c = Bt^{1/\beta}$	Applicable to primary stage; $\beta = 3$;
Lomnitz (1957)	$\varepsilon_c = A\ln(1 + at)$	Applicable to primary stage
Modified Lomnitz law	$\varepsilon_c = A + B\log(t) + Ct$	Primary and secondary stages
Norton's law (1929)	$\varepsilon_c = A\sigma_a^n t$ or $\dot{\varepsilon}_c = A\sigma_a^n$	Applicable to secondary stage and n = 4–5

3.4 Creep of rock joints

Since the pioneering works of Amadei & Curran (1980), very little research has been done into the creep behavior of rock joints. Dieterich & Kilgore (1994) established empirical state-variable formulations to predict frictional processes, including stick-slip instability, based on direct observation of frictional contacts. Rock joint creep susceptibility is highly correlated to the nature if the infill material and its water content. Recently, Pellet *et al.* (2013), presented shear test results performed on dry and saturated clay rock discontinuities; they showed that both the friction coefficient and the cohesion decrease when the discontinuity is saturated. Overall, the shear strength of the discontinuity is also substantially reduced. Other results have been presented by Xu *et al.* (2005) who carried out direct shear tests on unfilled rock joints. It was observed that the long term strength is smaller than the short term one. The shear behavior of rock joints is also rate-dependent; this was shown by Jafari *et al.* (2004) who performed cyclic tests at different loading rates.

Wang *et al.* (2015) analyzed the behavior of micro-contacts in rock joints under direct shear creep loading. Using CT and laser scanning images, they identified two distinct patterns of contacts in rock joints. As a result, they concluded that the long-term shear strength of fractured rocks is composed of the shear resistance built up by the interlocked micro-asperities at the scale of roughness and the frictional resistance produced between macro-asperities at the scale of waviness on the shear fractured rocks.

3.5 Time-dependent failure

Time-dependent failure materializes at the end of the tertiary creep phase (Figure 7). It appears when the rock material has been heavily damaged, resulting in crack coalescence. Despite the fact that it is a time-dependent process, the related theory is given in the chapter devoted to Rock Damage Mechanics.

762   Pellet

In summary, to predict time-to-failure at a meso scale, it is necessary to couple the viscoplastic constitutive model to the continuum damage theory (Nawrocki & Mroz, 1998). The latter theory is inspired by Fracture Mechanics since multiple modes of failure are usually involved (namely Mode I for tensile failure and mode II for shear failure). Brandut et al. (2013) gathered a lot of data in a comprehensive review on time-dependent cracking and brittle creep in crustal rocks. One of the most popular expressions to compute time-to-failure, t_f, is given by Costin (1987):

$$t_f = (1/\beta) \exp \left[E/RT + b \left(S^* - \sigma \right) \right] \qquad (20)$$

where: E is the activation energy, RT the product of the ideal gas constant and the temperature, S^* the stress corrosion limit, β and b material constants and σ the applied stress.

More generally, viscoplastic constitutive models should account for damage-induced anisotropy (Pellet et al., 2005). For practical purposes at a large scale, the classification proposed by Lauffer (1958) still allows one to assess the stand-up time of unsupported openings.

3.6 Parameter identification

The identification of constitutive model parameters from test data is an important step. Sometimes it is useful to restate the law as a logarithm of the strain rate, in order to proceed using linear regression. Boidy et al. (2002) proposed a procedure for the Lemaitre constitutive equation. Note that most tests are carried out in compression. In some cases the reverse creep that occurs consecutive to unloading may be more realistic, in particular for the analysis of rock excavations. In any case, accurate instrumentation that provides linear strains and volumetric strain will be very valuable to help identify the model parameters.

4 NUMERICAL MODELING OF CREEP IN ROCK ENGINEERING

In terms of numerical solutions, rate-dependent constitutive models offer a major advantage since the solution is regularized with time. Therefore, there is no loss of solution uniqueness when it comes to dealing with strain-softening.

Viscoplastic models have been used for many rock engineering problems such as rock slopes (Li and Li, 2013), foundations or underground cavities (Ladanyi, 1993). One of the major applications during the last two decades was the issue of underground disposal vaults for radioactive waste (Fairhurst, 2002). Here, the challenge is to predict the behavior of the host rock over exceptionally long periods of time (several centuries). This required comprehensive approaches including coupled thermal-hydrological-mechanical behavior of the rock mass (Selvadurai & Nguyen, 1997). Other issues, dealing with gas or CO_2 storage are also a current concern for the long term behavior of rock mass (Lux, 2007; Selvadurai, 2013). In the following we concentrate on time-dependent behavior of underground openings.

4.1 Specificities of underground openings

Viscoplastic constitutive models have been advantageously used in the design of underground works, especially for tunnels excavated in soft rocks prone to creep (Gioda & Cividini, 1996). In such a case, time is involved both in terms of the progression of the tunnel excavation and in the mechanical behavior of the rock. Several studies have presented in situ measurements of the time-dependent closures of galleries (Boidy *et al.*, 2002). In one study (Armand *et al.*, 2013) it was found that the stress orientations induced an anisotropic closure in relation to the direction of the major principal in situ stress.

Tunneling in difficult geological conditions, such swelling and squeezing rocks, can lead to large deformations which often require extra support (Einstein, 1996). Several studies have been published in relation of this problem. Barla *et al.* (2011), Vu *et al.* (2013), Sterpi & Gioda (2009), Pellet (2009), and Pellet *et al.* (2009) proposed countermeasures to account for the rock-support interaction, while Cantieni *et al.* (2011) focused on the core extrusion.

4.2 Time-dependent behavior of monitored galleries

Back analysis of the time-dependent behavior of a monitored tunnel excavated in Opalinus clay was performed by Boidy *et al.* (2002). This study was concerned with a specific section of the Mont-Terri reconnaissance gallery (Switzerland). From a theoretical point of view, Lemaitre's viscoplastic model (see section 3.2.2) was used with a finite-difference numerical code. The model parameters were well identified based on creep test results. Subsequently, complete numerical simulations gave good results when compared to the in situ convergence measurements. Based on the comparison of strain and stress measured in the lining support, the time-dependent behavior was extrapolated over a period of approximately 10 years. The scale effects between the laboratory tests results and in situ measurements were also accounted for.

In Figure 16, the gallery closures (convergences) and the excavation progress (working face position) are reported with respect to time, for different data sets. Computed values show good agreement with the measured magnitudes. For design purposes, the well used analytical Sulem's solution (Sulem *et al.*, 1987) is also shown.

4.3 Time-dependent rock-support interaction

In time-dependent geological formations, the rock-support interaction depends on construction sequences. In practice, the New Austrian Tunneling Method (NATM) aims to limit the size of the opening and thus allows displacements to mobilize the intrinsic rock mass strength. The counterpart is that the rate of excavation is slow and the rock mass properties may have time to deteriorate. On the other hand, when using the New Italian Tunneling Method (NITM) the excavation is performed in a full section with full face reinforcement. The rate of advancement is faster and therefore the displacements are limited. The consequence is that a stiff heavy support must be installed as quickly as possible. To summarize, the stand-up time without support, which is related to strength degradation of the rock, has to be longer than the time

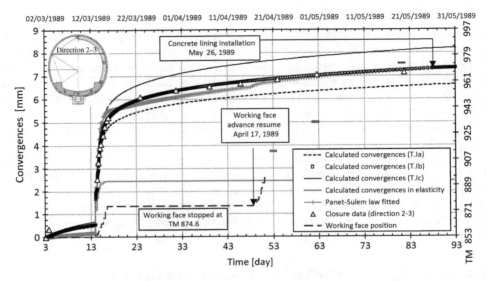

Figure 16 Measured and computed convergences for different sets of parameters in a section of the Mont-Terri, Switzerland, reconnaissance gallery (after Boidy et al., 2002).

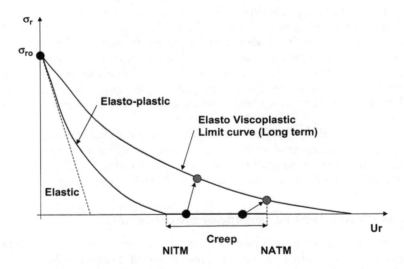

Figure 17 Stress decrease due to excavation versus wall displacements for different rock rheology and different types of excavation methods – NATM-NITM, (after Pellet & Roosefid, 2007).

required to mobilize the rock strength. Therefore, the appropriate method that should be used has to be a compromise to overcome these two phenomena.

These two tunneling options are illustrated in Figure 17. NITM allows a small convergence and provides a stiff support, whereas NATM provides a softer support reaction after larger convergences.

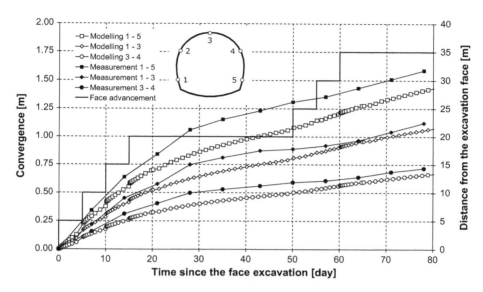

Figure 18 Convergence versus time in section C for different directions: comparison between numerical results and the measured data (after Pellet, 2009).

The contact and the time-dependent interaction between the tunnel lining support and the viscoplastic medium has been investigated by Pellet (2009). First, back analysis of the time-dependent behavior of a drift excavated across a carboniferous layer, which exhibited large delayed displacements, was undertaken. Drift closure was simulated using an elasto-viscoplastic constitutive model that included the strength limitation. This 3D numerical simulation was performed taking into account both stage construction sequences and the rate of excavation advancement. A comparison of the numerical results with the measured data allowed the calibration of the model parameters (Figure 18).

Subsequently, the installation of a concrete lining was simulated to account for the contact with the rock mass. This blind numerical simulation aimed to optimize the tunnel cross-section and to establish the dimensions of a suitable concrete supporting lining. Three months after installation, the stresses measured in the concrete lining were in agreement with the numerically predicted stresses. Figure 19 shows the radial stress decrease due to the excavation followed by an increase due to the contact with the lining support.

Changes in both the radial and tangential stress with time in the rock mass behind the concrete lining are presented in Figure 20. It should be pointed out that the maximum tangential stress in the rock mass progressively relaxes.

5 CONCLUSION

Creep deformations are a concern for many rock engineering activities. To address this problem, a comprehensive multi-scale approach is needed. Firstly, high-quality

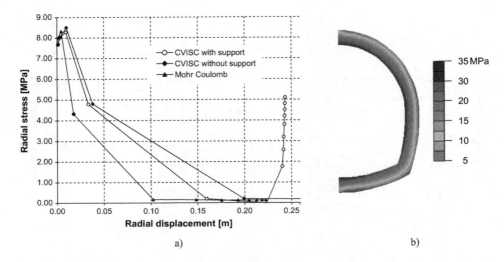

Figure 19 Radial stress at the contact between the concrete lining support and rock mass versus radial displacement for different situations, b) Normal stress distribution in the concrete lining support 80 days after installation (after Pellet, 2009).

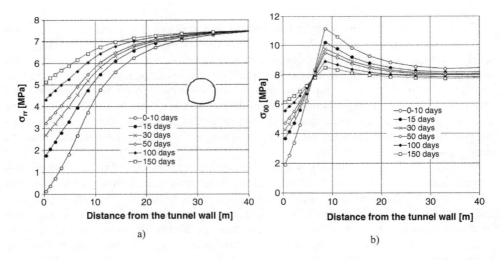

Figure 20 Changes over time for radial stress (a) and tangential stress (b) in the rock mass behind the lining.

investigations of the petrophysics properties of the rock under study will provide the insights necessary to assess creep susceptibility. This phase includes mineralogical identification and quantification at a micro-scale (possibly with the help of a Scanning Electron Microscope). It also requires the characterization of the in situ conditions in terms of temperature and moisture. Secondly, the time scale of the investigation has to be defined in a realistic way.

Creep is one of the manifestations of viscoplasticity. In practice, it is inextricably linked to stress relaxation. Therefore, to characterize viscoplastic behavior, it is necessary to perform creep test, relaxation test or monotonic compression tests with different rates of loading in appropriate ambient conditions of temperature and humidity. It must be pointed out that conventional tests are mostly performed in compression. Under unloading conditions, reverse creep tests (extension) would be more representative of the actual stress path.

For the purpose of numerical simulation, a suitable rate-dependent constitutive model has to be carefully selected. Indeed, some rocks exhibit a large transient deformation stage, due to strain hardening, whereas other materials are prone to stationary creep. Micro-cracks developed during the time-dependent damage stage (tertiary creep) could lead to the rock failure, depending on the level of the applied stress. Time-to-failure prediction requires the input of Fracture Mechanics and/or Damage Mechanics (see the chapter devoted to Rock Damage Mechanics).

The identification of the parameters for the selected constitutive models, based on tests data, is also a sensitive step. To cope with the inherent lack of representativeness of the lab specimen volume, the scale effect has to be accounted for. In any case, carefully instrumented specimens will provide valuable information.

The modeling of the behavior of underground structures over time needs a good reproduction of the stage construction sequences. This will allow a reliable description of the interaction between the ground and the support system when creep strain and stress relaxation are concomitantly developed. To mitigate the uncertainties of the constitutive parameters used and to account for the scale effects mentioned above, inverse analysis of the data collected from a monitored tunnel is a reliable tool.

REFERENCES

Amadei, B., Curran, J.H. (1980) Creep behavior of rock joints. *In: Proc. 13th Canadian Rock Mechanics Symposium, Underground Rock Engineering*, H.R. Rice (ed.), Toronto, Ont, pp. 146–150.

Andrade, E.N.D. (1910) On the viscous flow in metals and allied phenomena. *In: Proc. Royal Society, Section A – Mathematical and Physical Sciences*, London, 84, (567), 1–12.

Armand, G, Noiret, A, Zghondi, J, Seyedi, D.M. (2013) Short and long-term behaviors of drifts in the Callovo-Oxfordian claystone at the Meuse/Haute-Marne Underground Research Laboratory. *Journal of Rock Mechanics and Geotechnical Engineering*, 5, 221–230.

Aydan, Ö, Rassouli, F, Ito, T. (2011) Multi-parameter responses of Oya tuff during experiments on its time-dependent characteristics. *In: Proc. 45th US Rock Mechanics/Geomechanics Symposium, San Francisco*, ARMA 11–294.

Aydan, O., Ito, T., Özbay, U., Kwasniewski, M., Shariar, K., Okuno, T., Özgenoğlu, A., Malan, D.F., Okada, T. (2014) ISRM suggested methods for determining the creep characteristics of rock. *Rock Mechanics and Rock Engineering*, 47, (1), 275–290.

Barla, G., Debernardi, D., Sterpi, D. (2011) Time dependent modeling of tunnels in squeezing conditions. *International Journal of Geomechanics*, 12, 697–710.

Bažant, Z.P., Chern, J.C. (1984) Double power logarithmic law for concrete creep. *Cement & Concrete Research*, 14, 793–806.

Bérest, P., Blum, P., Charpentier, J., Gharbi, H., Vales, F. (2005) Very slow creep tests on rock samples. *International Journal of Rock Mechanics and Mining Sciences*, 42, 569–576.

Boidy, E., Pellet, F.L. (2000) Identification of mechanical parameters for modeling time-dependent behavior of shales. In: *Proc. International Workshop on Geomechanics, Behavior of Deep Argillaceous Rocks: Theory and Experiment*, Andra, Paris, pp. 11–22.

Boidy, E., Bouvard, A., Pellet, F.L. (2002) Back analysis of time-dependent behaviour of a test gallery in claystone. *Tunnelling and Underground Space Technology*, 17, (4), 415–424.

Brantut, N., Heap, M.J, Meredith, P.G., Baud, P. (2013) Time-dependent cracking and brittle creep in crustal rocks: A review. *Journal of structural Geology*, 52, (1), 17–43.

Butenuth, C. (2001) *Strength and weathering of rock as boundary layer problems*, Imperial College Press, London, UK.

Cantieni, L., Anagnostou, G., Hug, R. (2011) Interpretation of core extrusion measurements when tunnelling through squeezing ground. *Rock Mechanics and Rock Engineering*, 44, (6), 641–670.

Costin, L.S. (1987) *Time-dependent deformation and failure, in Fracture mechanics of rock*, Atkinson B.K. (ed.), Academic Press Geology Series, San Diego, pp. 167–215.

Cristescu, N.D., Cazacu, O. (1995) Viscoplasticity of anisotropic rocks. *In: Proc. 5th International Symposium on Plasticity and its Current Applications*, S. Tinimura and A.S. Khan (eds.), Gordon and Breach, New York, pp. 499–502.

Cristescu, N.D., Hunsche, U. (1998) *Time Effects in Rock Mechanics*, John Wiley & Sons, Chichester, UK.

Davis, G.H., Reynolds, S.J. (1996) *Structural Geology of Rocks and Regions*, John Wiley & Sons, New York.

Dieterich, J.H., Kilgore, B.D. (1994) Direct observation of frictional contacts: New insights for state-dependent properties. *Pure and Applied Geophysics*, 143, (1), 283–302.

Dusseault, M. B., Fordham, C.J. (1993) Time-dependent behavior of rocks. *Comprehensive Rock Engineering*, J. A. Hudson (ed.), Pergamon Press, 3, pp. 119–149.

Einstein, H.H. (1996) Tunnelling in difficult ground – Swelling behaviour and identification of swelling rocks. *Rock Mechanics and Rock Engineering*, 29, (3), 113–124.

Fabre, G., Pellet, F. (2002) Identification des caractéristiques visqueuses d'une roche argileuse. *In: Proc. International Symposium on Identification and Determination of Soil and Rock Parameters for Geotechnical Design*, PARAM, Paris – France, pp. 33–40.

Fabre, G., Pellet, F.L. (2006) Creep and time-dependent damage in argillaceous rocks. *International Journal of Rock Mechanics and Mining Sciences*, 43, (6), 950–960.

Fairhurst, C. (2002) Geomechanics issues related to long-term isolation of nuclear waste. *Comptes-Rendus Physique*, 3, (7–8), 961–974.

Field, J.E., Walley, S.M. (2013) Review of the dynamic properties of materials: History, techniques and results. In: *Proc. Int. Conf. on Rock Dynamics and Applications – State of the Art*, J. Zhao and J. Li (eds.), Lausanne, Switzerland, pp. 3–24.

Gasc-Barbier, M., Chanchole, S., Bérest, P. (2004) Creep behavior of Bure clayey rock. *Applied Clay Science*, 26, (1–4), 449–458.

Gioda, G., Cividini, A. (1996) Numerical methods for the analysis of tunnel performance in squeezing rocks. *Rock Mechanics and Rock Engineering*, 29, (4), 171–193.

Grgic, D., Amitrano, D. (2009) Creep of a porous rock and associated acoustic emission under different hydrous conditions. *Journal of Geophysical research – Solid Earth – Chemistry and Physics of Minerals and Rocks/Volcanology*, 114, (B10), 1–19.

Heim, A. (1878) *The mechanical transformation of rocks during mountain building*. English translation, Geologische Institute.

Hou, Z, Lux, K.H. (1999) A constitutive model for rock salt including structural damages as well as practice-oriented applications. *In: Proc. 5th Conference on the Mechanical. Behavior of Salt, Basic and Applied Salt Mechanics*, CRC Balkema, Bucharest, Romania, pp. 151–169.

Itasca Inc. (2006) *Flac2D 5.0, User's Manual*. Minneapolis, USA.

Jafari, M.K., Pellet, F.L., Boulon, M., Amini Hosseini, K. (2004) Experimental study of mechanical behaviour of rock joints under cyclic loading. *Rock Mechanics and Rock Engineering*, 37, (1), 3–23.

Keshavarz, M., Pellet, F.L., Loret, B. (2010) Damage and changes in mechanical properties of a gabbro thermally loaded up to 1000°C. *Pure and Applied Geophysics*, 167(12), 1511–1523.

Kharchafi, M., Descoeudres, F. (1995) Comportement différé des roches marneuses encaissant les tunnels. In: *C.R. Colloque Mandanum Craies et Schistes*, Groupement Belge de Mécanique des Roches, Bruxelles, pp. 58–67.

Ladanyi, B. (1993) Time-dependent response of rock around tunnels. *Comprehensive Rock Engineering*, J. A. Hudson (ed.), Pergamon Press, Oxford, 2, pp. 77–112.

Lauffer, H. (1958) Gebirgsklassifiezierung für den Stollenbau, *Geologie und Bauwesen*, 24, (1), 46–51.

Lemaitre, J., Chaboche, J.L. (1978) Aspect phénoménologique de la rupture par endommagement, *Journal de Mécanique Appliquée*, 2, (3), 317–365.

Li, L.C., Li, S.H. (2013) Numerical investigation on factors influencing the time-dependent stability of the rock slopes with Weak Structure Planes. *Applied Mechanics and Materials*, 353–356, 177–182.

Lomnitz, C. (1957) Creep measurements in igneous rocks, *The Journal of Geology*, 64, (5), 473–479.

Lux, K.H. (2007) Long-term behaviour of sealed brine-filled cavities in rock salt mass – A new approach for physical modelling and numerical simulation, *In: Proc. 6th Conference on the Mechanical Behavior of Salt 'SALTMECH6' – The Mechanical Behavior of Salt – Understanding of THMC Processes in Salt*, Hannover; Germany, pp. 435–444.

McClintock, F.A., Argon, A.S. (1966) *Mechanical Behavior of Materials*, Addison-Wesley Publishing Company, Reading, MA.

Munson, D.E. (1997) Constitutive model of creep in rock salt applied to underground room closure. *International Journal of Rock Mechanics and Mining Sciences*, 34, (2), 233–247.

Nawrocki, P.A., Mróz, M. (1998) A viscoplastic degradation model for rocks. *International Journal of Rock Mechanics and Mining Sciences*, 35, (7), 991–1000.

Norton, F.H. (1929) *Creep of steel at high temperatures*, McGraw Hill Book, New York.

Palandri J.L., Kharaka, Y.K. (2004) *A compilation of rate parameters of water- mineral interaction kinetics for applications to geochemical modeling*. U.S. Geological Survey: Open file report 2004–1068.

Pellet F. L. (2009) Contact between a tunnel lining and a damage-susceptible viscoplastic medium. *Computer Modeling in Engineering and Sciences*, 52, (3), 279–296.

Pellet, F.L. (2015) Micro-structural analysis of time-dependent cracking in shale. *Environmental Geotechnics*, 2, (2), 78–86.

Pellet, F.L., Fabre, G. (2007) Damage evaluation with P-wave velocity measurements during uniaxial compression tests on argillaceous rocks. *International Journal of Geomechanics*, 7, (6), 431–436.

Pellet, F.L, Roosefid, M. (2007) Time-dependent behaviour of rock and practical implications to tunnel design. In: *Proc. 11th International Congress on Rock Mechanic*, Lisbon, Portugal, pp. 1079–1082.

Pellet, F.L., Hajdu, A., Deleruyelle, F, Besnus, F. (2005) A viscoplastic constitutive model including anisotropic damage for the time-dependent mechanical behaviour of rock. *International Journal for Numerical and Analytical Methods in Geomechanics*, 29, (9), 941–970.

Pellet, F.L., Keshavarz, M., Boulon, M. (2013) Influence of humidity conditions on shear strength of clay rock discontinuities. *Engineering Geology*, 157, 33–38.

Pellet, F.L., Roosefid, M., Deleruyelle, F. (2009) On the 3D numerical modelling of the time-dependent development of the Damage Zone around underground galleries during and after excavation. *Tunnelling and Underground Space Technology*, 24, (6), 665–674.

Peng, S. (1973) Time-dependent aspects of rock behavior as measured by a servo-controlled hydraulic testing machine. *International Journal of Rock Mechanics and Mining Sciences and Geomechanics Abstract*, 10, 235–246.

Perzyna, P. (1966) Fundamental problems in viscoplasticity. *Advances in Applied Mechanics*, 9, 243–377.

Poirier, J.P. (1995) Plastic rheology of crystals. In: *Mineral physics & Crystallography, A Handbook of Physical Constants*, Ahrens T.J (ed.), American Geophysical Union, AGU, Reference Shelf 2, 237–247.

Püsch, R. (1993) Mechanisms and consequences of creep in crystalline rock, In: *Comprehensive Rock Engineering*. J. A. Hudson (ed.), Pergamon Press, Oxford, 1, pp. 227–241.

Selvadurai, A.P.S. (2013). Caprock breach: A potential threat to secure geologic sequestration, Chapter 5. In *Geomechanics of CO2 Storage Facilities*, G. Pijaudier-Cabot and J.-M. Pereira (eds.), John Wiley, Hoboken, NJ, pp.75–93.

Selvadurai, A.P.S., Nguyen, T.S. (1997) Scoping analyses of the coupled thermal-hydrological-mechanical behaviour of the rock mass around a nuclear fuel waste repository. *Engineering Geology*, 47, 379–400.

Sterpi, D., Gioda, G. (2009) Visco-plastic behaviour around advancing tunnels in squeezing rock. *Rock Mechanics and Rock Engineering*, 42, (2), 319–339.

Sulem, J., Panet, M., Guenot, A. (1987) An analytical solution for time-dependent displacements in a circular tunnel. *International Journal of Rock Mechanics and Mining Sciences & Geomechanic Abstract*, 24, 155–164.

Vu, T.M., Sulem, J., Subrin, D., Monin, N., Lascols, J. (2013) Anisotropic closure in squeezing rocks: The example of Saint-Martin-la-Porte access gallery. *Rock Mechanics and Rock Engineering*, 46, 231–224.

Wang, J.A, Wang, Y.X, Cao, Q.J., Yang Ju, Mao, L.T. (2015) Behavior of micro-contacts in rock joints under direct shear creep loading. *International Journal of Rock Mechanics and Mining Sciences*, 78, 217–229.

Xu, W.Y., Yang, S.Q., Shao, J.F., Xie, S.Y. (2005) Experimental and modeling study on shear creep behavior of joint rock. *In: Proc. International Symposium of the International Society for Rock Mechanics, Eurock 2005*, Brno, Czech Republic, pp. 697–704.

The five-volume set *Rock Mechanics and Engineering* consists of the following volumes

Volume 1: Principles
ISBN: 978-1-138-02759-6 (Hardback)
ISBN: 978-1-315-36426-1 (eBook)

Volume 2: Laboratory and Field Testing
ISBN: 978-1-138-02760-2 (Hardback)
ISBN: 978-1-315-36425-4 (eBook)

Volume 3: Analysis, Modeling and Design
ISBN: 978-1-138-02761-9 (Hardback)
ISBN: 978-1-315-36424-7 (eBook)

Volume 4: Excavation, Support and Monitoring
ISBN: 978-1-138-02762-6 (Hardback)
ISBN: 978-1-315-36423-0 (eBook)

Volume 5: Surface and Underground Projects
ISBN: 978-1-138-02763-3 (Hardback)
ISBN: 978-1-315-36422-3 (eBook)